ABBREVIATIONS

oz	ounce	in.	inch	Btu	British thermal unit
lb	pound	ft	foot	dB	decibel
cm	centimeter	yd	yard	ft-lb	foot-pound
m	meter	mi	mile	hp	horsepower
km	kilometer	ft/sec^2 or (ft/sec)/sec	feet per second per second	ppm	parts per million
g	gram			ppt	parts per trillion
L	liter				
sec	second	mi/hr	miles per hour	rpm	revolutions per minute
hr	hour				

METRIC EQUIVALENTS

LINEAR MEASURE

1 centimeter . . . . 0.3937 inch
1 inch . . . . 2.54 centimeters
1 decimeter . . . . . 3.937 inches . . . . 0.328 foot
1 foot . . . . 3.048 decimeters
1 meter . . . . . 39.37 inches . . . . . 1.0936 yards
1 yard . . . . 0.9144 meter
1 dekameter . . . . 1.9884 rods
1 rod . . . . 0.5029 dekameter
1 kilometer . . . . 0.62137 mile
1 mile . . . . 1.6094 kilometers

SQUARE MEASURE

1 sq. centimeter . . . . 0.1550 sq. inch
1 sq. inch . . . . 6.452 sq. centimeters
1 sq. decimeter . . . . 0.1076 sq. foot
1 sq. foot . . . . 9.2903 sq. decimeters
1 sq. meter . . . . 1.196 sq. yards
1 sq. yard . . . . 0.8361 sq. meter
1 hectare . . . . 2.471 acres
1 acre . . . . 0.4047 hectare
1 sq. kilometer . . . . 0.386 sq. mile
1 sq. mile . . . . 2.59 sq. kilometers

MEASURE OF VOLUME

1 cu. centimeter . . . . 0.061 cu. inch
1 cu. inch . . . . 16.39 cu. centimeters
1 cu. decimeter . . . . 0.0353 cu. foot
1 cu. foot . . . . 28.317 cu. decimeters
1 cu. yard . . . . 0.7646 cu. meter
1 cu. meter . . . . 0.2759 cord
1 cord . . . . 3.625 steres
1 liter . . 0.908 quart dry . . 1.0567 quart liquid
1 quart dry . . . . 1.101 liters
1 quart liquid . . . . 0.9463 liter
1 dekaliter . . . . 2.6417 gallons . . . . 1.135 pecks
1 gallon . . . . 0.3785 dekaliter
1 peck . . . . 0.881 dekaliter
1 hektoliter . . . . 2.8378 bushels
1 bushel . . . . 0.3524 hektoliter

WEIGHT

1 gram . . . . 0.03527 ounce
1 ounce . . . . 28.35 grams
1 kilogram . . . . 2.2046 pounds
1 pound . . . . 0.4536 kilogram
1 short ton (U.S.) . . . . 2000 pounds
or 0.907 metric ton
1 long ton (English) . . . . 2240 pounds
or 1.016 metric ton
1 metric ton . . . . 2204.6 pounds
or 1.102 short ton (U.S.)
or 0.98421 long ton (English

Intermediate
Algebra

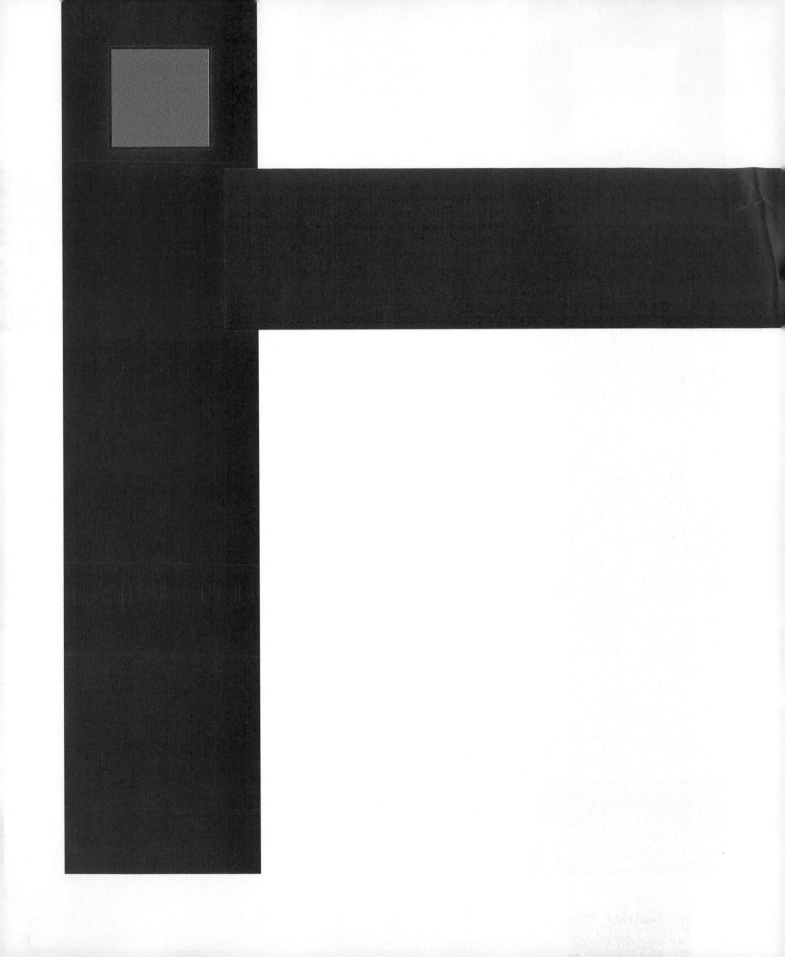

Intermediate Algebra

Ignacio Bello

Hillsborough Community College
Tampa, Florida

Brooks/Cole Publishing Company

I**T**P® An International Thomson Publishing Company

Pacific Grove • Albany • Belmont • Bonn • Boston • Cincinnati • Detroit • Johannesburg • London
Madrid • Melbourne • Mexico City • New York • Paris • Singapore • Tokyo • Toronto • Washington

PRODUCTION CREDITS

SPONSORING EDITORS Peter Marshall and Bob Pirtle
PRODUCTION EDITOR Sharon Kavanagh
PERMISSIONS EDITOR Angela Musey
DESIGN TECHarts
COVER DESIGN AND IMAGE Michelle Austin
ARTWORK Scientific Illustrators
COMPOSITION Dawna Fisher, G&S Typesetters; Typesetters Jan Mullis, Carol Sue Hagood, and Carolyn Briggs; Proofreaders Teri Gaus and Christine Gever
COPYEDITING Luana Richards
PROOFREADING Amy Mayfield

For more information, contact:

BROOKS/COLE PUBLISHING COMPANY
511 Forest Lodge Road
Pacific Grove, CA 93950
USA

International Thomson Publishing Europe
Berkshire House 168-173
High Holborn
London WC1V 7AA
England

Thomas Nelson Australia
102 Dodds Street
South Melbourne, 3205
Victoria, Australia

Nelson Canada
1120 Birchmount Road
Scarborough, Ontario
Canada M1K 5G4

International Thomson Editores
Seneca 53
Col. Polanco
México, D.F., México
C. P. 11560

International Thomson Publishing GmbH
Königswinterer Strasse 418
53227 Bonn
Germany

International Thomson Publishing Asia
221 Henderson Road
#05-10 Henderson Building
Singapore 0315

International Thomson Publishing Japan
Hirakawacho Kyowa Building, 3F
2-2-1 Hirakawacho
Chiyoda-ku, Tokyo 102
Japan

THIS BOOK IS PRINTED ON ACID-FREE RECYCLED PAPER

British Library Cataloguing-in-Publication Data. A catalogue record for this book is available from the British Library.

Printed in the United States of America
04 03 02 01 00 99 98 97   8 7 6 5ᐧ4 3 2

LIBRARY OF CONGRESS CATALOGING-IN-PUBLICATION DATA

LIBRARY OF CONGRESS CATALOGING-IN-PUBLICATION DATA
Bello, Ignacio.
  Intermediate algebra / Ignacio Bello.
    p. cm.
  Includes index.
  ISBN 0-314-06858-9 (alk. paper)
  1. Algebra. I. Title.
  QA152.2.B4517   1997
512.9—dc20

96-9902
CIP

Contents

v

4 RATIONAL EXPRESSIONS   202

SKIP

5 RATIONAL EXPONENTS AND RADICALS   278

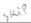

6 QUADRATIC EQUATIONS AND INEQUALITIES   332

7   GRAPHS AND FUNCTIONS   384

Review + ⑪

⑩

8   SOLVING SYSTEMS OF LINEAR EQUATIONS   458

9   QUADRATIC FUNCTIONS AND THE CONIC SECTIONS   524

10   INVERSE, EXPONENTIAL, & LOGARITHMIC FUNCTIONS   582

11   SEQUENCES AND SERIES   652

Preface:
A Guide to *Intermediate Algebra*

If you have never taken algebra, if you need to review the subject, or if you must meet a mathematics requirement that has algebra as a prerequisite, this is the book for you. This book is designed so that you can become familiar or reacquainted with the symbols, terminology, operations, and procedures used in solving equations and, more important, how to use algebra to solve problems.

We view this book as a complete learning system; it is the culmination of many years of teaching experience. Each chapter begins with a list of topics for easy reference, an overview of the material to be covered, and a discussion of the people who contributed to its development. In each lesson, we tell you the prerequisites you need to succeed and the objectives to be met. We then introduce the topic and explain it using definitions, procedures, rules, notes, and cautions carefully set off in boxes. You are then given examples to illustrate the topic and exercises to practice. In keeping with new technology, a special section called *Graph It* is included in every section. All exercise sets contain a *Using Your Knowledge,* a *Write On,* a *Skill Checker,* and a *Mastery Test.* When appropriate, *Applications* and a *Calculator Corner* are also included. Many sections also have a *Problem Solving* feature in a special two-column format for easy recognition and special emphasis.

We have followed the *Standards for Introductory College Mathematics* published by The American Mathematical Association of Two Year Colleges (AM-ATYC) as well as the *Curriculum and Evaluation Standards for School Mathematics* of the National Council of Teachers of Mathematics (NCTM). While all the traditional topics of Intermediate Algebra are included and discussed in clear-cut, easy-to-follow steps, we also show how new technologies, such as graphing calculators, can be used to study these topics. The learning process is carefully guided by these features:

1.1 **Numbers and Their Properties**

1.2 **Operations and Properties of Real Numbers**

1.3 **Properties of Exponents**

1.4 **Grouping Symbols and the Order of**

the human side of algebra

Who invented algebra, anyway? One of the earliest accounts of an algebra problem is in the Rhind papyrus, an ancient Egyptian document written by an Egyptian priest named Ahmes (ca. 1620 B.C.) and purchased by Henry Rhind. When Ahmes wished to find a number such that the number added to its seventh made 19, he symbolized the number by the sign we translate as "heap." Today, we write the problem as

$$x + \frac{x}{7} = 19$$

Can you find the answer? Ahmes says it is $16 + \frac{1}{2} + \frac{1}{8}$.

Western Europeans, however, learned their algebra from the works of Mohammed ibn Musa al-Khowarizmi (this translates as Mohammed the son of Moses of Khowarizmi; ca. A.D. 820), an astronomer and mathematician of Baghdad and author of the treatise *Hisab al-jabr w'al muqabalah,* the science of restoring (placing variables on one side of an equation) and reduction (collecting like terms). With the passage of time, the word *al-jabr* evolved into our present word *algebra,* a subject that we shall continue to study now.

Chapter Preview
Gives the list of topics to be covered in each section and provides an overview of the material to be studied in the chapter as well as the ways in which the topics are related.

The Human Side of Algebra
Discusses the persons who devised the material being studied or who contributed to its development. This section offers a historical perspective and conveys the message that mathematics is a growing body of knowledge and that all advances in algebra began as part of a problem-solving process.

SOLVING EQUATIONS BY FACTORING: APPLICATIONS

To succeed, review how to:

1. Factor polynomials (pp. 173–176).
2. Solve linear equations (pp. 69–73).
3. Evaluate expressions (pp. 42–44).

Objectives:

A Solve equations by factoring.

B Use the Pythagorean Theorem to find the length of one side of a right triangle when the lengths of the other two sides are given.

C Solve applications involving quadratic equations.

To Succeed, Review How to

Details the material you must review *before* you start the section.

Objectives

Specifies the objectives to be met in the lesson. Each of the *Objectives* is tied to the subsections. (Objective **A** goes with subsection **A**, objective **B** goes with subsection **B**, and so on.)

Getting Started

Appears at the beginning of every section. The *Getting Started* is an application that demonstrates how the material relates to the real world and the lesson being studied.

Definitions, Rules, Cautions, Notes

Boxed to emphasize important concepts, rules, procedures, and notes as well as to call your attention to potential pitfalls.

Titled Examples

Includes a wide range of computational, drill, and applied problems carefully selected to build confidence, competency, skill, and understanding.

Graph It

Appears at the end of each section and instructs you on the use of graphers in solving the examples covered in the section. Although this feature is optional, detailed instructions are given to enable you to obtain the solutions of most examples in the section using a TI-82 graphing calculator. (A correlation chart presenting equivalent keystrokes is available for other graphers.)

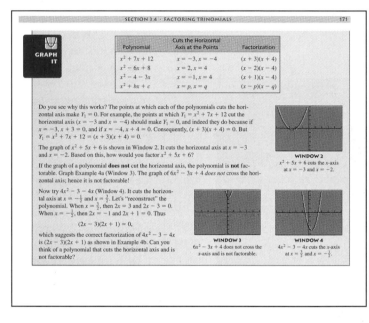

Problem Solving

Interspersed throughout each chapter. Since problem solving is presented as an on-going theme in this book, numerous *Problem Solving* examples are given and clearly formatted using two columns. The left column uses the **RSTUV** method (**R**ead, **S**elect, **T**hink, **U**se, and **V**erify) to guide you through the problem. In the right column, you will find a carefully developed and clearly laid-out solution.

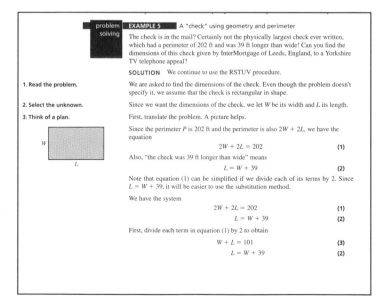

67. When you exercise, your pulse rate should be within a certain *target zone*. The *upper limit U* of your target zone when exercising is a function of age *a* (in years) and is given by

$$U(a) = -a + 190 \quad \text{(your pulse or heart rate)}$$

Find the highest safe heart rate for a person who is
a. 50 years old
b. 60 years old

68. The lower limit *L* of your target zone when exercising is a function of *a* (in years) and is given by

$$L(a) = -\frac{2}{3}a + 150$$

The target zone for a person *a* years old consists of all the heart rates between *L(a)* and *U(a)*, inclusive. Thus if a person's heart rate is *R*, that person's target zone is described by

---

WRITE ON ...

90. What is the main difference in the techniques used when solving equations as opposed to inequalities?

CALCULATOR CORNER — Using a Calculator with the Quadratic Formula

Any calculator that has a store ([STO]) and recall ([RCL]) keys can be extremely helpful in finding the solutions of a quadratic equation with the quadratic formula. Of course, the solutions you obtain are being approximated by decimals. It's especially convenient to start with the radical part in the solution of the quadratic equation and then store this value so you can evaluate both solutions without having to backtrack or write down any intermediate steps. Let's look at the equation of Example 1:

$$8x^2 + 7x + 1 = 0$$

Using the quadratic formula, one solution is obtained by entering

7 [x²] [−] 4 [×] 8 [×] 1 [=]
[√x] [STO] 7 [+/−] [+] [RCL] [=] [÷] 2 [÷] 8 [=]

The display shows −0.1798059 (which was given as

$$\frac{-7 + \sqrt{17}}{16}$$

in the example). To obtain the other solution, enter

7 [+/−] [−] [RCL] [=] [÷] 2 [÷] 8 [=]

which yields −0.6951194. In general, to solve the equation $ax^2 + bx + c = 0$ using your calculator, enter the following:

[b] [x²] [−] 4 [×] [a] [×] [c] [=] [√x] [STO] [b] [+/−] [+] [RCL] [=]
[÷] 2 [÷] [a] [=] [b] [+/−] [−] [RCL] [=] [÷] 2 [÷] [a] [=]

Note that if $b^2 − 4ac < 0$, the calculator will give you an error message when you press [√x]. In such cases, you will have to change the sign before pressing [√x] and supply the *i* in the final answer. (Try this in Example 6.)

MASTERY TEST — If you know how to do these problems, you have learned your lesson!

93. The number of bacteria *B* present in a laboratory culture after *t* minutes is given by $B = Ke^{0.05t}$. If the initial number of bacteria is 1000, how long would it take for there to be 20,000 bacteria present?

94. After a bactericide is introduced, the number of bacteria present in a laboratory culture after *t* minutes is given by $B = Ke^{-0.02t}$. If the initial number of bacteria is 50,000, how long would it be before this number is reduced to 10,000?

Exercises

Usually graded regarding increased difficulty. Note that each subsection is correlated with the *Exercises;* that is, the material covered by objective **A** is in subsection **A**, and the exercises corresponding to that objective appear in subsection **A** of the *Exercises.*

Applications

Included whenever possible in the *Exercises*. These problems are based on real data or real situations.

Skill Checker

Correlated with the *To Succeed, Review How to* items appearing in the following section. You will thus actually be practicing the skills you will need to master the next section.

Using Your Knowledge

Interesting applications and problems related to the material being studied. This feature will help you to generalize and apply the material you have just learned to real-life situations. Especially designed to answer that often-asked question, "Why do I have to learn this, and what good is it?"

Calculator Corner

Develops your ability to understand how and when technology can be utilized. Not all examples require a grapher for their solution; some will be better solved using a *scientific calculator*. The *Calculator Corner* feature provides essential background on how to solve problems requiring a scientific calculator.

Write On

Useful as a writing exercise or for class discussion. These brief questions provide an opportunity to think about and clarify ideas, concepts, and procedures. Especially helpful to those of you with satisfactory writing skills who may be more comfortable writing about, rather than doing, the algebra connected with a particular topic.

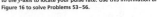

to the *y*-axis to locate your pulse rate. Use this information and Figure 16 to solve Problems 53–56.

53. What is the lower limit pulse rate for a 20-year-old person? Write the answer as an ordered pair (age, limit).

54. What is the upper limit pulse rate for a 20-year-old person? Write the answer as an ordered pair.

55. What is the upper limit pulse rate for a 45-year-old person? Write the answer as an ordered pair.

56. What is the lower limit pulse rate for a 50-year-old person? Write the answer as an ordered pair.

SKILL CHECKER

Find the quotients:

57. $\dfrac{3 - 6}{3 - 1}$ **58.** $\dfrac{6 - 3}{1 - 3}$

59. $\dfrac{3 - (-6)}{1 - 4}$ **60.** $\dfrac{4 - (-2)}{2 - 5}$

61. $\dfrac{4 - (-2)}{3 - (-6)}$ **62.** $\dfrac{2 - (-6)}{4 - (-8)}$

USING YOUR KNOWLEDGE — The Long, Hot Summer Exercise

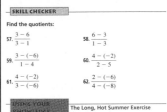

The ideas presented in this section are vital for understanding graphs. For example, do you exercise in the summer? To determine the risk of exercising in the heat, you must know how

---

Mastery Test

Tests every objective covered in each lesson in the *Mastery Test*. This test is useful in diagnosing any weaknesses in your knowledge so you can correct them before you go on to the next section.

Summary

Provides brief descriptions and examples for key topics in the chapter. Importantly, it also contains section references to encourage you to reread sections rather than memorizing definitions out of context.

SUMMARY

SECTION	ITEM	MEANING	EXAMPLE
4.1	Fraction	An expression denoting a division	$\frac{3}{4}, \frac{-2}{3},$ and $\frac{1}{x}$ are fractions.
	Rational expression	An expression of the form $\frac{P}{Q}$, where P and Q are polynomials, $Q \neq 0$	$\frac{1}{x}, \frac{x}{x+y}, \frac{x+y}{z},$ and $\frac{x^2 + 21x - 1}{x^3 + 3}$ are rational expressions.
4.1B	Fundamental property of fractions	If P, Q, and K are polynomials, $\frac{P}{Q} = \frac{P \cdot K}{Q \cdot K}$ for all values for which the denominator is not zero.	$\frac{3}{4} = \frac{3 \cdot 8}{4 \cdot 8}, \frac{x}{2} = \frac{x \cdot 5}{2 \cdot 5},$ and $\frac{3}{x+2} = \frac{3(x+5)}{(x+2)(x+5)}$
4.1C	Standard forms of a fraction	The forms $\frac{-a}{b}$ and $\frac{a}{-b}$ are the standard forms of a fraction.	$-\frac{5}{4}$ is written as $\frac{5}{-4}, -\frac{3}{4}$ is written as $\frac{-3}{4},$ and $-\frac{3}{4}$ is written as $\frac{-3}{4}.$
4.1D	Simplified (reduced) fraction	A fraction is simplified (reduced) if the numerator and denominator have no common factor.	$\frac{1}{4}, \frac{2}{3},$ and $\frac{4}{7}$ are reduced, but $\frac{3}{6}$ and $\frac{x+y}{x^2-y^2}$ are not.
4.2A	Multiplication of rational expressions	If a, b, c, and d are rational expressions $(b \neq 0, d \neq 0)$, $\frac{a}{b} \cdot \frac{c}{d} = \frac{a \cdot c}{b \cdot d}$	$\frac{3}{4} \cdot \frac{5}{7} = \frac{3 \cdot 5}{4 \cdot 7} = \frac{15}{28}$ $\frac{x}{x+y} \cdot \frac{3}{x-y} = \frac{3x}{(x+y)(x-y)} = \frac{3x}{x^2-y^2}$

Research Questions

Presents an additional opportunity to explore how various mathematical topics were developed and to reinforce the message given in the *Human Side of Algebra*—that the body of mathematical knowledge has evolved through human thought, experience, and communications. Combined with the *Write On,* this section adds a strong, interesting writing component to the course.

Review Questions

Can be used to prepare for the chapter test. Each of the questions has several parts for maximum practice. In addition, each question indicates the location of the material being reviewed.

Practice Test

Designed to help you check comprehension of the topics under discussion. A diagnostic component provides the answers, as well as the example(s), section, and page corresponding to the question under consideration.

26. [7.5B] If the temperature of a gas is held constant, the pressure P varies inversely as the volume V.
  a. Write an equation of variation using k for the constant of variation.
  b. A pressure of 1600 lb/in.^{2} is exerted by 2 ft^3 of air in a cylinder fitted with a piston. Find k.

27. [7.5B] The force F with which the Earth attracts an object above the Earth's surface varies inversely as the square of the distance d from the center of the Earth.
  a. Write an equation of variation using k for the constant of variation.
  b. An astronaut weighs 120 lb on the Earth's surface, and the radius of Earth is about 4000 mi. Find the value of k.

28. [7.5B] What would the astronaut of Problem 27 weigh if she is on a [space voyage] 1000 mi above the Earth's surface?

29. [7.5C] The horsepower h that a rotating shaft can safely transmit varies jointly as the cube of its diameter d and the number of revolutions r it makes per minute.
  a. Write an equation of variation using k for the constant of variation.
  b. A 2-in. shaft at a speed of 1000 rev/min can safely transmit 400 hp (horsepower). Find k for this shaft.

SUPPLEMENTS

Our supplements package includes materials for both *students* and *instructors.*

Annotated Instructor's Edition An *Annotated Instructor's Edition* containing the answer to every exercise in the text with the exception of the *Write On* exercises, the *Calculator Corners,* and the *Research Questions.*

Instructor's Test Manual The *Instructor's Test Manual* includes eight forms of the chapter tests for each chapter: four short-answer (free-response) and four multiple-choice versions as well as possible questions for a final examination. An extensive set of exercises for each textbook objective that can be used as an additional source of questions for practice, quizzes, tests, or review of difficult topics is provided as well.

WESTEST 4.01 for DOS *Westest 4.01 for DOS* has been completely rewritten, giving it a new look and feel and a simplified, streamlined interface based on the F1–F10 function keys. Instructors will be able to quickly add questions to an exam, arrange and rearrange the questions, and print the exam or send the file to a word processor such as Word-Perfect or Word for Windows.

WESTEST 3.2 for Windows	*Westest 3.2* prints directly to a word processing format for easy editing and printing. The package includes the user's manual, the *Westest* program disks, and the chapter disks containing questions from the text's test bank for easy selection and printing of tests. Questions can be edited or new ones added to better reflect the instructor's own teaching style.
Instructor's Solutions Manual	Detailed, worked-out solutions to all even-numbered exercises in the book including the *Write On* and the *Calculator Corner* exercises.
Student's Solutions Manual	Solutions to every odd-numbered exercise including the *Write On* and the *Calculator Corner.* Answers to all *Review Exercises* and all *Practice Test* questions are also included.
Activities Manual	Classroom activities that will enable students of *Intermediate Algebra* to use collaborative and hands-on learning experiences in the classroom.
Interactive Algebra Tutorial Software: Mac or Windows	A graphically driven tutorial software covering the major topics in elementary and intermediate algebra, as well as basic mathematics for those students who may need more review. An extremely intuitive program with carefully graded examples and exercises. The fifth section of each topic is designed as a motivational game. The management system provides a performance report to the student upon completion of each unit. The chapter and page of additional material is given whenever further study is recommended.
Videotapes	Separated into lessons for each section in the book, the videos cover all objectives, topics, and problem-solving techniques given in the text. Examples in the video are keyed to examples in the *Practice Test* at the end of each chapter.

OTHER BOOKS IN THIS SERIES

Other books in this series include *Intermediate Algebra: A Graphing Approach* and *Elementary Algebra.*

ACKNOWLEDGMENTS

The author would like to express his appreciation to the following persons:

Reviewers
Jeanne Baird, *Hillsborough Community College*
Carole Bauer, *Triton College*
Chuck Beals, *Hartnell College*
Barbara Burrows, *Santa Fe Community College*
Douglas Cameron, *The University of Akron*
Tom Carnevale, *Shawnee State University*
Robert Dodge, *Jackson Community College*
Arthur Dull, *Diablo Valley College*
Patricia Foard, *South Plains College*
Marie Franzosa, *Oregon State University*
Robert Hafer, *Brevard Community College*
Cheri Halcrow, *University of North Dakota*
Kay Haralson, *Austin Peay University*
Linda Harper, *Harrisburg Area Community College*
Karen Hay, *Mesa Community College*
Charles Hogue, *Pasadena City College*

Robert Horvath, *El Camino College*
Judy Jones, *Madison Area Technical College*
Mary Ann Justinger, *Erie Community College-South Campus*
Marie Kelly, *Kings River Community College*
George Kosan, *Hillsborough Community College*
Calvin Lathan, *Monroe Community College*
Wanda Long, *St. Charles County Community College*
James Magliano, *Union County College*
Sandra Orr, *West Virginia State University*
Joanne Peeples, *El Paso Community College*
Joseph Phillips, *Warren County Community College*
Larry Pontaski, *Pueblo Community College*
Janice Rech, *University of Nebraska at Omaha*
Amy Relyea, *Cuyahoga Community College-Eastern Campus*
Debbie Ritchie, *Moorepark College*
Donald Rose II, *McCloud Hospital*
Sharon Schwenk, *Columbia Basin College*
Richard Semmler, *Northern Virginia Community College*
Lynn Smith, *Gloucester County College*
Patricia Stanley, *Ball State University*
Ara Sullenberger, *Tarrant County Community College*
Lee Topham, *Kingwood College*
Cora West, *Florida Community College-Kent Campus*
Gail Wiltse, *St. John Community College*
Thomas Witten, *Southwest Virginia Community College*
Brenda Wood, *Florida Community College*

Special thanks go to Liana Fox, who offered many teaching tips and techniques for teaching and presenting Intermediate Algebra; to Jolene Rhodes, Judy Jones, and Debbie Garrison from Valencia Community College, who also offered many suggestions and finally came up with a set of activities to accompany the book; to Josephine Rinaldo, who is responsible for checking the material and working on many ancillaries and to Fran Hopf for her annotations.

Of course, this book would not have been possible without the wisdom of Peter Marshall and the help of his staff, Angie Vroom, Becky Stovall, and Angela Mussei, as well as the superb production staff he provided under the able leadership of Sharon Kavanagh, with the wisdom to assign an excellent staff to help me out. Luana Richards did a superb job of editing the book, Kathy Townes of TECHarts kept me informed of the designing process and provided Quark challenges to be solved, George Morris with Scientific Illustrators showed me that he cared about students, Brian Morris endured waiting for the equations while I looked for them, G & S composed the book swiftly and never complained about the many changes, Amy Mayfield did a fantastic job proofreading the book wondering all the time if we are really having fun and the great detectives in Ms. Kavanagh's office, Lisa Waalen, Mary Cybyske, and Gail Halpenny were always able to find her when I called. To all, my heartfelt thanks. Finally, thanks go to Professor Jack Britton, gentleman, scholar, and bon vivant, who has been trying to retire and make our books more readable for the last several years.

Intermediate Algebra

1

The Real Numbers

Why do we need algebra? Because we need to solve problems! If you have studied introductory algebra, this chapter is a review. If you haven't, you will acquire the necessary background for solving the equations and problems of Chapter 2. Algebra may be a new language to you, so we start by discussing the symbols, operations, and rules of this language. First we study the terminology, different sets of numbers, and their properties. We then learn about the operations we can perform with these numbers and the order in which these operations should be done.

The development of the number system used in algebra has been a multicultural undertaking. More than 20,000 years ago, our ancestors needed to count their possessions, their livestock, and the passage of days. Australian aborigines counted to two, South American Indians near the Amazon counted to six and the Bushmen of South Africa were able to count to ten ($10 = 2 + 2 + 2 + 2 + 2$).

The earliest technique for visibly expressing a number is tallying (from the French verb *tallier*, "to cut"). Tallying, a practice that reached its highest level of development in the British Exchequer tallies, used flat pieces of hazelwood about 6–9 inches long and about an inch thick, with notches of varying sizes and types. When a loan was made, the appropriate notches were cut and the stick split into two pieces, one for the debtor, one for the Exchequer. In this manner, transactions could easily be verified by fitting the two halves together and noticing whether the notches coincided, hence the expression "our accounts tallied."

The development of written numbers is due mainly to the Egyptians (about 3000 B.C.), the Babylonians (about 2000 B.C.), the early Greeks (about 400 B.C.), the Hindus (about 250 B.C.), and the Arabs (about 200 B.C). Here are the numbers these civilizations used:

Egyptian, about 3000 B.C.

❘	∩	⌒	𝔰	⧵	⌔	𝔜
1	10	100	1000	10,000	100,000	1,000,000

Babylonian, about 2000 B.C.

ξ	▼	⟨	⟨▮▮	⟨⟨	▼	▶━
0	1	10	12	20	60	600

Early Greek, about 400 B.C.

❘	Γ	△	Γ̄	Η	Γ̄	Γ̄
1	5	10	50	100	500	5000

1.1

NUMBERS AND THEIR PROPERTIES

To succeed, review how to:

1. Write the fraction $\frac{a}{b}$ as a decimal by dividing a by b.

2. Distinguish between a positive and a negative number.

NOTE: The *To succeed* section tells you what you need to know or review *before* you go on.

Objectives:

A Write a set of numbers using roster or set-builder notation.

B Write a rational number as a decimal.

C Classify a number as natural, whole, integer, rational, irrational, or real.

D Find the additive inverse of a number.

E Find the absolute value of a number.

F Given two numbers, use the correct notation to indicate equality or which is larger.

getting started

Algebra and Gender Wages

Look at the graph. What is the difference in wages between men and women with a high school diploma? The answer is $26,766 − $18,648, or $8118.

In arithmetic we are accustomed to expressions like

$$26 + 18, \quad 8 \times 32, \quad 26 - 18, \quad \text{and} \quad \frac{40}{5}$$

In algebra we use expressions like

$$26 + x, \quad 2\pi r, \quad 26 - w, \quad \text{and} \quad \frac{d}{t}$$

The letters x, r, w, d, and t are **variables** that stand for different numbers. When a letter stands for just *one* number, it's called a **constant**. For example, if in the graph a were to represent the median wage women with bachelor's degrees earned in 1992, a would be a *constant*. However, wages change from year to year, so in some other algebra problem w could be the **variable** representing women's wages for a given year.

What kind of numbers can we use in place of x, r, w, d, and t? In algebra we start with **real numbers** and then study **complex numbers**. We begin this section by discussing different sets of numbers.

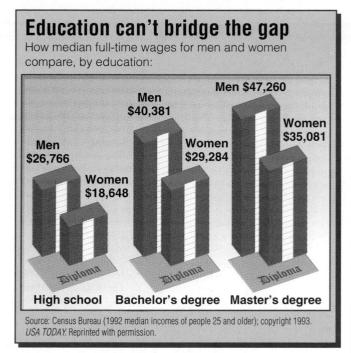

Education can't bridge the gap

How median full-time wages for men and women compare, by education:

Men $26,766
Women $18,648
Men $40,381
Women $29,284
Men $47,260
Women $35,081

High school   **Bachelor's degree**   **Master's degree**

Source: Census Bureau (1992 median incomes of people 25 and older); copyright 1993. *USA TODAY.* Reprinted with permission.

The idea of a **set** is familiar in everyday life. You may own a set of dishes, a tool set, or a set of books. In algebra we use sets of numbers.

A   Sets of Numbers

We use capital letters to denote sets and lowercase letters to denote elements (or members) of these sets. If possible, we "list" the elements of a set in braces { } and separate them by commas. Thus $A = \{1, 2, 3\}$ is the set A that has 1, 2, and 3 for its elements. This notation is known as **roster notation**. If a set has no elements, it is called the **empty**, or **null**, set and is denoted by { } or by the symbol $\varnothing$ (used without braces and read "the empty or null set").

> **NOTE** Note that { } is empty but the set $\{\varnothing\}$ is not the empty set because $\{\varnothing\}$ has one element, $\varnothing$!

Here are three sets of numbers frequently used in algebra.

NATURAL NUMBERS

The set of numbers used for counting.

$$N = \{1, 2, 3, \ldots\}$$

WHOLE NUMBERS

> The set of natural numbers and zero.
>
> $$W = \{0, 1, 2, 3, \ldots\}$$

INTEGERS

> The set of whole numbers and their opposites (negatives).
>
> $$I = \{\ldots, -2, -1, 0, 1, 2, \ldots\}$$
>
> The dots (called ellipsis points) mean that the pattern continues in the indicated direction.

The Greek letter $\in$ (epsilon) is used to indicate that an element *belongs* to a set. Thus $-5 \in I$ indicates that -5 is in the set *I;* that is, -5 *is an integer.* On the other hand, $0 \notin N$ indicates that 0 *is not a natural number.*

EXAMPLE 1   Roster notation

Use roster notation to write the following sets:
a. The natural numbers between 2 and 7.
b. The first three whole numbers.
c. The first two negative integers.
d. The only number that is neither positive nor negative.

SOLUTION
a. The set of natural numbers between 2 and 7 is $\{3, 4, 5, 6\}$. (Note that 2 and 7 are not included.)
b. The set of the first three whole numbers is $\{0, 1, 2\}$.
c. The set of the first two negative integers is $\{-1, -2\}$.
d. The set containing the only number that is neither positive nor negative is $\{0\}$. ∎

If a number can be written in the form $\frac{a}{b}$, where a and b are integers and b is not 0 ($b \neq 0$), the number is called a *rational number* (because it is a *ratio* of two integers). Thus

$$\frac{1}{5}, \qquad \frac{-4}{3}, \qquad \frac{4}{1} = 4, \qquad \text{and} \qquad \frac{-7}{1} = -7$$

are rational numbers. There is no obvious pattern with which we can list all the rational numbers, so we use a new notation, called **set-builder notation**, to define this set.

RATIONAL NUMBERS

> The set Q of **rational numbers** consists of all the numbers that can be written as the ratio of two integers. Thus
>
> $$Q = \left\{ r \,\middle|\, r = \frac{a}{b}, a \text{ and } b \text{ integers}, b \neq 0 \right\}$$
>
> This is read "Q equals the set of all r such that r equals a divided by b, a and b integers and b not equal to 0." Note that the symbol $|$ is read as "such that."

Using set-builder notation, the sets N, W, and I can be written as

$$N = \{x \mid x \text{ is a counting number}\}$$
$$W = \{x \mid x \text{ is a whole number}\}$$
$$I = \{x \mid x \text{ is an integer}\}$$

B Writing Rational Numbers as Decimals

The rational number $\frac{a}{b}$ can also be written as a decimal by dividing a by b. The result is either a **terminating decimal** (as in $\frac{1}{2} = 0.5$) or a **nonterminating, repeating decimal** (as in $\frac{1}{3} = 0.333 \ldots$). We often place a bar over the repeating digits in a nonterminating repeating decimal. Thus $\frac{1}{3} = 0.\overline{3}$, and $\frac{2}{11} = 0.181818 \ldots = 0.\overline{18}$.

EXAMPLE 2   Writing fractions as decimals

Write as a decimal:

a. $\dfrac{4}{5}$ **b.** $\dfrac{3}{11}$ **c.** $\dfrac{95}{30}$

SOLUTION

a. Dividing 4 by 5, we obtain $\frac{4}{5} = 0.8$, a terminating decimal.

b. Dividing 3 by 11, we have

$$\frac{3}{11} = 0.272727 \ldots = 0.\overline{27}$$

a **nonterminating**, **repeating decimal**.

c. Dividing 95 by 30, we have

$$\frac{95}{30} = 3.1666 \ldots = 3.1\overline{6}.$$ ∎

Since any rational number of the form $\frac{a}{b}$ ($b \neq 0$) is either a terminating or a repeating decimal, the set Q of rational numbers can also be defined as follows.

ALTERNATIVE DEFINITION FOR THE SET Q

$Q = \{x \mid x \text{ is a terminating or a repeating decimal}\}$

There are some numbers such as $\sqrt{2}$ (the square root of 2), π, and $\sqrt{10}$ that are *not* rational numbers. These are called *irrational* numbers because they cannot be written as the ratio of two integers. Note that when written as decimals, irrational numbers are nonterminating and nonrepeating. For example, $0.101001000 \ldots$ and $3.1234567 \ldots$ are irrational. Here is the definition of irrational numbers.

IRRATIONAL NUMBERS

Irrational numbers are numbers that *cannot* be written as ratios of two integers. The set of irrational numbers is

$$H = \{x \mid x \text{ is a number that is not rational}\}$$

The **real numbers** include both the rational numbers and the irrational numbers.

REAL NUMBERS

Numbers that are either rational or irrational are called **real numbers**. The set of real numbers R is defined by

$$R = \{x \mid x \text{ is a number that is rational or irrational}\}$$

Classifying Numbers

Here are some real numbers:

$$5, 17, -4, -9, 0, \frac{3}{5}, 0.6, \frac{1}{-10}, -0.\overline{1}, \frac{4}{-3}, \sqrt{3}, \pi, 0.345\ldots$$

EXAMPLE 3   Classifying numbers

Classify the given number by making a check mark (✓) in the appropriate row:

Set	0	$-\frac{4}{5}$	-4	$\sqrt{2}$	7	$-\pi$	$0.\overline{8}$	$0.01001000\ldots$
Natural number					✓			
Whole number	✓				✓			
Integer	✓		✓		✓			
Rational number	✓	✓	✓		✓		✓	
Irrational number				✓		✓		✓
Real number	✓	✓	✓	✓	✓	✓	✓	✓

Do you have a good idea of the relationship between the sets of numbers we have discussed? We can clarify the situation by using the idea of a *subset*. We say that A is a **subset** of B, denoted by $A \subseteq B$, when all the elements in A are also in B. Thus since all natural numbers N are whole numbers, $N \subseteq W$ (read "N is a subset of W"). Also, since all whole numbers are integers, $W \subseteq I$. Here is the complete picture:

$$N \subseteq W \subseteq I \subseteq Q \subseteq R$$

The diagram in Figure 1 shows the sets involved.

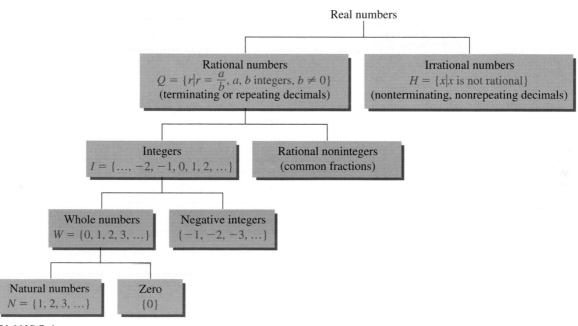

FIGURE 1

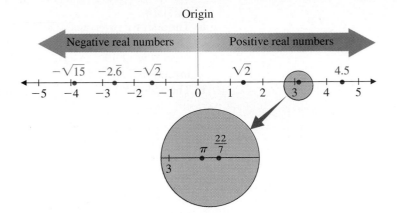

FIGURE 2

To make a picture or graph of some real numbers, we can think of a number line that is completely filled with all the real numbers. In this line the *positive* real numbers are to the *right* of zero, the *negative* real numbers are to the *left* of zero and zero itself, in the center of the number line, is called the *origin,* as shown in Figure 2.

The number corresponding to a point on the number line is the **coordinate** of the point, and the point is called the **graph** of the number. In Figure 2, some irrational numbers are shown above the number line, and some rational numbers are shown below the number line. Note that there is exactly one real number for each point graphed on the number line and one point for each real number.

D Additive Inverses (Opposites)

Each point on the number line has another point *opposite* it with respect to zero. The numbers corresponding to these two points are called the *additive inverses (opposites)* of each other. Thus 3 and -3 (read "the additive inverse of 3," or "negative 3") are additive inverses, as are -4 and 4. (See Figure 3.) Note that a number and its additive inverse are always the same distance from zero.

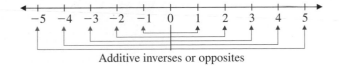

FIGURE 3

Here is the definition.

ADDITIVE INVERSE

> The **additive inverse (opposite)** of a is $-a$.

Note that the sum of a number and its additive inverse is always zero—that is, $a + (-a) = -a + a = 0$.

EXAMPLE 4   Finding additive inverses

Find the additive inverse:

a. 5   **b.** -3.5 **c.** $\dfrac{2}{3}$

SOLUTION

a. The additive inverse of 5 is -5. (See Figure 4.)

b. The additive inverse of -3.5 is 3.5. (See Figure 4.) In symbols, $-(-3.5) = 3.5$ (read "the additive inverse of negative 3.5 is 3.5," or "the additive inverse of the inverse of 3.5 is 3.5").

c. The additive inverse of $\frac{2}{3}$ is $-\frac{2}{3}$. (See Figure 4.)

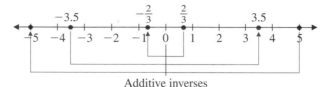

FIGURE 4   Additive inverses   ■

E **Absolute Values**

Now, let's go back to the number line. What is the distance between 3 and 0? The answer is 3 units. What about the distance between -3 and 0? The answer is *still* 3 units. The distance between any number a and 0 is called the *absolute value* of a and is denoted by $|a|$. Thus $|-3| = 3$ and $|3| = 3$. In general, we have the following definition.

ABSOLUTE VALUE OF A REAL NUMBER

The **absolute value** of a real number a, denoted by $|a|$, is defined as the distance between a and 0 on the real-number line. In general,

$$|a| = \begin{cases} a & \text{if } a \text{ is positive} \\ 0 & \text{if } a \text{ is zero} \\ -a & \text{if } a \text{ is negative} \end{cases}$$

$|11| = 11$ and $\left|\frac{1}{2}\right| = \frac{1}{2}$

$|0| = 0$

$|-5| = -(-5) = 5$ and $\left|-\frac{1}{3}\right| = -\left(-\frac{1}{3}\right) = \frac{1}{3}$

 CAUTION Since the absolute value of a number represents a distance and a distance is *never* negative, the absolute value of a number is *never* negative. It is always positive or zero. However, if a is not 0, $-|a|$ is *always* negative. Thus

$$-|-3| = -3, \quad -|4.2| = -4.2, \quad \text{and} \quad -|0.\overline{3}| = -0.\overline{3}$$

EXAMPLE 5   Finding absolute values

Find:

a. $|-8|$ b. $\left|\frac{1}{7}\right|$ c. $|0|$ d. $|4.2|$ e. $|0.\overline{3}|$ f. $-\left|-\frac{5}{9}\right|$

SOLUTION

a. $|-8| = 8$ -8 is 8 units from 0.   Note that $|-8| = -(-8) = 8$.

b. $\left|\frac{1}{7}\right| = \frac{1}{7}$ $\frac{1}{7}$ is $\frac{1}{7}$ units from 0.

c. $|0| = 0$ 0 is 0 units from 0.

d. $|4.2| = 4.2$ 4.2 is 4.2 units from 0.

e. $|0.\overline{3}| = 0.\overline{3}$ $0.\overline{3}$ is $0.\overline{3}$ units from 0.

f. $-\left|-\dfrac{5}{9}\right| = -\dfrac{5}{9}$ $-\dfrac{5}{9}$ is $\dfrac{5}{9}$ units from 0.   ∎

F Equality and Inequality

TRICHOTOMY LAW

If you are given any two real numbers a and b, only one of three things can be true:

1. a is **equal to** b, denoted by $a = b$, or

2. a is **less than** b, denoted by $a < b$, or

3. a is **greater than** b, denoted by $a > b$.

On the number line, numbers are shown in order; they *increase* as you move right and *decrease* as you move left. Thus

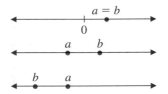

1. $a = b$ means that the graphs of a and b coincide.

2. $a < b$ means that a is to the left of b on the number line.

3. $a > b$ means that a is to the right of b on the number line.

The symbols $<$ and $>$ are called **inequality** signs and statements such as $a > b$ or $b < a$ are called **inequalities**. For example, $3 < 4$ (and $4 > 3$) because 3 is to the left of 4 on the number line, and $-3 < -2$ (and $-2 > -3$) because -3 is to the left of -2. Similarly, $3.14 > 3.13$ (and $3.13 < 3.14$) because 3.14 is to the right of 3.13 on the number line.

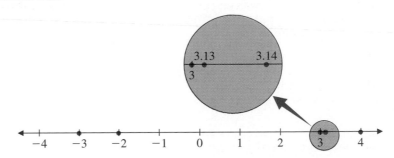

EXAMPLE 6   Determining relationships between numbers

Fill in the blank with $<$, $>$, or $=$ to make the resulting statement true:

a. $-3 \underline{\phantom{<}} -2$ **b.** $-5 \underline{\phantom{<}} -6$ **c.** $\dfrac{1}{2} \underline{\phantom{<}} 0$

d. $\dfrac{1}{5} \underline{\phantom{<}} \dfrac{1}{3}$ **e.** $-2.3 \underline{<} -2.2$ **f.** $\dfrac{1}{3} \underline{\phantom{<}} 0.\overline{3}$

SOLUTION

a.

number line from -4 to 4 with points at -3 and -2

$-3 < -2$

b.
number line from -8 to 0 with points at -6 and -5

$-5 > -6$

c.
number line from -2 to 2 with point at $\frac{1}{2}$

$\frac{1}{2} > 0$

d.
number line from -1 to 1 with points at $\frac{1}{5}$ and $\frac{1}{3}$

$\frac{1}{5} < \frac{1}{3}$ Note that $\frac{1}{5} = 0.2$ and $\frac{1}{3} = 0.333\ldots$. Thus $\frac{1}{5} < \frac{1}{3}$.

e.
-2.3 -2.2
number line from -3 to -1 with points at -2.3 and -2.2

$-2.3 < -2.2$

f.
$0.\overline{3}$
number line from -1 to 1 with point at $\frac{1}{3}$

$\frac{1}{3} = 0.\overline{3}$ Dividing 1 by 3, we obtain $\frac{1}{3} = 0.333\ldots = 0.\overline{3}$. Thus $\frac{1}{3}$ and $0.\overline{3}$ correspond to the same point.   ∎

GRAPH IT

If you have a grapher (graphing calculator), you can use it to do the numerical calculations in this section, but be aware that calculator procedures vary. (When in doubt, read the manual!) Let us start at the Home Screen* (see window). To do Example 2, press 4 ÷ 5 ENTER. The result is shown in the next line as .8. (Note that the answer in Example 2 is given as 0.8, not just .8.)

Additive inverses are found in most graphers by using the (−) key. (The gray (−) key and the blue − key in TI calculators perform very different functions. The (−) key finds the **additive inverse** of a number, while the − key indicates **subtraction**, just as in algebra!) To do Example 4a, simply press (−) 5 ENTER, and the answer −5 appears on the next line. To find the additive inverse of −3.5, enter (−) (−) 3.5 ENTER to obtain the answer 3.5. This suggests a property we shall study later—that is, $-(-a) = a$. To study absolute values with your grapher, follow this tip: On some graphers the absolute value of a number a—that is, $|a|$— is entered as 2nd ABS a, but other graphers use different notations. Let's try Example 5a. Key in 2nd ABS (−) 8 ENTER. The answer 8 appears on the next line.

Your grapher can also help you compare decimals. Thus to do Example 6f, note that $\frac{1}{3}$ means 1 divided by 3, so key in 1 ÷ 3 ENTER. The next line indicates that the answer is .3333333333. This is close to the meaning of $0.\overline{3}$, but you have to *know* that $\frac{1}{3}$ is nonterminating: The grapher cannot show you that! It can only give you an *approximation* to ten decimal places, and then you have to decide that $\frac{1}{3} = 0.\overline{3}$.

---

*Key terms shown in color appear in the *Keystroke Guide* accompanying this book. Consult this guide to find the equivalent keystrokes for your grapher.

EXERCISE 1.1

A **In Problems 1–10, use roster notation to write the indicated set.**

1. The first two natural numbers

2. The first six natural numbers

3. The natural numbers between 4 and 8

4. The natural numbers between 7 and 10

5. The first three negative integers

6. The negative integers between -4 and 7

7. The whole numbers between -3 and 4

8. The integers less than 0

9. The integers greater than 0

10. The nonnegative integers

In Problems 11–16, write the set using set-builder notation.

11. $\{1, 2, 3\}$ 12. $\{6, 7, 8\}$

13. $\{-2, -1, 0, 1, 2\}$ 14. $\{-8, -7, -6, -5\}$

15. The set of even numbers between 19 and 78

16. The set of all multiples of 3 between 8 and 78

In Problems 17–26, classify each statement as true or false. (Recall that N, W, I, Q, H, and R are the sets of natural, whole, integer, rational, irrational, and real numbers, respectively.)

17. $0 \in N$ 18. $0 \in W$

19. $-0.3 \notin I$ 20. $-0.\overline{8} \in Q$

21. $\sqrt{3} \in H$ 22. $8.101001000\ldots \in Q$

23. $8.112233 \notin H$ 24. $\sqrt{7} \in H$

25. $\dfrac{3}{8} \in R$ 26. $-0 \in W$

[B] **In Problems 27–34, write the given number as a decimal.**

27. $\dfrac{2}{3}$ 28. $\dfrac{1}{6}$

29. $\dfrac{7}{8}$ 30. $\dfrac{5}{6}$

31. $\dfrac{5}{2}$ 32. $\dfrac{4}{3}$

33. $\dfrac{7}{6}$ 34. $\dfrac{9}{8}$

[C] **In Problems 35–44, classify the given numbers by placing a check mark in the appropriate row.**

Set	35. $\frac{-3}{8}$	36. 0	37. $\sqrt{8}$	38. $\frac{3}{7}$	39. 0.3	40. -9	41. 0.9	42. -3.4	43. 3.1416	44. $3.141618\ldots$
Natural numbers		✓								
Whole numbers		✓								
Integers		✓				✓				
Rational numbers	✓	✓		✓	✓	✓	✓			
Irrational numbers	—		✓							
Real numbers	✓	✓	✓	✓	✓	✓	—			

[D] **In Problems 45–60, find the additive inverse of the given number.**

45. 8 46. -9

47. -7 48. 6

49. $\dfrac{3}{4}$ 50. $-\dfrac{1}{4}$

51. $-\dfrac{1}{5}$ 52. $\dfrac{2}{5}$

53. 0.5 54. -0.6

55. $0.\overline{2}$ 56. $-0.\overline{3}$

57. $-1.\overline{36}$ 58. $2.\overline{38}$

59. π 60. $-\pi$

[E] **In Problems 61–74, find each value.**

61. $|10|$ 62. $|-11|$

63. $|-17|$ 64. $|18|$

65. $\left|\dfrac{3}{5}\right|$ 66. $\left|-\dfrac{5}{7}\right|$

67. $|0.\overline{5}|$ 68. $|-0.\overline{7}|$

69. $|-3.\overline{61}|$ 70. $|2.\overline{48}|$

71. $-|\sqrt{2}|$ 72. $-|-\sqrt{3}|$

73. $|-\pi|$ 74. $|\pi|$

[F] **In Problems 75–84, fill in the blanks with $<$ or $>$ to make the resulting statement true.**

75. -5 _____ 2 76. 3 _____ -4

77. -6 _____ -8 78. -7 _____ -5

79. $\dfrac{1}{2}$ ——— $\dfrac{1}{4}$

80. $\dfrac{1}{3}$ ——— $\dfrac{1}{2}$

81. $-\dfrac{3}{5}$ ——— $-\dfrac{1}{4}$

82. $-\dfrac{1}{3}$ ——— $-\dfrac{1}{4}$

83. -3.5 ——— -3.4

84. -3.2 ——— -3.1

USING YOUR KNOWLEDGE   Hamburgers, Lawyers, and Baseball

In this section we learned how to compare integers and decimals; that is, we learned that $0.33 > 0.32$ and that $\frac{1}{3} < \frac{1}{2}$. To compare 0.33 and $\frac{1}{3}$, we write $\frac{1}{3}$ as a decimal by dividing the numerator by the denominator, obtaining $0.333\ldots$. We then write 0.33 as $0.33\mathbf{0}$ (note the extra zero) and write both numbers in a column with the decimal points aligned.

$0.333\ldots$

0.330

$\llcorner\rightarrow 3 > 0$, so $0.333\ldots > 0.330$

Thus $0.333\ldots > 0.330$.

We can use this knowledge in solving problems.

85. A McDonald's hamburger weighs 100 grams (g) and contains 11 g of fat; that is, $\frac{11}{100}$ is fat. A Burger King hamburger is 0.11009 fat. Write $\frac{11}{100}$ as a decimal and determine which hamburger has the larger percentage of fat.

86. Lawyers have to know about fractions and decimals, too. In a court case called the *U.S. v. Forty Barrels and Twenty Kegs of Coca-Cola*, a chemical analysis indicated that $\frac{3}{7}$ of the Coca-Cola was water. A second analysis showed that 0.41 was water. Which of the two analyses indicated more water in the Coke?

87. In a recent year the New York Yankees won the Eastern Division Championship with 100 wins in 160 games. The Los Angeles Dodgers had a 0.584 win-loss average. Write $\frac{100}{160}$ as a decimal and determine which of the two teams had the better average.

Here are the heights of three of the tallest people in the world:

Name	Feet	Inches
(a) Sulaiman Ali Nashnush	8	$\frac{1}{25}$
(b) Gabriel Estavao Monjane	8	$\frac{3}{4}$
(c) Constantine	8	0.8

88. Which one is the tallest of the three?

89. Which one is the second tallest?

90. Which one is the shortest?

WRITE ON . . .

91. Consider the statement, "$5 \subseteq N$, where N is the set of natural numbers." Is this statement true or false? Explain.

92. Consider the statement, "$N \in W$, where W is the set of whole numbers." Is this statement true or false? Explain.

93. Write in your own words a definition for the set of rational numbers and the set of irrational numbers using the idea of a fraction.

94. Write in your own words a definition of the set of rational numbers and the set of irrational numbers using the idea of a decimal.

95. Explain why every integer is a rational number but not every rational number is an integer.

MASTERY TEST   If you know how to do these problems, you have learned your lesson!

Write as a decimal:

96. $\dfrac{5}{7}$

97. $\dfrac{1}{8}$

Find the additive inverse of:

98. 0

99. $-|-8|$

100. $-|-\sqrt{17}|$

101. $|-x|$ if x is -2

Write the following using set-builder notation:

102. $\{3, 6, 9, \ldots\}$

103. $\{3, 6, 10, 15, \ldots\}$

104. Classify the numbers in the following set as rational or irrational.

$$\left\{ \frac{22}{7}, 3.1415, \pi, 3, -5, 0.\overline{77}, 34.010010001\ldots \right\}$$

Fill in the blank with $<$ or $>$ so that the result is a true statement:

105. $\dfrac{1}{3}$ ——— $0.331332333334\ldots$

106. 3.1416 ——— $\dfrac{22}{7}$

1.2

OPERATIONS AND PROPERTIES OF REAL NUMBERS

To succeed, review how to:

1. Find the additive inverse of a number (pp. 9–10).

2. Find the absolute value of a number (pp. 10–11).

3. Recognize positive and negative numbers (p. 9).

Objectives:

A Add, subtract, multiply, and divide signed numbers.

B Identify uses of the properties of the real numbers.

---

getting started

Signed Numbers and Population Changes

Look at the graph. What is the U.S. population change per hour? To find the answer, we have to add the births (+460), the deaths (−250), and the new immigrants (+100). The result is

$$460 + (-250) + 100 =$$

$$560 + (-250) = \text{Adding 460 and 100.}$$

$$+310 \quad \text{Subtracting 250 from 560.}$$

This answer means that the U.S. population is increasing by 310 persons every hour! You can use other operations to find out what the population change is every day, every week, or every year.

Note that

1. We wrote the addition of 460 and −250 as 460 + (−250), instead of 460 + −250, which is confusing.

2. To add 560 and −250, we can subtract 250 from 560; that is, subtracting 250 from 560 is the same as adding 560 and −250.

U.S. population changes per hour

+310 Net result

+460 Births

New immigrants +100

−250 Deaths

Source: Population Reference Bureau, Inc. (1992 data); copyright 1993. *USA TODAY.* Reprinted with permission.

In this section we shall learn to perform additions, subtractions, multiplications, and divisions using real numbers, and we shall study the rules for these four fundamental operations.

In algebra we use the idea of absolute value to define the addition and subtraction of "signed" numbers.

A **Operations with Signed Numbers**

PROCEDURE

To Add Two Numbers with the Same Sign

Add their absolute values and give the sum the common sign (+ if both numbers are positive, − if both numbers are negative).

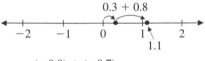

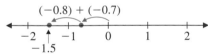

Thus

$0.3 + 0.8 = +(|0.3| + |0.8|) = 1.1$ The answer is positive, but we omit the + sign.

$(-0.7) + (-0.8) = -(|-0.7| + |-0.8|) = -1.5$

$$\frac{2}{3} + \frac{1}{6} = +\left(\left|\frac{2}{3}\right| + \left|\frac{1}{6}\right|\right) = +\left(\frac{2}{3} + \frac{1}{6}\right) = +\left(\frac{4}{6} + \frac{1}{6}\right) = \frac{5}{6}$$

We write $\frac{2}{3}$ as $\frac{4}{6}$ to add it to $\frac{1}{6}$.

$$-\frac{1}{2} + \left(-\frac{1}{4}\right) = -\left(\left|-\frac{1}{2}\right| + \left|-\frac{1}{4}\right|\right) = -\left(\frac{1}{2} + \frac{1}{4}\right) = -\left(\frac{2}{4} + \frac{1}{4}\right) = -\frac{3}{4}$$

And, we write $\frac{1}{2}$ as $\frac{2}{4}$ before adding.

PROCEDURE

> **To Add Two Numbers with Different Signs**
>
> **1.** Find the absolute value of the numbers.
>
> **2.** Subtract the number with the smaller absolute value from the one with the greater absolute value.
>
> **3.** Use the sign of the number with the greater absolute value for the result obtained in step 2.

Of course, if zero is added to any number a, the result is the number a. Thus $3 + 0 = 3$ and $0 + \frac{1}{3} = \frac{1}{3}$. The number 0 is called the **identity** for addition or simply the **additive identity**.

ADDITIVE IDENTITY

> For any real number a,
>
> $$a + 0 = a = 0 + a$$
>
> (Zero is the identity for addition.)

EXAMPLE 1   Adding signed numbers

Find:

a. $(-5) + (-11)$ **b.** $0.8 + (-0.5)$ **c.** $-0.7 + (0.4)$

d. $\frac{4}{5} + \left(-\frac{2}{5}\right)$ **e.** $-\frac{3}{8} + \left(\frac{1}{4}\right)$ **f.** $-0.2 + 0.6$

SOLUTION

a. $(-5) + (-11) = -(|-5| + |-11|) = -(5 + 11) = -16.$

b. $0.8 + (-0.5) = (|0.8| - |-0.5|) = 0.8 - 0.5 = 0.3$

c. $-0.7 + (0.4) = -(|-0.7| - |0.4|) = -(0.7 - 0.4) = -0.3$

d. $\frac{4}{5} + \left(-\frac{2}{5}\right) = \left(\left|\frac{4}{5}\right| - \left|-\frac{2}{5}\right|\right) = \left(\frac{4}{5} - \frac{2}{5}\right) = \frac{2}{5}$

e. $-\frac{3}{8} + \left(\frac{1}{4}\right) = -\frac{3}{8} + \frac{2}{8} = -\left(\left|-\frac{3}{8}\right| - \left|\frac{2}{8}\right|\right) = -\left(\frac{3}{8} - \frac{2}{8}\right) = -\frac{1}{8}$

(We write $\frac{1}{4}$ as $\frac{2}{8}$ first so we can compare it with $-\frac{3}{8}$ to determine the sign of the answer.)

f. $-0.2 + 0.6 = +(|0.6| - |0.2|) = +(0.6 - 0.2) = 0.4$ ■

Since we have developed a procedure for adding real numbers, it will be convenient to describe subtraction in terms of addition.

SUBTRACTION OF SIGNED NUMBERS

If a and b are real numbers, $a - b = a + (-b)$.

This means that we can *subtract* by adding the additive inverse. Thus

$$6 - (-3) = 6 + 3 = 9$$

$$-0.7 - 0.2 = -0.7 + (-0.2) = -0.9 \qquad \text{Subtract by adding the additive inverse.}$$

$$-\frac{1}{5} - \left(-\frac{4}{5}\right) = -\frac{1}{5} + \frac{4}{5} = \frac{3}{5}$$

EXAMPLE 2 Subtracting signed numbers

Find:

a. $-20 - 5$ **b.** $-0.6 - (-0.2)$ **c.** $-\frac{1}{7} - \left(-\frac{2}{7}\right)$

SOLUTION

a. $-20 - 5 = -20 + (-5) = -25$

b. $-0.6 - (-0.2) = -0.6 + 0.2 = -0.4$

c. $-\frac{1}{7} - \left(-\frac{2}{7}\right) = -\frac{1}{7} + \frac{2}{7} = \frac{1}{7}$ ■

In actual practice most of these operations are carried out mentally. Thus we write $6 + (-3) = 3$ and $-8 + 4 = -4$. Note that $-3 + 3 = 0$, $\frac{1}{2} + \left(-\frac{1}{2}\right) = 0$, and $-2.1 + 2.1 = 0$. Thus when **opposites (additive inverses)** are added, their sum is zero.

ADDITIVE INVERSES (OPPOSITES)

For any real number a, $a + (-a) = -a + a = 0$.

In arithmetic the product of a and b is written as $a \times b$. In algebra, however, the multiplication sign ($\times$) can be mistaken for the letter x; thus the product of a and b is written in one of the following ways.

PROCEDURE

How to Signify Multiplication	
Using a raised dot, ·	$a \cdot b$
Writing a and b next to each other	ab
Using parentheses	$(a)(b)$, $a(b)$, or $(a)b$

The numbers represented by a and b are called **factors**, and the result of the multiplication is called the **product** of a and b.

Now suppose you own four shares of stock and the price is *down* \$3, written as -3. Your loss that day would be the product $4 \cdot (-3)$, with factors 4 and -3. What is this product? This multiplication is just the repeated addition of -3.

$$4 \cdot (-3) = \underbrace{(-3) + (-3) + (-3) + (-3)}_{\text{4 negative threes}} = -12$$

Next, look at $(-3) \cdot 4$. In Section B, you will see that multiplication of real numbers has the commutative property. Thus

$$(-3) \cdot 4 = 4 \cdot (-3) = -12$$

We can generalize this idea to show that the product of any two numbers, one *positive* and the other *negative,* is a *negative number.* As in addition, we can state this result in terms of absolute values.

PROCEDURE

> **Multiplying Numbers with Opposite Signs**
>
> To multiply a *positive* number by a *negative* number, multiply their absolute values. Make the product *negative.*

Thus $3 \cdot (-2) = -6$, $-4 \cdot 8 = -32$, and $2 \cdot (-3.5) = -7$. Note that the factors in each case, 3 and -2, -4 and 8, and 2 and -3.5, have *different* (*unlike*) signs.

EXAMPLE 3   Multiplying signed numbers

Find:

a. $7 \cdot (-8)$ **b.** $-2.5 \cdot 4$

SOLUTION

a. $7 \cdot (-8) = -56$ **b.** $-2.5 \cdot 4 = -10$ ■

What about the product of two negative integers such as $-2 \cdot (-3)$? First, look for a pattern.

This number decreases by 1.   This number increases by 3.

$$2 \cdot (-3) = -6$$
$$1 \cdot (-3) = -3$$
$$0 \cdot (-3) = 0$$
$$-1 \cdot (-3) = 3$$
$$-2 \cdot (-3) = 6$$

Thus $-2 \cdot (-3) = 6$. Note that in this case -2 and -3 have the *same* sign $(-)$. When we multiply $2 \cdot 3$, 2 and 3 are both positive, so they also have the *same* sign $(+)$. We can summarize this discussion by the following rules.

SIGNS OF MULTIPLICATION PRODUCTS

WHEN MULTIPLYING TWO NUMBERS WITH	THE PRODUCT IS
Same (like) signs	Positive $(+)$
Different (unlike) signs	Negative $(-)$

Thus

$$(\overset{\frown}{-3}) \cdot (\overset{\frown}{-4.2}) = 12.6 \qquad \overset{\frown}{4 \cdot (-2.1)} = -8.4$$

Same signs	Positive answer	Different signs	Negative answer

Of course, multiplying by 1 leaves the number unchanged ($3 \cdot 1 = 3$ and $-7 \cdot 1 = -7$). Thus 1 is the identity for multiplication.

IDENTITY FOR MULTIPLICATION

> For any real number a,
>
> $$a \cdot 1 = 1 \cdot a = a.$$ (Note that 1 is the identity element for multiplication.)

What about fractions? To multiply fractions, we need the following definition.

MULTIPLICATION OF FRACTIONS

> $$\frac{a}{b} \cdot \frac{c}{d} = \frac{a \cdot c}{b \cdot d} \qquad (b, d \neq 0)$$

The same laws of signs apply. Thus,

$$\left(-\frac{9}{5}\right) \cdot \frac{3}{4} = -\frac{9 \cdot 3}{5 \cdot 4} = -\frac{27}{20}$$

Different signs	Negative answer

When multiplying fractions, it saves time if common factors are divided out before you multiply. Thus to multiply $\frac{5}{7} \cdot \left(-\frac{2}{5}\right)$, we write

$$\frac{\overset{1}{\cancel{5}}}{7} \cdot \left(-\frac{2}{\underset{1}{\cancel{5}}}\right) = -\frac{1 \cdot 2}{7 \cdot 1} = -\frac{2}{7} \qquad \text{Note that } \tfrac{5}{5} = 1.$$

EXAMPLE 4   Multiplying signed numbers

Find:

a. $\left(-\dfrac{3}{7}\right) \cdot \dfrac{7}{8}$ **b.** $\left(-\dfrac{5}{8}\right) \cdot \left(-\dfrac{4}{15}\right)$

SOLUTION

a. $\left(-\dfrac{3}{\underset{1}{\cancel{7}}}\right) \cdot \dfrac{\overset{1}{\cancel{7}}}{8} = -\dfrac{3 \cdot 1}{1 \cdot 8} = -\dfrac{3}{8}$ Different (unlike) signs; the answer is negative.

b. $\left(-\dfrac{\overset{1}{\cancel{5}}}{\underset{2}{\cancel{8}}}\right) \cdot \left(-\dfrac{\overset{1}{\cancel{4}}}{\underset{3}{\cancel{15}}}\right) = \dfrac{1 \cdot 1}{2 \cdot 3} = \dfrac{1}{6}$ Same (like) signs; the answer is positive. ∎

Just as we were able to define subtraction in terms of addition, we can also define division in terms of multiplication.

DIVISION OF REAL NUMBERS

If a and b are real numbers and b is not zero,

$$\frac{a}{b} = q \quad \text{means that} \quad a = b \cdot q$$

where a is called the **dividend**, b is the **divisor**, and q is the **quotient**.

Thus

$$\frac{48}{-6} = -8 \quad \text{means that} \quad 48 = -6 \cdot (-8)$$

and

$$\frac{-28}{-7} = 4 \quad \text{means that} \quad -28 = -7 \cdot (4)$$

When Dividing Two Numbers with	The Quotient Is
Same (like) signs	Positive (+)
Different (unlike) signs	Negative (−)

Because of this, the same rules of sign that apply to the multiplication of real numbers also apply to the division of real numbers; that is, the quotient of two numbers with the *same* sign is *positive,* and the quotient of two numbers with *different* signs is *negative,* as shown in the table. Here are some examples:

$$\frac{24}{6} = 4 \qquad \frac{3.2}{1.6} = 2$$

$$\frac{-18}{-9} = 2 \qquad \frac{-3.3}{-1.1} = 3$$

Same signs, positive answers

$$\frac{-32}{4} = -8 \qquad \frac{-6.3}{0.9} = -7$$

$$\frac{35}{-7} = -5 \qquad \frac{4.5}{-0.5} = -9$$

Different signs, negative answers

Note that since

$$\frac{-32}{4} = \frac{32}{-4} = -\frac{32}{4} = -8$$

the following holds true.

SIGNS OF A FRACTION

For any real number a and any nonzero real number b,

$$\frac{-a}{b} = \frac{a}{-b} = -\frac{a}{b}$$

Thus there are *three* signs associated with every fraction: the sign of the numerator, the sign of the denominator, and the sign of the fraction itself.

EXAMPLE 5   Dividing signed numbers

Find:

a. $48 \div 6$ **b.** $\dfrac{54}{-9}$ **c.** $\dfrac{-63}{-7}$

d. $-28 \div 4$ **e.** $5 \div 0$ **f.** $3.4 \div 1.7$

g. $\dfrac{4.8}{-1.2}$ **h.** $\dfrac{-5.6}{-0.8}$ **i.** $0 \div 3.5$

SOLUTION

a. $48 \div 6 = 8$ 48 and 6 have the same sign; the answer is positive.

b. $\dfrac{54}{-9} = -6$ 54 and -9 have different signs; the answer is negative.

c. $\dfrac{-63}{-7} = 9$ -63 and -7 have the same sign; the answer is positive.

d. $-28 \div 4 = 7$ -28 and 4 have different signs; the answer is negative.

e. $5 \div 0$ is not defined. Note that if you make $5 \div 0$ equal any number, say *a*, you will have

$\dfrac{5}{0} = a$,   which means $5 = a \cdot 0 = 0$ This says that $5 = 0$, which is impossible.

Thus $\frac{5}{0}$ is not defined.

f. $3.4 \div 1.7 = 2$ 3.4 and 1.7 have the same sign; the answer is positive.

g. $\dfrac{4.8}{-1.2} = -4$ 4.8 and -1.2 have different signs; the answer is negative.

h. $\dfrac{-5.6}{-0.8} = 7$ -5.6 and -0.8 have the same signs; the answer is positive.

i. In this case, $0 \div 3.5 = 0$. We can check this using the definition of division:
$0 \div 3.5 = \frac{0}{3.5} = 0$ means $0 = 3.5 \cdot 0$, which is true.   ■

Here are three rules to help you out.

ZERO IN DIVISION

For $a \neq 0$, $\dfrac{0}{a} = 0$ and $\dfrac{a}{0}$ is *not* defined. Moreover, $\dfrac{0}{0}$ is indeterminate.

⚠ **CAUTION** $\dfrac{0}{k}$ is okay but $\dfrac{n}{0}$ is a no-no!

Let's look at the division problem $2 \div 5$. We can write $2 \div 5 = \frac{2}{5} = 2 \cdot \frac{1}{5}$. Thus to *divide 2 by 5*, we *multiply 2 by $\frac{1}{5}$*. The numbers 5 and $\frac{1}{5}$ are *reciprocals* or *multiplicative inverses*. Here is the definition.

MULTIPLICATIVE INVERSE (RECIPROCAL)

Every nonzero real number a has a **reciprocal (multiplicative inverse)** $\frac{1}{a}$ such that

$$a \cdot \frac{1}{a} = 1$$

The reciprocal of 3 is $\frac{1}{3}$, the reciprocal of -6 is $\frac{1}{-6} = -\frac{1}{6}$, and the reciprocal of $\frac{2}{3}$ is $\frac{3}{2}$. Note that the reciprocal of a *positive* number is *positive* and the reciprocal of a *negative* number is *negative*.

EXAMPLE 6   Finding reciprocals

Find the reciprocal:

a. $\dfrac{2}{3}$ 　　　　　　　　　 **b.** $-\dfrac{4}{5}$ 　　　　　　　　　 **c.** 0.2

SOLUTION

a. The reciprocal of $\frac{2}{3}$ is $\frac{3}{2}$.

b. The reciprocal of $-\frac{4}{5}$ is $-\frac{5}{4}$. (Remember that the reciprocal of a negative number is negative.)

c. The reciprocal of 0.2 is

$$\frac{1}{0.2} = \frac{1}{\dfrac{2}{10}} = 1 \cdot \frac{10}{2} = 5. \qquad \blacksquare$$

Division of fractions is done in terms of reciprocals. Since

$$\frac{a}{b} = a \cdot \frac{1}{b}$$

to divide by a number (such as b) we multiply by its reciprocal $\frac{1}{b}$. Here is the general definition.

DIVISION OF FRACTIONS

$$\frac{a}{b} \div \frac{c}{d} = \frac{a}{b} \cdot \frac{d}{c} \qquad (b, c, \text{ and } d \text{ are not zero.})$$

Thus to divide by $\frac{c}{d}$, we multiply by the reciprocal $\frac{d}{c}$. For example, to divide $\frac{4}{5}$ by $\frac{2}{3}$, we multiply $\frac{4}{5}$ by the reciprocal of $\frac{2}{3}$, that is, by $\frac{3}{2}$. Thus

$$\frac{4}{5} \div \frac{2}{3} = \frac{4}{5} \cdot \frac{3}{2}$$

EXAMPLE 7   Dividing fractions

Find the following using reciprocals:

a. $\dfrac{2}{5} \div \left(-\dfrac{3}{4}\right)$ 　　　　 **b.** $\left(-\dfrac{5}{6}\right) \div \left(-\dfrac{7}{2}\right)$ 　　　　 **c.** $\left(-\dfrac{3}{7}\right) \div \dfrac{6}{7}$

SOLUTION

a. $\dfrac{2}{5} \div \left(-\dfrac{3}{4}\right) = \dfrac{2}{5} \cdot \left(-\dfrac{4}{3}\right) = -\dfrac{8}{15}$

b. $\left(-\dfrac{5}{6}\right) \div \left(-\dfrac{7}{2}\right) = \left(-\dfrac{5}{6}\right) \cdot \left(-\dfrac{2}{7}\right) = \dfrac{10}{42} = \dfrac{5}{21}$

Note that it is easier to "reduce" before multiplying like this

$$\left(-\dfrac{5}{\overset{}{\underset{3}{6}}}\right) \cdot \left(-\dfrac{\overset{1}{2}}{7}\right) = \dfrac{5}{21}$$

c. $\left(-\dfrac{3}{7}\right) \div \dfrac{6}{7} = \left(-\dfrac{3}{7}\right) \cdot \dfrac{7}{6} = -\dfrac{21}{42} = -\dfrac{1}{2}$

You can also "reduce" before multiplying by writing

$$\left(-\dfrac{\overset{1}{3}}{\underset{1}{7}}\right) \cdot \dfrac{\overset{1}{7}}{\underset{2}{6}} = -\dfrac{1}{2}$$ ∎

B Real-Number Properties

We have already mentioned two properties of the real numbers: **zero** (0) is the identity for addition, and **one** (1) is the identity for multiplication. We end this section by summarizing some additional real-number properties in Table 1 and then examining how they can help us in our work with algebra.

TABLE 1 Properties of Real Numbers (a, b, and c represent real numbers)

Property	Addition	Multiplication
Closure	$a + b$ is a real number. $-8 + \sqrt{5}$ is a real number.	$a \cdot b$ is a real number. $-8 \cdot \sqrt{5}$ is a real number.
Commutative	$a + b = b + a$ $\sqrt{2} + 8 = 8 + \sqrt{2}$	$ab = ba$ $3 \cdot \dfrac{1}{5} = \dfrac{1}{5} \cdot 3$
Associative	$a + (b + c) = (a + b) + c$ $\dfrac{1}{3} + \left(\dfrac{2}{5} + \dfrac{3}{4}\right) = \left(\dfrac{1}{3} + \dfrac{2}{5}\right) + \dfrac{3}{4}$	$a(bc) = (ab)c$ $-3 \cdot \left(\sqrt{5} \cdot \dfrac{1}{8}\right) = \left(-3 \cdot \sqrt{5}\right) \cdot \dfrac{1}{8}$
Identity	$a + 0 = 0 + a = a$ $-\dfrac{4}{5} + 0 = 0 + \left(-\dfrac{4}{5}\right) = -\dfrac{4}{5}$ (0 is called the additive identity.)	$1 \cdot a = a \cdot 1 = a$ $0.\overline{38} \cdot 1 = 1 \cdot 0.\overline{38} = 0.\overline{38}$ (1 is called the multiplicative identity.)
Inverse	$a + (-a) = (-a) + a = 0$ $-a$ is the additive inverse or opposite of a. $-\dfrac{1}{5}$ is the inverse (opposite) of $\dfrac{1}{5}$. Thus $\dfrac{1}{5} + \left(-\dfrac{1}{5}\right) = 0$.	$a \cdot \dfrac{1}{a} = \dfrac{1}{a} \cdot a = 1 \qquad (a \neq 0)$ $\dfrac{1}{a}$ is the multiplicative inverse (reciprocal) of a. 8 is the reciprocal of $\dfrac{1}{8}$.
Distributive	$a(b + c) = ab + ac \qquad 3(1 + 6) = 3 \cdot 1 + 3 \cdot 6$	$a(b - c) = ab - ac \qquad 4(1 - 6) = 4 \cdot 1 - 4 \cdot 6$

Note that the **commutative property** tells us that the *order* in which we add or multiply two numbers does not affect the result; thus for the sum

$$\begin{array}{r} 28 \\ + \ 39 \\ \hline \end{array}$$

you can add down (28 + 39) and then check by adding up (39 + 28). Similarly, it is easier to multiply

$$\begin{array}{r} 48 \\ \times \ 9 \\ \hline \end{array} \qquad \text{instead of} \qquad \begin{array}{r} 9 \\ \times \ 48 \\ \hline \end{array}$$

but by the commutative property, the result is the same. On the other hand, the **associative property** tells us that the *grouping* of the numbers does not affect the final answer. Thus if you wish to add 3 + 24 + 6 by adding 24 + 6 first—that is, by finding 3 + (24 + 6)—you will get the same answer as if you had added (3 + 24) + 6.

EXAMPLE 8   Properties of real numbers

Which property is illustrated in the following statements?

a. $(-3) + \dfrac{7}{5} = \dfrac{7}{5} + (-3)$

b. $3 \cdot (4 \cdot 7) = (4 \cdot 7) \cdot 3$

c. $(-4) \cdot \left(\dfrac{1}{2} \cdot \dfrac{1}{8} \right) = \left(-4 \cdot \dfrac{1}{2} \right) \cdot \dfrac{1}{8}$

d. $2 + (3 \cdot 4) = 2 + (4 \cdot 3)$

SOLUTION
a. Since we changed the *order,* the commutative property of addition applies.
b. Here again we changed the *order* (the 3 changed position). The commutative property of multiplication was used.
c. This time we *grouped* the numbers differently. The associative property of multiplication was used.
d. The commutative property of multiplication applies (the 3 and 4 changed position). ■

Note that the associative property can help you find the answer to a problem such as (3)(−5)(6). To use this property, you can write either of the following:

1. $(3)(-5)(6) = (-15)(6)$ Multiply (3)(−5) first and (−15)(6) next.
$\qquad\qquad\quad = -90$

2. $(3)(-5)(6) = (3)(-30)$ Multiply (−5)(6) first and (3)(−30) next.
$\qquad\qquad\quad = -90$

Of course, the answer is the same in both cases. We shall practice with this type of problem in the exercises.

We have not yet added or subtracted expressions like $2\sqrt{3} + 5\sqrt{3}$ or $8\sqrt{2} - 12\sqrt{2}$. We can do this using the commutative and distributive properties as follows:

$$2\sqrt{3} + 5\sqrt{3} = (\sqrt{3} \cdot 2) + (\sqrt{3} \cdot 5) \quad \text{By the commutative property of multiplication}$$

$$= \sqrt{3}(2 + 5) \qquad\qquad \text{By the distributive property}$$

$$= 7\sqrt{3} \qquad\qquad\quad \text{By the commutative property of multiplication}$$

In practice you can see that $2\sqrt{3}$ and $5\sqrt{3}$ are **like** terms that can be combined. Using this idea, $8\sqrt{2} - 12\sqrt{2} = -4\sqrt{2}$. We shall study the distributive property and how to use it to combine like terms in Section 1.4.

GRAPH IT

Your grapher can really help in this section. Examples 1, 2, and 3 are easy with a grapher, *except* when fractions are involved. (Some graphers have special keys to enter fractions). If in Example 1e you key in [(−)] 3 [÷] 8 [+] 1 [÷] 4 [ENTER], you get −.125. Is that the same as the $-\frac{1}{8}$ given in the text? Key in [(−)] 1 [÷] 8 [ENTER] and you also get −.125, the same answer! To change the −.125 to a fraction with a TI-82, press [MATH] 1 [ENTER] and the answer is given as $-\frac{1}{8}$ (Window 1). Can you do the same with Example 2c? For Example 7c, press [(] [(−)] 3 [÷] 7 [)] [÷] [(] 6 [÷] 7 [)] [ENTER]; this yields −.5, which is equivalent to $-\frac{1}{2}$ (Window 2). Check what happens if you try Example 5e! Even the grapher can't do that one!

```
-3/8+1/4
                    -.125
Ans▶Frac
                     -1/8
```
WINDOW 1

```
(-3/7)/(6/7)
                      -.5
```
WINDOW 2

EXERCISE 1.2

Ⓐ **In Problems 1–86, perform the indicated operations.**

1. $\dfrac{3}{5} + \left(-\dfrac{1}{5}\right)$

2. $-0.4 + 0.9$

3. $(-0.3) + 0.2$

4. $(-8) + 5$

5. $(-4) + 6$

6. $(-0.2) + 0.3$

7. $(-0.5) + (-0.3)$

8. $(-7) + (-11)$

9. $\left(-\dfrac{1}{5}\right) + \dfrac{1}{4} + \dfrac{3}{20}$

10. $\left(-\dfrac{4}{7}\right) + \dfrac{2}{9} + \dfrac{1}{63}$

11. $6 - 13$

12. $8 - 13$

13. $0.6 - 0.9$

14. $0.3 - 0.8$

15. $\dfrac{1}{7} - \dfrac{3}{8}$

16. $\dfrac{3}{8} - \dfrac{4}{8}$

17. $-8 - 4 - 2$

18. $-4 - 6 - 3$

19. $-0.4 - 0.2$

20. $-0.3 - 0.5$

21. $-\dfrac{3}{7} - \dfrac{2}{9}$

22. $-\dfrac{4}{9} - \dfrac{1}{9}$

23. $-6 - (-5)$

24. $-7 - (-9)$

25. $-8 - (-4)$

26. $-9 - (-2)$

27. $-0.7 - (-0.6)$

28. $-0.9 - (-0.3)$

29. $-\dfrac{2}{7} - \left(-\dfrac{4}{3}\right)$

30. $-\dfrac{3}{4} - \left(-\dfrac{5}{3}\right)$

31. $(-5)(8)$

32. $(-9)(6)$

33. $(4)(-3)$

34. $(6)(-8)$

35. $(-10)(-5)$

36. $(-6)(-9)$

37. $(-3)(4)(-5)$

38. $(-5)(2)(3)$

39. $(-4)(-2)(5)$

40. $(-2)(-5)(9)$

41. $(-3)(5)(-2)$

42. $(-3)(10)(-2)$

43. $(4)(-5)(2)$

44. $(10)(-3)(6)$

45. $-2.2(3.3)$

46. $-1.4(3.1)$

47. $-1.3(-2.2)$

48. $-1.5(-1.1)$

49. $\dfrac{5}{6}\left(-\dfrac{5}{7}\right)$

50. $\dfrac{3}{8}\left(-\dfrac{5}{7}\right)$

51. $-\dfrac{3}{5}\left(-\dfrac{5}{12}\right)$

52. $-\dfrac{4}{7}\left(-\dfrac{21}{8}\right)$

53. $-\dfrac{6}{7}\left(\dfrac{35}{8}\right)$

54. $-\dfrac{7}{5}\left(\dfrac{15}{28}\right)$

55. $\dfrac{-18}{9}$

56. $\dfrac{-32}{16}$

57. $\dfrac{20}{-5}$

58. $\dfrac{36}{-3}$

59. $\dfrac{-14}{-7}$

60. $\dfrac{-24}{-8}$

61. $\dfrac{0}{-3}$

62. $\dfrac{0}{-9}$

63. $\dfrac{4}{0}$　　**64.** $\dfrac{-7}{0}$

65. $-\left(\dfrac{-4}{-2}\right)$　　**66.** $-\left(\dfrac{-10}{-5}\right)$

67. $-\left(\dfrac{-27}{3}\right)$　　**68.** $-\left(\dfrac{-9}{3}\right)$

69. $-\left(\dfrac{15}{-5}\right)$　　**70.** $-\left(\dfrac{18}{-6}\right)$

71. $\dfrac{-3}{-3}$　　**72.** $\dfrac{-18}{-9}$

73. $\dfrac{-16}{4}$　　**74.** $\dfrac{-48}{6}$

75. $\dfrac{-56}{8}$　　**76.** $\dfrac{-54}{6}$

77. $\dfrac{3}{5} \div \left(-\dfrac{4}{7}\right)$　　**78.** $\dfrac{4}{9} \div \left(-\dfrac{1}{7}\right)$

79. $-\dfrac{2}{3} \div \left(-\dfrac{7}{6}\right)$　　**80.** $-\dfrac{5}{6} \div \left(-\dfrac{25}{18}\right)$

81. $-\dfrac{5}{8} \div \dfrac{7}{8}$　　**82.** $-\dfrac{4}{5} \div \dfrac{8}{15}$

83. $\dfrac{-3.1}{6.2}$　　**84.** $\dfrac{1.2}{-4.8}$

85. $\dfrac{-1.6}{-9.6}$　　**86.** $\dfrac{-9.8}{-1.4}$

B In Problems 87–94, indicate which property is illustrated in each statement (*a* and *b* represent real numbers).

87. $5.6 + 9.2 = 9.2 + 5.6$　　**88.** $-3 \cdot 4 = 4 \cdot (-3)$

89. $5 \cdot (-2) = -2 \cdot 5$

90. $\left(\dfrac{1}{5} + \dfrac{2}{7}\right) + \dfrac{1}{8} = \dfrac{1}{5} + \left(\dfrac{2}{7} + \dfrac{1}{8}\right)$

91. $(-3 \cdot a) \cdot 2 = 2 \cdot (-3 \cdot a)$　　**92.** $5(3a) = (3a)5$

93. $1 \cdot (3 + b) = 3 + b$　　**94.** $(a + b) \cdot 1 = a + b$

95. If the area of a rectangle is found by multiplying the length times the width, express the area of the rectangle in the figure in two ways to illustrate the distributive property for $a(b + c)$.

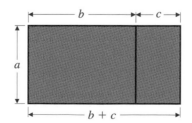

96. Express the shaded area of the rectangle in the figure in two ways to illustrate the distributive property for $a(b - c)$.

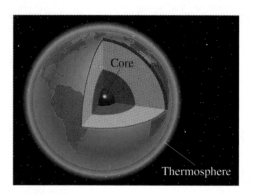

APPLICATIONS

97. The temperature in the center core of the Earth reaches $+5000°C$. In the thermosphere (a region in the upper atmosphere), the temperature is $+1500°C$. Find the difference in temperature between the center of the Earth and the thermosphere.

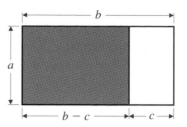

Core

Thermosphere

98. The record high temperature in Calgary, Alberta, is $+99°F$. The record low temperature is $-46°F$. Find the difference between these extremes.

99. The price of a certain stock at the beginning of the week was $47. Here are the changes in price during the week: $+1, +2, -1, -2, -1$. What is the price of the stock at the end of the week?

100. The price of a stock at the beginning of the week was $37. On Monday, the price went up $2; on Tuesday, it went down $3; and on Wednesday, it went down another $1. What was the price of the stock then?

Here are the temperature changes (in degrees Celsius, C) by the hour in a certain city:

　1 P.M., +2　　2 P.M., +1　　3 P.M., −1　　4 P.M., −3

101. If the temperature was initially $15°C$, what was it at 4 P.M.?

102. If the temperature was initially $-15°C$, what was it at 4 P.M.?

The *Skill Checker,* which appears periodically in the exercise sets, reviews skills previously studied and will help you maintain your mastery.

SKILL CHECKER

In Problems 103–106, fill in the blanks.

Number	Additive Inverse	Reciprocal
103. 7		
104. −2.8		
105. 0		
106. $-\dfrac{2}{3}$		

USING YOUR KNOWLEDGE

Would You Rather Be "Very Lovable" or "Extremely Good"? Use Signed Numbers to Decide.

Have you met anybody *nice* today? Have you had an *unpleasant* experience? Perhaps the person you met was *very nice* or your experience *very unpleasant.* Psychologists and linguists have a numerical way to indicate the difference between nice and very nice or between unpleasant and very unpleasant. Suppose you assign a positive number (+2, for example) to the adjective *nice,* and a negative number (say, −2) to *unpleasant,* and a positive number greater than 1 (say, +1.75) to *very.* Then, by definition, **"very nice"** means

very nice
↓   ↓
$(1.75) \cdot (2) = 3.50$

and "very unpleasant" means

very unpleasant
↓   ↓
$(1.75) \cdot (-2) = -3.50$

Here are some adverbs and adjectives and their average numerical values as rated by a panel of college students.

Adverbs		Adjectives	
Slightly	0.54	Wicked	−2.5
Rather	0.84	Disgusting	−2.1
Decidedly	0.16	Average	−0.8
Very	1.25	Good	3.1
Extremely	1.45	Lovable	2.4

Find the value of:

107. Slightly wicked

108. Decidedly average

109. Extremely disgusting

110. Rather lovable

111. Very good

By the way, if you got all the answers correct, you are 4.495!

WRITE ON . . .

112. Explain why division by zero is not defined.

113. The distributive property $a(b + c) = ab + ac$ is actually called the distributive property of multiplication over addition. Is addition distributive over multiplication? Explain and give examples to support your answer.

114. We mentioned that the set of real numbers is *closed* under addition and multiplication. Are the natural numbers closed under addition? What about under subtraction? Explain and give examples to support your answer.

MASTERY TEST

If you know how to do these problems, you have learned your lesson!

Perform the indicated operation:

115. $-\dfrac{2}{5} \div \left(-\dfrac{5}{8}\right)$

116. $-\dfrac{9}{5} - \left(-\dfrac{3}{4}\right)$

117. $-\left(\dfrac{3}{5}\right)\left(-\dfrac{10}{3}\right)$

118. $23.4 + (-29.7)$

119. $-\dfrac{5}{8} + \left(-\dfrac{3}{7}\right)$

120. $18.7 - (-13.2)$

121. $3.2(-4)$

122. $-3.4 \div 1.7$

123. $(-3 - 5)(-2) + 8(3 - 7 + 4)$

124. $-4(5 - 7) + (-3 - 5)(-2)$

Name the property used:

125. $(-8) + 8 = 0$

126. $(a + 3) = 1 \cdot (a + 3)$

127. $3(ab) = (3a)b$

128. $8 + (b + c) = (8 + b) + c$

129. $2 \cdot (b + 4) = (b + 4) \cdot 2$

130. $(a + 1) \cdot \dfrac{1}{a + 1} = 1, a \neq -1$

1.3

PROPERTIES OF EXPONENTS

To succeed, review how to:

Perform the four fundamental operations (addition, subtraction, multiplication, and division) using signed numbers (pp. 16–21).

Objectives:

A. Evaluate expressions containing natural numbers as exponents.

B. Write an expression containing negative exponents as a fraction.

C. Multiply and divide expressions containing exponents.

D. Raise a power to a power.

E. Raise a quotient to a power.

F. Convert between ordinary decimal notation and scientific notation and use scientific notation in computations.

getting started

Rubik's Cube, Areas, and Volumes

Have you ever seen a Rubik's cube? If the side of each of the squares in the cube is 1 centimeter (cm) long, can you find the **area** of the top of the cube? Can you find the **volume** of the whole cube? As you may recall, to find the area of a square, we multiply the length of two sides of the square. Since the top of the square is 3 cm long, the area of the top is

$$(3 \text{ cm}) \cdot (3 \text{ cm}) = 3^2 \text{ cm}^2 \quad \text{or} \quad 9 \text{ square centimeters}$$

Similarly, the volume of the cube is

$$(3 \text{ cm}) \cdot (3 \text{ cm}) \cdot (3 \text{ cm}) = 3^3 \text{ cm}^3 \quad \text{or} \quad 27 \text{ cubic centimeters}$$

Note that when calculating areas, we use **square** units, and when calculating volumes, we use **cubic** units.

The notation 3^2 means to multiply 3 by 3, and 3^3 means to multiply 3 by 3 by 3, that is,

$$3^2 = \overbrace{3 \times 3}^{2 \text{ times}} \quad \text{and} \quad 3^3 = \overbrace{3 \times 3 \times 3}^{3 \text{ times}}$$

This is **exponential** notation, one of the topics we shall study in this section.

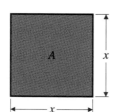

What is the area A of the square? The area is

$$A = x \cdot x = x^2 \qquad \text{Read "} x \text{ squared, or } x \text{ to the second power."}$$

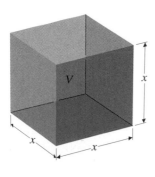

In the expression x^2, the *exponent* 2 indicates that the *base x* is to be used as a factor twice. What about the volume V of the cube? It is

$$V = x \cdot x \cdot x = x^3 \qquad \text{Read "} x \text{ cubed, or } x \text{ to the third power."}$$

This time the *exponent* 3 indicates that the *base x* is used as a factor three times. In general, we have the following definition.

EXPONENT AND BASE

If a is a real number and n is a natural number,

$$a^n = \underbrace{a \cdot a \cdot a \cdots a}_{n \text{ factors}}$$

where n is called the **exponent** and a is the **base**.

When $n = 1$, the exponent is usually omitted. Thus, $a^1 = a$, $b^1 = b$, and $c^1 = c$.

 Natural-Number Exponents

We can **evaluate** (find the value of) expressions containing natural-number exponents. Thus

$$4^2 = 4 \cdot 4 = 16$$

$$(-3)^2 = (-3)(-3) = 9$$

$$(-2)^3 = (-2)(-2)(-2)$$

$$= 4(-2) \qquad\qquad (-2)(-2) = 4$$

$$= -8$$

$$(-4)^4 = (-4)(-4)(-4)(-4)$$

$$= (16)(16)$$

$$= 256$$

Note that for the **even** exponents 2 and 4 in $(-3)^2 = 9$ and $(-4)^4 = 256$, respectively, we get the **positive** answers 9 and 256, but for the **odd** exponent 3 in $(-2)^3 = -8$, the answer is **negative**. This suggests the following.

PRODUCTS OF NEGATIVE NUMBERS

The product of an **odd** number of **negative** factors is **negative**. The product of an **even** number of **negative** factors is **positive**.

EXAMPLE 1   Evaluating expressions with natural-number exponents

Evaluate:

a. $(-4)^2$ **b.** -4^2 **c.** $(-4)^3$ **d.** -4^3

SOLUTION

a. $(-4)^2 = (-4)(-4) = 16$ Note that the exponent 2 is applied to the base (-4).

b. $-4^2 = -1 \cdot 4^2$ Here the exponent 2 applies to the base 4 only.

$\quad\quad = -1 \cdot 16$ Since $4^2 = 16$

$\quad\quad = -16$

c. $(-4)^3 = (-4)(-4)(-4) = -64$

d. $-4^3 = -1 \cdot 4^3$

$\quad\quad = -1 \cdot 64$ Since $4^3 = 4 \cdot 4 \cdot 4 = 64$

$\quad\quad = -64$

> ⚠️ **CAUTION** Note that $(-4)^2 \neq -4^2$ because $(-4)^2 = 16$, which is positive, while -4^2 is the additive inverse of 4^2, that is, $-4^2 = -16$. The placement of parentheses when using exponents is extremely important. Always interpret -4^2 as $-(4^2)$.

B Negative and Zero Exponents

In science and technology, negative numbers are used as exponents. For example, the diameter of a DNA molecule is 10^{-8} m, and the time it takes for an electron to go from source to screen in a TV tube is 10^{-6} sec. What do 10^{-8} and 10^{-6} mean? Look at the pattern obtained by dividing by 10 in each of the following:

$$10^3 = 1000$$
$$10^2 = 100$$
$$10^1 = 10$$
$$10^0 = 1$$

The exponents decrease by 1. ⟶   ⟶ The number decreases by a factor of 10.

$$10^{-1} = \frac{1}{10} = \frac{1}{10^1}$$

$$10^{-2} = \frac{1}{100} = \frac{1}{10^2}$$

$$10^{-3} = \frac{1}{1000} = \frac{1}{10^3}$$

As you can see, this procedure yields $10^0 = 1$. In general, we make the following definition.

ZERO EXPONENT

If a is a nonzero real number,
$$a^0 = 1$$
Moreover, 0^0 is *not* defined.

Thus $5^0 = 1$, $8^0 = 1$, and $9^0 = 1$. Note that 0^0 is **not** defined. Now look again at the numbers in the first box. Note that we obtained

$$10^{-1} = \frac{1}{10}, \qquad 10^{-2} = \frac{1}{10^2}, \qquad \text{and} \qquad 10^{-3} = \frac{1}{10^3}$$

Thus we make the following definition.

NEGATIVE EXPONENTS

If n is a positive integer,

$$a^{-n} = \frac{1}{a^n} \qquad (a \neq 0)$$

This definition says that a^{-n} and a^n are reciprocals since

$$a^{-n} \cdot a^n = \frac{1}{a^n} \cdot a^n = 1$$

By definition

$$5^{-2} = \frac{1}{5^2} = \frac{1}{5 \cdot 5} = \frac{1}{25}$$

and

$$(-2)^{-3} = \frac{1}{(-2)^3} = \frac{1}{(-2) \cdot (-2) \cdot (-2)} = \frac{1}{-8} = -\frac{1}{8}$$

Similarly,

$$\frac{1}{4^2} = 4^{-2}$$

and

$$\frac{1}{3^4} = 3^{-4}$$

EXAMPLE 2   From negative exponents to fractions

Write the following as a fraction without negative exponents:

a. 6^{-2} **b.** $(-4)^{-3}$ **c.** x^{-4} **d.** $-a^{-4}$ **e.** $3x^{-2}$

SOLUTION

a. $6^{-2} = \dfrac{1}{6^2} = \dfrac{1}{6 \cdot 6} = \dfrac{1}{36}$

b. $(-4)^{-3} = \dfrac{1}{(-4)^3} = \dfrac{1}{(-4) \cdot (-4) \cdot (-4)} = \dfrac{1}{-64} = -\dfrac{1}{64}$

c. $x^{-4} = \dfrac{1}{x^4}$

d. $-a^{-4} = -\dfrac{1}{a^4}$

e. $3x^{-2} = 3 \cdot x^{-2} = 3 \cdot \dfrac{1}{x^2} = \dfrac{3}{x^2}$ ∎

Multiplication and Division with Exponents

Suppose we wish to multiply $x^2 \cdot x^3$. We write

$$\underbrace{\overbrace{x^2}\cdot\overbrace{x^3}}_{x^5}$$
$$\underbrace{x \cdot x \cdot x \cdot x \cdot x}$$

Since we want to know the total number of factors, we simply add the exponents 2 and 3 of x^2 and x^3 to find the exponent 5 of the result. Similarly,

$$a^3 \cdot a^4 = a^{3+4} = a^7$$

$$b^2 \cdot b^4 = b^{2+4} = b^6$$

$$c^5 \cdot c^0 = c^{5+0} = c^5$$

Now, let's consider $x^5 \cdot x^{-3}$, where one of the exponents is negative. By the definition of exponents,

$$x^5 \cdot x^{-3} = x \cdot x \cdot x \cdot x \cdot x \cdot \frac{1}{x \cdot x \cdot x}$$

$$= \frac{x \cdot x \cdot \cancel{x} \cdot \cancel{x} \cdot \cancel{x}}{\cancel{x} \cdot \cancel{x} \cdot \cancel{x}}$$

$$= x^2$$

Adding exponents,

$$x^5 \cdot x^{-3} = x^{5+(-3)} = x^2 \qquad \text{Same answer}$$

Let us try $x^{-2} \cdot x^{-3}$. This time we have

$$x^{-2} \cdot x^{-3} = \frac{1}{x \cdot x} \cdot \frac{1}{x \cdot x \cdot x}$$

$$= \frac{1}{x \cdot x \cdot x \cdot x \cdot x} \qquad \text{Multiplying}$$

$$= x^{-5} \qquad \text{By the definition of negative exponents}$$

Adding exponents,

$$x^{-2} \cdot x^{-3} = x^{-2+(-3)} = x^{-5} \qquad \text{Same answer}$$

This suggests the following law.

FIRST LAW OF EXPONENTS (PRODUCT RULE)

If a is a real number and m and n are integers,

$$a^m \cdot a^n = a^{m+n}$$

This law tells us that to *multiply* expressions with the *same* base, we *add* the exponents. Note that this law *does not* apply to expressions such as $x^m \cdot y^n$ because the bases are *different*. [Note that $3^2 \cdot 2^4 = 3 \cdot 3 \cdot 2 \cdot 2 \cdot 2 \cdot 2$, so we have different bases (factors) that cannot be combined.] If the expressions involved have numerical coefficients, we multiply numbers by numbers and letters (variables) by letters using the commutative and associative properties that we have studied. Thus to multiply $(-3x^2)(5x^3)$, we write

$$(-3x^2)(5x^3) = (-3 \cdot 5)(x^2)(x^3)$$

$$= -15x^{2+3} = -15x^5$$

Similarly,

$$(-4x^{-2})(3x^5) = (-4 \cdot 3)(x^{-2+5})$$
$$= -12x^3 \qquad\qquad -2 + 5 = 3$$

and

$$(-2x^{-5})(-3x^2) = (-2)(-3)(x^{-5+2})$$
$$= 6x^{-3} \qquad\qquad -5 + 2 = -3$$
$$= 6 \cdot \frac{1}{x^3}$$
$$= \frac{6}{x^3}$$

 NOTE Note that we write the answer without using negative exponents.

EXAMPLE 3   Using the product rule of exponents

Multiply and simplify:

a. $(-4x^2)(5x^5)$ **b.** $(3x^7)(-2x^4)$ **c.** $(4x^3y)(-2x^{-8}y^6)$

SOLUTION

a. $(-4x^2)(5x^5) = (-4 \cdot 5)(x^2 \cdot x^5)$
$$= -20x^{2+5}$$
$$= -20x^7$$

b. $(3x^7)(-2x^4) = (3)(-2)(x^7 \cdot x^4)$
$$= -6x^{7+4}$$
$$= -6x^{11}$$

c. $(4x^3y)(-2x^{-8}y^6) = (4)(-2)(x^3 \cdot x^{-8})(y^1 \cdot y^6)$
$$= -8(x^{3+(-8)})(y^{1+6})$$
$$= -8(x^{-5})(y^7)$$
$$= -8\left(\frac{1}{x^5}\right)(y^7)$$
$$= -\frac{8y^7}{x^5} \qquad\qquad ■$$

Now consider the following division:

$$\frac{x^5}{x^3} = \frac{x \cdot x \cdot x \cdot x \cdot x}{x \cdot x \cdot x}$$
$$= x \cdot x \cdot \frac{x \cdot x \cdot x}{x \cdot x \cdot x}$$
$$= x^2 \cdot 1$$
$$= x^2$$
$$= x^{5-3}$$

As you can see, the exponent 2 in the answer can be obtained by *subtracting* $5 - 3 = 2$. Similarly,

$$\frac{x^5}{x^{-3}} = x^{5-(-3)} = x^8$$

This is because

$$\frac{x^5}{x^{-3}} = \frac{x^5}{\dfrac{1}{x^3}} = x^5 \cdot x^3 = x^8$$

This line of reasoning suggests the following law.

SECOND LAW OF EXPONENTS (QUOTIENT RULE)

If a is a real number and m and n are integers,

$$\frac{a^m}{a^n} = a^{m+n} \qquad (a \neq 0)$$

This law tells us that to *divide* expressions with the same base, we *subtract* the exponent of the denominator from that of the numerator. Thus

$$\frac{x^8}{x^5} = x^{8-5} = x^3$$

$$\frac{x^3}{x^8} = x^{3-8} = x^{-5} = \frac{1}{x^5}$$

$$\frac{x^{-3}}{x^{-8}} = x^{-3-(-8)} = x^{-3+8} = x^5$$

$$\frac{x^8}{x^8} = x^{8-8} = x^0 = 1$$

EXAMPLE 4   Using the quotient rule of exponents

Divide and simplify:

a. $\dfrac{4x^8}{2x^5}$ **b.** $\dfrac{-12x^4}{-3x^6}$ **c.** $\dfrac{5x^{-4}}{-15x^{-6}}$ **d.** $\dfrac{30x^3}{-15x^{-6}}$

SOLUTION

a. $\dfrac{4x^8}{2x^5} = \dfrac{4}{2} \cdot \dfrac{x^8}{x^5}$

$\qquad = 2x^{8-5}$

$\qquad = 2x^3$

b. $\dfrac{-12x^4}{-3x^6} = \dfrac{-12}{-3} \cdot \dfrac{x^4}{x^6}$

$\qquad = 4x^{4-6}$

$\qquad = 4x^{-2} \qquad 4 - 6 = -2$

$\qquad = \dfrac{4}{x^2}$

c. $\dfrac{5x^{-4}}{-15x^{-6}} = \dfrac{5}{-15} \cdot \dfrac{x^{-4}}{x^{-6}}$

$\qquad\qquad = -\dfrac{1}{3} \cdot x^{-4-(-6)}$

$\qquad\qquad = -\dfrac{1}{3} \cdot x^{-4+6} \qquad -4 - (-6) = -4 + 6$

$\qquad\qquad = -\dfrac{1}{3} \cdot x^{2}$

$\qquad\qquad = -\dfrac{x^{2}}{3}$

d. $\dfrac{30x^{3}}{-15x^{-6}} = \dfrac{30}{-15} \cdot \dfrac{x^{3}}{x^{-6}}$

$\qquad\qquad = -2x^{3-(-6)}$

$\qquad\qquad = -2x^{9} \qquad 3 - (-6) = 3 + 6 = 9$ ■

Raising a Power to a Power

Now let us consider $(5^3)^2$. By definition,

$$(5^3)^2 = (5^3)(5^3) = 5^{3+3} = 5^6$$

Notice that the exponent is $6 = 3 \cdot 2$, so we could have obtained the answer by multiplying exponents in the expression $(5^3)^2$. Similarly,

$$(3^{-2})^3 = \left(\frac{1}{3^2}\right)\left(\frac{1}{3^2}\right)\left(\frac{1}{3^2}\right) = \frac{1}{3^6} = 3^{-6}$$

Again, the exponent -6 could be obtained by multiplying the original exponents -2 and 3. Generalizing this result, we have the following law.

THIRD LAW OF EXPONENTS (POWER RULE)

If a is a real number and m and n are integers,

$$(a^m)^n = a^{m \cdot n} \qquad (a \neq 0)$$

Thus to raise a power to another power, we multiply the exponents, that is,

$$(x^3)^4 = x^{3 \cdot 4} = x^{12}$$

$$(y^6)^{-3} = y^{6 \cdot (-3)} = y^{-18} = \frac{1}{y^{18}}$$

$$(z^{-5})^4 = z^{-5 \cdot 4} = z^{-20} = \frac{1}{z^{20}}$$

Now, let's consider $(5x^4)^3$. By definition,

$$(5x^4)^3 = (5x^4)(5x^4)(5x^4)$$

$$= (5 \cdot 5 \cdot 5)(x^4 \cdot x^4 \cdot x^4)$$

$$= 5^3 \cdot x^{3 \cdot 4}$$

$$= 5^3 \cdot x^{12}$$

Note that the exponent 3 is applied to the 5 and the x^4; that is, if we raise several factors inside parentheses to a power, we raise each factor to the given power. In general, we have the following law.

RAISING A PRODUCT TO A POWER

If a and b are real numbers and m, n, and k are integers,
$$(a^m b^n)^k = a^{m \cdot k} b^{n \cdot k} \qquad (a \neq 0, b \neq 0)$$

This result can be generalized further to apply to any number of factors inside the parentheses.

EXAMPLE 5   Using the power rule of exponents

Simplify:

a. $(3x^5 y^3)^{-2}$ **b.** $(-2x^5 y^{-4})^3$

SOLUTION

a. $(3x^5 y^3)^{-2} = (3)^{-2} (x^5)^{-2} (y^3)^{-2}$

$$= \frac{1}{3^2} \cdot x^{5 \cdot (-2)} \cdot y^{3 \cdot (-2)} \qquad (3)^{-2} = \frac{1}{3^2}$$

$$= \frac{1}{3^2} \cdot x^{-10} \cdot y^{-6}$$

$$= \frac{1}{3^2} \cdot \frac{1}{x^{10}} \cdot \frac{1}{y^6} \qquad x^{-10} = \frac{1}{x^{10}},\ y^{-6} = \frac{1}{y^6}$$

$$= \frac{1}{9x^{10} y^6}$$

b. $(-2x^5 y^{-4})^3 = (-2)^3 (x^5)^3 (y^{-4})^3$

$$= -8x^{5 \cdot 3} y^{-4 \cdot 3}$$

$$= -8x^{15} y^{-12} \qquad y^{-12} = \frac{1}{y^{12}}$$

$$= -\frac{8x^{15}}{y^{12}}$$

■

E   Raising a Quotient to a Power

We have already raised a product to a power. Can we raise a quotient to a power? Let us try $(2^3/3^4)^2$. By the definition of exponents

$$\left(\frac{2^3}{3^4} \right)^2 = \frac{2^3}{3^4} \cdot \frac{2^3}{3^4} = \frac{2^{3+3}}{3^{4+4}} = \frac{2^6}{3^8} = \frac{2^{3 \cdot 2}}{3^{4 \cdot 2}}$$

Note that the same answer is obtained by multiplying each of the exponents in the numerator and denominator by 2. Here is the general rule.

RAISING A QUOTIENT TO A POWER

If a and b are real numbers and m, n, and k are integers,
$$\left(\frac{a^m}{b^n} \right)^k = \frac{a^{m \cdot k}}{b^{n \cdot k}} \qquad (a \neq 0, b \neq 0)$$

EXAMPLE 6   Raising a quotient to a power

Simplify:

a. $\left(\dfrac{x^4}{y^{-3}}\right)^{-2}$ **b.** $\left(\dfrac{3x^{-3}y^2}{2y^3}\right)^3$

SOLUTION

a. $\left(\dfrac{x^4}{y^{-3}}\right)^{-2} = \dfrac{(x^4)^{-2}}{(y^{-3})^{-2}} = \dfrac{x^{4\cdot(-2)}}{y^{-3\cdot(-2)}} = \dfrac{x^{-8}}{y^6}$ $x^{-8} = \dfrac{1}{x^8}$

$= \dfrac{1}{x^8 y^6}$

b. In this case, it is easier to do the operations inside the parentheses first. Thus

$$\left(\frac{3x^{-3}y^2}{2y^3}\right)^3 = \left(\frac{3}{2} \cdot x^{-3} \cdot y^{2-3}\right)^3 \qquad \frac{y^2}{y^3} = y^{2-3}$$

$$= \left(\frac{3y^{-1}}{2x^3}\right)^3 \qquad x^{-3} = \frac{1}{x^3}$$

$$= \left(\frac{3}{2x^3 y}\right)^3 \qquad y^{-1} = \frac{1}{y}$$

$$= \frac{3^3}{(2x^3 y)^3}$$

$$= \frac{27}{8x^9 y^3}$$ ∎

Since the reciprocal of $\dfrac{a}{b}$ is $\dfrac{b}{a}$,

$$\left(\frac{a}{b}\right)^{-1} = \left(\frac{b}{a}\right).$$

Thus

$$\left[\left(\frac{a}{b}\right)^{-1}\right]^n = \left(\frac{b}{a}\right)^n$$

and

PROCEDURE

> **Raising a Quotient to a Negative Power**
>
> $$\left(\frac{a}{b}\right)^{-n} = \left(\frac{b}{a}\right)^n \qquad (a \neq 0, b \neq 0)$$

This means that a fraction raised to the $-n$th power is equivalent to its *reciprocal* raised to the nth power. In Example 6a, we could write

$$\left(\frac{x^4}{y^{-3}}\right)^{-2} = \left(\frac{y^{-3}}{x^4}\right)^2 = \frac{y^{-6}}{x^8} = \frac{1}{x^8 y^6}$$

You can use this method when you work the exercise set.

F   Scientific Notation

In science and other areas of endeavor, very large or very small numbers occur frequently. For example, a red cell of human blood contains 270,000,000 hemoglobin molecules, and the mass of a single carbon atom is 0.000 000 000 000 000 000 000 019 9 gram. Numbers in this form are difficult to write and to work with, so they are written in scientific notation for which we have the following definition.

SCIENTIFIC NOTATION

A number is said to be in **scientific notation** if it is written in the form

$$m \times 10^n$$

where m is a number greater than or equal to 1 and less than 10 ($1 \le m < 10$) and n is an integer.

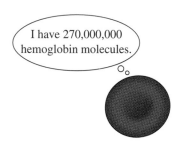

I have 270,000,000 hemoglobin molecules.

For any given number, the m is obtained by placing the decimal point so that there is exactly one nonzero digit to its left. The integer n is then the number of places that the decimal point must be moved from its position in m to its original position; it is positive if the point must be moved to the right and negative if the point must be moved to the left. Thus

$5.3 = 5.3 \times 10^0$	Decimal point in 5.3 must be moved 0 places.
$87 = 8.7 \times 10^1 = 8.7 \times 10$	Decimal point in 8.7 must be moved 1 place to the *right* to get 87.
$68,000 = 6.8 \times 10^4$	Decimal point in 6.8 must be moved 4 places to the *right* to get 68,000.
$0.49 = 4.9 \times 10^{-1}$	Decimal point in 4.9 must be moved 1 place to the *left* to get 0.49.
$0.072 = 7.2 \times 10^{-2}$	Decimal point in 7.2 must be moved 2 places to the *left* to get 0.072.
$0.0003875 = 3.875 \times 10^{-4}$	Decimal point in 3.875 must be moved 4 places to the *left* to get 0.0003875.

EXAMPLE 7   Writing numbers in scientific notation

Write in scientific notation:

a. 270,000,000

b. 0.000 000 000 000 000 000 000 019 9

SOLUTION

a. $270,000,000 = 2.7 \times 10^8$

b. $0.000\ 000\ 000\ 000\ 000\ 000\ 000\ 019\ 9 = 1.99 \times 10^{-23}$ ■

EXAMPLE 8   From scientific to standard decimal notation

Write in standard decimal notation:

a. 2.5×10^{10} **b.** 7.4×10^{-6}

SOLUTION

a. $2.5 \times 10^{10} = 25,000,000,000$

b. $7.4 \times 10^{-6} = 0.0000074$ ■

We can use the laws of exponents when working with numbers in scientific notation. The next example shows you how.

EXAMPLE 9   Calculations in scientific notation

Do the following calculations, and write the answers in scientific notation:

a. $(5 \times 10^4) \times (9 \times 10^{-7})$

b. $\dfrac{6 \times 10^5}{3 \times 10^{-4}}$

SOLUTION

a. $(5 \times 10^4) \times (9 \times 10^{-7}) = (5 \times 9) \times (10^4 \times 10^{-7})$

$$= 45 \times 10^{4-7}$$
$$= 45 \times 10^{-3}$$
$$= 4.5 \times 10^1 \times 10^{-3}$$
$$= 4.5 \times 10^{1-3}$$
$$= 4.5 \times 10^{-2}$$

b. $\dfrac{6 \times 10^5}{3 \times 10^{-4}} = \dfrac{6}{3} \times \dfrac{10^5}{10^{-4}}$

$$= 2 \times 10^{5-(-4)}$$
$$= 2 \times 10^9$$ ■

GRAPH IT

You can use a grapher to clarify the concepts in Example 1. Enter $(-4)^2$ by keying in ⟮(⟯ ⟮(−)⟯ 4 ⟮)⟯ ⟮^⟯ 2 ⟮ENTER⟯, where the ⟮^⟯ key raises the base 4 to the second power. (You can also enter ⟮(⟯ ⟮(−)⟯ 4 ⟮)⟯ ⟮x²⟯.) In either case, the answer is 16. To find -4^2—that is, the *additive inverse* of 4^2—key in ⟮(−)⟯ 4 ⟮^⟯ 2 ⟮ENTER⟯. This time the answer is -16, so $(-4)^2 \neq -4^2$. To verify the answer in Example 3b, use a standard window, one that covers from -10 to 10 in both the horizontal and vertical axes. (Press ⟮ZOOM⟯ 6 to set one up on the TI-82.) Now enter ⟮Y=⟯ and the original expression $(3x^7)(-2x^4)$ as $Y_1 =$ ⟮(⟯ 3 ⟮X,T,θ⟯ ⟮^⟯ 7 ⟮)⟯ ⟮(⟯ ⟮(−)⟯ 2 ⟮X,T,θ⟯ ⟮^⟯ 4 ⟮)⟯ ⟮GRAPH⟯. (The TI-81 has an ⟮X/T⟯ key, the TI-82 has an ⟮X,T,θ⟯ key.)

The graph you have just keyed in appears in the window. Now enter the answer $-6x^{11}$ as Y_2; that is, enter $Y_2 =$ ⟮(−)⟯ 6 ⟮X,T,θ⟯ ⟮^⟯ 11 ⟮GRAPH⟯. If you get the same graph for the original expression $(3x^7)(-2x^4)$ and for the answer $-6x^{11}$, your solution is correct! (Try entering 12 instead of 11 and see what happens.) To write numbers in scientific notation, press ⟮MODE⟯, use your ⟮>⟯ button to move the cursor from **Normal** to **Sci** and then press ⟮ENTER⟯. To do Example 7a, go to the Home Screen, key in 270 000 000 ⟮ENTER⟯, and you will get the answer 2.7 E 8, which means 2.7×10^8.

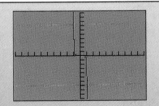

EXERCISE 1.3

A In Problems 1–10, evaluate.

1. $-4^2 =$ _____

2. $(-4)^2 =$ _____

3. $(-5)^2 =$ _____

4. $-5^2 =$ _____

5. $-5^3 =$ _____

6. $(-5)^3 =$ _____

7. $(-6)^4 =$ _____

8. $-6^4 =$ _____

9. $-2^5 =$ _____

10. $(-2)^5 =$ _____

B In Problems 11–20, write the expression given as a fraction in simplified form and without negative exponents.

11. 4^{-2}

12. 2^{-3}

13. 5^{-3}

14. 7^{-2}

15. 3^{-4}

16. 6^{-3}

17. x^{-6}

18. y^{-7}

19. a^{-8}

20. b^{-4}

C **In Problems 21–50, perform the indicated operations and simplify.**

21. $2^{-4} \cdot 2^{-2}$

22. $4^{-1} \cdot 4^{-2}$

23. $(3x^6) \cdot (4x^{-4})$

24. $(4y^7) \cdot (5y^{-3})$

25. $(-3y^{-3}) \cdot (5y^5)$

26. $(-5x^{-7}) \cdot (4x^8)$

27. $(-4a^3) \cdot (-5a^{-8})$

28. $(-2b^4) \cdot (-3b^{-7})$

29. $(3x^{-5}) \cdot (5x^2y)(-2xy^2)$

30. $(4y^{-6}) \cdot (5xy^4)(-2x^2y)$

31. $(-2x^{-3}y^2)(3x^{-2}y^3)(4xy)$

32. $(-3xy^{-5})(4x^2y)(2x^3y^2)$

33. $(4a^{-2} \cdot b^{-3})(5a^{-1}b^{-1})(-2ab)$

34. $(2a^{-5} \cdot b^{-2})(3a^{-1}b^{-1})(5ab)$

35. $(6a^{-3} \cdot b^3)(5a^2b^2)(-ab^{-5})$

36. $(7a^6 \cdot b^{-6})(2ab^5)(-a^{-7}b)$

37. $\dfrac{8x^7}{4x^3}$

38. $\dfrac{8a^3}{4a^2}$

39. $\dfrac{-8a^4}{-16a^2}$

40. $\dfrac{-9y^5}{-18y^2}$

41. $\dfrac{12x^5y^3}{-6x^2y}$

42. $\dfrac{18x^6y^2}{-9xy}$

43. $\dfrac{-6x^{-4}}{12x^{-5}}$

44. $\dfrac{8x^{-3}}{4x^{-4}}$

45. $\dfrac{-14a^{-5}}{-21a^{-2}}$

46. $\dfrac{-2a^{-6}}{-6a^{-3}}$

47. $\dfrac{-27a^{-4}}{-36a^{-4}}$

48. $\dfrac{-5x^{-3}}{10x^{-3}}$

49. $\dfrac{3a^{-2} \cdot b^5}{2a^4b^2}$

50. $\dfrac{x^{-3} \cdot y^6}{x^4 \cdot y^3}$

D **In Problems 51–60, simplify the expression given and write your answer without negative exponents.**

51. $(2x^3y^{-2})^3$

52. $(3x^2y^{-3})^2$

53. $(2x^{-2}y^3)^2$

54. $(3x^{-4}y^4)^3$

55. $(-3x^3y^2)^{-3}$

56. $(-2x^5y^4)^{-4}$

57. $(x^{-6}y^{-3})^2$

58. $(y^{-4}z^{-3})^5$

59. $(x^{-4}y^{-4})^{-3}$

60. $(y^{-5}z^{-3})^{-4}$

E **In Problems 61–70, simplify.**

61. $\left(\dfrac{a}{b^3}\right)^2$

62. $\left(\dfrac{a^2}{b}\right)^3$

63. $\left(\dfrac{-3a}{2b^2}\right)^{-3}$

64. $\left(\dfrac{-2a^2}{3b^0}\right)^{-2}$

65. $\left(\dfrac{a^{-4}}{b^2}\right)^{-2}$

66. $\left(\dfrac{a^{-2}}{b^3}\right)^{-3}$

67. $\left(\dfrac{x^5}{y^{-2}}\right)^{-3}$

68. $\left(\dfrac{x^6}{y^{-3}}\right)^{-2}$

69. $\left(\dfrac{x^{-4}y^3}{x^5y^5}\right)^{-3}$

70. $\left(\dfrac{x^{-2}y^0}{x^7y^2}\right)^{-2}$

F **In Problems 71–74, write the numbers in scientific notation.**

71. 268,000,000 (U.S. population in the year 2000)

72. 1,900,000,000 (dollars spent on waterbeds and accessories in 1 year)

73. 0.00024 (probability of four of a kind in poker)

74. 0.00000009 (wavelength in centimeters of an X-ray)

In Problems 75–78, write the numbers in decimal notation.

75. 8×10^6 (bagels eaten per day in the United States)

76. $\$6.85 \times 10^9$ (estimated wealth of the five wealthiest women)

77. 2.3×10^{-1} (kilowatts per hour used by your TV)

78. 4×10^{-11} (joules of energy released by splitting one uranium atom)

In Problems 79–84, write your answer in scientific notation.

79. The width of the asteroid belt is 2.8×10^8 kilometers (km). The speed of *Pioneer 10* in passing through this belt was 1.4×10^5 km/hr. Thus *Pioneer 10* took

$$\frac{2.8 \times 10^8}{1.4 \times 10^5} \text{ hr}$$

to go through the belt. How many hours was that?

80. The mass of Earth is 6×10^{21} tons. The Sun is about 300,000 times as massive. Thus the mass of the Sun is $(6 \times 10^{21}) \times$ 300,000 tons. How many tons is that?

81. The velocity of light can be measured by knowing the distance from the Sun to Earth (1.47×10^{11} meter, m) and the time it takes for sunlight to reach Earth (490 sec). Thus the velocity of light is

$$\frac{1.47 \times 10^{11}}{490} \text{ m/sec}$$

How many meters per second is that?

82. Oil reserves in the United States are estimated to be 3.5×10^{10} barrels. Production amounts to 3.2×10^9 barrels per year. At this rate, how long would U.S. oil reserves last? (Give your answer to the nearest year.)

83. The world's oil reserves are estimated to be 6.28×10^{11} barrels. Production is 2.0×10^{10} barrels per year. At this rate, how long would the world's oil reserves last? (Give your answer to the nearest year.)

84. Scientists have estimated that the total energy received from the Sun each minute is 1.02×10^{19} calories. Since the area of Earth is 5.1×10^8 km^2 (square kilometers), the amount of energy received per square centimeter of Earth's surface per minute (the solar constant) is

$$\frac{1.02 \times 10^{19}}{(5.1 \times 10^8) \times 10^{10}} \qquad (1 \text{ km}^2 = 10^{10} \text{ cm}^2)$$

How many calories per square centimeter is that?

SKILL CHECKER

Find:

85. a. $(-3.2)(-1.4)(-2.2)$ **b.** $(-1.1)(-1.2)(-2.1)$

86. a. $\dfrac{-3.2}{1.6}$ **b.** $\dfrac{-4.8}{-1.2}$

USING YOUR KNOWLEDGE **If You Have a Scientific Calculator**

If you have a scientific calculator, and you multiply 9,800,000 by 4,500,000, the display may show

| 4.41   13 |

This means that the answer is 4.41 × 10^{13}.

87. The display on a calculator shows

| 3.34   5 |

Write this number in scientific notation.

88. The display on a calculator shows

| −9.97   −6 |

Write this number in scientific notation.

89. To enter large or small numbers in a calculator with scientific notation, you must write the number using this notation first. Thus to enter the number 8,700,000,000 in the calculator, you must know that 8,700,000,000 is 8.7×10^9, *then* you can key in

    or   8 · 7 EXP 9

The calculator displays

| 8.7   09 |

a. What would the display read when you enter the number 73,000,000,000?

b. What would the display read when you enter the number 0.000000123?

WRITE ON . . .

Write an explanation for:

90. Why a^0 is defined as 1.

91. Why $0^n = 0$ for n not 0.

92. Why 0^0 is not defined.

93. Why $x^m x^n = x^{m+n}$, using the concept of a factor.

94. Why

$$\frac{x^m}{x^n} = x^{m-n}$$

using the concept of a factor.

95. Why $(x^m)^n = x^{m \cdot n}$, using the concept of a factor.

MASTERY TEST **If you know how to do these problems, you have learned your lesson!**

Multiply and simplify:

96. $(-3x^4y^2)(3x^{-7}y)$ **97.** $(-2x^{-5}y^{-3})(-4xy)$

Divide and simplify:

98. $\dfrac{45y^4}{-15y^{-7}}$ **99.** $\dfrac{-6x^{-8}}{30x^6}$

Evaluate:

100. $(-5)^{-4}$ **101.** -2^6

Simplify:

102. $(-2x^4y^{-5})^3$ **103.** $(-3x^{-4}y^5)^{-2}$

Write as a fraction:

104. $(-3)^{-5}$ **105.** $(-x)^{-5}$

Simplify:

106. $\left(\dfrac{2x^{-4}y^3}{3y^5}\right)^{-2}$ **107.** $\left(-\dfrac{5x^{-5}y^7}{7y^3}\right)^{-2}$

Write in scientific notation:

108. 387,000,000   **109.** $(4 \times 10^{-5}) \times (6 \times 10^2)$

1.4

GROUPING SYMBOLS AND THE ORDER OF OPERATIONS

To succeed, review how to:

1. Perform the four fundamental operations using signed numbers (pp. 16–21).
2. Calculate powers of integers (pp. 29–30).
3. Use the identity for multiplication (p.19).

Objectives:

A Evaluate numerical expressions that contain grouping symbols.

B Evaluate expressions using the correct order of operations.

C Use the distributive property to simplify expressions.

D Simplify expressions by combining like terms.

E Simplify expressions by removing grouping symbols and combining like terms.

getting started

Swimming and Your Heart Rate

How do you calculate your ideal heart rate when swimming? One way to do this is as follows:

Subtract your age from 205 and multiply by 0.70.

This means that if you are a years old, you would subtract your age a from 205 and multiply by 0.70.

Should you subtract your age from 205 first and then multiply by 0.70, or multiply your age by 0.70 first, and then subtract from 205? In algebra we can make our meaning clear by using parentheses. The formula is written as $(205 - a) \cdot 0.70$ to indicate that the subtraction should be done first. (Try it by substituting your age for a.)

A Grouping Symbols and Operations

In algebra and arithmetic *parentheses* ( ) are grouping symbols used to indicate which operations are to be performed first. Square brackets [ ] and braces { } are also grouping symbols; they can be used in the same manner as parentheses. Thus

$$4 \cdot (3 + 2), \qquad 4 \cdot [3 + 2], \qquad \text{and} \qquad 4 \cdot \{3 + 2\}$$

all mean that we must first add 3 and 2 and then multiply this sum by 4. Note that the expressions $4 \cdot (3 + 2)$ and $(4 \cdot 3) + 2$ have *different* meanings. In the first expression we add 3 and 2 first, while in the second we multiply 4 by 3 first. Thus $4 \cdot (3 + 2) = 4 \cdot (5) = 20$, but $(4 \cdot 3) + 2 = (12) + 2 = 14$. Hence, $4 \cdot (3 + 2) \neq (4 \cdot 3) + 2$.

EXAMPLE 1 Evaluating expressions containing parentheses

Evaluate:

a. $(-4 \cdot 5) + 6$ **b.** $-4 \cdot (5 + 6)$ **c.** $-48 \div (4 \cdot 3)$ **d.** $(-48 \div 4) \cdot 3$

SOLUTION   Perform the operations inside the parentheses *first*.

a. $(-4 \cdot 5) + 6 = -20 + 6$ Multiply -4 and 5 first.

$\qquad\qquad\qquad = -14$

b. $-4 \cdot (5 + 6) = -4 \cdot 11$ Add 5 and 6 first.

$\qquad\qquad\qquad = -44$

c. $-48 \div (4 \cdot 3) = -48 \div 12$ Multiply 4 and 3 first.

$\qquad\qquad\qquad = -4$

d. $(-48 \div 4) \cdot 3 = -12 \cdot 3$ Divide -48 by 4 first.

$\qquad\qquad\qquad = -36$ ■

If we have more than one set of grouping symbols, we perform the operations inside the innermost grouping symbols first. Thus

$[2 \cdot (88 + 14)] + 12 = [2 \cdot (102)] + 12$ Add 88 and 14 inside the parentheses first.

$\qquad\qquad\qquad\qquad = 204 + 12$

$\qquad\qquad\qquad\qquad = 216$

EXAMPLE 2   Evaluating expressions containing grouping symbols

Evaluate:

a. $[-5 \cdot (6 + 4)] + 9$ **b.** $[-10 \cdot (8 - 3)] - 9$

SOLUTION

a. $[-5 \cdot (6 + 4)] + 9 = [-5 \cdot 10] + 9$ Add 6 and 4 first.

$\qquad\qquad\qquad\qquad = -50 + 9$

$\qquad\qquad\qquad\qquad = -41$

b. $[-10 \cdot (8 - 3)] - 9 = [-10 \cdot 5] - 9$ Subtract 3 from 8 first.

$\qquad\qquad\qquad\qquad = -50 - 9$

$\qquad\qquad\qquad\qquad = -59$ ■

In some cases we have to evaluate expressions in which a bar is used to indicate division. For instance, if we wish to convert a temperature given in degrees Fahrenheit to degrees Celsius, we have to evaluate the expression

$$\frac{5 \cdot (F - 32)}{9}$$

where F represents the temperature in degrees Fahrenheit. Thus if the temperature is 77°F, the corresponding Celsius temperature is calculated as follows:

$$\frac{5 \cdot (77 - 32)}{9} = \frac{5 \cdot (45)}{9}$$

$$= \frac{225}{9}$$

$$= 25$$

You can also do this by dividing 45 by 9 first and then multiplying the result, 5, by 5, to obtain 25.

The corresponding temperature is 25°C.

EXAMPLE 3   An application using grouping symbols

In September 1933, a freak heat flash struck the city of Coimbra, in Portugal. On this day, the temperature rose to 158°F for 120 sec. How many degrees Celsius is that?

SOLUTION In this case $F = 158$; thus the Celsius temperature is given by

$$\frac{5 \cdot (158 - 32)}{9} = \frac{5 \cdot (126)}{9}$$

$$= \frac{630}{9}$$

$$= 70$$

You can also do this by dividing 126 by 9 first and then multiplying the result, 14, by 5, to obtain 70.

The corresponding Celsius temperature is 70°C. ∎

B   The Order of Operations

If an expression does not contain parentheses or brackets, we must establish the order in which operations are to be performed. For example, the expression $6^2 + 9 \div 3$ might be evaluated in two ways:

$6^2 + 9 \div 3$	Square 6 and divide 9 by 3.	$6^2 + 9 \div 3$	Square 6.
$36 + 3$	Add.	$36 + 9 \div 3$	Add $36 + 9$.
39		$45 \div 3$	Divide by 3.
		15	

To avoid this ambiguity, we agree to perform any sequence of operations from left to right and in the following order.

PROCEDURE

> **Order of Operations**
> 1. Do the operations inside the parentheses (or other grouping symbols) starting with the innermost grouping symbols and operations above and below fraction bars.
> 2. Evaluate all exponential expressions.
> 3. Perform multiplications and divisions as they occur from left to right.
> 4. Perform additions and subtractions as they occur from left to right.

EXAMPLE 4   Order of operations

Evaluate:

$$-6^2 + \frac{(4 - 8)}{2} + 10 \div 5$$

SOLUTION

$$-6^2 + \frac{(4 - 8)}{2} + 10 \div 5$$

$$= -6^2 + \frac{-4}{2} + 10 \div 5 \qquad \text{Perform the operation inside the parentheses.}$$

$$= -36 + \frac{-4}{2} + 10 \div 5 \qquad \text{Evaluate } -6^2 = -36.$$

$$= -36 + (-2) + 2 \qquad \text{Perform multiplications and divisions as they occur from left to right.}$$

$$= -38 + 2 \qquad -36 + (-2) = -38 \quad \text{Perform additions and subtractions as they occur from left to right.}$$

$$= -36 \qquad -38 + 2 = -36 \qquad ∎$$

Removing Parentheses

In algebra the distributive property, $a(b + c) = ab + ac$, is used to remove parentheses in expressions such as $3(x + 5)$ or $4(x - 7)$, where x is a real number. Thus

$$3(x + 5) = 3x + 3 \cdot 5 = 3x + 15$$

and

$$4(x - 7) = 4x - 4 \cdot 7 = 4x - 28$$

EXAMPLE 5   Removing parentheses

Remove the parentheses (simplify):

a. $-2(x + 8)$ **b.** $0.5(7 - y)$

SOLUTION

a. $-2(x + 8) = -2x + (-2 \cdot 8)$

$= -2x + (-16)$

$= -2x - 16$ Recall that $a - b = a + (-b)$.

b. $0.5(7 - y) = 0.5 \cdot 7 - 0.5y$

$= 3.5 - 0.5y$ ■

Expressions of the form $-(a + b)$ or $-(a - b)$, where a and b are called **terms**, require special consideration. We first recall the following.

PROCEDURE

Identity for Multiplication

For any real number a, $a = 1 \cdot a$.

Since any real number has an additive inverse and the additive inverse of a is $-a$, the additive inverse of $1 \cdot a$ is $-1 \cdot a$.

PROCEDURE

Additive Inverse

For any real number a, $-a = -1 \cdot a$.

Hence

$$-(a + b) = -1 \cdot (a + b)$$

$$= -1 \cdot a + (-1 \cdot b)$$

$$= -a - b$$

PROCEDURE

Additive Inverse of a Sum

$$-(a + b) = -a - b$$

Similarly,

$$-(a - b) = -1 \cdot (a - b)$$

$$= -1 \cdot [a + (-b)]$$

$$= -1 \cdot a + (-1)(-b)$$

$$= -a + b$$

PROCEDURE

> **Additive Inverse of a Difference**
> $$-(a - b) = -a + b$$

These rules tell us that to remove the parentheses in an expression preceded by a minus sign, we simply *change the sign of every term inside the parentheses* or, equivalently, *multiply each term inside the parentheses by* -1.

EXAMPLE 6   Removing parentheses

Remove the parentheses (simplify):

a. $-(x - 2)$ **b.** $-(ab + 3)$

SOLUTION

a. $-(x - 2) = -1 \cdot (x - 2)$
$$= -1 \cdot x + (-1)(-2)$$
$$= -x + 2$$

Note that changing the signs inside the parentheses in $-(x - 2)$ will immediately yield

$$-(\overset{\text{change sign}}{x} - 2) = \overset{}{-}x + 2$$

change sign

b. $-(ab + 3) = -1 \cdot (ab + 3) = -1 \cdot ab + -1 \cdot (3)$
$$= -ab + (-3)$$
$$= -ab - 3$$

Note that changing signs inside the parentheses will immediately yield the answer $-ab - 3$. ∎

We can summarize this discussion by the following two facts.

PROCEDURE

> **Removing Parentheses**
>
> 1. If the factor in front of the parentheses has no written sign, multiply each term inside the parentheses by this factor; that is,
> $$a(b - c + d - e) = ab - ac + ad - ae$$
>
> 2. If the factor in front of the parentheses is preceded by a minus sign, multiply this factor by each of the terms inside the parentheses and change the sign of each of these terms; that is,
> $$-a(b - c + d - e) = -ab + ac - ad + ae$$

EXAMPLE 7   Removing parentheses

Remove the parentheses (simplify):

a. $4(x - 2y + 3)$ **b.** $-5(2x + y - z)$

c. $0.4(-3x + 2y - 7z - 8)$ **d.** $0.5x(y + 3z - 5)$

SOLUTION
a. $4(x - 2y + 3) = 4x - 8y + 12$
b. $-5(2x + y - z) = -10x - 5y + 5z$
c. $0.4(-3x + 2y - 7z - 8) = -1.2x + 0.8y - 2.8z - 3.2$
d. $0.5x(y + 3z - 5) = 0.5xy + 1.5xz - 2.5x$ ■

 Combining Like Terms

Suppose we wish to simplify $3x + 2(x + 5)$. We start by simplifying $2(x + 5)$, to obtain

$$3x + 2(x + 5) = 3x + 2x + 10$$

The terms $3x$ and $2x$ are called *like terms*. They differ only in their numerical parts (coefficients). Similarly, $-3y$ and $5y$ are like terms, and $9z^2$ and $-3z^2$ are like terms. In general, we have the following definition.

LIKE TERMS

> Constant terms or terms with exactly the same variable factors are called **similar** or **like** terms.

 Like terms differ only in their *numerical* coefficients.

We can *combine* like terms by using a variation of the distributive property. As you recall, the distributive property states that, for any real numbers *a, b,* and *c,*

$$a(b + c) = ab + ac$$

and

$$a(b - c) = ab - ac$$

Using the commutative property of multiplication, we can rewrite the two distributive properties as follows.

PROCEDURE

Distributive Properties for Like Terms
$$(b + c)a = ba + ca$$
$$(b - c)a = ba - ca$$

Now

$$3x + 2x = (3 + 2)x = 5x$$

Similarly,

$$7z^2 - 5z^2 = (7 - 5)z^2 = 2z^2$$

and

$$8xy + 3xy - 2xy = (8 + 3 - 2)xy$$
$$= 9xy$$

Note that $x + x = 1 \cdot x + 1 \cdot x = (1 + 1)x = 2x$. Thus the coefficient of x is understood to be 1. Also, if an expression within parentheses is preceded by a plus sign, we can simply remove the parentheses and combine any like terms. Thus, using the commutative and associative properties,

$$3x + (2 + 5x) = 3x + 2 + 5x$$
$$= 8x + 2$$

> **NOTE** You can combine like terms by simply adding or subtracting their coefficients.

Use this idea in the next example.

EXAMPLE 8   Combining like terms

Simplify:

a. $5x + 2(x - 4)$ **b.** $-3(x + 5) - 2x$ **c.** $5x - 2(x + 1) + (x + 3)$

SOLUTION

a. $5x + 2(x - 4) = 5x + 2x - 8$
$$= 7x - 8$$

b. $-3(x + 5) - 2x = -3x - 15 - 2x$
$$= -5x - 15 \qquad -3x - 2x = (-3 - 2)x = -5x$$

c. $5x - 2(x + 1) + (x + 3) = 5x - 2x - 2 + x + 3$
$$= 4x - 2 + 3 \qquad 5x - 2x + x = 4x$$
$$= 4x + 1 \qquad\qquad\qquad\qquad \blacksquare$$

 Removing Other Grouping Symbols

To avoid confusion when parentheses occur within other parentheses, we do not write $((x + 5) + 3)$. Instead, we may use a different grouping symbol, the brackets [ ], and write $[(x + 5) + 3]$. To simplify (combine like terms) in such expressions, the innermost grouping symbols are removed first. This procedure is illustrated in the next example.

EXAMPLE 9   Removing other grouping symbols

Remove the grouping symbols and simplify:

$$[(4x^2 - 1) + (2x + 5)] - [(x - 2) + (3x^2 - 3)]$$

SOLUTION We first remove the innermost parentheses and then add like terms. Thus

$$[(4x^2 - 1) + (2x + 5)] - [(x - 2) + (3x^2 - 3)]$$

$$= [4x^2 - 1 + 2x + 5] - [x - 2 + 3x^2 - 3] \qquad \text{Remove parentheses.}$$

$$= [4x^2 + 2x + 4] - [3x^2 + x - 5] \qquad \text{Add like terms.}$$

$$= 4x^2 + 2x + 4 - 3x^2 - x + 5 \qquad \begin{array}{l}\text{Multiply by } -1 \text{ and}\\ \text{remove brackets.}\end{array}$$

$$= x^2 + x + 9 \qquad \text{Add like terms.} \quad \blacksquare$$

GRAPH IT

Graphers automatically follow the order of operations. Thus if you key in 3 $\boxed{\times}$ 4 $\boxed{+}$ 5, the grapher multiplies 3 by 4 *first,* and then adds 5 to obtain 17. On the other hand, if you enter 3 $\boxed{+}$ 4 $\boxed{\times}$ 5, the grapher does **not** add 3 and 4 first, but rather *multiplies* 4 by 5 and then adds 3 to obtain 23. As before, if you want to add 3 and 4 first, you have to key in $\boxed{(}$ 3 $\boxed{+}$ 4 $\boxed{)}$ $\boxed{\times}$ 5.

As you did in Section 1.3, you can check the accuracy of answers containing only one variable. In Example 8c, make a picture (graph) of the original expression $5x - 2(x + 1) + (x + 3)$ by first keying in $\boxed{Y=}$ and entering $Y_1 = 5 \boxed{X,T,\theta} \boxed{-} 2 \boxed{(} \boxed{X,T,\theta} \boxed{+} 1 \boxed{)} \boxed{+} \boxed{(} \boxed{X,T,\theta} \boxed{+} 3 \boxed{)} \boxed{GRAPH}$, then entering $Y_2 = 4 \boxed{X,T,\theta} \boxed{+} 1$. If the two pictures coincide, your answer is correct. You can confirm that there are two graphs (1 and 2) by pressing $\boxed{TRACE}$ and the $\boxed{\blacktriangle}$ and $\boxed{\blacktriangledown}$ keys. See the number at the top right-hand side of the screen change? (Note that you entered $5x$ as $5 \boxed{X,T,\theta}$ *without* the multiplication sign. As in algebra the grapher knows that $5x$ means 5 times x.)

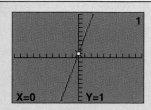

X=0   Y=1

This is the graph of
$5x - 2(x + 1) + (x + 3)$
and the graph of $4x + 1$. Both
graphs are the same, so the
expressions are equal.

EXERCISE 1.4

A **In Problems 1–30, evaluate the expression.**

1. a. $(-10 \cdot 3) + 4$ **2. a.** $(6 \cdot 4) + 6$

   b. $-10 \cdot (3 + 4)$ **b.** $6 \cdot (4 + 6)$

3. a. $(36 \div 4) \cdot 3$ **4. a.** $(-28 \div 7) \cdot 2$

   b. $36 \div (4 \cdot 3)$ **b.** $-28 \div (7 \cdot 2)$

5. $[-5 \cdot (8 + 2)] + 3$ **6.** $[7 \cdot (4 + 3)] + 1$

7. $-7 + [3 \cdot (4 + 5)]$ **8.** $-8 + [3 \cdot (4 + 1)]$

9. $[-6 \cdot (4 - 2)] - 3$ **10.** $[-2(7 - 5)] - 8$

11. $3 - [8 \cdot (5 - 3)]$ **12.** $7 - [3(4 - 5)]$

13. $-5 \cdot 6 - 6$ **14.** $-5 \cdot 2 - 2$

15. $-7 \cdot 3 \div 3 - 3$ **16.** $-36 \cdot 2 \div 18 - 4$

17. $(-20 - 5 + 3 \div 3) \div 6$ **18.** $(-10 - 2 + 10 \div 5) \cdot 4$

19. $\dfrac{8 + (-3)}{5} - 1$ **20.** $\dfrac{7 + (-3)}{2} - 4$

21. $\dfrac{4 \cdot (6 - 2)}{-8} - \dfrac{6}{-2}$ **22.** $\dfrac{5 \cdot (6 - 2)}{-4} - \dfrac{16}{-4}$

23. $-8[3 - 2(4 + 1)] + 1$ **24.** $6[7 - 2(5 - 7)] - 2$

25. $48 \div \{4(8 - 2[3 - 1])\}$ **26.** $-96 \div \{4(8 - 2[1 - 3])\}$

27. $\left[\dfrac{9 - (-3)}{8 - 6}\right]\left[\dfrac{3 + (-8)}{7 - 2}\right]$ **28.** $\left[\dfrac{6 + (-2)}{3 + (-7)}\right]\left[\dfrac{8 + (-12)}{2 - 4}\right]$

29. $\dfrac{3 - 5\left(\dfrac{4 + 2}{2 + 1}\right) - 2}{-4 + 3\left(\dfrac{4 - 2}{4 - 6}\right) - 2}$ **30.** $\dfrac{8 + 2\left(\dfrac{9 - 15}{3 - 1}\right) - 2}{-4 + 8\left(\dfrac{6 - 3}{1 - 4}\right) + 12}$

B **In Problems 31–40, use the correct order of operations and simplify.**

31. $4 \div 2 + 3 - 5^2$ **32.** $8 \div 4 + 7 - 2^2$

33. $4 + 6 \cdot 4 \div 2 - 2^3$ **34.** $6 + 6 \div 3 - 3^3$

35. $-5^2 + \dfrac{2 - 10}{4} + 12 \div 4$

36. $-4^2 + \dfrac{3-7}{2} + 18 \div 9$

37. $-3^3 + 4 - 6 \cdot 8 \div 4 - \dfrac{8-2}{-3}$

38. $-2^3 + 6 - 6 \div 3 \cdot 2 - \dfrac{9-3}{-6}$

39. $4 \cdot 9 \div 3 \cdot 10^3 - 2 \cdot 10^2$ **40.** $5 \cdot 8 \div 4 \cdot 10^3 - 2 \cdot 10^2$

C In Problems 41–70, remove the parentheses (simplify).

41. $4(x - y)$

42. $3(a - b)$

43. $-9(a - b)$

44. $-6(x - y)$

45. $0.3(4x - 2)$

46. $0.2(3a - 9)$

47. $-\left(\dfrac{3a}{2} - \dfrac{6}{7}\right)$

48. $-\left(\dfrac{2x}{3} - \dfrac{1}{5}\right)$

49. $-(2x - 6y)$

50. $-(3a - 6b)$

51. $-(2.1 + 3y)$

52. $-(5.4 + 4b)$

53. $-4(a + 5)$

54. $-6(x + 8)$

55. $-x(6 + y)$

56. $-y(2x + 3)$

57. $-8(x - y)$

58. $-9(a - b)$

59. $-3(2a - 7b)$

60. $-4(3x - 9y)$

61. $0.5(x + y - 2)$

62. $0.8(a + b - 6)$

63. $-\dfrac{6}{5}(a - b + 5)$

64. $-\dfrac{2}{3}(x - y + 4)$

65. $-2(x - y + 3z + 5)$

66. $-4(a - b + 2c + 8)$

67. $-0.3(x + y - 2z - 6)$

68. $-0.2(a + b - 3c - 4)$

69. $-\dfrac{5}{2}(a - 2b + c + 2d - 2)$

70. $-\dfrac{4}{7}(2a - b + 3c + 7d - 7)$

D In Problems 71–85, remove the parentheses and combine like terms.

71. $6x + 3(x - 2)$

72. $8y + 6(y - 3)$

73. $-4(x + 2) - 5x$

74. $-5(x + 3) - 6x$

75. $(5L - 3W) - (W - 6L)$

76. $(2ab - 2ac) - (ab - 4ac)$

77. $5x - (8x + 1) + (x + 1)$

78. $3x - (7x + 2) + (x + 2)$

79. $\dfrac{2x}{9} - \left(\dfrac{x}{9} - 2\right)$

80. $\dfrac{5x}{7} - \left(\dfrac{2x}{7} - 3\right)$

81. $4a - (a + b) + 3(b + a)$ **82.** $8x - 3(x + y) - (x - y)$

83. $7x - 3(x + y) - (x + y)$

84. $4(b - a) + 3(b + a) - 2(a + b)$

85. $-(x + y - 2) + 3(x - y + 6) - (x + y - 16)$

E In Problems 86–95, remove the grouping symbols and simplify.

86. $[(a^2 - 4) + (2a^3 - 5)] + [(4a^3 + a) + (a^2 + 9)]$

87. $(x^2 + 7 - x) + [-2x^3 + (8x^2 - 2x) + 5]$

88. $[(0.4x - 7) + 0.6x^2] - [(0.3x^2 - 2) - 0.8x]$

89. $\left[\left(\dfrac{5}{7}x^2 + \dfrac{1}{5}x\right) - \dfrac{1}{8}\right] - \left[\left(\dfrac{3}{7}x^2 - \dfrac{3}{5}x\right) + \dfrac{5}{8}\right]$

90. $[3(x + 2) - 10] + [5 + 2(5 + x)]$

91. $[3(2a - 4) + 5] - [2(a - 1) + 6]$

92. $[6(a - b) + 2a] - [3b - 4(a - b)]$

93. $[4a - (3 + 2b)] - [6(a - 2b) + 5a]$

94. $-[-(x + y) + 3(x - y)] - [4(x + y) - (3x - 5y)]$

95. $-[-(0.2x + y) + 3(x - y)] - [2(x + 0.3y) - 5]$

SKILL CHECKER

Find:

96. $[(-3)(-3)](-3)$ **97.** $[(-2)(-2)][(-2)(-2)]$

98. $(-2)[(-2)(-2)]$ **99.** $\dfrac{(-2)(-2)(-2)}{(-2)(-2)}$

100. $\dfrac{(-3)(-3)(-3)(-3)}{(-3)(-3)}$

USING YOUR KNOWLEDGE   Average Velocity, Momentum, and Kinetic Energy

The distributive property is helpful in solving problems in many areas. Use your knowledge of the distributive property to remove the parentheses in Problems 101–104.

101. If your car is accelerating at a constant rate, and v_1 is the initial velocity and v_2 is the final velocity, the *average* velocity is

$$v_a = \dfrac{1}{2}(v_1 + v_2)$$

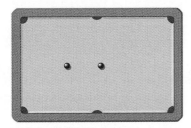

102. The momentum M of a billiard ball is the product of its mass m and its velocity v. If two billiard balls of equal mass m and moving in the same straight line, with velocities v_1 and v_2, respectively, collide, the total momentum M is given by

$$M = m(v_1 + v_2)$$

103. The total kinetic energy (KE) of the billiard balls in Problem 102 is given by

$$\text{KE} = \frac{1}{2}m(v_1^2 + v_2^2)$$

104. The length of a belt L needed to connect two pulleys of radius r_1 and r_2, respectively, with centers d units apart is

$$L = \pi(r_1 + r_2) + 2d$$

WRITE ON . . .

105. Explain why $(32 \div 4) \cdot 2$ is different from $32 \div (4 \cdot 2)$.

106. Write in your own words the definition for "like terms."

Evaluate:

107. $-3^2 + \left(\dfrac{4 - 8}{2}\right) + 48 \div 6$

108. $\left(\dfrac{2 - 14}{6}\right) + 16 \div 4 - 5^2$

Simplify:

109. $-2(x - 3y + 2z - 4)$

110. $-\dfrac{2}{3}(a + 6b - 9c - 12)$

111. $\dfrac{3}{8}x - \left(\dfrac{x}{8} - 5\right)$

112. $\dfrac{5}{7}x - \left(4 - \dfrac{5}{7}x\right)$

113. $[(a^2 - 5) + (2a^3 - 3)] - [(4a^3 + a) - (a^2 - 9)]$

114. $\left[\left(\dfrac{5}{9}x^2 + \dfrac{1}{5}x\right) - \dfrac{1}{3}\right] - \left[\left(\dfrac{2}{9}x^2 - \dfrac{2}{5}x\right) - \dfrac{4}{3}\right]$

Evaluate:

115. $\left[\dfrac{8 - (-4)}{8 - 10}\right]\left[\dfrac{5 + (-9)}{7 - 3}\right]$

116. $\left[\dfrac{7 + (-4)}{4 - 7}\right]\left[\dfrac{9 + (-14)}{3 - 5}\right]$

research questions

Sources of information for these questions can be found in the *Bibliography* at the end of this book.

1. There is a charming story about the long-accumulated used wooden tally sticks mentioned in the *Chapter Preview*. Find out how their disposal literally resulted in the destruction of the old Houses of Parliament in England.

2. Write a paper detailing the Egyptian number system and the base and symbols used, and enumerate the similarities and differences between the Egyptian and our (Hindu-Arabic) system of numeration.

3. Write a paper detailing the Greek number system and the base and symbols used, and enumerate the similarities and differences between the Greek and our system of numeration.

4. Find out about the development of the symbols we use in our present numeration system. Where was the symbol for zero invented and by whom?

5. When were negative numbers introduced, by whom were they introduced, and what were they first called?

6. Write a short paper about the Rhind, or Ahmes, papyrus. What is the significance of the names *Rhind* and *Ahmes?* What is the content of the papyrus and who discovered it?

7. Find out what "gematria" (not geometry!) is, the significance of 666, and the reason why many old editions of the Bible substitute the number 99 for *amen* at the end of a prayer.

SUMMARY

SECTION	ITEM	MEANING	EXAMPLE
1.1A	Empty or null set $\varnothing$	The set containing no elements	The set of natural numbers between 5 and 6 is $\varnothing$.
	Natural numbers	$N = \{1, 2, 3, \ldots\}$	2, 76, and 308 are natural numbers.
	Whole numbers	$W = \{0, 1, 2, \ldots\}$	0, 8, and 93 are whole numbers.
	Integers	$I = \{\ldots, -2, -1, 0, 1, 2, \ldots\}$	-7 and 23 are integers.
	Rational numbers	$Q = \left\{ r \mid r = \dfrac{a}{b}, a \text{ and } b \text{ are integers, and } b \neq 0 \right\}$	$\dfrac{1}{5}, -\dfrac{2}{3}, 0, 9, 1.4,$ and $0.\overline{3}$ are rational numbers.
1.1B	Rational numbers	The set Q of rational numbers is the same as the set of terminating or repeating decimals.	0.345 and $0.\overline{3}$ are rational numbers.
	Irrational numbers	$H = \{x \mid x \text{ is not rational}\}$	$\sqrt{2}$ and π are irrational numbers.
	Real numbers (R)	The set of all rationals and irrationals	$\dfrac{2}{7}, -\dfrac{2}{3}, 0, 9, 1.4, 0.\overline{3}, \sqrt{2},$ and π are real numbers.
1.1C	$N \subseteq W \subseteq I$ $\subseteq Q \subseteq R$	N is a subset of W, W is a subset of I, and so on.	Every natural number is a whole number, every whole number is an integer, and so on.
1.1D	Additive inverses (opposites)	a and $-a$ are additive inverses.	8 and -8 are additive inverses.
1.1E	Absolute value $\lvert a \rvert$	The distance from 0 to a on the number line. $\lvert a \rvert = \begin{cases} a & \text{when } a \geq 0 \\ -a & \text{when } a < 0 \end{cases}$	$\lvert -8 \rvert = 8, \left\lvert \dfrac{2}{3} \right\rvert = \dfrac{2}{3},$ and $\lvert -0.4 \rvert = 0.4$
1.1F	Trichotomy law	If a and b are real numbers, then 1. $a = b$, or 2. $a < b$, or 3. $a > b$	
1.2A	Adding signed numbers with the same sign	Add their absolute values and give the sum the common sign.	$-3 + (-7) = -(\lvert -3 \rvert + \lvert -7 \rvert)$ $= -10$
	Adding signed numbers with different signs	Subtract the smaller absolute value from the greater absolute value and use the sign of the number with the greater absolute value.	$3 + (-5) = -(5 - 3) = -2$ $-7 + 9 = +(9 - 7) = 2$
	Subtraction	If a and b are real numbers, $a - b = a + (-b)$.	$3 - (-4) = 3 + 4 = 7$
	$a \cdot b, ab, (a)(b),$ $a(b), (a)b$	The product of a and b	$5 \cdot 2 = (5)(2) = 5(2) = (5)2 = 10$
	Multiplying signed numbers with different signs	Multiply their absolute values; the product is negative.	$3 \cdot (-4) = -12$ $-7 \cdot 2 = -14$

SECTION	ITEM	MEANING	EXAMPLE
1.2A	Multiplying signed numbers with the same sign	Multiply their absolute values; the product is positive.	$3 \cdot 8 = 24$ and $(-9)(-2) = 18$
	Multiplication of fractions	$\dfrac{a}{b} \cdot \dfrac{c}{d} = \dfrac{a \cdot c}{b \cdot d}$ $\quad (b \neq 0, d \neq 0)$	$\left(-\dfrac{3}{4}\right) \cdot \dfrac{2}{7} = -\dfrac{3}{14}$
	Division	If a, b, c are real numbers, $\dfrac{a}{b} = c$ means $a = bc$ $\quad (b \neq 0)$	$\dfrac{6}{3} = 2$ means $6 = 3 \cdot 2$.
	Dividing signed numbers	The quotient of two real numbers with the same sign is positive, with different signs, negative.	$\dfrac{6}{-3} = -2, \dfrac{-6}{3} = -2,$ and $\dfrac{-8}{-2} = 4$
	Zero in division problems	$\dfrac{0}{a} = 0 \quad (a \neq 0)$	$\dfrac{0}{9} = 0, \dfrac{0}{-8} = 0,$ and $\dfrac{0}{-2.4} = 0$
		$\dfrac{a}{0}$ is not defined.	$\dfrac{9}{0}, \dfrac{-8}{0},$ and $\dfrac{-2.4}{0}$ are not defined.
		$\dfrac{0}{0}$ is indeterminate.	
	Reciprocal	The reciprocal of a is $\dfrac{1}{a}$ $\quad (a \neq 0)$	The reciprocal of -7 is $-\dfrac{1}{7}$.
	Division of fractions	$\dfrac{a}{b} \div \dfrac{c}{d} = \dfrac{a}{b} \cdot \dfrac{d}{c}$ $\quad (b \neq 0, c \neq 0, d \neq 0)$	$\dfrac{3}{4} \div \dfrac{9}{2} = \dfrac{3}{4} \cdot \dfrac{2}{9} = \dfrac{1}{6}$

1.2B   Real-Number Properties

If a, b, and c are real numbers:

NAME	ADDITION	MULTIPLICATION
Closure	$a + b$ is a real number.	$a \cdot b$ is a real number.
Commutative	$a + b = b + a$	$a \cdot b = b \cdot a$
Associative	$a + (b + c) = (a + b) + c$	$a \cdot (b \cdot c) = (a \cdot b) \cdot c$
Identity	$a + 0 = 0 + a = a$ (0 is the identity.)	$1 \cdot a = a \cdot 1 = a$ (1 is the identity.)
Inverse	For each real number a, there is a unique inverse $-a$ such that $a + (-a) = -a + a = 0$.	For each nonzero real number a, there is a unique real number $\dfrac{1}{a}$ such that $a \cdot \dfrac{1}{a} = \dfrac{1}{a} \cdot a = 1$ $\left(\dfrac{1}{a}$ is called the reciprocal of $a.\right)$

Distributive property of multiplication (over addition)	$a(b + c) = ab + ac$	
Distributive property of multiplication (over subtraction)	$a(b - c) = ab - ac$	

SECTION	ITEM	MEANING	EXAMPLE
1.3	Exponent	$a^n = a \cdot a \cdot a \cdots a$ (n factors) n is called the exponent.	$3^4 = 3 \cdot 3 \cdot 3 \cdot 3$
	Base	In the expression a^n, a is called the base.	In the expression 2^5, 2 is the base.
1.3B	Negative exponent	$a^{-n} = \dfrac{1}{a^n}$ $\quad (a \neq 0)$	$5^{-2} = \dfrac{1}{5^2} = \dfrac{1}{25}$
	Zero exponent	$a^0 = 1 \quad (a \neq 0)$	$2^0 = 1$ and $\left(\dfrac{-1}{4}\right)^0 = 1$
1.3C	First law of exponents	$a^m \cdot a^n = a^{m+n}$	$x^5 \cdot x^4 = x^9$
	Second law of exponents	$\dfrac{a^m}{a^n} = a^{m-n} \quad (a \neq 0)$	$\dfrac{x^8}{x^3} = x^5$
1.3D	Third law of exponents	$(a^m)^n = a^{m \cdot n}$	$(x^2)^5 = x^{10}$
	Raising powers to powers	$(a^m b^n)^k = a^{m \cdot k} b^{n \cdot k}$	$(x^3 y^5)^6 = x^{3 \cdot 6} y^{5 \cdot 6} = x^{18} y^{30}$
1.3E	Raising a quotient to a power	$\left(\dfrac{a^m}{b^n}\right)^k = \dfrac{a^{m \cdot k}}{b^{n \cdot k}}$	$\left(\dfrac{x^5}{y^3}\right)^4 = \dfrac{x^{5 \cdot 4}}{y^{3 \cdot 4}} = \dfrac{x^{20}}{y^{12}}$
1.3F	Scientific notation	A number is in scientific notation when it is written in the form $m \times 10^n$, where m is greater than or equal to 1 and less than 10 and n is an integer.	$352 = 3.52 \times 10^2$ is in scientific notation.
1.4B	Order of operations (from left to right)	Parentheses (or other grouping symbols) Division bars Exponentiation Multiplication Division Addition Subtraction	
1.4C	$-1 \cdot a$ $-(a + b)$ $-(a - b)$	$-1 \cdot a = -a$ $-(a + b) = -a - b$ $-(a - b) = -a + b$	$-1 \cdot 4 = -4$ $-(x + 7) = -x - 7$ $-(x - 3) = -x + 3$
1.4D	Similar or like terms	Two or more terms that differ only in their numerical coefficients	$-3a$ and $7a$ are similar or like terms.

REVIEW EXERCISES

(If you need help with these exercises, look in the section indicated in brackets.)

1. **[1.1A]** Use roster notation to list the natural numbers that fall between
   a. 3 and 9
   b. 4 and 8

2. **[1.1B]** Write as a decimal:
   a. $\dfrac{1}{5}$
   b. $\dfrac{2}{5}$

3. **[1.1B]** Write as a decimal:
   a. $\dfrac{1}{9}$
   b. $\dfrac{2}{9}$

4. **[1.1C]** Classify the given number by making a check mark (✓) in the appropriate place.

Set	0.3	0	$\frac{-3}{4}$	-5	$\sqrt{3}$
Natural number					
Whole number					
Integer					
Rational number					
Irrational number					
Real number					

5. **[1.1D]** Find the additive inverse:
   a. -3.5
   b. $\dfrac{3}{4}$

6. **[1.1E]** Find:
   a. $|-9|$
   b. $|4.2|$

7. **[1.1E]** Find:
   a. $\left|-\dfrac{1}{8}\right|$
   b. $|0.\overline{4}|$

8. **[1.1F]** Fill in the blank with $<$, $>$, or $=$ to make the resulting statement true:
   a. -8 _____ -7
   b. -4 _____ -3

9. **[1.1F]** Fill in the blank with $<$, $>$, or $=$ to make the resulting statement true:
   a. $\dfrac{1}{4}$ _____ $\dfrac{1}{5}$
   b. $\dfrac{3}{4}$ _____ 0.75

10. **[1.1F]** Fill in the blank with $<$, $>$, or $=$ to make the resulting statement true:
    a. $\dfrac{1}{5}$ _____ 0.25
    b. $0.\overline{6}$ _____ $\dfrac{2}{3}$

11. **[1.2A]** Find:
    a. $-3 + (-8)$
    b. $-5 + 2$

12. **[1.2A]** Find:
    a. $\dfrac{1}{7} - \dfrac{3}{7}$
    b. $-0.2 - 0.4$

13. **[1.2A]** Find:
    a. $8 - (-4)$
    b. $-3 - (-7)$

14. **[1.2A]** Find:
    a. $\dfrac{3}{4} - \left(-\dfrac{1}{5}\right)$
    b. $\dfrac{5}{6} - \left(-\dfrac{1}{4}\right)$

15. **[1.2A]** Find:
    a. $9 \cdot (-4)$
    b. $-2.4 \cdot 6$

16. **[1.2A]** Find:
    a. $\left(-\dfrac{3}{4}\right) \cdot \dfrac{7}{8}$
    b. $\left(-\dfrac{5}{6}\right) \cdot \left(-\dfrac{2}{7}\right)$

17. **[1.2A]** Find:
    a. $\dfrac{0}{7}$
    b. $\dfrac{8}{0}$

18. [1.2A] Find the reciprocal:

a. $-\dfrac{3}{5}$

b. 0.3

19. [1.2A] Find:

a. $-\dfrac{3}{5} \div \dfrac{4}{15}$

b. $\dfrac{3.6}{-1.2}$

20. [1.2B] Name the property illustrated in the statement:

a. $(4 + 9) + 5 = 5 + (4 + 9)$

b. $(3 + 5) + 8 = 3 + (5 + 8)$

21. [1.3A] Evaluate:

a. $(-3)^4$

b. -3^4

22. [1.3B] Evaluate:

a. 9^0

b. $\left(\dfrac{1}{7}\right)^0$

23. [1.3B] Write as a fraction:

a. $(-8)^{-3}$

b. x^{-10}

24. [1.3C] Multiply and simplify:

a. $(3x^4y)(-5x^{-8}y^9)$

b. $(4x^{-3}y^{-1})(-6x^{-8}y^{-7})$

25. [1.3C] Divide and simplify:

a. $\dfrac{48x^4}{16x^6}$

b. $\dfrac{8x^5}{-2x^{-6}}$

26. [1.3C] Divide and simplify:

a. $\dfrac{-5x^{-3}}{15x^{-4}}$

b. $\dfrac{8x^{-4}}{-4x^7}$

27. [1.3D] Simplify:

a. $(-2x^7y^{-6})^3$

b. $(-2x^{-6}y^{-6})^4$

28. [1.3E] Simplify:

a. $\left(\dfrac{x^6}{y^{-3}}\right)^{-4}$

b. $\left(\dfrac{x^{-5}}{y^3}\right)^{-5}$

29. [1.3F] Write in scientific notation:

a. $340,000$

b. 0.000047

30. [1.3F] Write in decimal notation:

a. 3.7×10^4

b. 7.8×10^{-3}

31. [1.4A] Evaluate:

a. $[-8 \cdot (9 + 2)] + 13$

b. $[-7(3 - 8)] + 15$

32. [1.4A] Evaluate:

a. $6^2 \div 3 - 9 \cdot 2 \div 3 + 3$

b. $\dfrac{5 \cdot (68 - 32)}{9}$

33. [1.4B] Evaluate:

a. $-3^2 + \dfrac{4 - 10}{2} + 15 \div 3$

b. $-4^3 + \dfrac{2 - 10}{2} - 25 \div 5$

34. [1.4C] Remove the parentheses (simplify):

a. $-3(x - 7)$

b. $3(x + 8) - (x + 7)$

35. [1.4E] Simplify:

a. $3(x + 2y - 2) - 2(2x - 2y + 5)$

b. $[(5x^2 - 3) + (4x + 5)] - [(x - 4) + (2x^2 - 2)]$

PRACTICE TEST

(Answers are on pages 58–59.)

1. Use roster notation to list the natural numbers between 5 and 9.

2. Write as a decimal:

   a. $\dfrac{3}{8}$ b. $\dfrac{2}{3}$

3. Classify the given number by making a check mark (✓) in the appropriate place.

Set	0.5	0	−6	$\frac{-2}{7}$	$\sqrt{5}$
Natural number					
Whole number					
Integer					
Rational number					
Irrational number					
Real number					

4. Find the additive inverse of $\dfrac{4}{5}$.

5. Find:

   a. $|-9|$ b. $|0.5|$

6. Fill in the blank with $<$, $>$, or $=$ to make the resulting statement true:

   a. $-\dfrac{1}{4}$ _____ $-\dfrac{1}{3}$ b. 0.4 _____ $\dfrac{2}{5}$

7. Find:

   a. $-9 + 5$ b. $-0.8 + (-0.7)$

8. Find:

   a. $-16 - 7$ b. $-0.6 - (-0.4)$

9. Find:

   a. $-\dfrac{1}{8} - \dfrac{3}{4}$ b. $-\dfrac{3}{4} - \left(-\dfrac{5}{6}\right)$

10. Find:

    a. $6 \cdot (-9)$ b. $-4 \cdot (-1.2)$

11. Find:

    a. $-\dfrac{1}{2} \cdot \dfrac{2}{9}$ b. $-\dfrac{3}{2} \div \dfrac{9}{8}$

12. Name the property illustrated in the statement:
    a. $(7 + 3) + 6 = (3 + 7) + 6$ b. $(2 + 9) + 4 = 2 + (9 + 4)$

13. Name the property illustrated in the statement:

    a. $3 \cdot \dfrac{1}{3} = 1$ b. $0.3 + (-0.3) = 0$

14. Evaluate:
    a. $(-3)^4$ b. -3^4

15. Write as a fraction:
    a. 7^{-2} b. x^{-8}

16. Perform the indicated operation and simplify:

    a. $(3x^4y)(-4x^{-8}y^8)$ b. $\dfrac{48x^4}{16x^{-8}}$

17. Simplify:

    a. $(-2x^8y^{-2})^3$ b. $\left(\dfrac{x^5}{y^{-3}}\right)^{-3}$

18. Write 6.5×10^{-3} as a decimal.

19. Write 8.5×10^5 as a whole number.

20. Perform the calculation and write your answer in scientific notation:

    $(7.1 \times 10^5) \times (4 \times 10^{-7})$

21. Evaluate:

    a. $[-7(4 + 3)] + 9$ b. $\dfrac{5 \cdot (131 - 32)}{9}$

22. Evaluate $-4^3 + \dfrac{6 - 12}{2} + 15 \div 3$.

23. Simplify:
    a. $-5(x + 7)$ b. $7x - (3x + 1) + (2x + 2)$

24. Simplify $[(5x^2 - 3) + (3x + 7)] - [(x - 3) + (2x^2 - 2)]$.

25. Simplify $[(8x^2 - 5) - (3x - 7)] - [(2 - 3x) + (8x^2 + 1)]$.

ANSWERS TO PRACTICE TEST

Answer	If you missed:	Review:		
	Question	Section	Examples	Page
1. $\{6, 7, 8\}$	1	1.1	1	6
2a. 0.375	2a	1.1	2	7
2b. $0.\overline{6}$	2b	1.1	2	7
3. (see table below)	3	1.1	3	8
4. $-\dfrac{4}{5}$	4	1.1	4	9–10
5a. 9	5a	1.1	5	10–11
5b. 0.5	5b	1.1	5	10–11
6a. $>$	6a	1.1	6	11–12
6b. $=$	6b	1.1	6	11–12
7a. -4	7a	1.2	1	16–17
7b. -1.5	7b	1.2	1	16–17
8a. -23	8a	1.2	2a	17
8b. -0.2	8b	1.2	2b	17
9a. $-\dfrac{7}{8}$	9a	1.2	2c	17
9b. $\dfrac{1}{12}$	9b	1.2	2c	17
10a. -54	10a	1.2	3	18
10b. 4.8	10b	1.2	3	18
11a. $-\dfrac{1}{9}$	11a	1.2	4	19

3.

Set	0.5	0	-6	$\frac{-2}{7}$	$\sqrt{5}$
N					
W		✓			
I		✓	✓		
Rat.	✓	✓	✓	✓	
Irr.					✓
R	✓	✓	✓	✓	✓

Answer	If you missed:		Review:	
	Question	Section	Examples	Page
11b. $-\dfrac{4}{3}$	11b	1.2	5, 6, 7	21−23
12a. Commutative property of addition	12a	1.2	8	24
12b. Associative property of addition	12b	1.2	8	24
13a. Inverse (reciprocal) property for multiplication	13	1.2	8	24
13b. Inverse (opposite) property for addition				
14a. 81	14a	1.3	1	29
14b. -81	14b	1.3	1	29
15a. $\dfrac{1}{49}$	15a	1.3	2	31
15b. $\dfrac{1}{x^8}$	15b	1.3	2	31
16a. $-\dfrac{12y^9}{x^4}$	16a	1.3	3	33
16b. $3x^{12}$	16b	1.3	4	34
17a. $-\dfrac{8x^{24}}{y^6}$	17a	1.3	5	36
17b. $\dfrac{1}{x^{15}y^9}$	17b	1.3	6	37
18. 0.0065	18	1.3	8	38
19. 850,000	19	1.3	8	38
20. 2.84×10^{-1}	20	1.3	7, 8, 9	38−39
21a. -40	21a	1.4	1, 2	42, 43
21b. 55	21b	1.4	3	43−44
22. -62	22	1.4	4	44
23a. $-5x - 35$	23a	1.4	5, 6, 7	45−47
23b. $6x + 1$	23b	1.4	8	48
24. $3x^2 + 2x + 9$	24	1.4	9	49−50
25. -1	25	1.4	9	49−50

2

Linear Equations and Inequalities

ou have probably heard that algebra is simply a generalized arithmetic. We shall show why this is true in this chapter where we cover the topics most students associate with algebra. In Section 2.1, we shall learn how to solve linear equations in one variable, and in Section 2.2, we shall use the techniques of Section 2.1 to solve for specified variables in formulas that you may have seen in some of your other courses. We then discuss how to solve word problems with equations. In Sections 2.3 and 2.4, we develop problem-solving techniques that will be used throughout the text. We generalize the methods for solving linear equations and then use them in Section 2.5 to solve linear and compound inequalities. We conclude this chapter by learning how to solve absolute-value equations and inequalities.

Who invented algebra, anyway? One of the earliest accounts of an algebra problem is in the Rhind papyrus, an ancient Egyptian document written by an Egyptian priest named Ahmes (ca. 1620 B.C.) and purchased by Henry Rhind. When Ahmes wished to find a number such that the number added to its seventh made 19, he symbolized the number by the sign we translate as "heap." Today, we write the problem as

$$x + \frac{x}{7} = 19$$

Can you find the answer? Ahmes says it is $16 + \frac{1}{2} + \frac{1}{8}$.

Western Europeans, however, learned their algebra from the works of Mohammed ibn Musa al-Khowarizmi (this translates as Mohammed the son of Moses of Khowarizmi; ca. A.D. 820), an astronomer and mathematician of Baghdad and author of the treatise *Hisab al-jabr w'al muqabalah,* the science of restoring (placing variables on one side of an equation) and reduction (collecting like terms). With the passage of time, the word *al-jabr* evolved into our present word *algebra,* a subject that we shall continue to study now.

2.1

LINEAR EQUATIONS IN ONE VARIABLE

To succeed, review how to:

1. Add, subtract, multiply, and divide positive and negative numbers (pp. 15–21).

2. Use the commutative, associative, and distributive properties (pp. 23–25).

3. Find the sum of opposites (additive inverses) (p. 17).

4. Find the product of two reciprocals (pp. 19, 22).

(See the *Skill Checker*, page 50.)

Objectives:

A Determine whether a number is a solution of a given equation.

B Solve linear equations using the properties of equality.

C Solve linear equations in one variable using the six-step procedure.

D Solve linear equations involving decimals.

---

getting started

Crickets, Ants, and Temperatures

Does temperature affect animal behavior? You certainly know about bears hibernating and languid students in the spring. But what about crickets and ants? Farmers claim that they can tell the temperature F in degrees Fahrenheit by listening to the number of chirps N a cricket makes in 1 minute! How? By using the formula

$$F = \frac{N}{4} + 40$$

Thus if a cricket is chirping 80 times a minute, the temperature is $F = \frac{80}{4} + 40 = 60°F$ (60 degrees Fahrenheit). Now suppose the temperature is 90°F. How fast is the cricket chirping? To find the answer, you must know how to solve the equation

$$90 = \frac{N}{4} + 40$$

Similarly, the speed S of an ant, in centimeters per second (cm/sec), is given by $S = \frac{1}{6}(C - 4)$, where C is the temperature in degrees Celsius. If an ant is moving at 4 cm/sec, what is the temperature? Here, you have to solve the equation $4 = \frac{1}{6}(C - 4)$. You will learn how to solve these and other similar equations in this chapter.

Finally, if you know the relationship between Fahrenheit and Celsius temperatures and you look at these two formulas, can you tell whether the crickets stop chirping before the ants stop crawling?

How do we express our ideas in algebra? We start with a collection of letters (**variables**) and real numbers (**constants**) and then perform the basic operations of addition, subtraction, multiplication, and division. The result is an **algebraic expression** such as

$$2x + 7, \qquad x^2 - 3x + 5, \qquad \text{or} \qquad \frac{x^5 y^7}{z^3}$$

 CAUTION [Note that in expressions such as

$$\frac{x^5 y^7}{z^3}$$

with variables in the denominator, the denominator cannot be zero.]

An **equation** is a sentence stating that two algebraic expressions are equal. Equalities have three important properties: reflexive, symmetric, and transitive.

PROPERTIES OF EQUALITIES

For all real numbers a, b, and c,

1. $a = a$ Reflexive property

2. If $a = b$, then $b = a$.   Symmetric property

3. If $a = b$ and $b = c$, then $a = c$.   Transitive property

Examples of the Reflexive Property

$$0.5 = 0.5$$

$$x + 7 = x + 7$$

$$x^2 + 5x - 7 = x^2 + 5x - 7$$

Examples of the Symmetric Property

If $x = 7$, then $7 = x$.

If $C = 2\pi r$, then $2\pi r = C$.

If $y = x^2 + 3x - 7$, then $x^2 + 3x - 7 = y$.

Examples of the Transitive Property

If $x = y$ and $y = 2a$, then $x = 2a$.

If $C = 2\pi r$ and $2\pi r = \pi d$, then $C = \pi d$.

If $x = \dfrac{a}{b}$ and $\dfrac{a}{b} = \dfrac{ac}{bc}$, then $x = \dfrac{ac}{bc}$.

In this section we consider *linear equations* involving only real numbers and *one* variable. Here are some examples of linear equations:

$$x = 8, \qquad 2x - 5 = 7, \qquad 3(y - 1) = 2y + 8, \qquad \text{and} \qquad 3k + 7 = 10$$

In general, we have the following definition.

LINEAR EQUATIONS

A **linear equation** in one variable is an equation that can be written in the form

$$ax + b = c$$

where a, b, and c are real numbers and $a \neq 0$.

Since the highest power of the variable in a linear equation is 1, linear equations are also called **first-degree equations**.

 Solutions of an Equation

Some equations are always *true* (identities: $2 + 2 = 4, 5 - 3 = 2$), some are always *false* (contradictions: $2 + 2 = 22, 5 - 3 = -2$), and some are neither true nor false. For example, the equation $x + 1 = 5$ is neither true nor false. It is a *conditional* equation, and its truth or falsity depends on the value of x.

In the equation $x + 1 = 5$, the **variable** x can be replaced by many numbers, but only one number will make the resulting statement true. This number is called the *solution* of the equation.

SOLUTIONS OF AN EQUATION

> The **solutions** (**roots**) of an equation are the replacements of the variable that make the equation a true statement. To **solve** an equation is to find all its solutions.

To determine whether a number is a solution of an equation, we replace the variable by the number. For example, 4 is a solution of $x + 1 = 5$ because replacing x with 4 in the equation yields $4 + 1 = 5$, a true statement, but -6 is *not* a solution because $-6 + 1 \neq 5$. Since 4 is the *only* number that yields a true statement, the solution set of $x + 1 = 5$ is $\{4\}$.

EXAMPLE 1　Determining when a number is a solution

Determine whether
a. 8 is a solution of $x - 5 = 3$.

b. 5 is a solution of $3 = 2 - y$.

c. 6 is a solution of $\frac{1}{3}z - 4 = 2z - 14$.

SOLUTION
a. Substituting 8 for x in $x - 5 = 3$, we have $8 - 5 = 3$, a true statement. Thus 8 is a solution of $x - 5 = 3$.

b. Substituting 5 for y in $3 = 2 - y$, we obtain $3 = 2 - 5$, which is false. Hence 5 is not a solution of $3 = 2 - y$.

c. If we replace z by 6 in $\frac{1}{3}z - 4 = 2z - 14$, we obtain

$$\frac{1}{3}(6) - 4 = 2(6) - 14$$

$$2 - 4 = 12 - 14$$

$$-2 = -2$$

which is a true statement. Thus 6 is a solution of $\frac{1}{3}z - 4 = 2z - 14$. ■

 Solving Equations Using the Properties of Equality

We have learned how to determine whether a number is a solution of an equation; now we will learn how to find these solutions—that is, how to *solve* the equation. The procedure is to find an *equivalent* equation whose solution is obvious. For example, the equations $x = 2$ and $x + 3 = 5$ are equivalent because $x = 2$, with the obvious solution 2, is the only solution of the equation $x + 3 = 5$.

EQUIVALENT EQUATIONS

> Two or more equations are **equivalent** if they have the same solution set.

To solve equations, we use the idea that adding or subtracting the same number on both sides of the equation and multiplying or dividing both sides of an equation by the same nonzero number produces an equivalent equation. Here is the principle.

PROPERTIES OF EQUALITY

If C is a real number, then the following equations are all equivalent.

$$A = B$$

Add C.	$A + C = B + C$
Subtract C.	$A - C = B - C$
Multiply by C.	$A \cdot C = B \cdot C$ $\quad (C \neq 0)$
Divide by C.	$\dfrac{A}{C} = \dfrac{B}{C}$ $\quad (C \neq 0)$

Now suppose we want to solve $x + 5 = 7$. Since we are trying to find a value of x that will satisfy the equation, we try to get x by itself on one side of the equation. To do this, we "undo" the addition of 5 by *adding* the inverse of 5 on both sides. Thus

$$x + 5 = 7 \qquad \text{Given.}$$
$$x + 5 + (-5) = 7 + (-5) \qquad \text{Add } (-5) \text{ to both sides.}$$
$$x + 0 = 2 \qquad 5 + (-5) = 0$$
$$x = 2$$

The solution is 2, and the solution set is $\{2\}$. Note that we can also solve the equation by subtracting 5 from both sides:

$$x + 5 = 7 \qquad \text{Given.}$$
$$x + 5 - 5 = 7 - 5 \qquad \text{Subtract 5 from both sides.}$$
$$x + 0 = 2 \qquad 5 - 5 = 0$$
$$x = 2$$

Using the same idea, we can solve $8 = 3x - 7$ by first adding 7 to both sides. Thus

$$8 = 3x - 7 \qquad \text{Given.}$$
$$8 + 7 = 3x - 7 + 7 \qquad \text{Add 7 to both sides.}$$
$$15 = 3x$$
$$\frac{15}{3} = \frac{3x}{3} \qquad \text{Divide both sides by 3.}$$
$$5 = x$$

The solution is 5, and the solution set is $\{5\}$.

EXAMPLE 2   Solving linear equations using the equality properties

Solve:

a. $2x - 4 = 6$ **b.** $\dfrac{2}{3}y - 3 = 9$

SOLUTION

a. To solve this equation, we want the variable x by itself on one side of the equation. We start by adding 4 to both sides.

$$2x - 4 = 6 \qquad \text{Given.}$$

$$2x - 4 + 4 = 6 + 4 \qquad \text{Add 4 to both sides.}$$

$$2x = 10$$

$$\frac{1}{2} \cdot 2x = \frac{1}{2} \cdot 10 \qquad \text{Multiply both sides by } \tfrac{1}{2}.$$

$$x = 5 \qquad \tfrac{1}{2} \cdot 2 = 1 \text{ and } \tfrac{1}{2} \cdot 10 = 5$$

Thus, the solution set is $\{5\}$.

Note that to solve $2x = 10$, we could also divide both sides by 2

$$\frac{2x}{2} = \frac{10}{2}$$

to obtain the same result, $x = 5$. To check this solution, we substitute 5 for x in the original equation and use the following diagram:

$$
\begin{array}{c|c}
\multicolumn{2}{c}{2x - 4 \overset{?}{=} 6} \\
\hline
2(5) - 4 & 6 \\
10 - 4 & \\
6 &
\end{array}
$$

Since both sides yield 6, we have a true statement and our result is correct.

b.

$$\frac{2}{3}y - 3 = 9 \qquad \text{Given.}$$

$$\frac{2}{3}y - 3 + 3 = 9 + 3 \qquad \text{Add 3 to both sides.}$$

$$\frac{2}{3}y = 12$$

$$\frac{3}{2} \cdot \frac{2}{3}y = \frac{3}{2} \cdot 12 \qquad \text{Multiply both sides by } \tfrac{3}{2}.$$

$$y = 18 \qquad \tfrac{3}{2} \cdot 12 = \tfrac{3}{2} \cdot \tfrac{12}{1} = 18$$

Thus the solution set is $\{18\}$. You can check this answer by substituting 18 for y in $\frac{2}{3}y - 3 = 9$ to obtain

$$
\begin{array}{c|c}
\multicolumn{2}{c}{\frac{2}{3}y - 3 \overset{?}{=} 9} \\
\hline
\frac{2}{3} \cdot 18 - 3 & 9 \\
12 - 3 & \\
9 &
\end{array}
$$

∎

EXAMPLE 3   Solving linear equations using the equality properties

Solve: $4a - 7 = a + 4$

SOLUTION   We start by adding 7 to both sides. We then add the inverse of a on both sides so that only variables are on the left. Here are the steps:

$4a - 7 = a + 4$	Given.
$4a - 7 + 7 = a + 4 + 7$	Add **7** to both sides.
$4a = a + 11$	
$4a + (-a) = a + (-a) + 11$	Add $(-a)$ to both sides so that all variables are on the left.
$3a = 11$	
$a = \dfrac{11}{3}$	Divide both sides by **3** $\left(\text{or multiply by } \tfrac{1}{3}\right)$.

Thus the solution is $\frac{11}{3}$, and the solution set is $\left\{\frac{11}{3}\right\}$, as can be checked by substituting $\frac{11}{3}$ for a in $4a - 7 = a + 4$.   ■

Sometimes we need to simplify an equation before solving it. For instance, to solve $x + 6 = 3(2x - 2)$, we use the distributive property to simplify the right-hand side of the equation and then solve for x. We do this next.

EXAMPLE 4   Simplifying equations before solving

Solve: $x + 6 = 3(2x - 2)$

SOLUTION

$x + 6 = 3(2x - 2)$	Given.
$x + 6 = 6x - 6$	Simplify the right-hand side.

Now we have two choices; we can isolate the variables on the right or on the left. To avoid negative expressions, this time we keep them on the right by adding (6) to both sides. We have

$x + 6 + (6) = 6x - 6 + (6)$	
$x + 12 = 6x$	
$x + (-x) + 12 = 6x + (-x)$	Add $(-x)$ so all variables are on the right.
$12 = 5x$	
$\dfrac{12}{5} = x$	Divide by **5** $\left(\text{or multiply by } \tfrac{1}{5}\right)$.

Thus the solution is $\frac{12}{5}$, and the solution set is $\left\{\frac{12}{5}\right\}$.   ■

If an equation involves fractions, we "clear" them by multiplying both sides of the equation by the *smallest* number that is a multiple of each denominator—that is, by the least common denominator (LCD). Thus to solve

$$\frac{x}{6} + \frac{x}{4} = 10$$

we have to find the LCD of $\frac{x}{6}$ and $\frac{x}{4}$. One way is to pick the larger of the two denominators and double it, triple it, and so on, until the other number divides into the result. Using this idea, we find that 12 is the LCD. Multiplying both sides of the equation by 12, we have

$$12 \cdot \left(\frac{x}{6} + \frac{x}{4} \right) = 12 \cdot 10$$

$$12 \cdot \frac{x}{6} + 12 \cdot \frac{x}{4} = 12 \cdot 10 \qquad \text{Use the distributive property.}$$

$$2x + 3x = 120$$

$$5x = 120 \qquad \text{Combine like terms.}$$

$$x = 24 \qquad \text{Divide both sides by 5.}$$

Thus the solution is 24, and the solution set is {24}. Here is the check:

$$\frac{x}{6} + \frac{x}{4} \stackrel{?}{=} 10$$

$$\begin{array}{c|c} \dfrac{24}{6} + \dfrac{24}{4} & 10 \\[2mm] 4 + 6 & \\[2mm] 10 & \end{array}$$

EXAMPLE 5   Clearing fractions in linear equations

Solve:

a. $\dfrac{x + 1}{3} + \dfrac{x - 1}{10} = 5$ **b.** $\dfrac{x + 1}{3} - \dfrac{x - 1}{8} = 4$

SOLUTION

a. The LCD of

$$\frac{x + 1}{3} \qquad \text{and} \qquad \frac{x - 1}{10}$$

is $3 \cdot 10 = 30$, since 3 and 10 do not have any common factors. Multiplying both sides by 30, we have

$$30\left(\frac{x + 1}{3} + \frac{x - 1}{10} \right) = 30 \cdot 5$$

$$30\left(\frac{x + 1}{3} \right) + 30\left(\frac{x - 1}{10} \right) = 30 \cdot 5 \qquad \text{By the distributive property}$$

$$10(x + 1) + 3(x - 1) = 150 \qquad \text{Since } \overset{10}{\cancel{30}}\left(\frac{x + 1}{\cancel{3}} \right) = 10\,(x + 1) \text{ and } \overset{3}{\cancel{30}}\left(\frac{x - 1}{\cancel{10}} \right) = 3(x - 1)$$

$$10x + 10 + 3x - 3 = 150 \qquad \text{Use the distributive property.}$$

$$13x + 7 = 150 \qquad \text{Add like terms.}$$

$$13x = 143 \qquad \text{Subtract 7.}$$

$$x = 11 \qquad \text{Divide by 13.}$$

Thus the solution set is {11}. The check is left to you.

b. Here the LCD is $3 \cdot 8 = 24$. Multiplying both sides by 24, we obtain

$$24\left(\frac{x+1}{3} - \frac{x-1}{8}\right) = 24 \cdot 4$$

$$24\left(\frac{x+1}{3}\right) - 24\left(\frac{x-1}{8}\right) = 24 \cdot 4 \qquad \text{By the distributive property}$$

$$8(x+1) - 3(x-1) = 96 \qquad \text{Since } \overset{8}{24}\left(\frac{x+1}{\underset{1}{\cancel{3}}}\right) = 8(x+1) \text{ and } \overset{3}{24}\left(\frac{x-1}{\underset{1}{\cancel{8}}}\right) = 3(x-1)$$

$$8x + 8 - 3x + 3 = 96 \qquad \text{Use the distributive property.}$$

$$5x + 11 = 96 \qquad \text{Add like terms.}$$

$$5x = 85 \qquad \text{Subtract 11.}$$

$$x = 17 \qquad \text{Divide by 5.}$$

The solution set is $\{17\}$. Be sure you check this answer in the original equation.

Solving Linear Equations

All the equations we have solved in this section can be written in the form $ax + b = c$ where a, b, and c are real numbers. Such equations are called **linear equations**. (You will see in Chapter 7 that the graph of $ax + b = y$ is a *straight line*.) Thus

$$2x - 4 = 6 \qquad \text{and} \qquad \frac{x}{10} + \frac{x}{8} = 9$$

are linear equations. In the equation $2x - 4 = 6$, $2x - 4$ and 6 are called **terms**: $2x$ is a variable term and -4 and 6 are constant terms. We will use this terminology to give you a general procedure used to solve linear equations.

PROCEDURE

Procedure for Solving Linear Equations

1. If there are fractions, multiply both sides of the equation by the LCD of the fractions.

2. Remove parentheses and collect like terms (simplify) if necessary.

3. Add or subtract the same quantity on both sides of the equation so that one side has only the terms containing variables.

4. Add or subtract the same quantity on both sides of the equation so that the other side has only a constant.

5. If the coefficient of the variable is not 1, divide both sides of the equation by this coefficient (or, equivalently, multiply by the reciprocal of the coefficient of the variable).

6. Be sure to check your answer in the original equation.

EXAMPLE 6   Solving linear equations using the six-step procedure

Solve:

a. $\dfrac{7}{24} = \dfrac{x}{8} + \dfrac{1}{6}$

b. $\dfrac{1}{5} - \dfrac{x}{4} = \dfrac{7(x + 3)}{10}$

SOLUTION   We use the six steps as follows.

a. Given: $\dfrac{7}{24} = \dfrac{x}{8} + \dfrac{1}{6}$

1. $24 \cdot \dfrac{7}{24} = 24\left(\dfrac{x}{8} + \dfrac{1}{6}\right)$ Multiply by **24**, the LCD.

$24 \cdot \dfrac{7}{24} = 24 \cdot \dfrac{x}{8} + 24 \cdot \dfrac{1}{6}$

2. $7 = 3x + 4$ Simplify.

3. $7 - 4 = 3x + 4 - 4$ Subtract **4** (or add −4).

4. $3 = 3x$ The left side has numbers only.

5. $\dfrac{3}{3} = \dfrac{3x}{3}$ Divide by **3** $\left(\text{or multiply by } \tfrac{1}{3}\right)$.

$1 = x$

$x = 1$

The solution set is $\{1\}$.

6. Since $\tfrac{1}{8} = \tfrac{3}{24}$ and $\tfrac{1}{6} = \tfrac{4}{24}$,   Check.

$\dfrac{7}{24} = \dfrac{1}{8} + \dfrac{1}{6} = \dfrac{3}{24} + \dfrac{4}{24}$

which is true.

b. Given: $\dfrac{1}{5} - \dfrac{x}{4} = \dfrac{7(x + 3)}{10}$

1. $20\left(\dfrac{1}{5} - \dfrac{x}{4}\right) = 20 \cdot \left[\dfrac{7(x + 3)}{10}\right]$ Multiply by **20**, the LCD.

$20 \cdot \dfrac{1}{5} - 20 \cdot \dfrac{x}{4} = 20 \cdot \dfrac{7(x + 3)}{10}$

2. $4 - 5x = 14(x + 3)$ Simplify.

$= 14x + 42$

3. $4 - 5x - 42 = 14x + 42 - 42$ Subtract **42** (or add −42).

$-38 - 5x = 14x$

4. $-38 - 5x + 5x = 14x + 5x$ Add **5**x.

$-38 = 19x$

5. $\dfrac{-38}{19} = \dfrac{19x}{19}$ Divide by **19** $\left(\text{or multiply by } \tfrac{1}{19}\right)$.

$-2 = x$

$x = -2$

The solution set is $\{-2\}$.

6. $\dfrac{1}{5} - \dfrac{x}{4} \overset{?}{=} \dfrac{7(x+3)}{10}$ for $x = -2$ Check.

$$
\begin{array}{c|c}
\dfrac{1}{5} - \dfrac{(-2)}{4} & \dfrac{7(-2+3)}{10} \\[2ex]
\dfrac{1}{5} + \dfrac{1}{2} & \dfrac{7(1)}{10} \\[2ex]
\dfrac{7}{10} & \dfrac{7}{10}
\end{array}
$$

■

So far all our equations have had a solution. They are called **conditional equations**. There are two other possibilities:

  1. Equations with *no* solution (*contradictions*)

  2. Equations with *infinitely many* solutions (*identities*)

Consider the following example.

$$-4x + 6 = 2(1 - 2x) + 3 \qquad \text{Given.}$$

$$-4x + 6 = 2 - 4x + 3 \qquad \text{Since } 2(1 - 2x) = 2 - 4x.$$

$$-4x + 6 = -4x + 5 \qquad \text{Collect like terms.}$$

$$-4x + 6 + (-6) = -4x + 5 + (-6) \qquad \text{Add } -6.$$

$$-4x = -4x + (-1)$$

$$-4x + 4x = -4x + 4x + (-1) \qquad \text{Add } 4x.$$

$$0 = (-1) \quad \text{False!} \qquad \text{Simplify.}$$

We cannot find a replacement for x to satisfy this equation. This equation is a **contradiction**; it has no solution. Its solution set is $\varnothing$ (the empty set). On the other hand, consider the following equation.

$$-4x + 6 = 2(1 - 2x) + 4$$

$$-4x + 6 = 2 - 4x + 4 \qquad \text{Since } 2(1 - 2x) = 2 - 4x.$$

$$-4x + 6 = -4x + 6 \qquad \text{Collect like terms.}$$

$$6 = 6 \quad \text{True!} \qquad \text{Add } 4x.$$

Here, any x will be a solution (try 0, 1, or any other number in the original equation). This equation is an **identity**. Any real number is a solution. Its solution set is R, the set of real numbers.

EXAMPLE 7   Solving linear equations using the six-step procedure

Solve:

a. $3x + 8 = 3(x + 1) + 5$ **b.** $3x + 8 = 3(x + 1) + 2$

SOLUTION   We use the six-step procedure we studied.

a. Given: $3x + 8 = 3(x + 1) + 5$ There are no fractions.

           $3x + 8 = 3x + 3 + 5$ Use the distributive property.

           $3x + 8 = 3x + 8$ Collect like terms.

We can stop here. This equation is always true. It is an identity, and its solution set is R.

b. Given: $3x + 8 = 3(x + 1) + 2$ There are no fractions.

$3x + 8 = 3x + 3 + 2$ Use the distributive property.

$3x + 8 = 3x + 5$ Collect like terms.

We can stop here. This equation is a contradiction; it is always false. (Try subtracting $3x$ from both sides.) It has no solution, and its solution set is $\varnothing$. ∎

> ⚠ **CAUTION** Do not write $\{\varnothing\}$ to represent the empty set because the set $\{\varnothing\}$ is **not** empty, it has one element, $\varnothing$! Always use $\varnothing$ or $\{\ \ \}$ to represent the empty set.

Linear Equations with Decimals

EXAMPLE 8   Solving equations involving decimals

Solve: $14.5 - 3.15x = 5.5$

SOLUTION

$$14.5 - 3.15x = 5.5 \qquad \text{Given.}$$

$$14.5 - 14.5 - 3.15x = 5.5 - 14.5 \qquad \text{Subtract 14.5.}$$

$$-3.15x = -9 \qquad \text{Simplify.}$$

$$\frac{-3.15x}{-3.15} = \frac{-9}{-3.15} \qquad \text{Divide by } -3.15.$$

$$x = \frac{9 \cdot 100}{3.15 \cdot 100} \qquad \text{Multiply numerator and denominator by 100.}$$

$$= \frac{900}{315} = \frac{20}{7}$$

The solution is $\frac{20}{7}$, and the solution set is $\left\{\frac{20}{7}\right\}$, which you can check by substitution in the original equation. ∎

> **NOTE** To "clear decimals" from the very beginning, you can multiply both sides of $14.5 - 3.15x = 5.5$ by a power of 10 that has as many zeros as the number in the coefficient with the most decimal places, that is, by 100.

Later in this chapter we translate word problems into equations such as $1.30P + 1.50(50 - P) = 72.50$. But, first let's solve this equation.

EXAMPLE 9   Solving equations involving decimals

Solve: $1.30P + 1.50(50 - P) = 72.50$

SOLUTION

$$1.30P + 1.50(50 - P) = 72.50 \qquad \text{Given.}$$

$$1.30P + 75 - 1.50P = 72.50 \qquad \text{Distributive property}$$

$$-0.20P + 75 = 72.50 \qquad \text{Simplify.}$$

$$-0.20P + 75 - 75 = 72.50 - 75 \qquad \text{Subtract 75.}$$

$$-0.20P = -2.50 \qquad \text{Simplify.}$$

$$\frac{-0.20P}{-0.20} = \frac{-2.50}{-0.20} \qquad \text{Divide by } -0.20.$$

$$P = \frac{2.50 \cdot 100}{0.20 \cdot 100} \qquad \text{Multiply numerator and denominator by 100.}$$

$$= \frac{250}{20} = \frac{25}{2}$$

The solution is $\frac{25}{2}$, and the solution set is $\left\{\frac{25}{2}\right\}$. The solution set can also be written as $\{12.5\}$ or $\left\{12\frac{1}{2}\right\}$. Make sure you check this! ■

GRAPH IT

We have already mentioned that the graph of $ax + b = y$ is a straight line. To illustrate this, select any two numbers for a and b, say 2 and 3, and graph $y = 2x + 3$ with a grapher. You will always get a line regardless of the numbers you select for a and b. As a matter of fact, all of the linear equations we have solved can be graphed by first writing an equivalent equation with all variables on the left and zero on the right—that is, an equation of the form $A(x) = 0$ (read "A of x is 0") and then graphing the line $y = A(x)$. The number at which the line crosses the x-axis (where $y = 0$) is the desired solution because at this point $y = A(x) = 0$.

Let us use a grapher to solve the equation in Example 6b:

$$\frac{1}{5} - \frac{x}{4} = \frac{7(x + 3)}{10}$$

$$\frac{1}{5} - \frac{x}{4} - \frac{7(x + 3)}{10} = 0 \qquad \begin{array}{l}\text{Subtract } 7(x + 3)/10 \\ \text{from both sides.}\end{array}$$

Next, graph the equivalent equation for Y_1 using

$$Y_1 = \frac{1}{5} - \frac{x}{4} - \frac{7(x + 3)}{10}$$

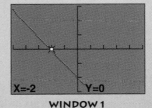

X=-2   Y=0

WINDOW 1

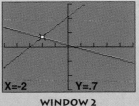

X=-2   Y=.7

WINDOW 2

The two lines meet at a point where $x = -2$.

using a decimal or an integer window. You **do not** have to multiply by the LCD here, but the equation would be easier to enter if you write

$$\frac{7(x + 3)}{10} \qquad \text{as} \qquad .7(x + 3)$$

What type of graph do you get? Where does the graph cross the x-axis? You should get a line crossing the x-axis at $x = -2$. Thus the solution set of the given equation is, as before, $\{-2\}$; see Window 1.

You can also find the solution set of the equation by graphing

$$Y_1 = \frac{1}{5} - \frac{x}{4} \qquad \text{and} \qquad Y_2 = .7(x + 3)$$

The x-coordinate at the point of intersection of the two lines, $x = -2$, is the solution (See Window 2). (*Note:* To verify that $x = -2$ is the solution, use your $\boxed{\text{TRACE}}$ feature with a decimal or integer window so that the cursor moves in x-increments of 0.1 or 1, respectively.)

What happens if you try to graph the equations in Example 7? Can you use your grapher to distinguish between an equation that has no solution and an equation that has infinitely many solutions? The moral here is, even with a grapher you have to know how to do your algebra!

EXERCISE 2.1

A In Problems 1–10, determine if the number in the box is a solution of the equation.

1. $2x + 8 = 14;$ $\boxed{3}$

2. $5x + 5 = 10;$ $\boxed{2}$

3. $-2x + 1 = 3;$ $\boxed{-1}$

4. $-3x + 4 = 10;$ $\boxed{-2}$

5. $2y - 5 = y - 2;$ $\boxed{3}$

6. $3y - 7 = y + 1;$ $\boxed{4}$

7. $\dfrac{4}{5}t - 1 = 5t;$ $\boxed{\dfrac{-1}{5}}$

8. $6t + 1 = t + \dfrac{2}{3};$ $\boxed{\dfrac{-1}{3}}$

9. $\dfrac{1}{2}x + 5 = 5 - \dfrac{1}{3}x;$ $\boxed{3}$

10. $\dfrac{-1}{3}x + 1 = x - 3;$ $\boxed{3}$

B In Problems 11–36, solve the equation.

11. $3x - 4 = 8$

12. $5a + 16 = 6$

13. $2y + 8 = 10$

14. $4b - 6 = 2$

15. $-3z - 6 = -12$

16. $-4r - 3 = 5$

17. $-5y + 2 = -8$

18. $-3x + 2 = -10$

19. $3x + 5 = x + 19$

20. $4x + 6 = x + 9$

21. $7(x - 1) - 3 + 5x = 3(4x - 3) + x$

22. $5(x + 1) + 3x + 2 = 8(x + 2) + 2x + 1$

23. $6v - 8 = 8v + 8$

24. $8t + 3 = 15t - 11$

25. $7m - 4m + 12 = 0$

26. $10k + 25 - 5k = 35$

27. $4(2 - z) + 8 = 8(2 - z)$

28. $4(3 - y) + 8 = 12(3 - y)$

29. $5(x + 3) = 3(x + 3) + 6$

30. $y - (5 - 2y) = 7(y - 1) - 2$

31. $5(4 - 3a) = 7(3 - 4a)$

32. $\dfrac{3}{4}y - 4 = \dfrac{1}{4}y - 2$

33. $-\dfrac{7}{8}c + 5 = -\dfrac{5}{8}c + 3$

34. $x + \dfrac{2}{3}x = 10$

35. $-2x + \dfrac{1}{4} = 2x + \dfrac{4}{5}$

36. $6x + \dfrac{1}{7} = 2x - \dfrac{2}{7}$

C In Problems 37–60, use the six-step procedure given in the text to solve the equation.

37. $\dfrac{t}{6} + \dfrac{t}{8} = 7$

38. $\dfrac{f}{9} + \dfrac{f}{12} = 14$

39. $\dfrac{x}{2} + \dfrac{x}{5} = \dfrac{7}{10}$

40. $\dfrac{a}{3} + \dfrac{a}{7} = \dfrac{20}{21}$

41. $\dfrac{c}{3} - \dfrac{c}{5} = 2$

42. $\dfrac{F}{4} - \dfrac{F}{7} = 3$

43. $\dfrac{W}{6} - \dfrac{W}{8} = \dfrac{5}{12}$

44. $\dfrac{m}{6} - \dfrac{m}{10} = \dfrac{4}{3}$

45. $\dfrac{x}{5} - \dfrac{3}{10} = \dfrac{1}{2}$

46. $\dfrac{3y}{7} - \dfrac{1}{14} = \dfrac{1}{14}$

47. $\dfrac{x + 4}{4} - \dfrac{x + 2}{3} = -\dfrac{1}{2}$

48. $\dfrac{w - 1}{2} + \dfrac{w}{8} = \dfrac{7w + 1}{16}$

49. $\dfrac{x + 1}{4} - \dfrac{2x - 2}{3} = 3$

50. $\dfrac{z + 4}{3} = \dfrac{z + 6}{4}$

51. $\dfrac{2h - 1}{3} = \dfrac{h - 4}{12}$

52. $\dfrac{5 - 6y}{7} - \dfrac{-7 - 4y}{3} = 2$

53. $\dfrac{2w + 3}{2} - \dfrac{3w + 1}{4} = 1$

54. $\dfrac{8x - 23}{6} + \dfrac{1}{3} = \dfrac{5}{2}x$

55. $\dfrac{7r + 2}{6} + \dfrac{1}{2} = \dfrac{r}{4}$

56. $\dfrac{x + 1}{2} + \dfrac{x + 2}{3} + \dfrac{x + 4}{4} = -8$

57. $\dfrac{x - 5}{2} - \dfrac{x - 4}{3} = \dfrac{x - 3}{2} - (x - 2)$

58. $\dfrac{x + 1}{2} + \dfrac{x + 2}{3} + \dfrac{x + 3}{4} = 16$

59. $4(x - 2) + 4 = 4x - 4$

60. $8 - (2 - 3x) = 3(x + 2)$

D In Problems 61–76, solve the equation.

61. $6.3x - 8.4 = 16.8$

62. $15.5a + 49.6 = 18.6$

63. $-12.6y - 25.2 = 50.4$

64. $6.4y - 19.2 = 32$

65. $2.1y + 3.5 = 0.7y + 83.3$

66. $2.4x + 3.6 = 0.6x + 5.4$

67. $3.5(x + 3) = 2.1(x + 3) + 4.2$

68. $7.2(3 - t) = 2.4(3 - t) + 4.8$

69. $0.40y + 0.20(32 - y) = 9.60$

70. $0.30x + 0.35(50 - x) = 16$

71. $0.65x + 0.40(50 - x) = 25.375$

72. $0.09y + 0.12(200 - y) = 19.20$

73. $0.06P + 0.08(2000 - P) = 130$

74. $0.15P + 0.10(6000 - P) = 660$

75. $0.30y + 1.80 = 0.20(y + 12)$

76. $0.10x + 4 = 0.30(x + 10)$

SKILL CHECKER

Simplify:

77. $3x - (4x + 7) - 2x$ **78.** $5x - 3 - (7 + 4x)$

79. $6x - 2(3x - 1)$ **80.** $8x - 4(3 - 2x) + 5x$

Evaluate:

81. $\dfrac{400 - 2 \cdot 150}{2}$ **82.** $\dfrac{300.48 - 2 \cdot 100}{2}$

83. $\dfrac{5}{9}(F - 32)$ when $F = 41$ **84.** $\dfrac{5}{9}(F - 32)$ when $F = 32$

USING YOUR KNOWLEDGE   If the Shoe Fits

Use the knowledge you have gained to solve the following problems about shoes.

The relationship between your shoe size S and the length of your foot L (in inches) is given by

$$S = 3L - 22 \qquad \text{(for men)}$$
$$S = 3L - 21 \qquad \text{(for women)}$$

85. If Tyrone wears a size 11, what is the length L of his foot?

86. If Maria wears a size 7, what is the length L of her foot?

87. Sam's size 7 tennis shoes fit Sue perfectly! What size women's tennis shoe does Sue wear?

88. The largest shoes ever sold was a pair of size 42 built for the giant Harley Davidson of Avon Park, Florida. How long is Mr. Davidson's foot?

89. How long a foot requires a size 14, the largest standard shoe size for men?

90. How long is your foot when your shoe size is the same as the length of your foot and you are
   a. A man?   **b.** A woman?

91. In 1951, Eric Shipton photographed a 23-in. footprint believed to be that of the Abominable Snowman.
   a. What size shoe does the Abominable Snowman need?
   b. If the Abominable Snowman turned out to be a woman, what size shoe would she need?

WRITE ON . . .

Write a paragraph:

92. Explaining the difference between a conditional equation and an identity.

93. Defining a contradictory equation.

94. Detailing the terminology used for equations that have no solution, exactly one solution, and infinitely many solutions, and give examples.

MASTERY TEST   If you know how to do these problems, you have learned your lesson!

Solve:

95. $\dfrac{x + 4}{8} - \dfrac{x + 2}{6} = -\dfrac{1}{4}$ **96.** $\dfrac{y}{8} - \dfrac{1}{4} = \dfrac{7y + 2}{12}$

97. $0.8t + 0.4(32 - t) = 19.2$

98. $0.60x + 3.6 = 0.40(x + 12)$

99. $7(x - 1) - x = 3 - 5x + 3(4x - 3)$

100. $8(x + 2) + 4x + 3 = 5x + 4 + 5(x + 1)$

101. Is -5 a solution of $\dfrac{-1}{5}x + 2 = -x + 4$?

102. Is $-\dfrac{1}{4}$ a solution of $4t - 1 = t + \dfrac{1}{4}$?

2.2

FORMULAS AND TOPICS FROM GEOMETRY

To succeed, review how to:

1. Solve linear equations (pp. 69–73).

2. Evaluate expressions using the correct order of operations (p. 44).

(See the *Skill Checker*, page 75.)

Objectives:

A Solve a formula for a specified variable and then evaluate the answer for given values of the variables.

B Write a formula for a given situation that has been described in words.

C Solve problems about angle measures.

---

getting started

Trucks, Axles, and Tolls

You are a truck driver traveling on a toll road in Florida. If your truck has 5 axles, how much do you have to pay? You can use the toll schedule and a linear equation to figure this out. The cost C for a truck with n axles is given by

$$C = 0.50 + 0.25(n - 3)$$

For a truck with 5 axles, $n = 5$ and the answer is

$$C = 0.50 + 0.25(5 - 3)$$
$$= 0.50 + 0.25(2)$$
$$= 0.50 + 0.50$$
$$= \$1.00$$

Many problems in algebra can be solved if the correct formula is used. Now suppose a trucker pays \$1.25 for the toll. How many axles does the truck have? Here, we are interested in n in the equation $1.25 = 0.50 + 0.25(n - 3)$. Thus we are *solving for a specified variable*. The steps that we use are similar to those we used in solving linear equations. To remind you that we are solving for n, the n is in color. Here is the procedure.

$1.25 = 0.50 + 0.25(n - 3)$	Given.
$125 = 50 + 25(n - 3)$	Multiply each term by 100.
$125 = 50 + 25n - 75$	Use the distributive property.
$125 = -25 + 25n$	Combine like terms.
$25 + 125 = -25 + 25 + 25n$	Add 25.
$150 = 25n$	Simplify.
$\dfrac{150}{25} = n$	Divide by 25.
$6 = n$	

Thus the truck has 6 axles.

Solving Formulas for a Specified Variable

Sometimes we are asked to solve for one of the variables in an equation called a **literal equation**. For example, the equation $C = 0.50 + 0.25(n - 3)$ is a literal equation. The procedure used to solve $C = 0.50 + 0.25(n - 3)$ for n is similar to that used for solving linear equations (page 69). Here are some suggestions to follow when solving for a *specified variable*.

PROCEDURE

> **Procedure to Solve for a Specified Variable**
>
> 1. Add or subtract the same quantity on both sides of the equation so that the terms containing the specified **variable** (the one for which you are solving) are isolated on one side. The terms that **do not** contain the specified variable are thus on the other side.
>
> 2. If necessary, use the distributive property to write the side containing the specified variable as a product of the variable and a sum (or difference) of terms.
>
> 3. In general, use the rules for solving linear equations.

EXAMPLE 1   Solving for a variable in a literal equation

Anthropologists know how to estimate the height of a man (in centimeters) using only a bone. They use the formula

$$H = 2.89h + 70.64$$

$\underbrace{}$ Height of the man $\underbrace{}$ Length of the humerus

a. Solve for h.
b. A man is 157.34 cm tall. How long is his humerus?

SOLUTION

a. We have to solve for h; that is, we isolate h on one side of the equation. Given

$$H = 2.89h + 70.64$$

$$H - 70.64 = 2.89h + 70.64 - 70.64 \qquad \text{Subtract } \textbf{70.64}.$$

$$H - 70.64 = 2.89h \qquad \text{Simplify.}$$

$$\frac{H - 70.64}{2.89} = h \qquad \text{Divide by } \textbf{2.89}.$$

Now we can find h for any given value of H.

b. Substitute 157.34 for H to get

$$h = \frac{157.34 - 70.64}{2.89} = \frac{86.7}{2.89} = 30$$

The man's humerus is 30 cm long.   ∎

EXAMPLE 2   Solving for a specified variable

The formula for converting degrees Fahrenheit (°F) to degrees Celsius (°C) is

$$C = \frac{5}{9}(F - 32)$$

a. Solve for F.
b. If the temperature is 35°C, what is the equivalent Fahrenheit temperature?

SOLUTION

a. We use the procedure for solving linear equations given on page 69.

Given: $\qquad C = \dfrac{5}{9}(F - 32)$

1. $\qquad 9 \cdot C = 9 \cdot \dfrac{5}{9}(F - 32) \qquad$ Clear fractions by multiplying by **9** and simplify.

$\qquad\qquad 9C = 5(F - 32)$

2. $\qquad\qquad 9C = 5F - 160 \qquad$ Use the distributive property.

3. $9C + 160 = 5F \qquad$ Add **160**.

4. $\dfrac{9C + 160}{5} = F \qquad$ Divide by **5**.

b. Substitute 35 for C. $\qquad \dfrac{9 \cdot 35 + 160}{5} = \dfrac{475}{5} = 95$

The equivalent Fahrenheit temperature is 95°F. ■

Many of the formulas we encounter in algebra come from geometry. In geometry, the distance around a polygon is called the **perimeter**, and the number of square units occupied by the figure is its **area**. Table 1 shows the formulas for the perimeters and areas of some common geometric figures.

TABLE 1 Perimeters and Areas

Name	Shape	Perimeter	Area
Square		$P = 4S$	$A = S^2$
Rectangle		$P = 2L + 2W$	$A = LW$
Triangle		$P = s_1 + s_2 + b$	$A = \dfrac{1}{2}bh$
Circle		$C = 2\pi r = \pi d$	$A = \pi r^2$

Circumference   Radius   Diameter

Note that $2r = d$ and $\pi \approx 3.14$.

✓ **EXAMPLE 3**   Solving geometric shapes using a specified variable

The perimeter P of a rectangle is given by

$$P = 2L + 2W$$

a. Solve for W.

b. The largest poster ever made was a rectangular greeting card 166 ft long and with a perimeter of 458.50 ft. How wide was this poster?

SOLUTION

a.

$P = 2L + 2W$	Given.
$P + (-2L) = 2L + (-2L) + 2W$	Add $-2L$.
$P - 2L = 2W$	Simplify.
$\dfrac{P - 2L}{2} = W$	Divide by 2.

b. Since the perimeter P was 458.50 ft and the length L was 166 ft, we substitute 458.50 for P and 166 for L, to obtain

$$W = \frac{458.50 - 2 \cdot 166}{2}$$

$$= \frac{458.50 - 332}{2} = 63.25 \text{ ft}$$

The poster was 63.25 ft wide.   ■

✓ **EXAMPLE 4**   Solving interest rate problems

If you invest P dollars at the simple interest rate r, the amount of money A you will receive at the end of t years is $A = P + Prt$.

a. Solve for P.

b. At the end of 5 yr, an investor receives \$2250 on her investment at 10% simple interest. How much money did she invest?

SOLUTION

a. This time, P appears twice on the right-hand side of the equation. Using the distributive property, we write $A = P(1 + rt)$. To get P by itself on the right, we need only divide by $(1 + rt)$. Here are all the steps:

$A = P + Prt$	Given.
$A = P(1 + rt)$	Use the distributive property.
$\dfrac{A}{1 + rt} = \dfrac{P(1 + rt)}{1 + rt}$	Divide by $1 + rt$.
$P = \dfrac{A}{1 + rt}$	

b. We substitute 2250 for A, $10\% = 0.10$ for r, and 5 for t, to obtain

$$P = \frac{2250}{1 + 0.10 \cdot 5} = \frac{2250}{1.5} = 1500$$

Thus she invested $1500. ∎

√ **EXAMPLE 5**   Solving for a specified variable

Do you know how to add $a_1 + a_2 + a_3 + \cdots + a_n$? If the difference between successive terms is a constant, the sum is

$$S_n = \frac{n(a_1 + a_n)}{2}$$

where n is the number of terms to be added. Note that the subscript n in a_n indicates the number of the term. (We show you why in Chapter 11.)
a. Solve for n.
b. $S_n = 2 + 4 + 6 + \cdots + 100 = 2550$. Thus $a_1 = 2$, $a_n = 100$, and $S_n = 2550$. Verify that we have added 50 terms.

SOLUTION
a. Since we want n by itself on the right, we multiply by 2 and divide by $a_1 + a_n$. Here are the steps:

$$S_n = \frac{n(a_1 + a_n)}{2} \qquad \text{Given.}$$

$$2 \cdot S_n = 2 \cdot \frac{n(a_1 + a_n)}{2} \qquad \text{Multiply by 2.}$$

$$2S_n = n(a_1 + a_n) \qquad \text{Simplify.}$$

$$\frac{2S_n}{a_1 + a_n} = \frac{n(a_1 + a_n)}{a_1 + a_n} \qquad \text{Divide by } a_1 + a_n.$$

$$\frac{2S_n}{a_1 + a_n} = n \qquad \text{Simplify.}$$

b. We substitute $a_1 = 2$, $a_n = 100$, and $S_n = 2550$ in the formula

$$n = \frac{2S_n}{a_1 + a_n} \qquad \text{to get} \qquad n = \frac{2 \cdot 2550}{2 + 100} = \frac{5100}{102} = 50 \qquad ∎$$

B   **Writing and Evaluating Formulas**

In the preceding examples you were asked to solve for a specified variable in a formula. There are many applications in which a formula is given in words and you have to translate it into algebra. For example, anthropologists know that the relationship between a female's height H (in centimeters) and the length f of her femur bone can be found by

$$\overbrace{\text{Multiplying 1.95 by } f}^{} \text{ and } \overbrace{\text{adding 28.68}}^{}$$

Algebraically this is $\quad\quad 1.95f \quad\quad + \quad 28.68 = H$

See how your translation skills work in Example 6.

√ **EXAMPLE 6**   Writing formulas from words

The cost of a Boston-to-Nantucket dial-direct call during business hours is 50¢ for the first minute and 23¢ for each additional minute or fraction thereafter.
a. Write a formula for the cost C of a Boston-to-Nantucket call.
b. If C is the cost and t is the number of minutes for a Boston-to-Nantucket call, find t.
c. Roberto's telephone bill showed a $2.80 charge for a Boston-to-Nantucket call. Use the formula found in part b to find out for how many minutes he was charged.

SOLUTION   Let's look at the costs for 2, 3, and 4 minutes and see whether we can see a pattern.

For 2 minutes: $\quad C = 0.50 + 0.23(1)$

For 3 minutes: $\quad C = 0.50 + 0.23(2)$

For 4 minutes: $\quad C = 0.50 + 0.23(3)$

For t minutes: $\quad C = 0.50 + 0.23(t - 1)$

Note that you pay 23¢ for each minute after the first, so you have to multiply 0.23 by $(t - 1)$.
a. In general, the formula for C is $C = 0.50 + 0.23(t - 1)$.
b. To solve for t, think of t as the unknown and C as a number, then follow the procedure for solving a linear equation.

$$C = 0.50 + 0.23(t - 1) \quad\quad \text{Given.}$$

$$C = 0.50 + 0.23t - 0.23 \quad\quad \text{Distributive property}$$

$$C = 0.27 + 0.23t \quad\quad \text{Simplify.}$$

$$C - 0.27 = 0.23t \quad\quad \text{Subtract 0.27.}$$

$$\frac{C - 0.27}{0.23} = t \quad\quad \text{Divide by 0.23.}$$

c. Substituting 2.80 for C in the preceding equation gives us

$$\frac{2.80 - 0.27}{0.23} = \frac{2.53}{0.23} = 11$$

Thus Roberto was charged for 11 min.   ■

See Note

Angle Measures

In geometry, angles are measured in **degrees**, and $1°$ is defined as $\frac{1}{360}$ of a complete revolution. Certain relationships between angles are shown in Table 2.

TABLE 2

Angle Relationships	Examples
If L_1 and L_2 are parallel lines crossed by a **transversal** as shown to the right, angles 1, 2, 7, and 8 are **exterior angles**, and angles 3, 4, 5, and 6 are **interior angles**. Angles 1 and 5 are **corresponding angles**, as are angles 3 and 7. The measures of angles 1 and 5 are equal; and the measures of angles 3 and 7 are equal.	
Alternate interior angles (the pairs of angles *between* the parallel lines and on *opposite* sides of the transversal) are equal; that is, the measures of angles 3 and 6 are equal. Similarly, the measures of angles 4 and 5 are equal.	
Vertical angles (angles *opposite* each other) are equal. That is, the following pairs of angles are equal: 1 and 4; 2 and 3; 5 and 8; and 6 and 7.	

Here is quick way of remembering all these facts: Look at angles 1 and 2. Angle 1 is *acute* (less than $90°$), and angle 2 is *obtuse* (more than $90°$). Think of the measure of angle 1 as "small" and that of angle 2 as "big."

REMEMBER:

All the acute ("small") angles in the figures have the same measure. Thus angles 1, 4, 5, and 8 have the same measure.

All the obtuse ("big") angles in the figures have the same measure. Thus angles 2, 3, 6, and 7 have the same measure.

* Nae   ✓   **EXAMPLE 7**   Finding the measure of angles

If lines L_1 and L_2 are parallel, find x and the measure of the two unknown angles.

SOLUTION   The angles whose measures are given are both "small" (alternate exterior angles), and hence they have equal measures. Thus

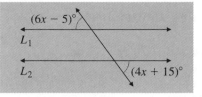

$$6x - 5 = 4x + 15$$

$$6x - 5 + 5 = 4x + 15 + 5 \qquad \text{Add 5.}$$

$$6x = 4x + 20 \qquad \text{Simplify.}$$

$$6x - 4x = 4x - 4x + 20 \qquad \text{Subtract } 4x.$$

$$2x = 20 \qquad \text{Divide by 2.}$$

$$x = 10$$

Substituting $x = 10$ in $6x - 5$, we find that the measure of one angle is $6 \cdot 10 - 5 = 55°$. Since the angles are equal, the other angle also measures $55°$. ■

GRAPH IT

Many students believe that if they have a grapher, they don't need to know any algebra. This is an appropriate time to dispel this notion. Suppose you want to graph the simple equation $2x + 5y = 10$ using a grapher. **You cannot do it unless you understand the concepts studied in this section!** Why? Because your grapher requires you to enter the expression representing y in $2x + 5y = 10$; that is, you have to *solve* for y. How do we do it? The way we have just learned, except that, to remember we want to isolate y, we place the variable y in a box:

$$2x + 5\boxed{y} = 10 \qquad \text{Given.}$$

$$2x - 2x + 5\boxed{y} = 10 - 2x \qquad \text{Subtract } 2x.$$

$$5y = 10 - 2x \qquad \text{Simplify.}$$

$$\boxed{y} = \frac{10 - 2x}{5} \qquad \text{Divide by 5.}$$

Now you can enter and graph

$$y = \frac{10 - 2x}{5}$$

Remember to enter this equation as $y = (10 - 2x) \div 5$, using parentheses. If you know a little bit of extra algebra and arithmetic, you can even shorten the work and enter $y = 2 - 0.4x$. Do you see why?

EXERCISE 2.2

A In Problems 1–12, solve the given formula for the indicated letter.

1. $V = \pi r^2 h$ for h

2. $V = \frac{1}{3}\pi r^2 h$ for h

3. $V = LWH$ for W

4. $V = LWH$ for H

5. $P = s_1 + s_2 + b$ for b

6. $P = s_1 + s_2 + b$ for s_2

7. $A = \pi(r^2 + rs)$ for s

8. $T = 2\pi(r^2 + rh)$ for h

9. $\dfrac{V_2}{V_1} = \dfrac{P_1}{P_2}$ for V_2

10. $\dfrac{V_2}{V_1} = \dfrac{P_1}{P_2}$ for P_1

11. $\dfrac{V_2}{V_1} = \dfrac{P_1}{P_2}$ for P_2

12. $\dfrac{V_2}{V_1} = \dfrac{P_1}{P_2}$ for V_1

13. The distance D traveled in time T by an object moving at rate R is given by $D = RT$.
  a. Solve for T.
  b. The distance between two cities A and B is 220 miles (mi). How long would it take a driver traveling at 55 mi/hr to go from A to B?

14. The ideal height H (in inches) of a man is related to his weight W (in pounds) by the formula $W = 5H - 190$.
  a. Solve for H.
  b. If a man weighs 160 lbs, how tall should he be?

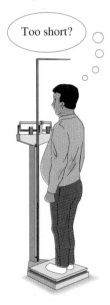

Too short?

15. The number of hours H a growing child should sleep is $H = 17 - \frac{A}{2}$, where A is the age of the child in years.
  a. Solve for A.
  b. The parents of an infant cannot wait until the child sleeps just 8 hr a day. At what age will that happen?

16. The efficiency energy rating (EER) of an air conditioner is given by $EER = \frac{Btu}{W}$, where Btu is the cooling capacity (per hour) in British thermal units and W is the watts of energy consumed.

  a. Solve for W.
  b. How many watts of electricity would an air conditioner with EER 9 and rated at 9000 Btu consume in 1 hr?

17. The operating profit margin (OPM) for a business is

$$OPM = \frac{CGS + OE}{NS}$$

where CGS is the cost of goods sold, OE is the operating expense, and NS is the net sales.
  a. Solve for CGS.

  b. If the operating expenses of a business amounted to \$18,500, the operating profit margin was $96\% = 0.96$, and the net sales were \$50,000, what was the cost of the goods sold?

18. The acid-test (AT) ratio for a business is given by

$$AT = \frac{C + R}{CL}$$

where C is the cash, R is the amount of receivables, and CL is the current liability.
  a. Solve for R.

  b. If the AT ratio for a business is 1, its current liability is \$7800, and the business has \$1200 cash on hand, what should the accounts receivable be?

19. The probability P of an event is

$$P = \frac{F}{F + U}$$

where F is the number of favorable outcomes for the event and U is the number of unfavorable outcomes for the event.
  a. Solve for U.

  b. The probability of throwing a 4 with a die (plural, dice) is $\frac{1}{6}$. If a die is thrown a number of times and 200 of the throws are 4's (favorable), how many unfavorable throws would you expect?

20. The area A of a trapezoid is given by $A = \frac{1}{2}h(a + b)$.
  a. Solve for b.

  b. If the area of a trapezoid is 60 square units, its height h is 10 units, and side a is 7 units, what is the length of side b?

Use the following information in Problems 21–23.

The relationship between the length L of the femur bone of a man and his height H is given by

$$H = 1.88L + 32 \tag{1}$$

where H and L are both measured in inches. The corresponding equation for a woman is

$$H = 1.95L + 29 \tag{2}$$

21. a. Solve equation (1) for L.

  b. Can a 20-in. femur belong to a man whose height was 6 ft?

  c. Can the bone belong to a man whose height was 69.6 in.?

  d. Solve equation (2) for L.

  e. Can a femur measuring 20 in. in length belong to a woman whose height was 5 ft, 8 in.?

22. A police pathologist wants to check the accuracy of equations (1) and (2). Use your calculator to find L for

  a. A man 5 ft, 6 in. in height.

  b. A man 5 ft, 8 in. in height.

  c. A woman 5 ft, 6 in. in height.

  d. A woman 5 ft, 10 in. in height.

23. Find the height to the nearest tenth of an inch of a woman whose femur is the same length as the femur of a 5 ft, 10 in. tall man.

B Translate the following word problems into formulas.

24. Anthropologists can estimate a person's height by using the person's skeletal remains.

  a. Find a formula for H if the height H (in inches) of a man with a femur bone of length x in. can be obtained by multiplying 1.88 by x and adding 32 to the result.

  b. Use the formula from part a to find the length of the femur for a man who was 65 in. tall.

25. The cost C of a Tampa-to-Los Angeles call during business hours is 36¢ for the initial minute and 28¢ for each additional minute t or fraction thereof.

  a. Write a formula for the cost C of a Tampa-to-Los Angeles call.

  b. Use the formula from part a to find the length of a phone call from Tampa-to-Los Angeles if the cost was $4.00.

26. A plumber charges $25 per hour ($h$) plus $30 for the service call.

  a. Find a formula for C if C is the total cost for the call.

  b. Use the formula from part a to find the number of hours worked by a plumber who charged $142.50.

27. A finance company will lend you $1000 for a finance charge of $20 plus simple interest at 1% per month. This means that you will pay interest of 1% of $1000—that is, $10 per month in addition to the $20 finance charge.

  a. Find a formula for F if F is the finance charge and m is the number of months before you repay the loan.

  b. Use the formula from part a to find the number of months outstanding on a $1000 loan that has already cost $140 in finance charges.

C In Problems 28–35, find x and then find the measure of each marked angle. In Problems 32–35, assume that L_1 and L_2 are parallel.

28.

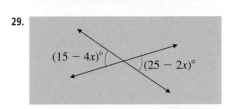

$(3x - 5)°$ $(2x + 25)°$

29.

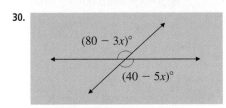

$(15 - 4x)°$ $(25 - 2x)°$

30.

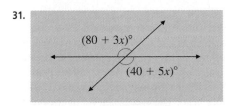

$(80 - 3x)°$

$(40 - 5x)°$

31.

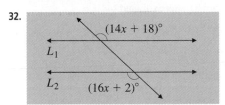

$(80 + 3x)°$

$(40 + 5x)°$

32.

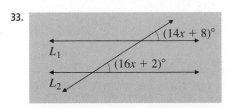

$(14x + 18)°$

L_1

L_2 $(16x + 2)°$

33.

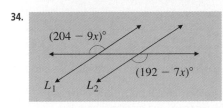

$(14x + 8)°$

L_1

$(16x + 2)°$

L_2

34.

$(204 - 9x)°$

$(192 - 7x)°$

L_1 L_2

35.

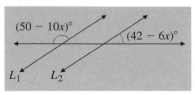

Hint: The two marked angles are supplementary. Their sum is 180°.

SKILL CHECKER

Simplify:

36. $n + (n + 1) + (n + 2)$ **37.** $n + (n + 2) + (n + 4)$

Solve:

38. $(p + 50) + p = 110$ **39.** $p + (p + 60,000) = 110,000$

40. $0.30m + 25 = 52$ **41.** $0.25m + 20 = 37.50$

42. $x + (90 - x) + 3x = 180$ **43.** $(90 - x) + x + 5x = 180$

USING YOUR KNOWLEDGE   **Global Warming and Sea Levels**

Do you know what global warming is? It is the theory that over the past century, the Earth has begun to warm significantly due to factors like pollution (from your car and power plants); deforestation (fewer trees, less shade, less rain, more heat); and the depletion of the ozone layer (ozone absorbs ultraviolet radiation; less ozone, more sun and more heat). The average global temperature C (in degrees Celsius) can be approximated by

$$C = 14.84 + 0.08t$$

where t is the number of years after 1960.

44. Solve for t in the equation.

45. In what year would the temperature be 20°C? Give your answer to the nearest year.

46. In what year would the temperature be 25°C? Give your answer to the nearest year.

47. The present average global temperature is about 15°C. An increase of 1°C causes a 1-ft rise in the sea level. What would the rise in sea level be in the year 2000?

WRITE ON . . .

48. In this section you have solved formulas for specified variables. Write a short paragraph explaining what that means.

49. A linear equation in one variable is an equation that can be written as $ax + b = c$ (a, b, and c are real numbers and $a \neq 0$). Write an explanation of how you would solve for x

in this equation. Use your explanation to solve the equation $-3x + 10 = 16$ for x.

50. The definition of linear equation in one variable (Problem 49) states that a cannot be zero ($a \neq 0$). Explain what happens when a is 0. Can $c = 0$?

MASTERY TEST   If you know how to do these problems, you have learned your lesson!

In Problems 51 and 52, find x and the measures of the marked angles:

51.

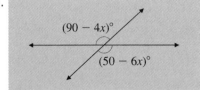

52. Assume L_1 and L_2 are parallel lines.

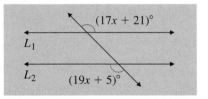

53. The cost C of mailing a first class letter is 32¢ for the first ounce (oz) and 23¢ for each additional ounce. If w is the weight of the letter in ounces:
   a. Write a formula for the cost C.
   b. Solve for w in the formula.
   c. What is the cost of mailing a letter weighing 10 oz?

54. The annual number N of cigarettes per person consumed in the United States is the difference between 4200 and the product of 70 and t, where t is the number of years after 1960.
   a. Write a formula for N.
   b. In what year would consumption drop to 2100 cigarettes per person?

55. The speed S (in centimeters per second) for a certain type of ant is $S = \frac{1}{6}(C - 4)$, where C is the temperature in degrees Celsius.
   a. Solve for C.
   b. If the ant is moving at a rate of 2 cm/sec, what is the temperature?

56. The number of chirps N a cricket makes per minute is $N = 4(F - 40)$, where F is the temperature in degrees Fahrenheit.
   a. Solve for F.
   b. If the cricket is chirping 80 times per minute, what is the temperature?

2.3

GENERAL, INTEGER, AND GEOMETRY PROBLEMS

To succeed, review how to:

1. Simplify expressions (pp. 45–49).
2. Solve linear equations involving decimals (pp. 72–73).

(See the *Skill Checker*, page 86.)

Objectives:

A Translate a word expression into a mathematical expression.

B Solve word problems of a general nature.

C Solve word problems about integers.

D Solve word problems about geometric formulas and angles.

getting started   RSTUV Procedure

In the preceding sections we learned how to solve certain kinds of equations. Now we are ready to apply this knowledge to solve problems. These problems will be stated in words and are consequently called **word**, or **story**, **problems**. Word problems frighten many students, but do not panic. We have a surefire method for tackling *any* word problem. Here is our five-step procedure:

1. **R**ead the problem. Not once or twice, but until you understand it.
2. **S**elect the unknown; that is, find out what the problem asks for.
3. **T**hink of a plan to solve the problem.
4. **U**se the techniques you are studying to carry out the plan.
5. **V**erify the answer.

Look at the first letter in each sentence. To help you remember the steps, we call it the **RSTUV** method. Later in this section we have additional tips on how to use this method, but first, we have to discuss the terminology we shall use.

You have probably noticed the frequent occurrence of certain words in the statements of word problems. Table 3 presents a brief mathematics dictionary to help you translate these words properly and consistently.

> **CAUTION** Don't confuse "7 less than x," that is, $x - 7$, with "7 is less than x," which is $7 < x$. Moreover, *do not* write "7 less than x" as $7 - x$, which means "x less than 7." A number x subtracted *from* y is always written as $y - x$, *not* $x - y$. For division, remember that the numerator is divided *by* the denominator. Thus x divided by y means
>
> $$\frac{x}{y} \begin{array}{l} \leftarrow \text{numerator} \\ \leftarrow \text{denominator} \end{array}$$
>
> Note that y divided *into* x also means $\frac{x}{y}$.

A   Translating Words into Mathematical Expressions

We are now ready to use our mathematics dictionary to translate words into mathematical expressions.

TABLE 3 Mathematics Dictionary

Words	Translation	Example	Translation
Add More than Sum Increased by Added to	+	Add n to 7 7 more than n The sum of n and 7 n increased by 7 7 added to n	$n + 7$
Subtract Less than Minus Difference Decreased by Subtracted from	−	Subtract 9 from x 9 less than x x minus 9 Difference of x and 9 x decreased by 9 9 subtracted from x	$x - 9$
Of The product Times Multiply by	×	$\frac{1}{2}$ of a number x The product of $\frac{1}{2}$ and x $\frac{1}{2}$ times a number x Multiply $\frac{1}{2}$ by x	$\frac{1}{2}x$
Divide Ratio Divided by The quotient	÷	Divide 10 by x Ratio of 10 to x 10 divided by x The quotient of 10 and x	$\dfrac{10}{x}$
The same, yields, gives, is, equals	=	The sum of x and 10 is the same as the sum of 10 and x.	$x + 10 = 10 + x$

EXAMPLE 1   Translating words into mathematical equations

Translate the sentences into equations.

Verbal Sentence	Translation
A number decreased by 7 yields 25.	$x - 7 = 25$
The quotient of a number plus 8 and the number minus 8 equals 22.	$\dfrac{x + 8}{x - 8} = 22$
Twice a number decreased by 9 equals the number divided into 9.	$2x - 9 = \dfrac{9}{x}$
If the product of a number and 10 is decreased by 3, the result is 47.	$10x - 3 = 47$
The difference of the consecutive integers n and $n + 1$ is always -1.	$n - (n + 1) = -1$

B   General Word Problems

The mathematics dictionary we have presented plays an important role in the solution of word problems. You will find that the words contained in the dictionary are often key words when you translate a problem. But there are other details that can help you. Here is a restatement of the **RSTUV** method with hints and tips you can use when you solve word problems.

| problem solving | **HINTS AND TIPS** |

1. Read the problem.

Mathematics is a language. As such you have to learn how to read it. You may not understand or even get through reading the problem the first time. Read it again and as you do, pay attention to key words or instructions such as *compute, draw, write, construct, make, show, identify, state, simplify, solve,* and *graph.* (Can you think of others?)

2. Select the unknown.

How can you answer a question if you don't know what the question is? One good way to look for the unknown is to look for the question mark (?) and read the material to its left. Try to determine what is given and what is missing.

3. Think of a plan.

Problem solving requires many skills and strategies: Some of them are *look for a pattern; examine a related problem; make tables, pictures, diagrams; write an equation; work backwards;* and *make a guess.* (Can you think of others?)

4. Use the techniques you are studying to carry out the plan.

If you are studying a mathematical technique, it is almost certain that you will have to use it to solve the given problem. Look for specific procedures that are given to solve certain problems.

5. Verify the answer.

Look back and check the results of the original problem. Is your answer reasonable? Can you find it some other way?

Now let's try a word problem. Do you know the name of the heaviest glider in the world? It is the space shuttle *Columbia,* and we are ready to solve a problem about it. This solution is presented in a two-column format. Cover the answers in the right-hand column (a 3-by-5 index card will do), and write *your own answers* so you can practice. After that, uncover the answers and check to see whether you are right. We will then give you another example and provide you with its solution.

| problem solving | **WORD PROBLEMS** |

When fully loaded, the space shuttle *Columbia* and its payload weigh 215,000 pounds (lb). The *Columbia* itself weighs 85,000 lb more than the payload. What is the weight of each?

1. Read the problem carefully and decide what it asks for.

The problem asks for the weight of each—that is, the weight of the *Columbia* and the weight of the payload.

2. Select a variable to represent the unknown.

Let p represent the weight of the payload in pounds. Since the *Columbia* weighs 85,000 lb more than the payload, the *Columbia* weighs $p + 85,000$.

3. Think of a plan.
Can you translate the information into an equation or inequality?

We translate the first sentence in the problem:

The *Columbia* and its payload weigh 215,000.

$$(p + 85,000) \quad + \quad p \quad = \quad 215,000$$

That is,

4. Use algebra to solve for the unknown.

$$p + 85,000 + p = 215,000$$
$$2p + 85,000 - 85,000 = 215,000 - 85,000$$
$$2p = 130,000$$
$$p = 65,000$$

Thus the payload weighs 65,000 lb, and the *Columbia* 65,000 + 85,000, or 150,000 lb.

5. Verify the answer.

To verify the answer, note that the combined weight of the *Columbia* and its payload is 150,000 + 65,000, or 215,000 lb, as stated in the problem.

problem solving

EXAMPLE 2   Car rental costs

Tom Jones rented an intermediate sedan at $25 per day plus 20¢ per mile. How many miles can Tom travel for $50?

SOLUTION   We use the RSTUV method.

1. Read the problem.

We are looking for the number of miles Tom can travel for $50.

2. Select the unknown.

Let m represent this number of miles.

3. Think of a plan.

Translate the problem into an equation. To do this, realize that Tom is paying 20¢ for each mile plus $25 for the day. Thus, if Tom travels

1 mi,	the cost is	$0.20(1) + 25$
2 mi,	the cost is	$0.20(2) + 25$
m mi,	the cost is	$0.20m + 25$

Because we want to know how many miles Tom can drive for $50, we put the cost for m mi equal to $50, which gives the equation $0.20m + 25 = 50$.

4. Use the procedure for solving linear equations given on page 69.

$0.20m + 25 = 50$	Given.
$0.20m = 25$	Subtract 25.
$20m = 2500$	Multiply by 100. This gets rid of the decimal.
$m = \dfrac{2500}{20} = 125$	Divide by 20.

Thus Tom can travel 125 mi for $50, as long as he does it in one day!

5. Verify the answer.

Verify that $(0.20)(125) + 25 = 50$. (We leave this to you.)   ■

Integer Word Problems

A popular algebra problem deals with integers. If you are given an integer, can you find the integer that comes right after it? For example, the integer that comes after 7 is $7 + 1 = 8$, and the one that comes after -6 is $-6 + 1 = -5$.

CONSECUTIVE INTEGERS

> If n is any integer, the next **consecutive** integer is $n + 1$.

On the other hand, if you are given an *even* integer such as 6, the next *even* integer is 8 (add 2 this time). The next even integer after 34 is $34 + 2 = 36$.

CONSECUTIVE EVEN (OR ODD) INTEGERS

> If n is an *even* (or *odd*) integer, the next *even* (or *odd*) integer is $n + 2$.

Thus, if $n = 34$, the next even integer is $n + 2 = 34 + 2 = 36$. Similarly, if $n = 21$, the next odd integer is $n + 2 = 21 + 2 = 23$.

We use this idea in the next example. (Don't forget to consult the dictionary when necessary.)

EXAMPLE 3   Integer problems ✓

The sum of three consecutive odd integers is 129. Find the integers.

SOLUTION   We use the RSTUV method.

1. Read the problem.   Note that we are asking for three consecutive odd *integers*.

2. Select the unknown.   Let n be the first of the integers. Since we want three consecutive odd integers, we need to find the next two consecutive odd integers. The next odd integer after n is $n + 2$ and the one after $n + 2$ is $n + 4$. Thus the three consecutive odd integers are n, $n + 2$, and $n + 4$.

3. Think of a plan.   Translate the problem into the language of algebra.

The sum of 3 consecutive odd integers is 129.

$$\overbrace{n + (n + 2) + (n + 4)}\quad \overbrace{= 129}$$

4. Use the procedure to solve linear equations given on page 69.

$$n + (n + 2) + (n + 4) = 129 \qquad \text{Given.}$$
$$n + n + 2 + n + 4 = 129 \qquad \text{Remove parentheses.}$$
$$3n + 6 = 129 \qquad \text{Combine like terms.}$$
$$3n + 6 - 6 = 129 - 6 \qquad \text{Subtract 6.}$$
$$3n = 123 \qquad \text{Simplify.}$$
$$\frac{3n}{3} = \frac{123}{3} \qquad \text{Divide by 3.}$$
$$n = 41$$

5. Verify the answer.   Thus the three consecutive odd integers are 41, 41 + 2 = 43, and 41 + 4 = 45. Verify that the sum of the three integers is 129. Since 41 + 43 + 45 = 129, our result is correct.   ∎

Geometry Problems

✓

As you recall, the *perimeter* of (distance around) a rectangle is $P = 2L + 2W$, where L is the length and W is the width of the rectangle. Here's a problem using this formula.

EXAMPLE 4   A geometry word problem ✓

The students at Osaka Gakun University made a rectangular poster whose length was 130 ft more than its width. If the perimeter of the poster was 416 ft, give its dimensions.

SOLUTION   We use the RSTUV method.

1. Read the problem.   The key words are "rectangular" and "perimeter."

2. Select the unknown.   We let L be the length and W the width of the rectangle.

3. Think of a plan.   Translate the problem. This time, we make a sketch to help us visualize the situation.

The length is 130 more than the width.

$$\overbrace{L} \quad \overbrace{=} \quad \overbrace{W + 130}$$

Now the perimeter is

$$P = 2L + 2W$$

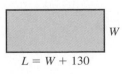

$L = W + 130$

Substituting 416 for P and $W + 130$ for L, $P = 2L + 2W$ becomes

$$416 = 2(W + 130) + 2W$$

4. Use the procedure to solve linear equations given on page 69.

$416 = 2(W + 130) + 2W$	Given.
$416 = 2W + 260 + 2W$	Remove parentheses.
$416 = 260 + 4W$	Collect like terms.
$156 = 4W$	Subtract 260.
$39 = W$	Divide by 4.

The length $L = W + 130$; thus $L = 39 + 130 = 169$. The dimensions of the poster are height, 169 ft, and width, 39 ft.

5. Verify the answer.

Is $P = 2L + 2W$? Is $416 = 2 \cdot 169 + 2 \cdot 39$?

$$416 = 338 + 78 \qquad \text{Yes.} \qquad ■$$

In Section 2.2, we studied vertical angles and angles formed by transversals that intersect parallel lines. Table 4 shows some other types of angles and their relationships.

TABLE 4

Angle	Relationship
A **right angle** is an angle whose measure is 90°. If the sum of the measures of two angles is 90°, the angles are **complementary angles** and they are **complements** of each other.	90° Right angle B A Complementary angles If the measure of one of the angles is $x°$, the measure of its complement is $(90 - x)°$.
A **straight angle** is an angle whose measure is 180°. If the sum of the measures of two angles is 180°, the angles are **supplementary angles** and they are **supplements** of each other.	180° Straight angle B A Supplementary angles If the measure of one of the angles is $x°$, the measure of its supplement is $(180 - x)°$.
The **sum** of the measures of the angles of a triangle is 180°. You can use this fact to find the measure of the third angle when the measures of the other two angles are given.	60° 35° The measure of the other angle, call it $x°$, is such that $$60 + 35 + x = 180$$ $$95 + x = 180$$ $$x = 85$$ Thus the third angle is 85°.

problem
solving

EXAMPLE 5   A geometry word problem involving angles

Two of the angles in a triangle are complementary. The third angle is twice the size of one of the complementary angles. What is the size of each of the angles?

SOLUTION   We use our RSTUV method.

1. Read the problem.

The key words are "complementary angles." Do you remember what that means?

2. Select the unknown.

Let $x°$ be the measure of one of the complementary angles.

3. Think of a plan.

We know that the sum of the three angles is 180°. We also know that two of the angles are complementary, so if one of the angles measures $x°$, the second angle measures $(90 - x)°$ and the third angle measures $(2x)°$. We can now translate the information:

The sum of all the angles is 180°.

$$x + (90 - x) + 2x = 180$$

4. Use the procedure to solve linear equations on page 69.

$x + (90 - x) + 2x = 180$	Given.
$x + 90 - x + 2x = 180$	Remove parentheses.
$90 + 2x = 180$	Since $x - x = 0$
$2x = 90$	Subtract 90.
$x = 45$	Divide by 2.

Thus one of the angles is 45°. Its complement is $90° - 45° = 45°$, and the third angle is $2 \cdot 45° = 90°$.

5. Verify the answer.

Since $45 + 45 + 90 = 180$, our answer is correct.   ■

GRAPH IT

How do you solve word problems with a grapher? The same way as without one! The grapher is simply an additional tool. Here are some suggestions for solving word problems with a grapher.

1. Use the RSTUV method until you reach step 3.

2. Graph the equation or inequality obtained in step 3 using your grapher. To do this, write the equation in the form $A(x) = 0$ and graph $Y_1 = A(x)$.

We must warn you, however, that in many cases it is *easier* to solve problems *algebraically* than it is to solve them *graphically*. Let's look at steps 3 and 4 in Examples 3 and 4.

Example 3

To graph $3x + 6 = 129$, subtract 129 from both sides to obtain $3x + 6 - 129 = 0$. Then graph

$$Y_1 = 3x + 6 - 129 = 3x - 123$$

But there's another problem. What window should you use? There are many answers to this question, but you may reason that since the sum of the three numbers is 129, each of the numbers must be about $\frac{129}{3} = 43$. (By the way, you almost got the answer by accident here!) Try a $[-5, 45]$ by $[-5, 5]$ window with Xscl = Yscl = 1. The resulting line (see Window 1) crosses the x-axis at $x = 41$, which represents the first integer as before. (You can show this by using the $\boxed{\text{ZOOM}}$ and $\boxed{\text{TRACE}}$ buttons repeatedly.) With a TI-82, you can find the solution (root) by pressing $\boxed{\text{2nd}}$ $\boxed{\text{CALC}}$ 2 and finding a lower bound (below the x-axis) and an upper bound (above the x-axis), and then giving a guess as prompted by the grapher. (See Window 2.)

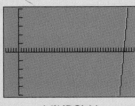

WINDOW 1

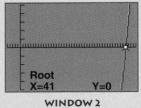

Root
X=41   Y=0

WINDOW 2

(continued on next page)

Example 4

From step 3, you can see that you have to graph

$$Y_1 = 2(x + 130) + 2x - 416$$

or

$$Y_1 = 2x + 260 + 2x - 416$$

WINDOW 3

Since the total perimeter is 416, each side is, on average, $\frac{416}{4} = 104$ ft. Thus select a window $[-10, 100]$ by $[-50, 100]$ with a scale of 10 units for both the x- and y-values (Xscl = 10 and Yscl = 10). Using the ⌐TRACE⌐ and ⌐ZOOM⌐ features, you can arrive at the answer, $x = 39$ (see Window 3). (The idea is to get the x-value when y is very close to zero. You actually obtained $x = 38.99723$ when $y = -0.0110803$.) But really, wouldn't you rather do it algebraically?

EXERCISE 2.3

A

1. Write an equation that is equivalent to the description: The product of 4 and a number m is the number increased by 18.

2. Given three consecutive, positive, even integers, write an inequality that is equivalent to the statement: The product of the smallest integer, say, $2n$, and the largest integer is always less than the square of the middle integer.

3. The units digit of a two-digit number is 3 less than the tens digit. The sum of the tens digit and the units digit is 7. Write an equation to find x, the tens digit.

4. The square of a number x, decreased by twice the number itself, is 10 more than the number. Write an equation expressing this statement.

In Problems 5–14, write the given statement as an equation, and then solve it.

5. If 4 times a number is increased by 5, the result is 29. Find the number.

6. Eleven more than twice a number is 19. Find the number.

7. The sum of 3 times a number and 8 is 29. Find the number.

8. If 6 is added to 7 times a number, the result is 69. Find the number.

9. If the product of 3 and a number is decreased by 2, the result is 16. Find the number.

10. Five times a certain number is 9 less than twice the number. What is the number?

11. Five times a certain number is the same as 12 increased by twice the number. What is the number?

12. If 5 is subtracted from half a number, the result is 1 less than the number itself. Find the number.

13. One-third of a number decreased by 2 yields 10. Find the number.

14. One-fifth of a certain number plus 2 times the number is 11. What is the number?

B **In Problems 15–20, use the RSTUV method to obtain the solution.**

15. The space shuttle *Columbia* consists of an orbiter, an external tank, two solid-fuel boosters, and fuel. At the time of lift-off, the weight of all these components is 4.16 million lb. The tank and the boosters weigh 1.26 million lb less than the orbiter and fuel. What is the weight of the orbiter and fuel?

16. The external tank and the boosters of the *Columbia* weigh 2.9 million lb, and the boosters weigh 0.34 million lb more than the external tank. Find the weight of the external tank and of the boosters.

17. Russia and Japan have the greatest number of merchant ships in the world. The combined total is 15,426 ships. If Japan has 2276 ships more than Russia, how many ships does each have?

18. Mary is 12 years old and her brother Joey is 2 years old. In how many years will Mary be just twice as old as Joey?

19. The cost of renting a car is $18 per day plus 20¢ per mile traveled. Margie rented a car and paid $44 at the end of the day. How many miles did Margie travel?

20. In baseball, the slugging average of a player is obtained by dividing his total bases (1 for a single, 2 for a double, 3 for a triple, and 4 for a home run) by his official number of times

at bat. José Cataña has 2 home runs, 1 triple, 2 doubles, and 9 singles. His slugging average is 1.2. How many times has he been at bat?

In Problems 21–32, consult your mathematics dictionary, if necessary, to solve the integer problems.

21. The sum of three consecutive even integers is 138. Find the integers.

22. The sum of three consecutive odd integers is 135. Find the integers.

23. The sum of three consecutive odd integers is −27. Find the integers.

24. The sum of three consecutive even integers is −24. Find the integers.

25. The sum of two numbers is 179, and one of them is 5 more than the other. Find the numbers.

26. The larger of two numbers is 6 times the smaller. Their sum is 147. Find the numbers.

27. The Beatles have 20 more Recording Industry Association of America awards than Paul McCartney. If the number of awards received by the Beatles and McCartney total 74, how many awards do the Beatles have?

28. The number of rooms in the Detroit Plaza Hotel exceeds the number of rooms in the Peachtree Plaza Hotel by 300. If the combined number of rooms in these hotels is 2500, how many rooms are there in each hotel?

29. Norway has 2811 merchant ships, 3 times as many as Sweden. How many merchant ships does Sweden have?

30. To find the weight of an object on the Moon, you can divide its weight on Earth by 6. The crew of *Apollo 16* collected lunar rocks and soil weighing 35.5 lb on the Moon. What is their weight on Earth?

31. The weight of an object on the Moon is obtained by dividing its earth-weight by 6. If an astronaut weighs 28 lb on the Moon, what is the corresponding earth-weight?

32. The greatest weight difference ever recorded in a major boxing bout is 140 lb in a match between John Fitzsimmons and Ed Punkhorst. If the combined weight of the contestants was 484 lb, find Fitzsimmons' weight. (He was the lighter of the two.)

Solve the geometry problems (33–46).

33. The largest painting in the world used to be the *Panorama of the Mississippi*, by John Banvard. If the length of this paint-

ing was 4988 ft more than its width and its perimeter was 10,024 ft, find the dimensions of this rectangular painting.

34. The largest painting now in existence is probably the *Battle of Gettysburg*. The length of this painting exceeds its width by 340 ft. If the perimeter of this rectangular painting is 960 ft, find its dimensions.

35. The scientific building with the greatest capacity is the Vehicle Assembly Building at the John F. Kennedy Space Center. The width of this building is 198 ft less than its length. If the perimeter of the rectangular building is 2468 ft, find its dimensions.

36. The largest fair hall is located in Hanover, Germany. The length of this hall exceeds its width by 295 ft. If the perimeter of this rectangular hall is 4130 ft, find its dimensions.

37. An angle has 4 times the measure of its complement. What is the measure of the angle?

38. The sum of the measures of an angle and one third a second angle is 32°. If the angles are complementary, what are their measures?

39. An angle has 3 times the measure of its supplement. What is the measure of the angle?

40. An angle is 5° less than twice another angle. Find their measures if they are supplementary angles.

41. Find x and the measure of each complementary angle.

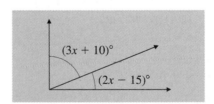

42. Find x and the measure of each complementary angle.

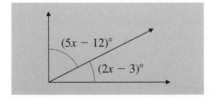

43. Find x and the measure of each supplementary angle.

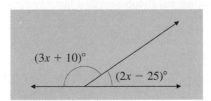

44. Find x and the measure of each supplementary angle.

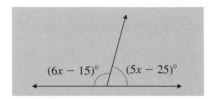

45. A triangle has two angles that are complementary. If the measure of the third angle is 3 times that of the first, what are the measures of the angles in this triangle?

46. A triangle has two angles that are complementary. If the measure of the third angle is 5 times that of the first, what are the measures of the angles in this triangle?

SKILL CHECKER

Solve:

47. $n + 0.30n = 26$ **48.** $n - 0.60n = 80$

49. $220 = 2(W + 70) + 2W$ **50.** $300 = 2(W + 60) + 2W$

51. $0.05x + 0.10(800 - x) = 20$

52. $0.10s + 0.20(6000 - s) = 1000$

53. $110T = 80(T + 3)$ **54.** $80T = 50(T + 3)$

USING YOUR KNOWLEDGE   Puzzles and Riddles

Here is a puzzle and a riddle that you can solve using the knowledge you've gained in these sections.

55. "What number," asked Professor Bourbaki, absentmindedly trying to find his algebra book, "must be added to the numerator and to the denominator of the fraction $\frac{1}{4}$ to obtain the fraction $\frac{2}{3}$?" "Well," he said after he found his book. Well?

56. Not much is known about Diophantus, sometimes called the father of algebra, except that he lived between A.D. 100 and 400. His age at death, however, is well known, because one of his admirers described his life using this algebraic riddle:

Diophantus' youth lasted $\frac{1}{6}$ of his life. He grew a beard after $\frac{1}{12}$ more of his life. After $\frac{1}{7}$ more of his life, he married. Five years later, he had a son. The son lived exactly $\frac{1}{2}$ as long as his father, and Diophantus died just 4 years after his son's death.

How many years did Diophantus live?

WRITE ON . . .

57. When reading a word problem, what is the first thing you should try to determine?

58. How can you verify the answer in a word problem?

59. Make your own integer word problem and show a detailed solution of it using the RSTUV method.

60. Make your own geometry word problem and show a detailed solution of it using the RSTUV method.

61. Step 3 of the RSTUV method calls for you to "**T**hink of a plan" to solve the problem. Some strategies you can use include *looking for a pattern* and *making a picture*. Can you think of three other strategies?

MASTERY TEST   If you know how to do these problems, you have learned your lesson!

62. The sum of two consecutive integers is 27. What are the integers?

63. The sum of three consecutive odd integers is 99. What are the three integers?

64. The square of a number x, increased by 4 times the number itself, is 2 less than the number. Write an equation expressing this statement.

65. The quotient of the sum of a number and 8, divided into the difference of the number and 5, yields 2. Write an equation expressing this statement.

66. The measure of an angle is 8 times that of another.
a. Find their measures if the angles are complementary.
b. Find their measures if the angles are supplementary.

67. Find x and the measure of each marked angle.

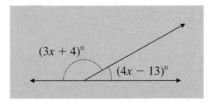

68. Denmark's cigarette tax tops the world. The tax on a pack of 20 cigarettes exceeds the price of the cigarettes by $3.03! If the total price of a pack of cigarettes in Denmark is $4.33, what is the tax?

69. The total number of students in U.S. schools in 1995 was 65 million. The number in secondary schools exceeded that of students in college by 2 million, while in elementary schools there were 3 times as many students as in college. What is the number (in millions) of students enrolled in college in 1995? Give your answer to two decimal places.

2.4

PERCENT, INVESTMENT, MOTION, AND MIXTURE PROBLEMS

To succeed, review how to:

1. Perform the four fundamental operations using decimals (pp. 15–21).

2. Solve linear equations involving decimals (pp. 72–73).

Objectives:

A Solve percent problems.

B Solve investment problems.

C Solve uniform motion problems.

D Solve mixture problems.

getting started   **Tax Forms**

Schedule Z—Use if your filing status is **Head of household**

If the amount on Form 1040, line 37, is: Over—	But not over—	Enter on Form 1040, line 38	of the amount over—
$ 0	$ 29,600	. . . . . .15%	$ 0
29,600	76,400	$4,440.00 + 28%	29,600
76,400	127,500	$17,544.00 + 31%	76,400
127,500	250,000	$33,385.00 + 36%	127,500
250,000	—	$77,485.00 + 39.6%	250,000

Suppose you are a "Head of household" and the amount on the IRS Form 1040, line 37 is $30,000. According to the table, the amount you have to enter on line 38 (the tax you owe) is

$4440.00 + 28% of the amount over $29,600

$= \$4440 + 0.28 \cdot (30{,}000 - 29{,}600)$ The amount over 29,600 is 30,000 − 29,600.

$= \$4440 + 0.28 \cdot 400$

$= \$4440 + 112$

$= \$4512$

Thus you owe $4512. Do you know what *percent* of the $30,000 is tax? We will learn how to find that in this section.

In algebra, rates, increases, decreases, and discounts are often written as percents (%). *Percent* means "by the hundred." Thus 28% means 28 parts of 100 or $\frac{28}{100}$. In most applications, however, percents are written as decimals. Thus

$$28\% = \frac{28}{100} = 0.28$$

$$30\% = \frac{30}{100} = 0.30$$

$$120\% = \frac{120}{100} = 1.2$$

and $$34.8\% = \frac{34.8}{100} = \frac{348}{1000} = 0.348$$

There are three basic types of percent problems.

TYPE 1 asks to find a percent of a number.

PROBLEM   30% of 80 is what number?

TRANSLATION $0.30 \times 80 = n$

SOLUTION $24 = n$

TYPE 2 asks what percent of a number is another given number.

PROBLEM   What percent of 40 is 8? Or 8 is what percent of 40?

TRANSLATION $n \times 40 = 8$ or $8 = n \times 40$

SOLUTION $n = \dfrac{8}{40} = \dfrac{1}{5} = 20\%$ or $\dfrac{8}{40} = \dfrac{1}{5} = 20\% = n$

TYPE 3 asks to find a number when a percent of the number is given.

PROBLEM   10 is 40% of what number?

TRANSLATION $10 = 0.40 \times n$

SOLUTION $\dfrac{10}{0.40} = \dfrac{1000}{40} = 25 = n$

Percent Problems

problem solving

EXAMPLE 1   Percent problems

Referring to the "Head of household" table, suppose the amount on line 37 (your taxable income) is $100,000.

a. What has to be entered on Form 1040, line 38 (the tax you owe)?

b. What percent of line 37 is line 38; that is, what percent of your taxable income is paid in taxes?

SOLUTION   We use the RSTUV method.

1. Read the problem.

a. Note that we have to refer to the table in the *Getting Started*.

2. Select the unknown.

We are asked to find the amount to be entered on line 38.

3. Think of a plan.

Since the amount on line 37 is $100,000, which is over $76,400 but not over $127,500, we use the third line in the table to find the amount to enter on line 38.

4. Use arithmetic to compute the total.

$$\$17,544 + 31\% \text{ of the amount over } \$76,400$$
$$= \$17,544 + 0.31 \cdot (100,000 - 76,400)$$
$$= \$17,544 + 0.31 \cdot 23,600$$
$$= \$17,544 + 7316$$
$$= \$24,860$$

Thus the amount to be entered on line 38 is $24,860.

5. Verify the solution.

The verification is left to you.

b. Again, we refer to the table in the *Getting Started*.

1. Read the problem.

We are asked to find what percent of the amount on line 37 is the amount on line 38.

2. Select the unknown.   Let n represent this percent.

3. Think of a plan. $$n \times 100{,}000 = 24{,}860$$

4. Use arithmetic to compute the total. $$n = \frac{24{,}860}{100{,}000}$$ Divide both sides by 100,000.

$$= 24.86\%$$

5. Verify the solution.   We leave the verification to you.   ■

problem solving   **EXAMPLE 2**   Solving percent increase problems

It is estimated that by the year 2000 the annual number of airline passengers at O'Hare airport in Chicago will reach 42 million, a 50% increase over this year's figures. How many passengers will use O'Hare this year?

SOLUTION   We use the RSTUV method.

1. Read the problem.   We are asked to find the number of passengers using O'Hare this year.

2. Select the unknown.   Let n millions be the number of passengers this year.

3. Think of a plan.   The number of passengers will increase by 50%. The increase is 50% of n, that is, $0.50n$. Now translate this into an equation.

A 50% increase	over	this year's figure	will reach	42 (million)
$0.50n$	$+$	n	$=$	42

4. Use algebra to solve for the unknown.   We now solve this equation

$$0.50n + n = 42 \qquad \text{Given.}$$

$$1.50n = 42 \qquad \text{Collect like terms.}$$

$$n = \frac{42}{1.5} = 28 \qquad \text{Divide by 1.5.}$$

5. Verify the answer.   The number of passengers this year will be 28 million. The answer is correct because $28 + 0.50(28) = 28 + 14 = 42$.   ■

B   Investment Problems

Investment problems also use percents. If you invest P dollars at a rate r, your **annual** interest is

$$I = Pr$$

When working this type of problem, it's helpful to enter all the information in a table. We do this in Example 3.

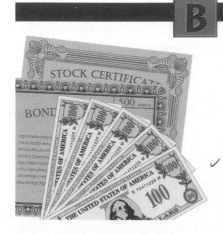

✓   **EXAMPLE 3**   Solving investment problems

A woman has some stocks yielding 5% annually and some bonds that yield 10%. If her investment totals $6000 and her annual income from the investment is $500, how much does she have invested in stocks and how much in bonds?

SOLUTION   We use the RSTUV method.

1. Read the problem.

We want to find how much she has invested in stocks and how much in bonds.

2. Select the unknown.

Let s be the amount she has invested in stocks. This makes the amount invested in bonds $(6000 - s)$.

3. Think of a plan.

We use a chart to help us visualize the problem. We enter the information in the chart:

	P $\times$	r $=$	I
Stocks	s	0.05	$0.05s$
Bonds	$6000 - s$	0.10	$0.10(6000 - s)$

The total interest is the sum of the entries in the last column, that is, $0.05s$ and $0.10(6000 - s)$. This amount must be $500, that is,

$$0.05s + 0.10(6000 - s) = 500$$

4. Use algebra to solve for the unknown.

Solve the equation.

$0.05s + 0.10(6000 - s) = 500$	Given.
$0.05s + 600 - 0.10s = 500$	Simplify.
$600 - 0.05s = 500$	Collect like terms.
$600 - 600 - 0.05s = 500 - 600$	Subtract 600.
$-0.05s = -100$	Simplify.
$\dfrac{-0.05s}{-0.05} = \dfrac{-100}{-0.05}$	Divide by -0.05.
$s = 2000$	

Thus the woman has $2000 in stocks and the rest, that is, $4000, in bonds.

5. Verify the answer.

Since 5% of $2000 = $100 and 10% of $4000 = $400, the total interest is indeed $500.   ■

Uniform Motion Problems

When traveling at a constant rate R, the distance D traveled in time T is given by

$$D = RT$$

Since this formula is similar to the one used for interest problems, we again use a table to enter the information. Here's how we do it.

√ **EXAMPLE 4**   Solving distance problems

A Supercruiser bus leaves Miami for San Francisco traveling at the rate of 40 mi/hr. Three hours later a car leaves Miami for San Francisco traveling at 55 mi/hr on the same route as the bus. How far from Miami does the car overtake the bus?

SOLUTION   We use the RSTUV method.

1. Read the problem.

Read the problem carefully. To find the distance, we need to know the time it takes the car to overtake the bus.

2. Select the unknown.

Let T represent this time.

3. Think of a plan.

We translate the given information and write it in a chart. Note that if the car travels for T hr, the bus travels for $T + 3$ hr (since it left 3 hr earlier).

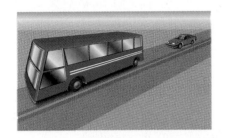

	R	$\times$	T	$=$	D
Car	55		T		$55T$
Bus	40		$T + 3$		$40(T + 3)$

When the car overtakes the bus, they will have traveled the same distance. According to the chart, the car has traveled $55T$ and the bus $40(T + 3)$ miles. Thus

$$55T = 40(T + 3)$$

4. Use algebra to solve for the unknown.

Solve the equation.

$55T = 40(T + 3)$	Given.
$55T = 40T + 120$	Simplify.
$55T - 40T = 40T - 40T + 120$	Subtract $40T$.
$15T = 120$	Simplify.
$\dfrac{15T}{15} = \dfrac{120}{15}$	Divide by 15.
$T = 8$	

It takes the car 8 hr to overtake the bus. In 8 hr, the car travels $8 \times 55 = 440$ mi, and overtakes the bus 440 mi from Miami.

5. Verify the answer.

The car has traveled for 8 hr at 55 mi/hr; thus it travels $55 \times 8 = 440$ mi, whereas the bus has traveled 11 hr at 40 mi/hr, a total of $40 \times 11 = 440$ mi.   ■

Mixture Problems

The last type of problem we discuss is the **mixture problem**, a type of problem in which two or more things are put together to form a mixture. Again, we use a chart to enter the information.

EXAMPLE 5   Solving mixture problems

How many ounces of a 50% acetic acid solution should a photographer add to 32 oz of a 5% acetic acid solution to obtain a 10% acetic acid solution?

SOLUTION We use the RSTUV method.

1. Read the problem.

We are asked to find the number of ounces of the 50% solution that should be added.

2. Select the unknown.

Select x to stand for the number of ounces of 50% solution to be added.

3. Think of a plan.

To translate the problem, we use a chart. In this case, the headings for the chart contain the percent of acetic acid and the amount to be mixed. The product of these two numbers will give us the amount of pure acetic acid.

50%   5%   10%

	%	×	Amount	=	Amount of Pure Acid
50% solution or 0.50			x		$0.50x$
5% solution or 0.05			32		1.60
10% solution or 0.10			$x + 32$		$0.10(x + 32)$

The percents have been converted to decimals. Note that you *should not* add the percents in this column.

Since we have x oz of one and 32 oz of the other, we have $(x + 32)$ ounces of the mixture.

Since the sum of the total amounts of pure acetic acid should be the same as the amount of pure acetic acid in the final mixture, we have

$$0.50x + 1.60 = 0.10(x + 32)$$

4. Use algebra to solve for the unknown.

Solve the equation.

$5x + 16 = x + 32$	Multiply by 10 to get rid of the decimals.
$5x + 16 - 16 = x + 32 - 16$	Subtract 16.
$5x = x + 16$	Simplify.
$5x - x = x - x + 16$	Subtract x.
$4x = 16$	Simplify.
$\dfrac{4x}{4} = \dfrac{16}{4}$	Divide by 4.
$x = 4$	

The photographer must add 4 oz of the 50% solution.

5. Verify the answer.

The verification of this fact is left to you.   ■

GRAPH IT

Suppose you want to use a grapher to solve Example 3. First, do steps 1 and 2 in the RSTUV method in the usual manner. Then go to step 3, write the equation

$$0.05s + 0.10(6000 - s) = 500$$

as

$$0.05s + 0.10(6000 - s) - 500 = 0$$

and graph

$$Y_1 = 0.05x + 0.10(6000 - x) - 500$$

Note: We use the variable x instead of the original s. (If you want to enter the original equation using s for the variable, use your grapher's ALPHA feature.)

What happens when you press the GRAPH button? Nothing, unless you have set your RANGE or WINDOW correctly. If you look at step 4 in the solution of the problem, you will note that the answer is $s = 2000$; thus *before* you graph the equation, you have to set your window at $[-5000, 5000]$ by $[-5000, 5000]$ with a scale of 1000. You will then see that the resulting line intersects the x-axis at $x = 2(000)$ as shown in Window 1. For a better view, adjust the y-scale to $[-1000, 1000]$ as shown in Window 2. If you have a TI-82, the "root" feature will find the answer. Press 2nd TRACE 2 to access this feature. Pick a point above the horizontal axis when the grapher asks for "Lower Bound?" and press ENTER. Move the cursor right and pick a point below the horizontal axis when the grapher asks for "Upper Bound?" then press ENTER. Press ENTER when asked for a "Guess?" The grapher will answer "Root X=2000, Y=0." Remember that the "root" of an equation means the "solution" of the equation. Thus the answer (root) is 2000 as in the problem.

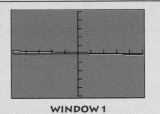

WINDOW 1

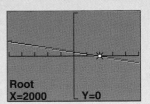

Root
X=2000   Y=0
WINDOW 2

EXERCISE 2.4

A Solve the following problems.

1. By weight, the average adult is composed of 43% muscle, 26% skin, 17.5% bone, 7% blood, and 6.5% organs. Suppose a person weighs 150 pounds.
   a. How many pounds of muscle does the person have?
   b. How many pounds of organs does the person have?

2. Refer to Problem 1.
   a. How many pounds of bone does the person have?
   b. How many pounds of skin does the person have?

3. A bicycle is priced at $196.50. If the sales tax rate is 5.5%, what is the tax?

4. The highest recorded shorthand speed was 300 words per minute for 5 min with 99.64% accuracy. How many errors were made? (Answer to the nearest whole number.)

5. In a recent year 41.2 million households with televisions watched the Super Bowl. This represents 41.7% of homes with television in major cities. How many homes with televisions are there in major cities?

6. On February 28, 1983, 125 million people watched the final episode of M*A*S*H. If the 125 million represents a 77% share of the viewing audience, what was the viewing audience that day?

In Problems 7–10, use the following formula:

Selling price = cost + markup

7. The cost of an article is $18.50. If the markup is 25% of the cost, what is the selling price of the article?

8. The selling price of an article is $30. If the markup is 20% of the cost, what is the cost of the article?

9. An article costing $30 is sold for $54. What is the markup and the percent of markup on selling price?

10. An article costing $60 is sold for $90. What is the markup and the percent of markup on cost?

11. If 25 is increased to 35, what is the percent increase?

12. If 32 is decreased to 24, what is the percent decrease?

13. If 40 is increased by 25% of itself, what is the result?

14. If 35 is decreased by 20% of itself, what is the result?

15. Luisa is a computer programmer. Her salary was increased from $25,000 to $30,000. What was her percent increase?

16. Tran was told that if he would switch to the next higher grade of gasoline, his mileage would increase to 120% of his present mileage of 25 mi/gal. What would be his increased mileage?

17. The average salary for pediatric doctors is $110,000. Surgeons make an astounding 250% more. On average, how much more do surgeons earn?

18. The number of beverages introduced in the market during two consecutive years decreased by 6% to 611 new beverages. How many beverages were introduced during the first year?

19. The total amount spent in corporate travel and entertainment increased by 21% in a 2-yr period. If the amount spent reached $115 billion at the end of the period, how much was spent at the beginning?

20. The annual amount spent on entertainment by the average person will increase by 33% to $1463. How much is the average person spending now on entertainment?

B **Solve the following problems.**

21. Two sums of money totaling $15,000 earn, respectively, 5% and 7% annual interest. If the interest from both investments amounts to $870, how much is invested at each rate?

22. An investor invested $20,000, part at 6% and the rest at 8%. Find the amount invested at each rate if the annual income from the two investments is $1500.

23. A woman invested $25,000, part at 7.5% and the rest at 6%. If her annual interest from these two investments amounted to $1620, how much money did she have invested at each rate?

24. A man has a savings account that pays 5% annual interest and some certificates of deposit paying 7% annually. His total interest from the two investments is $1100, and the total amount of money in the two investments is $18,000. How much money does he have in the savings account?

C **Solve the following problems.**

25. Two hours after a car leaves a certain town traveling at an average speed of 60 km/hr, a highway patrol officer leaves from the same starting point to overtake the car. If the average speed of the patrolman is 90 km/hr, how far from town does the officer overtake the car?

26. A group of smugglers cross the border in a car traveling in a straight line at 96 km/hr. An hour later, the border patrol starts after them in a light plane traveling 144 km/hr.

a. How long will it be before the border patrol reaches the smugglers?

b. At what distance from the border will the border patrol overtake the smugglers?

27. A freight train leaves the station traveling at 30 mi/hr. One hour later, a passenger train leaves the same station on a parallel track traveling at 60 mi/hr. How far from the station does the passenger train overtake the freight train?

28. A bus leaves the station traveling at 60 km/hr. Two hours later, the wife of one of the passengers shows up at the station with a briefcase belonging to an absentminded professor riding the bus. If she immediately starts after the bus at 90 km/hr, how far from the station is the briefcase reunited with the professor?

29. An accountant and her boss have to travel to a nearby town. The accountant catches a train traveling at 50 mi/hr while the boss leaves 1 hr later in a car traveling at 60 mi/hr. They have decided to meet at the train station and, strangely enough, they get there at exactly the same time! If the train and the car traveled in a straight line on parallel paths, how far is it from one town to the other?

30. The basketball coach at a local high school left for work on her bicycle, traveling at 15 mi/hr. Thirty minutes later, her husband noticed that she had left her lunch. He got in his car and took her lunch to her traveling at 60 mi/hr. Luckily, he got to school at exactly the same time as his wife. How far is it from the house to the school?

D **Solve the following problems.**

31. How many liters (L) of a 40% glycerin solution must be mixed with 10 L of an 80% glycerin solution to obtain a 65% solution?

32. How many parts of glacial acetic acid (99.5%) must be added to 100 parts of a 10% solution of acetic acid to give a 28% solution?

33. If the price of copper is 65¢ per pound and the price of zinc is 30¢ per pound, how many pounds of copper and zinc should be mixed to make 70 lb of brass selling for 45¢ per pound?

34. Oolong tea sells for $19 per pound. How many pounds of Oolong should be mixed with regular tea selling at $4 per pound to produce 50 lb of tea selling for $7 per pound?

35. You think the prices of coffee are high? You haven't seen anything yet! Jamaican Blue coffee sells for about $20 per pound! How many pounds of Jamaican Blue should be mixed with 80 lb of regular coffee selling at $8 per pound so that the result is a mixture selling for $10.40 per pound? (You can cleverly advertise this as "Containing the incomparable Jamaican Blue coffee.")

Fill in the blank with $<$ or $>$ to make the resulting statement true.

36. a. -3 _____ -1 b. -1.3 _____ -1.4 c. $\dfrac{1}{3}$ _____ $\dfrac{1}{2}$

37. a. $-\dfrac{1}{3}$ _____ $-\dfrac{1}{2}$ b. $-\dfrac{1}{5}$ _____ $-\dfrac{1}{2}$ c. $-\dfrac{1}{4}$ _____ $-\dfrac{1}{2}$

USING YOUR KNOWLEDGE   **We Owe, We Owe, and to the Deficit We Go.**

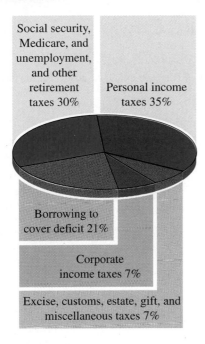

Social security, Medicare, and unemployment, and other retirement taxes 30%

Personal income taxes 35%

Borrowing to cover deficit 21%

Corporate income taxes 7%

Excise, customs, estate, gift, and miscellaneous taxes 7%

The last page of the IRS Instructions for Form 1040 indicates that the federal income this year was $1090.5 billion, while outlays (expenses) amounted to $1380.9 billion. (By the way, 1090.5 billion = 1.0905 _trillion_.)

38. What is the difference between income and expenses?

39. The pie chart indicates that 21% of the federal income was borrowed to cover the deficit. How much money was borrowed to cover the deficit?

40. The pie chart indicates that 35% of the federal income comes from personal taxes. How much money comes from personal taxes?

41. Formulate three different problems involving percents and show a detailed solution using the RSTUV method.

42. Formulate your own investment word problem and show a detailed solution using the RSTUV method.

43. Formulate your own motion word problem and show a detailed solution using the RSTUV method.

44. Ask your pharmacist or your chemistry instructor if he or she mixes products of different concentrations to make new mixtures. Write a paragraph on your findings.

45. All the problems in this text have precisely the information you need to solve them. In real life, however, some of the information given may be irrelevant. Such irrelevant information is called a _red herring_. Find some problems with red herrings and point them out.

MASTERY TEST   **If you know how to do these problems, you have learned your lesson!**

46. A man has two investments totaling $8000. One investment yields 5% and the other 10%. If the total annual interest is $650, how much is invested at each rate?

47. How many gallons of a 10% salt solution must be added to 15 gallons of a 20% salt solution to obtain a 16% solution?

48. The estimated number of passengers at the Atlanta airport in the year 2000 will reach 31.2 million, a 30% increase over this year's figure. How many passengers will use the Atlanta airport this year?

49. A bus leaves Los Angeles traveling at 50 mi/hr. An hour later a car traveling on the same road leaves at 70 mi/hr to try to overtake the bus. How far from Los Angeles does the car overtake the bus?

50. The number of students enrolled in higher education in the United States increased from 12 million in 1980 to 14.4 million in 1990. What was the percent increase?

2.5

LINEAR INEQUALITIES

To succeed, review how to:

1. Solve linear equations (pp. 69–73).

2. Add, subtract, multiply, and divide integers (pp. 15–21).

3. Properly use the symbols $>$ and $<$ when comparing numbers (pp. 11–12).

(See the *Skill Checker*, page 105.)

Objectives:

A   Graph linear inequalities.

B   Solve and graph linear inequalities.

C   Solve and graph compound inequalities.

D   Use the inequality symbols to translate sentences into inequalities.

---

getting started

Inequalities in Rental Cars

Suppose a rental car costs $25 each day and $0.20 per mile. If you drive m miles, the cost C for the day is

$$C = 0.20m + 25$$

 Cost for m miles   Cost per day

Now suppose your daily cost must be under $50. This means that $0.20m + 25 < 50$. How many miles can you drive? The expression $0.20m + 25 < 50$ is an example of a linear inequality and can be solved using the same techniques as those used to solve linear equations. Thus

$0.20m + 25 < 50$	Given.
$0.20m + 25 - 25 < 50 - 25$	Subtract 25.
$0.20m < 25$	Simplify.
$m < \dfrac{25}{0.20}$	Divide by 0.20.
$m < 125$	

Hence if you drive less than 125 mi, your cost will be less than $50 per day.

An **inequality** is a statement that two expressions are *not* equal. Thus $x + 2 < 5$, $x - 7 < 3$, and $3x - 8 \leq 15$ are *linear inequalities*.

LINEAR INEQUALITIES

> A **linear inequality** in one variable is an inequality that can be written in the form
>
> $$ax + b < c$$
>
> where a, b, and c are real numbers, and $a \neq 0$.

The inequality $ax + b < c$ is still a linear inequality if the $<$ (less than) symbol is replaced by $>$ (greater than), $\leq$ (is less than or equal to), or $\geq$ (is greater than or equal to). Thus, $2x + 5 > 8$, $-3x - 7 \geq 9$, $-\frac{1}{2}x + 8 < -\frac{3}{4}$, and $0.20m + 25 \leq 50$ are all linear inequalities. As with linear equations, we *solve* an inequality by finding all the replacements of the variable that make the inequality a true statement. This is done by finding an *equivalent* inequality whose solution is obvious. For example, the inequalities $x + 2 > 5$ and $x > 3$ are equivalent. They are both satisfied by all real numbers greater than 3. We write the *solution set* of $x + 2 > 5$ in set-builder notation as

$$\{x \mid x > 3\}$$

This is read "the set of all x's such that x is greater than 3."

Graphing Linear Inequalities

There are infinitely many numbers in the solution set of the inequality $x > 3$ (4, 7.5, $\sqrt{10}$, and $\frac{10}{3}$ are a few of them). Since we cannot list all these numbers, we show all solutions of $x > 3$ **graphically** by using the number line. This type of representation is called the **graph** of the inequality. The heavy line in the figure is the graph of $x > 3$.

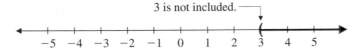

Note that the number 3 is *excluded* from the graph. This fact is shown by drawing a parenthesis at the point 3.

EXAMPLE 1   Graphing linear inequalities

Graph:

a. $x \geq -1$ **b.** $x < -2$

SOLUTION

a. The numbers that satisfy the inequality $x \geq -1$ are the numbers that are *greater than or equal to* -1, that is, the number -1 and all the numbers to the *right* of -1 (note that $\geq$ points to the right). The graph is shown in the following figure. The fact that -1 is included is shown by drawing a bracket at the point -1.

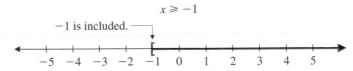

b. The numbers that satisfy the inequality $x < -2$ are the numbers that are *less than* -2, that is, the numbers to the *left* but *not including* -2 (note that $<$ points to the left). The graph of these points is

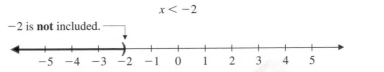

The inequalities $x > 3$, $x \geq -1$, and $x < -2$ resulted in graphs that did *not* have a finite length; they are called **unbounded** (or **infinite**) **intervals**. The four basic types of unbounded intervals are summarized in Table 5, where we use a parenthesis to denote the inequality symbols $<$ and $>$ and a bracket to indicate $\leq$ or $\geq$. Note that an **open** interval does not contain either of its end points.

TABLE 5 Unbounded Interval Notation

Set Notation	Interval Notation	Type of Interval	Graph	Inequality
$\{x \mid x > a\}$	$(a, +\infty)$	Open		$x > a$
$\{x \mid x < b\}$	$(-\infty, b)$	Open		$x < b$
$\{x \mid x \geq a\}$	$[a, +\infty)$	Half-open		$x \geq a$
$\{x \mid x \leq b\}$	$(-\infty, b]$	Half-open		$x \leq b$

In all these intervals a parenthesis is always used with the symbols ∞ (**positive infinity**) and $-\infty$ (**negative infinity**). Note that the symbols ∞ and $-\infty$ do not represent real numbers; they are used to indicate that the interval is unbounded.

B Solving Linear Inequalities

The inequalities $x + 2 > 5$ and $x > 3$ are equivalent. This is because we added (-2) to both sides of $x + 2 > 5$ to obtain the equivalent inequality

$$x + 2 + (-2) > 5 + (-2) \quad \text{or} \quad x > 3$$

We used the first of the following properties.

PROPERTIES OF INEQUALITIES: ADDITION AND SUBTRACTION

If C is a real number, then the following inequalities are all *equivalent:*

$$A < B$$

Add C. $A + C < B + C$

Subtract C. $A - C < B - C$

Now consider the inequality $2x - 3 < x + 1$. To solve this inequality, we need all the variables by themselves on one side. Thus we proceed as follows:

$$2x - 3 < x + 1 \qquad \text{Given.}$$
$$2x - 3 + 3 < x + 1 + 3 \qquad \text{Add 3.}$$
$$2x < x + 4 \qquad \text{Simplify.}$$
$$2x - x < x - x + 4 \qquad \text{Add } -x \text{ (or subtract } x).$$
$$x < 4 \qquad \text{Simplify.}$$

The solution set is $\{x \mid x < 4\}$ or, in interval notation, $(-\infty, 4)$.

The graph of this inequality is shown in the figure. You can check that this solution is correct by selecting a number from the graph (say 0) and replacing x with the number 0 in the original inequality to obtain $2(0) - 3 < 0 + 1$ or $-3 < 1$, a true statement. Of course, this is only a "partial" check since we did not try *all* the numbers in the graph.

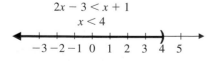

EXAMPLE 2   Solving and graphing inequalities

Solve and graph:

a. $3x - 2 < 2(x - 2)$ **b.** $4(x + 1) \geq 3x + 7$

SOLUTION

a.

$3x - 2 < 2(x - 2)$	Given.
$3x - 2 < 2x - 4$	Distributive property
$3x - 2 + 2 < 2x - 4 + 2$	Add 2.
$3x < 2x - 2$	Simplify.
$3x - 2x < 2x - 2x - 2$	Add $-2x$ (or subtract $2x$).
$x < -2$	Simplify.

The solution set is $\{x \mid x < -2\}$ or, in interval notation, $(-\infty, -2)$. The graph is

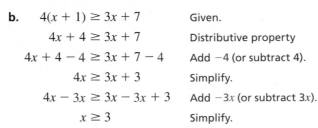

b.

$4(x + 1) \geq 3x + 7$	Given.
$4x + 4 \geq 3x + 7$	Distributive property
$4x + 4 - 4 \geq 3x + 7 - 4$	Add -4 (or subtract 4).
$4x \geq 3x + 3$	Simplify.
$4x - 3x \geq 3x - 3x + 3$	Add $-3x$ (or subtract $3x$).
$x \geq 3$	Simplify.

The solution set is $\{x \mid x \geq 3\}$ or, in interval notation, $[3, \infty)$. The graph of this inequality is

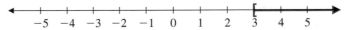

Before we state the multiplication and division properties of inequalities, let's see what happens if we multiply both sides of an inequality by a positive number. Consider these true inequalities:

$$
\begin{array}{lll}
1 < 3 & -1 < 3 & -3 < -1 \\
4 \cdot 1 < 4 \cdot 3 \quad & 4 \cdot (-1) < 4 \cdot 3 \quad & 4 \cdot (-3) < 4 \cdot (-1) \qquad \text{Multiply by 4.} \\
4 < 12 & -4 < 12 & -12 < -4
\end{array}
$$

Note that the resulting inequalities are all *true* and that the inequality symbol points in the *same* direction as the original. We say that multiplying both sides of an inequality by a *positive* number *preserves the sense,* or direction, of the inequality.

Now, let's multiply both sides of the original inequality by a *negative* number, say -4. We have

$$
\begin{array}{lll}
1 < 3 & -1 < 3 & -3 < -1 \\
-4 \cdot 1 \; ? \; -4 \cdot 3 \quad & -4 \cdot (-1) \; ? \; -4 \cdot 3 \quad & -4 \cdot (-3) \; ? \; -4 \cdot (-1) \qquad \text{Multiply by } -4. \\
-4 > -12 & 4 > -12 & 12 > 4
\end{array}
$$

This time, however, we had to reverse the direction of the inequalities to maintain a true statement. Thus multiplying both sides of an inequality by a *negative* number *reverses the sense* of the inequality. Note that since division is defined in terms of multiplication, these two properties apply to division as well. They can be stated as follows.

PROPERTIES OF INEQUALITIES: MULTIPLICATION AND DIVISION

If C is a real number, then the following inequalities are all equivalent:

$$A < B$$

$$A \cdot C < B \cdot C, \qquad \text{if } C \text{ is positive } (C > 0)$$

$$\frac{A}{C} < \frac{B}{C}, \qquad \text{if } C \text{ is positive } (C > 0)$$

$$A \cdot C > B \cdot C, \qquad \text{if } C \text{ is negative } (C < 0)$$

$$\frac{A}{C} > \frac{B}{C}, \qquad \text{if } C \text{ is negative } (C < 0)$$

CAUTION We can still multiply or divide both sides of an inequality by any nonzero number *as long as we remember to reverse the sense (direction) of the inequality if the number is negative.*

Thus to solve $-2x > 4$, we must divide both sides by -2 (or multiply by $-\frac{1}{2}$). When doing this, remember to reverse the sense (direction) of the inequality.

$$-2x > 4 \qquad \text{Given.}$$

$$\frac{-2x}{-2} < \frac{4}{-2} \qquad \text{Multiply by } -\tfrac{1}{2} \text{ (or divide by } -2\text{) and reverse the sign.}$$

$$x < -2 \qquad \text{Simplify.}$$

EXAMPLE 3   Solving and graphing inequalities

Solve and graph:

a. $4(x - 1) \le 6x + 2$ **b.** $2x + 9 \ge 5x + 3$

SOLUTION

a. We follow the same procedure as that for solving linear equations given on page 69.

Given: $4(x - 1) \le 6x + 2$

1.   There are no fractions.

2. $4x - 4 \le 6x + 2$ Use the distributive property.

3. $4x - 4 + 4 \le 6x + 2 + 4$ Add 4 to both sides.

     $4x \le 6x + 6$

4. $4x + (-6x) \le 6x + (-6x) + 6$ Add $-6x$ (or subtract $6x$).

    $-2x \le 6$

5. $\dfrac{-2x}{-2} \ge \dfrac{6}{-2}$ Multiply by $-\tfrac{1}{2}$ (or divide by -2) and *reverse* the sense of the inequality since -2 is negative.

    $x \ge -3$

Thus the solution set is $\{x \mid x \ge -3\}$ or, in interval notation, $[-3, \infty)$. The graph is

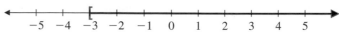

If we wish to avoid multiplying (or dividing) by negative numbers, we can add -6 in step 3 and $-4x$ in step 4 to obtain $-6 \leq 2x$. Then, dividing by 2, we get $-3 \leq x$, which is equivalent to $x \geq -3$.

b. Again, we isolate the x's on one side. This time, let's avoid multiplying or dividing by negative numbers. We do this by noting that there are more x's on the right-hand side of the inequality. Thus we isolate the x's on the right.

Given: $2x + 9 \geq 5x + 3$

1.   There are no fractions.

2.   The inequality is simplified.

3. $2x + 9 + (-3) \geq 5x + 3 + (-3)$ Add -3.

$\qquad\qquad 2x + 6 \geq 5x$

4. $(-2x) + 2x + 6 \geq 5x + (-2x)$ Add $-2x$.

$\qquad\qquad 6 \geq 3x$

5. $\qquad\qquad \dfrac{6}{3} \geq \dfrac{3x}{3}$ Multiply by $\frac{1}{3}$ (or divide by 3).

$\qquad\qquad 2 \geq x$

We can write this as $x \leq 2$, so the solution set is $\{x \mid x \leq 2\}$ or, in interval notation $(-\infty, 2]$. The graph is

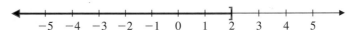

The last two inequalities contained no fractions. If fractions are present, we clear them by multiplying both sides of the inequality by the LCD of the fractions involved, as Example 4 shows.

EXAMPLE 4   Solving and graphing inequalities with fractions

Solve and graph: $\dfrac{x}{4} - \dfrac{x}{6} > \dfrac{x-3}{6}$

SOLUTION

Given: $\dfrac{x}{4} - \dfrac{x}{6} > \dfrac{x-3}{6}$

1. $\qquad 12\left(\dfrac{x}{4} - \dfrac{x}{6}\right) > 12\left(\dfrac{x-3}{6}\right)$ Multiply both sides by 12, the LCD of these three fractions.

2. $\quad 12\left(\dfrac{x}{4}\right) - 12\left(\dfrac{x}{6}\right) > 12\left(\dfrac{x-3}{6}\right)$ Use the distributive property.

3. $\qquad\qquad 3x - 2x > 2(x - 3)$ Simplify.

$\qquad\qquad\qquad x > 2x - 6$

4.   There are no constants on the left.

5. $\qquad x + (-2x) > 2x + (-2x) - 6$ Add $-2x$ (or subtract $2x$).

$\qquad\qquad\qquad -x > -6$

6. $\qquad\qquad \dfrac{-x}{-1} < \dfrac{-6}{-1}$ Multiply by -1 (or divide by the coefficient of x, -1), and reverse the sense of the inequality.

$\qquad\qquad\qquad x < 6$

The solution set is $\{x|x < 6\}$ or, in interval notation, $(-\infty, 6)$, and the graph is

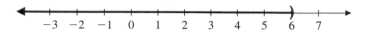

Solving Compound Inequalities

Sometimes we connect inequalities by using the word *or* as in

$$x < -2 \ \textbf{or} \ x > 1$$

The resulting inequality is a **compound** inequality and its graph is based on the idea of the *union* of two sets.

UNION OF TWO SETS
A AND B
$A \cup B$

If A and B are sets, the **union** of A and B, denoted by $A \cup B$, is the set of elements that are in either A *or* B.

To graph $\{x|x < -2 \ or \ x > 1\}$, first graph $x < -2$, then graph $x > 1$, and finally graph the *union*. The graph of $x < -2$ is

$(-\infty, -2)$

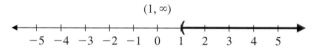

The graph of $x > 1$ is

$(1, \infty)$

The union is

$(-\infty, -2) \cup (1, \infty)$

Thus the solution set consists of all numbers less than -2 **or** greater than 1 or, in interval notation, $(-\infty, -2) \cup (1, \infty)$.

EXAMPLE 5   Solving a compound inequality with "or"

Graph: $\{x|x < -3 \ or \ x \geq 1\}$

SOLUTION   We do each of the graphs separately and then take the union of the two graphs. The graph of $x < -3$ is

$(-\infty, -3)$

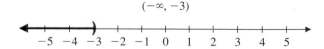

The graph of $x \geq 1$ is

$[1, \infty)$

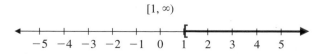

The union is

$$(-\infty, -3) \cup [1, \infty)$$

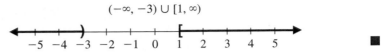

The inequality $3 \le s \le 5$ is also a compound inequality because it is equivalent to two other inequalities:

$$3 \le s \qquad \text{and} \qquad s \le 5$$

These inequalities use "and" as their connective. We can graph these inequalities using the idea of *intersection*.

INTERSECTION OF TWO SETS A **AND** B $A \cap B$	If A and B are two sets, the **intersection** of A and B, denoted by $A \cap B$, is the set of elements that are in both A *and* B.

The graph of $3 \le s$ is

$$[3, \infty)$$

The graph of $s \le 5$ is

$$(-\infty, 5]$$

The intersection $A \cap B$ is

$$[3, \infty) \cap (-\infty, -5] = [3, 5]$$

Thus the intersection is the line segment that the two graphs have in common. It is the graph of $\{s \mid 3 \le s \ and \ s \le 5\}$.

The interval $[3, 5]$ has a finite length, $5 - 3 = 2$. Because of this, $[3, 5]$ is called a **bounded** interval. In general, if a and b are real numbers and $a < b$, the intervals shown in Table 6 are bounded intervals and a and b are called the **end points** of each interval.

TABLE 6 Intervals

Set Notation	Interval Notation	Type of Interval	Graph	Inequality
$\{x \mid a \le x \le b\}$	$[a, b]$	Closed	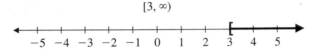 $a \qquad b$	$a \le x \le b$
$\{x \mid a < x < b\}$	(a, b)	Open	$a \qquad b$	$a < x < b$
$\{x \mid a \le x < b\}$	$[a, b)$	Half-open	$a \qquad b$	$a \le x < b$
$\{x \mid a < x \le b\}$	$(a, b]$	Half-open	$a \qquad b$	$a < x \le b$

EXAMPLE 6   Solving a compound inequality with "and"

Graph: $\{x \mid x > -2 \text{ and } x < 1\}$

SOLUTION   The graph of $x > -2$ is

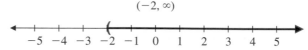

$(-2, \infty)$

The graph of $x < 1$ is

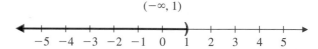

$(-\infty, 1)$

The intersection is

$$(-2, \infty) \cap (-\infty, 1) = (-2, 1)$$

■

Now suppose you wish to solve the compound inequality $2x + 7 < 9$ *and* $x + 3 \geq -1$. We solve each inequality, obtaining

$2x + 7 < 9$	Given.	and	$x + 3 \geq -1$	
$2x < 2$	Add -7.		$x \geq -4$	Add -3.
$x < 1$	Divide by 2.			

The graph of $x < 1$ is

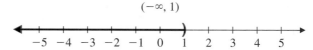

$(-\infty, 1)$

The graph of $x \geq -4$ is

$[-4, \infty)$

The intersection is

$$(-\infty, 1) \cap [-4, \infty) = [-4, 1)$$

Thus the intersection consists of all points such that $-4 \leq x$ *and* $x < 1$, which we can write as $-4 \leq x < 1$. A compound inequality that can be written in the form $a < x$ and $x < b$ (a and b real numbers) can be expressed more concisely as

$$a < x < b$$

The graph of $a < x < b$ is the solution set of $a < x$ *and* $x < b$, that is, (a, b). We use this idea to solve compound inequalities in Example 7.

EXAMPLE 7   Solving a compound inequality with "and"

Solve and graph:

a. $1 < x$ and $x < 3$ **b.** $5 \geq -x$ and $x \leq -3$ **c.** $x + 1 \leq 5$ and $-2x < 6$

SOLUTION

a. $1 < x$ and $x < 3$ is written as $1 < x < 3$. The graph is

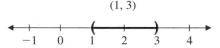

(1, 3)

b. Since we wish to write this inequality in the form $a < x < b$, we multiply the first inequality by -1 to obtain

$$-5 \leq x \qquad \text{and} \qquad x \leq -3$$

that is,

$$-5 \leq x \leq -3$$

The graph is

[−5, −3]

c. We solve the first inequality by subtracting 1 from both sides. We then have

$$x \leq 4 \qquad \text{and} \qquad -2x < 6$$

We then divide the second inequality by -2.

$$x \leq 4 \qquad \text{and} \qquad x > -3$$

Rearranging these inequalities gives

$$-3 < x \qquad \text{and} \qquad x \leq 4$$
$$-3 < x \leq 4$$

The graph is

(−3, 4]

Now suppose we want to solve $-5 < 2x + 3 \leq 9$. Since this inequality is equivalent to $-5 < 2x + 3$ and $2x + 3 \leq 9$, it is called a **double linear** inequality. We can solve this inequality by using the inequality properties we studied, keeping in mind that if we do any operations on the center expression ($2x + 3$), we must do the same operation on the outside expressions. As before, we wish to isolate x (this time in the middle). We start by adding (-3) to all three parts. We then have

$-5 <$	$2x + 3$	≤ 9	Given.
$-5 + (-3) < 2x + 3 + (-3) \leq 9 + (-3)$			Add (-3).
$-8 <$	$2x$	≤ 6	Simplify.
$\dfrac{-8}{2} <$	$\dfrac{2x}{2}$	$\leq \dfrac{6}{2}$	Multiply by $\frac{1}{2}$ (or divide by 2).
$-4 <$	x	≤ 3	Simplify.

The graph is

(−4, 3]

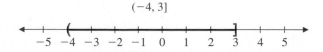

EXAMPLE 8   Solving and graphing double linear inequalities

Solve and graph: $-2 \le -3x - 5 < 4$

SOLUTION   We start by adding 5 to each part.

$$-2 + 5 \le -3x - 5 + 5 < 4 + 5$$

$3 \le$	$-3x$	< 9	Simplify.
$\dfrac{3}{-3} \ge$	$\dfrac{-3x}{-3}$	$> \dfrac{9}{-3}$	Divide by -3 and *reverse* the sense (direction).
$-1 \ge$	x	> -3	Simplify.
$-3 <$	x	≤ -1	Rewrite.

The graph is

$$(-3, -1]$$

Check your answer by substituting a number from the interval $(-3, -1]$ (say, -2) in the original inequality. ∎

> **CAUTION** When writing a double inequality such as $-2 < -3x - 5 \le 4$, make sure that the numbers are in the correct position. If you write $4 \le -3x - 5 < -2$, you are implying that $4 < -2$, which is *wrong*. Double inequalities must be written with the symbols pointing in the same direction toward the smaller number.
>
> Of course, no matter how hard you try, the inequality $\{x \mid x > 5 \text{ and } x < -5\}$ has no solution. (If you rewrite it as $5 < x < -5$, you can see that its solution set is $\varnothing$, the empty set).

At this point, you may have noticed that all linear inequalities we have solved have *unbounded* intervals for their graph, while all double inequalities have *bounded* intervals for their graph. Table 7 shows what happens in general.

TABLE 7 Equations, Inequalities, and Their Solution Sets

Type	Solution Set	Graph
Linear equation $ax + b = c$	$\{p\}$	
Linear inequality $ax + b < c$	$(-\infty, p)$ or (p, ∞)	
Compound inequality $x < p$ or $x > q$	$(-\infty, p) \cup (q, \infty)$	
Double inequality $c < ax + b < d$	(p, q)	

Translating Sentences into Inequalities

At the beginning of this section, we mentioned that the cost of renting a car must be under $50. We then translated this phrase by writing "< 50."

Here are some other phrases and their translations using inequalities.

Words	Translation	In Symbols
x is at least 10	x is 10 or more	$x \geq 10$
x is at most 20	x is 20 or less	$x \leq 20$
x is no more than 30	x is 30 or less	$x \leq 30$
x is no less than 40	x is 40 or more	$x \geq 40$

EXAMPLE 9   Translating sentences involving inequalities

Translate into an inequality:

a. The height h of a human (in feet) has never been known to exceed 9 ft.
b. The weight w of a human is at most 1400 lb.
c. The number n of puppies born in a single litter is no more than 23.
d. The cat population p in the United States (1993) is at least 64 million.

SOLUTION

a. $h \leq 9$ (Robert Wadlow was the tallest at 8 ft, 11.1 in.)
b. $w \leq 1400$ (Jon Browner Minnoch weighed 1400 lb.)
c. $n \leq 23$ (Lena, a foxhound, had 23 live puppies June 9, 1944.)
d. $p \geq 64$ million   ∎

GRAPH IT

Linear inequalities and linear equations are graphed similarly: Write the inequality as an equivalent one of the form $A(x) < 0$ [or $A(x) > 0$] and graph $y = A(x)$. The solution set for $A(x) < 0$ corresponds to all x-values for which the corresponding y-values are *below* the x-axis [*above* the x-axis for $A(x) > 0$].

Let's solve Example 2 using a grapher.

Example 2

To solve $3x - 2 < 2(x - 2)$, subtract $2(x - 2)$ and write

$$3x - 2 - 2(x - 2) < 0$$

Now, graph

$$Y_1 = 3x - 2 - 2(x - 2)$$

using a decimal viewing window (Window 1). Since the resulting line is *below* the x-axis for values of x that are less than -2, the solution set is $\{x \mid x < -2\}$ as before.

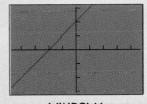

WINDOW 1
$Y_1 = 3x - 2 - 2(x - 2)$

Similarly, to graph $4(x + 1) \geq 3x + 7$, subtract $3x + 7$ and write

$$4(x + 1) - (3x + 7) \geq 0$$

Then graph

$$Y_1 = 4(x + 1) - (3x + 7)$$

This time (see Window 2), the resulting line is *above* or *on* the x-axis for values of x greater than or equal to 3. The solution set is $\{x \mid x \geq 3\}$.

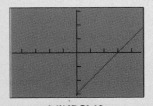

WINDOW 2
$Y_1 = 4(x + 1) - (3x + 7)$

(continued on next page)

Finally, if the inequality contains fractions, you **do not** have to clear the fractions. Let's illustrate this with Example 4.

Example 4

To solve

$$\frac{x}{4} - \frac{x}{6} > \frac{x-3}{6}$$

subtract

$$\frac{x-3}{6} \quad \text{to obtain} \quad \frac{x}{4} - \frac{x}{6} - \frac{x-3}{6} > 0$$

Now, graph

$$Y_1 = \frac{x}{4} - \frac{x}{6} - \frac{x-3}{6}$$

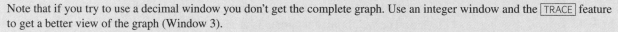

WINDOW 3

$$Y_1 = \frac{x}{4} - \frac{x}{6} - \frac{x-3}{6}$$

Note that if you try to use a decimal window you don't get the complete graph. Use an integer window and the TRACE feature to get a better view of the graph (Window 3).

The resulting line crosses the x-axis at $x = 6$ and is *above* the x-axis ($y > 0$) for values less than 6. You can check this better using the root feature on a TI-82. Thus the solution set is $\{x \mid x < 6\}$.

EXERCISE 2.5

A In Problems 1–10, graph the inequality and write the solution set using interval notation.

1. $x > 3$

2. $x > 4$

3. $x \leq -3$

4. $x \leq -4$

5. $2x \geq 6$

6. $3x \geq 8$

7. $-3x \leq 3$

8. $-5x \leq 10$

9. $-4x \geq -8$

10. $-6x \geq -12$

B In Problems 11–30, solve the inequality and graph the solution set.

11. $3x + 6 \leq 9$

12. $4y - 9 \leq 3$

13. $-2y - 4 \geq -10$

14. $-2z - 2 \geq -10$

15. $-3x + 1 \leq -14$

16. $-3x + 4 \leq -8$

17. $3a + 6 \leq a + 10$

18. $4b + 4 \leq b + 7$

19. $7z - 12 \geq 8z - 8$

20. $3z + 7 \geq 5z + 19$

21. $10 - 5x \leq 7 - 8x$

22. $6 - 4y \leq -14 + 6y$

23. $5(x + 2) \leq 3(x + 3) + 1$

24. $5(4 - 3x) < 7(3 - 4x) + 12$

25. $-4x + \dfrac{1}{2} \geq 4x + \dfrac{8}{5}$

26. $12x + \dfrac{2}{7} \geq 4x - \dfrac{4}{7}$

27. $\dfrac{x}{5} - \dfrac{x}{4} \leq 1$

28. $\dfrac{x}{3} - \dfrac{x}{2} \leq 1$

29. $\dfrac{7x + 2}{6} + \dfrac{1}{2} \geq \dfrac{3}{4}x$

30. $\dfrac{8x - 23}{6} + \dfrac{1}{3} \geq \dfrac{5}{2}x$

C In Problems 31–70, solve and graph the solution set. If the solution set is empty, write $\varnothing$ for the answer.

31. $\{x \mid x + 4 > 7 \text{ or } x - 2 < -6\}$

32. $\{x \mid x - 4 > 1 \text{ or } x + 2 < 1\}$

33. $3x + 4 > 10 \text{ or } 3x - 1 < 2$

34. $2x - 5 > 5 \text{ or } 2x < -4$

35. $2x \leq x + 4 \text{ or } x - 2 > 3$

36. $3x - 2 > 7 \text{ or } -5x \geq -5$

37. $\{x \mid -3x + 2 < -4 \text{ or } 2x - 1 < 3\}$

38. $\{x \mid -5x + 1 < -4 \text{ or } -2x + 1 > 5\}$

39. $-6x - 2 \geq -14 \text{ or } -7x + 2 < -19$

40. $-3x - 5 < 7 \text{ or } -2x + 1 > -5$

41. $\{x \mid x \leq 4 \text{ and } x \geq -2\}$

42. $\{x \mid x > 0 \text{ and } x \leq 5\}$

43. $x + 1 \leq 7 \text{ and } x > 2$

44. $x > -5 \text{ and } x - 1 < 0$

45. $\{x \mid 2x - 1 > 1 \text{ and } x + 1 < 4\}$

46. $\{x \mid x - 1 < 1 \text{ and } 3x - 1 > 11\}$

47. $x < -5 \text{ and } x > 5$

48. $x \geq 0 \text{ and } x < -2$

49. $\{x \mid x + 1 \geq 2 \text{ and } x \leq 4\}$

50. $\{x \mid x \leq 5 \text{ and } x > -1\}$

51. $\{x \mid x < 3 \text{ and } -x < -2\}$

52. $\{x \mid -x < 5 \text{ and } x < 2\}$

53. $\{x \mid x + 1 < 4 \text{ and } -x < -1\}$

54. $\{x \mid x - 2 < 1 \text{ and } -x < 2\}$

55. $\{x \mid x - 2 < 3 \text{ and } 2 > -x\}$

56. $\{x \mid x - 3 < 1 \text{ and } 1 > -x\}$

57. $\{x \mid x + 2 < 3 \text{ and } -4 < x + 1\}$

58. $\{x \mid x + 4 < 5 \text{ and } -1 < x + 2\}$

59. $\{x \mid x - 1 > 2 \text{ and } x + 7 < 11\}$

60. $\{x \mid x - 2 > 1 \text{ and } -x > -5\}$

61. $-3 < x - 1 < 3$

62. $-4 < x + 1 < 4$

63. $-8 < 2y + 4 < 6$

64. $-2 \leq 3x + 1 \leq 7$

65. $4 \leq 3y - 8 \leq 10$

66. $3 \leq 4z + 3 \leq 5$

67. $-1 < \dfrac{x}{2} < 2$

68. $-2 < \dfrac{y}{2} < 1$

69. $2 < 4 + \dfrac{2}{3}a < 6$

70. $1 < 5 + \dfrac{4}{5}b < 9$

D In Problems 71–76, translate the statement into an inequality.

71. The height h (in feet) of any mountain does not exceed that of Mt. Everest, 29,028 ft.

72. The number e of possible eclipses in a year is at most 7.

73. The number e of possible eclipses in a year is at least 2.

74. The altitude h (in feet) attained by the first liquid-fueled rocket was no more than 41 ft.

75. There are no less than 4×10^{25} nematode sea worms in the world. (Let n be the number of nematodes.)

76. There are at least 713 million people (p) that speak Mandarin Chinese.

77. When the variable cost per unit is $12 and the fixed cost is $160,000, the total cost for a certain product is $C = 12n + 160{,}000$ (n is the number of units sold). If the unit price is $20, the revenue R is $20n$. What is the minimum number of units that must be sold to make a profit? (You need $R > C$ to make a profit.)

78. The cost of first-class mail is 32¢ for the first ounce and 23¢ for each additional ounce. A delivery company will charge $4.00 for delivering a package weighing up to 2 lb (32 oz). When would the U.S. Post Office price, $P = 0.32 + 0.23(x - 1)$ (where x is the weight of the package in ounces), be cheaper than the delivery company's price?

79. The parking cost at a garage is $C = 1 + 0.75(h - 1)$, where h is the number of hours you park and C is the cost in dollars. When is the cost C less than $10?

SKILL CHECKER

Fill in the blank with < or > to make the result a true statement.

80. -7 _____ -8

81. $\dfrac{1}{3}$ _____ $\dfrac{1}{2}$

82. 0.34 _____ 0.342

83. -0.234 _____ -0.233

Find:

84. $|-9|$

85. $\left|-\dfrac{1}{5}\right|$

86. $|-0.34|$

87. $|\sqrt{2}|$

USING YOUR KNOWLEDGE

Inequalities and the Environment

88. Do you know why spray bottles use a pump rather than propellants? Because some of the propellants have chlorofluorocarbons (CFCs) and they deplete the ozone layer. In an international meeting in London, the countries represented agreed to stop producing CFCs by the year 2000. If the production of CFCs (in thousands of tons) is given by $P = 1260 - 110x$, where x is the number of years after 1980, approximate (to the nearest whole year) the number of years it would take for the amount of CFCs produced to be
   a. 50 thousand tons   **b.** Zero
   c. Will the production of CFCs be stopped by the year 2000?

89. Cigarette smoking also produces air pollution. The annual number of cigarettes per person consumed in the United States is given by $N = 4200 - 70x$, where x is the number of years after 1960. How many years after 1960 will consumption be less than 1000 cigarettes per person annually? (Answer to the nearest whole year.)

WRITE ON . . .

90. What is the main difference in the techniques used when solving equations as opposed to inequalities?

How would you describe in words the real numbers in the following intervals?

91. $(1, \infty)$

92. $(-\infty, 3]$

93. $(4, \infty)$

94. $[-2, \infty)$

95. Look at the following argument, where it is assumed that $a > b$:

$2 > 1$	Known.
$2(b - a) > 1(b - a)$	Multiply by $(b - a)$.
$2b - 2a > b - a$	Simplify.
$2b > b + a$	Add $2a$.
$b > a$	Subtract b.

But we assumed that $a > b$. Write a paragraph explaining what went wrong.

MASTERY TEST

If you know how to do these problems, you have learned your lesson!

Translate into an inequality:

96. x does not exceed 23.

97. p is at least 45.

Graph:

98. $x \leq -3$

99. $x > 4$

Solve and graph:

100. $-3 < -3x - 6 < 3$

101. $-2 \leq 2 + \dfrac{4}{5}x < 6$

102. $2(x - 1) \geq 3x + 1$

103. $3x + 10 < 6x + 4$

104. $\dfrac{1}{4}(x - 5) < \dfrac{1}{3}(x - 4)$

105. $2x + 3 > 1$ or $5x - 3 < -13$

106. $\{x \mid 2x + 9 > 11 \text{ or } x + 3 > -1\}$

107. $\{x \mid 2x + 3 < 5x - 3 \text{ and } 3x - 2 < 3 + 2x\}$

108. $\{x \mid -3x + 5 < 9 \text{ and } 3x - 7 < -3 + 2x\}$

2.6

ABSOLUTE-VALUE EQUATIONS AND INEQUALITIES

To succeed, review how to:

1. Find the absolute value of a number (p. 10).
2. Graph an inequality (pp. 108–116).

Objectives:

 A   Solve absolute-value equations.

B   Solve absolute-value inequalities of the form $|ax + b| < c$ or $|ax + b| > c$, where $c > 0$.

---

getting started

Budget Variances

Businesses and individuals usually try to predict how much money will be spent on certain items over a period of time. If you budget $120 for a month's utilities and a heat wave or cold snap makes your actual expenses jump to $150, the $30 difference might be an acceptable **variance**. Suppose b represents the budgeted amount for an item, a represents the actual expense, and you want to be within $10 of your estimate. The item will pass the variance test if the actual expenses a are within $10 of the budgeted amount b; that is, $b - a$ is between -10 and 10. In symbols, $-10 \leq b - a \leq 10$ or equivalently $|b - a| \leq 10$.

In general, if a and b are as before, a certain item will pass the variance test if $|b - a| \leq c$, where c is the variance. The quantity c can be a definite amount or a percent of the budget. For example, if you budget $50 for gas and you want to be within 10% of your budget, how much gas money can you spend and still be within your variance? Since 10% of $50 = 0.10 \cdot 50 = 5$, you can see that if you spend between $45 and $55, you will be within your 10% variance. Here $|b - a| \leq c$ becomes $|50 - a| \leq 5$ or equivalently $-5 \leq 50 - a \leq 5$.

$$-55 \leq -a \leq -45 \qquad \text{Subtract 50.}$$
$$55 \geq a \geq 45 \qquad \text{Multiply by } -1.$$
$$45 \leq a \leq 55 \text{ (as expected)} \qquad \text{Rewrite.}$$

Note that $|50 - a| \leq 5$ is equivalent to $-5 \leq 50 - a \leq 5$; we shall develop the rule for this concept in this section.

The absolute value of a, denoted by $|a|$, is the distance between a and zero on the number line. Thus, $|3| = 3$, $|-5| = 5$, and $|-0.7| = 0.7$. The equation $|x| = 2$ is read as "the distance between x and zero on the number line is 2." Since x is 2 units from zero, x can only be $+2$ or -2; that is, $x = \pm 2$.

 Absolute-Value Equations

To solve absolute-value equations, we need the following definition.

THE SOLUTIONS OF $|x| = a$ $(a \geq 0)$	If $a \geq 0$, the solutions of $|x| = a$ are $x = a$ and $x = -a$.

CAUTION Note that if $a < 0$, $|x| = a$ has no solution.

EXAMPLE 1   Solving absolute-value equations

Solve:

a. $|x| = 8$ **b.** $|y| = 2.5$ **c.** $|z| = -6$ **d.** $|w| = 0$

SOLUTION

a. Since the numbers 8 and -8 are 8 units from zero, the solutions of $|x| = 8$ are 8 and -8. The solution set of $|x| = 8$ is $\{-8, 8\}$.

b. The solutions of $|y| = 2.5$ are $y = 2.5$ and $y = -2.5$. Thus the solution set of $|y| = 2.5$ is $\{-2.5, 2.5\}$.

c. The absolute value of a number is never negative (it represents a distance). Thus the equation $|z| = -6$ has no solution. Its solution set is $\varnothing$ (the empty set).

d. The only solution is $w = 0$. ∎

The ideas of Example 1 can be generalized to more complicated equations. Look at the pattern:

$$|x| = 8 \text{ has solutions} \qquad x = 8 \qquad \text{and} \qquad x = -8$$

$$|y| = 2.5 \text{ has solutions} \qquad y = 2.5 \quad \text{and} \qquad y = -2.5$$

$$|x + 1| = 8 \text{ has solutions} \qquad x + 1 = 8 \quad \text{and} \quad x + 1 = -8$$

(Solve each equation by
adding (-1) on each side.) $\qquad x = 7 \qquad \text{and} \qquad x = -9$

We can check the solutions by substituting them into the equation $|x + 1| = 8$. Substituting 7 for x, we obtain $|7 + 1| = |8| = 8$, a true statement. If we substitute -9 for x, we have $|-9 + 1| = |-8| = 8$, also true.

EXAMPLE 2   Solving more complicated absolute-value equations

Solve:

a. $|2x + 1| = 6$ **b.** $\left|\dfrac{3}{4}x + 1\right| + 5 = 11$

SOLUTION

a. The solutions of $|2x + 1| = 6$ are

$$2x + 1 = 6 \qquad \text{and} \qquad 2x + 1 = -6$$

$$2x = 5 \qquad \text{and} \qquad 2x = -7 \qquad \text{Add } (-1) \text{ to both sides.}$$

$$x = \frac{5}{2} \qquad \text{and} \qquad x = -\frac{7}{2} \qquad \text{Multiply both sides by } \tfrac{1}{2}.$$

Thus the solution set of $|2x + 1| = 6$ is $\left\{-\frac{7}{2}, \frac{5}{2}\right\}$.

b. In order to use the definition of absolute value, we must isolate $\left|\frac{3}{4}x + 1\right|$. Thus we first add (-5) to both sides.

$$\left|\frac{3}{4}x + 1\right| + 5 = 11 \qquad \text{Given.}$$

$$\left|\frac{3}{4}x + 1\right| + 5 + (-5) = 11 + (-5) \qquad \text{Add } -5.$$

$$\left|\frac{3}{4}x + 1\right| = 6 \qquad \text{Simplify.}$$

The solutions of this equation are

$$\frac{3}{4}x + 1 = 6 \qquad \text{and} \qquad \frac{3}{4}x + 1 = -6$$

$$\frac{3}{4}x = 5 \qquad \text{and} \qquad \frac{3}{4}x = -7 \qquad \text{Add } -1.$$

$$\frac{4}{3} \cdot \frac{3}{4}x = \frac{4}{3} \cdot 5 \qquad \text{and} \qquad \frac{4}{3} \cdot \frac{3}{4}x = \frac{4}{3} \cdot (-7) \qquad \text{Multiply by } \frac{4}{3}.$$

$$x = \frac{20}{3} \qquad \text{and} \qquad x = -\frac{28}{3} \qquad \text{Simplify.}$$

The solution set of $\left|\frac{3}{4}x + 1\right| + 5 = 11$ is $\left\{-\frac{28}{3}, \frac{20}{3}\right\}$. ∎

EXAMPLE 3   Solving absolute-value equations

Solve: $|x - 3| = |x - 5|$

SOLUTION   The equation $|x - 3| = |x - 5|$ is true if $x - 3$ and $x - 5$ are equal to each other, or if they are additive inverses. The additive inverse of $(x - 5)$ is $-(x - 5)$. Here are the two cases.

Equal	**Opposites**	
$x - 3 = x - 5$	$x - 3 = -(x - 5)$	
$(-x) + x - 3 = (-x) + x - 5$ Add $(-x)$.	$x - 3 = -x + 5$	
$-3 = -5$	$x = -x + 8$	Add 3.
This is a contradiction. There is *no* solution for this case.	$2x = 8$	Add x.
	$x = 4$	Divide by 2.

The solution set is $\{4\}$. We can verify this by replacing x by 4 in $|x - 3| = |x - 5|$ to obtain

$$|4 - 3| = |4 - 5|$$
$$|1| = |-1|$$
$$1 = 1$$

This is a true statement, so our solution is correct. ∎

Absolute-Value Inequalities

As you recall, $|x| = 2$ means that the distance from 0 to x on the number line is 2 units. What would $|x| < 2$ and $|x| > 2$ mean? Here are the statements, their translations, their graphs, and the meaning in terms of the compound inequalities we just studied.

STATEMENT TRANSLATION

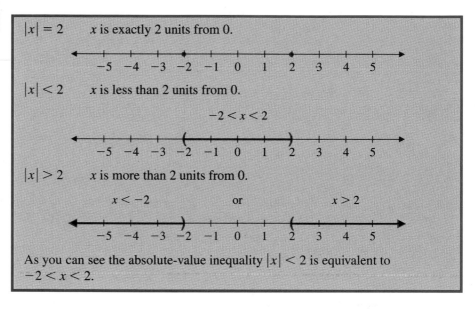

$|x| = 2$ x is exactly 2 units from 0.

$|x| < 2$ x is less than 2 units from 0.

$$-2 < x < 2$$

$|x| > 2$ x is more than 2 units from 0.

$x < -2$ or $x > 2$

As you can see the absolute-value inequality $|x| < 2$ is equivalent to $-2 < x < 2$.

In general, the following holds.

 $|x| < a$
IS EQUIVALENT TO
$-a < x < a$

The expression $|x| < a$ is equivalent to $-a < x < a$, where $a > 0$. A similar relationship exists if $\leq$ replaces $<$.

 CAUTION If $a < 0$, $|x| < a$ has no solution.

$|x| < 4$ is equivalent to $-4 < x < 4$

$|y| \leq 2.5$ is equivalent to $-2.5 \leq y \leq 2.5$

$|z| < \dfrac{3}{4}$ is equivalent to $-\dfrac{3}{4} < z < \dfrac{3}{4}$

What does $|x + 1| < 2$ mean? Since $|x| < a$ is equivalent to $-a < x < a$, $|x + 1| < 2$ is equivalent to $-2 < x + 1 < 2$.

To solve $-2 < x + 1 < 2$, add -1 and simplify like this:

$$-2 + (-1) < x + 1 + (-1) < 2 + (-1) \qquad \text{Add } -1.$$

$$-3 < \quad x \quad < 1 \qquad \text{Simplify.}$$

The solution is $(-3, 1)$, and the graph is

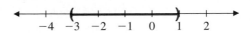

In general, $|ax + b| < c$ is equivalent to $-c < ax + b < c$.

EXAMPLE 4   Absolute-value inequalities of the form $|ax + b| \le c$

Solve and graph: $|3x + 4| \le 8$

SOLUTION $|3x + 4| \le 8$ is equivalent to $-8 \le 3x + 4 \le 8$.

$$-8 + (-4) \le 3x + 4 + (-4) \le 8 + (-4) \qquad \text{Add } -4.$$

$$-12 \le \qquad 3x \qquad \le 4 \qquad \text{Simplify.}$$

$$\frac{-12}{3} \le \qquad \frac{3x}{3} \qquad \le \frac{4}{3} \qquad \text{Divide by 3.}$$

$$-4 \le \qquad x \qquad \le \frac{4}{3} \qquad \text{Simplify.}$$

The solution is thus $\left[-4, \frac{4}{3}\right]$, and the graph is

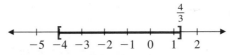

Let's look at $|x| > 2$. Its graph consists of all points such that $x < -2$ or $x > 2$. Thus $|x| > 2$ is equivalent to $x < -2$ or $x > 2$. If we generalize this result, we obtain the following.

$\|x\| > a$ **IS EQUIVALENT TO** $x < -a$ **OR** $x > a$	The expression $\|x\| > a$ is equivalent to $x < -a$ or $x > a$, where $a > 0$. A similar relationship applies if $\ge$ replaces $>$.

$$|x| > 5 \qquad \text{is equivalent to} \qquad x < -5 \text{ or } x > 5$$

$$|x| > 3.4 \qquad \text{is equivalent to} \qquad x < -3.4 \text{ or } x > 3.4$$

$$|x| \ge \frac{1}{5} \qquad \text{is equivalent to} \qquad x \le -\frac{1}{5} \text{ or } x \ge \frac{1}{5}$$

What does $|x - 1| > 3$ mean? Since $|x| > a$ is equivalent to $x < -a$ or $x > a$, $|x - 1| > 3$ means

$$x - 1 < -3 \qquad \text{or} \qquad x - 1 > 3$$

$$x < -2 \qquad \text{or} \qquad x > 4 \qquad \text{Solve each inequality by adding 1 to both sides.}$$

The solution is $(-\infty, -2) \cup (4, \infty)$, and the graph of the solution set is

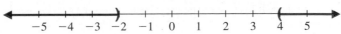

In general, $|ax + b| > c$ is equivalent to $ax + b < -c$ or $ax + b > c$.

EXAMPLE 5   Absolute-value inequalities of the form $|ax + b| > c$

Solve and graph: $|-2x + 3| > 1$

SOLUTION   The inequality $|-2x + 3| > 1$ is equivalent to

$$-2x + 3 < -1 \qquad\qquad \text{or} \qquad\qquad -2x + 3 > 1$$

Now add (-3) on both sides.

$$-2x + 3 + (-3) < -1 + (-3) \qquad \text{or} \qquad -2x + 3 + (-3) > 1 + (-3)$$

$$-2x < -4 \qquad\qquad \text{or} \qquad\qquad -2x > -2 \qquad \text{Simplify.}$$

Next divide by -2 and reverse the inequality signs.

$$\frac{-2x}{-2} > \frac{-4}{-2} \qquad \text{or} \qquad \frac{-2x}{-2} < \frac{-2}{-2}$$

$$x > 2 \qquad \text{or} \qquad x < 1 \qquad \text{Simplify.}$$

The solution is $(2, \infty) \cup (-\infty, 1)$, and the graph is

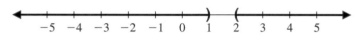

Table 8 shows a summary that will help with the exercises.

TABLE 8 Summary for Solving Absolute-Value Equations and Inequalities

1. The solutions of the equation $|ax + b| = c$ are obtained by solving the two equations
    $$ax + b = c \qquad \text{and} \qquad ax + b = -c$$
    Note: If c is negative ($c < 0$), there is no solution.

2. The solutions of the equation $|ax + b| = |cx + d|$ are obtained by solving the equations
    $$ax + b = cx + d \qquad \text{and} \qquad ax + b = -(cx + d)$$

3. The solution set of the inequality $|ax + b| < c$ is obtained by solving the inequality $-c < ax + b < c$. The graph is *bounded*. (If c is negative, there is no solution.)

4. The solution set of the inequality $|ax + b| > c$ is obtained by solving the inequalities $ax + b > c$ or $ax + b < -c$. The graph is the union of two *unbounded* intervals. (If c is negative, $|ax + b| > c$ is always true, and the solution set is the set of real numbers.)

GRAPH IT

To graph the solution set of an equation of the form $|ax + b| = c$, you must be able to find an equivalent equation of the form $|A(x)| - c = 0$, where $A(x)$ is a linear expression and then graph

$$Y_1 = |A(x)| - c$$

The x-value at which the graph crosses the x-axis is the solution for the equation.

First, find out how to enter $|ax + b|$ in your grapher. With a TI-82, enter [2nd] [ABS] $(ax + b)$. The resulting graph is always V-shaped. Make sure you experiment with your viewing window and the range you use so that you can see the complete graph.

(*continued on facing page*)

Let's do Example 2b with a grapher.

Example 2b

To solve $\left|\frac{3}{4}x + 1\right| + 5 = 11$, write an equivalent equation with zero on the right-hand side by subtracting 11 from both sides to obtain $\left|\frac{3}{4}x + 1\right| - 6 = 0$. (By the way, it is easier to enter $\frac{3}{4}x$ as .75x.) Now graph $y = \left|\frac{3}{4}x + 1\right| - 6$. If you use a viewing rectangle $[-5, 5]$ by $[-5, 5]$ you don't get the complete graph, so try a $[-10, 10]$ by $[-10, 10]$ standard rectangle. As predicted, the graph is V-shaped (Window 1), and it crosses the x-axis at $x = -\frac{28}{3} = -9\frac{1}{3}$ and $x = \frac{20}{3} = 6\frac{2}{3}$. You can verify this by using the $\boxed{\text{TRACE}}$ and $\boxed{\text{ZOOM}}$ buttons repeatedly to get more accurate values for x. (With a TI-82, graph $Y_2 = 0$ and press $\boxed{\text{2nd}}$ $\boxed{\text{CALC}}$ 5 to get $X = 6.6666667$ as shown or use your root feature $\boxed{\text{2nd}}$ $\boxed{\text{CALC}}$ 2). You can get the other root by using $\boxed{\text{TRACE}}$ and $\boxed{\text{ZOOM}}$ repeatedly near

the point at which the curve crosses the x-axis.

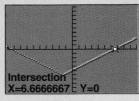

WINDOW 1
$y = |0.75x + 1| - 6$

The second type of absolute-value problem is of the form $|A(x)| = |B(x)|$ as in Example 3. Let's try this example with a grapher.

Example 3

To solve this problem, first obtain the equivalent equation $|x - 3| - |x - 5| = 0$ and then graph $y = |x - 3| - |x - 5|$ using a $[-10, 10]$ by $[-10, 10]$ viewing rectangle. Note that this time the graph (Window 2) is **not** V-shaped ($y = |A(x)| - a$ is V-shaped, but this equation is of the form $y = |A(x)| - |B(x)|$). The answer is $x = 4$ as before.

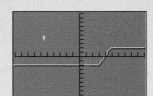

A grapher can also be used to do Examples 4 and 5. Here's how.

1. The solution set of $|ax + b| < c \ (c > 0)$ is obtained by graphing $Y_1 = |ax + b| - c$ and selecting the interval in which the x-coordinates have corresponding y-values *below* the x-axis ($Y_1 < 0$).

2. The solution set of $|ax + b| > c \ (c > 0)$ is obtained by graphing $Y_1 = |ax + b| - c$ and selecting the interval in which the x-coordinates have corresponding y-values *above* the x-axis ($Y_1 > 0$).

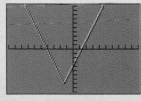

WINDOW 2
$y = |x - 3| - |x - 5|$

Now let's proceed to Examples 4 and 5.

Example 4

1. Since $|3x + 4| \leq 8$ is equivalent to $|3x + 4| - 8 \leq 0$ (subtract 8 from both sides of $|3x + 4| \leq 8$), we find the graph of $Y_1 = |3x + 4| - 8$. See Window 3.

2. The values that make $Y_1 \leq 0$ are *below* the x-axis and have x-coordinates between -4 and about $1\frac{1}{3} = \frac{4}{3}$.

3. Confirm that $x = -4$ and $x = \frac{4}{3}$ are the points at which the graph crosses the x-axis. (You can do this using the $\boxed{\text{TRACE}}$ and $\boxed{\text{ZOOM}}$ buttons repeatedly.) Thus the solution set is, as before, $\left[-4, \frac{4}{3}\right]$.

WINDOW 3
$y = |3x + 4| - 8$

Example 5

1. Since $|-2x + 3| > 1$ is equivalent to $|-2x + 3| - 1 > 0$ (subtract 1 from both sides of $|-2x + 3| > 1$), we find the graph of $Y_1 = |-2x + 3| - 1$, using the viewing rectangle $[-5, 5]$ by $[-1, 5]$. See Window 4.

2. The values that make $y > 0$ are *above* the x-axis ($Y_1 > 0$) and have x-coordinates that are less than 1 or greater than 2.

3. Confirm that $x = 1$ and $x = 2$ are the points at which the graph crosses the x-axis, noting that 1 and 2 are **not** included in the solution set. Thus the solution set is, as before, $(-\infty, 1) \cup (2, \infty)$.

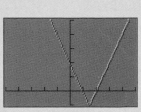

WINDOW 4
$Y_1 = |-2x + 3| - 1$

EXERCISE 2.6

A In Problems 1–30, solve the equation.

1. $|x| = 13$

2. $|y| = 17$

3. $|y| - 2.3 = 0$

4. $|x| - 3.7 = 0$

5. $|x| = 0$

6. $|y| = -3$

7. $|z| = -4$

8. $|x + 1| = 10$

9. $|x + 7| = 2$

10. $|x + 9| = 3$

11. $|2x - 4| = 8$

12. $|3x - 6| = 9$

13. $|5a - 2| - 8 = 0$

14. $|6b - 3| - 9 = 0$

15. $\left|\dfrac{1}{2}x + 4\right| = 6$

16. $\left|\dfrac{1}{3}x + 2\right| = 7$

17. $\left|\dfrac{2}{3}z - 3\right| = 9$

18. $\left|\dfrac{2}{5}x - 6\right| = 4$

19. $|x + 2| = |x + 4|$

20. $|y + 6| = |y + 2|$

21. $|2y - 4| = |4y + 6|$

22. $|3x - 2| = |6x + 4|$

23. $2|a + 1| - 3 = 9$

24. $2|3a + 1| + 5 = 13$

25. $3|2x + 1| - 4 = -6$

26. $5|x - 1| - 6 = -8$

27. $|x - 4| = |4 - x|$

28. $|2x - 2| = |2 - 2x|$

29. $|5x - 10| = |10 - 5x|$

30. $|6x - 3| = |3 - 6x|$

B In Problems 31–68 solve and graph the inequality.

31. $|x| < 4$

32. $|y| \le 1.5$

33. $|z| \le 2.4$

34. $|x| - 3 < 1$

35. $|a| - 2 \le 2$

36. $|b| + 2 \le 5$

37. $|x - 1| < 2$

38. $|x - 3| \le 1$

39. $|x + 3| < -2$

40. $|2x - 3| < -4$

41. $|2x + 3| \le 1$

42. $|3x + 2| < 5$

43. $|4x + 2| - 4 < 2$

44. $|3x + 3| - 2 < 4$

45. $|x| > 2$

46. $|y| \ge 2.5$

47. $|z| \ge 1.4$

48. $|x| - 2 > 1$

49. $|a| - 1 \ge 2$

50. $|b| + 1 \ge 5$

51. $|x - 1| > 1$

52. $|x - 3| \ge 2$

53. $|x + 3| > -1$

54. $|2x - 3| > -1$

55. $|2x + 3| \ge 1$

56. $|3x + 2| > 5$

57. $|3 - 4x| > 7$

58. $|5 - 3x| > 11$

59. $|4x + 2| - 4 \ge 2$

60. $|3x + 3| - 2 \ge 4$

61. $\left|2 - \dfrac{1}{2}a\right| > 1$

62. $\left|1 + \dfrac{2}{3}b\right| > 2$

63. $|2x + 3| > -2$

64. $|-2x - 4| > -3$

65. $|-2x + 4| < -4$

66. $|-2x - 3| < -5$

67. $|3x - 2| + 3 > -4$

68. $|-2x + 3| - 1 > -2$

In Problems 69–72, use $|b - a| \le c$, where b is the budgeted amount, a is the actual expense, and c is the variance.

69. A company budgets $500 for office supplies. If their variance is $50, write an inequality giving the amounts between which the actual expense a must fall.

70. A company budgets $800 for maintenance. If their acceptable variance is 5% of their budgeted amount, write an inequality giving the amounts between which the actual expense a must fall.

71. George budgets $300 for miscellaneous monthly expenses. His actual expenses for 1 month amounted to $290. Was he within a 5% budget variance?

72. If George in Problem 71 spent $310, was he within a 5% budget variance?

SKILL CHECKER

Evaluate:

73. $-16t^2 + 10t - 15$ when $t = 2$

74. $x^2 - 3x$ when $x = -3$

Simplify:

75. $(-6x^2 + 3x + 5) + (7x^2 + 8x + 2)$

76. $(3x - 2x^2 + 4) + (5 - 7x + 4x^2)$

77. $(3x - 7x^2 + 4) - (3x^2 + 5x + 1)$

78. $(5 - 3x^2 - 2x) - (8x^2 + 4x - 2)$

USING YOUR KNOWLEDGE   **The Boundaries of Inequalities**

The inequality $|-2x + 3| > 1$ can be solved by finding its *boundary numbers,* the solutions resulting when the inequality sign ($>$ in this case) is replaced by an equal sign. Thus we write $|-2x + 3| = 1$ with solutions

$-2x + 3 = 1$	or	$-2x + 3 = -1$
$-2x = -2$		$-2x = -4$	Subtract 3.
$x = 1$		$x = 2$	Divide by -2.

Thus the boundary numbers are 1 and 2.

Since the graph of an inequality of the form $|ax + b| > c$ is the union of two unbounded line segments, the graph is

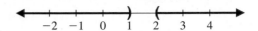

as in Example 5. Try doing Problems 37–42 and 51–60 using this technique.

WRITE ON . . .

79. Write your own definition for the absolute value of a number using the idea of distance between two numbers.

80. Write your own explanation of why the equation $|x| = a$ has no solution if a is negative.

81. Explain how to solve the equation $|A(x)| = a$, where $A(x)$ is a linear equation. When is the solution set of this equation empty?

82. Write in your own words the reasons why
  a. The inequality $|x| < a$ ($a < 0$) has no solution.
  b. The inequality $|x| > a$ ($a < 0$) is always true.

MASTERY TEST   **If you know how to do these problems, you have learned your lesson!**

83. $|x - 1| = |x - 8|$

84. $|3 - x| = |7 + x|$

85. $|3x + 1| = 5$

86. $\left| z \right| - \dfrac{3}{4} = 0$

87. $\left| \dfrac{2}{3}x + 1 \right| + 3 = 9$

88. $|y| = 3.4$

Solve and graph:

89. $|-2x + 1| - 1 > 4$

90. $\left| x - \dfrac{2}{3} \right| \le 5$

91. $|-2 + 3x| \ge 5$

92. $|2x - 3| - 5 < 0$

research questions

Sources of information for these questions can be found in the *Bibliography* at the end of this book.

1. Write a short paragraph about the contents of the Rhind papyrus with special emphasis on the algebraic method called "the method of false position."

2. Try to find three problems that appeared on the Rhind papyrus and explain the techniques used to solve them. Then try to solve them yourself using the techniques you've learned in this chapter. (*Hint:* Problems 25, 26, and 27 are easy to solve.)

3. Find out about al-Khowarizmi and then write a few paragraphs about him and his life. (Include what academy he belonged to, the titles of the books he wrote, and the types of problems that appeared in these books.)

4. A Latin corruption of the name "al-Khowarizmi" meant "the art of computing with Hindu Arabic numerals." What is this word and what does it mean today?

5. The traditional explanation of the word *jabr* is "the setting of a broken bone." Write a paragraph relating how this word reached Spain as *algebrista* and the context in which it was used.

6. Diophantus of Alexandria wrote a book described "as the earliest treatise devoted to algebra." Write a few paragraphs about Diophantus and, in particular, the problems contained in his book.

7. Write a short paragraph detailing the types of symbols used in Diophantus' book.

SUMMARY

SECTION	ITEM	MEANING	EXAMPLE
2.1	Equation	A sentence using $=$ as its verb	$x + 1 = 5$ and $3 - y = 7$ are equations.
2.1A	Solutions of an equation	The replacements of the variable that make the equation a true statement	5 is a solution of $x + 3 = 8$.
2.1B	Equivalent equations	Two or more equations are equivalent if they have the same solution set.	$x + 2 = 5$ and $x = 3$ are equivalent.
	Properties of equality	You can *add* or *subtract* the same quantity on both sides and *multiply* or *divide* both sides of an equation by the same nonzero quantity, and the result will be an equivalent equation.	If C is a real number and $A = B$, $A + C = B + C$ $A - C = B - C$ $A \cdot C = B \cdot C \, (C \neq 0)$ $\dfrac{A}{C} = \dfrac{B}{C} \quad (C \neq 0)$ are equivalent equations.
2.1C	Linear equation	An equation that can be written in the form $ax + b = c$, where a, b, and c are real numbers and $a \neq 0$.	$3x + 7 = 9$ and $-x - 3 = \frac{2}{3}$ are linear equations.
2.2A	Perimeter	The distance around a polygon (figure)	The perimeter P of a rectangle is $P = 2L + 2W$, where L is the length and W is the width.

SECTION	ITEM	MEANING	EXAMPLE
2.2C	Angles and transversals	All the acute angles (1, 4, 5, and 8) are equal. All the obtuse angles (2, 3, 6, and 7) are equal.	L_1 and L_2 are parallel.
2.3B	Word problem	The RSTUV procedure to solve a word problem is **R**ead. **S**elect a variable. **T**hink of a plan. **U**se algebra. **V**erify.	
2.3C	Consecutive integers	n and $n + 1$ are consecutive integers.	8 and 9; 28 and 29 are consecutive integers.
2.3D	Right angle	An angle whose measure is 90°	
	Complementary angles	Two angles whose measures add to 90° (A and B are complementary.)	
	Straight angle	An angle whose measure is 180°	
	Supplementary angles	Two angles whose measures add to 180° (A and B are supplementary.)	
2.4B	$I = Pr$	The annual interest I is the product of the principal P and the rate r.	The annual interest on a $3000 principal at a 5% rate is $I = 3000 \cdot 0.05 = \$150$.
2.4C	$D = RT$	When moving at a constant rate R, the distance D traveled in time T is $D = RT$.	A car moving at 55 mi per hour for 3 hr will travel $D = 55 \cdot 3 = 165$ mi.
2.5	Linear inequality	An inequality that can be written in the form $ax + b > c$, where a, b, and c are real numbers	$-2x + 3 > 5$ and $3 - 2x > 7$ are linear inequalities.
2.5B	Properties of inequalities	If $A < B$, then $A + C < B + C$.	The number C can be added to both sides to obtain an equivalent inequality.
		If $A < B$, then $A - C < B - C$.	The number C can be *subtracted* from both sides to obtain an equivalent inequality.
		If $A < B$, then $A \cdot C < B \cdot C$ when $C > 0$.	Both sides can be *multiplied* by $C > 0$ to obtain an equivalent inequality.
		If $A < B$, then $A \cdot C > B \cdot C$ when $C < 0$.	Both sides can be *multiplied* by $C < 0$ to obtain an equivalent inequality provided you change the sense (direction) of the inequality.

SECTION	ITEM	MEANING	EXAMPLE
		If $A < B$, then $\dfrac{A}{C} < \dfrac{B}{C}$ when $C > 0$.	Both sides can be *divided* by $C > 0$ to obtain an equivalent inequality.
		If $A < B$, then $\dfrac{A}{C} > \dfrac{B}{C}$ when $C < 0$.	Both sides can be *divided* by $C < 0$ to obtain an equivalent inequality provided you change the sense (direction) of the inequality.
2.5C	Union of sets	If A and B are sets, the union of A and B, denoted by $A \cup B$, is the set of elements that are in either A or B.	
	Intersection of sets	If A and B are sets, the intersection of A and B, denoted by $A \cap B$, is the set of elements that are in both A and B.	
2.6A	Absolute-value equation	The solutions of $|x| = a$ are $x = a$ or $x = -a$ for $a \geq 0$.	The solutions of $|x| = 2$ are 2 or -2.
2.6B	$|x| < a$ $|x| > a$	$|x| < a$ is equivalent to $-a < x < a$ $(a \geq 0)$. $|x| > a$ is equivalent to $x > a$ or $x < -a$.	$|x| < 3$ is equivalent to $-3 < x < 3$. $|x| > 3$ is equivalent to $x > 3$ or $x < -3$.

REVIEW EXERCISES

(If you need help with these exercises, look in the section indicated in brackets.)

1. [2.1A] Does -3 satisfy the equation?

  a. $7 = 8 - x$ **b.** $9 = 8 + x$ **c.** $4 = 1 - x$

2. [2.1B] Solve.

  a. $\dfrac{2}{3}y - 3 = 5$ **b.** $\dfrac{2}{3}y - 5 = 5$ **c.** $\dfrac{2}{3}y - 7 = 5$

3. [2.1B] Solve.

  a. $x + 2 = 2(2x - 2)$ **b.** $x + 3 = 3(2x - 4)$

  c. $x + 4 = 4(2x - 6)$

4. [2.1B] Solve.

  a. $\dfrac{x+4}{3} - \dfrac{x-4}{5} = 4$ **b.** $\dfrac{x+6}{3} - \dfrac{x-6}{5} = 6$

  c. $\dfrac{x+8}{3} - \dfrac{x-8}{5} = 8$

5. [2.1C] Solve.

  a. $\dfrac{x}{4} - \dfrac{x}{3} = \dfrac{x-4}{4}$ **b.** $\dfrac{x}{5} - \dfrac{x}{3} = \dfrac{x-5}{5}$ **c.** $\dfrac{x}{7} - \dfrac{x}{3} = \dfrac{x-7}{7}$

6. [2.1D] Solve.

  a. $0.05P + 0.10(2000 - P) = 175$

  b. $0.08P + 0.10(5000 - P) = 460$

  c. $0.06P + 0.10(10{,}000 - P) = 840$

7. [2.2A] Solve for h and evaluate when $H = 82.48$.

  a. $H = 2.5h + 72.48$

  b. $H = 2.5h + 77.48$

  c. $H = 2.5h + 84.98$

8. [2.2A] Solve for A.

  a. $B = \dfrac{2}{7}(A - 7)$ **b.** $B = \dfrac{3}{7}(A - 7)$ **c.** $B = \dfrac{4}{7}(A - 7)$

9. [2.2A] The perimeter of a rectangle is $P = 2L + 2W$, where L is the length and W is the width. Solve for L and give the dimensions when

  a. The perimeter is 180 ft and the length is 10 ft more than the width.

  b. The perimeter is 220 ft and the length is 10 ft more than the width.

  c. The perimeter is 260 ft and the length is 10 ft more than the width.

10. [2.2C]

  a. If L_1 and L_2 are parallel lines, find x and the measure of the unknown angles.

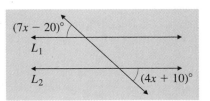

  b. Find x and the measure of the unknown angles if the measure of one angle is $(3x + 12)°$ and the measure of the other angle is $(4x - 3)°$.

c. Find x and the measure of the unknown angles if the measure of one angle is $(2x + 18)°$ and the measure of the other angle is $(3x - 2)°$.

11. [2.3B] Juan rented a car for \$30 per day plus 15¢ per mile. How many miles did Juan travel if he paid
a. \$45 for 1 day?   b. \$52.50 for 1 day? c. \$60 for 1 day?

12. [2.3C] Find three consecutive odd integers whose sum is
a. 153   b. 159   c. 207

13. [2.4A] A woman's salary is increased by 20%. What was her salary before the increase if her new salary is
a. \$24,000   b. \$36,000   c. \$18,000

14. [2.4B] An investor bought some municipal bonds yielding 5% annually and some certificates of deposit (CDs) yielding 10% annually. If his total investment amounts to \$20,000, find how much money is invested in bonds and how much in certificates of deposit if his annual interest is
a. \$1750   b. \$1150   c. \$1500

15. [2.4C] A car leaves a town traveling at 40 mi/hr. An hour later another car leaves the same town in the same direction traveling at 50 mi/hr.
a. How far from town does the second car overtake the first?
b. Repeat the problem where the first car is traveling at 50 mi/hr and the second one, 60 mi/hr.
c. Repeat the problem where the first car is traveling at 40 mi/hr and the second one, 60 mi/hr.

16. [2.4D] How many liters of a 40% salt solution must be mixed with 50 L of a 10% salt solution to obtain
a. A 30% solution? b. A 20% solution? c. A 10% solution?

17. [2.5A] Graph:
a. $x \geq -2$

b. $x \geq -3$

c. $x \geq -4$

18. [2.5B] Solve and graph:
a. $2(x - 1) \leq 4x + 4$

b. $3(x - 1) \leq 6x + 3$

c. $4(x - 1) \leq 8x + 4$

19. [2.5B] Solve and graph:
a. $\dfrac{x}{4} - \dfrac{x}{3} < \dfrac{x - 4}{4}$

b. $\dfrac{x}{5} - \dfrac{x}{3} < \dfrac{x - 5}{5}$

c. $\dfrac{x}{7} - \dfrac{x}{3} < \dfrac{x - 7}{7}$

20. [2.5C] Solve and graph:
a. $\{x \mid x > -1 \text{ and } x < 2\}$

b. $\{x \mid x > -2 \text{ and } x < 3\}$

c. $\{x \mid x > -3 \text{ and } x < 4\}$

21. [2.5C] Solve and graph:
a. $\{x \mid x < -2 \text{ or } x \geq 3\}$

b. $\{x \mid x < -3 \text{ or } x \geq 2\}$

c. $\{x \mid x < -4 \text{ or } x \geq 1\}$

22. [2.5C] Solve and graph:
a. $x + 1 \leq 3 \text{ and } -4x < 8$

b. $x + 1 \leq 4 \text{ and } -3x < 9$

c. $x + 1 \leq 5 \text{ and } -2x < 4$

23. [2.5C] Solve and graph:
a. $-4 \leq -2x - 6 < 4$

b. $-3 \leq -2x - 5 < 3$

c. $-6 \leq -2x - 4 < 2$

24. [2.6A] Solve:
a. $\left| \dfrac{2}{7}x + 2 \right| + 5 = 9$

b. $\left| \dfrac{2}{7}x + 2 \right| + 3 = 9$

c. $\left| \dfrac{2}{7}x + 2 \right| + 1 = 9$

25. [2.6A] Solve:
a. $|x - 1| = |x - 3|$

b. $|x - 3| = |x - 5|$

c. $|x - 5| = |x - 7|$

26. [2.6B] Solve and graph:
a. $|3x - 1| \leq 2$

b. $|4x - 1| \leq 3$

c. $|5x - 1| \leq 4$

27. [2.6B] Solve and graph:
a. $|3x - 1| \geq 2$

b. $|4x - 1| \geq 3$

c. $|5x - 1| \geq 4$

PRACTICE TEST

(Answers on pages 134–135.)

1. Does 4 satisfy the equation $5 = 9 - x$?

2. Solve $\dfrac{4}{5}y - 3 = 9$.

3. Solve $x + 1 = 2(3x - 3)$.

4. Solve $\dfrac{x + 4}{3} - \dfrac{x - 4}{5} = 4$.

5. Solve $\dfrac{6}{5} - \dfrac{x}{15} = \dfrac{2(x + 4)}{25}$.

6. Solve $0.06P + 0.07(1500 - P) = 96$.

7. $H = 2.75h + 71.48$
   a. Solve for h.   b. Find h if $H = 140.23$.

8. Solve for A in $B = \dfrac{3}{4}(A - 8)$.

9. The perimeter of a rectangle is $P = 2L + 2W$, where L is the length and W is the width.
   a. Solve for L.
   b. If the perimeter is 100 ft and the length is 20 ft more than the width, what are the dimensions of the rectangle?

10. If L_1 and L_2 are parallel lines, find x and the measure of the unknown angles.

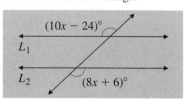

11. The bill for repairing an appliance totaled $72.50. If the repair shop charges $35 for the service call, plus $25 for each hour of labor, how many hours of labor did the repair take?

12. The sum of three consecutive odd integers is 117. What are the three integers?

13. A woman's salary is increased by 20% to $24,000. What was her salary before the increase?

14. An investor bought some municipal bonds yielding 5% annually and some certificates of deposit yielding 7%. If his total investment amounts to $20,000 and his annual interest is $1100, how much money is invested in bonds and how much in certificates of deposit?

15. A freight train leaves a station traveling at 40 mi/hr. Two hours later, a passenger train leaves the same station traveling in the same direction at 60 mi/hr. How far from the station does the passenger train overtake the freight train?

16. How many gallons of a 30% salt solution must be mixed with 40 gal of a 12% salt solution to obtain a 20% solution?

17. Solve and graph $4(x - 1) \le 8x + 4$.

18. Solve and graph $\dfrac{x}{7} - \dfrac{x}{3} < \dfrac{x - 7}{7}$.

19. Solve and graph $\{x \mid x < -1 \text{ or } x \ge 2\}$.

20. Solve and graph $x + 1 \le 4$ and $-2x < 6$.

21. Solve and graph $-4 \le -2x - 6 < 0$.

22. Solve $\left| \dfrac{3}{4}x + 2 \right| + 4 = 9$.

23. Solve $|x - 3| = |x - 7|$.

24. Solve and graph $|2x - 1| \le 5$.

25. Solve and graph $|2x + 1| > 3$.

ANSWERS TO PRACTICE TEST

Answer	If you missed:		Review:	
	Question	Section	Examples	Page
1. Yes	1	2.1	1	64
2. 15	2	2.1	2	65
3. $\dfrac{7}{5}$	3	2.1	3, 4	67
4. 14	4	2.1	5	68
5. 6	5	2.1	6	70
6. 900	6	2.1	8, 9	72
7a. $h = \dfrac{H - 71.48}{2.75}$	7a	2.2	1	77

Answer	If you missed: Question	Review: Section	Examples	Page
7b. 25	7b	2.2	1	77
8. $A = \dfrac{4B + 24}{3}$	8	2.2	2	77
9a. $L = \dfrac{P - 2W}{2}$	9a	2.2	3	79
9b. 15 ft by 35 ft	9b	2.2	3	79
10. $x = 15$, $126°$	10	2.2	7	83
11. 1.5 hr	11	2.3	2	90
12. 37, 39, 41	12	2.3	3	91
13. $20,000	13	2.4	2	99
14. $15,000 bonds; $5000 CD	14	2.4	3	99
15. 240 mi	15	2.4	4	100
16. 32	16	2.4	5	101

17. $x \geq -2$

	17	2.5	2, 3	109–110

18. $x > 3$

	18	2.5	4	111

19. $x < -1$ or $x \geq 2$

	19	2.5	5	112

20. $-3 < x \leq 3$

	20	2.5	6, 7	114

21. $-3 < x \leq -1$

	21	2.5	8	116

22. $4, -\dfrac{28}{3}$

	22	2.6	1, 2	122

23. 5

	23	2.6	3	123

24. $-2 \leq x \leq 3$

	24	2.6	4	125

25. $-2 < x$ or $x > 1$

	25	2.6	5	126

Polynomials

In this chapter we generalize the ideas of Chapter 1 by examining some of the fundamental operations for obtaining polynomials. In doing so, we make extensive use of the properties of exponents studied in Chapter 1. In Section 3.1, we discuss the basic terminology associated with polynomials as well as the techniques for adding and subtracting them. We continue this study in Section 3.2 where we learn how to multiply polynomials. We then reverse the process and "undo" these multiplications by using a process called factoring. We start by factoring out the greatest common factor of a polynomial (Section 3.3) and then factor polynomials using grouping. Next we factor trinomials (Section 3.4) and learn to recognize special polynomials that can be factored by reversing some of the multiplication formulas we learned (Section 3.5). We conclude our study of polynomials by discussing a general factoring strategy (Section 3.6) and then looking at some applications that are solved by factoring quadratic equations (Section 3.7).

Algebra has gone through three stages: rhetorical, in which statements and equations were written in ordinary language; syncopated, in which familiar terms were abbreviated; and symbolic, in which every part of an expression is written in symbols. At the time of Euclid, letters were used to represent quantities entered into equations until Diophantus introduced "the syncopation of algebra," using his own shorthand to express quantities and operations. (For example, Diophantus wrote the square of the unknown as Δ^Y.)

Francois Vieta (1540–1603), a French lawyer and member of parliament, used vowels to designate unknown quantities and consonants to represent constants. Vieta, however, retained part of the verbal algebra by writing *A quadratus* for x^2, *A cubus* for x^3, and so on. Vieta's contribution was a significant step toward a more abstract mathematics, but one of his most interesting contributions was his motto, *Leave no problem unsolved.* See if you can apply this motto to your study of algebra.

3.1

POLYNOMIALS: ADDITION AND SUBTRACTION

To succeed, review how to:

1. Define base and exponent (p. 29).
2. Evaluate expressions involving exponents (p. 29).
3. Use the properties of real numbers (p. 23).
4. Collect like terms (pp. 47–48).
5. Remove parentheses in an expression preceded by a minus sign (p. 45).

Objectives:

A. Classify polynomials.

B. Find the degree of a polynomial.

C. Rewrite a polynomial using exponents.

D. Evaluate a polynomial.

E. Add or subtract polynomials.

F. Solve applications involving sums or differences of polynomials.

getting started   Diving and Robberies

A man dives from an altitude of 118 ft. Do you know how high above sea level he will be after falling for t seconds? This height is given by

$$H = -16t^2 + 118$$

The right-hand side of this formula is an algebraic expression called a *polynomial*.

Polynomials have many applications. For example, do you think that the crime problem is getting worse? According to FBI data, the annual number of robberies (per 100,000 population) can be approximated by

$$R(t) = 1.76t^2 - 17.24t + 251$$

where $R(t)$ is read "R of t" (R to remind you of robberies and t for the number of years after 1980). Can you predict how many robberies (per 100,000) will occur annually by the year 2000? Since the year 2000 is 20 yr after 1980, we let $t = 20$ in $R(t)$; that is, we find

$$R(20) = 1.76(20)^2 - 17.24(20) + 251$$
$$= 610.2$$

or, to the nearest whole number, 610 robberies per 100,000. In 1980, this figure was 251 robberies per 100,000!

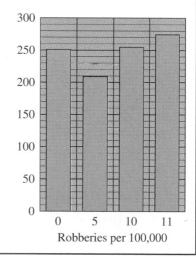

Robberies per 100,000

A   Monomials, Binomials, and Trinomials

An expression consisting of a constant or a constant times a product of variables with whole-number exponents is called a *monomial*. For example,

$$2x, \qquad -5x^2y, \qquad 10x^5yz^2, \qquad -\frac{2}{3}x^4, \qquad \text{and} \qquad -7y^3$$

are all monomials.

A **polynomial** is just a sum or difference of monomials. Thus

$$2x^2 - 4xy - 7y^3$$

is a polynomial. The individual monomials in a polynomial are called the **terms** of the polynomial. In the term ax^k, a is the **coefficient** and k is the **degree** of the term.

In the polynomial $2x^2 + xy - 7y^3$, the terms are $2x^2$, with coefficient 2; xy, with coefficient 1 (since $xy = 1xy$); and $-7y^3$, with coefficient -7. Note that

$$\frac{5}{x}, \qquad \frac{x+y}{z}, \qquad \text{and} \qquad 2 + \sqrt{x}$$

are *not* polynomials. (In the first two expressions we are dividing by a variable. The third expression involves taking the square root of the variable.)

Polynomials can be classified according to the number of terms they have. Polynomials with one term are called **monomials**, polynomials with two *unlike* terms are called **binomials**, and polynomials with three *unlike* terms are called **trinomials**. Note that polynomials with more than three unlike terms are simply called polynomials, as shown in the table.

Type	Example
Monomials	$-8x$, $\quad 3x^2y$, $\quad 9m^{10}$, $\quad -3$
Binomials	$x - y$, $\quad -3x^2 + xy$, $\quad -16t^2 + 118$
Trinomials	$x + y - z^5$, $\quad 2x^2 - 3x + \sqrt{2}$, $\quad z^3 + 2z - 12$
Polynomials	$t^5 + 3t^3 - 2t^2 - 8$, $\quad -3m^4 + 3n^2 + mn - 4m^3 + 2$

EXAMPLE 1   Classifying polynomials

Classify each of the following polynomials as a monomial, binomial, trinomial, or simply polynomial:

a. $x + x^2$ **b.** $-9x$ **c.** $3x^2 - y + 3xyz$ **d.** $x^5 - 3x^3 + 5x^2 - 7$

SOLUTION
a. Since $x + x^2$ has two terms, it is a binomial.
b. Since $-9x$ has one term, it is a monomial.
c. Since $3x^2 - y + 3xyz$ has three terms, it is a trinomial.
d. Since $x^5 - 3x^3 + 5x^2 - 7$ has more than three terms, it is just called a polynomial.   ■

The Degree of a Polynomial

In Example 1, the polynomials $x + x^2$ and $-9x$ contain only one variable, x. Single-variable polynomials can also be classified according to the *greatest exponent* of the variable. This number is called the *degree* of the polynomial. (Recall that in the monomial ax^k, k is the degree.) In general, we have the following definition.

DEGREE OF A POLYNOMIAL IN ONE VARIABLE

> The **degree** of a polynomial in one variable is the *greatest* exponent of that variable.

Thus $-8x^5$ is of the *fifth* degree, $-3x^2 + 8x^4 - 2x$ is of the *fourth* degree, and $0.5x$ is of the *first* degree. (Note that $x = x^1$.) Since $x^0 = 1$, $-3 = -3 \cdot 1 = -3x^0$. Similarly, $9 = 9x^0$. Thus the degree of nonzero numbers such as -3 and 9 is 0. The number 0 is called the **zero polynomial** and is not assigned a degree. These ideas can be extended to include polynomials in more than one variable. For example, the expression $3x^2 - y$ is a polynomial in *two* variables (x and y), and $3x^2 - 2xy - 3xyz^2$ is a polynomial in *three* variables (x, y, and z). To find the degree of these polynomials, we look at the degree of each term. Here is the definition.

DEGREE OF A POLYNOMIAL IN SEVERAL VARIABLES

> The degree of a polynomial in several variables is the greatest sum of the exponents of the variables in any one term of the polynomial.

Thus the degree of $3x^2 - y$ is 2 (the degree of the first term, $3x^2$). To find the degree of $3x^2 - 2xy - 3xyz^2$, find the degree of each term:

$$\underbrace{3x^2}_{} \quad - \quad \underbrace{2x^1y^1}_{} \quad - \quad \underbrace{3x^1y^1z^2}_{}$$

Degree: 2 $1 + 1 = 2$ $1 + 1 + 2 = 4$

Since the greatest sum of exponents is 4, the degree of $3x^2 - 2xy - 3xyz^2$ is 4.

EXAMPLE 2   Finding the degree of a polynomial

Find the degree of the given polynomials:

a. $-5x^2 + 3x^5 + 9$ **b.** $-x^2 + xy^2z^3 - x^5$ **c.** 3   **d.** 0

SOLUTION
a. The degree of $-5x^2 + 3x^5 + 9$ is 5.
b. The degree of $-x^2 + x^1y^2z^3 - x^5$ is $1 + 2 + 3 = 6$, the sum of the exponents in $x^1y^2z^3$
c. The degree of 3 is 0.
d. 0 has no degree.   ∎

Rewriting Polynomials Using Exponents

If we have a polynomial in several variables, it is easier to determine its degree if the multiplications indicated by exponents are written out. Thus, if $3x^2 - 2xy - 3xyz^2$ is written as $3 \cdot x \cdot x - 2 \cdot x \cdot y - 3 \cdot x \cdot y \cdot z \cdot z$, you can see that the last term involves variables as factors $1 + 1 + 2 = 4$ times; thus the degree of the polynomial is 4. We use this idea in the next example.

EXAMPLE 3   Finding the degree using exponents

Rewrite each of the given polynomials using exponents and state the degree of the polynomial.

a. $(x \cdot x) + (x \cdot y \cdot y) + (x \cdot x \cdot y \cdot y)$ **b.** $8xxxyy - 2xxyyzzz$

SOLUTION
a. $(x \cdot x) + (x \cdot y \cdot y) + (x \cdot x \cdot y \cdot y)$ can be written as $x^2 + xy^2 + x^2y^2$. The degree of this polynomial is $2 + 2 = 4$.
b. $8xxxyy - 2xxyyzzz$ can be written as $8x^3y^2 - 2x^2y^2z^3$. The degree of this polynomial is $2 + 2 + 3 = 7$.   ∎

Evaluating Polynomials

In mathematics, polynomials in one variable are sometimes represented by using symbols such as $P(t)$ (read "P of t"), $Q(x)$, and $D(y)$, *where the symbol in parentheses indicates the variable being used.* For example, we may have

$$P(t) = -16t^2 + 10t - 15 \qquad t \text{ is the variable.}$$

$$Q(x) = x^2 - 3x \qquad x \text{ is the variable.}$$

$$D(y) = -3y - 9 \qquad y \text{ is the variable.}$$

With this notation, it's easy to indicate the value of a polynomial for specific values of the variable. Thus $P(2)$ represents the value of the polynomial $P(t)$ when 2 is substituted for t in the polynomial. Similarly, $Q(3)$ represents the value of $Q(x)$ for $x = 3$. Thus

$$P(t) = -16t^2 + 10t - 15$$

$$P(2) = -16(2)^2 + 10(2) - 15$$

$$= -64 + 20 - 15 = -59$$

$P(2)$ represents the value of the polynomial $P(t)$ when $t = 2$. We also say that we are *evaluating* $P(t)$ at $t = 2$.

Similarly, to evaluate $Q(x)$ at $x = 3$ in

$$Q(x) = x^2 - 3x$$

we find

$$Q(3) = 3^2 - 3(3) = 0$$

 CAUTION Always enclose the substituted number in parentheses.

Using these ideas, we can find the height above sea level of the diver in the *Getting Started* section. As you recall, his altitude after t seconds was given as $H(t) = -16t^2 + 118$. Thus after 1 sec, his altitude will be

$$H(1) = -16(1)^2 + 118 = 102 \text{ ft}$$ Here we are evaluating $H(t)$ when $t = 1$.

After 2 sec, it will be

$$H(2) = -16(2)^2 + 118 = -64 + 118 = 54 \text{ ft}$$ Here we are evaluating $H(t)$ when $t = 2$.

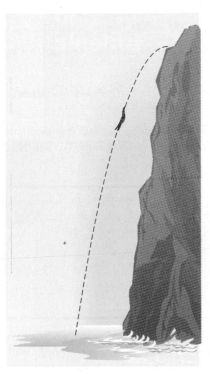

EXAMPLE 4   Evaluating polynomials using a formula

Find the altitude of the diver in the *Getting Started* after 3 sec.

SOLUTION   After 3 sec, the formula tells us that his altitude will be $P(3) = -16(3)^2 + 118 = -26$ ft—that is, 26 ft below sea level. Since the water is only 12 ft deep at this point, this is impossible. Moreover, divers cannot continue to free-fall after they hit the surface. In other words, the formula doesn't apply. ■

EXAMPLE 5   Evaluating polynomials

Let $P(x) = x^2 - 2x + 3$ and $Q(x) = x^2 + 3x - 5$.

a. Find: $P(0)$ **b.** Find: $Q(-1)$ **c.** Find: $P(0) + Q(-1)$

SOLUTION

a. To find $P(0)$, we substitute 0 for x in $P(x)$.

$$P(x) = x^2 - 2x + 3$$

$$P(0) = 0^2 - 2 \cdot 0 + 3$$

$$= 0 - 0 + 3$$

$$= 3$$

Hence $P(0) = 3$.

b. To find $Q(-1)$, we substitute -1 for x in $Q(x)$.

$$Q(x) = x^2 + 3x - 5$$

$$Q(-1) = (-1)^2 + 3(-1) - 5$$

$$= 1 - 3 - 5$$

$$= -7$$

Hence $Q(-1) = -7$.

c. Since $P(0) = 3$ and $Q(-1) = -7$,

$$P(0) + Q(-1) = 3 + (-7) = -4 \qquad\blacksquare$$

 Adding and Subtracting Polynomials

In the *Getting Started*, we mentioned that the annual number of robberies (per 100,000 population) can be approximated by

$$R(t) = 1.76t^2 - 17.24t + 251$$

If the corresponding number of aggravated assaults is given by

$$A(t) = -0.2t^3 + 4.7t^2 - 15t + 300$$

can we find the annual number of robberies and aggravated assaults? Yes, we do it by adding the two polynomials.

$$R(t) + A(t) = (1.76t^2 - 17.24t + 251) + (-0.2t^3 + 4.7t^2 - 15t + 300)$$

We shall add these polynomials in Example 7.

The procedure we use to add polynomials is dependent on the fact that the same properties used in the addition of numbers also apply to polynomials. We list these properties here for your convenience.

PROPERTIES FOR ADDING POLYNOMIALS

If P, Q, and R are polynomials,

$$P + Q = Q + P \qquad \text{Commutative property of addition}$$

$$P + (Q + R) = (P + Q) + R \qquad \text{Associative property of addition}$$

$$\left.\begin{array}{l} P(Q + R) = PQ + PR \\[6pt] (Q + R)P = QP + RP \end{array}\right\} \quad \text{Distributive property}$$

Now suppose we wish to add $(5x^2 + 3x + 9) + (7x^2 + 2x + 1)$. For our solution, we use the commutative, associative, and distributive properties just mentioned. With these facts, the terms in the expression $(5x^2 + 3x + 9) + (7x^2 + 2x + 1)$ can be added as follows:

$$(5x^2 + 3x + 9) + (7x^2 + 2x + 1) \qquad \text{Given.}$$

$$= (5x^2 + 7x^2) + (3x + 2x) + (9 + 1) \qquad \text{Group like terms.}$$

$$= (5 + 7)x^2 + (3 + 2)x + (9 + 1) \qquad \text{Use the distributive property.}$$

$$= 12x^2 + 5x + 10$$

This addition is done more efficiently by writing the terms of the polynomials in order of descending (or ascending) degree and then placing like terms in a column:

$$
\begin{array}{r}
5x^2 + 3x + \;\;9 \\
(+)\;\underline{7x^2 + 2x + \;\;1} \\
12x^2 + 5x + 10
\end{array}
$$

As we have already seen, $a - b = a + (-b)$. The signs in a polynomial are always taken to indicate positive or negative coefficients, and the operation involved is assumed to be addition. Thus

$$(6x^2 - 9x - 8) + (x^2 + 3x - 1)$$

$$= [6x^2 + (-9x) + (-8)] + [x^2 + 3x + (-1)]$$

$$= (6x^2 + x^2) + (-9x + 3x) + [-8 + (-1)]$$

$$= (6 + 1)x^2 + (-9 + 3)x + [-8 + (-1)]$$

$$= 7x^2 + (-6)x + (-9)$$

$$= 7x^2 - 6x - 9$$

Using the column method, this problem can be done more efficiently by writing

$$
\begin{array}{r}
6x^2 - 9x - 8 \\
(+)\;\underline{x^2 + 3x - 1} \\
7x^2 - 6x - 9
\end{array}
$$

EXAMPLE 6   Adding polynomials using the column method

Add: $10x^3 + 8x^2 - 7x - 3$ and $9 - 4x + x^2 - 5x^3$

SOLUTION   We write $9 - 4x + x^2 - 5x^3$ in descending order: $-5x^3 + x^2 - 4x + 9$. We then place like terms in a column and add.

$$
\begin{array}{r}
10x^3 + 8x^2 - \;\;7x - 3 \\
(+)\;\underline{-5x^3 + \;\;x^2 - \;\;4x + 9} \\
5x^3 + 9x^2 - 11x + 6
\end{array}
$$

EXAMPLE 7   Adding polynomials with missing terms

Add: $R(t) = 1.76t^2 - 17.24t + 251$ and $A(t) = -0.2t^3 + 4.7t^2 - 15t + 300$

SOLUTION   As before, we place like terms in a column, leaving space for any missing terms, and then add as follows:

$$
\begin{array}{r}
1.76t^2 - 17.24t + 251 \\
(+)\;\underline{-0.2t^3 + 4.7\;t^2 - 15\;\;\;t + 300} \\
-0.2t^3 + 6.46t^2 - 32.24t + 551
\end{array}
$$

By the way, this sum $R(t) + A(t)$ represents the annual number of robberies *and* aggravated assaults (per 100,000) t years after 1980.

To subtract polynomials, we first recall the following.

SUBTRACTING POLYNOMIALS

$$a - (b + c) = a - b - c \qquad \text{By the distributive property}$$

For example, the difference between the revenue $R(p) = 60 - 0.3p^2$ and the cost $C(p) = 4000 - 20p$ is

$$(60p - 0.3p^2) - (4000 - 20p) = 60p - 0.3p^2 - 4000 + 20p$$
$$= -0.3p^2 + 80p - 4000$$

Similarly,

$$(3x^2 + 4x - 5) - (5x^2 + 2x + 3) = 3x^2 + 4x - 5 - 5x^2 - 2x - 3$$
$$= -2x^2 + 2x - 8$$

Note that to subtract $(5x^2 + 2x + 3)$ from $(3x^2 + 4x - 5)$, we changed the sign of each term in $(5x^2 + 2x + 3)$ and then added. This procedure can also be done in columns:

$$
\begin{array}{r}
3x^2 + 4x - 5 \\
(-)\underline{5x^2 + 2x + 3}
\end{array}
\quad \xrightarrow{\text{is written}} \quad
\begin{array}{r}
3x^2 + 4x - 5 \\
(+)\underline{-5x^2 - 2x - 3} \\
-2x^2 + 2x - 8
\end{array}
$$

EXAMPLE 8   Subtracting polynomials

Subtract: $(9x^3 - 7x^2 + 3x - 5)$ from $(6x^3 + 2x^2 + 5)$

SOLUTION

> **NOTE** Subtracting *a from b* means to find $b - a$. Thus we find $(6x^3 + 2x^2 + 5) - (9x^3 - 7x^2 + 3x - 5)$ in column form.

$$
\begin{array}{r}
6x^3 + 2x^2 \quad\ + 5 \\
(-)\underline{9x^3 - 7x^2 + 3x - 5}
\end{array}
\quad \xrightarrow{\text{is written}} \quad
\begin{array}{r}
6x^3 + 2x^2 \quad\ + 5 \\
(+)\underline{-9x^3 + 7x^2 - 3x + 5} \\
-3x^3 + 9x^2 - 3x + 10
\end{array}
$$

Of course, the same result can be obtained by combining like terms, using the usual rules of signs. Thus

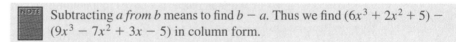

$$(6x^3 + 2x^2 + 5) - (9x^3 - 7x^2 + 3x - 5)$$
$$= (6 - 9)x^3 + [2 - (-7)]x^2 - 3x + [5 - (-5)]$$
$$= -3x^3 + 9x^2 - 3x + 10$$

∎

Applications

The profit P derived from selling x units of a product is related to the cost C and the revenue R, and is given by $P = R - C$. (Thus the profit P is the revenue minus the cost.)

EXAMPLE 9   Solving a word problem with polynomials

A company produces x video cassettes at a weekly cost $C = 2x + 1000$ (dollars). What is their weekly profit P if their revenue R is given by $R = 50x - 0.1x^2$ and they produce and sell 300 cassettes a week?

SOLUTION   We need to find

$$P = \overbrace{R} - \overbrace{C}$$
$$= (50x - 0.1x^2) - (2x + 1000)$$
$$= 50x - 0.1x^2 - 2x - 1000$$
$$= -0.1x^2 + 48x - 1000$$

If they produced and sold 300 cassettes, $x = 300$ and

$$P = -0.1(300)^2 + 48(300) - 1000$$
$$= -9000 + 14{,}400 - 1000$$
$$= 4400 \text{ (dollars)}$$

Their profit P when they sell 300 cassettes is \$4400.   ■

GRAPH IT

In Example 7, we added $R(t)$ and $A(t)$. To *graph* $R(t) + A(t)$, we need to find a suitable window. Since t represents the number of years after 1980, set the minimum for t (or x if you prefer) at 0 and the maximum at 20 with a scale of 1. What do we need for the y-values representing the number of robberies and assaults? First, note that $R(0) = 251$ and $A(0) = 300$, so we let Ymin = 100, Ymax = 500, and Yscl = 100. We now have a [0, 20] by [100, 500] window. Press $\boxed{Y=}$ and enter $Y_1 = 1.76t^2 - 17.24t + 251$ and $Y_2 = -0.2t^3 + 4.7t^2 - 15t + 300$. (Remember, you can use x for the variable instead of t.) Press $\boxed{\text{GRAPH}}$.

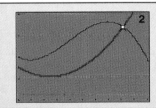

You can now answer questions that are not evident when you only have the polynomials defined algebraically. For example, it seems that the number of robberies is increasing but the number of aggravated assaults peaks at a point and then begins to decrease. Can you find in what year the number of aggravated assaults $A(t)$ peaks? You can use the $\boxed{\text{TRACE}}$ key and decide, or better yet, you can let a TI-82 decide for you. Press $\boxed{\text{2nd}}\boxed{\text{CALC}}$ and 4 for maximum. Now, use the $\boxed{\blacktriangleleft}$ key to trace around the curve $A(t)$ to a point to the *left* of the maximum and press $\boxed{\text{ENTER}}$. Use the $\boxed{\blacktriangleright}$ key to trace along the curve to a point to the *right* of the maximum and press $\boxed{\text{ENTER}}$. When the grapher asks for a "Guess," press $\boxed{\text{ENTER}}$. The calculator gives the x- and y-values of the maximum point. Can you find when the number of aggravated assaults and robberies will be the same? Use the $\boxed{\text{TRACE}}$ key to find out or use the intersection feature on a TI-82.

Some calculators (TI-82) can even evaluate polynomials. If you have graphed $R(t)$ and $A(t)$, press $\boxed{\text{2nd}}\boxed{\text{CALC}}$ 1, which prompts you for an x-value by showing "X=" on your screen. Enter 16 and press $\boxed{\text{ENTER}}$. The grapher shows $y = 425.72$, so $R(16) = 425.72$. Note that at the top right-hand side of the screen, a little 1 appears. This means that you have found the value of Y_1. To find the value of Y_2 at 16, press the $\boxed{\blacktriangle}$ key. The y-value is 444; thus $A(16) = 444$. (Note the 2 at the top right-hand side of the screen.)

EXERCISE 3.1

A **B** In Problems 1–10, classify the given polynomial as a monomial, binomial, or trinomial and give its degree.

1. xyz^2

2. u^2vw^3

3. $x^2 + yz^2$

4. $x^2y + z^3 - x^6$

5. $x + y^2 + z^3$

6. $xy + y^3$

7. $x^2yz - xy^3 - u^2v^3$

8. 8

9. 0

10. $3xyz - uv^2 + v^7$

B **C** In Problems 11–16, use exponents to rewrite each expression as a polynomial and give its degree.

11. $x \cdot x \cdot x + y \cdot y \cdot y \cdot y$

12. $(ab)(ab) - (xy)(xy)$

13. $xxxx + yyyy + zzzz$

14. $2aabb - 4(xy)(xy)$

15. $3xx - 4yyy + 5(xy)(xy)$

16. $9x^2 - 6(xyz)(xyz)$

D In Problems 17–26, evaluate the polynomial for the specified values of the variables.

17. z^3 for $z = -2$

18. $(xy)^3$ for $x = 2, y = -1$

19. $(x - 2y + z)^2$ for $x = 2, y = -1, z = -2$

20. $(x - y)(x - z)$ for $x = 2, y = -1, z = -2$

21. If $P(x) = 4x^2 + 4x - 1$, find $P(-1)$.

22. If $P(x) = -3x^2 + 3x + 2$, find $P(0)$.

23. If $Q(y) = y^2 - 7y - 2$, find $Q(-2)$.

24. If $R(t) = t^2 - 2t + 7$, find $R(-2)$.

25. If $S(u) = -16u^2 + 120$, find $S(4)$.

26. If $V(t) = -16t^2 + 80t$, find $V(3)$.

27. $P(x) = 2x^2 + 3x$ and $Q(y) = -3y^2 - 7y + 1$.
   a. Find $P(0)$.   b. Find $Q(1)$.   c. Find $P(0) + Q(1)$.

28. $P(x) = 3x^2 - 2x + 5$ and $Q(y) = -2y^2 + 3y - 1$.
   a. Find $P(-2)$.   b. Find $Q(0)$.   c. Find $P(-2) + Q(0)$.

29. $P(x) = x^2 - 2x + 5$ and $Q(y) = -2y^2 + 5y - 1$.
   a. Find $P(-1)$.   b. Find $Q(1)$.   c. Find $P(-1) - Q(1)$.

30. $P(x) = x^2 - 3x + 5$ and $Q(y) = -2y^2 + 5y - 1$.
   a. Find $P(-2)$.   b. Find $Q(2)$.   c. Find $P(-2) - Q(2)$.

E In Problems 31–55, perform the indicated operations.

31. $(x^2 + 4x - 8) + (5x^2 - 4x + 3)$

32. $(3x^2 + 2x + 1) + (8x^2 - 7x + 5)$

33. $\begin{array}{r} 5x^2 + 3x + 4 \\ (+) \underline{-4x^2 - 5x - 8} \end{array}$

34. $\begin{array}{r} -5x^2 + 4x - 3 \\ (+) \underline{6x^2 + 4x - 7} \end{array}$

35. $(4x^2 + 7x - 5) - (3x + x^2 + 4)$

36. $(8x^2 - 6x + 3) - (4x + 2x^2 - 6)$

37. $\begin{array}{r} -3y^2 + 6y - 5 \\ (-) \underline{8y^2 + 7y - 2} \end{array}$

38. $\begin{array}{r} -4y^2 + 5y - 2 \\ (-) \underline{5y^2 - 3y + 6} \end{array}$

39. $(x^3 - 6x^2 + 4x - 2) + (3x^3 - 6x^2 + 5x - 4)$

40. $(-6x^3 - 3x + 2x^2 + 2) + (2x^3 - 6x^2 + 8x - 4)$

41. Add $(-8y^3 + 5y + 7y^2 - 5)$ and $(8y^3 + 7y - 6)$.

42. Add $(5y^3 + 3y - 8)$ and $(-9y^3 - 6y + 2y^2 + 3)$.

43. Subtract $(3v^3 + v - v^2 + 2)$ from $(6v^3 - 3v^2 + 2v - 5)$.

44. Subtract $(5v^3 + 3v^2 - 6v + 4)$ from $(3v^3 - 7v^2 + 3v - 1)$.

45. $(4u^3 - 5u^2 - u + 3) - (2u + 9u^3 - 7)$

46. $(x^3 + y^3 - 8xy + 3) + (10xy - y^3 + 2x^3 - 6)$

47. $(x^3 + y^3 - 6xy + 7) + (3x^3 - y^3 + 8xy - 8)$

48. $(2x^3 - y^3 + 3xy - 5) - (x^3 + y^3 - 3xy + 9)$

49. $(x^3 - y^3 + 5xy - 2) - (x^2 - y^3 + 5xy + 2)$

50. $(4x^2 + y^2 - 3x^2y^2) - (x^3 + 3y^2 - 3x^2y^2)$

51. $(a + a^2) + (9a - 4a^2) + (a^2 - 5a)$

52. Subtract $(x + x^2)$ from $(2x - 5x^2) + (7x - x^2)$.

53. $2y + (x + 3y) - (x + y)$

54. $8y - (y + 3x) + 7y$

55. $(3x^2 + y) - (x^2 - 3y) + (3y + x^2)$

In Problems 56–60, let $P(x) = x^2 - 2x + 3$ and $Q(x) = 2x^2 + 3x - 1$ and find the solution.

56. $P(x) - Q(x)$ **57.** $P(0) + Q(0)$

58. $P(1) - Q(-1)$ **59.** $P(x) - P(x)$

60. $[P(x) + Q(x)] + P(x)$

In Problems 61–70, justify each of the equalities by using one of the three properties given in the text.

61. $3x^2 + 9x = 9x + 3x^2$

62. $(-y^3) + 7y = 7y + (-y)^3$

63. $(8x + 9)4 = 32x + 36$

64. $(7 - 2x)(-3) = -21 + 6x$

65. $x^2 + (x + 5) = (x^2 + x) + 5$

66. $(y^3 + y^2) + 4 = y^3 + (y^2 + 4)$

67. $x^7 + (3 + x) = x^7 + (x + 3)$

68. $(y + 7) + y^3 = y^3 + (y + 7)$

69. $3(x^2 + 5) = 3x^2 + 15$

70. $8(x^3 - 5x) = 8x^3 - 40x$

F Applications

71. The height $H(t)$ (in feet) of an object thrown straight up with an initial velocity of 64 ft/sec after t seconds (sec) is given by

$$H(t) = -16t^2 + 64t$$

a. Find the height of the object after 1 sec.

b. Find the height of the object after 2 sec.

72. The total dollar cost $C(x)$ of manufacturing x units of a certain product each week is given by

$$C(x) = 10x + 400$$

a. Find the cost of manufacturing 500 units.

b. Find the cost of manufacturing 1000 units.

73. The dollar revenue obtained by selling x units of a certain product each week is given by

$$R(x) = 100x - 0.03x^2$$

a. Find the revenue when 500 units are sold.

b. Find the revenue when 1000 units are sold.

74. In business, you can calculate your gross profit by subtracting the cost from the revenue. If the cost C and revenue R (in dollars) when x units are sold are given by

$$C(x) = 10x + 400 \quad \text{and} \quad R(x) = 100x - 0.03x^2$$

find the gross profit when

a. 500 units are sold.

b. 1000 units are sold.

75. If a \$100,000 computer depreciates 10% each year, its value $V(t)$ after t years is given by

$$V(t) = 100,000 - 0.10t(100,000)$$
$$= 100,000(1 - 0.10t)$$

a. Find the value of the computer after 5 yr.

b. Find the value of the computer after 10 yr.

76. The total number N of units of output per day when the number of employees m is given by $N = 20m - \frac{1}{2}m^2$. Find how many units are produced per day in a company with

a. 10 employees.   **b.** 20 employees.

In Problems 77–81 $P = R - C$, where P is the profit, R is the revenue, and C is the cost.

77. The cost C in dollars of producing x items is given by $C = 100 + 0.3x$. If the revenue $R = 1.50x$, find the profit P when 100 items are produced and sold.

78. The cost C in dollars of producing x pairs of jogging shoes is given by $C = 2000 + 60x$. If the revenue $R = 90x$, find the profit when 200 pairs are produced and sold.

79. The cost C in dollars of producing x pairs of jeans is given by $C = 1500 + 20x$. If the revenue

$$R = 50x - \frac{x^2}{20}$$

find the profit when 100 pairs of jeans are produced and sold.

80. The cost C in dollars of producing x pairs of sunglasses is given by $C = 30,000 + 60x$. Find the profit when 300 pairs of sunglasses are manufactured and sold if the revenue

$$R = 200x - \frac{x^2}{30}$$

81. The cost C in dollars of producing x pairs of shoes is given by $C = 100,000 + 30x$. Find the profit when 500 pairs of shoes are manufactured and sold if the revenue

$$R = 300x - \frac{x^2}{50}$$

SKILL CHECKER

Multiply:

82. $6x^2y \cdot 2x$ **83.** $6x^2y \cdot 3xy$

84. $-2x^3y \cdot (-2y^2)$ **85.** $-2x^3y \cdot (-7xy)$

86. $(2x)^2$

Sums of Areas

The addition of polynomials can be used to find the sum of the areas of several rectangles. Thus to find the total area of the rectangles, add the individual areas as shown.

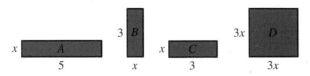

The total area is

$$\underbrace{\text{Area of A}}_{} + \underbrace{\text{area of B}}_{} + \underbrace{\text{area of C}}_{} + \underbrace{\text{area of D}}_{}$$

$$\underbrace{5x \quad + \quad 3x \quad + \quad 3x}_{11x} \quad + \quad \underbrace{(3x)^2}_{9x^2}$$

or, in descending order, $9x^2 + 11x$.

Find the sum of the areas of the rectangles.

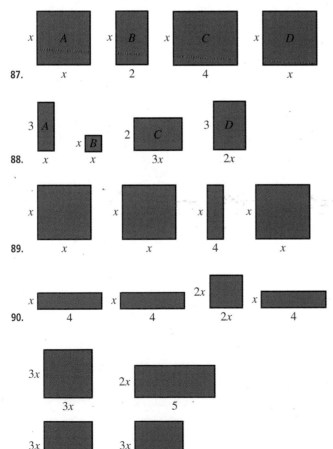

87.

88.

89.

90.

91.

92. Write your own definition of a polynomial.

93. Describe the procedure you use to find the degree of a polynomial.

94. If $P(x)$ and $Q(x)$ are polynomials and you subtract $Q(x)$ from $P(x)$, what happens to the signs of all the terms of $Q(x)$?

If you know how to do these problems, you have learned your lesson!

95. A company produces x units of a product at a cost $C = 3x + 100$ (dollars).
   a. If their revenue R is given by $R = 50x - 0.2x^2$ and their profit is $P = R - C$, find P.
   b. What is their profit when they produce and sell 100 units?

96. The height of an object thrown straight up with an initial velocity of 96 ft/sec after t seconds is given by

$$H(t) = -16t^2 + 96t$$

What is the height of the object after 3 sec?

Find the degree of:

97. $-2x^2 + 9 - 7x^4$

98. $-3xy^2z + x^2 - x^3$

99. Classify as monomials, binomials, trinomials, or simply polynomials:
   a. $x^2 - x$
   b. $t^5 - 4t^2 + 3t - 2t^3 + 5$
   c. $x^2y^3z^5$
   d. $\dfrac{2}{3}x^2 - 8 - 3x$

If $P(x) = x^2 - 3x + 2$ and $Q(x) = x^2 + 2x - 5$, find:

100. $P(2)$ and $Q(2)$

101. $P(2) - Q(2)$

102. $P(x) + Q(x)$

103. $P(x) - Q(x)$

104. Subtract $(8x^3 - 6x^2 + 5x - 3)$ from $(4x^2 + 4x + 3)$.

Rewrite each polynomial using exponents and give its degree.

105. $(y \cdot y \cdot y) + 5 \cdot x \cdot y - x \cdot y \cdot z \cdot z$

106. $7 \cdot x \cdot x \cdot x \cdot y - 4 \cdot x \cdot x \cdot y \cdot y \cdot y$

3.2

MULTIPLICATION OF POLYNOMIALS

To succeed, review how to:

1. Use the distributive property to simplify expressions (pp. 45–47).
2. Use the properties of exponents (pp. 31–37).
3. Combine like terms (pp. 47–48).

Objectives:

A Multiply a monomial by a polynomial.

B Multiplying two polynomials

C Use the FOIL method to multiply two binomials.

D Square a binomial sum or difference.

E Find the product of the sum and the difference of two terms.

F Use the ideas discussed to solve applications.

getting started

Building Bridges with Algebra

How much does a bridge beam bend (deflect) when a car or truck goes over the bridge? There's a formula that can tell us! For a certain beam of length L at a distance x from one end, the deflection is given by

$$(x - L)(x - 2L)$$

To multiply these two binomials, we first learn how to do several types of related multiplications. We show you how to multiply $(x - L)(x - 2L)$ in Example 3 of this section.

A Multiplying Monomials by Polynomials

When multiplying polynomials such as $6x^2$ and $2x + 3xy$, we use the commutative, associative, and distributive properties of multiplication for polynomials. These properties are generalizations of the properties discussed in Chapter 1. We state them here for polynomials.

MULTIPLICATION PROPERTIES FOR POLYNOMIALS

If P, Q, and R are polynomials,

$$P \cdot Q = Q \cdot P \qquad \text{Commutative property of multiplication}$$

$$P \cdot (Q \cdot R) = (P \cdot Q) \cdot R \qquad \text{Associative property of multiplication}$$

$$\left.\begin{array}{l} P(Q + R) = PQ + PR \\ (Q + R)P = QP + RP \end{array}\right\} \quad \text{Distributive property}$$

To multiply $6x^2y(2x + 3xy)$, we proceed as follows:

$$6x^2y(2x + 3xy) = 6x^2y(2x) + 6x^2y(3xy) \qquad \text{Use the distributive property.}$$

$$= 12x^3y + 18x^3y^2 \qquad \text{Multiply and rearrange using the commutative and associative properties.}$$

Note that we multiplied the coefficients but added the exponents. We can multiply $4x^2(3x^3 + 7x^2 - 4x + 8)$ in a similar way. Thus

$$4x^2(3x^3 + 7x^2 - 4x + 8) = 4x^2(3x^3) + 4x^2(7x^2) + 4x^2(-4x) + 4x^2(8)$$
$$= 12x^5 + 28x^4 - 16x^3 + 32x^2$$

EXAMPLE 1   Multiplying a monomial by a polynomial

Multiply:

a. $5x^2(3x^3 + 3x^2 - 2x - 3)$ **b.** $-2x^3y(x^2 + 7xy - 2y^2)$

SOLUTION

a. $5x^2(3x^3 + 3x^2 - 2x - 3) = 15x^5 + 15x^4 - 10x^3 - 15x^2$

b. $-2x^3y(x^2 + 7xy - 2y^2) = -2x^5y - 14x^4y^2 + 4x^3y^3$ ■

B   **Multiplying Two Polynomials**

To multiply $(3x + 2)(x + 3)$, we use the distributive property

$$a(b + c) = ab + ac$$

and multiply $(3x + 2)$ by each term of $(x + 3)$ to obtain

$$\overbrace{a}^{} \overbrace{(b + c)}^{} = \overbrace{a}^{} \overbrace{b}^{} + \overbrace{a}^{} \overbrace{c}^{}$$
$$(3x + 2)(x + 3) = (3x + 2)(x) + (3x + 2)(3)$$
$$= 3x^2 + 2x + 9x + 6$$
$$= 3x^2 + 11x + 6$$

This multiplication can also be done by arranging the work as in ordinary multiplication and placing the resulting like terms of the products in the same column. The procedure looks like this:

$$\begin{array}{r} x + 3 \\ 3x + 2 \\ \hline 2x + 6 \\ 3x^2 + 9x \\ \hline 3x^2 + 11x + 6 \end{array}$$

Multiply by 2: $2(x + 3) = 2x + 6$.
Multiply by $3x$: $3x(x + 3) = 3x^2 + 9x$.
Add like terms: $2x + 9x = 11x$.

Now let's use this same technique to multiply two polynomials like $(x + 5)$ and $(x^2 + x - 2)$. We call this technique the **vertical scheme**.

PROCEDURE

Vertical Scheme for Multiplying Polynomials

STEP 1	STEP 2	STEP 3
$\begin{array}{r} x^2 + x - 2 \\ x + 5 \\ \hline 5x^2 + 5x - 10 \end{array}$	$\begin{array}{r} x^2 + x - 2 \\ x + 5 \\ \hline 5x^2 + 5x - 10 \\ x^3 + x^2 - 2x \end{array}$	$\begin{array}{r} x^2 + x - 2 \\ x + 5 \\ \hline 5x^2 + 5x - 10 \\ x^3 + x^2 - 2x \\ \hline x^3 + 6x^2 + 3x - 10 \end{array}$
Multiply.	Multiply.	Add like terms.
$5(x^2 + x - 2)$ $= 5x^2 + 5x - 10$	$x(x^2 + x - 2)$ $= x^3 + x^2 - 2x$	

Of course, we could have obtained the same result by using the distributive property and multiplying $(x + 5)$ by x^2, $(x + 5)$ by x, and $(x + 5)$ by -2. The result would look like this:

$$(x + 5)(x^2 + x - 2) = (x + 5)(x^2) + (x + 5)x + (x + 5)(-2)$$
$$= x^3 + 5x^2 + x^2 + 5x - 2x - 10$$
$$= x^3 + (5x^2 + x^2) + (5x - 2x) - 10$$
$$= x^3 + 6x^2 + 3x - 10$$

EXAMPLE 2 Using the vertical scheme to multiply polynomials

Multiply:

a. $(x - 3)(x^2 - 2x - 4)$ **b.** $(x + 2)(x^2 - 2x + 4)$

SOLUTION We use the vertical scheme.

a.
$$
\begin{array}{r}
x^2 - 2x - 4 \\
x - 3 \\
\hline
-3x^2 + 6x + 12 \\
x^3 - 2x^2 - 4x \\
\hline
x^3 - 5x^2 + 2x + 12
\end{array}
$$
Multiply by -3.
Multiply by x.
Add like terms.

b.
$$
\begin{array}{r}
x^2 - 2x + 4 \\
x + 2 \\
\hline
2x^2 - 4x + 8 \\
x^3 - 2x^2 + 4x \\
\hline
x^3 \qquad\quad + 8
\end{array}
$$
Multiply by 2.
Multiply by x.
Add like terms. ■

The result of multiplying $(x - 3)(x^2 - 2x - 4)$ is $x^3 - 5x^2 + 2x + 12$. Since $(b - c)a = ba - ca$, you can also do this problem by multiplying $x(x^2 - 2x - 4)$ first and then multiplying $-3(x^2 - 2x - 4)$ to obtain the following result:

$$
\overbrace{(x - 3)}^{(b-c)}\ \overbrace{(x^2 - 2x - 4)}^{a} = \overbrace{x}^{b}\overbrace{(x^2 - 2x - 4)}^{a} \overbrace{-\ 3}^{-c}\overbrace{(x^2 - 2x - 4)}^{a}
$$
$$= x^3 - 2x^2 - 4x - 3x^2 + 6x + 12$$
$$= x^3 + (-2x^2 - 3x^2) + (-4x + 6x) + 12$$
$$= x^3 - 5x^2 + 2x + 12$$

Note that since $(x - 3)(x^2 - 2x - 4) = (x^2 - 2x - 4)(x - 3)$, the same result is obtained in both cases. Here is the rule we used.

RULE TO MULTIPLY ANY TWO POLYNOMIALS

> To multiply two polynomials, multiply each term of one by every term of the other and combine like terms.

Multiplying Two Binomials

Let's use the rule we just mentioned to multiply $(x + 4)$ and $(x - 7)$. We have to multiply each term of $x - 7$ by every term of $x + 4$, so we write

$$(x + 4)(x - 7) = x \cdot x + x \cdot (-7) + 4 \cdot x + 4 \cdot (-7)$$
$$= x^2 - 7x + 4x - 28$$
$$= x^2 - 3x - 28$$

If you look at the second line, you can see what the procedure for multiplying two binomials is.

The first term, x^2, is the product of the first terms.

$$(x + 4)(x - 7) \qquad x^2 \qquad \text{First terms} \qquad \text{F}$$

The second term, $-7x$, is the product of the outside terms.

$$(x + 4)(x - 7) \qquad -7x \qquad \text{Outside terms} \qquad \text{O}$$

The third term, $+4x$, is the product of the inside terms.

$$(x + 4)(x - 7) \qquad 4x \qquad \text{Inside terms} \qquad \text{I}$$

The last term, -28, is the product of the last two terms.

$$(x + 4)(x - 7) \qquad -28 \qquad \text{Last terms} \qquad \text{L}$$

Thus to multiply $(x + 4)(x - 7)$, we simply write

$$\begin{array}{cccc} \text{First} & \text{Outside} & \text{Inside} & \text{Last} \\ \text{F} & \text{O} & \text{I} & \text{L} \end{array}$$
$$(x + 4)(x - 7) = x^2 \quad - 7x \quad + 4x \quad - 28$$
$$= x^2 - 3x - 28$$

Here is the general rule for multiplying two binomials using the FOIL method. This is the first of several special products.

PROCEDURE

Using FOIL to Multiply the Two Binomials $(x + a)(x + b)$

To find the product of two binomials, multiply the terms in this order:

$$\begin{array}{cccc} \text{First} & \text{Outside} & \text{Inside} & \text{Last} \\ \text{F} & \text{O} & \text{I} & \text{L} \end{array}$$
$$(x + a)(x + b) = x^2 \ + \ bx \ + \ ax \ + \ ab \qquad \text{Special product (1)}$$
$$= x^2 + (b + a)x + ab$$

Let's do one more example, step by step, to give you more practice.

$$\text{F} \qquad (x + 7)(x - 4) \qquad x^2$$

$$\text{O} \qquad (x + 7)(x - 4) \qquad x^2 - 4x$$

$$\text{I} \qquad (x + 7)(x - 4) \qquad x^2 - 4x + 7x$$

$$\text{L} \qquad (x + 7)(x - 4) = x^2 - 4x + 7x - 28$$

Thus

$$(x + 7)(x - 4) = x^2 + 3x - 28$$

EXAMPLE 3   Using the FOIL method

Multiply:

a. $(5x + 2y)(2x + 3y)$ **b.** $(3x - y)(4x - 3y)$ **c.** $(x - L)(x - 2L)$

SOLUTION

$$\overset{\text{F}}{} \quad \overset{\text{O}}{} \quad \overset{\text{I}}{} \quad \overset{\text{L}}{}$$

a. $(5x + 2y)(2x + 3y) = (5x)(2x) + (5x)(3y) + (2y)(2x) + (2y)(3y)$

$$= 10x^2 + 15xy + 4xy + 6y^2$$

$$= 10x^2 + 19xy + 6y^2$$

$$\overset{\text{F}}{} \quad \overset{\text{O}}{} \quad \overset{\text{I}}{} \quad \overset{\text{L}}{}$$

b. $(3x - y)(4x - 3y) = (3x)(4x) + (3x)(-3y) + (-y)(4x) + (-y)(-3y)$

$$= 12x^2 - 9xy - 4xy + 3y^2$$

$$= 12x^2 - 13xy + 3y^2$$

$$\overset{\text{F}}{} \quad \overset{\text{O}}{} \quad \overset{\text{I}}{} \quad \overset{\text{L}}{}$$

c. $(x - L)(x - 2L) = x \cdot x + (x)(-2L) + (-L)(x) + (-L)(-2L)$

$$= x^2 - 2Lx - Lx + 2L^2$$

$$= x^2 - 3Lx + 2L^2 \qquad\blacksquare$$

Squaring Sums or Differences of Binomials

Now suppose we want to find $(x + 7)^2$. The exponent 2 means that we must multiply $(x + 7)(x + 7)$. Using FOIL, we write

$$(x + 7)(x + 7) = x^2 + 7x + 7x + 7 \cdot 7$$

$$= x^2 + 14x + 49$$

The result, $x^2 + 14x + 49$, is called a perfect square trinomial. Did you see how the middle term was calculated? It's the sum of $7x$ and $7x$, that is, $2 \cdot 7x$. Also, the last term is 7^2. In general, we have the following.

PROCEDURE

> **Square of a Binomial Sum $(x + a)^2$**
>
> $$\overset{\text{F}}{} \quad \overset{\text{O}}{} \quad \overset{\text{I}}{} \quad \overset{\text{L}}{}$$
> $$(x + a)^2 = (x + a)(x + a) = x^2 + ax + ax + a \cdot a \qquad \text{Special product (2)}$$
> $$= x^2 + 2ax + a^2$$

The same pattern applies to the difference of two binomials.

PROCEDURE

> **Square of a Binomial Difference $(x - a)^2$**
>
> $$\overset{\text{F}}{} \quad \overset{\text{O}}{} \quad \overset{\text{I}}{} \quad \overset{\text{L}}{}$$
> $$(x - a)^2 = (x - a)(x - a) = x^2 - ax - ax + a \cdot a \qquad \text{Special product (3)}$$
> $$= x^2 - 2ax + a^2$$

Thus to find the square of a binomial, add the square of the first term, twice the product of the two terms, and the square of the second term. The sign of the middle term is + for binomial sums and − for binomial differences. Note that the sign of the last term is always +.

EXAMPLE 4   Square of a binomial

Multiply:

a. $(2x + 3y)^2$ **b.** $(3x - 2y)^2$

c. $-3x(2x - 3y)^2$ **d.** $[(2x + 1) + y]^2$

SOLUTION

a. $(2x + 3y)^2 = (2x)^2 + 2 \cdot 3y \cdot 2x + (3y)^2$
$$= 4x^2 + 12xy + 9y^2$$

b. $(3x - 2y)^2 = (3x)^2 - 2 \cdot 2y \cdot 3x + (-2y)^2$
$$= 9x^2 - 12xy + 4y^2$$

c. $-3x(2x - 3y)^2 = -3x[(2x)^2 - 2 \cdot (3y)(2x) + (3y)^2]$
$$= -3x[4x^2 - 12xy + 9y^2]$$
$$= -12x^3 + 36x^2y - 27xy^2$$

d. $[(2x + 1) + y]^2 = (2x + 1)^2 + 2 \cdot y \cdot (2x + 1) + y^2$
$$= (4x^2 + 4x + 1) + 4xy + 2y + y^2$$
$$= 4x^2 + 4xy + y^2 + 4x + 2y + 1 \qquad ∎$$

Product of a Sum and a Difference

We have one more special product, and this one is really special. Suppose we multiply the sum of two terms by the difference of the same two terms; that is, suppose we want to find the product of

$$(x - 7)(x + 7)$$

Using FOIL, we get

$$
\overset{\text{F} \quad\ \text{O} \quad\ \text{I} \quad\ \text{L}}{(x - 7)(x + 7) = x^2 + 7x - 7x - 7^2}
$$
$$= x^2 + 0x - 7^2$$
$$= x^2 - 49$$

Since multiplication is commutative, $(x - 7)(x + 7) = (x + 7)(x - 7)$; thus

$$(x + 7)(x - 7) = x^2 - 49$$

In general, we have the following.

PROCEDURE

Product of the Sum and Difference of Two Monomials

$$(x - a)(x + a) = x^2 - a^2 \qquad \text{Special product (4)}$$

$$(x + a)(x - a) = x^2 - a^2$$

Note that the FOIL method is still operative here.

EXAMPLE 5   Finding the product of the sum
                             and difference of two monomials

Multiply:

a. $(x + 10)(x - 10)$ **b.** $(2x + y)(2x - y)$

c. $-(3x - 5y)(3x + 5y)$ **d.** $[2x - (3y + 1)][2x + (3y + 1)]$

SOLUTION

a. $(x + 10)(x - 10) = x^2 - 10^2$

$$= x^2 - 100$$

b. $(2x + y)(2x - y) = (2x)^2 - y^2$

$$= 4x^2 - y^2$$

c. $-(3x - 5y)(3x + 5y) = -[(3x)^2 - (5y)^2]$

$$= -9x^2 + 25y^2 \qquad \text{Since } -(a - b) = -a + b$$

$$= 25y^2 - 9x^2 \qquad \text{By the commutative property}$$

d. $[2x - (3y + 1)][2x + (3y + 1)] = (2x)^2 - (3y + 1)^2$

$$= 4x^2 - (9y^2 + 6y + 1)$$

$$= 4x^2 - 9y^2 - 6y - 1 \qquad ■$$

F Applications

We have mentioned that the profit P is the revenue R minus the cost C; that is, $P = R - C$. Do you know how revenue is calculated? Suppose you have 10 skateboards and sell them for $50 each. Your revenue is $R = 10 \cdot 50 = \$500$. In general,

$$R = \binom{\text{number of}}{\text{items sold}} \cdot \binom{\text{price of}}{\text{each item}}$$

or $\qquad\qquad R = xp$

EXAMPLE 6 Supply, demand, and skateboards

The research department of a company determines that the number of skateboards that are selling (the demand) is given by $x = 1000 - 10p$, where p is the price of each skateboard.

a. Write a formula for the revenue R.

b. Find the revenue obtained from selling the skateboards for $50 each.

SOLUTION

a. The revenue is

$$R = xp$$

$$R = (1000 - 10p)p \qquad \text{Substitute } x = 1000 - 10p.$$

$$= 1000p - 10p^2$$

b. When $p = 50$,

$$R = 1000(50) - 10(50)^2$$

$$= 50{,}000 - 10(2500)$$

$$= 50{,}000 - 25{,}000$$

$$= 25{,}000$$

The revenue is $25,000. $\qquad$ ■

A grapher is the perfect tool to verify products of polynomials in one variable. The procedure is simple: Graph the original problem as Y_1 and the answer (the product) as Y_2. If the two graphs are identical, you have done the multiplication correctly. Before you do this, however, find out how exponents are entered in your grapher. With a TI-82, press the exponent key $\boxed{\wedge}$ and then enter the exponent. (Some graphers also have x^2 and x^3 keys.) Use a standard $[-10, 10]$ by $[10, 10]$ window (to do this, press $\boxed{\text{ZOOM}}$ $\boxed{6}$ on the TI-82). Now try Example 1a (Window 1). Graph

$$Y_1 = 5x^2(3x^3 + 3x^2 - 2x - 3)$$

and

$$Y_2 = 15x^5 + 15x^4 - 10x^3 - 15x^2$$

(*Hint:* If you don't get the same graph, check your work and make sure you entered Y_1 and Y_2 correctly!)

What happens when you encounter a wrong answer? For example, suppose you obtained $(x + 3)^2 = x^2 + 9$. If you let $Y_1 = (x + 3)^2$ and $Y_2 = x^2 + 9$, you get the two graphs in Window 2, which shows that the multiplication is incorrect!

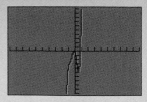

WINDOW 1

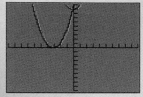

WINDOW 2

EXERCISE 3.2

A In Problems 1–10, do the indicated multiplications.

1. $3x(4x - 2)$ 2. $4x(x - 6)$

3. $-3x^2(x - 3)$ 4. $-5x^3(x^2 - 8)$

5. $-8x(3x^2 - 2x + 1)$ 6. $-4x^2(3x^2 - 5x - 1)$

7. $-3xy^2(6x^2 + 3y^2 - 7)$ 8. $-2x^2y^3(6xy^3 - 2x^2y + 9)$

9. $2xy^3(3x^2y^3 - 5xy^2 + xy)$ 10. $3x^4y(6x^3y^2 - 10x^2y + xy)$

B In Problems 11–20, do the indicated multiplications.

11. $(x + 3)(x^2 + x + 5)$ 12. $(x + 2)(x^2 + 5x + 6)$

13. $(x + 4)(x^2 - x + 3)$ 14. $(x + 5)(x^2 - x + 2)$

15. $x^2 - x - 2$ 16. $x^2 - x - 3$
    $\underline{\quad x + 3}$ $\underline{\quad x + 4}$

17. $(x - 2)(x^2 + 2x + 4)$ 18. $(x - 3)(x^2 + x + 1)$

19. $x^2 - x + 2$ 20. $x^2 - 2x + 1$
    $\underline{\quad x^2 - 1}$ $\underline{\quad x^2 - 2}$

C In Problems 21–38, do the indicated multiplications.

21. $(3x + 2)(3x + 1)$ 22. $(x + 5)(2x + 7)$

23. $(5x - 4)(x + 3)$ 24. $(2x - 1)(x + 5)$

25. $(3a - 1)(a + 5)$ 26. $(3a - 2)(a + 7)$

27. $(y + 5)(2y - 3)$ 28. $(y + 1)(5y - 1)$

29. $(x - 3)(x - 5)$ 30. $(x - 6)(x - 1)$

31. $(2x - 1)(3x - 2)$ 32. $(3x - 5)(x - 1)$

33. $(2x - 3a)(2x + 5a)$ 34. $(5x - 2a)(x + 5a)$

35. $(x + 7)(x + 8)$ 36. $(x + 1)(x + 9)$

37. $(2a + b)(2a + 4b)$ 38. $(3a + 2b)(3a + 5b)$

D **E** In Problems 39–74, use the special products to do the indicated multiplications.

39. $(4u + v)^2$ 40. $(3u + 2v)^2$

41. $(2y + z)^2$ 42. $(4y + 3z)^2$

43. $(3a - b)^2$ 44. $(4a - 3b)^2$

45. $(a + b)(a - b)$ 46. $(a + 4)(a - 4)$

47. $(5x - 2y)(5x + 2y)$ 48. $(2x - 7y)(2x + 7y)$

49. $-(3a - b)(3a + b)$ 50. $-(2a - 5b)(2a + 5b)$

51. $3x(x + 1)(x + 2)$ 52. $3x(x + 2)(x + 3)$

53. $-3x(x - 1)(x - 3)$ 54. $-2x(x - 5)(x - 1)$

55. $x(x + 3)^2$ 56. $3x(x + 7)^2$

57. $-2x(x - 1)^2$ 58. $-5x(x - 3)^2$

59. $(2x + y)(2x - y)y^2$ 60. $(3x + y)(3x - y)x^2$

61. $\left(x + \dfrac{3}{4}\right)^2$ 62. $\left(x + \dfrac{2}{5}\right)^2$

63. $\left(2y - \dfrac{1}{5}\right)^2$ 64. $\left(3y - \dfrac{3}{4}\right)^2$

65. $\left(\dfrac{3}{4}p + \dfrac{1}{5}q\right)^2$ **66.** $\left(\dfrac{2}{5}p + \dfrac{1}{4}q\right)^2$

67. $[(3x + 1) + 4y]^2$ **68.** $[(2x + 1) + 3y]^2$

69. $[(3x - 1) - 4y]^2$ **70.** $[(2x - 1) - 4y]^2$

71. $[2y + (3x - 1)]^2$ **72.** $[3y + (2x - 1)]^2$

73. $[4p - (3q - 1)]^2$ **74.** $[2p - (3q - 1)]^2$

F **Applications. In Problems 75 and 76, $R = xp$, where x is the number of items sold and p is the price of the item.**

75. The demand x for a certain product is given by $x = 1000 - 30p$.
  a. Write a formula for the revenue R.

  b. What is the revenue when the price is $20?

76. A company manufactures and sells x jogging suits at p dollars every day. If $x = 3000 - 30p$, write a formula for the daily revenue R and use it to find the revenue on a day in which the suits were selling for $40.

In Problems 77–79, multiply the expression given.

77. The heat transmission between two objects of temperature T_2 and T_1 involves the expression

$$(T_1^2 + T_2^2)(T_1^2 - T_2^2)$$

where T_1^2 means the square of T_1 and T_2^2 means the square of T_2.

78. The deflection of a certain beam involves the expression

$$w(l^2 - x^2)^2$$

Multiply this expression.

79. The heat output from a natural draught convector is given by

$$K(t_n - t_a)^2$$

Multiply this expression.

SKILL CHECKER

The distributive property can be written as $ab + ac = a(b + c)$. Use this rule to rewrite each expression:

80. $3x + 3y$ **81.** $5x + 5y$

82. $2xz + 2xy$ **83.** $3ab + 3ac$

84. $6bc + 6bd$

USING YOUR KNOWLEDGE   Avoiding Multiplication Mistakes

A common fallacy (mistake) when multiplying binomials is to assume that

$$(x + y)^2 = x^2 + y^2$$

Here are some arguments that should convince you otherwise.

85. Let $x = 1$, $y = 2$.
  a. What is $(x + y)^2$?   **b.** What is $x^2 + y^2$?
  c. Does $(x + y)^2 = x^2 + y^2$?

86. Let $x = 2$, $y = 1$.
  a. What is $(x - y)^2$?   **b.** What is $x^2 - y^2$?
  c. Does $(x - y)^2 = x^2 - y^2$?

87. Look at the large square. Its area is $(x + y)^2$. The square is divided into four smaller areas numbered 1, 2, 3, and 4.
  a. What is the area of square 1?
  b. What is the area of rectangle 2?
  c. What is the area of square 4?
  d. What is the area of rectangle 3?

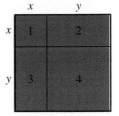

88. The total area of the square is $(x + y)^2$. It's also the sum of the areas of the regions numbered 1, 2, 3, and 4. What is the sum of these four areas? (Simplify your answer.)

89. From your answer to Problem 88, what can you say about $x^2 + 2xy + y^2$ and $(x + y)^2$?

90. If $x^2 + y^2$ is the sum of the areas of squares 1 and 4, does $x^2 + y^2 = (x + y)^2$?

WRITE ON . . .

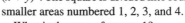

91. Write in your own words the procedure for multiplying two monomials with integer coefficients.

92. Explain why $(x - y)^2$ is different from $x^2 - y^2$.

93. Describe in your own words the procedure for multiplying two polynomials.

MASTERY TEST   If you know how to do these problems, you have learned your lesson!

Multiply:

94. $(5x - y)(2x - 3y)$ **95.** $(4x + 3y)(3x + 2y)$

96. $(3x + 5y)^2$ **97.** $(5x - 3y)^2$

98. $(5x - 3y)(5x + 3y)$ **99.** $(3x + y)(3x - y)$

100. $(2x - 3)(x^2 - 3x + 5)$ **101.** $(x - 2)(x^2 - 4x - 3)$

102. $-3xy^2(x^3 - 7xy - 2x^2)$ **103.** $4x^3(5x^3 + 3x^2 - 2x - 5)$

104. $[3x - (2y - 1)]^2$ **105.** $[(3x + 1) - 4y]^2$

106. The research department of a company determined that the number of skateboards to be sold is given by $x = 1000 - 20p$, where p is the price of each skateboard.
  a. If the revenue is the number of skateboards to be sold times the price of each skateboard, write a formula for the revenue R.
  b. Find the revenue obtained by selling the skateboards for $50 each.

3.3

THE GREATEST COMMON FACTOR AND FACTORING BY GROUPING

To succeed, review how to:

1. Use the distributive property to multiply expressions (pp. 149–151).

2. Use the properties of exponents (pp. 31–37).

Objectives:

A Factor out the greatest common factor in a polynomial.

B Factor a polynomial with four terms by grouping.

getting started   **Factoring with Interest**

If you buy a 1-yr, $1000 certificate of deposit, what amount A would you get at the end of the year? You will get $1000 plus interest. Since the annual interest is $1000 \cdot 0.0834$, the amount will be

$$A = 1000 + 1000 \cdot (0.0834)$$

In general, if you invest P dollars at an annual rate r, the amount A dollars that you get is

$$A = P + Pr$$

The expression $P + Pr$ can be written in a simpler way if we *factor* it, that is, if we write it as the product of its factors. Factoring is the reverse of multiplication. Thus if we tell you to multiply P by $1 + r$, you will write $P(1 + r) = P + Pr$. If we tell you to factor $P + Pr$, you should write

$$P + Pr = P(1 + r) \qquad \text{Use the distributive property to factor } P + Pr.$$

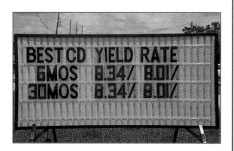

Using the distributive property to **factor** an expression changes a *sum of terms* into a *product of factors.* Here are some examples of factoring. Check the results by multiplying the *factors* on the right-hand side of the equation to obtain the *terms* on the left-hand side.

$$5x + 15 = 5(x + 3)$$
$$6x^2 + 6xy = 6x(x + y)$$
$$-12x^3 - 42x^2y = -6x^2(2x + 7y)$$

A The Greatest Common Factor

To factor the preceding expressions, you must know how to find the **greatest common factor (GCF)** of their terms. In arithmetic, we factor integers into a product of **primes**. (A **prime number** is a number that has exactly two factors: itself and 1.) The prime numbers are 2, 3, 5, 7, 11, 13, 17, and so on. We can write 18 and 12 as a product of primes by using successive divisions by primes:

$$
\begin{array}{cc}
2\,|\,\underline{18} & 2\,|\,\underline{12} \\
3\,|\,\underline{9} & 2\,|\,\underline{6} \\
3\,|\,\underline{3} & 3\,|\,\underline{3} \\
1 & 1
\end{array}
$$

Thus $18 = 2 \cdot 3 \cdot 3 = 2 \cdot 3^2$ and $12 = 2 \cdot 2 \cdot 3 = 2^2 \cdot 3$. To obtain the GCF, we write

Pick the factor with the smallest exponent in each column.

$$18 = \begin{vmatrix} 2 & \cdot 3^2 \\ 2^2 & \cdot 3 \end{vmatrix}$$
$$12 = $$

with all the primes in columns as shown, and then we select the prime with the smallest exponent from each column.

1. In the first column, we have 2 and 2^2; select 2.

2. In the second column, we have 3^2 and 3; select 3.

3. The GCF is the product of the numbers selected: $2 \cdot 3 = 6$.

Note that the common factors in $18 = 2 \cdot 3 \cdot 3$ and $12 = 2 \cdot 2 \cdot 3$ are 2 and 3. Thus the GCF of 18 and 12 is $2 \cdot 3 = 6$. Similarly, the GCF of xy^2 and x^2y is xy, and the GCF of x^3y^5 and x^2y^7 is x^2y^5. In general, we have the following.

GREATEST COMMON FACTOR

The term ax^n is the **greatest common monomial factor** (GCF) of a polynomial in x (with integer coefficients) if

1. a is the *greatest* integer that divides each of the coefficients in the polynomial.

2. n is the *smallest* exponent of x in all the terms of the polynomial.

Thus to factor $12x^3 + 18x^5$, we could write

$$12x^3 + 18x^5 = 2x(6x^2 + 9x^4) \qquad \text{Check by multiplying!}$$

But this is *not* completely factored. The greatest number that divides 12 and 18 is 6, and the power of x with the smallest exponent in all the terms of the polynomial is x^3. Thus the GCF is $6x^3$. The complete factorization is

$$12x^3 + 18x^5 = 6x^3(2 + 3x^2)$$

It may help your accuracy and understanding if you write an intermediate step showing the greatest common factor present in each term. Thus when factoring $12x^3 + 18x^5$, you can write

$$12x^3 + 18x^5 = 6x^3 \cdot 2 + 6x^3 \cdot 3x^2$$
$$= 6x^3(2 + 3x^2)$$

We have one more convention. When factoring expressions such as $-6x + 18$, we have two choices:

$$-6(x - 3) \qquad \text{and} \qquad 6(-x + 3)$$

Both are correct, but the factorization $-6(x - 3)$ is the *preferred* choice because the first term of the binomial $x - 3$ has a positive sign.

EXAMPLE 1   Factoring binomials

Factor:

a. $8x - 24$ 　　　　　　**b.** $-6y^2 + 12y$ 　　　　　**c.** $10x^2 - 25x^3$

SOLUTION

a. $8x - 24 = 8 \cdot x - 8 \cdot 3$ 　　　　　　　　8 is the GCF of 8 and 24.

$\qquad\qquad = 8(x - 3)$

b. $-6y^2 + 12y = -6y \cdot y - 6y \cdot (-2)$ $-6y$ is the GCF of $-6y^2$ and $12y$.

$= -6y(y - 2)$

c. $10x^2 - 25x^3 = 5x^2 \cdot 2 - 5x^2 \cdot 5x$ $5x^2$ is the GCF of $10x^2$ and $25x^3$.

$= 5x^2(2 - 5x)$

Check your answers by multiplying out the expressions to the right of the equal sign. ∎

We can also factor polynomials with more than two terms, as shown in Example 2.

EXAMPLE 2   Factoring polynomials

Factor:

a. $6x^3 + 12x^4 + 18x^2$ **b.** $10x^6 - 15x^5 + 20x^7 + 30x^2$

SOLUTION

a. $6x^3 + 12x^4 + 18x^2 = 6x^2 \cdot x + 6x^2 \cdot 2x^2 + 6x^2 \cdot 3$

$= 6x^2(x + 2x^2 + 3)$

$= 6x^2(2x^2 + x + 3)$

b. $10x^6 - 15x^5 + 20x^7 + 30x^2 = 5x^2 \cdot 2x^4 - 5x^2 \cdot 3x^3 + 5x^2 \cdot 4x^5 + 5x^2 \cdot 6$

$= 5x^2(2x^4 - 3x^3 + 4x^5 + 6)$

$= 5x^2(4x^5 + 2x^4 - 3x^3 + 6)$

Check by multiplying. ∎

EXAMPLE 3   Factoring polynomials with fractional coefficients

Factor: $\dfrac{3}{4}x^2 - \dfrac{1}{4}x^4 + \dfrac{5}{4}x^5$

SOLUTION   Here we have to be careful because the procedure for finding the GCF only works with polynomials with integer coefficients. However, we can see that the greatest common factor is $\frac{1}{4}x^2$. Thus

$$\frac{3}{4}x^2 - \frac{1}{4}x^4 + \frac{5}{4}x^5 = \frac{1}{4}x^2(3 - x^2 + 5x^3)$$

Check by multiplying. ∎

Factoring by Grouping

How do we factor expressions when there seems to be no common factor except 1? We group and factor the first two terms and also the last two terms and then use the distributive property. Here is our three-step procedure.

PROCEDURE

Factoring by Grouping

1. Group terms with common factors using the associative property.

2. Factor each resulting binomial.

3. Factor out the binomial using the GCF, by the distributive property.

Let's factor $x^3 + 2x^2 + 3x + 6$. As you can see, this expression doesn't appear to have a common factor other than 1. We use our three-step procedure:

1. Group terms. $\qquad\qquad\qquad x^3 + 2x^2 + 3x + 6 = (x^3 + 2x^2) + (3x + 6)$

2. Factor binomials. $\qquad\qquad\qquad\qquad\qquad\qquad = x^2(x + 2) + 3(x + 2)$

3. Use GCF to factor out binomials. $\qquad\qquad\qquad = (x + 2)(x^2 + 3)$

Thus $x^3 + 2x^2 + 3x + 6 = (x + 2)(x^2 + 3)$. Note that by the commutative property of multiplication, either $(x + 2)(x^2 + 3)$ or $(x^2 + 3)(x + 2)$ is a correct factorization for $x^3 + 2x^2 + 3x + 6$. Check by multiplying.

EXAMPLE 4 Factoring by grouping

Factor:

a. $3x^3 + 6x^2 + 4x + 8$ $\qquad\qquad\qquad$ **b.** $6x^3 - 3x^2 - 4x + 2$

SOLUTION

a. We proceed by steps.

1. Group terms. $\qquad\qquad 3x^3 + 6x^2 + 4x + 8 = (3x^3 + 6x^2) + (4x + 8)$

2. Factor binomials. $\qquad\qquad\qquad\qquad\qquad\qquad = 3x^2(x + 2) + 4(x + 2)$

3. Use GCF to factor out binomials. $\qquad\qquad\qquad = (x + 2)(3x^2 + 4)$

Note that you can write $3x^3 + 6x^2 + 4x + 8$ as $(3x^2 + 4)(x + 2)$ in step 3. Since $(x + 2)(3x^2 + 4) = (3x^2 + 4)(x + 2)$, both answers are correct! Do you know why?

b. Again, we proceed by steps.

1. Group terms. $\qquad\qquad 6x^3 - 3x^2 - 4x + 2 = (6x^3 - 3x^2) + (-4x + 2)$

2. Factor binomials. $\qquad\qquad\qquad\qquad\qquad\qquad = 3x^2(2x - 1) + (-2)(2x - 1)$

3. Use GCF to factor out binomials. $\qquad\qquad\qquad = (2x - 1)(3x^2 - 2)$

Thus $6x^3 - 3x^2 - 4x + 2 = (2x - 1)(3x^2 - 2)$. We leave the verification to you. ∎

You might save some time if in step 1 you realize that $-4x + 2 = -(4x - 2)$ and write

$$6x^3 - 3x^2 - 4x + 2 = (6x^3 - 3x^2) - (4x - 2)$$

$$= 3x^2(2x - 1) - 2(2x - 1)$$

$$= (2x - 1)(3x^2 - 2)$$

If the third term is preceded by a minus sign, write $-(\quad)$ with the appropriate signs for the expression inside the parentheses.

EXAMPLE 5 Factoring out the common factor

Factor:

a. $2x^4 - 4x^3 - x^2 + 2x$ $\qquad\qquad\qquad$ **b.** $6x^6 - 9x^4 + 4x^3 - 6x$

SOLUTION

a. First, factor out the common factor x and write

$$2x^4 - 4x^3 - x^2 + 2x = x(2x^3 - 4x^2 - x + 2) \qquad \text{Then factor } 2x^3 - 4x^2 - x + 2.$$

We again use our three-step procedure. Can you tell what is happening without the words? If not, check the procedure box on the preceding page.

1. $2x^4 - 4x^3 - x^2 = x[(2x^3 - 4x^2) - (x - 2)]$

2. $= x[2x^2(x - 2) - 1(x - 2)]$

3. $= x[(x - 2)(2x^2 - 1)]$

Thus $2x^4 - 4x^3 - x^2 + 2x = x(x - 2)(2x^2 - 1)$.

b. Factor out x to obtain

$$6x^6 - 9x^4 + 4x^3 - 6x = x(6x^5 - 9x^3 + 4x^2 - 6) \qquad \text{Then factor}$$
$$\hspace{9.5cm} 6x^5 - 9x^3 + 4x^2 - 6.$$

Now we use the procedure.

1. $6x^6 - 9x^4 + 4x^3 - 6x = x[(6x^5 - 9x^3) + (4x^2 - 6)]$

2. $= x[3x^3(2x^2 - 3) + 2(2x^2 - 3)]$

3. $= x(2x^2 - 3)(3x^3 + 2)$

Thus

$$6x^6 - 9x^4 + 4x^3 - 6x = x(2x^2 - 3)(3x^3 + 2)$$
$$\uparrow$$
Don't forget the x!

We leave the verification to you.   ∎

GRAPH IT

In Section 3.2, we used the grapher as a tool to verify that the products we obtained were correct. We can do the same for factorization. Let's look at Example 1c. To verify that $10x^2 - 25x^3 = 5x^2(2 - 5x)$, graph $Y_1 = 10x^2 - 25x^3$ and $Y_2 = 5x^2(2 - 5x)$. If the graphs are identical, the factorization is probably correct. The graphs for Y_1 and Y_2 are shown in Window 1.

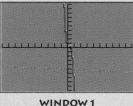

WINDOW 1
$10x^2 - 25x^3 = 5x^2(2 - 5x)$

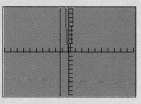

WINDOW 2
$10x^6 - 15x^5 + 20x^7 + 30x^2$

Similarly, we can check our factorization in Example 2b.
Graph $Y_1 = 10x^6 - 15x^5 + 20x^7 + 30x^2$ and $Y_2 = 5x^2(4x^5 + 2x^4 - 3x^3 + 6)$. Window 2 shows the same graph for Y_1 and Y_2. But how do we know that portions of the graph that may not be shown are the same? What we need is a more complete graph. To do this, adjust the vertical scale by selecting Ymin = -100 and Ymax = 100. A more complete graph is shown in Window 3, and Y_1 and Y_2 appear to be identical.

To do Example 4a, let $Y_1 = 3x^3 + 6x^2 + 4x + 8$ and $Y_2 = (x + 2)(3x^2 + 4)$. If we graph Y_1 and Y_2 in the standard window, we get a small portion of the graph as shown in Window 4. But we need a more complete graph, so we change the vertical scale to $[-100, 100]$. Window 5 shows that both graphs are identical and our factorization is probably correct.

WINDOW 3
$10x^6 - 15x^5 + 20x^7 + 30x^2$
using a $[-100, 100]$ vertical scale

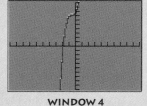

WINDOW 4
$Y_1 = 3x^3 + 6x^2 + 4x + 8$
in the standard window

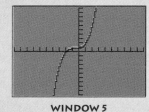

WINDOW 5
$Y_1 = 3x^3 + 6x^2 + 4x + 8$
with a $[-100, 100]$ vertical scale

EXERCISE 3.3

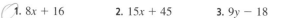

A In Problems 1–26, factor.

1. $8x + 16$

2. $15x + 45$

3. $9y - 18$

7. $-8x - 24$

8. $-6x - 36$

9. $4x^2 + 36x$

4. $11y - 88$

5. $-5y + 25$

6. $-4y + 28$

10. $5x^3 + 20x$

11. $6x - 42x^3$

12. $7x - 14x^5$

13. $-5x^2 - 35x^4$

14. $-8x^3 - 16x^6$

15. $3x^3 + 6x^2 + 39x$

16. $8x^3 + 4x^2 - 36x$

17. $63y^3 - 18y^2 + 27y$

18. $10y^3 - 5y^2 + 20y$

19. $36x^6 + 12x^5 - 18x^4 + 30x^2$

20. $15x^7 - 15x^6 + 30x^3 - 20x^2$

21. $48y^8 + 16y^5 - 24y^4 + 8y^3$

22. $12y^9 - 4y^6 + 6y^5 + 8y^4$

23. $\dfrac{4}{7}x^3 + \dfrac{3}{7}x^2 - \dfrac{9}{7}x + \dfrac{3}{7}$

24. $\dfrac{2}{5}x^3 + \dfrac{3}{5}x^2 - \dfrac{2}{5}x + \dfrac{4}{5}$

25. $\dfrac{7}{8}y^9 + \dfrac{3}{8}y^6 - \dfrac{5}{8}y^4 + \dfrac{5}{8}y^2$

26. $\dfrac{4}{3}y^7 - \dfrac{1}{3}y^5 + \dfrac{2}{3}y^4 - \dfrac{5}{3}y^3$

B **In Problems 27–52, factor by grouping.**

27. $x^3 + 2x^2 + x + 2$

28. $x^3 + 3x^2 + x + 3$

29. $y^3 - 3y^2 + y - 3$

30. $y^3 - 5y^2 + y - 5$ extra H.W

31. $4x^3 + 6x^2 + 2x + 3$

32. $6x^3 + 3x^2 + 2x + 1$

33. $6x^3 - 2x^2 + 3x - 1$

34. $6x^3 - 9x^2 + 2x - 3$

35. $4y^3 + 8y^2 + y + 2$

36. $2y^3 + 6y^2 + y + 3$

37. $2a^6 + 3a^4 + 2a^2 + 3$

38. $3a^6 + 2a^4 + 3a^2 + 2$

39. $3x^5 + 12x^3 + x^2 + 4$

40. $2x^5 + 2x^3 + x^2 + 1$

41. $6y^5 + 9y^3 + 2y^2 + 3$

42. $12y^5 + 8y^3 + 3y^2 + 2$

43. $4y^7 + 12y^5 + y^4 + 3y^2$

44. $2y^7 + 2y^5 + y^4 + y^2$

45. $3a^7 - 6a^5 - 2a^4 + 4a^2$

46. $4a^7 - 12a^5 - 3a^4 + 9a^2$

47. $8a^5 - 12a^4 - 10a^3 + 15a^2$

48. $3y^7 - 21y^4 + y^5 - 7y^2$

49. $x^6 - 2x^5 + 2x^4 - 4x^3$

50. $v^6 + 3v^5 - 7v^4 - 21v^3$

51. $(x - 4)(x + 2) + (x - 4)(x + 3)$

52. $(y + 2)(y - 7) + (2y + 3)(y - 7)$

53. Factor $\alpha L t_2 - \alpha L t_1$, where α is the coefficient of linear expansion, L is the length of the material, and t_2 and t_1 are the temperatures in degrees Celsius.

54. Factor the expression $-kx - kl$, which represents the restoring force of a spring stretched an amount l to its equilibrium position and then an additional x units.

55. When solving for the equivalent resistance of two circuits, we have to factor the expression $R^2 - R - R + 1$. Factor this expression by grouping.

56. The bending moment of a cantilever beam of length L, at x inches from its support, involves the expression $L^2 - Lx - Lx + x^2$. Factor this expression by grouping.

SKILL CHECKER

Multiply:

57. $(x + 3)(x + 4)$

58. $(x + 7)(x + 2)$

59. $(x + 5)(x - 2)$

60. $(x - 5)(x + 3)$

61. $(5x + 2y)^2$

62. $(3x + 4y)^2$

63. $(5x - 2y)^2$

64. $(3x - 4y)^2$

65. $(u + 6)(u - 6)$

66. $(2a + 7b)(2a - 7b)$

USING YOUR KNOWLEDGE   **General Formulas**

There are many formulas that can be simplified by factoring. Here are a few.

67. The vertical shear at any section of a cantilever beam of uniform cross section is

$$-wl + wz$$

Factor this expression.

68. The bending moment of any section of a cantilever beam of uniform cross section is

$$-Pl + Px$$

Factor this expression.

69. The surface area of a square pyramid is

$$a^2 + 2as$$

Factor this expression.

70. The energy of a moving object is given by

$$800m - mv^2$$

Factor this expression.

71. The height (in feet after t seconds) of a rock thrown from the roof of a certain building is given by

$$-16t^2 + 80t + 240$$

Factor this expression. (*Hint:* -16 is a common factor.)

72. What should the first step be in factoring a polynomial?

73. Write your own definition for the greatest common factor of a list of integers.

74. The procedure we outlined to find the greatest common monomial factor of a polynomial doesn't apply to Example 3. Explain why.

75. Describe the relationship between multiplying and factoring a polynomial.

If you know how to do these problems, you have learned your lesson!

Factor:

76. $3x^4 - 6x^3 - x^2 + 2x$

77. $6x^6 - 9x^4 + 2x^3 - 3x$

78. $2x^3 + 2x^2 + 3x + 3$

79. $6x^3 - 9x^2 - 2x + 3$

80. $7x^3 + 14x^4 - 49x^2$

81. $3x^6 - 6x^5 + 12x^7 + 27x^2$

82. $4x^2 - 32x^3$

83. $6x - 48$

84. $-3y^2 + 21y$

85. $\dfrac{2}{5}x^2 - \dfrac{4}{5}x^4 - \dfrac{1}{5}x^5$

3.4

FACTORING TRINOMIALS

To succeed, review how to:

1. Multiply two binomials using the FOIL method (pp. 151–153).

2. Add and multiply signed numbers (pp. 15–19).

Objectives:

A Factor a trinomial of the form $x^2 + bx + c$ (b and c are integers).

B Factor a trinomial of the form $ax^2 + bx + c$ using trial and error.

C Factor a trinomial of the form $ax^2 + bx + c$ using the ac test.

getting started

Applying Factoring to Fire Fighting

How many hundred gallons of water per minute can this fire truck pump? If it has a hose 100 ft long, you can find the answer by solving $2g^2 + g - 36 = 0$. The expression $2g^2 + g - 36$ is factorable. How do we factor it? By using reverse multiplication, but let's start with a simpler problem.

 Factoring Trinomials of the Form $x^2 + bx + c$ (b and c integers)

In Section 3.2, we multiplied two binomials using the FOIL method:

$$\overset{\text{F} \quad \text{O} \quad \text{I} \quad \text{L}}{(x + 4)(x - 7) = x^2 - 7x + 4x - 28}$$

and

$$(x + 7)(x - 4) = x^2 - 4x + 7x - 28$$

To factor $x^2 - 3x - 28$, we recall that

$$(x + a)(x + b) = x^2 + bx + ax + ab$$
$$= x^2 + (b + a)x + ab$$

Rewriting this trinomial, we have the F-1 factoring form.

PROCEDURE

> **Factoring Trinomials of the Form $x^2 + (b + a)x + ab$**
>
> **F-1** $\quad x^2 + (b + a)x + ab = (x + a)(x + b)$

This means that to factor a trinomial with a leading coefficient of 1, we need two integers a and b whose product is the last term and whose sum is the coefficient of the middle term. Since

$$x^2 + (b + a)x + ab = (x + a)(x + b)$$

we write

$$x^2 - \underbrace{3x} - \underbrace{28} = (x + a)(x + b)$$

We need two integers a and b whose product ab is -28 and whose sum is -3. Since the product is negative, one number must be positive and the other negative, with the larger number being negative, since the sum is -3. The numbers are -7 and 4. [*Check:* $-7 \cdot 4 = -28$ and $-7 + 4 = -3$, the coefficient of the middle term.] Thus

$$x^2 - 3x - 28 = (x - 7)(x + 4)$$

Since the multiplication of polynomials is commutative,

$$x^2 - 3x - 28 = (x + 4)(x - 7)$$

Now suppose we want to factor $x^2 - 8x + 12$. This time, we need two integers whose product is 12 and whose sum is -8 (the coefficient of the middle term). The numbers are -6 and -2. [*Check:* $(-6)(-2) = 12$ and $(-6) + (-2) = -8$.] Thus

$$x^2 - 8x + 12 = (x - 6)(x - 2)$$

You can check this by multiplying $(x - 6)$ by $(x - 2)$.

EXAMPLE 1   Factoring trinomials with one variable

Factor:

a. $x^2 + 7x + 12$ **b.** $x^2 - 6x + 8$ **c.** $x^2 - 4 - 3x$

SOLUTION

a. To factor $x^2 + 7x + 12$, we need two integers whose product is 12 and whose sum is 7. The numbers are 3 and 4. Thus

$$x^2 + 7x + 12 = (x + 3)(x + 4)$$

b. To factor $x^2 - 6x + 8$, we need two integers whose product is 8 and whose sum is -6. Since the product is positive, both numbers must be negative. They are -2 and -4. [*Check:* $(-2)(-4) = 8$ and $(-2) + (-4) = -6$.] Thus

$$x^2 - 6x + 8 = (x - 2)(x - 4)$$

c. First, rewrite $x^2 - 4 - 3x$ in descending order as $x^2 - 3x - 4$. This time, we need integers with product -4 and sum -3. The numbers are 1 and -4. Thus

$$x^2 - 4 - 3x = x^2 - 3x - 4 = (x + 1)(x - 4)$$

You can check all these results by multiplying. For example, $(x + 1)(x - 4) = x^2 - 4x + x - 4 = x^2 - 3x - 4$. ∎

EXAMPLE 2   Factoring trinomials with two variables

Factor:

a. $x^2 - 8xy + 7y^2$ **b.** $x^2 + 3xy + 7y^2$

SOLUTION

a. We need two integers whose product is 7 and whose sum is -8. Since the product is positive, both numbers must be negative. The numbers are -1 and -7. Thus

$$x^2 - 8xy + 7y^2 = (x - 1y)(x - 7y)$$

$$= (x - y)(x - 7y) \qquad \text{Check by multiplication.}$$

b. We need two integers whose product is 7 and whose sum is 3. There are no such numbers. The polynomial $x^2 + 3xy + 7y^2$ is not factorable using integer coefficients. A polynomial that is not factorable using integer coefficients is called a **prime polynomial**. ∎

B   Factoring Trinomials of the Form $ax^2 + bx + c$

To factor the polynomial $2g^2 + g - 36$ mentioned in the *Getting Started* at the beginning of this section, we can rely on our experience with FOIL. To obtain $2g^2 + 1g - 36$, we multiply

$$(2g + \underline{})(g - \underline{})$$

We need to fill the blanks with two integers that have a product of -36 and that give a middle term of $1g$. The possibilities are as follows.

Trial Factors	Middle Term
$(2g + \underline{1})(g - \underline{36})$	$-71g$
$(2g + \underline{2})(g - \underline{18})$	$-34g$
Note that $(2g + 2) = 2(g + 1)$, so we won't use even numbers in both terms of a factor.	
$(2g + \underline{3})(g - \underline{12})$	$-21g$
$(2g + \underline{9})(g - \underline{4})$	$1g$

Thus $2g^2 + g - 36 = (2g + 9)(g - 4)$.

EXAMPLE 3   Factoring trinomials of the form $ax^2 + bx + c$

Factor: $6x^2 + 17x + 12$

SOLUTION   The factors of 6 are 6 and 1 or 3 and 2. The possible combinations are

$$(6x + \underline{})(x + \underline{}) \qquad \text{or} \qquad (3x + \underline{})(2x + \underline{})$$

Trial Factors	
$(6x + 12)(x + 1)$	$(6x + 1)(x + 12)$
$(6x + 6)(x + 2)$	$(6x + 2)(x + 6)$
$(6x + 4)(x + 3)$	$(6x + 3)(x + 4)$
$(3x + 12)(2x + 1)$	$(3x + 1)(2x + 12)$
$(3x + 6)(2x + 2)$	$(3x + 2)(2x + 6)$
$(3x + 4)(2x + 3)$	$(3x + 3)(2x + 4)$
$(3x + 3)(2x + 4)$	$(3x + 4)(2x + 3)$

The only combination yielding the correct middle term is

$$(3x + 4)(2x + 3) = 6x^2 + 9x + 8x + 12 = 6x^2 + 17x + 12 \qquad ■$$

The *ac* Test

The method used in Example 3 is not efficient when the coefficients are large. Moreover, we don't even know whether the polynomial is factorable. We remedy this situation with the following test.

THE *ac* TEST

> The polynomial $ax^2 + bx + c$ is factorable only if there are two integers whose product is ac and whose sum is b.

Thus to find if $3x^2 + 2x + 5$ is factorable, we first look at $3 \cdot 5 = 15$. If we can find two integers whose product is 15 and whose sum is 2, we can factor the polynomial. No such integers exist, so $3x^2 + 2x + 5$ is *prime*. On the other hand, the polynomial $5x^2 + 11x + 2$ is factorable since we can find two integers (10 and 1) whose product is $5 \cdot 2 = 10$ and whose sum is 11. The number ac plays such an important part in the procedure used to factor $ax^2 + bx + c$ that we call it the **key number**.

We now give you a method that uses the key number to factor polynomials. For example, suppose we want to factor $2x^2 - 7x - 4$. We proceed as follows.

1. Find the key number $2 \cdot (-4) = -8$. $2x^2 - 7x - 4$ $\boxed{-8}$

2. Find the factors of the key number and use the appropriate ones to rewrite the middle term. $2x^2 - 8x + 1x - 4$ $-8, 1$

3. Group the terms into pairs (as in Section 3.3). $(2x^2 - 8x) + (1x - 4)$

4. Factor each pair. $2x(x - 4) + 1(x - 4)$

5. Note that $(x - 4)$ is the GCF. $(x - 4)(2x + 1)$

Thus $2x^2 - 7x - 4 = (x - 4)(2x + 1)$. You can check that this is the correct factorization by multiplying $(x - 4)$ by $(2x + 1)$.

Suppose you want to factor the trinomial $5x^2 + 7x + 2$. Here is one way of doing it.

1. Find the key number $(5 \cdot 2 = 10)$. $5x^2 + \underline{7x} + 2$ ⑩

2. Find the factors of the key number and use them to rewrite the middle term. $5x^2 + \underline{5x + 2x} + 2$ 5, 2

3. Group the terms into pairs. $(5x^2 + 5x) + (2x + 2)$

4. Factor each pair. $5x(x + 1) + 2(x + 1)$

5. Note that $(x + 1)$ is the GCF. $(x + 1)(5x + 2)$

Thus $5x^2 + 7x + 2 = (x + 1)(5x + 2)$.

Here is another way to proceed.

1. Find the key number. $5x^2 + \underline{7x} + 2$ ⑩

2. Find the factors of the key number and use them to rewrite the middle term. $5x^2 + \underline{2x + 5x} + 2$ 2, 5

3. Group the terms into pairs. $(5x^2 + 2x) + (5x + 2)$

4. Factor each pair. $x(5x + 2) + 1(5x + 2)$

5. Note that $(5x + 2)$ is the GCF. $(5x + 2)(x + 1)$

In this case, we found that

$$5x^2 + 7x + 2 = (5x + 2)(x + 1)$$

Which is the correct factorization, $(5x + 2)(x + 1)$ or $(x + 1)(5x + 2)$? Both are correct! The multiplication of polynomials is commutative, and the order in which the product is written makes no difference. You can write the factorization of $ax^2 + bx + c$ in *two* ways.

EXAMPLE 4   Factoring trinomials using the key number

Factor:

a. $6x^2 - 3x + 4$ **b.** $4x^2 - 3 - 4x$

SOLUTION

a. We proceed by steps:

1. Find the key number $(6 \cdot 4 = 24)$. $6x^2 - 3x + 4$ ㉔

2. Find the factors of the key numbers and use them to rewrite the middle term. Unfortunately, it is impossible to find two integers with product 24 and sum -3. This trinomial is *not* factorable.

b. We first rewrite the polynomial (in descending order) as $4x^2 - 4x - 3$ and then proceed by steps.

1. Find the key number $[4 \cdot (-3) = -12]$. $4x^2 - \underline{4x} - 3$ ⊝12

2. Find the factors of the key number and use them to rewrite the middle term. $4x^2 \underline{- 6x + 2x} - 3$ $-6, 2$

3. Group the terms into pairs. $(4x^2 - 6x) + (2x - 3)$

4. Factor each pair. $2x(2x - 3) + 1(2x - 3)$

5. Note that $(2x - 3)$ is the GCF. $(2x - 3)(2x + 1)$

Thus $4x^2 - 4x - 3 = (2x - 3)(2x + 1)$, as can be easily verified by multiplication.   ■

EXAMPLE 5   Factoring trinomials with two variables using the key number

Factor: $6x^2 + xy - y^2$.

SOLUTION

1. Find the key number
   $[6 \cdot (-1) = -6]$. $6x^2 + \underline{xy} - y^2$ $\widehat{-6}$

2. Find the factors of the key
   number and use them to
   rewrite the middle term. $6x^2 + \underline{3xy - 2xy} - y^2$ $3, -2$

3. Group the terms into pairs. $(6x^2 + 3xy) - (2xy + y^2)$ Note that
                                                                             $-(2xy + y^2) =$
4. Factor each pair. $3x(2x + y) - y(2x + y)$ $-2xy - y^2$.

5. Note that $(2x + y)$ is
   the GCF. $(2x + y)(3x - y)$

Thus $6x^2 + xy - y^2 = (2x + y)(3x - y)$.   ■

If the terms of the trinomial have a common factor, we factor it out first, as in the next example.

EXAMPLE 6   Factoring out common factors

Factor: $12x^3y^2 + 14x^2y^3 - 6xy^4$

SOLUTION   The greatest common factor of these three terms is $2xy^2$.
Thus $12x^3y^2 + 14x^2y^3 - 6xy^4 = 2xy^2(6x^2 + 7xy - 3y^2)$. We then factor
$6x^2 + 7xy - 3y^2$.

1. The key number is -18. $6x^2 + \underline{7xy} - 3y^2$

2. The factors of -18 with a sum of
   7 are 9 and -2. Rewrite the middle term. $6x^2 + \underline{9xy - 2xy} - 3y^2$

3. Group the terms into pairs. $(6x^2 + 9xy) - (2xy + 3y^2)$

4. Factor each pair. $3x(2x + 3y) - y(2x + 3y)$

5. Note that $(2x + 3y)$ is the GCF. $(2x + 3y)(3x - y)$

Thus

$$12x^3y^2 + 14x^2y^3 - 6xy^4 = 2xy^2(6x^2 + 7xy - 3y^2)$$

$$= 2xy^2(2x + 3y)(3x - y)$$

You can check this by multiplying all the factors on the right-hand side of the equation.   ■

We have factored polynomials of the form $ax^2 + bx + c$, where $a > 0$. If a is negative, it's helpful to factor out -1 first and proceed as before.

EXAMPLE 7   Factoring $ax^2 + bx + c$ where a is negative

Factor: $-2x^2 - 7x - 5$

SOLUTION

$$-2x^2 - 7x - 5 = -1(2x^2 + 7x + 5) \qquad \text{Factor out } -1.$$

1. Find the key number $[2 \cdot (5) = 10]$. $2x^2 + \underline{7x} + 5$

2. Find the factors of the key number
(5 and 2) that add up to the middle
term and rewrite the middle term. $-1[2x^2 + \underline{5x + 2x} + 5]$

3. Group the terms into pairs. $-1[(2x^2 + 5x) + (2x + 5)]$

4. Factor each pair. $-1[x(2x + 5) + 1(2x + 5)]$

5. Note that $(2x + 5)$ is the GCF. $-1[(2x + 5)(x + 1)]$

Since $-1 \cdot a = -a$,

$$2x^2 + 7x + 5 = -(2x + 5)(x + 1) \qquad \blacksquare$$

Finally, some polynomials that look complicated can be factored if we make a substitution. For example, to factor $6(y - 1)^2 - 5(y - 1) - 4$, we can think of $(y - 1)$ as A, and write $6A^2 - 5A - 4$, factor this polynomial, and substitute $(y - 1)$ for A in the final answer. We illustrate the procedure in the next example.

EXAMPLE 8   Factoring by substitution

Factor: $6(y - 1)^2 - 5(y - 1) - 4$

SOLUTION   Let $A = (y - 1)$; that is, $6(y - 1)^2 - 5(y - 1) - 4 = 6A^2 - 5A - 4$.

1. The key number is $6(-4) = -24$. $6A^2 - \underline{5A} - 4$

2. The factors of -24 adding to -5
are -8 and 3. $6A^2 - \underline{8A + 3A} - 4$

3. Group the terms into pairs. $(6A^2 - 8A) + (3A - 4)$

4. Factor each pair. $2A(3A - 4) + 1(3A - 4)$

5. Factor out the GCF, $(3A - 4)$. $(3A - 4)(2A + 1)$

Substitute $(y - 1)$ for A and simplify:

$$6A^2 - 5A - 4 = [3(y - 1) - 4][2(y - 1) + 1]$$
$$= [3y - 3 - 4][2y - 2 + 1]$$
$$= (3y - 7)(2y - 1) \qquad \blacksquare$$

GRAPH IT

In the preceding sections we used the grapher as a *tool* to verify multiplication or factoring results. Now we use it as a *resource* to tell us how to factor expressions.

In Example 1, note where each polynomial cuts the horizontal axis and the factorization of the polynomial. We graphed $x^2 + 7x + 12$ using a $[-5, 5]$ by $[-5, 5]$ viewing window (Window 1). Now try $x^2 - 6x + 8$ and $x^2 - 4 - 3x$. Do you see the pattern?

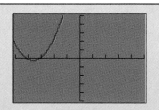

WINDOW 1
$x^2 + 7x + 12$ cuts the x-axis
at $x = -3$ and $x = -4$.

(continued on facing page)

Polynomial	Cuts the Horizontal Axis at the Points	Factorization
$x^2 + 7x + 12$	$x = -3, x = -4$	$(x + 3)(x + 4)$
$x^2 - 6x + 8$	$x = 2, x = 4$	$(x - 2)(x - 4)$
$x^2 - 4 - 3x$	$x = -1, x = 4$	$(x + 1)(x - 4)$
$x^2 + bx + c$	$x = p, x = q$	$(x - p)(x - q)$

Do you see why this works? The points at which each of the polynomials cuts the horizontal axis make $Y_1 = 0$. For example, the points at which $Y_1 = x^2 + 7x + 12$ cut the horizontal axis ($x = -3$ and $x = -4$) should make $Y_1 = 0$, and indeed they do because if $x = -3$, $x + 3 = 0$, and if $x = -4$, $x + 4 = 0$. Consequently, $(x + 3)(x + 4) = 0$. But $Y_1 = x^2 + 7x + 12 = (x + 3)(x + 4) = 0$.

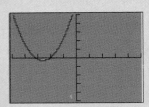

The graph of $x^2 + 5x + 6$ is shown in Window 2. It cuts the horizontal axis at $x = -3$ and $x = -2$. Based on this, how would you factor $x^2 + 5x + 6$?

WINDOW 2
$x^2 + 5x + 6$ cuts the x-axis at $x = -3$ and $x = -2$.

If the graph of a polynomial **does not** cut the horizontal axis, the polynomial is **not** factorable. Graph Example 4a (Window 3). The graph of $6x^2 - 3x + 4$ *does not* cross the horizontal axis; hence it is not factorable!

Now try $4x^2 - 3 - 4x$ (Window 4). It cuts the horizontal axis at $x = -\frac{1}{2}$ and $x = \frac{3}{2}$. Let's "reconstruct" the polynomial. When $x = \frac{3}{2}$, then $2x = 3$ and $2x - 3 = 0$. When $x = -\frac{1}{2}$, then $2x = -1$ and $2x + 1 = 0$. Thus

$$(2x - 3)(2x + 1) = 0,$$

which suggests the correct factorization of $4x^2 - 3 - 4x$ is $(2x - 3)(2x + 1)$ as shown in Example 4b. Can you think of a polynomial that cuts the horizontal axis and is not factorable?

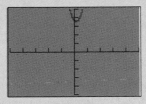

WINDOW 3
$6x^2 - 3x + 4$ does not cross the x-axis and is not factorable.

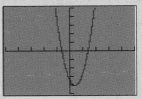

WINDOW 4
$4x^2 - 3 - 4x$ cuts the x-axis at $x = \frac{3}{2}$ and $x = -\frac{1}{2}$.

EXERCISE 3.4

A In Problems 1–16, factor the expressions.

1. $x^2 + 5x + 6$
2. $x^2 + 15x + 56$
3. $a^2 + 7a + 10$
4. $a^2 + 10a + 24$
5. $x^2 + x - 12$
6. $x^2 + 5x - 6$
7. $x^2 - 2 + x$
8. $x^2 - 18 + 7x$
9. $x^2 - x - 2$
10. $x^2 - 5x - 14$
11. $x^2 - 3x - 10$
12. $x^2 - 4x - 21$
13. $a^2 - 16a + 63$
14. $a^2 - 4a + 3$
15. $y^2 + 22 - 13y$
16. $y^2 + 11 - 12y$

A B In Problems 17–46, factor the expressions if possible.

17. $9x^2 + 37x + 4$
18. $2x^2 + 5x + 2$
19. $3a^2 - 5a - 2$
20. $8a^2 - 2a - 21$
21. $2y^2 - 3y - 20$
22. $6y^2 - 13y - 5$
23. $4x^2 - 11x + 6$
24. $16x^2 - 16x + 3$
25. $6x^2 + x - 12$
26. $20y^2 + y - 1$
27. $21a^2 + 11a - 2$
28. $18x^2 - 3x - 10$
29. $6x^2 + 7xy - 3y^2$
30. $3x^2 + 13xy - 10y^2$
31. $7x^4 - 10x^3y + 3x^2y^2$
32. $6x^4 - 17x^3y + 5x^2y^2$

33. $15x^2y^3 - xy^4 - 2y^5$

34. $5x^2y^3 - 6xy^4 - 8y^5$

35. $15x^3y^2 - 2x^2y^3 - 2xy^4$

36. $4x^3y^2 - 13x^2y^3 - 3xy^4$

37. $-2b^2 + 13b - 20$

38. $-3b^2 - 16b + 12$

39. $-12y^2 - 7y + 12$

40. $-12y^2 - 8y + 15$

41. $2(y + 2)^2 + (y + 2) - 3$

42. $3(y + 3)^2 - 11(y + 3) + 6$

43. $2(x + 1)^2 - 13(x + 1) + 20$

44. $3(u - 1)^2 + 16(u - 1) - 12$

45. $-(a^2 + 2a)^2 - 2(a^2 + 2a) - 1$

46. $-(y^2 - 6y)^2 - 18(y^2 - 6y) - 81$

In Problems 47–50, factor the given expression.

47. The flow g (in hundreds of gallons per minute) in 100 ft of $2\frac{1}{2}$-in. rubber-lined hose when the friction loss is 21 lb/in.^{2} is given by

$$2g^2 + g - 21$$

48. The flow g (in hundreds of gallons per minute) in 100 ft of $2\frac{1}{2}$-in. rubber-lined hose when the friction loss is 55 lb/in.^{2} is given by

$$2g^2 + g - 55$$

49. The equivalent resistance R of two electric circuits is given by

$$2R^2 - 3R + 1$$

50. The time t at which an object thrown upward at 12 m/sec will be 4 m above the ground is given by

$$5t^2 - 12t + 4$$

SKILL CHECKER

Multiply:

51. $(2a + b)^2$　　　52. $(3a + 2b)^2$

53. $(a - 2b)^2$　　　54. $(2a - 3b)^2$

55. $(a + b)(a - b)$　　56. $(a - 2b)(a + 2b)$

57. $(2x - 3y)(2x + 3y)$　58. $(5x + 7y)(5x - 7y)$

USING YOUR KNOWLEDGE　　Factoring Applications

The ideas presented in this section are important in many fields. Use your knowledge to factor the expressions given in Problems 59 and 60.

59. The deflection of a beam of length L at a distance of 3 ft from its end is given by

$$2L^2 - 9L + 9$$

60. If the distance from the end of the beam in Problem 59 is x feet, then the deflection is given by

$$2L^2 - 3xL + x^2$$

61. The distance (in meters) traveled in t seconds by an object thrown upward at 12 m/sec is

$$-5t^2 + 12t$$

To determine the time at which the object will be 7 m above ground, we must solve the equation

$$5t^2 - 12t + 7 = 0$$

Factor the trinomial on the left-hand side of this equation.

WRITE ON . . .

62. The factors of a polynomial are $3x^3$, $x + 5$, and $2x - 1$. Write a procedure that can be used to find the polynomial.

63. If you multiply $(2x - 2)(x - 1)$, you get $2x^2 - 4x + 2$. However, $(2x - 2)(x - 1)$ is *not* the complete factorization for $2x^2 - 4x + 2$. Why not?

64. Without using trial and error, explain why the polynomial $x^2 + 3x - 17$ is prime.

MASTERY TEST　　If you know how to do these problems, you have learned your lesson!

Factor completely if possible:

65. $6x^2 + 13x + 6$　　　66. $6x^2 - 17x + 5$

67. $x^2 - 3x - 10$　　　68. $x^2 - 5x - 6$

69. $x^2 - 2xy + 5y^2$　　70. $x^2 - 7xy + 10y^2$

71. $12x^4y + 2x^3y^2 - 4x^2y^3$　72. $2x^2 + xy - 3y^2$

73. $5x^2 - 2x + 2$　　　74. $3x^2 - 4 - 4x$

75. $-8x^2 + 2x + 21$　　76. $-21x^2 - 11x + 2$

77. $2(y - 3)^2 + 5(y - 3) + 2$

78. $-9(y - 2)^2 - 37(y - 2) - 4$

3.5

SPECIAL FACTORING

To succeed, review how to:

1. Square a binomial (p. 153).

2. Multiply a binomial sum by a binomial difference (p. 154).

3. Multiply a binomial and a trinomial (p. 150).

Objectives:

A Factor a perfect square trinomial.

B Factor the difference of two squares.

C Factor the sum or difference of two cubes.

getting started   Applying Factoring to Bending Moments

At x feet from its support, the bending moment (the product of a quantity and the distance from its perpendicular axis) for the crane in the photograph involves the expression

$$\frac{w}{2}(x^2 - 20x + 100)$$

where w is the weight of the crane in pounds per foot. The expression $x^2 - 20x + 100$ is the result of expanding the binomial $(x - 10)^2$ and is called a *perfect square trinomial*. We learned that

$$(a + b)^2 = a^2 + 2ab + b^2 \qquad \text{and} \qquad (a - b)^2 = a^2 - 2ab + b^2$$
$$\text{Perfect square trinomial} \qquad\qquad\qquad \text{Perfect square trinomial}$$

The trinomials on the right-hand side of the equations are perfect square trinomials.

The expression $x^2 - 20x + 100$ is also a perfect square trinomial. Let's factor it using the fact that $(a - b)^2 = a^2 - 2ab + b^2$. Write $x^2 - 20x + 100$ as $x^2 - 2 \cdot x \cdot 10 + 10^2$, where $a = x$ and $b = 10$. We then have $x^2 - 2 \cdot x \cdot 10 + 10^2 = (x - 10)^2$.

A   Factoring Perfect Square Trinomials

To factor perfect square trinomials, we rewrite $(x + a)^2 = x^2 + 2ax + a^2$ and $(x - a)^2 = x^2 - 2ax + a^2$ as follows.

PROCEDURE

> **Factoring Perfect Square Trinomials**
>
> **F-2** $x^2 + 2ax + a^2 = (x + a)^2$
> **F-3** $x^2 - 2ax + a^2 = (x - a)^2$

In a perfect square trinomial the following applies:

1. The first and last terms (x^2 and a^2) are perfect squares.

2. There are no minus signs before x^2 or a^2.

3. The middle term is twice the product of the two terms in the binomial being squared ($2ax$), or it is the additive inverse of this product ($-2ax$).

If you can write the trinomial in the form shown on the left of F-2, then you may factor it as shown on the right. Thus to factor $x^2 + 6x + 9$, note that the first and last terms are perfect squares [$(x)^2 = x^2$ and $3^2 = 9$], there are no minus signs before x^2 or 9, and the middle term is $2 \cdot 3 \cdot x = 6x$. Hence

$$x^2 + 6x + 9 = x^2 + 2 \cdot 3x + 3^2 = (x + 3)^2$$

Similarly, $x^2 - 8x + 16$ is a perfect square trinomial whose middle term is the negative of twice the product of the square root of the first and last terms; that is, $-2 \cdot 4 \cdot x = -8x$. Thus

$$x^2 - 8x + 4^2 = x^2 - 2 \cdot 4 \cdot x + 4^2 = (x - 4)^2$$

You can also factor $x^2 - 8x + 16$ by finding two factors whose product is 16 and whose sum is -8, as we did in the previous section. Of course, the factors are -4 and -4. Thus $x^2 - 8x + 16 = (x - 4)(x - 4) = (x - 4)^2$.

EXAMPLE 1   Factoring perfect square trinomials in one variable

Factor if possible:

a. $x^2 - 10x + 25$ **b.** $x^2 + 12x + 36$ **c.** $x^2 + 7x + 49$

SOLUTION

a. $x^2 - 10x + 25 = x^2 - 2 \cdot 5 \cdot x + 5^2 = (x - 5)^2$

b. $x^2 + 12x + 36 = x^2 + 2 \cdot 6 \cdot x + 6^2 = (x + 6)^2$

You can verify that parts a and b are correct by expanding $(x - 5)^2$ and $(x + 6)^2$.

c. $x^2 + 7x + 49$ has perfect squares for the first (x^2) and last (7^2) terms. However, the middle term is *not* $2 \cdot 7 \cdot x$. Thus $x^2 + 7x + 49$ is not a perfect square; it isn't even factorable. (We cannot find two integers whose product is 49 and whose sum is 7, so, by the *ac* test, $x^2 + 7x + 49$ is prime.) ∎

We can use the same idea to factor trinomials in two variables. Thus to factor $25x^2 + 20xy + 4y^2$, we write

$$25x^2 + 20xy + 4y^2 = (5x)^2 + 2 \cdot (2y)(5x) + (2y)^2$$
$$= (5x + 2y)^2$$

EXAMPLE 2   Factoring perfect square trinomials in two variables

Factor if possible:

a. $9x^2 - 12xy + 4y^2$ **b.** $4x^2 - 10xy + 25y^2$ **c.** $16x^2 + 24xy + 9y^2$

SOLUTION

a. $9x^2 - 12xy + 4y^2 = (3x)^2 - 2 \cdot 3x \cdot 2y + (2y)^2$
$$= (3x - 2y)^2$$

b. Even though the first term, $(2x)^2$, and last term, $(5y)^2$, are perfect squares, the middle term is *not* $2 \cdot 2x \cdot 5y$. Thus $4x^2 - 10xy + 25y^2$ is not a perfect square. The *ac* test shows that it is prime.

c. $16x^2 + 24xy + 9y^2 = (4x)^2 + 2 \cdot 4x \cdot 3y + (3y)^2$
$$= (4x + 3y)^2$$ ∎

Ⓑ Factoring the Difference of Two Squares

As you recall from Section 3.2,

$$(x + a)(x - a) = x^2 - a^2$$

where $x^2 - a^2$ is called the **difference of two squares**. The corresponding factoring formula is as follows.

PROCEDURE

> **Factoring the Difference of Two Squares**
>
> F-4 $\quad x^2 - a^2 = (x + a)(x - a)$

 $x^2 + a^2$ is not factorable using real numbers.

Thus, to factor $x^2 - 9$, we write

$$x^2 - 9 = x^2 - 3^2 = (x + 3)(x - 3)$$

Similarly,

$$25x^2 - 16y^2 = (5x)^2 - (4y)^2 = (5x + 4y)(5x - 4y)$$

EXAMPLE 3   Factoring the difference of two squares

Factor:

a. $9x^2 - 1$ 　　　　　　　　　　　　　**b.** $81x^4 - 16y^4$

SOLUTION

a. $9x^2 - 1 = (3x)^2 - 1^2 = (3x + 1)(3x - 1)$

b. $81x^4 - 16y^4 = (9x^2)^2 - (4y^2)^2$ 　　　　　　　　Difference of two squares

$$= (9x^2 + 4y^2)\overline{(9x^2 - 4y^2)}$$

$$= \underbrace{(9x^2 + 4y^2)}_{\text{Not factorable}}\underbrace{(3x + 2y)(3x - 2y)}_{\text{Factored}}$$　■

The polynomial $(x + y)^2 - 4$ is also the difference of two squares. If you think of $(x + y)$ as A and 4 as 2^2, you are factoring

$$A^2 - 2^2 = (A + 2)(A - 2)$$

or equivalently

$$(x + y)^2 - 2^2 = (x + y + 2)(x + y - 2)$$

If $(x + y)^2$ appears as $x^2 + 2xy + y^2$, factor it first. The procedure is

$$x^2 + 2xy + y^2 - 4 \qquad \text{Given.}$$

$$= (x + y)^2 - 2^2 \qquad \text{Factor the trinomial.}$$

$$= (x + y + 2)(x + y - 2) \qquad \text{Factor the difference.}$$

EXAMPLE 4   Factoring by grouping

Factor:

a. $x^2 - 6x + 9 - y^2$ 　　　　　　　　**b.** $x^2 + 6xy + 9y^2 - 4$

SOLUTION

a. Note that if you try to group the terms as $(x^2 - 6x) + (9 - y^2)$ and factor each binomial, you get $x(x - 6) + (3 + y)(3 - y)$ but there is no GCF. However, the first three terms are a perfect square trinomial. Thus

$$x^2 - 6x + 9 = (x - 3)^2$$

and

$$x^2 - 6x + 9 - y^2 = (x^2 - 6x + 9) - y^2$$
$$= (x - 3)^2 - y^2 \qquad \text{The difference of two squares}$$
$$= [(x - 3) + y][(x - 3) - y]$$
$$= (x + y - 3)(x - y - 3)$$

b. Since $x^2 + 6xy + 9y^2 = (x + 3y)^2$,

$$x^2 + 6xy + 9y^2 - 4 = (x^2 + 6xy + 9y^2) - 4$$
$$= (x + 3y)^2 - 4 \qquad \text{The difference of two squares}$$
$$= [(x + 3y) + 2][(x + 3y) - 2]$$
$$= (x + 3y + 2)(x + 3y - 2) \qquad ■$$

The Sum and Difference of Two Cubes

We know how to factor the difference of two squares. Can we factor the difference of two cubes—that is, $x^3 - a^3$? Yes! As a matter of fact, we can also factor $x^3 + a^3$, the sum of two cubes. Here are the formulas.

PROCEDURE

> **Factoring the Sum and Difference of Two Cubes**
>
> F-5 $\quad x^3 + a^3 = (x + a)(x^2 - ax + a^2)$
>
> F-6 $\quad x^3 - a^3 = (x - a)(x^2 + ax + a^2)$

Since we did not give corresponding product formulas, we verify these results:

$$
\begin{array}{r}
x^2 - ax + a^2 \\
x + a \\
\hline
ax^2 - a^2x + a^3 \\
x^3 - ax^2 + a^2x \\
\hline
x^3 \qquad\qquad\quad + a^3
\end{array}
$$

Multiply $a(x^2 - ax + a^2)$.
Multiply $x(x^2 - ax + a^2)$.

Similarly,

$$
\begin{array}{r}
x^2 + ax + a^2 \\
x - a \\
\hline
-ax^2 - a^2x - a^3 \\
x^3 + ax^2 + a^2x \\
\hline
x^3 \qquad\qquad\quad - a^3
\end{array}
$$

Multiply $-a(x^2 + ax + a^2)$.
Multiply $x(x^2 + ax + a^2)$.

Now that we have verified the formulas, we can factor sums and differences of cubes. For example,

$$8 - x^3 = 2^3 - x^3$$
$$= (2 - x)(2^2 + 2x + x^2)$$
$$= (2 - x)(\underbrace{4 + 2x + x^2})$$
$$\qquad\qquad\qquad \text{Not factorable}$$

and

$$x^3 + 125 = x^3 + 5^3$$
$$= (x + 5)(x^2 - 5x + 5^2)$$
$$= (x + 5)(\underbrace{x^2 - 5x + 25}_{\text{Not factorable}})$$

EXAMPLE 5 Factoring the sum and difference of two cubes

Factor:

a. $27 - \dfrac{1}{64}x^3$ **b.** $125x^3 + 64y^3$ **c.** $(x + y)^3 - 8$

SOLUTION

a. $27 - \dfrac{1}{64}x^3 = 3^3 - \left(\dfrac{1}{4}x\right)^3$

$$= \left(3 - \dfrac{1}{4}x\right)\left[3^2 + 3 \cdot \dfrac{1}{4}x + \left(\dfrac{1}{4}x\right)^2\right]$$

$$= \left(3 - \dfrac{1}{4}x\right)\left(9 + \dfrac{3}{4}x + \dfrac{1}{16}x^2\right)$$

b. $125x^3 + 64y^3 = (5x)^3 + (4y)^3$

$$= (5x + 4y)[(5x)^2 - 5x \cdot 4y + (4y)^2]$$

$$= (5x + 4y)(25x^2 - 20xy + 16y^2)$$

c. $(x + y)^3 - 8$ is the difference of two cubes. As before, think of $(x + y)$ as A and 8 as 2^3. Thus you are factoring

$$A^3 - 2^3 = (A - 2)(A^2 + 2A + 2^2)$$

or

$$[(x + y)^3 - 2^3] = (x + y - 2)[(x + y)^2 + 2(x + y) + 4]$$

$$= (x + y - 2)(x^2 + 2xy + y^2 + 2x + 2y + 4) \quad \blacksquare$$

EXAMPLE 6 Factoring the difference of two cubes

Factor completely: $x^6 - 64$

SOLUTION

$$x^6 - 64 = x^6 - 2^6$$
$$= (x^3)^2 - (2^3)^2$$
$$= (x^3 + 2^3)(x^3 - 2^3)$$
$$= [(x + 2)(x^2 - 2x + 4)][(x - 2)(x^2 + 2x + 4)]$$
$$= (x + 2)(x - 2)(x^2 + 2x + 4)(x^2 - 2x + 4)$$

You can also factor $x^6 - 64$ by writing it as the difference of two cubes; that is

$$x^6 - 64 = (x^2)^3 - (2^2)^3$$
$$= (x^2 - 2^2)[(x^2)^2 + 2^2x^2 + (2^2)^2]$$
$$= (x^2 - 4)(x^4 + 4x^2 + 16)$$
$$= (x + 2)(x - 2)(x^4 + 4x^2 + 16)$$

But the expression $(x^4 + 4x^2 + 16)$ is **not** completely factored! You can factor it by first *adding* and *subtracting* $4x^2$ to obtain

$$x^4 + 4x^2 + 16 = x^4 + 4x^2 + 4x^2 + 16 - 4x^2$$

$$= (x^4 + 8x^2 + 16) - 4x^2 \qquad \text{By the associative property}$$

$$= (x^2 + 4)^2 - 4x^2 \qquad \text{Factor } x^4 + 8x^2 + 16.$$

$$= [(x^2 + 4) + 2x][(x^2 + 4) - 2x] \qquad \text{Factor the difference of two squares.}$$

$$= (x^2 + 2x + 4)(x^2 - 2x + 4) \qquad \text{By the commutative property}$$

Thus, as before,

$$x^6 - 64 = (x^2)^3 - (2^2)^3 = (x + 2)(x - 2)(x^2 + 2x + 4)(x^2 - 2x + 4) \qquad ■$$

NOTE As you can see, it's easier to factor $x^6 - 64$ when it is *first* written as the difference of two squares as we did here.

GRAPH IT

The grapher is a perfect tool for factoring perfect square trinomials. In Example 1a, we factored $x^2 - 10x + 25$. Look at its graph (Window 1). It touches the horizontal axis at $x = 5$ and the factorization is $(x - 5)^2$. Clearly, if the graph of a perfect square trinomial just touches the horizontal axis at $x = a$, the factorization is $(x - a)^2$. Of course, if you know your algebra, you noticed that when the middle term of a perfect square trinomial is preceded by a *minus* sign, the factorization is of the form $(x - a)^2$.

Can you use your grapher to help you factor $x^2 + 7x + 49$? No! The graph does not cut the horizontal axis (Window 2). (If you want to graph this polynomial, you need to set your viewing window so that Ymin = 0 and Ymax = 100.) You have to know the *ac* test to discover that $x^2 + 7x + 49$ is *prime*.

What about $x^3 + 125$? Since the graph (Window 3) cuts the horizontal axis at $x = -5$, you know that $x + 5$ is one of the factors. You can verify that $x^3 + 125 = (x + 5)(x^2 - 5x + 25)$ by graphing $x^3 + 125$ and $(x + 5)(x^2 - 5x + 25)$ and making sure you get the same graph. However, only algebra can help you find the other factor $(x^2 - 5x + 25)$. If you divide $x^3 + 125$ by $x + 5$, you get $x^2 - 5x + 25$. We will show you how to divide polynomials in Chapter 4.

The moral continues to be, even with a grapher, you need algebra!

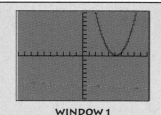
WINDOW 1
$x - 10x + 25$ touches the
x-axis at $x = 5$.

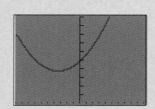
WINDOW 2
$x + 7x + 49$ does not cut
the horizontal axis.

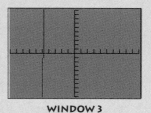
WINDOW 3
$x + 125$ cuts the x-axis at
$x = -5$.

EXERCISE 3.5

⚠ **In Problems 1–24, factor the expressions.**

1. $x^2 + 2x + 1$

2. $x^2 + 20x + 100$

3. $y^2 + 22y + 121$

4. $y^2 + 14x + 49$

5. $1 + 4x + 4x^2$

6. $1 + 6x + 9x^2$

7. $9x^2 + 30xy + 25y^2$

8. $25x^2 + 30xy + 9y^2$

9. $36a^2 + 48a + 16$

10. $9a^2 + 60a + 100$

11. $y^2 - 2y + 1$

12. $25 - 10y + y^2$

13. $49 - 14x + x^2$

14. $x^2 - 100x + 2500$

15. $49a^2 - 28ax + 4x^2$

16. $4a^2 - 12ax + 9x^2$

17. $16x^2 - 24xy + 9y^2$

18. $9x^2 - 42xy + 49y^2$

19. $9x^4 + 12x^2 + 4$

20. $25y^4 + 20y^2 + 4$

21. $16x^4 - 24x^2 + 9$

22. $4y^4 - 20y^2 + 25$

23. $1 + 2x^2 + x^4$

24. $4 + 12x^2 + 9x^4$

B In Problems 25–44, factor the expressions.

25. $y^2 - 64$

26. $y^2 - 121$

27. $a^2 - \dfrac{1}{9}$

28. $x^2 - \dfrac{1}{16}$

29. $64 - b^2$

30. $81 - b^2$

31. $36a^2 - 49b^2$

32. $36a^2 - 25b^2$

33. $\dfrac{x^2}{9} - \dfrac{y^2}{16}$

34. $\dfrac{x^2}{16} - \dfrac{9y^2}{25}$

35. $a^2 + 4ab + 4b^2 - c^2$

36. $9a^2 + 6ab + b^2 - 1$

37. $4x^2 - 4xy + y^2 - 1$

38. $9x^2 - 30xy + 25y^2 - 9$

39. $9y^2 - 12xy + 4x^2 - 25$

40. $16y^2 - 40xy + 25x^2 - 36$

41. $16a^2 - (x^2 + 6xy + 9y^2)$

42. $25a^2 - (4x^2 - 4xy + y^2)$

43. $y^2 - a^2 + 2ab - b^2$

44. $9y^2 - 9x^2 + 6xz - z^2$

C In Problems 45–70, factor the expressions.

45. $x^3 + 125$

46. $x^3 + 64$

47. $1 + a^3$

48. $343 + a^3$

49. $8x^3 + y^3$

50. $125x^3 + 8y^3$

51. $x^3 - 1$

52. $x^3 - 216$

53. $125a^3 - 8b^3$

54. $216a^3 - 125b^3$

55. $x^6 - 64$

56. $y^6 - 1$

57. $x^6 - \dfrac{1}{64}$

58. $y^6 - 729$

59. $\dfrac{x^6}{64} - 1$

60. $\dfrac{y^6}{729} - 1$

61. $(x - y)^3 + 1$

62. $(x + 2y)^3 + 8$

63. $1 + (x + 2y)^3$

64. $27 + (x + y)^3$

65. $(y - 2x)^3 - 1$

66. $(y - 4x)^3 - 1$

67. $27 - (x + 2y)^3$

68. $8 - (y - 4x)^3$

69. $64 + (x^2 - y^2)^3$

70. $27 + (y^2 - x^2)^3$

SKILL CHECKER

Multiply:

71. $(x + 3)(x - 5)$

72. $(x - 5)(x + 7)$

73. $(x - 8)(x + 2)$

74. $(x + 2y)^2$

75. $(2x - 3y)^2$

76. $(x + 2y)(x - 2y)$

USING YOUR KNOWLEDGE   Is There Demand for the Supply?

Have you heard of supply and demand? In business the supply and demand of a product can be expressed by using a polynomial. In Problems 77–79, factor the given expression.

77. A business finds that when x units of an item are demanded by consumers, the price per unit is given by

$$D(x) = 100 - x^2 \qquad \text{Factor } 100 - x^2.$$

78. When x units are supplied by sellers, the price per unit of an item is given by

$$S(x) = x^3 + 216 \qquad \text{Factor } x^3 + 216.$$

79. When x units of a certain item are produced, the cost is given by

$$C(x) = 8x^3 + 1 \qquad \text{Factor } 8x^3 + 1.$$

WRITE ON . . .

80. In Example 6, $(x + 2)(x - 2)(x^2 - 2x + 4)(x^2 + 2x + 4)$ is the "preferred" factorization of $x^6 - 64$. Explain why you think this is true.

81. Write a procedure to determine whether a quadratic trinomial is a perfect square trinomial.

82. Find two different values of k that will make $9x^2 + kx + 4$ a perfect square trinomial. Explain the procedure you used to find your answer.

83. Find a value for k that will make $4x^2 + 16x + k$ a perfect square trinomial. Describe the procedure you used to find your answer.

84. Explain why $a^2 + b^2$ cannot be factored.

85. Can $a^4 + 64$ be factored? (*Hint:* Add and subtract $16a^2$.)

$$a^4 + 64 = a^4 + 16a^2 + 64 - 16a^2$$
$$= (a^4 + 16a^2 + 64) - 16a^2$$

Can you factor it now?

If you know how to do these problems, you have learned your lesson!

Factor completely if possible:

86. $64 - \dfrac{1}{27}x^3$

87. $8x^3 + 27y^3$

88. $x^6 - 1$

89. $x^6 + 1$

90. $16x^2 - 1$

91. $16x^4 - 81y^4$

92. $x^2 + 2xy + y^2 + 16$

93. $x^2 + 10xy + 25y^2 - 16$

94. $4x^2 - 4xy + 9y^2$

95. $4x^2 - 12xy + 9y^2$

96. $x^2 - 16x + 64$

97. $x^2 + 18x + 81$

98. $(x - y)^4 - (x - y)^2$

99. $(x + y)^2 - (y - x)^2$

100. $(x - y)^3 + 125$

101. $125 + (x + y)^3$

3.6

GENERAL METHODS OF FACTORING

To succeed, review how to:	**Objectives:**
Use the factoring formulas F-1–F-6.	Factor a polynomial using the procedure given in the text.

getting started

Applying Factoring to Your Arteries

Have you heard of blocked arteries? The velocity of the blood inside a blocked artery (see photo) depends on the diameter of the inside wall (r) and outside wall (R) and is given by

$$CR^2 - Cr^2$$

where C is a constant. How do we factor this expression? We shall follow a general pattern that uses one or more of the techniques we've learned.

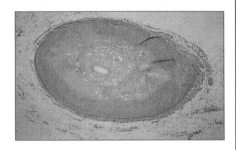

One of the most important skills when factoring polynomials is to know when the polynomial is completely factored. Here are the guidelines you need.

COMPLETELY FACTORED POLYNOMIAL

A polynomial is **completely factored** when

1. The polynomial is written as the product of prime polynomials with **integer** coefficients.

2. All the polynomial factors are prime, *except* that monomial factors need not be factored completely. Thus $6x^3(2x + 1)$ is the factored form of $12x^4 + 6x^3$. You don't need to write $6x^3$ as $2 \cdot 3 \cdot x \cdot x \cdot x$.

Now that we know when a polynomial is completely factored, let's learn how to factor any polynomial completely.

PROCEDURE

A General Factoring Strategy

1. Factor out the GCF, if there is one.

2. Look at the number of terms in the given polynomial (or inside the parentheses if the GCF was factored out).

   A. If there are *two terms*, check for
   - Difference of two squares

   $$x^2 - a^2 = (x + a)(x - a)$$

   - Difference of two cubes

   $$x^3 - a^3 = (x - a)(x^2 + ax + a^2)$$

   - Sum of two cubes

   $$x^3 + a^3 = (x + a)(x^2 - ax + a^2)$$

   Note that the sum of two squares, $x^2 + a^2$, is not factorable.

   B. If there are *three terms*, check for
   - Perfect square trinomial

   $$x^2 + 2ab + b^2 = (x + a)^2$$
   $$x^2 - 2ab + b^2 = (x - a)^2$$

   - Trinomials of the form

   $$ax^2 + bx + c \qquad (a > 0)$$

   Use the *ac* method or trial and error. If $a < 0$, factor out -1 first.

   C. If there are *four terms*, factor by grouping.

3. Check the result by multiplying the factors.

Thus to factor $CR^2 - Cr^2$ (see the *Getting Started*) we follow these steps.

1. Factor out the GCF, C. $\qquad\qquad CR^2 - Cr^2 = C(R^2 - r^2)$

2. Factor the difference of two squares inside
   the parentheses. $\qquad\qquad\qquad\qquad = C(R + r)(R - r)$

3. We leave the check to you.

Now, if you were the doctor you could tell the patient that when R and r are very close ($R - r$ close to 0) the artery will be almost blocked!

EXAMPLE 1   Using the general factoring strategy to factor a binomial

Factor:

a. $8x^5 - x^2y^3$ $\qquad\qquad\qquad\qquad\qquad$ **b.** $6x^5 + 24x^3$

SOLUTION   We use the steps in our general factoring strategy.

a. $8x^5 - x^2y^3$

1. Factor out the GCF, x^2. $\qquad 8x^5 - x^2y^3 = x^2(8x^3 - y^3)$

2. Factor the difference of two
   cubes inside the parentheses. $\qquad\qquad = x^2(2x - y)(4x^2 + 2xy + y^2)$

3. We leave the check to you.

b. $6x^5 + 24x^3$

   1. Factor out $6x^3$, the GCF: $6x^5 + 24x^3 = 6x^3(x^2 + 4)$.

   2. Since $x^2 + 4$ is the sum of two squares, it is not factorable.

   3. Check this!

Thus the complete factorization of $6x^5 + 24x^3$ is $6x^3(x^2 + 4)$.   ■

EXAMPLE 2   Using the general factoring strategy to factor a trinomial

Factor:

a. $12x^5 + 12x^4y + 3x^3y^2$ **b.** $36x^2y^2 - 24xy^3 + 4y^4$

SOLUTION   As usual, factor out the GCFs, $3x^3$ in part a and $4y^2$ in part b, first. Then note that we have a perfect square trinomial inside the parentheses. Here are the steps.

a. $12x^5 + 12x^4y + 3x^3y^2$

   1. Factor out the GCF, $3x^3$. $12x^5 + 12x^4y + 3x^3y^2 = 3x^3\underbrace{(4x^2 + 4xy + y^2)}$

   2. Factor the perfect square trinomial. $= 3x^3 \quad (2x + y)^2$

   3. We leave the check to you.

b. $36x^2y^2 - 24xy^3 + 4y^4$

   1. Factor out the GCF, $4y^2$. $36x^2y^2 - 24xy^3 + 4y^4 = 4y^2\underbrace{(9x^2 - 6xy + y^2)}$

   2. Factor the perfect square trinomial inside the parentheses. $= 4y^2 \quad (3x - y)^2$

   3. Don't forget to check this!   ■

EXAMPLE 3   Using the general factoring strategy to factor a polynomial

Factor:

a. $4x^3y - 10x^2y^2 - 6xy^3$ **b.** $2x^5 + x^4y + x^3y^2$

SOLUTION

a. The GCF is $2xy$. After factoring out this GCF, we have three terms inside the parentheses. We can use the *ac* method or trial and error to finish the problem. The steps are as follows.

   1. Factor out the GCF, $2xy$. $4x^3y - 10x^2y^2 - 6xy^3 = 2xy\,(2x^2 - 5xy - 3y^2)$

   2. Use the *ac* method or trial and error to factor $2x^2 - 5xy - 3y^2$. $= 2xy(2x + y)(x - 3y)$

   3. Check the answer by multiplying the factors. $2xy(2x + y)(x - 3y) = 4x^3y - 10x^2y^2 - 6xy^3$

b. Factor out the GCF, x^3. $2x^5 + x^4y + x^3y^2 = x^3(2x^2 + xy + y^2)$

Note that the expression inside the parentheses is *not* factorable. According to the *ac* method, the key number is 2. We need two integers whose product is 2 and whose sum is 1; no such integers exist.   ■

EXAMPLE 4   Finding the complete factorization

Factor: $2x^5 - x^4y + 4x^3y^2$

SOLUTION We start by factoring out the GCF, which is x^3.

$$2x^5 - x^4y + 4x^3y^2 = x^3(2x^2 - xy + 4y^2)$$

Although $2x^2 - xy + 4y^2$ is a trinomial, it is *not* factorable. (The *ac* method gives $2 \cdot 4 = 8$, and there are no factors whose product is 8 and whose sum is -1.) Thus the factorization shown is the complete factorization. ∎

EXAMPLE 5   Factoring by grouping

Factor: $4x^3 - 12x^2 - x + 3$

SOLUTION In this case, there is no common factor. Since the polynomial has four terms, we factor by grouping.

$$4x^3 - 12x^2 - x + 3 = 4x^2(x - 3) - (x - 3) \qquad \text{Group into pairs and factor.}$$

$$= (x - 3)\underbrace{(4x^2 - 1)} \qquad \text{Factor out the GCF, } (x - 3).$$

$$= (x - 3)(2x + 1)(2x - 1) \qquad \text{Factor the difference of two squares, } (4x^2 - 1). \quad ∎$$

EXAMPLE 6   Factoring by grouping

Factor: $x^2 - 6x + 9 - 9y^2$

SOLUTION Since the polynomial has four terms, you may try to factor it by grouping. If we try groups of two, we have

$$x^2 - 6x + 9 - 9y^2 = (x^2 - 6x) + (9 - 9y^2)$$

$$= x(x - 6) + 9(1 - y^2)$$

But there is no GCF, so we can't factor any further using this technique. However, note that the first three terms form a perfect square trinomial, so we write

$$x^2 - 6x + 9 - 9y^2 = (x^2 - 6x + 9) - 9y^2$$

$$= (x - 3)^2 - (3y)^2 \qquad \text{The difference of two squares}$$

$$= [(x - 3) + 3y][(x - 3) - 3y]$$

$$= (x + 3y - 3)(x - 3y - 3) \quad ∎$$

EXAMPLE 7.   General factoring strategies

Factor: $-2x^2 - 9x + 5$

SOLUTION Since this is a polynomial of the form $ax^2 + bx + c$ with $a = -2$, which is negative, we first factor out -1 to obtain

Factor out -1. $-2x^2 - 9x + 5 = -1(2x^2 + 9x - 5)$

1. The key number is $2(-5) = -10$. $= -1(2x^2 + 10x - 1x - 5)$

2. We write the middle term as $10x - 1x$.

3. Grouping by twos. $= -1[(2x^2 + 10x) + (-1x - 5)]$

4. Factoring each group. $= -1[2x(x + 5) - 1(x + 5)]$

5. Factoring the GCF. $= -1(x + 5)(2x - 1)$

Since $-1 \cdot a = -a$, $-2x^2 - 9x + 5 = -(x + 5)(2x - 1)$ ∎

GRAPH IT

If you are using a grapher, you need a general factoring strategy too. Here is one.

A General Factoring Strategy

1. Factor out the GCF if there is one. After you do this algebraically, graph the original polynomial and the factored one. You must get the same graph.

2. Look at the number of terms in the given polynomial (or inside the parentheses if the GCF was factored out).

- If there are *two* terms, graph the polynomial and determine the points at which the graph cuts the horizontal axis.

  If you have the difference of two squares and the points are $x = \pm p$, the factorization is

  $$(x - p)(x + p)$$

  If you have the sum or difference of two cubes and the point is $x = \pm p$, the factorization is

  $$(x \pm p)(x^2 \mp px + p^2)$$

- If there are *three* terms and the degree is 2, graph the polynomial and determine the point at which the graph touches the horizontal axis.

  If the point is $x = p$, the factorization is

  $$(x - p)^2$$

  If the points are

  $$x = \frac{p}{q} \quad \text{and} \quad x = \frac{r}{s}$$

  the factorization is $(qx - p)(sx - r)$. (If the leading coefficient is negative, factor out a -1 first.)

- If there are *four* terms, factor by grouping and check by graphing the original and the factored polynomial. The graphs should be identical.

- If you graph the polynomial and nothing shows, try different windows. Press $\boxed{\text{TRACE}}$ to give you an idea of the x- and y-values so you can adjust your window. If still nothing shows and you graphed correctly, the polynomial is not factorable. (Try the ac test for polynomials of the form $ax^2 + bx + c$.)

EXERCISE 3.6

A **In Problems 1–78, factor the expressions.**

1. $3x^4 - 3x^3 - 18x^2$ **2.** $4x^5 - 12x^4 - 16x^3$

3. $5x^4 + 10x^3y - 40x^2y^2$ **4.** $6x^7 + 18x^6y - 60x^5y^2$

5. $-3x^6 - 6x^5 - 21x^4$ **6.** $-6x^5 - 18x^4 - 12x^3$

7. $2x^6y - 4x^5y^2 - 10x^4y^3$ **8.** $3x^8y - 12x^7y^2 - 9x^6y^3$

9. $-4x^6 - 12x^5y - 18x^4y^2$ **10.** $-5x^6 - 25x^5 - 30x^4$

11. $6x^3y^2 + 12x^2y^2 + 2xy^2 + 4y^2$

12. $6x^3y^2 + 24x^2y^2 + 3xy^2 + 12y^2$

13. $-9x^4y - 9x^3y - 6x^2y - 6xy$

14. $-8x^4y - 16x^3y - 6x^2y - 12xy$

15. $-4x^4 - 4x^3y + 2x^2y + 2xy^2$

16. $-9x^4 - 18x^3y + 3x^2y + 6xy^2$

17. $3x^2y^2 + 24xy^3 + 48y^4$ **18.** $8x^2y^2 + 24xy^3 + 18y^4$

19. $-18kx^2 - 24kxy - 8ky^2$ **20.** $-12kx^2 - 60kxy - 75ky^2$

21. $16x^3y^2 - 48x^2y^3 + 36xy^4$ **22.** $45x^3y^2 - 60x^2y^3 + 20xy^4$

23. $kx^2 - 12kx + 36$ **24.** $kx^2 - 20kx + 25$

25. $3x^5 + 12x^4y + 12x^3y^2$ **26.** $2x^5 + 16x^4y + 32x^3y^2$

27. $18x^6 + 12x^5y + 2x^4y^2$ **28.** $12x^6 + 12x^5y + 3x^4y^2$

29. $12x^4y^2 - 36x^3y^3 + 27x^2y^4$

30. $18x^4y^2 - 24x^3y^3 + 8x^2y^4$

31. $6x^3 + 12x^2 - 6x - 12$ **32.** $4x^3 + 16x^2 - 16x - 64$

33. $7x^4 - 7y^4$ **34.** $9x^4 - 9z^4$

35. $2x^6 - 32x^2y^4$ **36.** $x^7 - 81x^3y^4$

37. $-2x^2 - 12x - 18$ **38.** $-2x^2 - 20x - 50$

39. $-3x^2 - 12x - 12$

40. $-4x^2 - 24x - 36$

41. $-4x^4 - 4x^3y - x^2y^2$

42. $-9x^4 - 6x^3y - x^2y^2$

43. $-9x^2y^2 - 12xy^3 - 4y^4$

44. $-4x^2y^2 - 12xy^3 - 9y^4$

45. $-8x^2y^2 + 24xy^3 - 18y^4$

46. $-18x^4 + 24x^3y - 8x^2y^2$

47. $-18x^3 - 24x^2y - 8xy^2$

48. $-12x^3 - 36x^2y - 27xy^2$

49. $-18x^3 - 60x^2y - 50xy^2$

50. $-12x^3 - 60x^2y - 75xy^2$

51. $-x^3 + xy^2$

52. $-x^3 + 9xy^2$

53. $-x^4 + 4x^2y^2$

54. $-x^4 + 16x^2y^2$

55. $-4x^4 + 9x^2y^2$

56. $-9x^4 + 4x^2y^2$

57. $-8x^3 + 18xy^2$

58. $-12x^3 + 3x$

59. $-18x^4 + 8x^2y^2$

60. $-12x^4 + 27x^2y^2$

61. $27x^2 - x^5$

62. $64x^3 - x^6$

63. $x^7 - 8x^4$

64. $8x^{10} - \dfrac{1}{27}x^7$

65. $27x^4 + 8x^7$

66. $8x^5 + 27x^8$

67. $27x^7 + 64x^4y^3$

68. $8x^8 + 27x^5y^3$

69. $x^2 + 4x + 4 - y^2$

70. $x^2 + 8x + 16 - y^2$

71. $x^2 + y^2 - 6y + 9$

72. $x^2 + y^2 - 8x + 16$

73. $x^2 - y^2 - 4y - 4$

74. $x^2 - y^2 - 8y - 16$

75. $-9x^2 + 30xy - 25y^2$

76. $-9x^2 + 12xy - 4y^2$

77. $18x^3 - 60x^2y + 50xy^2$

78. $-12x^3 + 60x^2y - 72xy^2$

SKILL CHECKER

Factor:

79. $6x^2 - x - 2$

80. $6x^2 - 7x - 3$

81. $12x^2 - x - 1$

82. $10x^2 - 17x + 3$

USING YOUR KNOWLEDGE

Technical Applications

Many of the ideas presented in this section are used by engineers and technicians. Use your knowledge to factor the expressions in Problems 83–86.

83. The bend allowance needed to bend a piece of metal of thickness t through an angle A when the inside radius of the bend is R is given by

$$\frac{2\pi A}{360}R + \frac{2\pi A}{360}Kt \qquad K \text{ is a constant.}$$

Factor this expression.

84. The change in kinetic energy of a moving object of mass m with initial velocity v_1 and terminal velocity v_2 is given by

$$\frac{1}{2}mv_1^2 - \frac{1}{2}mv_2^2$$

Factor this expression.

85. The parabolic distribution of shear stress on the cross section of a certain beam is given by

$$\frac{3Sd^2}{2bd^3} - \frac{12Sz^2}{2bd^3}$$

Factor this expression.

86. The polar moment of inertia, J, of a hollow round shaft of inner diameter d_1 and outer diameter d is given by

$$\frac{\pi d^4}{32} - \frac{\pi d_1^4}{32}$$

Factor this expression.

WRITE ON . . .

87. Explain why the expression $x^2(x^2 - 5) + y^2(x^2 - 5)$ is not completely factored.

88. Explain the difference between the statements "a polynomial is factored" and "a polynomial is completely factored." Give examples of polynomials that are factored but are not completely factored.

89. Is the statement "$9(x^2 - x)$ is factored" true or false? Explain.

90. Suppose you factor $x^2 - 2$ as $(x + \sqrt{2})(x - \sqrt{2})$. Is $x^2 - 2$ completely factored according to the guidelines we mentioned at the beginning of this section? Explain.

MASTERY TEST

If you know how to do these problems, you have learned your lesson!

Factor completely if possible:

91. $9x^3 - 18x^2 - x + 2$

92. $4x^3 - 12x^2 - x + 3$

93. $2x^4 + x^3y + 2x^2y^2$

94. $3x^5 + x^4y + x^3y^2$

95. $x^2 - 10x + 25 - y^2$

96. $y^2 - 9x^2 + 12x - 4$

97. $-8y^2 + 2y + 21$

98. $-2 - 5y - 2y^2$

99. $36x^5 + 24x^4y + 4x^3y^2$

100. $12x^2y^2 - 12xy^3 + 3y^4$

101. $8x^5 + 72x^3$

102. $27x^5 - x^2y^3$

3.7

SOLVING EQUATIONS BY FACTORING: APPLICATIONS

To succeed, review how to:

1. Factor polynomials (pp. 173–176).
2. Solve linear equations (pp. 69–73).
3. Evaluate expressions (pp. 42–44).

Objectives:

 Solve equations by factoring.

 Use the Pythagorean Theorem to find the length of one side of a right triangle when the lengths of the other two sides are given.

 Solve applications involving quadratic equations.

getting started

Quadratic Equations in Fire Fighting

How much water is the fire truck pumping if the friction loss is 36 lb/in.^{2}? You can find out by solving the equation

$$2g^2 + g - 36 = 0 \qquad (g \text{ in hundreds of gallons per minute})$$

This equation is a *quadratic equation in standard form*. In this section we shall study how to solve equations like this one.

We have already studied linear equations, equations that can be written in the form $ax + b = c$, where a, b, and c are real numbers and $a \neq 0$. We are now ready to study *quadratic equations*. These equations can be written in standard form and then solved by the factoring methods we have studied. Here is the definition we need.

STANDARD FORM OF A QUADRATIC EQUATION

> An equation that can be written in the **standard form**
> $$ax^2 + bx + c = 0$$
> where a, b, and c are constants and $a \neq 0$ is a **quadratic equation**.

Here are some quadratic equations:

$$x^2 = 5, \qquad 3x^2 - 8x + 7 = 0, \qquad \text{and} \qquad x^2 - 2x = 4$$

Of these, only $3x^2 - 8x + 7 = 0$ is in standard form, with $a = 3$, $b = -8$, and $c = 7$.

 Solving Equations by Factoring

The equation $2g^2 + g - 36 = 0$ can be solved by factoring. As you recall from Section 3.5, $2g^2 + g - 36$ can be factored as $(2g + 9)(g - 4)$. We then write

$$2g^2 + g - 36 = 0 \qquad \text{Given.}$$
$$(2g + 9)(g - 4) = 0 \qquad \text{Factor.}$$

Note that the product of the factors is zero. The only way this can happen is if at least one of the factors is zero. (Try getting zero for a product without having any zero factors!) Here is the property we need.

ZERO-FACTOR PROPERTY

> For all real numbers a and b, $a \cdot b = 0$ means that $a = 0$ or $b = 0$ (or both).

Thus

$$(2g + 9)(g - 4) = 0$$

means that

$$
\begin{array}{lll}
2g + 9 = 0 & \text{or} & g - 4 = 0 \\
2g = -9 & \text{or} & g = 4 \qquad \text{Solve the linear equations.} \\
g = -\dfrac{9}{2} & \text{or} & g = 4
\end{array}
$$

The two possible solutions are $g = -\frac{9}{2}$ and $g = 4$. Since g is the flow of water in hundreds of gallons per minute, g must be positive, so we discard the negative solution $g = -\frac{9}{2}$. Thus the fire truck can pump 400 gal/min.

EXAMPLE 1   Solving equations using the zero-factor property

Solve by factoring:

a. $x^2 - 9 = 0$ **b.** $x^2 + 8x = 0$

SOLUTION

a. $x^2 - 9 = 0$

$(x + 3)(x - 3) = 0$ Factor.

$x + 3 = 0$ or $x - 3 = 0$ Use the zero-factor property, with $a = x + 3$ and $b = x - 3$.

$x = -3$ or $x = 3$ Solve the equations $x + 3 = 0$ and $x - 3 = 0$.

We can check the solutions by substituting in the original equation, $x^2 - 9 = 0$.

CHECK

$$
\begin{array}{c|c}
x^2 - 9 = 0 & x^2 - 9 = 0 \\
\hline
(-3)^2 - 9 \mid 0 & 3^2 - 9 \mid 0 \\
9 - 9 & 9 - 9 \\
0 & 0
\end{array}
$$

In both cases the result is true. The solution set is $\{3, -3\}$.

b. $x^2 + 8x = 0$

$x(x + 8) = 0$ Factor.

$x = 0$ or $x + 8 = 0$ Use the zero-factor property, with $a = x$ and $b = x + 8$.

$x = 0$ or $x = -8$ Solve the equation $x + 8 = 0$.

CHECK

$$
\begin{array}{c|c}
x^2 + 8x = 0 & x^2 + 8x = 0 \\
\hline
(0)^2 + 8(0) \mid 0 & (-8)^2 + 8(-8) \mid 0 \\
0 + 0 & 64 - 64 \\
0 & 0
\end{array}
$$

Since both results check, the solution set is $\{0, -8\}$.   ■

The equation $x^2 = -x + 6$ is *not* in standard form. To write it in standard form, we add x and subtract 6 on both sides to obtain $x^2 + x - 6 = 0$, which can be solved by factoring. As you recall, to factor $x^2 + x - 6$, we need to find integers whose sum is 1 and whose product is -6. These integers are 3 and -2. Thus we have

$$x^2 = -x + 6 \qquad \text{Given.}$$

$$x^2 + x - 6 = 0 \qquad \text{Write in standard form (add } x \text{ and subtract 6).}$$

$$(x + 3)(x - 2) = 0 \qquad \text{Factor.}$$

$$x + 3 = 0 \qquad \text{or} \qquad x - 2 = 0 \qquad \text{Use the zero-factor property.}$$

$$x = -3 \qquad \text{or} \qquad x = 2 \qquad \text{Solve } x + 3 = 0 \text{ and } x - 2 = 0.$$

CHECK

$x^2 = -x + 6$		$x^2 = -x + 6$	
$(2)^2$	$-(2) + 6$	$(-3)^2$	$-(-3) + 6$
4	4	9	$3 + 6$
			9

The solution set is $\{2, -3\}$.

EXAMPLE 2   Writing and solving equations in standard form

Solve by factoring:

a. $x^2 = 6x - 8$ **b.** $x^2 + x = 2$

SOLUTION

a. We first write the equation in standard form by subtracting $6x$ and adding 8 to obtain $x^2 - 6x + 8 = 0$. To factor $x^2 - 6x + 8$, we must find two integers whose sum is -6 and whose product is 8. These numbers are -4 and -2. Here is the procedure,

$$x^2 = 6x - 8 \qquad \text{Given.}$$

$$x^2 - 6x + 8 = 0 \qquad \text{Subtract } 6x \text{ and add 8.}$$

$$(x - 4)(x - 2) = 0 \qquad \text{Factor.}$$

$$x - 4 = 0 \qquad \text{or} \qquad x - 2 = 0 \qquad \text{Use the zero-factor property.}$$

$$x = 4 \qquad \text{or} \qquad x = 2 \qquad \text{Solve } x - 4 = 0 \text{ and } x - 2 = 0.$$

The solution set is $\{2, 4\}$. Check this!

b. The equation $x^2 + x = 2$ is *not* in standard form. To solve an equation by factoring, we must first write the equation in the standard form $ax^2 + bx + c = 0$. Thus subtracting 2 from both members of $x^2 + x = 2$, we have

$$x^2 + x - 2 = 0$$

$$(x + 2)(x - 1) = 0 \qquad \text{Factor.}$$

$$x + 2 = 0 \qquad \text{or} \qquad x - 1 = 0 \qquad \text{Use the zero-factor property.}$$

$$x = -2 \qquad \text{or} \qquad x = 1 \qquad \text{Solve } x + 2 = 0 \text{ and } x - 1 = 0.$$

The solution set is $\{1, -2\}$. The check is left to you. ∎

To solve the equation $6x^2 - x - 2 = 0$, we first factor $6x^2 - x - 2$ by trial and error or by the *ac* method (shown). For $6x^2 - x - 2 = 0$, the key number is $6 \cdot (-2) = -12$. Thus we have to find integers whose product is -12 and whose sum is -1 (that is, -4 and 3) and use these numbers to rewrite the middle term x. We have

$6x^2 - x - 2 = 0$	Given.
$6x^2 - 4x + 3x - 2 = 0$	Write the middle term $-x$ as $-4x + 3x$.
$2x(3x - 2) + 1(3x - 2) = 0$	Factor the first and last pair of terms.
$(3x - 2)(2x + 1) = 0$	Factor out the common factor, $3x - 2$.
$3x - 2 = 0 \quad$ or $\quad 2x + 1 = 0$	Use the zero-factor property.
$3x = 2 \quad$ or $\quad 2x = -1$	Solve $3x - 2 = 0$ and $2x + 1 = 0$.
$x = \dfrac{2}{3} \quad$ or $\quad x = \dfrac{-1}{2}$	

The solution set is $\left\{ \frac{2}{3}, -\frac{1}{2} \right\}$. We check this for $x = \frac{2}{3}$.

$$\textbf{CHECK} \quad \frac{6x^2 - x - 2 = 0}{}$$

$$6\left(\frac{2}{3}\right)^2 - \frac{2}{3} - 2 \quad \bigg| \quad 0$$

$$6\left(\frac{4}{9}\right) - \frac{2}{3} - 2$$

$$\frac{8}{3} - \frac{2}{3} - 2$$

$$2 - 2$$

$$0$$

The check that $x = -\frac{1}{2}$ is a solution is left to you. ∎

EXAMPLE 3   Solving quadratic equations by factoring using the ac method

a. $12x^2 + 5x - 3 = 0$ **b.** $6x^2 - x = 1$ **c.** $x(x + 2) = (3x - 1)x + 1$

SOLUTION

a. You can factor $12x^2 + 5x - 3$ by trial and error or note that the key number for $12x^2 + 5x - 3 = 0$ is $(12)(-3) = -36$. Thus

$12x^2 + 5x - 3 = 0$	Given.
$12x^2 + 9x - 4x - 3 = 0$	Write $5x$ using coefficients whose sum is 5 and whose product is -36; that is, $5x = 9x - 4x$.
$3x(4x + 3) - 1(4x + 3) = 0$	Factor the first and last pairs of terms.
$(4x + 3)(3x - 1) = 0$	Factor out the common factor, $4x + 3$.
$4x + 3 = 0 \quad$ or $\quad 3x - 1 = 0$	Use the zero-factor property.
$4x = -3 \quad$ or $\quad 3x = 1$	Solve $4x + 3 = 0$ and $3x - 1 = 0$.
$x = \dfrac{-3}{4} \quad$ or $\quad x = \dfrac{1}{3}$	

The solution set is $\left\{ \frac{1}{3}, -\frac{3}{4} \right\}$. Check this!

b. $6x^2 - x = 1$ is *not* in standard form. Subtracting 1 from both sides of the equation, we have

$$6x^2 - x - 1 = 0$$ The key number is -6.

$$6x^2 - 3x + 2x - 1 = 0$$ The integers whose product is -6 with sum -1 are -3 and 2.

$$3x(2x - 1) + (2x - 1) = 0$$

$$(2x - 1)(3x + 1) = 0$$ Factor out $(2x - 1)$.

$$2x - 1 = 0 \quad \text{or} \quad 3x + 1 = 0$$ Use the zero-factor property.

$$2x = 1 \quad \text{or} \quad 3x = -1$$ Solve the equations $2x - 1 = 0$ and $3x + 1 = 0$.

$$x = \frac{1}{2} \quad \text{or} \quad x = \frac{-1}{3}$$

The solution set is $\left\{\frac{1}{2}, -\frac{1}{3}\right\}$. Check this!

c. $x(x + 2) = (3x - 1)x + 1$ is *not* in standard form. So we first simplify both sides and then write the result in standard form.

$$x(x + 2) = (3x - 1)x + 1$$ Given.

$$x^2 + 2x = 3x^2 - x + 1$$ Distributive property

$$0 = 3x^2 - x^2 - 2x - x + 1$$ Subtract $x^2 + 2x$.

$$0 = 2x^2 - 3x + 1$$ Simplify.

$$2x^2 - 3x + 1 = 0$$ Symmetric property

$$2x^2 - 2x - x + 1 = 0$$ Since the key number is 2, rewrite $-3x = -2x - x$.

$$(2x^2 - 2x) + (-x + 1) = 0$$ Group.

$$2x(x - 1) - 1 \cdot (x - 1) = 0$$ Factor.

$$(x - 1)(2x - 1) = 0$$ Since the GCF is $(x - 1)$

$$x - 1 = 0 \quad \text{or} \quad 2x - 1 = 0$$ Zero-factor property

$$x = 1 \quad \text{or} \quad x = \frac{1}{2}$$ Solve $x - 1 = 0$ and $2x - 1 = 0$.

The solution set is $\left\{1, \frac{1}{2}\right\}$. Check this! ∎

EXAMPLE 4   Solving equations by grouping and the zero-factor property

Solve: $x^3 + 3x^2 - 4x - 12 = 0$

SOLUTION   Since the polynomial has four terms, we try to factor it by grouping. Here are the steps.

$$x^3 + 3x^2 - 4x - 12 = 0$$ Given.

$$x^2(x + 3) - 4(x + 3) = 0$$ Group for factoring.

$$(x + 3)(x^2 - 4) = 0$$ Factor out the GCF.

$$(x + 3)(x + 2)(x - 2) = 0$$ Factor $x^2 - 4$.

$$x + 3 = 0 \quad \text{or} \quad x + 2 = 0 \quad \text{or} \quad x - 2 = 0$$

$$x = -3 \quad \text{or} \quad x = -2 \quad \text{or} \quad x = 2$$

The solution set is $\{2, -2, -3\}$. Check this! ∎

Using the Pythagorean Theorem

Quadratic equations can be used to find the lengths of the sides of right triangles using the **Pythagorean Theorem**, which we state here.

THE PYTHAGOREAN THEOREM

In any right triangle (a triangle with a 90° angle), the square of the longest side (hypotenuse) is equal to the sum of the squares of the other two sides (the legs). In symbols, this is

$$c^2 = a^2 + b^2$$

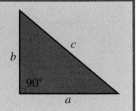

problem solving   **EXAMPLE 5**   Using the Pythagorean Theorem to solve word problems

The length of the three sides of a right triangle are consecutive integers. What are these lengths?

SOLUTION   We use the RSTUV method.

1. Read the problem.

We need to find the lengths of the sides of the right triangle.

2. Select the unknown.

If x is an integer, what are the next two consecutive integers?

Let the length of the shortest side be x. Since the lengths of the sides are consecutive integers, we have

x	Length of the shortest side
$x + 1$	Length of the next side
$x + 2$	Length of the hypotenuse (the longest side)

3. Think of a plan.

Make a diagram and use the Pythagorean Theorem to obtain the equation

$$(x + 2)^2 = (x + 1)^2 + x^2$$

4. Use algebra to solve the resulting equation.

$x^2 + 4x + 4 = x^2 + 2x + 1 + x^2$	Multiply.
$x^2 + 4x + 4 = 2x^2 + 2x + 1$	Simplify.
$0 = x^2 - 2x - 3$	Subtract x^2, $4x$, and 4 from both sides.
$x^2 - 2x - 3 = 0$	Write in standard form.
$(x - 3)(x + 1) = 0$	Factor.
$x - 3 = 0$ or $x + 1 = 0$	Use the zero-factor property.
$x = 3$ or $x = -1$	Solve $x - 3 = 0$ and $x + 1 = 0$.

Since the lengths of the sides must be positive, we discard the negative answer, -1. Thus the shortest side is 3 units, so the other two sides are 4 and 5 units.

5. Verify the solution.

The verification is left to you.   ■

problem solving   **EXAMPLE 6**   An application of the Pythagorean Theorem

A ladder is leaning against the side of a house with its bottom x feet away from the wall and resting against the wall of the house $x + 7$ feet above the ground. If the length of the ladder is 1 ft more than its height above the ground, how long is the ladder?

SOLUTION   As before, we use the RSTUV method.

1. Read the problem.

Form a mental picture of the dimensions.

2. Select the unknown.

Let x feet be the distance from the foot of the ladder to the wall.

3. Think of a plan.

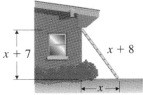

Make a diagram of the situation. The distance from the bottom of the ladder to the house is x feet. The ladder is resting on the side of the house $x + 7$ feet from the ground, and the length of the ladder is 1 ft more than the height from the ground. Thus the length of the ladder is $x + 7 + 1$ or $x + 8$ feet. By the Pythagorean Theorem, we have the equation

$$(x + 8)^2 = x^2 + (x + 7)^2$$

4. Use algebra to solve the resulting equation.

$(x + 8)^2 = x^2 + (x + 7)^2$	Given.
$x^2 + 16x + 64 = x^2 + x^2 + 14x + 49$	Simplify.
$0 = x^2 - 2x - 15$	Subtract $x^2 + 16x + 64$ from both sides.
$x^2 - 2x - 15 = 0$	By the symmetric property
$(x - 5)(x + 3) = 0$	Factor. (We need two numbers whose product is $- 15$ and whose sum is -2 (-5 and 3).
$x - 5 = 0$ or $x + 3 = 0$	Use the zero-factor property.
$x = 5$ or $x = -3$	Solve each equation.

Since x represents a length, $x = -3$ must be discarded as an answer. Thus the foot of the ladder is 5 ft from the wall and leans against the wall $x + 7 = 5 + 7 = 12$ ft from the ground. Its length is $x + 8 = 5 + 8 = 13$ ft.

5. Verify the solution.

We note that $13^2 = 5^2 + 12^2$.   ■

Applications

Many problems can be studied by using quadratic equations. For example, do you use hair spray containing chlorofluorocarbons (CFCs) for propellants? A U.N.-sponsored conference negotiated an agreement to stop producing CFCs by the year 2000 because they harm the ozone layer. Can we check to see whether the goal will be reached if current levels of production continue? We shall see in Example 7.

EXAMPLE 7   An application of quadratic equations

The production of CFCs for use as aerosol propellants (in thousands of tons) can be represented by $P(t) = -0.4t^2 + 22t + 120$, where t is the number of years after 1960. When will production be stopped?

SOLUTION   Production will be stopped when $P(t) = 0$; that is, we need to solve the equation

$P(t) = -0.4t^2 + 22t + 120 = 0$	
$-4t^2 + 220t + 1200 = 0$	Multiply by 10 (to clear decimals).
$t^2 - 55t - 300 = 0$	Divide by -4 (to obtain t^2).
$(t - 60)(t + 5) = 0$	We need two numbers whose product is -300 and whose sum is -55 (-60 and 5).
$t - 60 = 0$ or $t + 5 = 0$	By the zero-factor property
$t = 60$ or $t = -5$	Solve each equation.

Since t represents the number of years *after* 1960, production will be zero (stopped) 60 years after 1960 or in 2020 (not in the year 2000 as promised!). The answer $t = -5$ has to be discarded since it represents 5 years *before* 1960, but the equation applies only to years *after* 1960. ■

GRAPH IT

The problems in this section are easily solved with a grapher and you don't even have to factor the given polynomials to solve them! Consider Example 1a. To solve $x^2 - 9 = 0$, graph $Y_1 = x^2 - 9$ with a standard window (Window 1). The points at which the graph cuts the horizontal axis, 3 and -3, are the solutions of $x^2 - 9 = 0$. Similarly, to do Example 2a, write the equation in the standard form, $x^2 - 6x + 8 = 0$, and graph. Since the graph cuts the horizontal axis at $x = 2$ and $x = 4$, these are the solutions. (We leave these for you to graph.) To do Example 3c, first rewrite the equation as $x(x + 2) - (3x - 1)x - 1 = 0$ (you don't have to multiply) and graph it using a decimal window; then use TRACE to find the solutions (Window 2).

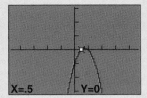

WINDOW 1
The solutions of $x^2 - 9 = 0$ are $x = -3$ and $x = 3$.

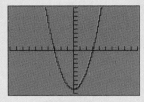

WINDOW 2
One of the solutions of $x(x + 2) - (3x - 1)x - 1 = 0$ is 0.5.

Some graphers (the TI-82) have an intersect and a solve feature. To find the point at which the curve intersects the horizontal axis, graph the horizontal axis $Y_2 = 0$. Get the cursor to a point near the intersection you want to find. Press 2nd CALC 5. The grapher asks "First Curve?" "Second Curve?" and "Guess?" Press ENTER each time, and the grapher gives the intersection $x = 0.5$, $y = 0$. To use the solve feature, press MATH 0. The screen shows "solve(." Enter the expression, say $x(x + 2) - (3x - 1)x - 1$, press , and enter the variable, which is x, press , and then enter a guess for the solution, say 1/4. Now close the parentheses and press ENTER. The answer .5 appears (Window 3). If you want the answer as a fraction, press MATH 1 ENTER; the answer is shown as 1/2. To guess the other answer, change your guess to 2. Press 2nd ENTER and *then* enter the 2.

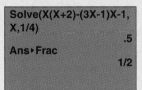

WINDOW 3

Finally, let's find the solutions of $x^3 + 3x^2 - 4x - 12 = 0$ in Example 4. First graph $Y_1 = x^3 + 3x^2 - 4x - 12$ in a standard window. The points at which the graph cuts the horizontal axis, $x = -3$, $x = -2$, and $x = 2$, are the solutions of the equation, as shown in Window 4.

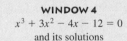

WINDOW 4
$x^3 + 3x^2 - 4x - 12 = 0$ and its solutions

EXERCISE 3.7

A In Problems 1–46, solve the equations.

1. $(x + 1)(x + 2) = 0$

2. $(x + 3)(x + 4) = 0$

3. $(x - 1)(x + 4)(x + 3) = 0$

4. $(x + 5)(x - 3)(x + 2) = 0$

5. $\left(x - \dfrac{1}{2}\right)\left(x - \dfrac{1}{3}\right) = 0$

6. $\left(x - \dfrac{1}{4}\right)\left(x - \dfrac{1}{7}\right) = 0$

7. $y(y - 3) = 0$

8. $y(y - 4) = 0$

9. $y^2 - 64 = 0$

10. $y^2 - 1 = 0$

11. $y^2 - 81 = 0$

12. $y^2 - 100 = 0$

13. $x^2 + 6x = 0$

14. $x^2 + 2x = 0$

15. $x^2 - 3x = 0$

16. $x^2 - 8x = 0$

17. $y^2 - 12y = -27$

18. $y^2 - 10y = -21$

19. $y^2 = -6y - 5$

20. $y^2 = -3y - 2$

21. $x^2 = 2x + 15$

22. $x^2 = 4x + 12$

23. $3y^2 + 5y + 2 = 0$

24. $3y^2 + 7y + 2 = 0$

25. $2y^2 - 3y + 1 = 0$

26. $2y^2 - 3y - 20 = 0$

27. $2y^2 - y - 1 = 0$

28. $2y^2 - y - 15 = 0$

Hint for Problems 29–34: multiply each term by the LCD *first*.

29. $\dfrac{x^2}{12} + \dfrac{x}{3} - 1 = 0$

30. $\dfrac{x^2}{2} - \dfrac{x}{12} - 1 = 0$

31. $\dfrac{x^2}{3} - \dfrac{x}{2} = \dfrac{-1}{6}$ **32.** $\dfrac{x^2}{6} + \dfrac{x}{3} = \dfrac{1}{2}$

33. $\dfrac{x^2}{12} + \dfrac{x}{2} = \dfrac{-2}{3}$ **34.** $\dfrac{x^2}{3} + \dfrac{x}{3} = \dfrac{1}{4}$

35. $(2x - 1)(x - 3) = 3x - 5$

36. $(3x + 1)(x - 2) = x + 7$

37. $(2x + 3)(x + 4) = 2(x - 1) + 4$

38. $(5x - 2)(x + 2) = 3(x + 1) - 7$

39. $(2x - 1)(x - 1) = x - 1$

40. $(3x - 2)(3x - 1) = 1 - 3x$

41. $x^3 + 4x^2 - 4x - 16 = 0$

42. $x^3 - 4x^2 - 4x + 16 = 0$

43. $x^3 - 5x^2 - 9x + 45 = 0$

44. $x^3 + 5x^2 - 9x - 45 = 0$

45. $3x^3 + 3x^2 = 12x + 12$

46. $2x^3 - 2x^2 - 18x + 18 = 0$

B **Use the Pythagorean Theorem to solve Problems 47–50.**

47. The sides of a right triangle are consecutive even integers. Find their lengths.

48. The hypotenuse of a right triangle is 4 cm longer than the shortest side and 2 cm longer than the remaining side. Find the dimensions of the triangle.

49. The hypotenuse of a right triangle is 16 in. longer than the shortest side and 2 in. longer than the remaining side. Find the dimensions of the triangle.

50. The hypotenuse of a right triangle is 8 in. longer than the shortest side and 1 in. longer than the remaining side. Find the dimensions of the triangle.

C **In Problems 51–54, use**

$$h = 5t^2 + V_0 t$$

where h is the distance (in meters) traveled in t seconds by an object thrown downward with an initial velocity V_0 (in meters per second).

51. An object is thrown downward with an initial velocity of 5 m/sec from a height of 10 m. How long does it take the object to hit the ground?

52. An object is thrown downward from a height of 28 m with an initial velocity of 4 m/sec. How long does it take the object to reach the ground?

53. An object is thrown downward from a building 15 m high with an initial velocity of 10 m/sec. How long does it take the object to hit the ground?

54. How long would it take a package thrown downward from a plane with an initial velocity of 10 m/sec to hit the ground 175 m below?

55. It costs a business $(0.1x^2 + x + 50)$ dollars to serve x customers. How many customers can be served if $250 is the cost?

56. The cost of serving x customers is given by $(x^2 + 10x + 100)$ dollars. If $1300 is spent serving customers, how many customers are served?

57. A manufacturer will produce x units of a product when its price is $(x^2 + 25x)$ dollars per unit. How many units will be produced when the price is $350 per unit?

58. When the price of a ton of raw materials is $(0.01x^2 + 5x)$ dollars, a supplier will produce x tons of it. How many tons will be produced when the price is $5000 per ton?

59. To attract more students, the campus theater decides to reduce ticket prices by x dollars from the current $5.50 price.
   a. If the number of tickets sold is $100 + 100x$, what is the new price after the x dollar reduction?
   b. If the revenue is the number of tickets sold times the price of each ticket, what is the revenue?
   c. If the theater wishes to have $750 in revenue, how much is the price reduction?
   d. If the reduction must be less than $1, what is the reduction?

60. An apartment owner wants to increase the monthly rent from the current $250 in n increases of $10.
   a. If the number of apartments rented is $70 - 2n$, what is the new price after the n increases of $10?
   b. If the revenue is the number of apartments rented times the rent, what is the revenue?
   c. If the owner wants to receive $17,980 per month, how many $10 increases can the owner make?
   d. What will be the monthly rent?

SKILL CHECKER

Simplify:

61. $\dfrac{3x^5}{15x^7}$ **62.** $\dfrac{10x^{-3}}{5x^6}$

63. $\dfrac{20x^7}{10x^{-3}}$ **64.** $\dfrac{8x^{-4}}{16x^{-5}}$

65. $\dfrac{18x^{-10}}{9x^2}$ **66.** $\dfrac{4x^{-5}}{2x^{-5}}$

USING YOUR
KNOWLEDGE   **Play Ball!**

Have you been to a baseball game lately? Did anybody hit a home run? The trajectory of a baseball is usually very complicated, but we can get help from *The Physics of Baseball,* by Robert Adair. According to Mr. Adair, after t seconds, starting 1 second after the ball leaves the bat, the height of a ball hit at a $35°$ angle rotating with an initial backspin of 2000 revolutions per minute (rpm) and hit at about 110 mi/hr is given by

$$H(t) = -80t^2 + 340t - 260 \text{ (in feet)}$$

67. How many seconds will it be before the ball hits the ground?

The distance traveled by the ball is given by

$$D(t) = -5t^2 + 115t - 110 \text{ (in feet)}$$

68. How far will the ball travel before it hits the ground?

69. How far will the ball travel in 6 sec, the time it takes a high fly ball to hit the ground?

WRITE ON . . .

70. Write an explanation of the difference between quadratic and linear equations.

71. Explain why the zero-factor property works for more than two numbers whose product is zero.

72. Explain the differences in the procedure for solving $3(x - 1)(x + 4) = 0$ and $3x(x - 1)(x + 4) = 0$.

73. Write a word problem that uses the Pythagorean theorem in its solution.

MASTERY
TEST   If you know how to do these problems, you have learned your lesson!

Solve:

74. $x^3 + 2x^2 - 9x - 18 = 0$ **75.** $8x^2 - 2x = 1$

76. $12x^2 + 13x - 4 = 0$ **77.** $x^2 = 6x - 5$

78. $x^2 + x = 6$ **79.** $x^2 - 16 = 0$

80. $x^2 + 9x = 0$

81. The lengths of one leg of a right triangle and its hypotenuse are consecutive integers. If the shortest leg is 7 units shorter than the longer leg, find the lengths of the three sides.

research
questions

Sources of information for these questions can be found in the *Bibliography* at the end of this book.

1. Who was the first person to use the notation x, xx, x^3, x^4, . . . for exponents, and when did this notation first occur?

2. In Section 3.1, we mentioned that the height of a diver is given by $H(t) = -16t^2 + 118$. An eminent mathematician, born in Pisa in 1564, performed experiments from the leaning tower of Pisa. Find out the name of this mathematician, the nature of his experiments, and his conclusions.

3. How old do you think quadratic equations are? There are clay tablets indicating that the Babylonians of 2000 B.C. were familiar with formulas for solving quadratic equations. Write a paper about Babylonian mathematics with special emphasis on the solution of quadratic equations.

4. About 4000 years ago, the Egyptians used trained surveyors, the *harpedonaptae*. Find out what the word *harpedonaptae* means. Then write a paragraph explaining what these surveyors did and the ways in which they used the Pythagorean theorem in their work.

5. If you are looking for multicultural discovery, the Pythagorean theorem is it! Write a paper about the proofs of Pythagoras' theorem by

  a. The ancient Chinese

  b. Bhaskara

  c. The early Greeks

  d. Euclid

  e. Garfield (not the cat, the twentieth U.S. president!)

  f. Pappus

SUMMARY

SECTION	ITEM	MEANING	EXAMPLE
3.1A	Monomial	A constant times a product of variables with whole-number exponents	$3x^2y$, $-7x$, $0.5x^3$, $\frac{2}{3}x^2yz^4$
	Polynomial	A sum or difference of monomials	$3x^2 - 7x + 8$, $x^2y + y^3$
	Terms	The individual monomials in a polynomial	The terms of $3x^2 - 7x + 8$ are $3x^2$, $-7x$, and 8.
	Coefficient	The numerical factor of a term	The coefficient of $3x^2$ is 3.
	Binomial	A polynomial with two terms	$5x^2 - 7$ is a binomial.
	Trinomial	A polynomial with three terms	$-3 + x^2 + x$ is a trinomial.
3.1B	Degree of a polynomial	Largest sum of the exponents in any term	The degree of $x^3 + 7x$ is 3. The degree of $-2x^3yz^2$ is 6.
3.1E	Commutative property of addition	If P and Q are polynomials, $P + Q = Q + P$.	$x + 3x^2 = 3x^2 + x$
	Associative property of addition	If P, Q, and R are polynomials, $P + (Q + R) = (P + Q) + R$.	$x^2 + (3 + 7x) = (x^2 + 3) + 7x$
	Distributive properties	If P, Q, and R are polynomials, $P(Q + R) = PQ + PR$ $(Q + R)P = QP + RP$	$2x(x^2 + 3) = 2x^3 + 6x$
3.2A	Commutative property of multiplication	If P and Q are polynomials, $P \cdot Q = Q \cdot P$.	$x \cdot 3x^2 = 3x^2 \cdot x$
	Associative property of multiplication	If P, Q, and R are polynomials, $P \cdot (Q \cdot R) = (P \cdot Q) \cdot R$.	$x^2 \cdot (3 \cdot 7x) = (x^2 \cdot 3) \cdot 7x$
3.2C	Product of two binomials	$(x + a)(x + b) = x^2 + (b + a)x + ab$	$(x + 5)(x - 7) = x^2 - 2x - 35$
3.2D	Square of a binomial sum	$(x + a)^2 = x^2 + 2ax + a^2$	$(x + 5y)^2 = x^2 + 10xy + 25y^2$
	Square of a binomial difference	$(x - a)^2 = x^2 - 2ax + a^2$	$(x - 5y)^2 = x^2 - 10xy + 25y^2$
3.2E	Product of a sum and a difference	$(x + a)(x - a) = x^2 - a^2$	$(x + 2y)(x - 2y) = x^2 - 4y^2$
3.3A	Greatest common factor (GCF) of a polynomial in x (ax^n)	1. a is the greatest integer that divides each of the coefficients in the polynomial. 2. n is the smallest exponent of x in all terms of the polynomial.	The GCF of $3x^6 + 6x^3$ is $3x^3$.
3.4A	Factoring $x^2 + (b + a)x + ab$	$x^2 + (b + a)x + ab = (x + a)(x + b)$	$x^2 + 5x + 4 = (x + 1)(x + 4)$

SECTION	ITEM	MEANING	EXAMPLE
3.4C	The *ac* test	$ax^2 + bx + c$ is factorable only if there are two integers whose product is ac and whose sum is b.	$3x^2 + 5x + 2$ is factorable, since there are two integers with product 6 and sum 5.
3.5A	Factoring perfect square trinomials	$x^2 + 2ax + a^2 = (x + a)^2$ $x^2 - 2ax + a^2 = (x - a)^2$	$x^2 + 10x + 25 = (x + 5)^2$ $x^2 - 14x + 49 = (x - 7)^2$
3.5B	Factoring the difference of two squares	$x^2 - a^2 = (x + a)(x - a)$	$16x^2 - 9y^2 = (4x + 3y)(4x - 3y)$
3.5C	Factoring the sum or difference of two cubes	$x^3 + a^3 = (x + a)(x^2 - ax + a^2)$ $x^3 - a^3 = (x - a)(x^2 + ax + a^2)$	$8x^3 + 27 = (2x + 3)(4x^2 - 6x + 9)$ $8x^3 - 27 = (2x - 3)(4x^2 + 6x + 9)$
3.6	General factoring strategy	1. Factor out the GCF. 2. Check for: 　Difference of two squares 　Difference of two cubes 　Sum of two cubes 　Perfect square trinomials 　Trinomials of the form 　　$ax^2 + bx + c$ 　Four terms (grouping) 3. Check by multiplying factors.	
3.7	Quadratic equation	An equation that can be written in the form $ax^2 + bx + c = 0\ (a \neq 0)$	$3x^2 + 5x = -6$ is a quadratic equation.
3.7A	Zero-factor property	For all real numbers a and b, $a \cdot b = 0$ means that $a = 0$ or $b = 0$, or both.	$(x + 1)(x + 2) = 0$ means $x + 1 = 0$ or $x + 2 = 0$.
3.7B	Pythagorean theorem	In any right triangle, the square of the longest side is equal to the sum of the squares of the other two sides: $c^2 = a^2 + b^2$.	If the sides of a right triangle are of lengths 3 and 4 and the hypotenuse is 5, $3^2 + 4^2 = 5^2$.

REVIEW EXERCISES

(If you need help with these exercises, look in the section indicated in brackets.)

1. **[3.1A, B]** Classify as a monomial, binomial, or trinomial and give the degree.
   a. $x^3 + x^2y^3z$ b. $x^3y^2z^3$ c. $x^4 - 5x^2y^3 + xyz$

2. **[3.1C]** Rewrite using exponents and give the degree.
   a. $8xxxyy - 3xxxyy + 7zzz$
   b. $3xxyyzzz + 4xxyy - 4zzzzz$
   c. $6xxyyzz + 3xyzzz - 4xxyyyzzz$

3. **[3.1D]** Let $P(x) = x^2 - 3x + 5$ and find.
   a. $P(-1)$ b. $P(2)$ c. $P(-3)$

4. **[3.1E]** Add $2x^3 + 5x^2 - 3x - 1$ and
   a. $8 - 7x + 3x^2 - x^3$ b. $9 - 8x^2 + 3x^3$
   c. $7 - 4x + x^3$

5. **[3.2E]** Subtract $7x^3 - 5x^2 + 3x - 1$ from
   a. $4x^3 + 2x^2 + 2$ b. $7x^3 + 5x^2 + 4x - 7$
   c. $8x^2 - 9x + 3$

6. **[3.2A]** Multiply.
   a. $-2x^2y(x^2 + 3xy - 2y^3)$ b. $-3x^2y^2(x^2 + 3xy - 2y^3)$
   c. $-4xy^2(x^2 + 3xy - 2y^3)$

7. **[3.2B]** Multiply.
   a. $(x - 1)(x^2 - 3x - 2)$ b. $(x - 2)(x^2 - 3x - 2)$
   c. $(x - 3)(x^2 - 3x - 2)$

8. **[3.2C]** Multiply.
   a. $(2x + 3y)(4x - 5y)$ b. $(2x + 3y)(3x - 2y)$
   c. $(2x + 3y)(5x - 3y)$

9. **[3.2D]** Multiply.
   a. $(2x + 7y)^2$ b. $(3x + 7y)^2$ c. $(4x + 7y)^2$

10. **[3.2D]** Multiply.
    a. $(3x - 7y)^2$ b. $(4x - 7y)^2$ c. $(5x - 7y)^2$

11. **[3.2E]** Multiply.
    a. $(3x + 2y)(3x - 2y)$ b. $(4x + 3y)(4x - 3y)$
    c. $(5x + 3y)(5x - 3y)$

12. **[3.3A]** Factor.
    a. $15x^5 - 20x^4 + 10x^3 + 25x^2$
    b. $9x^5 - 12x^4 + 6x^3 + 15x^2$
    c. $6x^5 - 8x^4 + 4x^3 + 10x^2$

13. **[3.3B]** Factor.
    a. $6x^6 - 2x^4 + 15x^3 - 5x$ b. $6x^6 - 8x^4 + 15x^3 - 20x$
    c. $6x^6 - 4x^4 + 9x^3 - 6x$

14. **[3.4A]** Factor.
    a. $x^2 - 3xy - 18y^2$ b. $x^2 - 4xy - 12y^2$
    c. $x^2 - 5xy - 6y^2$

15. **[3.4B]** Factor.
    a. $2x^2 - 7xy - 30y^2$ b. $2x^2 - 3xy - 20y^2$
    c. $2x^2 - 5xy - 25y^2$

16. **[3.4C]** Factor.
    a. $-18x^4y - 3x^3y^2 + 6x^2y^3$
    b. $-30x^4y - 5x^3y^2 + 10x^2y^3$
    c. $-36x^4y - 6x^3y^2 + 12x^2y^3$

17. **[3.5A]** Factor.
    a. $4x^2 - 28xy + 49y^2$ b. $9x^2 - 42xy + 49y^2$
    c. $16x^2 - 56xy + 49y^2$

18. **[3.5A]** Factor.
    a. $9x^2 + 24xy + 16y^2$ b. $9x^2 + 30xy + 25y^2$
    c. $9x^2 + 36xy + 36y^2$

19. **[3.5B]** Factor.
    a. $81x^4 - y^4$ b. $x^4 - 16y^4$
    c. $81x^4 - 16y^4$

20. **[3.5B]** Factor.
    a. $x^2 - 4x + 4 - y^2$ b. $x^2 - 6x + 9 - y^2$
    c. $x^2 - 8x + 16 - y^2$

21. **[3.5C]** Factor.
    a. $27x^3 + 8y^3$ b. $27x^3 + 64y^3$
    c. $64x^3 + 27y^3$

22. **[3.5C]** Factor.
    a. $27x^3 - 8y^3$ b. $27x^3 - 64y^3$
    c. $64x^3 - 27y^3$

23. **[3.6]** Factor.
    a. $27x^6 - 8x^3y^3$ b. $27x^7 - 64x^4y^3$
    c. $64x^8 - 27x^5y^3$

24. **[3.6]** Factor.
    a. $27x^6 + 3x^4$ b. $4x^6 + 64x^4$ c. $2x^6 + 18x^4$

25. **[3.6]** Factor.
  a. $27x^4 + 36x^3y + 12x^2y^2$ **b.** $36x^4 + 48x^3y + 16x^2y^2$
  c. $45x^4 + 60x^3y + 20x^2y^2$

26. **[3.6]** Factor.
  a. $27x^4 - 36x^3y + 12x^2y^2$ **b.** $36x^4 - 48x^3y + 16x^2y^2$
  c. $45x^4 - 60x^3y + 20x^2y^2$

27. **[3.6]** Factor.
  a. $12x^3y - 44x^2y^2 - 16xy^3$ **b.** $15x^3y - 55x^2y^2 - 20xy^3$
  c. $18x^3y - 66x^2y^2 - 24xy^3$

28. **[3.6]** Factor.
  a. $2x^3 - x^2 - 2x + 1$ **b.** $18x^3 - 9x^2 - 2x + 1$
  c. $32x^3 - 16x^2 - 2x + 1$

29. **[3.7A]** Solve.
  a. $x^2 = -x + 12$ **b.** $x^2 = -x + 20$ **c.** $x^2 = -2x + 24$

30. **[3.7A]** Solve.
  a. $6x^2 + x = 1$ **b.** $8x^2 + 2x = 1$
  c. $10x^2 + 3x = 1$

31. **[3.7A]** Solve.
  a. $x^3 + 2x^2 - x - 2 = 0$ **b.** $x^3 + 4x^2 - x - 4 = 0$
  c. $x^3 + 2x^2 - 9x - 18 = 0$

32. **[3.7B]** Find the dimensions of a right triangle whose sides are
  a. x, $x + 2$, and $x + 4$ units long
  b. x, $x + 3$, and $x + 6$ units long
  c. x, $x + 4$, and $x + 8$ units long

PRACTICE TEST

(Answers on pages 200–201.)

1. Classify as a monomial, binomial, or trinomial and give the degree of $xy^3z^4 - x^7$.

2. Rewrite $3xxxyyyy - 8xxyyyy - 4zzzzz$ using exponents and give its degree.

3. Let $P(x) = x^2 - 3x + 2$. Find $P(-2)$.

4. Add $6x^3 + 8x^2 - 6x - 4$ and $6 - 3x + x^2 - 3x^3$.

5. Subtract $8x^3 - 6x^2 + 5x - 3$ from $5x^3 + 3x^2 + 3$.

6. Multiply $-3x^2y(x^2 + 5xy - 3y^3)$.

7. Multiply $(x - 2)(x^2 - 4x - 5)$.

8. Multiply $(3x + 5y)(4x - 7y)$.

9. Multiply.
  a. $(2x + 3y)^2$ **b.** $(3x - 4y)^2$

10. Multiply $(3x + 4y)(3x - 4y)$.

11. Factor $12x^6 - 16x^5 + 8x^4 + 20x^3$.

12. Factor $6x^7 + 6x^5 + 15x^4 + 15x^2$.

13. Factor.
  a. $x^2 - 3xy - 18y^2$ **b.** $2x^2 + xy - 10y^2$

14. Factor $36x^4y + 12x^3y^2 - 8x^2y^3$.

15. Factor.
  a. $16x^2 - 24xy + 9y^2$ **b.** $9x^2 + 30xy + 25y^2$

16. Factor $x^4 - 16y^4$.

17. Factor $x^2 - 10x + 25 - y^2$.

18. Factor.
  a. $27x^3 + 8y^3$ **b.** $8y^3 - 27x^3$

19. Factor.
  a. $8x^7 - x^4y^3$ **b.** $6x^8 + 24x^6$

20. Factor.
  a. $8x^4 + 24x^3y + 18x^2y^2$ **b.** $48x^2y^2 - 72xy^3 + 27y^4$

21. Factor $9x^3y - 33x^2y^2 - 12xy^3$.

22. Factor $16x^3 - 12x^2 - 4x + 3$.

23. Solve.
  a. $x^2 = -3x + 10$ **b.** $6x^2 + 7x = 3$

24. Solve $x^3 - x^2 - 4x + 4 = 0$.

25. The sides of a right triangle are x, $x + 2$, and $x + 4$ units long. Find the dimensions of the triangle.

ANSWERS TO PRACTICE TEST

Answer	If you missed: Question	Review: Section	Examples	Page
1. Binomial: 8	1	3.1A, B	1, 2	139–140
2. $3x^3y^4 - 8x^2y^4 - 4z^5$; 7	2	3.1C	3	140
3. 12	3	3.1D	4, 5	141–142
4. $3x^3 + 9x^2 - 9x + 2$	4	3.1E	6, 7	143
5. $-3x^3 + 9x^2 - 5x + 6$	5	3.1E	8	144
6. $-3x^4y - 15x^3y^2 + 9x^2y^4$	6	3.2A	1	150
7. $x^3 - 6x^2 + 3x + 10$	7	3.2B	2	151
8. $12x^2 - xy - 35y^2$	8	3.2C	3	153
9a. $4x^2 + 12xy + 9y^2$	9a	3.2D	4	154
9b. $9x^2 - 24xy + 16y^2$	9b	3.2D	4	154
10. $9x^2 - 16y^2$	10	3.2E	5	154–155
11. $4x^3(3x^3 - 4x^2 + 2x + 5)$	11	3.3A	1, 2, 3	159–160
12. $3x^2(x^2 + 1)(2x^3 + 5)$	12	3.3B	4, 5	161–162
13a. $(x + 3y)(x - 6y)$	13a	3.4A	1, 2	165–166
13b. $(x - 2y)(2x + 5y)$	13b	3.4B, C	3, 4, 5	166–167, 168–169
14. $4x^2y(3x + 2y)(3x - y)$	14	3.4C	6	169
15a. $(4x - 3y)^2$	15a	3.5A	1, 2	174
15b. $(3x + 5y)^2$	15b	3.5A	1, 2	174
16. $(x^2 + 4y^2)(x + 2y)(x - 2y)$	16	3.5B	3	175
17. $(x - 5 + y)(x - 5 - y)$ or $(x + y - 5)(x - y - 5)$	17	3.5B	4	175–176
18a. $(3x + 2y)(9x^2 - 6xy + 4y^2)$	18a	3.5C	5	177
18b. $(2y - 3x)(4y^2 + 6xy + 9x^2)$	18b	3.5C	5	177
19a. $x^4(2x - y)(4x^2 + 2xy + y^2)$	19a	3.6	1	181–182
19b. $6x^6(x^2 + 4)$	19b	3.6	1	181–182
20a. $2x^2(2x + 3y)^2$	20a	3.6	2	182
20b. $3y^2(4x - 3y)^2$	20b	3.6	2	182

Answer	If you missed:	Review:		
	Question	Section	Examples	Page
21. $3xy(3x + y)(x - 4y)$	21	3.6	3	182
22. $(4x - 3)(2x + 1)(2x - 1)$	22	3.6	5	183
23a. $2, -5$	23a	3.7A	1, 2, 3	187–190
23b. $\dfrac{1}{3}, -\dfrac{3}{2}$	23b	3.7A	1, 2, 3	187–190
24. $1, 2, -2$	24	3.7A	4	190
25. 6, 8, and 10 units	25	3.7B	5	191

4

Rational Expressions

When you studied arithmetic, you first studied the whole numbers and then formed quotients called rational numbers. We follow the same pattern in algebra: After studying polynomials, we form quotients called rational expressions. We familiarize you with the terminology regarding rational expressions in Section 4.1 and show you how to add, subtract, multiply, and divide them in Sections 4.2 and 4.3. We work with a special type of rational expression called a complex fraction in Section 4.4, and then in Section 4.5, we divide one polynomial by another using long division and a shortened type of division called synthetic division. We solve equations involving rational expressions in Section 4.6 and finish the chapter in Section 4.7 by using our knowledge of rational expressions to solve number, work, and distance problems, and to solve for specific variables in a literal equation.

One of the most spectacular mathematical achievements of the sixteenth century was the discovery of an algebraic solution to cubic and quartic equations. The story of the discovery rivals the plots of contemporary novels. It starts with Scipione del Ferro's formula for solving the cubic $x^3 + mx = n$, a formula he passed on to his pupil Antonio Maria Fior. Enter Nicolo de Brescia, cruelly known as Tartaglia, "the stammerer," because of a speech impediment acquired during the sacking of Brescia by the French, which left his father dead and Nicolo with a sabre cut that cleft his jaw and palate. So poor was his mother that she could only pay his tutor for a meager 15 days, and even then he was relegated to using tombstones as slates on which to work his exercises! Tartaglia announced in 1535 the discovery of a more general formula to solve $x^3 + mx^2 = n$. Fior, believing the announcement to be a bluff, challenged him to a problem-solving contest. Who won? Tartaglia and his two formulas handily defeated Fior, who failed to solve a single problem, thus entering the annals of mathematical history in ignominious defeat.

4.1

RATIONAL EXPRESSIONS

To succeed, review how to:

1. Find the GCF of two expressions (p. 159).
2. Use the properties of exponents (pp. 31–37).
3. Factor polynomials (pp. 165–170, 173–178).

Objectives:

A   Find the numbers that make a rational expression undefined.

B   Write a given fraction with the indicated denominator.

C   Write a fraction in the standard forms.

D   Reduce a fraction to lowest terms.

getting started

An Application to Recycling and Waste Recovery

Do you recycle bottles, newspaper, and plastic? What fraction of the waste generated in the United States is recovered? According to the Environmental Protection Agency the waste generated (in millions of tons) is: $G(t) = 0.04t^2 + 2.34t + 90$, where t is the number of years after 1960. On the other hand, the waste recovered (in millions of tons) is $R(t) = 0.04t^2 - 0.59t + 7.42$. Thus the fraction of the waste recovered is

$$P(t) = \frac{R(t)}{G(t)} = \frac{0.04t^2 - 0.59t + 7.42}{0.04t^2 + 2.34t + 90}$$

The expression

$$\frac{R(t)}{G(t)}$$

is called a *rational expression* or *rational function*. Rational expressions are the topic of this section. By the way, to find out what fraction of our waste is expected to be recovered in the year 2000, 40 years after 1960, find

$$P(4) = \frac{R(40)}{G(40)} = \frac{47.82}{247.6} = 19.31\%$$

Can you compare this with the percent recovered in 1960?

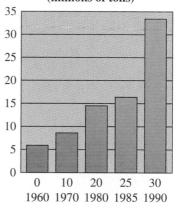

Materials recovered (millions of tons)

0	10	20	25	30
1960	1970	1980	1985	1990

The word *fraction* is derived from the Latin word *fractio,* which means "to break" or "to divide." Any **fraction** of the form $\frac{a}{b}$, where a and b are integers and $b \neq 0$, is a rational number. We extend this idea to polynomial expressions as follows:

RATIONAL EXPRESSION

If P and Q are polynomials, the algebraic expression

$$\frac{P}{Q} \qquad (Q \neq 0)$$

is a **rational expression**. If $Q = 0$, the expression is **undefined**.

Thus

$$\frac{x^2 - 2x + 1}{x + 2}$$

is undefined when $x + 2 = 0$, that is, when $x = -2$. Similarly,

$$\frac{1}{y^2 - 1}$$

is undefined when $y^2 - 1 = (y + 1)(y - 1) = 0$, that is, when $y = 1$ or $y = -1$. On the other hand,

$$\frac{x - 2}{3}$$

is *never* undefined, since the denominator 3 is *never* zero.

 Finding the Values That Make a Rational Expression Undefined

The values that make a rational expression undefined are the values that make the denominator of the fraction zero. For instance,

$$\frac{1}{x + 3} \qquad \text{is undefined for} \qquad x = -3$$

EXAMPLE 1   Finding values that make a rational expression undefined

For what values is the following rational expression

$$\frac{x^2 + 3x + 8}{x^2 + 2x - 3}$$

undefined?

SOLUTION   The rational expression is undefined for values of x that make the denominator zero. We find these values by setting the denominator equal to zero and solving the resulting equation.

$$x^2 + 2x - 3 = 0 \qquad \text{Set the denominator equal to zero.}$$
$$(x - 1)(x + 3) = 0 \qquad \text{Factor.}$$
$$x - 1 = 0 \quad \text{or} \quad x + 3 = 0 \qquad \text{Use the zero-factor property.}$$
$$x = 1 \quad \text{or} \quad x = -3 \qquad \text{Solve each equation.}$$

Thus the numbers 1 and -3 make the rational expression undefined.   ■

To avoid mentioning over and over that the denominators of algebraic fractions must not be zero, we make the following rule.

UNDEFINED RATIONAL EXPRESSIONS

> The variables in a rational expression may not be replaced by values that will make the denominator zero.

 Writing Fractions with an Indicated Denominator

The fraction $\frac{4}{5}$ can be written as an equivalent fraction with a denominator of 10 by multiplying both numerator and denominator by 2 to obtain

$$\frac{4}{5} = \frac{4 \cdot 2}{5 \cdot 2} = \frac{8}{10}$$

We can generalize this idea to rational expressions as follows.

FUNDAMENTAL PROPERTY OF FRACTIONS

If P, Q, and K are polynomials,

$$\frac{P}{Q} = \frac{P \cdot K}{Q \cdot K}$$

for all values for which the denominator is not zero.

Thus to write

$$\frac{5x^3}{3y^2}$$

with a denominator of $6y^7$, first write the new equivalent fraction with the old denominator factored out:

$$\frac{5x^3}{3y^2} = \frac{?}{3y^2(2y^5)}$$

$\quad\quad\quad$ Multiply by $2y^5$ to obtain $6y^7$.

Since the multiplier is $2y^5$, we have

Multiply by $2y^5$.

$$\frac{5x^3}{3y^2} = \frac{5x^3(2y^5)}{3y^2(2y^5)} = \frac{10x^3y^5}{6y^7}$$

We are multiplying the denominator by $2y^5$, so we have to multiply the numerator by $2y^5$.

Thus

$$\frac{5x^3}{3y^2} = \frac{10x^3y^5}{6y^7}$$

EXAMPLE 2   Writing fractions with a specified denominator

Write:

a. $\dfrac{5}{8}$ with a denominator of 16

b. $\dfrac{2x^2}{9y^3}$ with a denominator of $18y^8$

c. $\dfrac{3x + 1}{x - 1}$ with a denominator of $x^2 + 2x - 3$

SOLUTION

a. $\dfrac{5}{8} = \dfrac{?}{16}$

$\quad$ Multiply by 2.

Multiply by 2.

$$\frac{5}{8} = \frac{10}{16}$$

Thus $\dfrac{5}{8} = \dfrac{10}{16}$.

b. Since $18y^8 = 9y^3(2y^5)$,

$$\frac{2x^2}{9y^3} = \frac{?}{9y^3(2y^5)}$$

└─ Multiply by $2y^5$. ─┘

┌─ Multiply by $2y^5$. ─┐

$$\frac{2x^2}{9y^3} = \frac{2x^2(2y^5)}{9y^3(2y^5)} = \frac{4x^2y^5}{18y^8}$$

c. We first note that $x^2 + 2x - 3 = (x - 1)(x + 3)$. Thus

$$\frac{3x + 1}{x - 1} = \frac{?}{(x - 1)(x + 3)}$$

└─ Multiply by $(x + 3)$. ─┘

┌─ Multiply by $(x + 3)$. ─┐

$$\frac{3x + 1}{x - 1} = \frac{(3x + 1)(x + 3)}{(x - 1)(x + 3)} = \frac{3x^2 + 10x + 3}{x^2 + 2x - 3}$$ ∎

Writing a Fraction in the Standard Forms

It's important to know that there are three signs associated with a fraction:

1. The sign before the fraction

2. The sign of the numerator

3. The sign of the denominator

Using our definition of a quotient and the fundamental property of fractions, we can conclude that

$$\frac{-a}{b} = \frac{a}{-b} = -\frac{a}{b} = -\frac{-a}{-b} \quad \text{and} \quad \frac{a}{b} = \frac{-a}{-b} = -\frac{a}{-b} = -\frac{-a}{b}$$

The forms $\frac{-a}{b}$ and $\frac{a}{b}$, in which the sign of the fraction and that of the denominator are positive, are called the **standard forms** of the fractions. Thus $\frac{-2}{9}$ and $\frac{4}{7}$ are in standard form, but $\frac{2}{-9}$ and $\frac{-4}{-7}$ are not. Of course, in expressions with more than one term in the numerator or denominator, there are alternative standard forms. For example,

$$\underbrace{\frac{-1}{x - y}} = \underbrace{\frac{-1}{-(y - x)}} = \frac{1}{y - x} \qquad \text{Recall that } x - y = -(y - x), \text{ since } -(y - x) = -y + x = x - y.$$

$\underbrace{}_{\text{same}}$

Thus either

$$\frac{-1}{x - y} \qquad \text{or} \qquad \frac{1}{y - x}$$

can be used as the standard form. We prefer

$$\frac{1}{y - x}$$

because it has only one minus sign, whereas

$$\frac{-1}{x - y}$$

has two.

EXAMPLE 3   Writing fractions in standard form

Write in standard form:

a. $\dfrac{x}{-2}$ **b.** $-\dfrac{-3}{y}$ **c.** $-\dfrac{x-y}{5}$

SOLUTION

a. $\dfrac{x}{-2} = \dfrac{-x}{2}$ **b.** $-\dfrac{-3}{y} = \dfrac{3}{y}$ **c.** $-\dfrac{x-y}{5} = \dfrac{-(x-y)}{5}$, or $\dfrac{y-x}{5}$ ■

D Reducing Fractions to Lowest Terms

The fundamental property of fractions can also be used to **simplify** (**reduce**) fractions—that is, to write fractions as equivalent ones in which no integers other than 1 can be divided exactly into both the numerator and denominator. For example, the fraction $\frac{14}{21}$ can be simplified by writing the numerator and denominator in factored form and using the fundamental principle of fractions. Thus

$$\frac{14}{21} = \frac{2 \cdot \overset{1}{\cancel{7}}}{3 \cdot \underset{1}{\cancel{7}}} = \frac{2}{3}$$

Here we are dividing the numerator and denominator by the common factor 7. We usually write

$$\frac{\overset{2}{\cancel{14}}}{\underset{3}{\cancel{21}}}$$

and say that the fraction $\frac{2}{3}$ is in **lowest terms**.

   The fraction

$$\frac{(x + 3)(x^2 - 4)}{3(x + 2)(x^2 + x - 6)}$$

can also be written in lowest terms by using the fundamental property of fractions. We do it by steps.

PROCEDURE

> **Procedure for Reducing Fractions**
>
> 1. Write the numerator and denominator of the fraction in factored form.
> 2. Find the greatest common factor (GCF) of the numerator and denominator.
> 3. Replace the quotient of the common factors by the number 1, since $\frac{a}{a} = 1$.
> 4. Rewrite the fraction in lowest terms.

We are now ready to simplify

$$\frac{(x + 3)(x^2 - 4)}{3(x + 2)(x^2 + x - 6)}$$

Here are the steps:

1. Write the numerator and denominator in factored form.

$$\frac{(x + 3)(x + 2)(x - 2)}{3(x + 2)(x + 3)(x - 2)}$$

2. Find the GCF of the numerator and denominator (we rearranged factors so the same ones are in columns).

$$\frac{(x + 2)(x + 3)(x - 2)}{3(x + 2)(x + 3)(x - 2)}$$

3. Replace the quotient of the common factors by the number 1.

$$\frac{\overset{1}{\cancel{(x + 2)}}\cancel{(x + 3)}\cancel{(x - 2)}}{3\cancel{(x + 2)}\cancel{(x + 3)}\cancel{(x - 2)}}$$

4. Rewrite the fraction in lowest terms.

$$\frac{1}{3}$$

The whole procedure can be written as

$$\frac{(x + 3)(x^2 - 4)}{3(x + 2)(x^2 + x - 6)} = \frac{\overset{1}{\cancel{(x + 3)}}\overset{1}{\cancel{(x + 2)}}\overset{1}{\cancel{(x - 2)}}}{3\cancel{(x + 2)}\cancel{(x + 3)}\cancel{(x - 2)}} = \frac{1}{3}$$

> ⚠ **CAUTION** Only common factors can be divided out. It is incorrect to write
>
> $$\frac{\cancel{x} + 8}{\cancel{x} + 4} = 2$$
>
> The x is *not* a factor.

EXAMPLE 4   Reducing fractions to lowest terms

Reduce each fraction to lowest terms:

a. $\dfrac{x^3 y^4}{xy^6}$ **b.** $\dfrac{xy - y^2}{x^2 - y^2}$ **c.** $\dfrac{2x + xy}{x}$

SOLUTION

a. $\dfrac{x^3 y^4}{xy^6} = \dfrac{x^2 \cdot \overset{1}{\cancel{xy^4}}}{y^2 \cdot \cancel{xy^4}}$ Factor the numerator and denominator using the GCF as a factor.

$\phantom{\dfrac{x^3 y^4}{xy^6}} = \dfrac{x^2}{y^2}$ Divide out xy^4, the GCF.

b. $\dfrac{xy - y^2}{x^2 - y^2} = \dfrac{y \cdot \cancel{(x - y)}}{(x + y)\cancel{(x - y)}}$ Factor the numerator and denominator using the GCF as a factor.

$\phantom{\dfrac{xy - y^2}{x^2 - y^2}} = \dfrac{y}{x + y}$ Divide out $(x - y)$, the GCF.

c. $\dfrac{2x + xy}{x} = \dfrac{\overset{1}{\cancel{x}}(2 + y)}{\cancel{x}}$ Factor the numerator and denominator using the GCF as a factor.

$\phantom{\dfrac{2x + xy}{x}} = 2 + y$ Divide out x, the GCF. ■

CAUTION In Example 4b, the final answer is

$$\frac{y}{x+y}.$$

The answer cannot be reduced further. A common mistake is to try to divide out the y. This is not correct! You can only divide out *factors*, and y is not a factor of both the numerator and denominator.

Now, look at the expression

$$\frac{a-b}{b-a}$$

Are there any common factors in the numerator and denominator? Since $-(b-a) = -b + a = a - b$, we write

$$\frac{a-b}{b-a} = \frac{-(b-a)}{b-a} = -1$$

Thus

Quotient of Additive Inverses

$$\frac{a-b}{b-a} = -1$$

We use this idea in Example 5.

EXAMPLE 5   Quotients involving additive inverses

Simplify:

a. $\dfrac{x^3 - y^3}{y - x}$ **b.** $\dfrac{x^2 - y^2}{y^3 - x^3}$

SOLUTION

a. $\dfrac{x^3 - y^3}{y - x} = \dfrac{\overset{-1}{\cancel{(x - y)}}(x^2 + xy + y^2)}{\cancel{y - x}}$ Factor the numerator and denominator.

$\phantom{\dfrac{x^3 - y^3}{y - x}} = -1(x^2 + xy + y^2)$ $\dfrac{x - y}{y - x} = -1$

$\phantom{\dfrac{x^3 - y^3}{y - x}} = -(x^2 + xy + y^2)$

b. $\dfrac{x^2 - y^2}{y^3 - x^3} = \dfrac{(x + y)(x - y)}{(y - x)(y^2 + xy + x^2)}$ Factor the numerator and denominator.

$\phantom{\dfrac{x^2 - y^2}{y^3 - x^3}} = \dfrac{\overset{-1}{\cancel{(x - y)}}(x + y)}{\cancel{(y - x)}(y^2 + xy + x^2)}$ $\dfrac{x - y}{y - x} = -1$

$\phantom{\dfrac{x^2 - y^2}{y^3 - x^3}} = \dfrac{-(x + y)}{y^2 + xy + x^2}$ ∎

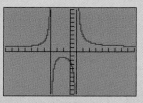

WINDOW 1

$$Y_1 = \frac{(x^2 + 3x + 8)}{(x^2 + 2x - 3)}$$

using the connected mode

Your grapher can help you determine the points at which a rational fraction is undefined. For instance, in Example 1, enter

$$Y_1 = \frac{x^2 + 3x + 8}{x^2 + 2x - 3}$$

and graph. (Don't forget to enter the parentheses in the numerator and denominator!) You will notice two vertical lines crossing the horizontal axis at $x = -3$ and $x = 1$, the two points at which the fraction is undefined (see Window 1). The reason for this is that the grapher is trying to connect all the dots to produce a smoother picture. The graph of a rational fraction with a zero denominator at $x = a$ will show a vertical line at $x = a$.

We can verify this by using the dot mode. With a TI-82, press MODE, go to line 5, press ▶ to move right and select "DOT." Now press ENTER GRAPH. The new graph, using dots that are *not* connected, shows that there is *nothing* at $x = 1$ and $x = -3$ (see Window 2). To further confirm your suspicions, press TRACE and note that as you approach $x = -3$ from the left, the value of y increases. If you continue to TRACE the graph, x becomes greater than -3, and y becomes negative. However, x is never -3. You can do the same for $x = 1$.

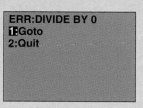

WINDOW 2

$$Y_1 = \frac{(x^2 + 3x + 8)}{(x^2 + 2x - 3)}$$

using the dot mode

Another way to verify the undefined point is to evaluate the function at $x = -3$. To do this with a TI-82, go to the home screen (2nd MODE) store the -3 in memory by pressing ─ 3 STO X,T,θ ENTER. Since Y_1 is already entered, evaluate Y_1 at -3—that is, $Y_1(-3)$— by pressing 2nd Y-VARS ENTER ENTER and entering the -3 in parentheses. When you press ENTER again, the grapher gives you an error message (see Window 3).

Finally, the grapher can also help with reducing fractions. Enter

$$Y_1 = \frac{(x + 3)(x^2 - 4)}{3(x + 2)(x^2 + x - 6)}$$

You will get a horizontal line. To see it better, change the window to $[-1, 1]$ by $[-1, 1]$. The trace cursor will verify that $y = 0.33333333$, which is an approximation for $y = \frac{1}{3}$, the simplified value of the given rational fraction.

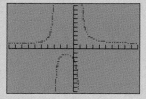

ERR:DIVIDE BY 0
1:Goto
2:Quit

WINDOW 3

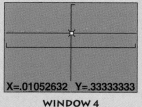

X=.01052632 Y=.33333333

WINDOW 4

$$\frac{(x + 3)(x^2 - 4)}{3(x + 2)(x^2 + x - 6)}$$

in simplified form is $\frac{1}{3}$.

EXERCISE 4.1

A In Problems 1–12, find the values (if any exist) that make the rational expression undefined.

1. $\dfrac{x}{x + 3}$

2. $\dfrac{y}{y - 4}$

3. $\dfrac{5m - 5}{5}$

4. $\dfrac{4q + 4}{4}$

5. $\dfrac{m + 7}{m^2 - m - 2}$

6. $\dfrac{u - 9}{u^2 + 6u + 5}$

7. $\dfrac{p^2 - 2p}{p^2 - 9}$

8. $\dfrac{z^2 - 3z - 4}{9z^2 - 4}$

9. $\dfrac{a^2 + 4}{2a^2 - 11a - 6}$

10. $\dfrac{b^2 + 9}{6a^2 - 5a - 6}$

11. $\dfrac{4v^2 - 9}{4v^2 + 9}$

12. $\dfrac{9y^2 - 16}{16 + y^2}$

B In Problems 13–30, write the given rational expression with the indicated denominator.

13. $\dfrac{2x}{3y}$;   denominator $6y^3$

14. $\dfrac{-3y}{2x}$;   denominator $8x^2$

15. $\dfrac{x}{x + y}$;   denominator $x^2 - y^2$

16. $\dfrac{-y}{x - y}$; denominator $x^2 - y^2$

17. $\dfrac{-x}{y - x} = \dfrac{?}{y^2 - x^2}$ 18. $\dfrac{-4x}{y - x} = \dfrac{?}{y^2 - x^2}$

19. $\dfrac{-x}{2x - 3y}$; denominator $4x^2 - 9y^2$

20. $\dfrac{-x}{2x - y}$; denominator $4x^2 - y^2$

21. $\dfrac{4x}{x + 1} = \dfrac{?}{x^2 - x - 2}$ 22. $\dfrac{5y}{y - 1} = \dfrac{?}{y^2 + 2y - 3}$

23. $\dfrac{-5x}{x + 3}$; denominator $x^2 + x - 6$

24. $\dfrac{-3y}{y - 4}$; denominator $y^2 - 2y - 8$

25. $\dfrac{3}{x + y}$; denominator $x^3 + y^3$

26. $\dfrac{-4}{x + y}$; denominator $x^3 + y^3$

27. $\dfrac{x}{x - y} = \dfrac{?}{x^3 - y^3}$

28. $\dfrac{-y}{x - y} = \dfrac{?}{x^3 - y^3}$

29. $\dfrac{x}{x^2 - xy + y^2} = \dfrac{?}{x^3 + y^3}$

30. $\dfrac{x}{x^2 + xy + y^2} = \dfrac{?}{x^3 - y^3}$

C In Problems 31–40, write each fraction in standard form.

31. $-\dfrac{y}{-2}$ 32. $-\dfrac{x - 3}{y}$ 33. $-\dfrac{x}{x - 5}$

34. $\dfrac{2x - y}{-x}$ 35. $-\dfrac{-2x}{-5y}$ 36. $\dfrac{-x}{-y}$

37. $\dfrac{-(x + y)}{-(x - y)}$ 38. $-\dfrac{-(3x + y)}{-(x - 5y)}$ 39. $\dfrac{-1}{-(x - 2)}$

40. $\dfrac{-y}{-(x + 1)}$

D In Problems 41–70, simplify each fraction.

41. $\dfrac{x^4 y^2}{xy^5}$ 42. $\dfrac{x^5 y^3 c^2}{x^2 y^6 c^4}$ 43. $\dfrac{3x - 3y}{x - y}$

44. $\dfrac{4x^2}{4x - 4y}$ 45. $\dfrac{3x - 2y}{9x^2 - 4y^2}$ 46. $\dfrac{4x^2 - 9y^2}{2x + 3y}$

47. $\dfrac{(x - y)^3}{x^2 - y^2}$ 48. $\dfrac{x^2 - y^2}{(x + y)^3}$ 49. $\dfrac{ay^2 - ay}{ay}$

50. $\dfrac{a^3 + 2a^2 + a}{a}$ 51. $\dfrac{x^2 + 2xy + y^2}{x^2 - y^2}$ 52. $\dfrac{x^2 + 3x + 2}{x^2 + 2x + 1}$

53. $\dfrac{y^2 - 8y + 15}{y^2 + 3y - 18}$ 54. $\dfrac{y^2 + 7y - 18}{y^2 - 3y + 2}$ 55. $\dfrac{2 - y}{y - 2}$

56. $\dfrac{3(x - y)}{4(y - x)}$ 57. $\dfrac{9 - x^2}{x - 3}$ 58. $\dfrac{25 - 9x^2}{3x - 5}$

59. $\dfrac{y^3 - 8}{2 - y}$ 60. $\dfrac{2 + x}{x^3 + 8}$ 61. $\dfrac{3x - 2y}{2y - 3x}$

62. $\dfrac{5y - 2x}{2x - 5y}$ 63. $\dfrac{x^2 + 4x - 5}{1 - x}$ 64. $\dfrac{x^2 - 2x - 15}{5 - x}$

65. $\dfrac{x^2 - 6x + 8}{4 - x}$ 66. $\dfrac{x^2 - 8x + 15}{3 - x}$ 67. $\dfrac{2 - x}{x^2 + 4x - 12}$

68. $\dfrac{3 - x}{x^2 + 3x - 18}$ 69. $-\dfrac{3 - x}{x^2 - 5x + 6}$ 70. $-\dfrac{4 - x}{x^2 - 3x - 4}$

SKILL CHECKER

Simplify:

71. $\dfrac{4x^3}{2x}$ 72. $\dfrac{8x^4}{2x}$ 73. $\dfrac{-16x^4}{8x^2}$ 74. $\dfrac{-48x^5}{12x^3}$

Factor:

75. $x^2 + x - 12$ 76. $x^2 - 2x - 15$

77. $9x^2 - 4y^2$ 78. $4x^2 - 9y^2$

79. $x^3 - 1$ 80. $x^3 - 8$

81. $8x^3 + 1$ 82. $27x^3 + 8$

Multiply:

83. $\dfrac{3}{7} \cdot \dfrac{14}{9}$ 84. $-\dfrac{5}{8} \cdot \dfrac{16}{15}$ 85. $\left(-\dfrac{4}{9}\right)\left(-\dfrac{27}{8}\right)$

Divide:

86. $\left(-\dfrac{3}{4}\right) \div \left(-\dfrac{3}{8}\right)$ 87. $-\dfrac{4}{5} \div \dfrac{8}{15}$

88. $\left(\dfrac{6}{7}\right) \div \left(-\dfrac{3}{14}\right)$

Buying Pollution Permits

89. Can you buy a permit to pollute the air? Unfortunately, yes! On March 29, 1993, the Chicago Board of Trade sold a permit for about 21 million dollars to emit 150,000 tons of sulfur dioxide. (By the way, this amount represents 1% of the yearly sulfur dioxide emissions.) If the price (in millions) for removing p% of the sulfur dioxide is given by

$$\frac{2100p}{100 - p}$$

find the price that should be paid to remove
a. 20%   **b.** 40%   **c.** 60%

d. Can we afford to remove 100% of the sulfur dioxide? Explain.

90. The demand for a product is given by

$$N(p) = \frac{2p + 100}{10p + 10}$$

where $N(p)$ represents the number of units people are willing to buy when the price is p dollars ($1 \le p \le 10$).
a. Reduce this expression to lowest terms.

b. Find the demand when the price is $3.

c. What happens to the demand as the price increases?

91. A vendor's profit (in dollars) for the sale of x sunglasses is

$$\frac{5x^2 - 5}{x + 1} \qquad (1 \le x \le 100)$$

a. Reduce this expression to lowest terms.

b. What is the profit when 10 sunglasses are sold?

c. What is the maximum profit possible?

WRITE ON . . .

We have already mentioned that when reducing rational expressions, you may divide out only factors. Explain what is wrong with the simplifications in Problems 92 and 93.

92. $\dfrac{\cancel{x} + y}{\cancel{x}} = 1 + y$ **93.** $\dfrac{\cancel{y}}{x + \cancel{y}} = \dfrac{1}{x + 1}$

94. In this section, we used the fundamental property of fractions in two different ways. Write a paragraph explaining these two ways.

95. Just before Example 4, we showed that

$$\frac{(x + 3)(x^2 - 4)}{3(x + 2)(x^2 + x - 6)} = \frac{1}{3}$$

Is this equation always true? Explain why or why not.

96. Explain that since

$$\frac{a - b}{b - a} = -1, \qquad \text{then} \qquad \frac{a - b}{-a + b} = -1$$

97. Explain how you would reduce

$$\frac{4x^2 - 7x + 1}{-4x^2 + 7x - 1}$$

and then explain how to reduce a rational expression where the numerator and denominator differ only in sign.

If you know how to do these problems, you have learned your lesson!

Reduce to lowest terms:

98. $\dfrac{y^3 - x^3}{y - x}$ **99.** $\dfrac{x^5 y^7}{xy^3}$ **100.** $\dfrac{3y + xy}{y}$

101. $\dfrac{x^2 - y^2}{y^3 - x^3}$ **102.** $\dfrac{x^2 - xy}{x^2 - y^2}$ **103.** $\dfrac{4y^2 - xy^2}{y^2}$

104. For what values of x is $\dfrac{x - 1}{x^2 + 2x - 3}$ undefined?

105. For what values of x is $\dfrac{x + 1}{x^2 - 1}$ undefined?

Write in standard form:

106. $\dfrac{x}{-7}$ **107.** $-\dfrac{-4}{x}$ **108.** $-\dfrac{a - b}{8}$

Write:

109. $\dfrac{7}{8}$ with a denominator of 16

110. $\dfrac{3x^2}{8y^3}$ with a denominator of $24y^6$

111. $\dfrac{4x + 1}{x + 2}$ with a denominator of $x^2 - x - 6$

4.2

MULTIPLICATION AND DIVISION OF RATIONAL EXPRESSIONS

To succeed, review how to:

1. Simplify quotients using the properties of exponents (p. 34).

2. Multiply polynomials (pp. 151–158).

Objectives:

A Multiply rational expressions.

B Divide rational expressions.

C Use multiplication and division together.

getting started   It's All in the Cards

A regular deck of 52 cards has 4 kings. The probability that you pick two kings when drawing two cards is

$$\frac{4}{52} \cdot \frac{3}{51} = \frac{4}{4 \cdot 13} \cdot \frac{3}{3 \cdot 17} \qquad \text{Factor the denominators.}$$

$$= \frac{\overset{1}{\cancel{4}}}{\cancel{4} \cdot 13} \cdot \frac{\overset{1}{\cancel{3}}}{\cancel{3} \cdot 17} \qquad \text{Simplify.}$$

$$= \frac{1}{221} \qquad \text{Multiply.}$$

We can save time by simplifying in the first step:

$$\frac{4}{52} \cdot \frac{3}{51} = \frac{\overset{1}{\cancel{4}}}{\underset{13}{\cancel{52}}} \cdot \frac{\overset{1}{\cancel{3}}}{\underset{17}{\cancel{51}}}$$

In this section, we shall learn how to multiply rational expressions using rules involving the three steps we have here: factor, simplify, and multiply.

A   ### Multiplying Rational Expressions

The multiplication of rational expressions follows the same procedure as the multiplication of rational numbers. Here is the definition from Section 1.2.

MULTIPLICATION OF RATIONAL EXPRESSIONS

If a, b, c, and d are real numbers,

$$\frac{a}{b} \cdot \frac{c}{d} = \frac{a \cdot c}{b \cdot d} \qquad (b \neq 0, d \neq 0)$$

PROCEDURE

> **Procedure to Multiply Rational Expressions**
> 1. **Factor** the numerators and denominators completely.
> 2. **Simplify** each rational expression completely.
> 3. **Multiply** the remaining factors in the numerator and denominator.
> 4. Make sure the final product is in **lowest terms**.

In the case of

$$\frac{6x^3}{2y} \cdot \frac{4y^5}{9x^2}$$

both numerators and denominators are to be simplified and reduced, so we proceed as follows:

$$\frac{6x^3}{2y} \cdot \frac{4y^5}{9x^2} = \frac{24x^3y^5}{18x^2y} = \frac{4xy^4 \cdot 6x^2y}{3 \cdot 6x^2y} = \frac{4xy^4}{3}$$

Note that, as we did in the *Getting Started,* the common factors in the numerator and denominator can be divided out *before* doing the multiplications in the numerator and denominator. The procedure goes like this:

$$\frac{6x^3}{2y} \cdot \frac{4y^5}{9x^2} = \frac{2x \cdot 3x^2}{2y} \cdot \frac{2y^4 \cdot 2y}{3 \cdot 3x^2} = \frac{4xy^4}{3}$$

Remember to divide out common factors before you do the multiplications!

$$\frac{x^2 + 2x - 8}{x^2 + 5x + 6} \cdot \frac{x + 2}{x + 4} = \frac{(x - 2)(x + 4)}{(x + 2)(x + 3)} \cdot \frac{(x + 2)}{(x + 4)} \qquad \text{Factor.}$$

$$= \frac{(x - 2)(x + 4)}{(x + 2)(x + 3)} \cdot \frac{(x + 2)}{(x + 4)} \qquad \text{Simplify.}$$

$$= \frac{x - 2}{x + 3}$$

EXAMPLE 1   Multiplying rational expressions

Multiply:

a. $\dfrac{x - 1}{2x - 3y} \cdot \dfrac{4x^2 - 9y^2}{2x^2 - x - 1}$

b. $\dfrac{x^2 - 4}{4x^2 - 9y^2} \cdot \dfrac{2x^2 - 3xy}{2x + 4}$

SOLUTION

a. $\dfrac{x - 1}{2x - 3y} \cdot \dfrac{4x^2 - 9y^2}{2x^2 - x - 1} = \dfrac{(x - 1)}{(2x - 3y)} \cdot \dfrac{(2x + 3y)(2x - 3y)}{(x - 1)(2x + 1)}$ Factor.

$$= \frac{(x - 1)}{(2x - 3y)} \cdot \frac{(2x + 3y)(2x - 3y)}{(x - 1)(2x + 1)} \qquad \text{Simplify.}$$

$$= \frac{2x + 3y}{2x + 1}$$

b. $\dfrac{x^2 - 4}{4x^2 - 9y^2} \cdot \dfrac{2x^2 - 3xy}{2x + 4} = \dfrac{(x + 2)(x - 2)}{(2x - 3y)(2x + 3y)} \cdot \dfrac{x(2x - 3y)}{2(x + 2)}$ Factor.

$= \dfrac{\cancel{(x + 2)}(x - 2)}{\cancel{(2x - 3y)}(2x + 3y)} \cdot \dfrac{x\cancel{(2x - 3y)}}{2\cancel{(x + 2)}}$ Simplify.

$= \dfrac{x(x - 2)}{2(2x + 3y)}$ Multiply.

$= \dfrac{x^2 - 2x}{4x + 6y}$ ■

EXAMPLE 2   More practice at multiplying rational expressions

Multiply:

a. $\dfrac{2 - x}{x + 1} \cdot \dfrac{x^2 + 3x + 2}{x^2 - 4}$ **b.** $\dfrac{x^2 + 3x + 2}{x^2 + 5x + 4} \cdot \dfrac{x^2 + 2x - 3}{x^2 + x - 2}$

SOLUTION

a. $\dfrac{2 - x}{x + 1} \cdot \dfrac{x^2 + 3x + 2}{x^2 - 4} = \dfrac{(2 - x)}{(x + 1)} \cdot \dfrac{(x + 1)(x + 2)}{(x + 2)(x - 2)}$ Factor.

$= \dfrac{\overset{-1}{\cancel{(2 - x)}}}{\cancel{(x + 1)}} \cdot \dfrac{\cancel{(x + 1)}\cancel{(x + 2)}}{\cancel{(x + 2)}\cancel{(x - 2)}}$ Simplify. Recall that

$\dfrac{2 - x}{x - 2} = -1.$

$= -1$

b. $\dfrac{x^2 + 3x + 2}{x^2 + 5x + 4} \cdot \dfrac{x^2 + 2x - 3}{x^2 + x - 2} = \dfrac{(x + 2)(x + 1)}{(x + 4)(x + 1)} \cdot \dfrac{(x - 1)(x + 3)}{(x - 1)(x + 2)}$ Factor.

$= \dfrac{\cancel{(x + 2)}\cancel{(x + 1)}}{(x + 4)\cancel{(x + 1)}} \cdot \dfrac{\cancel{(x - 1)}(x + 3)}{\cancel{(x - 1)}\cancel{(x + 2)}}$ Simplify.

$= \dfrac{x + 3}{x + 4}$ ■

B   Dividing Rational Expressions

To divide one rational expression by another, we use the definition of division from Section 1.2:

DIVISION OF REAL NUMBERS

If a, b, c, and d are real numbers,

$$\frac{a}{b} \div \frac{c}{d} = \frac{a}{b} \cdot \frac{d}{c} \qquad (b, d, \text{ and } c \neq 0)$$

Thus the quotient of two rational expressions is the product of the first and the reciprocal of the second. For example,

$$\frac{2}{5} \div \frac{3}{8} = \frac{2}{5} \cdot \frac{8}{3} = \frac{16}{15}$$ The reciprocal of $\frac{3}{8}$ is $\frac{8}{3}$.

Reciprocal

and

$$\frac{3x^2}{2y} \div \frac{6x}{4y^2} = \frac{3x^2}{2y} \cdot \frac{4y^2}{6x}$$

Reciprocal

$$= \frac{\overset{1xy}{\cancel{12x^2y^2}}}{\cancel{12xy}}$$ Answers should be given in simplified form.

$$= xy$$

EXAMPLE 3   Dividing rational expressions

Divide:

a. $\dfrac{x^3y}{z} \div \dfrac{x^2y}{z^4}$ **b.** $\dfrac{5x^2 - 5}{3x + 6} \div \dfrac{x + 1}{3}$ **c.** $\dfrac{x + 3}{x - 3} \div (x^2 + 6x + 9)$

SOLUTION

a. $\dfrac{x^3y}{z} \div \dfrac{x^2y}{z^4} = \dfrac{x^3y}{z} \cdot \dfrac{z^4}{x^2y}$

Reciprocal

$$= \frac{\overset{xz^3}{\cancel{x^3yz^4}}}{\cancel{x^2yz}}$$

$$= xz^3$$

b. $\dfrac{5x^2 - 5}{3x + 6} \div \dfrac{x + 1}{3} = \dfrac{5(x + 1)(x - 1)}{3(x + 2)} \cdot \dfrac{3}{x + 1}$ Factor.

Reciprocal

$$= \frac{5\cancel{(x + 1)}(x - 1)}{\cancel{3}(x + 2)} \cdot \frac{\cancel{3}}{\cancel{x + 1}}$$ Simplify.

$$= \frac{5x - 5}{x + 2}$$ Multiply.

c. We first write $x^2 + 6x + 9$ as $\dfrac{x^2 + 6x + 9}{1}$.

$$\frac{x + 3}{x - 3} \div (x^2 + 6x + 9) = \frac{x + 3}{x - 3} \div \frac{x^2 + 6x + 9}{1}$$

$$= \frac{x + 3}{x - 3} \cdot \frac{1}{x^2 + 6x + 9}$$

$$= \frac{(x + 3)}{(x - 3)} \cdot \frac{1}{(x + 3)(x + 3)}$$ Factor.

$$= \frac{\cancel{(x + 3)}}{(x - 3)} \cdot \frac{1}{\cancel{(x + 3)}(x + 3)}$$ Simplify.

$$= \frac{1}{x^2 - 9}$$ Multiply.   ■

Using Multiplication and Division

Sometimes both multiplications and divisions are involved, as shown in Example 4.

EXAMPLE 4　Operations involving multiplications and divisions

Perform the indicated operations:

$$\frac{3-x}{x+2} \div \frac{x^3-27}{x+4} \cdot \frac{x^3+8}{x+4}$$

SOLUTION　We first rewrite the division as a multiplication by the reciprocal:

$$\frac{3-x}{x+2} \div \frac{x^3-27}{x+4} \cdot \frac{x^3+8}{x+4}$$

$$= \frac{3-x}{x+2} \cdot \frac{x+4}{x^3-27} \cdot \frac{x^3+8}{x+4}$$　Change the division to multiplication by the reciprocal.

$$= \frac{(3-x)}{(x+2)} \cdot \frac{(x+4)}{(x-3)(x^2+3x+9)} \cdot \frac{(x+2)(x^2-2x+4)}{x+4}$$　Factor.

$$= \frac{\overset{-1}{(3-x)}}{(x+2)} \cdot \frac{(x+4)}{(x-3)(x^2+3x+9)} \cdot \frac{(x+2)(x^2-2x+4)}{(x+4)}$$　Simplify. Recall that $\frac{3-x}{x-3} = -1$.

$$= \frac{-(x^2-2x+4)}{x^2+3x+9}$$　Multiply $(-1 \cdot a = -a)$.　■

EXAMPLE 5　More practice at using multiplication and division

Perform the indicated operations:

$$\frac{x^3+2x^2-x-2}{x+3} \div \frac{x^3+8}{x^2-9} \cdot \frac{1}{x^2-1}$$

SOLUTION　We first rewrite the division as a multiplication by the reciprocal to obtain

$$\frac{x^3+2x^2-x-2}{x+3} \div \frac{x^3+8}{x^2-9} \cdot \frac{1}{x^2-1}$$

$$= \frac{x^3+2x^2-x-2}{x+3} \cdot \frac{x^2-9}{x^3+8} \cdot \frac{1}{x^2-1}$$

Now we have the polynomial $x^3 +2x^2 - x - 2$ with four terms in the first numerator. We factor it by grouping

$$x^3 + 2x^2 - x - 2 = x^2(x+2) - 1 \cdot (x+2)$$

$$= (x+2)(x^2-1)$$

$$= (x+2)(x+1)(x-1)$$

and factor as much as possible to get

$$\frac{(x + 2)(x + 1)(x - 1)}{(x + 3)} \cdot \frac{(x + 3)(x - 3)}{(x + 2)(x^2 - 2x + 4)} \cdot \frac{1}{(x + 1)(x - 1)} \qquad \text{Factor.}$$

$$= \frac{(x + 2)(x + 1)(x - 1)}{(x + 3)} \cdot \frac{(x + 3)(x - 3)}{(x + 2)(x^2 - 2x + 4)} \cdot \frac{1}{(x + 1)(x - 1)} \qquad \text{Simplify.}$$

$$= \frac{x - 3}{x^2 - 2x + 4} \qquad \text{Multiply.} \quad \blacksquare$$

GRAPH IT

A grapher can be used to check the multiplication of rational expressions involving one variable. In Example 2a, use a $[-2, 2]$ by $[-2, 2]$ window and enter

$$Y_1 = \frac{(2 - x)}{(x + 1)} \cdot \frac{(x^2 + 3x + 2)}{(x^2 - 4)}$$

Note the parentheses! The result is $Y_1 = -1$ as before (see Window 1). For part 2b, enter

$$Y_1 = \frac{(x^2 + 3x + 2)}{(x^2 + 5x + 4)} \cdot \frac{(x^2 + 2x - 3)}{(x^2 + x - 2)}$$

and

$$Y_2 = \frac{(x + 3)}{(x + 4)}$$

Note the parentheses again! The graphs of Y_1 and Y_2 are identical (see Window 2). Some graphers (TI-82) give additional support for this fact by using the [TRACE] and [▲] [▼] keys. When you press [▲] and then [▼], the little number at the top right-hand side of your window changes from 1 to 2, telling you that there are really two curves being graphed. Note that the [TRACE] key gives the same x- and y-coordinates for both graphs. Of course, division problems can be verified similarly. For Example 3b, you need to be extra careful with the parentheses. Note the extra set when entering Y_1:

$$Y_1 = \left(\frac{(5x^2 - 5)}{(3x + 6)} \right) \div \left(\frac{(x + 1)}{3} \right) \qquad \text{and} \qquad Y_2 = \frac{(5x - 5)}{(x + 2)}$$

The graphs for Y_1 and Y_2 are identical (Window 3). You can use the trace feature to confirm this.

To do Example 5, we use a new technique to avoid multiple sets of parentheses. Enter

$$Y_1 = \frac{(x^3 + 2x^2 - x - 2)}{(x + 3)}, \qquad Y_2 = \frac{(x^3 + 8)}{(x^2 - 9)} \qquad \text{and} \qquad Y_3 = \frac{1}{(x^2 - 1)}$$

Now for Y_4, enter $Y_1 \div Y_2 \cdot Y_3$ by pressing [2nd] [Y-VARS] 1 1 [÷] [2nd] [Y-VARS] 1 2 [×] [2nd] [Y-VARS] 1 3. Turn Y_1, Y_2 and Y_3 **off** (so their graphs don't clutter the window) by pressing [Y=], going to Y_1 and placing the cursor on top of the = sign. Then press [ENTER]. The Y_1 is now off. Do the same for Y_2 and Y_3. Finally, let

$$Y_5 = \frac{(x - 3)}{(x^2 - 2x + 4)}$$

The graphs of Y_4 and Y_5 are identical (Window 4). (If you want to see this better, make the vertical scale go from -2 to 2.)

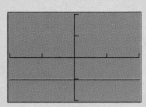

WINDOW 1

$$\frac{(2 - x)}{(x + 1)} \cdot \frac{(x^2 + 3x + 2)}{(x^2 - 4)}$$

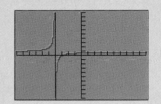

WINDOW 2

$$Y_2 = \frac{(x + 3)}{(x + 4)}$$

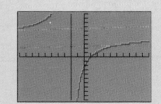

WINDOW 3

$$Y_2 = \frac{(5x - 5)}{(x + 2)}$$

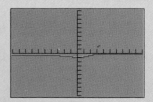

WINDOW 4

$$Y_5 = \frac{(x - 3)}{(x^2 - 2x + 4)}$$

EXERCISE 4.2

A In Problems 1–20, perform the indicated multiplications.

1. $\dfrac{3}{4} \cdot \dfrac{2}{5}$

2. $\dfrac{-9}{10} \cdot \dfrac{2}{3}$

3. $\dfrac{14x^2}{15} \cdot \dfrac{5}{7x}$

4. $\dfrac{-5x^3}{7y} \cdot \dfrac{4y^3}{9x^6}$

5. $\dfrac{-2xy^4}{9z^5} \cdot \dfrac{-3z}{7x^3y^3}$

6. $\dfrac{-35x^5z}{24x^3y^9} \cdot \dfrac{84x^3y^8}{15x^4y^7z}$

7. $\dfrac{10x + 50}{6x + 6} \cdot \dfrac{12}{5x + 25}$

8. $\dfrac{x + y}{xy - y^2} \cdot \dfrac{y^2}{x^2 - y^2}$

9. $\dfrac{6y + 3}{2y^2 - 3y - 2} \cdot \dfrac{y^2 - 4}{3y + 6}$

10. $\dfrac{y^2 + 9y + 18}{y - 2} \cdot \dfrac{2y - 4}{5y + 15}$

11. $\dfrac{y - x}{x^2 + 2xy} \cdot \dfrac{5x + 10y}{x^2 - y^2}$

12. $\dfrac{2 - 2x}{9x^2 - 25} \cdot \dfrac{6x - 10}{x^2 - 1}$

13. $\dfrac{3y^2 - 17y + 10}{y^2 - 4y - 5} \cdot \dfrac{y^2 + 3y + 2}{y^2 + y - 2}$

14. $\dfrac{y^2 + 2y - 3}{y^2 - 4y - 5} \cdot \dfrac{y^2 - 3y - 10}{y^2 + 5y - 6}$

15. $\dfrac{y^2 + 2y - 8}{y^2 + 7y + 12} \cdot \dfrac{y^2 + 2y - 3}{y^2 - 3y + 2}$

16. $\dfrac{y^2 + 2y - 15}{y^2 - 7y + 10} \cdot \dfrac{y^2 - 6y + 8}{y^2 - y - 12}$

17. $\dfrac{x^3 - 8}{4 - x^2} \cdot \dfrac{x^2 + x - 2}{x^2 + 2x + 4}$

18. $\dfrac{x^3 + y^3}{y^2 - x^2} \cdot \dfrac{x - y}{x^2 - xy + y^2}$

19. $\dfrac{a^3 + b^3}{a^3 - b^3} \cdot \dfrac{a^2 + ab + b^2}{a^2 - ab + b^2}$

20. $\dfrac{a^3 - 8}{a^2 + 2a + 4} \cdot \dfrac{a^2 + 3a + 9}{a^3 - 27}$

B In Problems 21–40, perform the indicated division.

21. $\dfrac{3}{5} \div \dfrac{10}{9}$

22. $\dfrac{3}{7} \div \dfrac{-9}{14}$

23. $\dfrac{4}{5x^2} \div \dfrac{12}{25x^3}$

24. $\dfrac{6x^2}{7} \div \dfrac{30x}{28}$

25. $\dfrac{24a^2b}{7c^2d} \div \dfrac{8ab}{21cd^2}$

26. $\dfrac{16a^3b}{15a} \div \dfrac{12ab^2}{20b^4}$

27. $\dfrac{3x - 3}{x} \div \dfrac{x^2 - 1}{x^2}$

28. $\dfrac{5x^2 - 45}{x^3} \div \dfrac{x + 3}{x}$

29. $\dfrac{y^2 - 25}{y^2 - 4} \div \dfrac{3y - 15}{4y - 8}$

30. $\dfrac{y^2 + y - 12}{y^2 - 1} \div \dfrac{3y + 12}{4y^2 + 4y}$

31. $\dfrac{a^3 + b^3}{a^3 - b^3} \div \dfrac{a^2 - ab + b^2}{a^2 + ab + b^2}$

32. $\dfrac{a^2 + ab + b^2}{a^3 + b^3} \div \dfrac{a^2 + ab + b^2}{a^2 - ab + b^2}$

33. $\dfrac{8a^3 - 1}{6u^4w^3} \div \dfrac{1 - 2a}{3u^2w}$

34. $\dfrac{-b^2c}{27a^3 - 1} \div \dfrac{b^3c^2}{3a - 1}$

35. $\dfrac{x - x^3}{2x^2 + 6x} \div \dfrac{5x^2 - 5x}{2x + 6}$

36. $\dfrac{121y - y^3}{y^2 - 49} \div \dfrac{y^2 - 11y}{y + 7}$

37. $\dfrac{y^2 + y - 12}{y^2 - 8y + 15} \div \dfrac{3y^2 + 7y - 20}{2y^2 - 7y - 15}$

38. $\dfrac{3y^2 + 11y + 6}{4y^2 + 16y + 7} \div \dfrac{3y^2 - y - 2}{2y^2 - y - 28}$

39. $\dfrac{4x^2 - 12x + 9}{25 - 4x^2} \div \dfrac{6x^2 - 5x - 6}{6x^2 + 19x + 10}$

40. $\dfrac{4x^2 + 12x + 9}{9 - 4x^2} \div \dfrac{10x^2 + 27x + 18}{8x^2 - 2x - 15}$

C In Problems 41–54, perform the indicated operations.

41. $\dfrac{x^2 + 2x - 3}{x - 5} \div \dfrac{x^2 + 6x + 9}{x^2 - 2x - 15} \cdot \dfrac{1}{x^2 - 1}$

42. $\dfrac{x^2 - 3x + 2}{x^2 - 5x + 6} \div \dfrac{x^2 - 5x + 4}{x^2 - 7x + 12} \cdot \dfrac{x^3 + 1}{x^2 - 1}$

43. $\dfrac{x^2 - 1}{x^2 + 3x - 10} \div \dfrac{x^2 - 3x - 4}{x^2 - 25} \cdot \dfrac{x - 2}{x - 5}$

44. $\dfrac{x^2 - 25}{x^2 - 49} \cdot \dfrac{x^2 - 4x - 21}{x^2 - 10x + 25} \div \dfrac{x^2 + 2x - 3}{x^2 - 6x + 5}$

45. $\dfrac{x - 3}{3 - x} \cdot \dfrac{x^2 + 3x - 4}{x^2 + 7x + 12} \div \dfrac{x^2 + x - 2}{x^2 + 5x + 6}$

46. $\dfrac{x^3 - 125}{x^3 - 8} \cdot \dfrac{x^2 + x - 2}{x^2 + 6x - 7} \div \dfrac{x^2 - 3x - 10}{x^2 + 5x - 14}$

47. $\dfrac{x^2 - y^2}{x^2 - 2xy} \div \dfrac{x^2 + xy - 2y^2}{x^2 - 4y^2} \cdot \dfrac{x^2}{(x + y)^2}$

48. $\dfrac{x^2 + xy - 2y^2}{x^2 - 4y^2} \div \dfrac{x^2 - y^2}{x^2 - 2xy} \cdot \dfrac{(x + y)^2}{x^2}$

49. $\dfrac{x^2 + 2xy - 3y^2}{y^2 - 7y + 10} \div \dfrac{x^2 - 3xy + 2y^2}{y^2 - 3y - 10} \cdot \dfrac{x^2 - 4y^2}{x^2 - 9y^2}$

50. $\dfrac{x^2 + 2xy - 8y^2}{x^2 + 7xy + 12y^2} \div \dfrac{x^2 - 3xy + 2y^2}{x^2 + 2xy - 3y^2} \cdot \dfrac{x^2 - 9}{9 - x^2}$

51. $\dfrac{x^3 + x^2 - x - 1}{x + 2} \div \dfrac{x^3 + 1}{x + 3} \cdot \dfrac{x + 2}{x - 1}$

52. $\dfrac{x^3 + 3x^2 - 9x - 27}{x + 4} \div \dfrac{x^3 + 27}{x + 4} \cdot \dfrac{x + 1}{x - 3}$

53. $\dfrac{x - 2}{x^2 - 9} \div \dfrac{x^3 - 8}{x + 3} \cdot \dfrac{x - 3}{x}$

54. $\dfrac{x - 1}{x^2 - 25} \div \dfrac{x - 3}{x^3 + 125} \cdot \dfrac{(x - 5)^2}{x - 1}$

55. If the price for x units of a product is

$$\frac{3x + 9}{4}$$

and the demand for these is

$$\frac{600}{x^2 + 3x}$$

find the product of the price and the demand.

56. If the price for x units of a product is given by

$$\frac{4x + 8}{5}$$

and the demand is

$$\frac{200}{x^2 + 2x}$$

what is the product of the price and the demand?

57. In a simple electrical circuit, the current I is the quotient of the voltage E and the resistance R. If the current changes with the time t according to

$$R = \frac{t^2 + 9}{t^2 + 6t + 9}$$

and the voltage changes according to the formula

$$E = \frac{4t}{t + 3}$$

find the current I.

58. If in Problem 57 the resistance is

$$R = \frac{t^2 + 5}{t^2 + 4t + 4}$$

and the voltage is

$$E = \frac{5t}{t + 2}$$

find the current I.

Write the rational expression with the indicated denominator:

59. $\dfrac{x - 1}{x - 3}$;   denominator $x^2 - x - 6$

60. $\dfrac{x + 4}{x - 3}$;   denominator $x^2 - 9$

61. $\dfrac{1}{x^2 + 2x + 4}$;   denominator $x^3 - 8$

62. $\dfrac{1}{x^2 - 3x + 9}$;   denominator $x^3 + 27$

More Applications

63. When studying parallel resistors, the expression

$$R \cdot \frac{R_T}{R - R_T}$$

occurs, where R is a known resistance and R_T is a required resistance. Perform the indicated multiplication.

64. The molecular model predicts that the pressure of a gas is given by

$$\frac{2}{3} \cdot \frac{mv^2}{2} \cdot \frac{N}{v}$$

where m is the mass, N is a constant, and v is the velocity. Perform the indicated multiplication.

65. Suppose a store orders 3000 items each year. If it orders x units at a time, the number N of reorders is

$$N = \frac{3000}{x}$$

If there is a fixed $20 reorder fee and a $3 charge per item, the cost of each order is

$$C = 20 + 3x$$

The yearly reorder cost R is then given by

$$R = N \cdot C$$

Find R.

The formula for the area A of a rectangle is $A = LW$, where L is the length and W is the width of the rectangle. In Problems 66 and 67, find the area of the shaded rectangle.

66.

	$\frac{3x+2}{3}$	$\frac{3x+2}{3}$	$\frac{3x+2}{3}$
$\frac{x}{2}$			
$\frac{x}{2}$			

67.

	$\frac{2W-L}{2}$	$\frac{2W-L}{2}$
$\frac{w}{3}$		
$\frac{w}{3}$		
$\frac{w}{3}$		

68. Write in your own words the procedure you use to multiply two rational expressions.

69. Write in your own words the procedure you use to divide two rational expressions.

70. When multiplying $(x + 2)(x + 3)$, we get $x^2 + 5x + 6$. When multiplying

$$\frac{x+2}{2} \cdot \frac{x+3}{3}$$

most textbooks write the answer as

$$\frac{(x+2)(x+3)}{6}$$

Do you agree? Why or why not?

MASTERY TEST   If you know how to do these problems, you have learned your lesson!

Perform the indicated operations:

71. $\dfrac{2-x}{x+3} \div \dfrac{x^3-8}{x-5} \cdot \dfrac{x^3+27}{x-5}$

72. $\dfrac{x^5 y}{z} \div \dfrac{x^3 y}{z^6}$

73. $\dfrac{3x^2-3}{5x+10} \div \dfrac{x+1}{5}$

74. $\dfrac{x+2}{x-2} \div (x^2+4x+4)$

75. $\dfrac{3-x}{x+2} \cdot \dfrac{x^2+5x+6}{x^2-9}$

76. $\dfrac{x^2+4x+3}{x^2+5x+6} \cdot \dfrac{x^2-2x-8}{x^2-2x-3}$

77. $\dfrac{x^3+x^2-x-1}{x+4} \div \dfrac{x^3+1}{x^2-16} \cdot \dfrac{1}{x^2-1}$

78. $\dfrac{x-2}{3x-2y} \cdot \dfrac{9x^2-4y^2}{2x^2-3x-2}$

79. $\dfrac{x^2-9}{9x^2-4y^2} \cdot \dfrac{3x^2-2xy}{3x+9}$

4.3

ADDITION AND SUBTRACTION OF RATIONAL EXPRESSIONS

To succeed, review how to:

1. Add and subtract real numbers (pp. 15–16).
2. Write a fraction with a given denominator (p. 205).
3. Factor polynomials (pp. 165–170, 173–178).

Objectives:

A Add or subtract rational expressions with the same denominator.

B Add or subtract rational expressions with different denominators.

getting started

What Type of Materials Will Be Recovered in the Year 2000?

In the *Getting Started* for Section 4.1, we mentioned that the fraction of the waste recovered is

$$P(t) = \frac{R(t)}{G(t)} = \frac{0.04t^2 - 0.59t + 7.42}{0.04t^2 + 2.34t + 90}$$

where *t* is number of years after 1960. Most of this waste is paper and paperboard. As a matter of fact, the fraction of paper and paperboard recovered is given by

$$Q(t) = \frac{0.02t^2 - 0.25t + 6}{0.04t^2 + 2.34t + 90}$$

Can you find the fraction of the waste recovered that is *not* paper and paperboard? To do this, you need to find $P(t) - Q(t)$. Fortunately, $P(t)$ and $Q(t)$ have the same denominator, so you only need to subtract the numerators and keep the denominator. Make sure you understand that the subtraction sign must be distributed to every term in the numerator of the fraction that follows it, like this:

$$= \frac{0.04t^2 - 0.59t + 7.42}{0.04t^2 + 2.34t + 90} - \frac{0.02t^2 - 0.25t + 6}{0.04t^2 + 2.34t + 90}$$

$$= \frac{0.04t^2 - 0.59t + 7.42 - (0.02t^2 - 0.25t + 6)}{0.04t^2 + 2.34t + 90}$$

$$= \frac{0.04t^2 - 0.59t + 7.42 - 0.02t^2 + 0.25t - 6}{0.04t^2 + 2.34t + 90}$$

$$= \frac{0.02t^2 - 0.34t + 1.42}{0.04t^2 + 2.34t + 90}$$

If you let $t = 40$ in this expression, you can find out what fraction of the waste will *not* be paper or paperboard in the year 2000. You should come out with 0.08 or 8%. Thus most of the material recovered (92%) is paper or paperboard.

A **Adding and Subtracting Rational Expressions with the Same Denominator**

In general, if *a, b,* and *c* are real numbers ($b \neq 0$),

$$\frac{a}{b} + \frac{c}{b} = \frac{a + c}{b}$$ ⟵ Add numerators.
⟵ Keep the denominator.

$$\frac{a}{b} - \frac{c}{b} = \frac{a - c}{b}$$ ⟵ Subtract numerators.
⟵ Keep the denominator.

The important thing to remember is that to add (or subtract) rational expressions with the *same* denominators, we add (or subtract) the numerators and *keep* the denominator. Thus

$$\frac{2}{x} + \frac{6}{x} = \frac{2+6}{x} = \frac{8}{x} \qquad \begin{array}{l} \longleftarrow \text{ Add numerators.} \\ \longleftarrow \text{ Keep the denominator.} \end{array}$$

Similarly,

$$\frac{5x}{x^2+1} + \frac{2x}{x^2+1} = \frac{5x+2x}{x^2+1} = \frac{7x}{x^2+1}$$

$$\frac{3x}{7(x-1)^2} + \frac{4x}{7(x-1)^2} = \frac{3x+4x}{7(x-1)^2} = \frac{7x}{7(x-1)^2} = \frac{x}{(x-1)^2}$$

and

$$\frac{8x}{x^2+5} - \frac{2x}{x^2+5} = \frac{8x-2x}{x^2+5} = \frac{6x}{x^2+5}$$

$$\frac{10x}{9(x-3)^2} - \frac{x}{9(x-3)^2} = \frac{10x-x}{9(x-3)^2} = \frac{9x}{9(x-3)^2} = \frac{x}{(x-3)^2}$$

Note that we write the final answer in simplified form.

EXAMPLE 1   Adding and subtracting with the same denominator

Perform the indicated operations:

a. $\dfrac{8x}{3(x-2)} + \dfrac{x}{3(x-2)}$
 b. $\dfrac{7x}{5(x+4)^2} + \dfrac{3x}{5(x+4)^2}$

c. $\dfrac{x}{x^2-1} - \dfrac{1}{x^2-1}$
 d. $\dfrac{4x}{x+2} - \dfrac{3x-2}{x+2}$

SOLUTION

a. $\dfrac{8x}{3(x-2)} + \dfrac{x}{3(x-2)} = \dfrac{8x+x}{3(x-2)} = \dfrac{\overset{3}{\cancel{9}}x}{\cancel{3}(x-2)} = \dfrac{3x}{x-2}$

b. $\dfrac{7x}{5(x+4)^2} + \dfrac{3x}{5(x+4)^2} = \dfrac{7x+3x}{5(x+4)^2} = \dfrac{\overset{2}{\cancel{10}}x}{\cancel{5}(x+4)^2} = \dfrac{2x}{(x+4)^2}$

c. Since both expressions have the same denominator, we subtract numerators and use the same denominator. However, the answer can be simplified. You can see that only when the denominator is completely factored. Make sure your denominators are in factored form so you can determine all the possible ways to simplify the answer.

$$\frac{x}{x^2-1} - \frac{1}{x^2-1} = \frac{x-1}{x^2-1} \qquad \begin{array}{l} \longleftarrow \text{ Subtract numerators.} \\ \longleftarrow \text{ Keep denominators.} \end{array}$$

$$= \frac{\cancel{x-1}}{(x+1)(\cancel{x-1})} \qquad \text{Divide out the common factor } (x-1).$$

$$= \frac{1}{x+1}$$

d. We indicate the subtraction of the *numerators* by using parentheses. Be careful with the signs when removing the parentheses.

$$\frac{4x}{x+2} - \frac{3x-2}{x+2} = \frac{4x - (3x-2)}{x+2}$$

$$= \frac{4x - 3x + 2}{x+2} \qquad \text{Recall that } -(3x-2) = -1(3x-2)$$
$$\qquad\qquad\qquad\qquad\qquad = -3x+2$$

$$= \frac{x+2}{x+2} = 1 \qquad \text{Combine like terms and simplify.} \quad \blacksquare$$

B Adding and Subtracting Rational Expressions with Different Denominators

To add or subtract fractions with different denominators, we must first find a common denominator. It is most convenient to use the smallest one available, called the **Least Common Denominator (LCD)**—that is, the smallest multiple of the denominators. Thus to add

$$\frac{5}{12} + \frac{7}{18}$$

we start by writing 12 and 18 as a product of *primes*. (See Section 3.3.) Thus we have

$$12 = 2 \cdot 2 \cdot 3 \quad = 2^2 \cdot 3 \qquad \text{Note that the 2s are in one column and the 3s are}$$
$$18 = \quad 2 \cdot 3 \cdot 3 = 2 \cdot 3^2 \qquad \text{written in another column.}$$

Since we need the *smallest* number that is a multiple of 12 and 18, we select the factors raised to the *greatest* power in each column. The product of these factors is the LCD. Thus the LCD of 12 and 18 is $2^2 \cdot 3^2 = 4 \cdot 9 = 36$. We then write each fraction with a denominator of 36 and add:

$$\frac{5}{12} = \frac{5 \cdot 3}{12 \cdot 3} = \frac{15}{36} \qquad \text{We multiply the denominator of } \frac{5}{12} \text{ by 3 (to get 36),}$$
$$\text{so we do the same to the numerator.}$$

$$\frac{7}{18} = \frac{7 \cdot 2}{18 \cdot 2} = \frac{14}{36} \qquad \text{We also multiply the numerator and denominator of}$$
$$\frac{7}{18} \text{ by 2 to get 36 as the denominator.}$$

$$\frac{5}{12} + \frac{7}{18} = \frac{15}{36} + \frac{14}{36} = \frac{29}{36}$$

The procedure used to find the LCD of two or more rational expressions is similar to that used to find the LCD of two numbers.

PROCEDURE

Finding the LCD of Two or More Rational Expressions

1. Factor each denominator. (Place identical factors in columns.)

2. From each column, select the factor with the greatest exponent.

3. The product of all the factors obtained in step 2 is the LCD.

If the denominators involved have no common factors, the LCD is the product of the denominators. For example, the denominators in $\frac{3}{5}$ and $\frac{1}{7}$ have no common factors.

The LCD is $5 \cdot 7$. To subtract $\frac{1}{7}$ from $\frac{3}{5}$, we first write each fraction with a denominator of 35 and then subtract. Here are the steps.

1. The LCD is $5 \cdot 7 = 35$.

2. Write each fraction with 35 as denominator.

$$\frac{3}{5} = \frac{3 \cdot 7}{5 \cdot 7} = \frac{21}{35} \quad \text{and} \quad \frac{1}{7} = \frac{1 \cdot 5}{7 \cdot 5} = \frac{5}{35}$$

3. Subtract: $\dfrac{3}{5} - \dfrac{1}{7} = \dfrac{21}{35} - \dfrac{5}{35} = \dfrac{16}{35}$.

EXAMPLE 2   Adding and subtracting with different denominators

Perform the indicated operation:

$$\frac{2x}{x + 1} - \frac{x}{x + 2}$$

SOLUTION $(x + 1)$ and $(x + 2)$ do not have any common factors. Thus the LCD in

$$\frac{2x}{x + 1} - \frac{x}{x + 2} \quad \text{is} \quad (x + 1)(x + 2)$$

Now rewrite

$$\frac{2x}{x + 1} \quad \text{and} \quad \frac{x}{x + 2}$$

with the LCD as the denominator.

$$\frac{2x}{x + 1} = \frac{2x(x + 2)}{(x + 1)(x + 2)} \quad \text{and} \quad \frac{x}{x + 2} = \frac{x(x + 1)}{(x + 2)(x + 1)}$$

$$\frac{2x}{x + 1} - \frac{x}{x + 2} = \frac{2x(x + 2)}{(x + 1)(x + 2)} - \frac{x(x + 1)}{(x + 1)(x + 2)}$$

$$= \frac{2x(x + 2) - x(x + 1)}{(x + 1)(x + 2)} \qquad \text{Subtract.}$$

$$= \frac{2x^2 + 4x - x^2 - x}{(x + 1)(x + 2)} \qquad \text{Use the distributive property,}$$
$$\qquad\qquad\qquad\qquad\qquad\qquad -x(x + 1) = -x^2 - x.$$

$$= \frac{x^2 + 3x}{(x + 1)(x + 2)} \qquad \text{Combine like terms in the numerator.}$$

$$= \frac{x(x + 3)}{(x + 1)(x + 2)} \qquad \text{Simplify.} \qquad\blacksquare$$

In general, to add or subtract fractions with different denominators, we use the following procedure.

PROCEDURE

Procedure to Add (or Subtract) Fractions with Different Denominators

1. Find the LCD.

2. Write all fractions as equivalent ones with the LCD as the denominator.

3. Add (or subtract) numerators; keep the LCD as denominator.

4. Simplify if possible.

EXAMPLE 3   Using the LCD to add and subtract with different denominators

Perform the indicated operations:

a. $\dfrac{x+1}{x^2+x-2} + \dfrac{x+3}{x^2-1}$

b. $\dfrac{x-1}{x^2-x-6} - \dfrac{x+4}{x^2-9}$

SOLUTION

a. We use the four-step procedure to add fractions.

1. We first find the LCD of the denominators. Write the denominators in factored form with the same factors in a column.

$$x^2 + x - 2 = \ (x+2) \ (x-1)$$
$$x^2 - 1 = \qquad\quad (x-1) \ (x+1)$$

Select the factors with the greatest exponents in each column, $(x+2)$ in column 1, $(x-1)$ in column 2, and $(x+1)$ in column 3. The LCD is the product of these factors, that is,

$$(x+2)(x-1)(x+1)$$

2. We then write each fraction as an equivalent one with the LCD as the denominator.

$$\frac{x+1}{x^2+x-2} = \frac{x+1}{(x+2)(x-1)} = \frac{(x+1)(x+1)}{(x+2)(x-1)(x+1)}$$

$$\frac{x+3}{x^2-1} = \frac{x+3}{(x+1)(x-1)} = \frac{(x+3)(x+2)}{(x+1)(x-1)(x+2)}$$

$$= \frac{(x+3)(x+2)}{(x+2)(x-1)(x+1)}$$

3. $\dfrac{x+1}{x^2+x-2} + \dfrac{x+3}{x^2-1} = \dfrac{(x+1)(x+1)}{(x+2)(x-1)(x+1)} + \dfrac{(x+3)(x+2)}{(x+2)(x-1)(x+1)}$

$$= \frac{(x^2 + 2x + 1) + (x^2 + 5x + 6)}{(x+2)(x-1)(x+1)} \qquad \text{Add the numerators; keep the denominator.}$$

$$= \frac{2x^2 + 7x + 7}{(x+2)(x-1)(x+1)} \qquad \text{Combine like terms in the numerator.}$$

4. The answer cannot be simplified; that is, $2x^2 + 7x + 7$ is *not* factorable.

b. To subtract fractions we again use the four-step procedure.

1. To find the LCD, we factor the denominators keeping the same factors in a column:

$$x^2 - x - 6 = \qquad (x-3)(x+2)$$
$$x^2 - 9 = (x+3)(x-3)$$

Thus the LCD is $(x+3)(x-3)(x+2)$.

2. Write each fraction as an equivalent one with the LCD as the denominator:

$$\frac{x-1}{x^2-x-6} = \frac{(x-1)(x+3)}{(x-3)(x+2)(x+3)} = \frac{(x-1)(x+3)}{(x+3)(x-3)(x+2)}$$

$$\frac{x+4}{x^2-9} = \frac{(x+4)(x+2)}{(x+3)(x-3)(x+2)}$$

3. $\dfrac{x-1}{x^2-x-6} - \dfrac{x+4}{x^2-9} = \dfrac{(x-1)(x+3)}{(x+3)(x-3)(x+2)} - \dfrac{(x+4)(x+2)}{(x+3)(x-3)(x+2)}$

$$= \frac{(x^2+2x-3)-(x^2+6x+8)}{(x+3)(x-3)(x+2)}$$ Subtract the numerators; keep the denominator.

$$= \frac{x^2+2x-3-x^2-6x-8}{(x+3)(x-3)(x+2)}$$ Remember that $-(x^2+6x+8) = -x^2-6x-8.$

$$= \frac{-4x-11}{(x+3)(x-3)(x+2)}$$ Combine like terms in the numerator.

4. The answer is not reducible. ■

How would you add $\frac{6}{12} + \frac{1}{8}$? You can start by finding the LCD, 24. However, it is easier to reduce $\frac{6}{12}$ to $\frac{1}{2}$ first. Then we simply add $\frac{1}{2} + \frac{1}{8} = \frac{4}{8} + \frac{1}{8} = \frac{5}{8}$. We illustrate a similar problem in Example 4.

EXAMPLE 4 Simplifying before adding and subtracting

Perform the indicated operations:

a. $\dfrac{x+y}{x^2+2xy+y^2} + \dfrac{x-y}{x^2-2xy+y^2}$ **b.** $\dfrac{x}{(x+2)(x-2)} - \dfrac{2}{(2-x)(x+2)}$

SOLUTION

a. $\dfrac{x+y}{x^2+2xy+y^2} = \dfrac{x+y}{(x+y)^2} = \dfrac{1}{x+y}$ Factor and simplify the first fraction.

$\dfrac{x-y}{x^2-2xy+y^2} = \dfrac{x-y}{(x-y)^2} = \dfrac{1}{x-y}$ Factor and simplify the second fraction.

Since $x+y$ and $x-y$ have no common factors, the LCD is $(x+y)(x-y)$. We have

$$\frac{x+y}{x^2+2xy+y^2} + \frac{x-y}{x^2-2xy+y^2} = \frac{1}{x+y} + \frac{1}{x-y}$$

$$= \frac{x-y}{(x+y)(x-y)} + \frac{x+y}{(x+y)(x-y)}$$ Write each fraction with $(x+y)(x-y)$ as the denominator.

$$= \frac{x-y+x+y}{(x+y)(x-y)}$$ Combine like terms in the numerator.

$$= \frac{2x}{(x+y)(x-y)}$$

b. We note that $x - 2 = -(2 - x)$, so we start by multiplying the numerator and denominator of the second fraction by -1.

$$\frac{x}{(x + 2)(x - 2)} - \frac{2}{(2 - x)(x + 2)}$$

$$= \frac{x}{(x + 2)(x - 2)} - \frac{-1 \cdot (2)}{-1 \cdot (2 - x)(x + 2)} \qquad \text{Multiply the numerator and denominator by } -1.$$

$$= \frac{x}{(x + 2)(x - 2)} - \frac{-2}{(x - 2)(x + 2)} \qquad \begin{array}{l} -1 \cdot 2 = -2 \text{ and } -1(2 - x) = \\ -2 + x = x - 2 \end{array}$$

$$= \frac{x - (-2)}{(x + 2)(x - 2)}$$

$$= \frac{x + 2}{(x + 2)(x - 2)} \qquad \text{Divide out } x + 2.$$

$$= \frac{1}{x - 2}$$

■

GRAPH IT

As we have been doing for several sections, the results of adding or subtracting rational expressions can be verified by graphing the original problem and the answer. Thus to verify Example 4b, graph

$$Y_1 = \frac{x}{(x + 2)(x - 2)} - \frac{2}{(2 - x)(x + 2)} \quad \text{and} \quad Y_2 = \frac{1}{x - 2}$$

WINDOW 1
$P(t) - Q(t)$

and make sure that both graphs are identical. Note, however, that Y_1 is not defined for $x = -2$ but Y_2 is. The grapher didn't indicate this! You have to know some algebra to recognize it.

A more challenging problem would be to verify the results obtained in the *Getting Started.* Does

$$P(t) - Q(t) = \frac{0.02t^2 - 0.34t + 1.42}{0.04t^2 + 2.34t + 90}$$

when simplified?

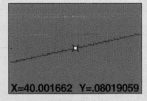

WINDOW 2
When the trace feature is used, we can find that for $x = 40, y = 0.21005654.$

The problem here is to select the correct viewing window. If we start with the standard window, no part of the graph will show (see Window 1). The problem tells us that t represents the number of years after 1960; thus it makes sense to let Xmin = 0 and Xmax = 50 with a scale of 10, that is, Xscl = 10. The values of the resulting rational expression are expressed as percents; thus we let Ymin = 0, Ymax = 1, and Yscl = 0.01. The part of the graph we are interested in is shown in Window 2. (If we want to see a complete graph, we let Xmin = -50 and Ymin = -1, but note that in this case, it isn't reasonable to predict what happened when $t = -50$—that is, 50 years *before* 1960! Moreover, answers representing negative percents, $y = -1$, for example, make no sense in this case.)

Now let's check our prediction regarding the percent of waste that is *not* paper or paperboard. If we use the trace feature to find what happens when $x = 40$, we find $y = 0.08019059$. If the value you get in your grapher isn't close enough to $x = 40$, you can zoom in several times by pressing $\boxed{\text{ZOOM}}$ 2 $\boxed{\text{ENTER}}$. If the value for x is still not close enough to 40, continue to press $\boxed{\text{ENTER}}$ and $\boxed{\text{TRACE}}$, and your grapher will continue to allow you to get values closer to 40. One such value is $x = 40.001662$, yielding $y = 0.08019059$, which is very close to our original answer of 8%.

EXERCISE 4.3

A In Problems 1–10, perform the indicated operations.

1. $\dfrac{x}{5} + \dfrac{2x}{5}$

2. $\dfrac{x+1}{3x} + \dfrac{2x+7}{3x}$

3. $\dfrac{7x}{3} - \dfrac{2x}{3}$

4. $\dfrac{2x-1}{5x} - \dfrac{x+1}{5x}$

5. $\dfrac{3}{5x+10} + \dfrac{2x}{5(x+2)}$

6. $\dfrac{2x+1}{3(x+2)} + \dfrac{3x+1}{3x+6}$

7. $\dfrac{2x+1}{2(x+1)} - \dfrac{x-1}{2x+2}$

8. $\dfrac{3x-1}{4(x-1)} - \dfrac{4x-1}{4x-4}$

9. $\dfrac{2x+1}{3(x-1)} + \dfrac{x+3}{3x-3} - \dfrac{x-1}{3(x-1)}$

10. $\dfrac{3x-1}{5(x+1)} - \dfrac{x+1}{5x+5} + \dfrac{2x-5}{5(x+1)}$

B In Problems 11–50, perform the indicated operations.

11. $\dfrac{x}{x^2+3x-4} + \dfrac{x}{x^2-16}$

12. $\dfrac{x-2}{x^2-9} + \dfrac{x+1}{x^2-x-12}$

13. $\dfrac{3x}{x^2+3x-10} + \dfrac{2x}{x^2+x-6}$

14. $\dfrac{x+3}{x^2-x-2} + \dfrac{x-1}{x^2+2x+1}$

15. $\dfrac{1}{x^2-y^2} + \dfrac{5}{(x+y)^2}$

16. $\dfrac{3}{(x+y)^2} + \dfrac{5}{(x+y)}$

17. $\dfrac{2}{x-5} - \dfrac{3x}{x^2-25}$

18. $\dfrac{x+3}{x^2-x-2} - \dfrac{x-1}{x^2+2x+1}$

19. $\dfrac{x-1}{x^2+3x+2} - \dfrac{x+7}{x^2+5x+6}$

20. $\dfrac{2}{x^2+3xy+2y^2} - \dfrac{1}{x^2-xy-2y^2}$

Hint: **For Problems 21–29, first simplify the fractions.**

21. $\dfrac{x+2}{x^2-4} + \dfrac{x+3}{x^2-9}$

22. $\dfrac{x-3}{x^2-9} + \dfrac{x+3}{x^2-9}$

23. $\dfrac{x-3}{x^2-9} + \dfrac{x+3}{x^2+6x+9}$

24. $\dfrac{a-4}{a^2-16} + \dfrac{a+3}{a^2+5a+6}$

25. $\dfrac{a+3}{a^2+5a+6} + \dfrac{a+2}{a^2+6a+8}$

26. $\dfrac{a+3}{a^2+5a+6} - \dfrac{a-4}{a^2-16}$

27. $\dfrac{3a+3}{a^2+5a+4} - \dfrac{a-3}{a^2+a-12}$

28. $\dfrac{2a}{5a-7b} - \dfrac{5a+7b}{25a^2-49b^2}$

29. $\dfrac{5a-15}{a^2+2a-15} - \dfrac{a^2+5a}{a^2+8a+15}$

30. $\dfrac{3}{y^2-9} + \dfrac{2y}{y-3}$

31. $\dfrac{y}{y^2-1} + \dfrac{y}{y-1}$

32. $\dfrac{3y}{y^2-4} - \dfrac{y}{y+2}$

33. $\dfrac{3y+1}{y^2-16} - \dfrac{2y-1}{y-4}$

34. $\dfrac{3x-5y}{2x-3y} + \dfrac{2x-3y}{2x+3y}$

35. $\dfrac{5x+2y}{5x-2y} + \dfrac{5x-2y}{5x+2y}$

36. $\dfrac{x+3y}{x-5y} - \dfrac{x+5y}{x-3y}$

37. $\dfrac{3x-y}{2x-y} - \dfrac{2x+y}{3x+y}$

38. $\dfrac{a+3}{a^2+a-6} + \dfrac{a-2}{a^2+3a-10}$

39. $\dfrac{x+3}{x^2-x-2} + \dfrac{x-1}{x^2+2x+1}$

40. $\dfrac{8x}{x^2-4y^2} - \dfrac{2x}{x^2-5xy+6y^2}$

41. $\dfrac{x+1}{x^2-x-2} - \dfrac{x}{x^2-5x+4}$

42. $\dfrac{3}{x^2-4} + \dfrac{1}{2-x} - \dfrac{1}{2+x}$

43. $\dfrac{2}{5+x} + \dfrac{5x}{x^2-25} + \dfrac{7}{5-x}$

44. $\dfrac{1}{x^2+x-12} + \dfrac{2}{x^2+2x-15} + \dfrac{3}{x^2+9x+20}$

45. $\dfrac{x}{(x-y)(2-x)} - \dfrac{y}{(y-x)(2-x)} + \dfrac{y}{(x-y)(x-2)}$

46. $\dfrac{a}{(b-a)(c-a)} - \dfrac{b}{(b-c)(a-b)} + \dfrac{c}{(a-c)(b-c)}$

47. $\dfrac{4a^2-9b^2}{4a^2-12ab+9b^2} + \dfrac{12a+18b}{4a^2+12ab+9b^2} - \dfrac{2a+3b}{2a+3b}$

48. $\dfrac{x+2y}{x^3+8y^3} + \dfrac{5}{x+2y} + \dfrac{2x-3y}{x^2-2xy+4y^2}$

49. $\dfrac{x+5}{x^3+125} + \dfrac{x-5}{x^2-25} - \dfrac{1}{x+5}$

50. $\dfrac{a}{a+3} + \dfrac{a-2}{a^2-3a+9} + \dfrac{5a^2-13a}{a^3+27}$

APPLICATIONS

51. The moment M of a cantilever beam of length L x units from the end is given by

$$-\frac{w_0 x^3}{6L} + \frac{w_0 Lx}{2} - \frac{w_0 L^2}{3}$$

Write this expression as a single rational expression in reduced form.

52. The deflection d of the beam of Problem 51 involves the expression

$$\frac{-x^4}{24L} + \frac{Lx^2}{4} - \frac{L^2 x}{3}$$

Write this expression as a single rational expression in reduced form.

53. In astronomy, planetary motion is given by

$$\frac{p^2}{2mr^2} - \frac{gmM}{r}$$

Write this expression as a single rational expression in reduced form.

54. The motion of a pendulum is given by

$$\frac{P_1^2 + P_2^2}{2(h_1 + h_2)} + \frac{P_1^2 - P_2^2}{2(h_1 - h_2)}$$

Write this expression as a single rational expression in reduced form.

SKILL CHECKER

Multiply:

55. $9\left(2 + \dfrac{2}{9}\right)$

56. $4\left(60 - \dfrac{15}{2}\right)$

57. $12xy\left(\dfrac{2}{y} + \dfrac{3}{2x}\right)$

58. $6ab\left(\dfrac{3}{a} - \dfrac{4}{b}\right)$

59. $x^2\left(1 - \dfrac{1}{x^2}\right)$

60. $x^3\left(1 - \dfrac{1}{x^3}\right)$

USING YOUR KNOWLEDGE

Looking Ahead to Calculus

In calculus, the derivative of a polynomial P is defined as the limiting value of

$$\frac{P(x+h) - P(x)}{h}$$

as h approaches zero.

Let $P(x) = x^2$.

61. Find $P(x + h)$ and write it in expanded form.

62. Find $P(x + h) - P(x)$ and simplify it.

63. Find $\dfrac{P(x+h) - P(x)}{h}$ in simplified form.

64. Find $\dfrac{P(x+h) - P(x)}{h}$ for $P(x) = x^2 + x$.

WRITE ON ...

65. Write in your own words the procedure you use to find the LCD of two or more rational expressions.

66. Write in your own words the procedure you use to add or subtract two rational expressions.

67. When adding two or more rational expressions, do you *always* have to find the LCD or can you use *any* common denominator? What is the advantage of using the LCD?

MASTERY TEST

If you know how to do these problems, you have learned your lesson!

68. $\dfrac{x+1}{(x+3)(x-1)} + \dfrac{x+4}{(x^2-1)}$

69. $\dfrac{x-3}{x^2-x-2} - \dfrac{x+3}{x^2-4}$

70. $\dfrac{x-y}{x^2-2xy+y^2} + \dfrac{x-y}{x^2-y^2}$

71. $\dfrac{x}{(x+3)(x-3)} + \dfrac{3}{(3-x)(x+3)}$

72. $\dfrac{5x}{x+1} - \dfrac{3x}{x+3}$

73. $\dfrac{4x}{5(x-2)} + \dfrac{6x}{5(x-2)}$

74. $\dfrac{11x}{3(x+2)^2} + \dfrac{4x}{3(x+2)^2}$

75. $\dfrac{x}{x^2-9} - \dfrac{3}{x^2-9}$

76. $\dfrac{5x}{x+3} - \dfrac{4x-3}{x+3}$

4.4

COMPLEX FRACTIONS

To succeed, review how to:

1. Find the LCD of two or more rational expressions (p. 225).

2. Remove parentheses using the distributive property (pp. 149–151).

3. Add, subtract, multiply, and divide rational expressions (pp. 214–217, 223–229).

Objective:

Write a complex fraction as a simple fraction in reduced form.

getting started

Rocking Along with Fractions

Suppose a disc jockey devotes $7\frac{1}{2}$ min each hour to commercials, leaving $60 - 7\frac{1}{2}$ min for music. If the songs she plays last an average of $3\frac{1}{4}$ min and it takes her about $\frac{1}{2}$ min to get a song going, how many songs can she play each hour? The answer is

$$\frac{60 - 7\frac{1}{2}}{3\frac{1}{4} + \frac{1}{2}}$$ ← Time allowed for music

← Time it takes to play each song

but you have to know how to simplify it to find out how many songs she plays each hour!

The fraction

$$\frac{60 - 7\frac{1}{2}}{3\frac{1}{4} + \frac{1}{2}}$$

← Numerator fraction

← Main fraction bar

← Denominator fraction

contains other fractions in its numerator and denominator. A fraction whose numerator or denominator (or both) contains other fractions is called a **complex fraction**. A fraction that is not complex is called a **simple fraction**. Thus,

$$\frac{\frac{1}{2}}{\frac{3}{4} + \frac{1}{5}}, \qquad \frac{\frac{3x}{5} - \frac{1}{8}}{\frac{1x}{7}}, \qquad \frac{\frac{1}{3}}{\frac{1}{9}}, \qquad \text{and} \qquad \frac{\frac{x}{7}}{\frac{8}{}}$$

are all complex fractions, but $\frac{1}{7}$, $\frac{3}{5}$, and $\frac{x}{9}$ are simple fractions.

To simplify a complex fraction, it's necessary to recall that the main fraction bar indicates that the numerator of the fraction is to be divided by the denominator of the fraction. Thus

$$\frac{60 - 7\frac{1}{2}}{3\frac{1}{4} + \frac{1}{2}} \quad \text{means} \quad \left(60 - 7\frac{1}{2}\right) \div \left(3\frac{1}{4} + \frac{1}{2}\right)$$

$\uparrow$

Use the ÷ sign
instead of the bar.

Here are the procedures we use to simplify complex fractions.

PROCEDURE

Procedures for Simplifying Complex Fractions

Method 1. Multiply the numerator and denominator of the complex fraction by the LCD of all the simple fractions appearing; or

Method 2. Perform the operations indicated in the numerator and denominator of the given complex fraction, and then divide the numerator by the denominator.

We now simplify

$$\frac{60 - 7\frac{1}{2}}{3\frac{1}{4} + \frac{1}{2}} = \frac{60 - \frac{15}{2}}{\frac{13}{4} + \frac{1}{2}}$$

using each of these methods.

METHOD 1   The LCD of $\frac{15}{2}$, $\frac{13}{4}$, and $\frac{1}{2}$ is 4, so we have

$$\frac{60 - \frac{15}{2}}{\frac{13}{4} + \frac{1}{2}} = \frac{4 \cdot \left(60 - \frac{15}{2}\right)}{4 \cdot \left(\frac{13}{4} + \frac{1}{2}\right)}$$

Multiply numerator and denominator by **4**, the LCD of $\frac{15}{2}$, $\frac{13}{4}$, and $\frac{1}{2}$.

$$= \frac{240 - 30}{13 + 2}$$

$\longleftarrow 4\left(60 - \frac{15}{2}\right) = 240 - \frac{60}{2}$

$\longleftarrow 4\left(\frac{13}{4} + \frac{1}{2}\right) = 4 \cdot \frac{13}{4} + \frac{4}{2}$

$= 13 + 2$

$$= \frac{210}{15}$$

$\longleftarrow 240 - 30 = 210$

$\longleftarrow 13 + 2 = 15$

$$= 14$$

Divide.

The disc jockey plays 14 songs each hour.

METHOD 2

$$\frac{60 - \dfrac{15}{2}}{\dfrac{13}{4} + \dfrac{1}{2}} = \frac{\dfrac{120}{2} - \dfrac{15}{2}}{\dfrac{13}{4} + \dfrac{2}{4}}$$ ⟵ Write 60 as $\frac{120}{2}$, so we can subtract $\frac{15}{2}$.

⟵ Write $\frac{1}{2}$ as $\frac{2}{4}$, so we can add it to $\frac{13}{4}$.

$$= \frac{\dfrac{105}{2}}{\dfrac{15}{4}}$$ ⟵ $\dfrac{120}{2} - \dfrac{15}{2} = \dfrac{105}{2}$

⟵ $\dfrac{13}{4} + \dfrac{2}{4} = \dfrac{15}{4}$

$$= \frac{105}{2} \div \frac{15}{4}$$ Replace the bar by the division sign, ÷.

$$= \frac{\overset{7}{\cancel{105}}}{2} \cdot \frac{\overset{2}{\cancel{4}}}{\cancel{15}}$$ Multiply by the reciprocal of $\frac{15}{4}$, which is $\frac{4}{15}$, and simplify.

$$= 14$$ As before

EXAMPLE 1   Simplifying complex fractions using method 1

Use method 1 to write the following as a simple fraction in simplified form:

$$\frac{\dfrac{3}{a} - \dfrac{4}{b}}{\dfrac{1}{2a} + \dfrac{2}{3b}}$$

SOLUTION   The LCD of

$$\frac{3}{a}, \qquad \frac{4}{b}, \qquad \frac{1}{2a}, \qquad \text{and} \qquad \frac{2}{3b}$$

is $6ab$. Therefore, we multiply the numerator and denominator of the given fraction by $6ab$ to obtain

$$\frac{6ab \cdot \left(\dfrac{3}{a} - \dfrac{4}{b}\right)}{6ab \cdot \left(\dfrac{1}{2a} + \dfrac{2}{3b}\right)} = \frac{6\cancel{a}b \cdot \dfrac{3}{\cancel{a}} - 6a\cancel{b} \cdot \dfrac{4}{\cancel{b}}}{\overset{3}{\cancel{6}}\cancel{a}b \cdot \dfrac{1}{\overset{}{2\cancel{a}}} + \overset{}{\cancel{6}}a\cancel{b} \cdot \dfrac{2}{3\cancel{b}}}$$ Use the distributive property and simplify.

$$= \frac{18b - 24a}{3b + 4a}$$ ⟵ $6b \cdot 3 = 18b$ and $6a \cdot 4 = 24a$

⟵ $3b \cdot 1 = 3b$ and $2a \cdot 2 = 4a$ ∎

EXAMPLE 2 More practice using method 1

Use method 1 to simplify:

$$\frac{x - \dfrac{1}{x^3}}{x + \dfrac{1}{x^2}}$$

SOLUTION Here the LCD is x^3. Thus

$$\frac{x - \dfrac{1}{x^3}}{x + \dfrac{1}{x^2}} = \frac{x^3 \cdot \left(x - \dfrac{1}{x^3}\right)}{x^3 \cdot \left(x + \dfrac{1}{x^2}\right)}$$

$$= \frac{x^3 \cdot x - x^3 \cdot \dfrac{1}{x^3}}{x^3 \cdot x + x^3 \cdot \dfrac{1}{x^2}}$$

$$= \frac{x^4 - 1}{x^4 + x} = \frac{x^4 - 1}{x(x^3 + 1)} \quad \text{Factor the numerator and denominator.}$$

$$= \frac{(x^2 + 1)(x + 1)(x - 1)}{x(x + 1)(x^2 - x + 1)} \quad \text{Divide out } x + 1.$$

$$= \frac{(x^2 + 1)(x - 1)}{x(x^2 - x + 1)}$$

or $\dfrac{x^3 - x^2 + x - 1}{x^3 - x^2 + x}$

EXAMPLE 3 Comparing method 1 with method 2

Simplify:

$$\frac{\dfrac{x}{x - 2} + x}{2 + \dfrac{1}{x^2 - 4}}$$

a. Use method 1.
b. Use method 2.

SOLUTION
a. We first write $x^2 - 4$ as $(x + 2)(x - 2)$, x as $\frac{x}{1}$, and 2 as $\frac{2}{1}$, to obtain

$$\frac{\dfrac{x}{x - 2} + x}{2 + \dfrac{1}{x^2 - 4}} = \frac{\dfrac{x}{x - 2} + \dfrac{x}{1}}{\dfrac{2}{1} + \dfrac{1}{(x + 2)(x - 2)}}$$

The LCD of the fractions is $(x + 2)(x - 2)$. Multiply numerator and denominator by this LCD.

$$\dfrac{\dfrac{(x + 2)(x - 2)}{1} \cdot \left(\dfrac{x}{x - 2} + x\right)}{\dfrac{(x + 2)(x - 2)}{1} \cdot \left[2 + \dfrac{1}{(x + 2)(x - 2)}\right]}$$

$$= \dfrac{x(x + 2) + x(x + 2)(x - 2)}{2(x + 2)(x - 2) + 1} \qquad \text{Simplify.}$$

$$= \dfrac{x^2 + 2x + x^3 - 4x}{2x^2 - 8 + 1} \qquad \text{Remove parentheses.}$$

$$= \dfrac{x^3 + x^2 - 2x}{2x^2 - 7} \qquad \text{Collect like terms.}$$

$$= \dfrac{x(x^2 + x - 2)}{2x^2 - 7} \qquad \text{Factor out } x.$$

$$= \dfrac{x(x - 1)(x + 2)}{2x^2 - 7} \qquad \text{Factor } x^2 + x - 2 = (x - 1)(x + 2).$$

 NOTE $\dfrac{x^3 + x^2 - 2x}{2x^2 - 7}$ is also an acceptable answer.

b. Using method 2, we first perform the operations in the numerator and denominator, and then we divide.

$$\dfrac{\dfrac{x}{x - 2} + x}{2 + \dfrac{1}{x^2 - 4}} = \dfrac{\dfrac{x}{x - 2} + \dfrac{x(x - 2)}{x - 2}}{\dfrac{2(x^2 - 4)}{x^2 - 4} + \dfrac{1}{x^2 - 4}} \qquad \longleftarrow \text{Rewrite } x \text{ as } \dfrac{x(x - 2)}{x - 2}.$$

$$\qquad\qquad\qquad\qquad\qquad \longleftarrow \text{Rewrite 2 as } \dfrac{2(x^2 - 4)}{x^2 - 4}.$$

$$= \dfrac{\dfrac{x + x(x - 2)}{x - 2}}{\dfrac{2(x^2 - 4) + 1}{x^2 - 4}} \qquad \text{Add.}$$

$$= \dfrac{\dfrac{x + x^2 - 2x}{x - 2}}{\dfrac{2x^2 - 8 + 1}{x^2 - 4}} \qquad \text{Remove parentheses.}$$

$$= \dfrac{x^2 - x}{x - 2} \div \dfrac{2x^2 - 7}{x^2 - 4} \qquad \text{Use the division sign, } \div \text{, instead of the bar.}$$

$$= \dfrac{x(x - 1)}{x - 2} \cdot \dfrac{(x + 2)(x - 2)}{2x^2 - 7} \qquad \text{Multiply by the reciprocal of } \dfrac{2x^2 - 7}{x^2 - 4} \text{ and factor } x^2 - 4.$$

$$= \dfrac{x(x - 1)(x + 2) \cdot \overset{1}{\cancel{(x - 2)}}}{\cancel{(x - 2)}(2x^2 - 7)} \qquad \text{Multiply.}$$

$$= \dfrac{x(x - 1)(x + 2)}{2x^2 - 7} \qquad \text{Divide out } x - 2.$$

or $$\dfrac{x^3 + x^2 - 2x}{2x^2 - 7}$$ ∎

EXAMPLE 4   Simplifying a complex fraction using method 2

Simplify:

$$1 + \cfrac{a}{1 + \cfrac{1}{1 + a}}$$

SOLUTION   We start by working on the denominator of the fraction.

$$\cfrac{a}{1 + \cfrac{1}{1 + a}}$$

Since we have to add 1 to

$$\frac{1}{1 + a}$$

we rewrite 1 as

$$\frac{1 + a}{1 + a}$$

$$1 + \cfrac{a}{\boxed{1 + \cfrac{1}{1 + a}}} = 1 + \cfrac{a}{\boxed{\cfrac{1 + a}{1 + a} + \cfrac{1}{1 + a}}}$$ Work on the denominator.

$$= 1 + \cfrac{a}{\boxed{\cfrac{2 + a}{1 + a}}}$$ $\dfrac{1 + a}{1 + a} + \dfrac{1}{1 + a} = \dfrac{1 + a + 1}{1 + a} = \dfrac{2 + a}{1 + a}$

$$= 1 + a \div \boxed{\frac{2 + a}{1 + a}}$$ Use the ÷ sign instead of the bar.

$$= 1 + a \cdot \frac{1 + a}{2 + 1}$$ To divide a by $\dfrac{2 + a}{1 + a}$, multiply by the reciprocal $\dfrac{1 + a}{2 + a}$.

$$= 1 + \frac{a(1 + a)}{2 + a}$$

$$= \frac{2 + a}{2 + a} + \frac{a(1 + a)}{2 + a}$$ Write 1 as $\dfrac{2 + a}{2 + a}$.

$$= \frac{2 + a + a + a^2}{2 + a}$$ Add the numerator and keep the denominator.

$$= \frac{2 + 2a + a^2}{2 + a}$$

$$= \frac{a^2 + 2a + 2}{a + 2}$$ Write the numerator and denominator in descending order. ∎

Sometimes complex fractions are written using negative exponents. To simplify such expressions, we begin by rewriting them using the definition of a negative exponent, that is,

$$a^{-n} = \frac{1}{a^n}$$

Using this definition,

$$(x - 1)^{-1} = \frac{1}{x - 1}, \qquad (x + 2)^{-3} = \frac{1}{(x + 2)^3}, \qquad \text{and} \qquad (x + y)^{-5} = \frac{1}{(x + y)^5}$$

We shall use these ideas in Example 5. Pay close attention to what we do in each step of the problem.

EXAMPLE 5   Simplifying a complex fraction involving negative exponents

Simplify: $\dfrac{x(x - 3)^{-1} + x}{x(3 - x)^{-1} - x}$

SOLUTION

$$\frac{x(x - 3)^{-1} + x}{x(3 - x)^{-1} - x} = \frac{\dfrac{x}{x - 3} + \dfrac{x}{1}}{\dfrac{x}{3 - x} - \dfrac{x}{1}}$$

Rewrite $x(x - 3)^{-1}$ as $\dfrac{x}{x - 3}$,

$x(3 - x)^{-1}$ as $\dfrac{x}{3 - x}$,

and x as $\dfrac{x}{1}$.

$$= \frac{\dfrac{x}{x - 3} + \dfrac{x}{1}}{\dfrac{x}{-(x - 3)} - \dfrac{x}{1}}$$

Since $3 - x$ and $x - 3$ are opposites, that is, $3 - x = -(x - 3)$, we substitute $-(x - 3)$ for $3 - x$.

$$= \frac{(x - 3)\left(\dfrac{x}{x - 3} + \dfrac{x}{1}\right)}{(x - 3)\left[\dfrac{x}{-(x - 3)} - \dfrac{x}{1}\right]}$$

The LCD of all the denominators is $x - 3$. Multiply the numerator and denominator by $(x - 3)$.

$$= \frac{\dfrac{(x - 3)x}{x - 3} + \dfrac{(x - 3)x}{1}}{\dfrac{(x - 3)x}{-(x - 3)} - \dfrac{(x - 3)x}{1}}$$

Use the distributive property to distribute the $(x - 3)$.

$$= \frac{x + (x - 3)x}{-x - (x - 3)x}$$

Divide out the $(x - 3)$ terms and note that $\dfrac{(x - 3)x}{1} = (x - 3)x$.

$$= \frac{x + x^2 - 3x}{-x - (x^2 - 3x)}$$

Multiply $(x - 3)x$ in the numerator and denominator.

$$= \frac{x^2 - 2x}{-x^2 + 2x}$$

Collect like terms in the numerator and denominator.

$$= \frac{x^2 - 2x}{-(x^2 - 2x)}$$

Rewrite the denominator $-x^2 + 2x$ as $-(x^2 - 2x)$.

$$= -1$$

Since $\dfrac{a}{-a} = -1$, the answer is -1. ■

In step 2, the numerator is

$$\frac{x}{x-3} + \frac{x}{1}$$

and the denominator is

$$\frac{x}{-(x-3)} - \frac{x}{1} = -\left(\frac{x}{x-3} + \frac{x}{1}\right)$$

which is the additive inverse of the numerator! Thus

$$\frac{\dfrac{x}{x-3} + \dfrac{x}{1}}{\dfrac{x}{-(x-3)} - \dfrac{x}{1}} = \frac{\dfrac{x}{x-3} + \dfrac{x}{1}}{-\left(\dfrac{x}{x-3} + \dfrac{x}{1}\right)} = -1$$

These two expressions are additive inverses of each other.

GRAPH IT

In the last few sections we have been verifying the results of multiplying, dividing, adding, and subtracting polynomials by making sure that the graph of the original problem and the graph of the final answer are identical. This, of course, can also be done with complex fractions. However, we want to explore the numerical capabilities of your grapher. In Example 3, we found that

```
                                    4
(X/(X-2)+X)/(2+1
/(X^2-4))
                                 2.88
X(X-1)(X+2)/(2X^
2-7)
                                 2.88
```

$$\frac{\dfrac{x}{x-2} + x}{2 + \dfrac{1}{x^2-4}} = \frac{x(x-1)(x+2)}{2x^2-7}$$

To check the answer, substitute any convenient number for x and see if both sides are equal. Of course, you cannot choose numbers that will give you a zero denominator (2 and -2 will yield zero denominators). A simple number to use is $x = 4$. Now, go to the home screen and store the value 4 at x. With a TI-82, press 4 [STO▶] [X,T,θ] [ENTER]. The grapher confirms the entry by showing 4→X and the answer 4. Now enter the original complex fraction as

$$\left(\frac{x}{(x-2)} + x\right) \div \left(2 + \frac{1}{(x^2-4)}\right)$$

Be extremely careful with the parentheses! Press [ENTER]. The answer is 2.88, as shown in the window. Next, enter the simplified complex fraction as $x(x-1)(x+2) \div (2x^2-7)$. Press [ENTER] again. You should get the same answer as before, 2.88. The final confirmation is shown on the screen. You can check the rest of the problems involving one variable using this method. By the way, if you are checking Example 4, you can enter an x instead of an a when writing the complex fraction involved, but we must warn you that there are so many parentheses involved that it's probably easier to do the problem without the grapher. Try it if you don't believe this!

EXERCISE 4.4

A **In Problems 1–40, perform the indicated operation and give the answer in simplified form.**

1. $\dfrac{50 - 5\dfrac{1}{2}}{7\dfrac{3}{4} + \dfrac{1}{2}}$

2. $\dfrac{70 - 17\dfrac{1}{2}}{2\dfrac{1}{4} + 1\dfrac{1}{2}}$

3. $\dfrac{\dfrac{a}{b}}{\dfrac{c}{b}}$

4. $\dfrac{\dfrac{-a^2}{c}}{\dfrac{-b^2}{c}}$

5. $\dfrac{\dfrac{x}{y}}{\dfrac{x^2}{z}}$

6. $\dfrac{\dfrac{x^2}{y^2}}{\dfrac{x}{z}}$

7. $\dfrac{\dfrac{3x}{5y}}{\dfrac{3x}{2z}}$

8. $\dfrac{\dfrac{7x}{3y}}{\dfrac{14x}{5y}}$

9. $\dfrac{\dfrac{1}{2}}{2 - \dfrac{1}{2}}$

33. $\dfrac{y + 3 - \dfrac{16}{y + 3}}{y - 6 + \dfrac{20}{y + 3}}$

34. $\dfrac{w + 2 - \dfrac{18}{w - 5}}{w - 1 - \dfrac{12}{w - 5}}$

10. $\dfrac{\dfrac{1}{4}}{3 - \dfrac{1}{4}}$

11. $\dfrac{a - \dfrac{a}{b}}{1 + \dfrac{a}{b}}$

12. $\dfrac{1 - \dfrac{1}{a}}{1 + \dfrac{1}{a}}$

35. $\dfrac{\dfrac{8x}{3x + 1} - \dfrac{3x - 1}{x}}{\dfrac{x}{3x + 1} - \dfrac{2x - 2}{x}}$

36. $\dfrac{\dfrac{3}{m - 4} - \dfrac{16}{m - 3}}{\dfrac{2}{m - 3} - \dfrac{15}{m + 5}}$

13. $\dfrac{y + \dfrac{2}{x}}{y^2 - \dfrac{4}{x^2}}$

14. $\dfrac{y - \dfrac{3}{x}}{y^2 - \dfrac{9}{x^2}}$

15. $\dfrac{\dfrac{x}{y^2} - \dfrac{y}{x^2}}{x^2 + xy + y^2}$

37. $\dfrac{\dfrac{c}{d} - \dfrac{d}{c}}{\dfrac{c}{d} - 2 + \dfrac{d}{c}}$

38. $\dfrac{\dfrac{a^2}{b^2} + 4 + \dfrac{4b^2}{a^2}}{\dfrac{a}{b} + \dfrac{2b}{a}}$

16. $\dfrac{\dfrac{x}{y^2} + \dfrac{y}{x^2}}{x^2 - xy + y^2}$

17. $3 - \dfrac{3}{3 - \dfrac{1}{2}}$

18. $2 - \dfrac{2}{2 - \dfrac{1}{2}}$

39. $\dfrac{\dfrac{a^2 - b^2}{a^2 + b^2} - \dfrac{a^2 + b^2}{a^2 - b^2}}{\dfrac{a - b}{a + b} - \dfrac{a + b}{a - b}}$

40. $\dfrac{1 + \dfrac{4uv}{(u - v)^2}}{1 + \dfrac{uv - 3v^2}{(u - v)^2}}$

19. $a - \dfrac{a}{a + \dfrac{1}{2}}$

20. $a + \dfrac{a}{a + \dfrac{1}{2}}$

21. $x - \dfrac{x}{1 - \dfrac{x}{1 - x}}$

22. $2x - \dfrac{x}{2 - \dfrac{x}{2 - x}}$

23. $\dfrac{1}{1 + \dfrac{1}{2 + \dfrac{1}{3 + \dfrac{1}{4}}}}$

24. $\dfrac{1}{1 - \dfrac{1}{2 - \dfrac{1}{3 - \dfrac{1}{4}}}}$

25. $\dfrac{\dfrac{x - 1}{x + 1} + \dfrac{x + 1}{x - 1}}{\dfrac{x - 1}{x + 1} - \dfrac{x + 1}{x - 1}}$

26. $\dfrac{\dfrac{x - 1}{x + 1} - \dfrac{x + 1}{x - 1}}{\dfrac{x - 1}{x + 1} + \dfrac{x + 1}{x - 1}}$

27. $\dfrac{\dfrac{1}{x - y} + \dfrac{1}{x + y}}{\dfrac{1}{x - y} - \dfrac{1}{x + y}}$

28. $\dfrac{\dfrac{v + 1}{v - 1} + \dfrac{v - 1}{v + 1}}{\dfrac{v + 1}{v - 1} - \dfrac{v - 1}{v + 1}}$

29. $\dfrac{x(x - 2)^{-1} - x}{x(2 - x)^{-1} + x}$

30. $\dfrac{x + x(4 - x)^{-1}}{x(x - 4)^{-1}}$

31. $\dfrac{\dfrac{1}{x^2} + \dfrac{3}{x} - 4}{\dfrac{1}{x^2} + \dfrac{5}{x} + 4}$

32. $\dfrac{\dfrac{6}{v^2} - \dfrac{11}{v} - 10}{\dfrac{2}{v^2} + \dfrac{1}{v} - 15}$

APPLICATIONS

41. When connected in parallel, the combined resistance R of two electrical resistances R_1 and R_2 is given by

$$R = \dfrac{1}{\dfrac{1}{R_1} + \dfrac{1}{R_2}}$$

Simplify this expression.

42. When connected in parallel, the combined resistance R of three electrical resistances R_1, R_2, and R_3 is given by

$$R = \dfrac{1}{\dfrac{1}{R_1} + \dfrac{1}{R_2} + \dfrac{1}{R_3}}$$

Simplify this expression.

43. The formula for the Doppler effect in light is

$$f = f_{static} \sqrt{\dfrac{1 + \dfrac{v}{c}}{1 - \dfrac{v}{c}}}$$

where f and f_{static} are frequencies, v is the velocity of the moving body, and c is the speed of light.

Simplify this expression.

44. Balmer's formula for the wavelength λ (lambda) of the hydrogen spectrum light is given by

$$\lambda = \dfrac{1}{\dfrac{1}{m^2} - \dfrac{1}{n^2}}$$

Simplify this expression.

Simplify:

45. $\dfrac{8x^4}{2x^3}$

46. $\dfrac{28x^7}{7x^5}$

47. $\dfrac{-30x^6}{6x^4}$

48. $\dfrac{-10x^2}{20x^4}$

49. $\dfrac{-30x}{10x^3}$

50. $\dfrac{-50}{10x^3}$

Multiply:

51. $x^2(5x + 5)$

52. $6x^2(x - 2)$

53. $3x^4(3x - 5)$

Factor:

54. $6x^2 + x - 2$

55. $6x^2 + 7x - 3$

56. $20x^2 - 7x - 6$

57. Subtract $6x^3 + 24x^2$ from $6x^3 + 25x^2 + 2x - 8$

58. Subtract $5x^2 + 15x$ from $5x^2 + 9x - 18$

Interest Rates and Planetary Orbits

Do you have monthly payments on any type of loan? Do you know what your **Annual Percentage Rate (APR)** is? If you financed P dollars to be paid in N monthly payments of M dollars, your APR is

$$\dfrac{\dfrac{24(NM - P)}{N}}{P + \dfrac{NM}{12}}$$

59. Simplify the APR formula.

60. Use the simplified version of the APR formula to determine which will give you a better APR on a $20,000 loan:
   a. $500 a month for 4 years?
   b. $400 a month for 5 years?

In the seventeenth century, the Dutch mathematician and astronomer Christian Huygens made a model of the solar system and found out that Saturn takes

$$29 + \dfrac{1}{2 + \dfrac{2}{9}}$$

years to go around the Sun. Now,

$$\dfrac{1}{2 + \dfrac{2}{9}} = \dfrac{9 \cdot 1}{9 \cdot \left(2 + \dfrac{2}{9}\right)}$$

$$= \dfrac{9}{18 + 2}$$

$$= \dfrac{9}{20}$$

Thus it takes Saturn $29 + \frac{9}{20} = 29\frac{9}{20}$ yr to go around the Sun. Use your knowledge to find the number of years it takes each of the following planets to go around the Sun by simplifying the fraction.

61. Mercury: $\dfrac{1}{4 + \dfrac{1}{6}}$ yr

62. Venus: $\dfrac{1}{1 + \dfrac{2}{3}}$ yr

63. Jupiter: $11 + \dfrac{1}{1 + \dfrac{7}{43}}$ yr

(Write your answer as a mixed number.)

64. Mars: $1 + \dfrac{1}{1 + \dfrac{3}{22}}$ yr

65. Write in your own words the definition of a complex fraction.

66. We have given two methods to simplify a complex fraction. Which method do you think is simpler and why?

67. Can you explain when one should use method 1 to simplify a complex fraction and what types of fractions are easier to simplify using this method?

68. Can you explain when one should use method 2 to simplify a complex fraction and what types of fractions are easier to simplify using this method?

If you know how to do these problems, you have learned your lesson!

Simplify:

69. $2 + \dfrac{a}{2 + \dfrac{2}{2 + a}}$

70. $3 + \dfrac{a}{3 + \dfrac{3}{3 + a}}$

71. $\dfrac{\dfrac{x}{x+2}+x}{1-\dfrac{5}{x^2-4}}$

72. $\dfrac{x+\dfrac{x}{x+3}}{\dfrac{1}{x^2-9}+1}$

75. $\dfrac{x-\dfrac{1}{x^3}}{x-\dfrac{1}{x^2}}$

76. $\dfrac{\dfrac{1}{x^2}-x}{\dfrac{1}{x^3}-x}$

73. $\dfrac{\dfrac{2}{b}-\dfrac{3}{a}}{\dfrac{1}{2b}+\dfrac{3}{4a}}$

74. $\dfrac{\dfrac{3}{b}+\dfrac{2}{a}}{\dfrac{1}{2a}-\dfrac{3}{4b}}$

77. $\dfrac{x(x-4)^{-1}+x}{x(4-x)^{-1}-x}$

78. $\dfrac{y-y(y-4)^{-1}}{y+y(4-y)^{-1}}$

4.5

DIVISION OF POLYNOMIALS AND SYNTHETIC DIVISION

To succeed, review how to:

1. Simplify quotients using properties of exponents (p. 34).

2. Multiply polynomials (pp. 151–153).

3. Evaluate expressions using the order of operations (pp. 44–45).

Objectives:

A Divide a polynomial by a monomial.

B Use long division to divide one polynomial by another.

C Completely factor a polynomial when one of the factors is known.

D Use synthetic division to divide one polynomial by a binomial.

E Use the remainder theorem to verify that a number is a solution of a given equation.

getting started   **Efficiency Quotients**

How efficient is your car engine? The efficiency E of an engine is given by

$$E=\dfrac{Q_1-Q_2}{Q_1}$$

where Q_1 is the horsepower rating of the engine and Q_2 is the horse-power delivered to the transmission. Can you do the indicated division in the rational expression

$$\dfrac{Q_1-Q_2}{Q_1}?$$

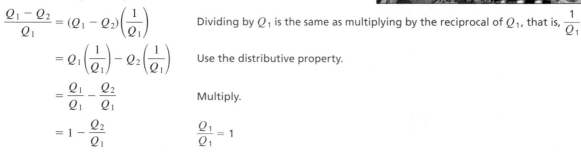

Follow the steps in the procedure.

$\dfrac{Q_1-Q_2}{Q_1}=(Q_1-Q_2)\left(\dfrac{1}{Q_1}\right)$ Dividing by Q_1 is the same as multiplying by the reciprocal of Q_1, that is, $\dfrac{1}{Q_1}$.

$=Q_1\left(\dfrac{1}{Q_1}\right)-Q_2\left(\dfrac{1}{Q_1}\right)$ Use the distributive property.

$=\dfrac{Q_1}{Q_1}-\dfrac{Q_2}{Q_1}$ Multiply.

$=1-\dfrac{Q_2}{Q_1}$ $\dfrac{Q_1}{Q_1}=1$

We shall learn next how to divide a polynomial by a monomial.

Dividing a Polynomial by a Monomial

To divide the trinomial $4x^4 - 8x^3 + 12x^2$ by $2x^2$, we proceed as in the *Getting Started*. Thus

$$\frac{4x^4 - 8x^3 + 12x^2}{2x^2} = (4x^4 - 8x^3 + 12x^2)\left(\frac{1}{2x^2}\right)$$

$$= 4x^4\left(\frac{1}{2x^2}\right) - 8x^3\left(\frac{1}{2x^2}\right) + 12x^2\left(\frac{1}{2x^2}\right)$$

$$= \frac{4x^4}{2x^2} - \frac{8x^3}{2x^2} + \frac{12x^2}{2x^2}$$

$$= 2x^2 - 4x + 6$$

$$\frac{4x^4 - 8x^3 + 12x^2}{2x^2} = \frac{4x^4}{2x^2} - \frac{8x^3}{2x^2} + \frac{12x^2}{2x^2} = 2x^2 - 4x + 6$$

To avoid doing all these steps, we can state the following rule.

PROCEDURE

Rule for Dividing a Polynomial by a Monomial

To divide a polynomial by a monomial, divide each term in the polynomial by the monomial.

EXAMPLE 1   Dividing a polynomial by a monomial

Divide:

a. $\dfrac{28x^5 - 14x^4 + 7x^3}{7x^2}$

b. $\dfrac{20x^4 - 15x^3 + 10x^2 - 30x + 50}{10x^3}$

SOLUTION

a. $\dfrac{28x^5 - 14x^4 + 7x^3}{7x^2} = \dfrac{28x^5}{7x^2} - \dfrac{14x^4}{7x^2} + \dfrac{7x^3}{7x^2}$

$$= 4x^3 - 2x^2 + x$$

b. $\dfrac{20x^4 - 15x^3 + 10x^2 - 30x + 50}{10x^3} = \dfrac{20x^4}{10x^3} - \dfrac{15x^3}{10x^3} + \dfrac{10x^2}{10x^3} - \dfrac{30x}{10x^3} + \dfrac{50}{10x^3}$

$$= 2x - \frac{3}{2} + \frac{1}{x} - \frac{3}{x^2} + \frac{5}{x^3} \qquad \blacksquare$$

Note that in this case, the answer is *not* a polynomial.

 Dividing One Polynomial by Another Polynomial

If we wish to divide a polynomial (called the **dividend**) by another polynomial (called the **divisor**), we proceed very much as we did in long division in arithmetic. To show you that this is so, we shall perform the division of 337 by 16 and $(x^3 + 3x^2 + 3x + 1)$ by $x^2 + x + 1$ side by side.

Think $\dfrac{x^3}{x^2} = x$.

1. $16\overline{)337}^{\,2}$ Divide 33 by 16. It goes twice. Write **2** over the 33.

$$x^2 + x + 1\overline{)x^3 + 3x^2 + 3x + 1}^{\;\;\;x}$$ Write x over the $3x$.

2. $16\overline{)\;337}^{\,2}$
$\;\;\;\;\underline{-32}$
$\;\;\;\;\;\;\;1$ Multiply 16 by 2 and subtract the product **32** from 33, obtaining 1.

$$x^2 + x + 1\overline{)x^3 + 3x^2 + 3x + 1}^{\;\;\;x}$$
$$(-)\underline{x^3 +\;\; x^2 +\;\; x}$$
$$0 + 2x^2 + 2x$$

Multiply $x(x^2 + x + 1)$. Subtract $x^3 + x^2 + x$ from $x^3 + 3x^2 + 3x + 1$. You can omit the zero.

Think $\dfrac{2x^2}{x^2} = 2$.

3. $16\overline{)\;337}^{\,21}$
$\;\;\;\;\underline{-32}$
$\;\;\;\;\;17$ "Bring down" the 7. Now, divide 17 by 16. It goes once. Write **1** after the 2.

$$x^2 + x + 1\overline{)x^3 + 3x^2 + 3x + 1}^{\;\;\;x + 2}$$
$$(-)\underline{x^3 +\;\; x^2 +\;\; x}$$
$$0 + 2x^2 + 2x + 1$$

"Bring down" the 1. Write $+ 2$ after the x.

4. $16\overline{)\;337}^{\,21}$
$\;\;\;\;\underline{-32}$
$\;\;\;\;\;17$
$\;\;\;\underline{-16}$
$\;\;\;\;\;\;\;1$ Multiply 16 by 1 and subtract the result from 17. The remainder is 1.

$$x^2 + x + 1\overline{)x^3 + 3x^2 + 3x + 1}^{\;\;\;x + 2}$$
$$(-)\underline{x^3 +\;\; x^2 +\;\; x}$$
$$0 + 2x^2 + 2x + 1$$
$$(-)\underline{2x^2 + 2x + 2}$$
$$-1$$

Multiply $2(x^2 + x + 1)$. Subtract $2x^2 + 2x + 2$ from $2x^2 + 2x + 1$. The remainder is -1.

5. The answer (**quotient**) can be written as 21 R 1 (read "21 remainder 1") or as $21 + \frac{1}{16}$, which is $21\frac{1}{16}$.

The answer (**quotient**) can be written as $x + 2$ R $- 1$ (read "$x + 2$ remainder $- 1$"), or you can write the result more completely as

$$\underbrace{\dfrac{\overbrace{x^3 + 3x^2 + 3x + 1}^{\text{Dividend}}}{\underbrace{x^2 + x + 1}_{\text{Divisor}}}} = x + 2 - \dfrac{\overbrace{1}^{\text{Remainder}}}{\underbrace{x^2 + x + 1}_{\text{Divisor}}}$$

Quotient

6. You can check this answer by multiplying 21 by 16 (336) and adding the remainder 1 to obtain 337, which is the dividend.

You can check the answer by multiplying $(x + 2)(x^2 + x + 1) = x^3 + 3x^2 + 3x + 2$ and adding the remainder -1 to get the dividend $x^3 + 3x^2 + 3x + 1$.

EXAMPLE 2   Dividing polynomials using long division

Divide: $(x^3 + 2x^2 - 17x) \div (x^2 + x - 3)$

SOLUTION

Think $\dfrac{x^3}{x^2} = x$.

Think $\dfrac{x^2}{x^2} = 1$.

$$x^2 + x - 3\overline{)x^3 + 2x^2 - 17x}^{\;\;\;x + 1}$$
$$(-)\underline{x^3 +\;\; x^2 -\;\; 3x}\quad\quad\longleftarrow x(x^2 + x - 3) = x^3 + x^2 - 3x$$
$$x^2 - 14x\quad\quad\text{Subtract.}$$
$$(-)\underline{x^2 +\;\;\;\; x - 3}\quad\quad\longleftarrow 1(x^2 + x - 3) = x^2 + x - 3$$
$$-15x + 3\quad\quad\longleftarrow \text{Remainder}$$

Thus

$$\frac{x^3 + 2x^2 - 17x}{x^2 + x - 3} = x + 1 + \frac{-15x + 3}{x^2 + x - 3}$$ ∎

If there are missing terms in the polynomial being divided, we insert zero coefficients, as shown in the next example.

EXAMPLE 3   Using long division when there are missing terms

Divide: $(4x^3 - 4 - 8x) \div (4 + 4x)$

SOLUTION   We write the polynomials in **descending** order, inserting $0x^2$ in the dividend.

Think $\dfrac{4x^3}{4x} = x^2$.

Think $\dfrac{-4x^2}{4x} = -x$.

Think $\dfrac{-4x}{4x} = -1$.

$$
\begin{array}{r}
x^2 - x - 1 \\
4x + 4 \overline{\smash{)}\, 4x^3 + 0x^2 - 8x - 4} \\
(-)\underline{4x^3 + 4x^2} \\
0 - 4x^2 - 8x \\
(-)\underline{-4x^2 - 4x} \\
-4x - 4 \\
(-)\underline{-4x - 4} \\
0
\end{array}
$$

⟵ $x^2(4x + 4) = 4x^3 + 4x^2$

Subtract; bring down $-8x$.

⟵ $-x(4x + 4) = -4x^2 - 4x$

Subtract; bring down -4.

⟵ $-1(4x + 4) = -4x - 4$

The remainder is zero.

 NOTE When the remainder is zero, the denominator divides *exactly* into the numerator.

Thus

$$\frac{4x^3 - 4 - 8x}{4 + 4x} = x^2 - x - 1$$

You can check this by multiplying $(4 + 4x)(x^2 - x - 1)$, obtaining $4x^3 - 4 - 8x$. ∎

Factoring When One of the Factors Is Known

Suppose we wish to factor the polynomial

$$6x^3 + 23x^2 + 9x - 18$$

None of the methods we have studied so far will work here (try them!), so we need some more information. If we know that $x + 3$ is one of the factors, we can write $6x^3 + 23x^2 + 9x - 18 = (x + 3)P$, where P is a polynomial. Dividing both sides by $x + 3$ gives

$$\frac{6x^3 + 23x^2 + 9x - 18}{x + 3} = P$$

Let's do the division:

$$\text{Think } \frac{6x^3}{x} = 6x^2.$$

$$\text{Think } \frac{5x^2}{x} = 5x.$$

$$\text{Think } \frac{-6x}{x} = -6.$$

$$
\begin{array}{r}
6x^2 + 5x - 6 \\
x + 3 \,\overline{)\, 6x^3 + 23x^2 + 9x - 18} \\
(-)\,\underline{6x^3 + 18x^2} \\
0 + 5x^2 + 9x \\
(-)+\,\underline{5x^2 + 15x} \\
0 - 6x - 18 \\
(-)-\,\underline{6x - 18} \\
0
\end{array}
$$

⟵ $6x^2(x + 3) = 6x^3 + 18x^2$

Subtract; bring down the $9x$.

⟵ $5x(x + 3) = 5x^2 + 15x$

Subtract; bring down the -18.

⟵ $-6(x + 3) = -6x - 18$

Subtract.

Thus the polynomial P is $6x^2 + 5x - 6$, and we have

$$\frac{6x^3 + 23x^2 + 9x - 18}{x + 3} = 6x^2 + 5x - 6$$

Now $6x^2 + 5x - 6 = (3x - 2)(2x + 3)$, which gives

$$\frac{6x^3 + 23x^2 + 9x - 18}{x + 3} = (3x - 2)(2x + 3)$$

Multiplying both sides by $x + 3$ yields

$$6x^3 + 23x^2 + 9x - 18 = (x + 3)(3x - 2)(2x + 3)$$

Thus if we wish to factor a polynomial and one of the factors is given, we can divide by this factor. The product of the quotient obtained (factored, if possible) and the given factor gives the complete factorization of the polynomial.

EXAMPLE 4   Factoring when one factor is known

Factor $6x^3 + 25x^2 + 2x - 8$ if $x + 4$ is one of its factors.

SOLUTION   We start by dividing $6x^3 + 25x^2 + 2x - 8$ by $x + 4$.

$$\text{Think } \frac{6x^3}{x} = 6x^2.$$

$$\text{Think } \frac{x^2}{x} = x.$$

$$\text{Think } \frac{-2x}{x} = -2.$$

$$
\begin{array}{r}
6x^2 + x - 2 \\
x + 4 \,\overline{)\, 6x^3 + 25x^2 + 2x - 8} \\
(-)\,\underline{6x^3 + 24x^2} \\
0 + x^2 + 2x \\
(-)\,\underline{x^2 + 4x} \\
0 - 2x - 8 \\
(-)\,\underline{-2x - 8} \\
0
\end{array}
$$

⟵ $6x^2(x + 4) = 6x^2 + 24x^2$

Subtract; bring down the $2x$.

⟵ $x(x + 4) = x^2 + 4x$

Subtract; bring down the -8.

⟵ $-2(x + 4) = -2x - 8$

Now we know that the polynomial we are trying to factor is the product of $(x + 4)$ and the quotient we obtained, $6x^2 + x - 2$. To factor completely, we have to factor $6x^2 + x - 2$ as $(3x + 2)(2x - 1)$. Thus

$$6x^3 + 25x^2 + 2x - 8 = (x + 4)(6x^2 + x - 2)$$

$$= (x + 4)(3x + 2)(2x - 1) \quad\blacksquare$$

 Using Synthetic Division to Divide a Polynomial by a Binomial

The factorization process we have discussed can be made faster and more efficient if we can find a simpler method for doing the division. Look at the long division (left) and its simplified version (right):

$$
\begin{array}{r}
3x^2 + 11x + 15 \\
x - 2\,\overline{)\,3x^3 + 5x^2 - 7x + 10} \\
\underline{3x^3 - 6x^2} \\
11x^2 \\
\underline{11x^2 - 22x} \\
15x \\
\underline{15x - 30} \\
+ 40
\end{array}
$$

Quotient · Dividend · Divisor · Remainder

$$
\begin{array}{r}
3 + 11 + 15 \\
1 - 2\,\overline{)\,3 + 5 - 7 + 10} \\
\underline{3\ (-6)} \\
+11 \\
\underline{+11\ (-22)} \\
+15 \\
\underline{+15\ (-30)} \\
+ 40
\end{array}
$$

Quotient · Dividend · Divisor · Remainder

First, notice that the diagram on the right omits the variables. Moreover, all the numbers in the boxes are repeated. Finally, the only numbers that we have to use to do the division are the circled ones and the ones in the answer (the quotient). We use these facts to write the division in an even shorter version using only three lines:

$$
\begin{array}{r}
-2\,\overline{)\,3 + 5 - 7 + 10} \\
- 6 - 22 - 30 \\
\hline
3 + 11 + 15 + 40
\end{array}
$$

The first line has the divisor with the coefficient 1 omitted, followed by the coefficients of the dividend; the second line shows the numbers we circled, and the third line the coefficients of the quotient (3, 11, and 15) and the remainder 40.

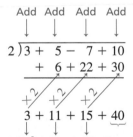

We do one more modification. To obtain the third line in the preceding form, we subtracted the second line from the first. To make it easier, we replace this subtraction with an addition by changing the sign of the indicated divisor; that is, we use $+2$ in place of -2 so that we can add at each step.

Quotient: $3x^2 + 11x + 15$ Remainder
Note that the degree (2) of the quotient is one less than the degree (3) of the dividend.

SYNTHETIC DIVISION

Synthetic division is a procedure used when dividing a polynomial by a binomial of the form $x - k$. The degree of the quotient is one less than the degree of the dividend.

EXAMPLE 5   Using synthetic division to divide a polynomial by a binomial

Use synthetic division to divide: $2x^4 - 3x^2 + 5x - 7 \div x + 3$

SOLUTION   Since synthetic division works only when dividing by binomials of the form $x - k$, we write $x + 3$ as $x - (-3)$ to obtain the indicated divisor. The zero in the first line of the division is in place of the missing x^3 term.

$$-3 \overline{)2 \quad 0 - 3 + 5 - \quad 7}$$
$$ - 6 + 18 - 45 + 120$$

$$2 - 6 + 15 - 40 + 113$$

Quotient: $2x^3 - 6x^2 + 15x - 40$ R 113

The answer is read from the bottom row, $2x^3 - 6x^2 + 15x - 40$ with 113 as the remainder, that is,

$$\frac{2x^4 - 3x^2 + 5x - 7}{x + 3} = 2x^3 - 6x^2 + 15x - 40 + \frac{113}{x + 3} \qquad \blacksquare$$

Using the Remainder and the Factor Theorems

Is there a quick way of checking at least part of the division process in Example 5? Amazingly enough, we can find the remainder in the division by using the *Remainder Theorem*, whose proof is given in more advanced courses.

THE REMAINDER THEOREM

> If the polynomial $P(x)$ is divided by $x - k$, then the remainder is $P(k)$.

This theorem says that when $P(x)$ is divided by $x - k$, the remainder can be found by evaluating P at k, that is, by finding $P(k)$.

Thus we can find the remainder in Example 5 by finding

$$P(-3) = 2(-3)^4 - 3(-3)^2 + 5(-3) - 7$$
$$= 2(81) - 27 - 15 - 7$$
$$= 162 - 27 - 15 - 7$$
$$= 113 \qquad \text{Remainder}$$

But there is a more important application of the Remainder Theorem. If you divide a polynomial $P(x)$ by $x - k$ and the remainder is zero, then $P(k) = 0$, which means that k is a solution of the equation $P(k) = 0$. Thus one way to show that $k = -3$ is a solution of $x^4 - 4x^3 - 5x^2 + 36x - 36 = 0$ is to substitute -3 for x in the equation. Another way is to use synthetic division and show that the remainder is zero, that is, $P(k) = 0$. We do this in Example 6.

EXAMPLE 6   Using the Remainder Theorem with synthetic division

Use synthetic division to show that -3 is a solution of

$$P(x) = x^4 - 4x^3 - 5x^2 + 36x - 36 = 0$$

SOLUTION We divide $x^4 - 4x^3 - 5x^2 + 36x - 36$ by $x - (-3)$.

$$
\begin{array}{r}
-3\,)\overline{1 - 4 - \ 5 + 36 - 36} \\
-3 + 21 - 48 + 36 \\
\hline
1 - 7 \quad 16 - 12 \qquad 0
\end{array}
$$

Since the remainder is zero, $P(-3) = 0$, so -3 is a solution of the given equation. ∎

Note that since the remainder is zero when the given polynomial is divided by $x + 3$, $x + 3$ must be a factor of the polynomial. Thus

$$x^4 - 4x^3 - 5x^2 + 36x - 36 = (x + 3)(x^3 - 7x^2 + 16x - 12)$$

where the second factor is the quotient polynomial found in the last row of the synthetic division. Thus the Remainder Theorem, used together with synthetic division, can help us evaluate a polynomial (by finding the remainder) and, if the remainder is zero, can even help us factor the polynomial. This last important result, which can be derived from the Remainder Theorem, is called the *Factor Theorem,* and we state it here.

THE FACTOR THEOREM

When a polynomial $P(x)$ has a factor $(x - k)$, it means that $P(k) = 0$.

GRAPH IT

Division problems can be checked with your grapher. To check the results of Example 2, graph the original problem

$$Y_1 = \frac{(x^3 + 2x^2 - 17x)}{(x^2 + x - 3)}$$

and the answer

$$Y_2 = x + 1 + \frac{(-15x + 3)}{(x^2 + x - 3)}$$

using the standard window. (Note the parentheses when entering the expression in your grapher.) The same graph is obtained in both cases (Window 1).

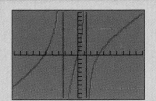
WINDOW 1
$$Y_1 = \frac{(x^3 + 2x^2 - 17x)}{(x^2 + x - 3)}$$

Example 4 can be checked by graphing the original expression $Y_1 = 6x^3 + 25x^2 + 2x - 8$ and the factored answer $Y_2 = (x + 4)(3x + 2)(2x - 1)$. (We leave this to you!)

Finally, we can do some exploring with the Remainder Theorem. In Example 6, we found out that -3 is a solution of $x^4 - 4x^3 - 5x^2 + 36x - 36 = 0$. How can we verify this graphically? We graph $Y_1 = x^4 - 4x^3 - 5x^2 + 36x - 36$ (see Window 2). If we use a standard window, it seems that the curve cuts the horizontal axis at $x = -3$, $x = 2$, and $x = 3$, but we cannot see the complete graph. To see more of the vertical axis, let Ymin $= -100$ and Ymax $= 100$. We can clearly see that $x = -3$ is a solution.

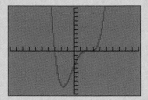
WINDOW 2
$$Y_1 = x^4 - 4x^3 - 5x^2 + 36x - 36$$

If you have a TI-82, you can check the two other possible solutions, $x = 2$ and $x = 3$. Start with $x = 2$. Press [2nd] [CALC] 2 and use the [◄] button to move the cursor to the left of 2. Press [ENTER]. Now move the cursor to the right of 2 (but less than 3) and press [ENTER]. When the grapher asks "Guess?" press [ENTER]. The root (solution) is $x = 2.00000002$. Do the same for 3. You can also verify that -3, 2, and 3 are the solutions of the equation by using the Remainder Theorem and checking that $P(-3) = 0$, $P(2) = 0$, and $P(3) = 0$. Here the grapher gives you the hint and the algebra confirms it.

Can we use all this information to factor the polynomial? Since -3, 2, and 3 are solutions of the equation, $x - (-3)$, $x - 2$ and $x - 3$ are factors of the polynomial; thus $x^4 - 4x^3 - 5x^2 + 36x - 36 = (x + 3)(x - 2)(x - 3)Q(x)$. To find $Q(x)$, divide $x^4 - 4x^3 - 5x^2 + 36x - 36$ by $(x + 3)(x - 2)(x + 3)$—that is, by $x^3 - 2x^2 - 9x + 18$. Using long division, we get the answer, $x - 2$. Substituting $x - 2$ for $Q(x)$, we find $x^4 - 4x^3 - 5x^2 + 36x - 36 = (x + 3)(x - 2)(x - 3)(x - 2)$, which means that $x - 2$ is a solution *twice*. The number 2 is called a **double root** of the equation.

EXERCISE 4.5

A In Problems 1–10, perform the indicated divisions.

1. $\dfrac{3x^3 + 9x^2 - 6x}{3x}$

2. $\dfrac{6x^3 + 8x^2 - 4x}{2x}$

3. $\dfrac{10x^3 - 5x^2 + 15x}{-5x}$

4. $\dfrac{24x^3 - 12x^2 + 6x}{-6x}$

5. $\dfrac{8y^4 - 32y^3 + 12y^2}{-4y^2}$

6. $\dfrac{9y^4 - 45y^3 + 18y^2}{-3y^2}$

7. $\dfrac{10x^5 + 8x^4 - 16x^3 + 6x^2}{2x^3}$

8. $\dfrac{12x^4 + 18x^3 + 16x^2}{4x^3}$

9. $\dfrac{15x^3y^2 - 10x^2y + 15x}{5x^2y}$

10. $\dfrac{18x^4y^4 - 24x^2y^3 + 6xy^2}{3x^2y^2}$

B In Problems 11–36, divide using long division.

11. $x^2 + 5x + 6$ by $x + 2$

12. $x^2 + 9x + 20$ by $x + 5$

13. $y^2 + 3y - 10$ by $y - 2$

14. $y^2 + 2y - 15$ by $y - 3$

15. $2x^3 - 4x - 2$ by $2x + 2$

16. $2x^3 + 5x^2 - x - 2$ by $2x - 1$

17. $3x^3 + 14x^2 + 13x - 6$ by $3x - 1$

18. $2x^3 - 5x^2 - 14x + 3$ by $2x - 1$

19. $2x^3 - 10x - 7x^2 + 24$ by $2x - 3$

20. $3x^3 + 8x + 13x^2 - 12$ by $3x - 2$

21. $2x^3 + 2x + 7x^2 - 2$ by $-1 + 2x$

22. $3x^3 + 3x + 8x^2 - 2$ by $-1 + 3x$

23. $y^4 - y^2 - 2y - 1$ by $y^2 + y + 1$

24. $y^4 - y^2 - 4y - 4$ by $y^2 + y + 2$

25. $8x^3 - 6x^2 + 5x - 9$ by $2x - 3$

26. $2x^4 - x^3 + 7x - 2$ by $2x + 3$

27. $x^3 + 8$ by $x + 2$

28. $x^3 + 64$ by $x + 4$

29. $8y^3 - 64$ by $2y - 4$

30. $27x^3 - 8$ by $3x - 2$

31. $a^4 - a^2 - 2a - 1$ by $a^2 + a + 1$

32. $b^4 - b^2 - 2b - 1$ by $b^2 - b - 1$

33. $x^5 - 5x + 12x^2$ by $x^2 + 5 - 2x$

34. $y^5 - y^4 + 10 - 27y + 7y^2$ by $y^2 + 5 - y$

35. $4x^4 - 13x^2 + 4x^3 - 3x - 21$ by $2x + 5$

36. $8y^4 - 75y^2 - 18y^3 + 46y + 121$ by $4y + 5$

C In Problems 37–42, factor completely.

37. $x^3 - 4x^2 + x + 6$ if $x + 1$ is one of the factors

38. $x^3 - 4x^2 + x + 6$ if $x - 3$ is one of the factors

39. $x^4 - 4x^3 + 3x^2 + 4x - 4$ if $x^2 - 4x + 4$ is one of the factors

40. $x^4 - 2x^3 - 13x^2 + 14x + 24$ if $x^2 - 6x + 8$ is one of the factors

41. $x^4 + 6x^3 + 3x + 140$ if $x^2 - 3x + 7$ is one of the factors

42. $x^4 - 22x^2 - 75$ if $x^2 + 3$ is one of the factors

D In Problems 43–52, use synthetic division to find the quotient and the remainder.

43. $(v^3 - 8v - 3) \div (v - 3)$

44. $(y^3 - 4y^2 - 25) \div (y - 5)$

45. $(x^3 + 4x^2 - 7x + 5) \div (x - 2)$

46. $(4w^3 - w^2 + 92) \div (w + 3)$

47. $(z^3 - 32z + 24) \div (z + 6)$

48. $(2y^4 - 3y^3 + y^2 - 3y) \div (y - 2)$

49. $(3y^4 - 41y^2 - 13y - 8) \div (y - 4)$

50. $(v^5 - 4v^3 + 5v^2 - 5) \div (v + 1)$

51. $(2y^4 - 13y^3 + 6y^2 + 5y - 30) \div (y - 6)$

52. $(4w^4 + 20w^3 - w^2 - 2w + 15) \div (w + 5)$

E In Problems 53–60, use synthetic division to show that the given number is a solution of the equation.

53. 4; $z^3 + 6z^2 - 6z - 136 = 0$

54. −7; $3y^3 + 13y^2 - 57y - 7 = 0$

55. −4; $5y^3 + 18y^2 - y + 28 = 0$

56. 6; $7x^3 - 39x^2 - 26x + 48 = 0$

57. 5; $3v^4 - 14v^3 - 7v^2 + 21v - 55 = 0$

58. −8; $8w^4 + 62w^3 - 15w^2 + 10w + 16 = 0$

59. −1; $y^5 + y^4 + 2y^3 + 5y^2 - 2y - 5 = 0$

60. 1; $3z^6 - z^5 + 5z^4 - 3z - 4 = 0$

APPLICATIONS

In business, the average cost per unit, $\bar{C}$, is given by

$$\bar{C} = \frac{C}{x}$$

where C is the total cost and x is the number of units.

61. Find the average cost when $C = 500 + 4x$.

62. Find the average cost when $C = 200 + 2x^2$.

SKILL CHECKER

Solve:

63. $4x + 8 = 6x$ **64.** $5x + 10 = 7x$

65. $x(x + 2) - (x - 3)(x - 4) = 4x + 3$

66. $x(x + 1) - (x - 1)(x - 2) = 2x + 2$

USING YOUR KNOWLEDGE
Finding Possible Factors

In this section we factored some polynomials by using a given first-degree factor. How did we find that one factor? Here is one way to do it.

If a polynomial of the form $c_0 x^n + c_1 x^{n-1} + \cdots + c_n$ with integer coefficients has a factor $ax + b$, where a and b are integers and $a > 0$, then a must divide c_0 and b must divide c_n. For example, to find a and b for the polynomial $2x^3 + 3x^2 - 8x + 3$, we consider the positive divisors of 2 and all the divisors of 3. Thus the possibilities for a are 1 and 2 and for b are 1, -1, 3, and -3. This means that only the binomials $x + 1$, $x - 1$, $x + 3$, $x - 3$, $2x + 1$, $2x - 1$, $2x + 3$, and $2x - 3$ have to be checked. It turns out that $x - 1$ is a factor and that division gives

$$2x^3 + 3x^2 - 8x + 3 = (x - 1)(2x^2 + 5x - 3)$$

Since $2x^2 + 5x - 3 = (2x - 1)(x + 3)$, we have

$$2x^3 + 3x^2 - 8x + 3 = (x - 1)(2x - 1)(x + 3)$$

67. What binomials should be checked as possible factors for the polynomial $x^3 + 3x^2 + 5x + 6$ if a and b are positive?

68. What binomials should be checked as possible factors for the polynomial $3x^3 + 5x^2 - 3x - 2$ if a and b are positive?

WRITE ON . . .

69. Describe the procedure you use to divide a polynomial by a binomial.

70. How can you check the result when dividing one polynomial by another? Use your answer to check that $(2x^3 + 5x^2 - 8x + 6) \div (x + 2) = 2x^2 + x - 10$ R 26.

71. Describe two different procedures that can be used to determine that there is a zero remainder when dividing a polynomial by a binomial of the form $x - k$.

MASTERY TEST
If you know how to do these problems, you have learned your lesson!

Use synthetic division to find the quotient and remainder when dividing:

72. $x^3 - 3x^2 + 3$ by $x - 3$

73. $2x^4 - 13x^3 + 16x^2 - 9x + 20$ by $x - 5$

Use synthetic division to show that the given number is a solution of the equation:

74. 2; $2x^3 + 5x^2 - 8x - 20 = 0$

75. -2; $3x^4 + 5x^3 - x^2 + x - 2 = 0$

Factor:

76. $2x^3 - x^2 - 18x + 9$ if $x + 3$ is one of the factors

77. $z^3 - 3z^2 - 4z + 12$ if $z + 2$ is one of the factors

Use long division to divide:

78. $6x^3 + 3x - 9$ by $x - 2$

79. $3x^3 - 2x^2 - x - 6$ by $x + 2$

Divide:

80. $\dfrac{24x^5 - 18x^4 + 12x^3}{6x^2}$

81. $\dfrac{16x^4 - 4x^3 + 8x^2 - 16x + 40}{8x^3}$

4.6

EQUATIONS INVOLVING RATIONAL EXPRESSIONS

To succeed, review how to:	Objective:
1. Find the LCD of two or more fractions (p. 225).	Solve equations involving rational expressions.
2. Factor polynomials (pp. 165–170, 173–178).	
3. Solve linear and quadratic equations (pp. 69, 186–190).	

getting started   **Play Ball with Equations Involving Rational Expressions**

How many two-sports superstars do you know? You can probably name Bo Jackson and Deion Sanders who starred in baseball and football. In 1994, Michael Jordan, the basketball superstar, decided to try baseball. His batting average in spring training (to three decimals) is found by using the formula:

$$\text{Average} = \frac{\text{number of hits}}{\text{number of times at bat}} = \frac{3}{20} = 0.150$$

How many consecutive hits h would he need to bring his average to 0.320? Here is the information we have:

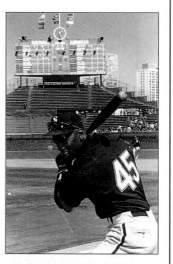

	Actual	New
Number of hits	3	$3 + h$
Times at bat	20	$20 + h$
New average $= 0.320 = \dfrac{32}{100} = \dfrac{3 + h}{20 + h}$		

Thus to find the answer, we must solve the equation. The first step in solving equations containing rational expressions is to multiply both sides of the equation by the LCD to clear the rational expressions; then we use the distributive property to clear parentheses and solve as usual. Here is the solution.

1. Multiply by the LCD, $100(20 + h)$.

$$100(20 + h)\,\frac{32}{100} = 100(20 + h)\,\frac{3 + h}{20 + h}$$

2. Reduce and use the distributive property.

$$640 + 32h = 300 + 100h$$

3. Subtract 300 from both sides.

$$340 + 32h = 100h$$

4. Subtract $32h$ from both sides.

$$340 = 68h$$

5. Divide by 68.

$$h = \frac{340}{68} = 5$$

Thus he would need only five consecutive hits to obtain a very respectable 0.320 average!

What is the difference between the equations we have solved up to now and an equation such as the following?

$$\frac{32}{100} = \frac{3 + h}{20 + h}$$

This equation has a variable in the *denominator,* so we have to avoid values of the variable that make the denominator zero. Such equations are solved with a procedure similar to the one given on page 69, except that when multiplying by the LCD of the fractions involved, we may end up with a quadratic equation, an equation that can be written in the form $ax^2 + bx + c = 0$. Here is the procedure we need.

PROCEDURE

Procedure for Solving Equations Containing Rational Expressions

1. Factor all denominators and multiply both sides of the equation by the LCD of all fractional expressions in the equation.

2. Write the result in reduced form and use the distributive property to remove parentheses.

3. Determine whether the equation is *linear* (can be written in the form $ax + b = 0$) or *quadratic* ($ax^2 + bx + c = 0$) and solve accordingly. (For linear equations see page 69; for quadratic equations, factor them and use the zero-factor property to solve.)

4. Check that the *proposed* or *trial* solution thus obtained satisfies the original equation. If it does not, discard it as an **extraneous solution**.

EXAMPLE 1   Solving equations containing rational expressions

Solve:

a. $\dfrac{4}{x} = \dfrac{6}{x + 2}$ **b.** $\dfrac{1}{x + 1} = \dfrac{2}{x + 2}$

SOLUTION

a. We use the four-step procedure.

 1. The LCD of

$$\frac{4}{x} \quad \text{and} \quad \frac{6}{x + 2}$$

 is $x(x + 2)$. Multiplying both sides of the equation by the LCD,

$$x(x + 2) = \frac{x(x + 2)}{1}$$

 gives

$$\frac{\cancel{x}(x + 2)}{1} \cdot \frac{4}{\cancel{x}} = \frac{x\cancel{(x + 2)}}{1} \cdot \frac{6}{\cancel{x + 2}}$$

 2. Reduce and remove parentheses.

$$(x + 2) \cdot 4 = x \cdot 6 \qquad \text{Divide out } x \text{ and } x + 2.$$
$$4x + 8 = 6x \qquad \text{Remove parentheses.}$$

 3. Solve.

$$8 = 2x \qquad \text{Subtract } 4x \text{ from both sides.}$$
$$4 = x \qquad \text{Divide by 2.}$$

4. The proposed solution is 4. To check the answer, we substitute 4 for x in the original equation to obtain

$$\frac{4}{4} = \frac{6}{4 + 2}$$

or $1 = 1$. Therefore, the solution 4 is correct, and the solution set is $\{4\}$.

b. Again, we use the four-step procedure.

1. The LCD of

$$\frac{1}{x + 1} \quad \text{and} \quad \frac{2}{x + 2}$$

is $(x + 1)(x + 2)$. Multiplying both sides of the equation by

$$\frac{(x + 1)(x + 2)}{1}$$

we obtain

$$\frac{(x + 1)(x + 2)}{1} \cdot \frac{1}{x + 1} = \frac{(x + 1)(x + 2)}{1} \cdot \frac{2}{x + 2}$$

2. $x + 2 = (x + 1) \cdot 2$ Simplify.

     $x + 2 = 2x + 2$ Use the distributive property.

3. $x = 2x$ Subtract 2.

     $0 = x$ Subtract x.

4. The proposed solution is zero. The verification that zero is the actual solution is left to you. ∎

As we mentioned, when variables occur in the denominator, it's possible to multiply both sides of the equation by the LCD of the fractions involved and obtain a solution of the resulting equation that does *not* satisfy the original equation. This points out the necessity of *checking,* by direct substitution in the original equation, any proposed solutions obtained after multiplying both sides of an equation by factors containing the unknown. If the proposed solution does not satisfy the original equation, it is called an *extraneous solution.*

CAUTION Note that, by definition, an extraneous solution does *not* satisfy the given equation and must *not* be listed as a solution.

 EXAMPLE 2   More practice solving equations containing rational expressions

Solve:

a. $\dfrac{1}{x - 4} - \dfrac{1}{x - 2} = \dfrac{2x}{x^2 - 6x + 8}$ **b.** $\dfrac{x}{x - 3} - \dfrac{x - 4}{x + 2} = \dfrac{4x + 3}{x^2 - x - 6}$

SOLUTION

a. We use our four-step procedure.

1. First factor the denominator of the right-hand side of the equation to obtain

$$\frac{1}{x-4} - \frac{1}{x-2} = \frac{2x}{(x-4)(x-2)}$$

Since the LCD of the fractions involved is

$$\frac{(x-4)(x-2)}{1}$$

we multiply each side of the equation by this LCD to get

$$\frac{(x-4)(x-2)}{1} \cdot \left[\frac{1}{x-4} - \frac{1}{x-2}\right] = \frac{(x-4)(x-2)}{1} \cdot \left[\frac{2x}{(x-4)(x-2)}\right]$$

2. Reduce and remove parentheses.

$$\frac{(x-4)(x-2)}{1} \cdot \frac{1}{x-4} - \frac{(x-4)(x-2)}{1} \cdot \frac{1}{x-2}$$

$$= \frac{(x-4)(x-2)}{1} \cdot \frac{2x}{(x-4)(x-2)}$$

3. Solve.
$$(x-2) - (x-4) = 2x$$
$$x - 2 - x + 4 = 2x$$
$$2 = 2x \qquad \text{Collect like terms.}$$
$$x = 1 \qquad \text{Divide by 2.}$$

4. The proposed solution is 1. Substituting 1 for x in the original equation gives

$$\frac{1}{1-4} - \frac{1}{1-2} \overset{?}{=} \frac{2 \cdot 1}{1^2 - 6 \cdot 1 + 8}$$

$$\frac{1}{-3} - \frac{1}{-1} \overset{?}{=} \frac{2}{3}$$

$$-\frac{1}{3} + 1 = \frac{2}{3}$$

which is a true statement. Thus our result is correct. The actual solution is 1 and the solution set is {1}.

b. We use our four-step procedure.

1. We first write the right-hand side of the equation with the denominator factored.

$$\frac{x}{x-3} - \frac{x-4}{x+2} = \frac{4x+3}{(x-3)(x+2)}$$

The LCD is

$$\frac{(x-3)(x+2)}{1}$$

We multiply both sides by the LCD.

$$\frac{(x-3)(x+2)}{1}\left[\frac{x}{x-3} - \frac{x-4}{x+2}\right] = \frac{(x-3)(x+2)}{1}\left[\frac{4x+3}{(x-3)(x+2)}\right]$$

2. Reduce and remove parentheses.

$$\frac{\cancel{(x-3)}(x+2)}{1} \cdot \frac{x}{\cancel{x-3}} - \frac{(x-3)\cancel{(x+2)}}{1} \cdot \frac{x-4}{\cancel{x+2}}$$

$$= \frac{\cancel{(x-3)}\cancel{(x+2)}}{1} \cdot \frac{4x+3}{\cancel{(x-3)}\cancel{(x+2)}}$$

$$(x+2)\cdot x - (x-3)(x-4) = 4x+3$$

$$x^2 + 2x - (x^2 - 7x + 12) = 4x + 3$$

3. Solve. $\qquad\qquad\qquad 9x - 12 = 4x + 3$ Combine like terms.

$$5x = 15 \qquad \text{Add 12 and subtract } 4x.$$

$$x = 3 \qquad \text{Divide by 5.}$$

4. The proposed solution is 3. However, if x is replaced by 3 in the original equation, the term

$$\frac{x}{x-3}$$

yields $\frac{3}{0}$, which is meaningless. Consequently, the equation

$$\frac{x}{x-3} - \frac{x-4}{x+2} = \frac{4x+3}{x^2 - x - 6}$$

has no solution. Its solution set is $\varnothing$. Thus 3 is an extraneous solution.   ■

Finally, we must point out that the equations that result from clearing denominators are not *always* linear equations. For example, to solve the equation

$$\frac{x^2}{x+3} = \frac{9}{x+3}$$

we first multiply by the LCD $(x + 3)$ to obtain

$$(x+3)\frac{x^2}{x+3} = (x+3)\frac{9}{x+3} \qquad \text{or} \qquad x^2 = 9$$

In this equation, the variable x has 2 as an exponent; thus it is a quadratic equation and can be solved when written in standard form—that is, by writing the equation as

$$x^2 - 9 = 0$$

$$(x+3)(x-3) = 0 \qquad \text{Factor.}$$

$$x + 3 = 0 \quad \text{or} \quad x - 3 = 0 \qquad \text{Use the zero-factor property.}$$

$$x = -3 \quad \text{or} \qquad x = 3 \qquad \text{Solve each equation.}$$

Obviously, 3 is a solution, since

$$\frac{3^2}{3+3} = \frac{9}{3+3}$$

However, for -3, the denominator $x + 3$ becomes zero. Thus -3 is an extraneous solution. The only actual solution is 3, and the solution set is $\{3\}$.

EXAMPLE 3   Solving equations containing extraneous solutions

Solve:

$$1 + \frac{3}{x-2} = \frac{12}{x^2-4}$$

SOLUTION   Since $x^2 - 4 = (x+2)(x-2)$, the LCD is $(x+2)(x-2)$. We then write the equation with the denominator $x^2 - 4$ in factored form and multiply each term by the LCD, as before. Here are the steps.

1. Multiply each term by the LCD.

$$(x+2)(x-2) \cdot 1 + (x+2)(x-2) \cdot \frac{3}{x-2} = (x+2)(x-2) \cdot \frac{12}{x^2-4}$$

2. Reduce and remove parentheses.

$$(x^2 - 4) + 3(x+2) = 12$$
$$x^2 - 4 + 3x + 6 = 12$$

3. Solve by factoring.

$$x^2 + 3x + 2 = 12$$
$$x^2 + 3x - 10 = 0 \qquad \text{Subtract 12 to write in standard form.}$$
$$(x+5)(x-2) = 0 \qquad \text{Factor.}$$
$$x + 5 = 0 \quad \text{or} \quad x - 2 = 0 \qquad \text{Use the zero-factor property.}$$
$$x = -5 \quad \text{or} \qquad x = 2 \qquad \text{Solve each equation.}$$

4. Since 2 makes the denominator $x - 2$ equal to zero, it is an extraneous solution. The only actual solution is -5, and the solution set is $\{-5\}$. This solution can be checked in the original equation.   ■

EXAMPLE 4   More practice with extraneous solutions

Solve:

$$\frac{x-3}{x^2-4x} = \frac{2}{x^2-16}$$

SOLUTION   As usual, we solve by steps.

1. Write all expressions in factored form and then multiply by the LCD,

$$\frac{x(x-4)(x+4)}{1}$$

2. Reduce and remove parentheses.

$$\frac{\cancel{x}(\cancel{x-4})(x+4)}{1} \cdot \frac{x-3}{\cancel{x}(\cancel{x-4})} = \frac{x(\cancel{x-4})(\cancel{x+4})}{1} \cdot \frac{2}{(\cancel{x+4})(\cancel{x-4})}$$

$$(x+4)(x-3) = 2x$$

3. Solve by factoring.

$$x^2 + x - 12 = 2x$$

$$x^2 - x - 12 = 0 \qquad \text{Subtract } 2x.$$

$$(x+3)(x-4) = 0 \qquad \text{Factor.}$$

$$x + 3 = 0 \quad \text{or} \quad x - 4 = 0 \qquad \text{Use the zero-factor property.}$$

$$x = -3 \quad \text{or} \qquad x = 4 \qquad \text{Solve.}$$

4. The proposed solutions are -3 or 4. If we substitute -3 for x in the original equation, we get a true statement. On the other hand, if we substitute 4 for x in the original equation, the denominators on both sides are zero, so the fractions are not defined. Thus the only actual solution is -3, and 4 is an extraneous solution. The solution set is $\{-3\}$. ■

EXAMPLE 5　Solving equations involving negative exponents

Solve: $3x(x+2)^{-1} + 8(x-3)^{-1} = 4$

SOLUTION　At first, it seems that there are *no* rational expressions involved. However, since $a^{-1} = \frac{1}{a}$,

$$3x(x+2)^{-1} = 3x \cdot \frac{1}{x+2} = \frac{3x}{x+2} \qquad \text{and} \qquad 8(x-3)^{-1} = 8 \cdot \frac{1}{x-3} = \frac{8}{x-3}$$

Thus $3x(x+2)^{-1} + 8(x-3)^{-1} = 4$ becomes

$$\frac{3x}{x+2} + \frac{8}{x-3} = 4$$

Using our four-step procedure, we have the following:

1. Since the only denominators are $x+2$ and $x-3$, the LCD is $(x+2)(x-3)$. Multiply both sides of the equation by the LCD.

$$(x+2)(x-3)\left(\frac{3x}{x+2} + \frac{8}{x-3}\right) = (x+2)(x-3)4$$

2. Reduce and remove parentheses.

$$(x+2)(x-3) \cdot \frac{3x}{x+2} + (x+2)(x-3) \cdot \frac{8}{x-3} = (x+2)(x-3)4$$

$$(x-3)3x + (x+2)8 = (x^2 - x - 6)4$$

$$3x^2 - 9x + 8x + 16 = 4x^2 - 4x - 24$$

3. Solve by factoring.

$$3x^2 - x + 16 = 4x^2 - 4x - 24$$

Subtract $3x^2 - x + 16$ from both sides. $\qquad 0 = x^2 - 3x - 40$

Factor. $\qquad 0 = (x+5)(x-8)$

$$x + 5 = 0 \quad \text{or} \quad x - 8 = 0 \qquad \text{Solve each equation.}$$

$$x = -5 \quad \text{or} \qquad x = 8$$

4. Neither of these proposed solutions makes a denominator zero. Thus the solutions are -5 and 8 and the solution set is $\{-5, 8\}$. Check this! ■

GRAPH IT

Now we shall discuss four techniques for solving rational equations: **graphing, finding roots (solutions), finding intersections,** and **using solve.** Let's select a typical equation, the one from Example 3:

$$1 + \frac{3}{(x-2)} = \frac{12}{x^2-4}$$

To solve this equation by graphing, first subtract $\frac{12}{x^2-4}$ from both sides to obtain the

equivalent rational expression $R(x) = 0$. To find the values of x for which $R(x) = 0$, that is, the values at which $R(x)$ crosses the horizontal axis, we graph

$Y_1 = 1 + \frac{3}{(x-2)} - \frac{12}{(x^2-4)}$ (Window 1). The graph seems to cross the horizontal

axis at $x = -5$. You can confirm this by using the $\boxed{\text{TRACE}}$ and $\boxed{\text{ZOOM}}$ keys as shown in Window 2. Note that at $x = -5$, $y = -3\text{E}^{-10} = -3 \times 10^{-10}$, which is nearly zero.

A second way of solving the equation is to find the roots (solutions) of $1 + \frac{3}{(x-2)} = \frac{12}{x^2-4}$

or equivalently, the roots of $1 + \frac{3}{(x-2)} - \frac{12}{(x^2-4)} = 0$. To do this on a TI-82, graph

$Y_1 = 1 + \frac{3}{(x-2)} - \frac{12}{(x^2-4)}$ and press $\boxed{\text{2nd}}$ $\boxed{\text{CALC}}$ 2. Since the graph crosses the hori-

zontal axis at about -5, move the cursor to the left of $x = -5$. When the grapher asks "Lower bound?" press $\boxed{\text{ENTER}}$. Now move the cursor to the right of $x = -5$. When the grapher asks "Upper bound?" and "Guess?" press $\boxed{\text{ENTER}}$ each time. The grapher responds with "Root $X = -5$" as shown in Window 3.

The third technique is to graph the left-hand side as $Y_1 = 1 + \frac{3}{(x-2)}$ and the right-hand

side as $Y_2 = \frac{12}{(x^2-4)}$ then find the point at which the two graphs intersect. The x-value is

the solution of the equation. Look for solutions in three regions: points to the left of $x = -2$, points between -2 and 2, and points to the right of $x = 2$. First, move the cursor to the left of $x = -2$ and press $\boxed{\text{2nd}}$ $\boxed{\text{CALC}}$ 5. Press $\boxed{\text{ENTER}}$ after the questions "First Curve?" "Second Curve?" and "Guess?" The grapher tells you "Intersection $X = -5$" (Window 4).

Finally, if your grapher has a **solve** feature, press $\boxed{\text{MATH}}$ 0 and enter $1 + \frac{3}{(x-2)} - \frac{12}{(x^2-4)}$

(Window 5). Again, note the parentheses in the denominators. Then enter a comma $\boxed{,}$; the variable you are using, which is $\boxed{\text{X,T,}\theta}$; another comma; and your guess for the answer. Now close the parentheses by pressing $\boxed{)}$. If you enter a number greater than 2 for your guess (say, 3), you will get an error message. If you enter a guess between -2 and 2 (say, 0), you will also get an error message. Finally, try a number less than -2 for your guess, say, -3. After you press $\boxed{\text{ENTER}}$, the grapher shows the answer -5. Note that to get to the final answer, it is essential that you understand what the graph looks like and where it is likely to have a solution. (This is your "guess.")

Now you can practice by verifying the rest of the examples using any one of the four techniques that is available on your grapher.

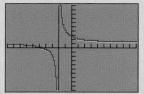

WINDOW 1

$$1 + \frac{3}{(x-2)} - \frac{12}{(x^2-4)} = 0$$

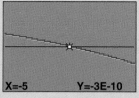

X=-5 Y=-3E-10

WINDOW 2

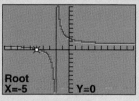

Root
X=-5 Y=0

WINDOW 3

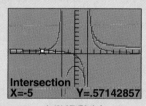

Intersection
X=-5 Y=.57142857

WINDOW 4

$$Y_1 = 1 + \frac{3}{(x-2)}$$

$$Y_2 = \frac{12}{(x^2-4)}$$

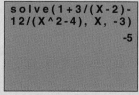

```
solve(1+3/(X-2)-
12/(X^2-4), X, -3)
                    -5
```

WINDOW 5

Note the parentheses when entering the expression. The solution -5 is shown in the last line.

EXERCISE 4.6

In Problems 1–36, solve the given equation.

1. $\dfrac{x}{3} + \dfrac{x}{6} = 3$

2. $\dfrac{x}{2} + \dfrac{x}{4} = \dfrac{3}{8}$

3. $\dfrac{x}{5} - \dfrac{3x}{10} = \dfrac{1}{2}$

4. $\dfrac{x}{6} - \dfrac{x}{5} = \dfrac{1}{15}$

5. $\dfrac{1}{y} + \dfrac{4}{3y} = 7$

6. $\dfrac{10}{3y} - \dfrac{9}{2y} = \dfrac{7}{30}$

7. $\dfrac{2}{y - 8} = \dfrac{1}{y - 2}$

8. $\dfrac{2}{y - 4} = \dfrac{3}{y - 2}$

9. $\dfrac{3}{3z + 4} = \dfrac{2}{5z - 6}$

10. $\dfrac{2}{4z - 1} = \dfrac{3}{2z + 1}$

11. $\dfrac{-2}{2x + 1} = \dfrac{3}{3x - 1}$

12. $\dfrac{-5}{2x + 3} = \dfrac{2}{3x - 1}$

13. $\dfrac{-1}{x + 1} = \dfrac{-2}{2x - 1}$

14. $\dfrac{-5}{5x - 2} = \dfrac{-3}{3x + 1}$

15. $\dfrac{2}{3x + 1} = \dfrac{4}{6x + 2}$

16. $\dfrac{3}{2x - 1} = \dfrac{6}{4x - 5}$

17. $\dfrac{2}{x^2 - 4} + \dfrac{5}{x + 2} = \dfrac{7}{x - 2}$

18. $\dfrac{3}{x^2 - 9} + \dfrac{5}{x + 3} = \dfrac{8}{x - 3}$

19. $\dfrac{t + 2}{t^2 - 3t + 2} = \dfrac{3}{t - 1} - \dfrac{1}{t - 2}$

20. $\dfrac{t + 3}{t^2 + 4t + 3} = \dfrac{4}{t + 3} - \dfrac{1}{t + 1}$

21. $\dfrac{x^2}{x^2 - 1} = 1 + \dfrac{1}{x + 1}$

22. $\dfrac{x^2}{x^2 - 9} = 1 + \dfrac{1}{x - 3}$

23. $\dfrac{1}{x^2 - 4x + 3} + \dfrac{1}{x^2 - 2x - 3} = \dfrac{1}{x^2 - 1}$

24. $\dfrac{1}{x^2 + 3x + 2} + \dfrac{1}{x^2 + x - 2} = \dfrac{1}{x^2 - 1}$

25. $\dfrac{x + 2}{3x^2 + 4x + 1} = \dfrac{x + 1}{3x^2 + 7x + 2}$

26. $\dfrac{x + 2}{2x^2 + x - 1} = \dfrac{x - 2}{2x^2 + x - 1}$

27. $\dfrac{2z + 13}{2z^2 + 5z - 3} + \dfrac{3}{z + 3} = \dfrac{4}{2z - 1}$

28. $\dfrac{z - 14}{2z^2 - 3z - 2} + \dfrac{3}{z - 2} = \dfrac{4}{2z + 1}$

29. $\dfrac{3 - x}{5x^2 - 4x - 1} + \dfrac{2}{5x + 1} = \dfrac{1}{x - 1}$

30. $\dfrac{16 - x}{4x^2 - 11x - 3} + \dfrac{5}{4x + 1} = \dfrac{2}{x - 3}$

31. $4x^{-1} + 2 = 7$ $\left(\text{Recall that } x^{-1} = \dfrac{1}{x}.\right)$

32. $3 + 6x^{-1} = 5$

33. $4x^{-1} + 6x^{-1} = 15(x + 1)^{-1}$

34. $6x^{-1} + 9x^{-1} = 25(x + 2)^{-1}$

35. $2(x - 8)^{-1} = (x - 2)^{-1}$

36. $3(3y + 4)^{-1} = 2(5y - 6)^{-1}$

SKILL CHECKER

Solve using the RSTUV method:

37. The sum of three consecutive odd integers is 69. What are the integers?

38. An investor bought some municipal bonds yielding 5% annually and some certificates of deposit yielding 8%. If the total investment amounts to $10,000 and the annual interest is $680, how much is invested in bonds and how much in certificates of deposit?

39. How many gallons of a 20% salt solution must be mixed with 40 gal of a 15% solution to obtain an 18% solution?

40. A car leaves a town traveling at 50 mi/hr. Two hours later, another car traveling at 60 mi/hr leaves on the same road in the same direction. How far from the town does the second car overtake the first?

USING YOUR KNOWLEDGE Solving for Letters

There are many instances in which a given formula must be changed to an equivalent form. For example, the formula

$$\dfrac{P}{R} = \dfrac{T}{V}$$

is frequently discussed in chemistry. Suppose you know P, R, and T. Can you find V? As before, we proceed by steps to solve for V.

1. Since the LCD is RV, we multiply each term by RV to obtain

$$RV \cdot \frac{P}{R} = \frac{T}{V} \cdot RV$$

2. Simplify.

$$VP = TR$$

3. Divide by P.

$$V = \frac{TR}{P}$$

Thus

$$V = \frac{TR}{P}$$

In Problems 41–45, use your knowledge of fractional equations to solve for the indicated variable.

41. The area A of a trapezoid is

$$A = \frac{h(b_1 + b_2)}{2}$$

Solve for h.

42. In an electrical circuit we have

$$\frac{1}{R} = \frac{1}{R_1} + \frac{1}{R_2}$$

Solve for R.

43. In refrigeration we find the formula

$$\frac{Q_1}{Q_2 - Q_1} = P$$

Solve for Q_1.

44. When studying the expansion of metals, we use the formula

$$\frac{L}{1 + at} = L_0$$

Solve for t.

45. Students of photography use the formula

$$\frac{1}{f} = \frac{1}{a} + \frac{1}{b}$$

Solve for f.

WRITE ON . . .

46. Consider the expression

$$\frac{x}{2} + \frac{x}{3}$$

and the equation

$$\frac{x}{2} + \frac{x}{3} = 5$$

a. What is the first step in simplifying the expression?

b. What is the first step in solving the equation?

c. What is the difference in the use of the LCD in adding two rational expressions as contrasted with solving an equation containing rational expressions?

47. Write your definition of an extraneous root.

48. Do you have to check the solutions to

$$\frac{x}{2} + \frac{x}{3} = 5$$

for extraneous solutions? Why or why not?

49. Do you have to check the solutions of

$$\frac{6}{x} + \frac{3}{x} = 4$$

for extraneous solutions? Why or why not?

50. In general, when would you check the solutions of an equation for extraneous solutions?

MASTERY TEST   If you know how to do these problems, you have learned your lesson!

Solve:

51. $\dfrac{x - 7}{x^2 - 8x} = \dfrac{2}{x^2 - 64}$

52. $1 - \dfrac{4}{x^2 - 1} = \dfrac{-2}{x - 1}$

53. $\dfrac{3}{x} = \dfrac{5}{x + 2}$

54. $\dfrac{2}{x + 1} = \dfrac{3}{x + 2}$

55. $4(x + 3)^{-1} + 3x(x - 2)^{-1} = -2$

56. $2x(x - 3)^{-1} + 6(x + 1)^{-1} = 6$

57. $\dfrac{1}{x - 6} - \dfrac{1}{x - 4} = \dfrac{6}{(x - 6)(x - 2)}$

58. $\dfrac{x}{x - 4} - \dfrac{x - 5}{x + 3} = \dfrac{3x + 16}{x^2 - x - 12}$

4.7

APPLICATIONS: PROBLEM SOLVING

To succeed, review how to:

1. Solve equations involving rational expressions (pp. 253–258).

2. Use the RSTUV procedure to solve word problems (p. 89).

Objectives:

A Solve integer problems.

B Solve work problems.

C Solve distance problems.

D Solve for a specified variable.

---

getting started

Getting the Golden

The unfinished canvas pictured here was painted by Leonardo da Vinci and is entitled *St. Jerome*. A Golden Rectangle (black overlay) fits so neatly around St. Jerome that experts conjecture that da Vinci painted the figure to conform to those proportions. For many years it has been said that the Golden Rectangle is one of the most visually satisfying of all geometric forms. Do you know how to construct a Golden Rectangle? Such a rectangle has a special ratio of length to width of about 8 to 5 and can be described by writing

$$\frac{\text{Length of rectangle}}{\text{Width of rectangle}} = \frac{8}{5}$$

Now suppose you want to make a Golden Rectangle of your own, but you want the length to be 6 in. longer than the width. What are the dimensions of your rectangle?

To solve this problem, you need to review the RSTUV procedure we used in Section 2.3.

---

problem solving   THE GOLDEN RECTANGLE

1. Read the problem.

You have to find the dimensions of the rectangle.

2. Select the unknown.

Let w be the width.

3. Think of a plan.

Since you want a Golden Rectangle,

$$\frac{\text{Length}}{\text{Width}} = \frac{8}{5}$$

$$\frac{w + 6}{w} = \frac{8}{5} \qquad \longleftarrow \text{Length is 6 in. more than the width.}$$
$$\qquad\qquad\qquad \longleftarrow \text{Width is } w.$$

4. Use the procedure for solving equations with rational expressions.

As before, find the LCD, which is $5w$, and multiply both sides by $5w$.

$$5w\left(\frac{w + 6}{w}\right) = 5w \cdot \frac{8}{5}$$

$$5w + 30 = 8w \qquad \text{Simplify.}$$

$$30 = 3w \qquad \text{Subtract } 5w.$$

$$10 = w \qquad \text{Divide by 3.}$$

Thus the width is 10 in. and the length is 6 in. more than that, or 16 in.

5. Verify the answer.

The rectangle is 10 in. by 16 in., so the ratio of length (16) to width (10) is $\frac{16}{10}$, or $\frac{8}{5}$, as desired.

Solving Integer Problems

We now discuss problems involving consecutive integers and other number properties. You can solve these problems using the RSTUV procedure.

problem
solving

EXAMPLE 1   Consecutive integers

There are two consecutive even integers such that the reciprocal of the first added to the reciprocal of the second is $\frac{3}{4}$. What are the integers?

SOLUTION   We use the RSTUV method.

1. Read the problem.

Make sure you understand the meaning of "consecutive even integers." For example, 2, 4 are consecutive even integers, as are 78, 80.

2. Select the unknown.

Let n be the first integer. The next even integer is $n + 2$.

3. Think of a plan.

First, translate the problem:

The reciprocal of the first	added to	the reciprocal of the second	is $\frac{3}{4}$.
$\dfrac{1}{n}$	$+$	$\dfrac{1}{n+2}$	$=\dfrac{3}{4}$

4. Use the procedure for solving equations with rational expressions.

Start by finding the LCD, which is $4n(n + 2)$. Multiply both sides by this LCD.

$$4n(n+2)\left(\frac{1}{n}+\frac{1}{n+2}\right)=4n(n+2)\cdot\frac{3}{4}$$

$$4n(n+2)\cdot\frac{1}{n}+4n(n+2)\cdot\frac{1}{n+2}=4n(n+2)\cdot\frac{3}{4}$$

$$4n+8+4n=3n^2+6n \qquad \text{Remove parentheses.}$$

$$0=3n^2-2n-8 \qquad \begin{array}{l}\text{Subtract } 8n \text{ and } 8 \\ \text{from both sides.}\end{array}$$

$$0=(3n+4)(n-2) \qquad \text{Factor.}$$

By the zero-factor property,

$$3n+4=0 \quad \text{or} \quad n-2=0$$

$$n=-\frac{4}{3} \quad \text{or} \qquad n=2 \qquad \text{Solve } 3n+4=0, n-2=0.$$

Since n was assumed to be an integer, we discard the answer $-\frac{4}{3}$. Thus the first even integer is 2 and the next one is 4.

5. Verify the answer.

Verify that the sum of the reciprocals is $\frac{3}{4}$. Since $\frac{1}{2}+\frac{1}{4}=\frac{3}{4}$, our result is correct.

B   Solving Work Problems

Have you ever wished for help with your taxes? The Internal Revenue Service estimates that it takes about 7 hours to file your 1040A form. (This includes record-keeping, familiarizing yourself with the form, and preparing and sending it.)

problem solving   **EXAMPLE 2**   Work problems

A couple is about to file their form 1040A. One of them can complete it in 8 hr, and the other can do it in 6 hr. How long would it take if they work on it together?

SOLUTION   Again, we use the RSTUV method.

1. Read the problem.   We need to find the total time it takes them when they work together.

2. Select the unknown.   Let t be the time it takes the couple to complete the form working together.

3. Think of a plan.   We concentrate on what happens each hour. Since one person can fill the form in 8 hr and the second can do it in 6 hr, the first person will complete $\frac{1}{8}$ of the form and the second will complete $\frac{1}{6}$ of the form each hour. Since they are working together and it takes t hours to do the whole thing, they complete $\frac{1}{t}$ of the form each hour. Here is what we have in one hour:

$$\underbrace{\text{Work done by first person}} + \underbrace{\text{work done by second person}} = \underbrace{\text{work done together}}$$

$$\frac{1}{8} + \frac{1}{6} = \frac{1}{t}$$

4. Use the procedure for solving equations with rational expressions.   First, find the LCD of $\frac{1}{8}$, $\frac{1}{6}$ and $\frac{1}{t}$. The LCD of these fractions is $24t$.

$$24t\left(\frac{1}{8} + \frac{1}{6}\right) = 24t \cdot \frac{1}{t} \qquad \text{Multiply by } 24t.$$

$$24t \cdot \frac{1}{8} + 24t \cdot \frac{1}{6} = 24t \cdot \frac{1}{t}$$

$$3t + 4t = 24 \qquad \text{Simplify.}$$

$$7t = 24$$

$$t = \frac{24}{7} = 3\frac{3}{7} \text{ hr}$$

It takes $3\frac{3}{7}$ hr (about 3 hr 26 min) to complete the job.

5. Verify the answer.   The verification is left to you! ■

Another type of problem can also be thought of as a work problem; this is the tank or pool problem. The idea is that the pipes filling or emptying a tank or pool are doing the *work* to fill or empty the pool. Here is how we solve these problems.

problem solving   **EXAMPLE 3**   Pool problems

A pool is filled by an intake pipe in 4 hr and is emptied by a drain pipe in 5 hr. How long will it take to fill the pool with both pipes open?

SOLUTION   As before, we use the RSTUV method.

1. Read the problem.   We are asked for the time it takes to fill the pool.

2. Select the unknown.   Let this time be T hours.

3. Think of a plan.   In 1 hr, the intake pipe fills $\frac{1}{4}$ of the pool, the drain pipe empties $\frac{1}{5}$, and together they fill $\frac{1}{T}$ of the pool. Thus, in 1 hr

Amount filled by intake pipe	−	amount emptied by drain pipe	=	amount filled by both
$\dfrac{1}{4}$	−	$\dfrac{1}{5}$	=	$\dfrac{1}{T}$

4. Use algebra to solve the equation.   The LCD is $20T$.

$$20T \cdot \frac{1}{4} - 20T \cdot \frac{1}{5} = 20T \cdot \frac{1}{T}$$

$$5T - 4T = 20$$

$$T = 20$$

It takes 20 hr to fill the pool if the intake and drain pipes are both open.

5. Verify the answer.   The intake pipe can fill the pool in 4 hr. It can then fill the pool five times in 20 hr. The drain pipe can empty the pool in 5 hr, so it can empty the pool four times in 20 hr. Since the intake can fill the pool five times and the drain can empty it four times in 20 hr, the pool would be filled once at the end of 20 hr. ■

Solving Distance Problems

The ideas we have studied can be used to solve uniform motion problems like the ones discussed in Section 2.4. As you recall, when traveling at a constant rate R, the distance D traveled in time T is given by $D = RT$. We use this information in Example 4.

problem solving   **EXAMPLE 4**   Uniform motion problems

One of the world's strongest currents is the Saltstraumen in Norway, reaching as much as 18 mi/hr. A speed boat can travel 48 mi downstream in this current in the same time it takes to go 12 mi upstream. What is the speed of the boat in still water?

SOLUTION   Once again, we use the RSTUV method.

1. Read the problem.   We want to find the speed (rate) of the boat in still water.

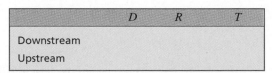

2. Select the unknown.   Let R be the rate of the boat in still water.

3. Think of a plan.   Make a chart with D, R, and T as headings.

	D	R	T
Downstream			
Upstream			

Speed downstream: $R + 18$ Current helps, add 18 to R.

Speed upstream: $R - 18$ Current hinders, subtract 18 from R.

The time T is given by

$$T = \frac{D}{R}$$

(solving for T in $D = RT$). We then have

Time downstream: $\dfrac{48}{R + 18}$

Time upstream: $\dfrac{12}{R - 18}$

Enter this information in the chart.

	D	R	T
Downstream	48	$R + 18$	$\dfrac{48}{R + 18}$
Upstream	12	$R - 18$	$\dfrac{12}{R - 18}$

Since it takes the same time to go upstream as downstream, we have

$$T_{\text{up}} = T_{\text{down}}$$

$$\frac{48}{R + 18} = \frac{12}{R - 18}$$

4. Use algebra to solve the resulting equation.

The LCD is $\dfrac{(R + 18)(R - 18)}{1}$ so we multiply both sides of the equation by this LCD

$$\frac{(R + 18)(R - 18)}{1} \cdot \frac{48}{R + 18} = \frac{(R + 18)(R - 18)}{1} \cdot \frac{12}{R - 18}$$

$$48(R - 18) = 12(R + 18)$$

$$4(R - 18) = R + 18 \qquad \text{Divide by 12.}$$

$$4R - 72 = R + 18 \qquad \text{Simplify.}$$

$$4R = R + 90 \qquad \text{Add 72.}$$

$$3R = 90 \qquad \text{Subtract } R.$$

$$R = 30$$

The speed of the boat in still water is 30 mi/hr.

5. Verify the answer.

The verification is left to you.

■

Solving for Specified Variables

Many students get the impression that the RSTUV procedure should only be used when solving word problems. This procedure, however, is very general and can be used in other situations, as we shall see in Example 5.

problem solving

EXAMPLE 5   Medicine dosages

There are many formulas that determine the medicine dosage c for children of age A when the adult dosage d is known. One rule, applicable to children aged $3-12$ yr, is Young's rule:

$$c = \frac{A}{A + 12} d$$

Solve for d in Young's rule and find out what the adult dose is if a 6-yr-old child is supposed to take 2 tablets every 12 hr.

SOLUTION   Solve for d in

$$c = \frac{A}{A + 12} d$$

1. Read the problem.

We are asked to solve for d and find the adult dose if a 6-yr-old child is supposed to take $c = 2$ tablets every 12 hr.

2. Select the unknown.

The unknown is d.

3. Think of a plan.

Since we want to solve for d, we place the d in a box to isolate it:

$$c = \frac{A}{A + 12} \boxed{d}$$

4. Use the procedure for solving equations involving rational expressions.

$$(A + 12)c = (A + 12)\frac{A}{A + 12} \boxed{d} \qquad \text{Multiply both sides by } (A + 12) \text{ and reduce.}$$

$$(A + 12)c = A\boxed{d}$$

$$\frac{(A + 12)c}{A} = \boxed{d} \qquad \text{Divide by } A.$$

$$d = \frac{(A + 12)c}{A} \qquad \text{By the symmetric property.}$$

When the child is 6 years old, $A = 6$, and we know that $c = 2$. Substituting in the equation, we get

$$d = \frac{(6 + 12)2}{6} = 6$$

The adult dosage is 6 tablets every 12 hr.

5. Verify the solution.

The verification is left to you. ∎

We have more problems for you in the *Using Your Knowledge* at the end of this section in which we ask you to solve for a variable.

GRAPH IT

By this time you should be convinced that graphing is a powerful tool in solving algebra problems. But is that all a grapher does? No, we haven't even touched on an important feature of some graphers—the construction of tables listing values that may be solutions. This would be useful in the problem in the *Getting Started*, which required us to construct a Golden Rectangle whose length is 6 in. more than its width. As before, we need to solve the equation

$$\frac{w + 6}{w} = \frac{8}{5} = 1.6$$

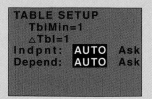

WINDOW 1

Now we will show you how to set up a table using a TI-82 that will solve the problem for us. First, let's agree to use x instead of w in the solution. Start by pressing $\boxed{\text{2nd}}$ $\boxed{\text{TBLSET}}$. The grapher wants to know the minimum at which you wish to start x. Let TblMin = 1. It then asks for ΔTbl=. This sets up the increments for x in your table. Let ΔTbl = 1. This means that the first x is 1, the next one is increased by 1 to 2, the next one is 3, and so on. For the "Indpnt" and "Depend," leave the setting in auto mode (see Window 1). Now, press $\boxed{\text{Y=}}$ and enter

$$Y_1 = \frac{(x + 6)}{x}$$

X	Y₁
1	7
2	4
3	3
4	2.5
5	2.2
6	2
7	1.8571

X=1

WINDOW 2

then press $\boxed{\text{2nd}}$ $\boxed{\text{TABLE}}$. The first column shows successive values for x, while the second shows the values of

$$Y_1 = \frac{(x + 6)}{x}$$

X	Y₁
4	2.5
5	2.2
6	2
7	1.8571
8	1.75
9	1.6667
10	1.6

X=10

WINDOW 3

(see Window 2). Press the $\boxed{\blacktriangledown}$ key while you are in the first column. It will show successive values of x and corresponding Y_1 values. You only need to move down until

$$Y_1 = \frac{8}{5} = 1.6$$

(see Window 3). This occurs when $x = 10$, the same answer as before, but it is more fun letting the grapher do the work for you!

Let's see if you can do Example 4 using this technique. (Remember, we are using x instead of R.) This time, let TblMin = 0 and ΔTbl = 10. Let

$$\text{Time downstream: } Y_1 = \frac{48}{R + 18}$$

$$\text{Time upstream: } Y_2 = \frac{12}{R - 18}$$

Now, look in your table. For what value of x is $Y_1 = Y_2$? That is your answer! Try solving some other problems using grapher tables.

EXERCISE 4.7

A In Problems 1–10, solve the problems.

1. The sum of an integer and its reciprocal is $\frac{65}{8}$. Find the integer.

2. The sum of an integer and its reciprocal is $\frac{50}{7}$. What is the integer?

3. One number is twice another. The sum of their reciprocals is $\frac{3}{10}$. Find the numbers.

4. One number is three times another. The sum of their reciprocals is $\frac{1}{3}$. Find the numbers.

5. Find two consecutive even integers the sum of whose reciprocals is $\frac{7}{24}$.

6. Find two consecutive odd integers such that the sum of their reciprocals is $\frac{16}{63}$.

7. The denominator of a fraction is 5 more than the numerator. If 3 is added to both numerator and denominator, the resulting fraction is $\frac{1}{2}$. Find the fraction.

8. The numerator of a certain fraction is 4 less than the denominator. If the numerator is increased by 8 and the denominator by 35, the resulting fraction is $\frac{1}{2}$. Find the fraction.

9. The current ratio of your business is defined by

$$\text{Current ratio} = \frac{\text{current assets}}{\text{current liabilities}}$$

By how much can you increase your current liabilities if they are $40,000 right now, your current assets amount to $90,000, and you wish to maintain the current ratio at $\frac{3}{2}$?

10. Repeat Problem 9 where you want the current ratio to be 2.

B In Problems 11–23, solve the work problems.

11. If one typist can finish a job in 3 hr while another typist can finish in 5 hr, how long will it take both of them working together to finish the job?

12. A carpenter can finish a job in 8 hr, and another one can do it in 10 hr. How long will it take them to finish the job working together?

13. The world record for riveting is 11,209 rivets in 9 hr, by J. Mair of Ireland. If another person can rivet 11,209 rivets in 10 hr, how long will it take both of them working together to rivet the 11,209 rivets?

14. Mr. Gerry Harley of England shaved 130 men in 60 min. If another barber can shave all these men in 5 hr, how long will it take both of them working together to shave the 130 men?

15. A printing press can print the evening paper in half the time another press takes to print it. Together, the presses can print the paper in 2 hr. How long will it take each of them to print the paper?

16. A computer can do a job in 4 hr. With the help of a newer computer, the job can be completed in 1 hr. How long will it take the newer computer to do the job alone?

17. A tank can be filled by an intake pipe in 9 hr and drained by another pipe in 21 hr. If both pipes are open, how long will it take to fill the tank?

18. A faucet fills a tank in 12 hr, and the drain pipe empties it in 18 hr. If the faucet and the drain pipe are both open, how long does it take to fill the tank?

19. A pipe fills a pool in 7 hr, and another one fills it in 21 hr. How long will it take to fill the pool using both pipes?

20. One pipe fills a tank in 6 hr, and another fills it in 4 hr. How long will it take both pipes together to fill the tank?

21. The main engine of a rocket burns for 60 sec on the fuel in the rocket's tank, while the auxiliary engine burns for 90 sec on the same amount of fuel. How long do both engines burn if they are operated together on the single tank of fuel?

22. An in-flow pipe fills a pool in 12 hr, and another pipe drains it in 4 hr. How long does it take to empty the pool if both pipes are open simultaneously? (Assume that the pool is full at the start.)

23. A pipe fills a tank in 9 hr, but the drain empties it in 6 hr. How long does it take to empty the tank if both pipes are open simultaneously? (Assume that the tank is full at the start.)

C In Problems 24–30, solve the distance problems.

24. A water skier travels 30 mi downstream in the same time it takes him to go 20 mi upstream. If the river current flows at 5 mi/hr, what is the skier's speed in still water?

25. A small plane flies 240 mi against the wind in the same time it takes it to fly 360 mi with a tail wind. If the wind velocity is 30 mi/hr, find the plane's speed in still air.

26. A jet plane flies 700 mi against the wind in the same time it takes it to fly 900 mi with a tail wind. If the wind velocity is 50 mi/hr, what is the plane's speed in still air?

27. A small plane cruises at 120 mi/hr in still air. It takes this plane the same time to travel 270 mi against the wind as it does to travel 450 mi with a tail wind. What is the wind velocity?

28. A small plane can travel 200 mi against the wind in the same time it takes it to travel 260 mi with a tail wind. If the plane's speed in still air is 115 mi/hr, find the wind velocity.

29. An automobile travels 200 mi in the same time in which a small plane travels 1000 mi. Find their rates of speed if the airplane is 100 mi/hr faster than the automobile.

30. Janice ran 1000 m in the same time that Paula ran 950 m. If Paula's speed was $\frac{1}{4}$ m/sec less than Janice's, what was Janice's speed?

SKILL CHECKER

Simplify:

31. $\dfrac{x^{-5}}{x^3}$ **32.** $\dfrac{x^5}{x^{-3}}$ **33.** $(x^{-4})^{-5}$

34. $(x^{-4})^5$ **35.** $(-2xy^2)^3$ **36.** $(-2x^2y)^{-3}$

37. $x^{-9} \cdot x^7$ **38.** $x^9 \cdot x^{-11}$ **39.** $\left(\dfrac{a^{-4}}{b^3}\right)^2$

40. $\left(\dfrac{a^4}{b^{-3}}\right)^{-2}$

USING YOUR KNOWLEDGE   **Formulas, Formulas, and More Formulas**

In Section 2.2, we learned how to solve a formula for a specified variable. You will find that many formulas involve rational expressions. Use your knowledge of solving equations involving rational expressions to solve for specified variables in the following equations.

41. To find the focal length of lenses, lens makers use the formula

$$\frac{1}{F} = \frac{1}{f_1} + \frac{1}{f_2}$$

Solve for F.

42. To find the radius of curvature R of a sphere, we use the formula

$$R = \frac{2AS}{L - 2S}$$

Solve for A.

43. The electric current i in a simple series circuit is given by

$$i = \frac{2E}{R + 2r}$$

Solve for R.

44. Cowling's rule states that a child's dose c for a child A years old, where A is between 2 and 13, is given by

$$c = \frac{A + 1}{24}d$$

where d is the adult dose. Solve for d.

45. In Problem 44, what would the adult dose be if a 5-yr-old child's dose for aspirin is 3 tablets a day?

46. Is there an integer A for which the dosages are the same for Cowling's rule and Young's rule?

$$\text{Young's rule: } c = \frac{A}{A + 12}d$$

WRITE ON . . .
---

47. There's another way to solve Example 2. If you assume that t is the time it takes for the couple to complete the form working together, then in t hr the first person will do $\frac{t}{6}$ of the job and the second person will do $\frac{t}{8}$ of the job. Working together, they will do $\frac{t}{6} + \frac{t}{8}$ of the job. To what should this sum be equal? Explain.

48. Using the method in Problem 47, what equation would you use to solve Example 3?

MASTERY TEST   **If you know how to do these problems, you have learned your lesson!**
---

49. A speed boat can travel 36 mi downstream in the same time it takes it to go 12 mi upstream. If the current is moving at 18 mi/hr, what is the speed of the boat in still water?

50. A pool is filled by an intake pipe in 5 hr and emptied by a drain pipe in 6 hr. How long would it take to fill the pool with both pipes open?

51. The sum of the reciprocals of two consecutive even integers is $\frac{5}{12}$. What are the integers?

52. According to Clark's rule, the dose c for a child weighing W pounds is

$$c = \frac{W}{150}d$$

where d is the adult dose. Solve for d.

research questions

Sources of information for these questions can be found in the *Bibliography* at the end of this book.

1. As we mentioned in *The Human Side of Algebra,* Tartaglia discovered new methods for solving cubics. But there is more to the story. Unfortunately, "an unprincipled genius who taught mathematics and practiced medicine in Milan, upon giving a solemn pledge of secrecy, wheedled the key to the cubic from Tartaglia." What was the name of this dastardly man, what publication did Tartaglia's work appear in, and what was the subject of the publication?

2. This "dastardly man" had a pupil who argued that his teacher received his information from del Ferro through a third party; the pupil then accused Tartaglia of plagiarism. What was the name of the pupil? Write a short paragraph about this pupil's contributions to mathematics.

3. Quadratic, cubic, and quartic equations were solved by formulas formed from the coefficients of the equation by using the four fundamental operations and taking radicals (which we shall study in the next chapter) of various sorts. Write a short paper detailing the struggles to solve quintic equations by these methods, which have culminated in the proof of the impossibility of obtaining such solutions by algebraic methods.

4. In the book *La Geometrie,* the author obtains the Factor Theorem as a major result. Who is the author of *La Geometrie,* and according to historians, what are the mathematical implications of this theorem?

SUMMARY

SECTION	ITEM	MEANING	EXAMPLE
4.1	Fraction	An expression denoting a division	$\frac{3}{4}, \frac{-8}{7},$ and $\frac{1}{2}$ are fractions.
	Rational expression	An expression of the form $\frac{P}{Q}$, where P and Q are polynomials, $Q \neq 0$	$\frac{1}{x}, \frac{x}{x+y}, \frac{x+y}{z}$, and $\frac{x^2 + 21x - 1}{x^3 + 3}$ are rational expressions.
4.1B	Fundamental property of fractions	If P, Q, and K are polynomials, $$\frac{P}{Q} = \frac{P \cdot K}{Q \cdot K}$$ for all values for which the denominator is not zero.	$\frac{3}{4} = \frac{3 \cdot 8}{4 \cdot 8}, \frac{x}{2} = \frac{x \cdot 5}{2 \cdot 5}$, and $$\frac{3}{x+2} = \frac{3(x+5)}{(x+2)(x+5)}$$
4.1C	Standard forms of a fraction	The forms $\frac{-a}{b}$ and $\frac{a}{b}$ are the standard forms of a fraction.	$-\frac{5}{4}$ is written as $\frac{5}{4}, -\frac{-8}{x}$ is written as $\frac{-8}{x}$, and $\frac{5}{-4}$ is written as $\frac{-5}{4}$.
4.1D	Simplified (reduced) fraction	A fraction is simplified (reduced) if the numerator and denominator have no common factor.	$\frac{3}{4}, \frac{9}{8}$, and $\frac{x}{7}$ are reduced, but $\frac{3}{6}$ and $\frac{x+y}{x^2 - y^2}$ are not.
4.2A	Multiplication of rational expressions	If a, b, c, and d are rational expressions $(b \neq 0, d \neq 0)$, $$\frac{a}{b} \cdot \frac{c}{d} = \frac{a \cdot c}{b \cdot d}$$	$\frac{3}{4} \cdot \frac{5}{7} = \frac{3 \cdot 5}{4 \cdot 7} = \frac{15}{28}$ $$\frac{x}{x+y} \cdot \frac{3}{x-y} = \frac{3x}{(x+y)(x-y)} = \frac{3x}{x^2 - y^2}$$

SECTION	ITEM	MEANING	EXAMPLE
4.2B	Division of rational expressions	If a, b, c, and d are rational expressions ($b \neq 0$, $c \neq 0$, $d \neq 0$), $$\frac{a}{b} \div \frac{c}{d} = \frac{a}{b} \cdot \frac{d}{c}$$	$$\frac{3}{4} \div \frac{7}{5} = \frac{3}{4} \cdot \frac{5}{7}$$ $$\frac{x}{x+y} \div \frac{x-y}{3} =$$ $$\frac{x}{x+y} \cdot \frac{3}{x-y}$$
4.3A	Addition and subtraction of rational expressions with the same denominator	If a, b, and c are rational expressions and $b \neq 0$, $$\frac{a}{b} + \frac{c}{b} = \frac{a+c}{b} \text{ and } \frac{a}{b} - \frac{c}{b} = \frac{a-c}{b}.$$	$\dfrac{3}{5} + \dfrac{1}{5} = \dfrac{4}{5}$ and $\dfrac{2}{x} + \dfrac{1}{x} = \dfrac{3}{x}$ $\dfrac{3}{5} - \dfrac{1}{5} = \dfrac{2}{5}$ and $\dfrac{2}{x} - \dfrac{1}{x} = \dfrac{1}{x}$
4.3B	Addition and subtraction of rational expressions with different denominators	$$\frac{a}{b} + \frac{c}{d} = \frac{ad+bc}{bd}$$ $$\frac{a}{b} - \frac{c}{d} = \frac{ad-bc}{bd}$$	$\dfrac{3}{4} + \dfrac{1}{7} = \dfrac{21+4}{28} = \dfrac{25}{28}$ $\dfrac{4}{5} - \dfrac{1}{3} = \dfrac{12-5}{15} = \dfrac{7}{15}$
	Prime number	A natural number greater than 1 with exactly two distinct divisors, itself and 1	2, 17, and 41 are prime.
4.4	Complex fraction	A fraction whose numerator or denominator (or both) contain other fractions	$\dfrac{\frac{1}{2}}{x+1}$, $\dfrac{3}{\frac{1}{5}+x}$, and $\dfrac{\frac{x}{2}}{x+\frac{1}{2}}$ are complex fractions.
	Simple fractions	A fraction that is not complex	$\dfrac{1}{2}$, $\dfrac{2x}{4}$, and $\dfrac{x+y}{x-y}$ are simple fractions.
4.5	Remainder Theorem	If the polynomial $P(x)$ is divided by $x - k$, then the remainder is $P(k)$.	If $P(x) = x^2 + 2x + 5$ is divided by $x - 3$, the remainder is $P(3) = 3^2 + 2(3) + 5 = 20$.
	Factor Theorem	When $P(x)$ has a factor $(x - k)$, $P(k) = 0$.	$P(x) = x^2 + x - 6$ has $x - 2$ as a factor, thus $P(2) = 0$.
4.6	Extraneous solution	A trial solution that does not satisfy the equation	3 is an extraneous solution of $$3 + \frac{1}{x-3} = \frac{1}{x-3}.$$
4.7	RSTUV method for solving word problems	**R**ead the problem. **S**elect a variable for the unknown. **T**hink of a plan. **U**se algebra to solve. **V**erify the answer.	

REVIEW EXERCISES

(If you need help with these exercises, look in the section indicated in brackets.)

1. [4.1A] Write the fraction with the indicated denominator.

a. $\dfrac{2x^2}{9y^4}$;   denominator $36y^7$

b. $\dfrac{2x^2}{9y^4}$;   denominator $45y^8$

c. $\dfrac{2x^2}{9y^4}$;   denominator $54y^9$

2. [4.1A] Write the fraction with the indicated denominator.

a. $\dfrac{2x+1}{x+1}$;   denominator $x^2 + 5x + 4$

b. $\dfrac{2x+1}{x+1}$;   denominator $x^2 + 6x + 5$

c. $\dfrac{2x+1}{x+1}$;   denominator $x^2 + 7x + 6$

3. [4.1B] Write in standard form.

a. $-\dfrac{-6}{y}$ b. $-\dfrac{-7}{y}$ c. $-\dfrac{-8}{y}$

4. [4.1B] Write in standard form.

a. $-\dfrac{x-y}{6}$ b. $-\dfrac{x-y}{7}$ c. $-\dfrac{x-y}{8}$

5. [4.1C] Reduce to lowest terms.

a. $\dfrac{x^4y^7}{xy^2}$ b. $\dfrac{x^4y^8}{xy^2}$ c. $\dfrac{x^4y^9}{xy^2}$

6. [4.1C] Simplify.

a. $\dfrac{xy^2 + y^3}{x^2 - y^2}$ b. $\dfrac{xy^3 + y^4}{x^2 - y^2}$ c. $\dfrac{xy^4 + y^5}{x^2 - y^2}$

7. [4.1C] Simplify.

a. $\dfrac{4y^2 - x^2}{x^3 + 8y^3}$ b. $\dfrac{4y^2 - x^2}{x^3 - 8y^3}$ c. $\dfrac{9y^2 - x^2}{x^3 - 27y^3}$

8. [4.2A] Multiply.

a. $\dfrac{x-2}{3x-2y} \cdot \dfrac{9x^2 - 4y^2}{3x^2 - 5x - 2}$

b. $\dfrac{x-2}{3x-2y} \cdot \dfrac{9x^2 - 4y^2}{4x^2 - 7x - 2}$

c. $\dfrac{x-2}{3x-2y} \cdot \dfrac{9x^2 - 4y^2}{5x^2 - 9x - 2}$

9. [4.2B] Divide.

a. $\dfrac{x+4}{x-2} \div (x^2 + 8x + 16)$

b. $\dfrac{x+5}{x-2} \div (x^2 + 10x + 25)$

c. $\dfrac{x+6}{x-2} \div (x^2 + 12x + 36)$

10. [4.2C] Perform the indicated operations.

a. $\dfrac{2-x}{x+3} \div \dfrac{x^3 - 8}{x+6} \cdot \dfrac{x^3 + 27}{x+6}$

b. $\dfrac{2-x}{x+3} \div \dfrac{x^3 - 8}{x+7} \cdot \dfrac{x^3 + 27}{x+7}$

c. $\dfrac{2-x}{x+3} \div \dfrac{x^3 - 8}{x+8} \cdot \dfrac{x^3 + 27}{x+8}$

11. [4.3A] Perform the indicated operations.

a. $\dfrac{x}{x^2 - 4} + \dfrac{2}{x^2 - 4}$

b. $\dfrac{x}{x^2 - 9} + \dfrac{3}{x^2 - 9}$

c. $\dfrac{x}{x^2 - 16} + \dfrac{4}{x^2 - 16}$

12. [4.3A] Perform the indicated operations.

a. $\dfrac{x}{x^2 - 9} - \dfrac{3}{x^2 - 9}$

b. $\dfrac{x}{x^2 - 16} - \dfrac{4}{x^2 - 16}$

c. $\dfrac{x}{x^2 - 25} - \dfrac{5}{x^2 - 25}$

13. [4.3B] Perform the indicated operations.

a. $\dfrac{x+1}{x^2 + x - 2} + \dfrac{x+5}{x^2 - 1}$

b. $\dfrac{x+1}{x^2 + x - 2} + \dfrac{x+6}{x^2 - 1}$

c. $\dfrac{x+1}{x^2 + x - 2} + \dfrac{x+7}{x^2 - 1}$

14. [4.3B] Perform the indicated operations.

a. $\dfrac{x-4}{x^2 - x - 6} - \dfrac{x+1}{x^2 - 9}$

b. $\dfrac{x-3}{x^2-x-6} - \dfrac{x+1}{x^2-9}$

c. $\dfrac{x-1}{x^2-x-6} - \dfrac{x+1}{x^2-9}$

15. [4.4] Simplify.

a. $\dfrac{\dfrac{1}{x}+\dfrac{1}{x^4}}{\dfrac{1}{x}-\dfrac{1}{x^5}}$

b. $\dfrac{\dfrac{1}{x^2}+\dfrac{1}{x^5}}{\dfrac{1}{x^2}-\dfrac{1}{x^6}}$

c. $\dfrac{\dfrac{1}{x^3}+\dfrac{1}{x^6}}{\dfrac{1}{x^3}-\dfrac{1}{x^7}}$

16. [4.4] Simplify.

a. $4 + \dfrac{a}{4+\dfrac{4}{4+a}}$

b. $5 + \dfrac{a}{5+\dfrac{5}{5+a}}$

c. $6 + \dfrac{a}{6+\dfrac{6}{6+a}}$

17. [4.5A] Divide.

a. $\dfrac{18x^5-12x^3+6x^2}{6x^2}$

b. $\dfrac{18x^5-12x^3+6x^2}{6x^3}$

c. $\dfrac{18x^5-12x^3+6x^2}{6x^4}$

18. [4.5B] Divide.

a. $2x^3 - 8 - 4x$ by $2 + 2x$

b. $2x^3 - 9 - 4x$ by $2 + 2x$

c. $2x^3 - 10 - 4x$ by $2 + 2x$

19. [4.5C] Factor $x^3 - 6x^2 + 11x - 6$ if

a. $x - 1$ is one of its factors

b. $x - 2$ is one of its factors

c. $x - 3$ is one of its factors

20. [4.5D] Use synthetic division to divide $x^4 + 10x^3 + 35x^2 + 50x + 28$ by

a. $x + 1$

b. $x + 2$

c. $x + 3$

21. [4.5E] If $P(x) = x^4 + 10x^3 + 35x^2 + 50x + 24$, use synthetic division to show that

a. -1 is a solution of $P(x) = 0$

b. -2 is a solution of $P(x) = 0$

c. -3 is a solution of $P(x) = 0$

22. [4.6] Solve.

a. $\dfrac{x}{x+4} - \dfrac{x}{x-4} = \dfrac{x^2+16}{x^2-16}$

b. $\dfrac{x}{x+5} - \dfrac{x}{x-5} = \dfrac{x^2+25}{x^2-25}$

c. $\dfrac{x}{x+6} - \dfrac{x}{x-6} = \dfrac{x^2+36}{x^2-36}$

23. [4.6] Solve.

a. $1 + \dfrac{4}{x-5} = \dfrac{40}{x^2-25}$

b. $1 + \dfrac{5}{x-6} = \dfrac{60}{x^2-36}$

c. $1 + \dfrac{6}{x-7} = \dfrac{84}{x^2-49}$

24. [4.7A] Find two consecutive even integers such that the sum of their reciprocals is

a. $\dfrac{11}{60}$ b. $\dfrac{13}{84}$ c. $\dfrac{15}{112}$

25. [4.7B] Jack can paint a room in 4 hr. Find how long it would take to paint the room if he is helped by Jill, who can paint the same room in

a. 5 hr   b. 6 hr   c. 7 hr

26. [4.7C] A plane traveled 1200 mi with a 25-mi/hr tail wind. What is the plane's speed in still air if it took the plane the same time to travel the given mileage against the wind?

a. 960 mi   b. 1000 mi   c. 1040 mi

27. [4.7C] If $A = \dfrac{a+2b+3c}{2}$,

a. Solve for a.

b. Solve for b.

c. Solve for c.

PRACTICE TEST

(Answers on pages 276–277.)

1. Write the fraction with the indicated denominator.

   a. $\dfrac{2x^2}{9y^4}$; denominator $36y^7$

   b. $\dfrac{2x + 1}{x + 1}$; denominator $x^2 + 4x + 3$

2. Write in standard form.

   a. $-\dfrac{-5}{y}$

   b. $-\dfrac{x - y}{5}$

3. Reduce to lowest terms.

   a. $\dfrac{x^4 y^6}{xy^2}$

   b. $\dfrac{xy + y^2}{x^2 - y^2}$

4. Reduce $\dfrac{y^2 - x^2}{x^3 - y^3}$ to lowest terms.

5. Multiply $\dfrac{x - 2}{3x - 2y} \cdot \dfrac{9x^2 - 4y^2}{2x^2 - x - 6}$.

6. Divide $\dfrac{x + 3}{x - 2}$ by $(x^2 + 6x + 9)$.

7. Perform the indicated operations.

   $$\frac{2 - x}{x + 3} \div \frac{x^3 - 8}{x + 5} \cdot \frac{x^3 + 27}{x + 5}$$

8. Perform the indicated operations.

   $$\frac{x}{x^2 - 1} + \frac{1}{x^2 - 1}$$

9. Perform the indicated operations.

   $$\frac{x}{x^2 - 4} - \frac{2}{x^2 - 4}$$

10. Perform the indicated operations.

    $$\frac{x + 1}{x^2 + x - 2} + \frac{x + 4}{x^2 - 1}$$

11. Perform the indicated operations.

    $$\frac{x - 5}{x^2 - x - 6} - \frac{x + 1}{x^2 - 9}$$

12. Simplify $\dfrac{x + \dfrac{1}{x^2}}{x - \dfrac{1}{x^3}}$.

13. Simplify $2 + \dfrac{a}{2 + \dfrac{2}{2 + a}}$.

14. Divide $\dfrac{28x^5 - 14x^3 + 7x^2}{7x^3}$.

15. Divide $2x^3 - 6 - 4x$ by $2 + 2x$.

16. Factor $2x^3 + 3x^2 - 23x - 12$ if $x - 3$ is one of its factors.

17. Use synthetic division to divide $x^4 - 4x^3 - 7x^2 + 22x + 25$ by $x + 1$.

18. Use synthetic division to show that -1 is a solution of $x^4 - 4x^3 - 7x^2 + 22x + 24 = 0$.

19. Solve $\dfrac{x}{x + 3} - \dfrac{x}{x - 3} = \dfrac{x^2 + 9}{x^2 - 9}$.

20. Solve $1 + \dfrac{3}{x - 4} = \dfrac{24}{x^2 - 16}$.

21. Find two consecutive even integers such that the sum of their reciprocals is $\frac{7}{24}$.

22. Jack can mow the lawn in 4 hr and Jill can mow it in 3. How long would it take them to mow the lawn if they work together?

23. A plane traveled 990 mi with a 30-mi/hr tail wind in the same time it took to travel 810 mi against the wind. What is the plane's speed in still air?

24. The area A of a trapezoid is given by

    $$A = \frac{(B + b)h}{2}$$

    where B is the length of one base, b is the length of the other base, and h is the height of the trapezoid. Solve for h.

25. The area of a triangle is

    $$A = \frac{bh}{2}$$

    If the area is 42 square units, and the base is 7 units, what is the height h?

ANSWERS TO PRACTICE TEST

Answer	If you missed:		Review:		
	Question		Section	Examples	Page
1a. $\dfrac{8x^2y^3}{36y^7}$	1a		4.1A	1	205
1b. $\dfrac{2x^2 + 7x + 3}{x^2 + 4x + 3}$	1b		4.1A	1	205
2a. $\dfrac{5}{y}$	2a		4.1B	2	206–207
2b. $\dfrac{y - x}{5}$	2b		4.1B	2	206–207
3a. x^3y^4	3a		4.1C	3	208
3b. $\dfrac{y}{x - y}$	3b		4.1C	3	208
4. $\dfrac{-(x + y)}{x^2 + xy + y^2}$	4		4.1D	4	209
5. $\dfrac{3x + 2y}{2x + 3}$	5		4.2A	1, 2	215–216
6. $\dfrac{1}{x^2 + x - 6}$	6		4.2B	3	217
7. $\dfrac{-(x^2 - 3x + 9)}{x^2 + 2x + 4}$	7		4.2C	4, 5	218–219
8. $\dfrac{1}{x - 1}$	8		4.3A	1	224–225
9. $\dfrac{1}{x + 2}$	9		4.3A	1	224–225
10. $\dfrac{2x^2 + 8x + 9}{(x + 2)(x + 1)(x - 1)}$	10		4.3B	2, 3	226–228
11. $\dfrac{-5x - 17}{(x + 2)(x + 3)(x - 3)}$	11		4.3B	2, 3	226–228
12. $\dfrac{x(x^2 - x + 1)}{(x^2 + 1)(x - 1)}$	12		4.4	1, 2	234–235
13. $\dfrac{a^2 + 6a + 12}{2a + 6}$	13		4.4	3, 4	235–237

Answer	If you missed:	Review:		
	Question	Section	Examples	Page
14. $4x^2 - 2 + \dfrac{1}{x}$	14	4.5	1	243
15. $x^2 - x - 1$ R -4	15	4.5	2, 3	244–245
16. $(x - 3)(x + 4)(2x + 1)$	16	4.5	4	246–247
17. $x^3 - 5x^2 - 2x + 24$ R 1	17	4.5	5	248
18. When the division is done, the remainder is zero.	18	4.5	6	248–249
19. No solution	19	4.6	1, 2	253–256
20. -7	20	4.6	3, 4	257–258
21. 6 and 8	21	4.7	1	263
22. $1\dfrac{5}{7}$ hr	22	4.7	2	264
23. 300 mi/hr	23	4.7	3	264–265
24. $h = \dfrac{2A}{B + b}$	24	4.7	5	267
25. 12 units	25	4.7	5	267

5

Rational Exponents and Radicals

For many years mathematics dealt only with rational numbers. Pythagoras, a Greek mathematician and scholar, changed all that with the discovery of *irrational* numbers such as $\sqrt{2}$ (read "the square root of 2"). In this chapter we shall study square, cube, fourth, and higher roots. Roots can be written using radicals as $\sqrt{2}$ or using rational exponents as $2^{1/2}$. In Section 5.1, we show you how to change from radicals to rational exponents and how to perform different operations using either notation. In Section 5.2, we discuss the properties of radicals and how they can be used to make the denominator of a fraction rational or to reduce the order of a radical. In Section 5.3, we study the operations that can be performed with these radicals and then discuss how to solve equations involving these radicals in Section 5.4. However, the equation $x^2 = -1$ cannot be solved using real numbers. To remedy this, we introduce the complex numbers in Section 5.5 and expand our knowledge of the operations that we can perform with complex numbers.

Imagine a universe in which only whole numbers, some classified as "perfect" and "amicable," existed. Even numbers were "feminine" and odd numbers "masculine," while 1 was the generator of all other numbers. This was the universe of the Pythagoreans, an ancient Greek secret society (ca. 540–500 B.C.). And then, disaster! In the midst of these charming fantasies, a new type of number was discovered, a type so unexpected that the brotherhood tried to suppress its discovery—the set of irrational numbers!

Pythagoras himself was born between 580 and 569 B.C., but accounts of his eventual demise differ. One claims that he died in about 501 B.C. when popular revolt erupted and the meeting house of the Pythagoreans was set afire, which resulted in his death. A more dramatic ending claims that his disciples made a bridge over the fire with their bodies enabling the master to escape to Metapontum. In the ensuing fight, Pythagoras was caught between freedom and a field of sacred beans. Rather than trampling the plants, he valiantly chose to die at the hands of his enemies.

5.1

RATIONAL EXPONENTS AND RADICALS

To succeed, review how to:

1. Use the laws of exponents (pp. 31–37).

2. Do operations involving signed numbers (pp. 15–19).

Objectives:

A Find the nth root of a number, if it exists.

B Evaluate expressions containing rational exponents.

C Simplify expressions involving rational exponents.

getting started

Speeding Along to Exponents

Have you noticed that the speed limit on a curve is lower than on a straight road? The velocity v (in miles per hour) that a car can travel on a curved concrete highway of radius r feet without skidding is $v = \sqrt{9r}$. If the radius r of the curve is 100 ft, the velocity is $\sqrt{900}$ (read "the square root of 900"). A *square root* of 900 is a number whose square is 900. Since $(30)^2 = 900$, a square root of 900 is 30. Similarly, a square root of 25 is 5 (since $5^2 = 25$), and a square root of 36 is 6 (since $6^2 = 36$).

There are many numbers that are not square roots of perfect squares—for example, $\sqrt{2}, \sqrt{3}, \sqrt{10}$, and $\sqrt{24}$. These real numbers are irrational; that is, they cannot be written as the quotient of two integers. The first irrational number was probably discovered by the Pythagoreans, an ancient Greek society of mathematicians who believed that everything was based on the whole numbers and that harmony consisted of numerical ratios. In this section we shall study rational exponents and radical expressions. Many of the important applications of mathematics use these two concepts. (See Problems 71–74.) We start by giving the definition of an nth root.

nTH ROOT

If a and x are real numbers and n is a positive integer, then x is an nth root of a if $x^n = a$.

For example,

1. A square (second) root of 4 is 2 because $2^2 = 4$.

2. Another square root of 4 is -2 because $(-2)^2 = 4$.

3. A cube (third) root of 27 is 3 because $(3)^3 = 27$.

4. A cube (third) root of -64 is -4 because $(-4)^3 = -64$.

5. A fourth root of $\frac{16}{81}$ is $\frac{2}{3}$ because $\left(\frac{2}{3}\right)^4 = \frac{16}{81}$.

6. Another fourth root of $\frac{16}{81}$ is $-\frac{2}{3}$ because $\left(-\frac{2}{3}\right)^4 = \frac{16}{81}$.

Note that 4 has two square roots, 2 and -2, and $\frac{16}{81}$ has two real fourth roots, $-\frac{2}{3}$ and $\frac{2}{3}$. To avoid this situation and make our work more precise, we introduce the idea of the principal nth root.

Radicals

PRINCIPAL nTH ROOT

If n is a positive integer, then $\sqrt[n]{a}$ denotes the **principal nth root** a, and

1. If a is positive $(a > 0)$, $\sqrt[n]{a}$ is the *positive nth root* of a.
2. If a is negative $(a < 0)$ and n is odd, $\sqrt[n]{a}$ is the *negative nth root* of a.
3. If a is negative and n is even, there is no real nth root.
4. $\sqrt[n]{0} = 0$.

In this definition, $\sqrt[n]{a}$ is called a **radical expression**, $\sqrt{}$ is called the **radical sign**, a is the **radicand**, and n is the **index** that tells you what root is being considered. By convention, the index 2 for square root is understood but not written, that is, $\sqrt{x}$ means $\sqrt[2]{x}$. Thus $\sqrt{9}$ means the principal square root of 9 and $\sqrt{25}$ means the principal square root of 25.

Now let's look at part 1 of our definition. It tells us that whenever the number under the radical sign is positive, the resulting root is also positive; that is, "If $a > 0$, $\sqrt{a} > 0$." Thus,

$$\sqrt{64} = 8 \quad \text{because} \quad 8^2 = 64$$

$$\sqrt[3]{8} = 2 \quad \text{because} \quad 2^3 = 8$$

$$\sqrt[4]{81} = 3 \quad \text{because} \quad 3^4 = 81$$

> **NOTE** A common mistake is to assume that $\sqrt{64}$ has two values. This is not correct. By our definition of principal root, $\sqrt{64} = 8$. If we wish to refer to the negative nth root, we write $-\sqrt[n]{a}$. Thus $-\sqrt{64} = -8$, $-\sqrt{16} = -4$, and $-\sqrt{25} = -5$.

Part 2 of our definition tells us that if a is *negative* and n is *odd,* $\sqrt[n]{a}$ is negative, that is, "If $a < 0$, $\sqrt[n]{a} < 0$. Thus,

$$\text{negative} \downarrow$$
$$\text{odd} \longrightarrow \sqrt[3]{\boxed{-27}} = -3 \quad \text{because} \quad (-3)^3 = -27 \qquad \text{Negative answer}$$
$$\text{odd} \longrightarrow \sqrt[5]{\boxed{-32}} = -2 \quad \text{because} \quad (-2)^5 = -32 \qquad \text{Negative answer}$$
$$\uparrow \text{negative}$$

What about $\sqrt{-16}$? There is no real number whose square is -16 because the square of a nonzero number is positive. Thus $\sqrt{-16}$ is not a real number. Similarly, $\sqrt[4]{-81}$ is not a real number since there is no real number whose fourth power is -81. Note that in both cases, the index n is even and the radicand is negative. This situation is not covered in the definition.

EXAMPLE 1   Finding the roots of numbers

Find if possible:

a. $\sqrt[3]{-64}$ 　　　　　　**b.** $\sqrt{-64}$ 　　　　　　**c.** $\sqrt[3]{\left(\dfrac{-1}{8}\right)}$

SOLUTION

a. $\sqrt[3]{-64} = -4$, since $(-4)^3 = -64$.

b. $\sqrt{-64}$ is not a real number. Note that $\sqrt{-64} \neq -8$, since $(-8)(-8) = 64$ and not -64.

c. $\sqrt[3]{\left(\dfrac{-1}{8}\right)} = \dfrac{-1}{2}$, since $\left(\dfrac{-1}{2}\right)^3 = \dfrac{-1}{8}$.   ∎

B   From Rational Exponents to Radicals

We have used radicals to define the nth root of a number. We can also define nth roots using rational exponents. For example, what do you think $a^{1/3}$ means? To find out, let

$$x = a^{1/3}$$

$$x^3 = (a^{1/3})^3 \qquad \text{Cube both sides.}$$

$$= a^{(1/3) \cdot (3)} \qquad \text{Assume } (a^{1/n})^n = a^{(1/n) \cdot (n)}.$$

$$= a^1$$

$$= a$$

Since $x^3 = a$, x must be the cube root of a; that is, $x = \sqrt[3]{a}$. But $x = a^{1/3}$, so

$$a^{1/3} = \sqrt[3]{a}$$

Similarly, if $\sqrt[4]{a}$ is defined,

$$a^{1/4} = \sqrt[4]{a}$$

In general, the following definition establishes the relationship between rational exponents and roots.

RATIONAL EXPONENTS AND THEIR ROOTS

> If n is a positive integer and $\sqrt[n]{a}$ is a real number, then
> $$a^{1/n} = \sqrt[n]{a}$$

Note that when we write

$$\overset{\text{same}}{a^{1/n} = \sqrt[n]{a}}$$

the denominator of the rational exponent is the index of the radical:

$$16^{1/2} = \sqrt[2]{16} = 4 \qquad \text{Since } 4^2 = 16. \text{ Remember } \sqrt{16} \text{ means } \sqrt[2]{16}.$$

$$(-8)^{1/3} = \sqrt[3]{-8} = -2 \qquad \text{Since } (-2)^3 = -8$$

$$\left(\dfrac{1}{81}\right)^{1/4} = \sqrt[4]{\dfrac{1}{81}} = \dfrac{1}{3} \qquad \text{Since } \left(\dfrac{1}{3}\right)^4 = \dfrac{1}{81}$$

EXAMPLE 2   Evaluating expressions containing rational exponents

Evaluate:

a. $9^{1/2}$ **b.** $(-125)^{1/3}$ **c.** $\left(\dfrac{1}{16}\right)^{1/4}$

SOLUTION

a. $9^{1/2} = \sqrt{9} = 3$ Since $3^2 = 9$

b. $(-125)^{1/3} = \sqrt[3]{-125} = -5$ Since $(-5)^3 = -125$

c. $\left(\dfrac{1}{16}\right)^{1/4} = \sqrt[4]{\left(\dfrac{1}{16}\right)} = \dfrac{1}{2}$ Since $\left(\dfrac{1}{2}\right)^4 = \dfrac{1}{16}$ ■

Table 1 shows a summary of our work so far.

TABLE 1 Rational Exponents and Radicals

If a is a real number greater than 1 and n is a positive integer		
	n even	n odd
For a positive a ($a > 0$)	$\sqrt[n]{a} = a^{1/n}$ is positive.	$\sqrt[n]{a} = a^{1/n}$ is positive.
For a negative a ($a < 0$)	$\sqrt[n]{a} = a^{1/n}$ is not a real number.	$\sqrt[n]{a} = a^{1/n}$ is negative.
For $a = 0$	$\sqrt[n]{a} = a^{1/n} = 0.$	$\sqrt[n]{a} = a^{1/n} = 0.$

We have already defined rational exponents of the form $1/n$. How do we define $a^{m/n}$, where m and n are positive integers when $n > 1$ and $\sqrt[n]{a}$ is a real number? If we assume that $(a^m)^n = a^{m \cdot n}$, then

$$a^{m/n} = (a^{1/n})^m = (a^m)^{1/n} = (\sqrt[n]{a})^m = \sqrt[n]{a^m}$$

From this we arrive at the following definition.

RADICAL EXPRESSION WITH AN m/n EXPONENT

$$a^{m/n} = (\sqrt[n]{a})^m = \sqrt[n]{a^m}$$

provided m and n are positive integers with no common factors and $\sqrt[n]{a}$ is a real number.

Note that the numerator of the exponent m/n is the exponent of the radical expression, and the denominator is the index of the radical—that is,

$$a^{m/n} = (\overset{\text{Exponent}}{\underset{\text{Index}}{\sqrt[n]{a}}})^m$$

For example,

$$a^{1/5} = \sqrt[5]{a} \quad \text{and} \quad a^{2/5} = (\sqrt[5]{a})^2$$

EXAMPLE 3   Evaluating expressions containing rational exponents

Evaluate:

a. $8^{2/3}$ **b.** $(-27)^{2/3}$ **c.** $(-25)^{3/2}$

SOLUTION

a. $8^{2/3}$ can be evaluated in two ways.

     1. $8^{2/3} = (\sqrt[3]{8})^2 = (2)^2 = 4$ Since $\sqrt[3]{8} = 2$

     2. $8^{2/3} = \sqrt[3]{8^2} = \sqrt[3]{64} = 4$

b. We can evaluate $(-27)^{2/3}$ in a similar manner.

1. $(-27)^{2/3} = (\sqrt[3]{-27})^2$

$\qquad\qquad = (-3)^2 \qquad$ Since $\sqrt[3]{-27} = -3$

$\qquad\qquad = 9$

2. $(-27)^{2/3} = \sqrt[3]{(-27)^2} = \sqrt[3]{729} = 9$

c. $(-25)^{3/2} = \sqrt[2]{(-25)^3} = \sqrt{-15,625}$, which is not a real number.   ■

Finally, to define negative rational exponents, we first note that if m and n are positive integers with no common factors,

$$-\frac{m}{n} = \frac{-m}{n}$$

Thus if $(a^m)^n = a^{m \cdot n}$,

$$a^{-m/n} = (a^{1/n})^{-m}$$

$$= \frac{1}{(a^{1/n})^m} \qquad \text{Since } m \text{ is negative}$$

Thus we make the following definition.

DEFINITION OF $a^{-m/n}$

$$a^{-m/n} = \frac{1}{(a^{m/n})}$$

where m and n are positive integers, $a^{1/n}$ a real number, $a \neq 0$.

Using this definition, we have the following.

1. $a^{-1/2} = \dfrac{1}{\sqrt{a}}$

2. $32^{-3/5} = \dfrac{1}{32^{3/5}} = \dfrac{1}{(\sqrt[5]{32})^3} = \dfrac{1}{2^3} = \dfrac{1}{8}$

3. $1000^{-2/3} = \dfrac{1}{1000^{2/3}} = \dfrac{1}{(\sqrt[3]{1000})^2} = \dfrac{1}{10^2} = \dfrac{1}{100}$

EXAMPLE 4   Evaluating expressions containing negative rational exponents

Evaluate:

a. $16^{-3/4}$ **b.** $(-8)^{-4/3}$ **c.** $125^{-2/3}$

SOLUTION

a. $16^{-3/4} = \dfrac{1}{16^{3/4}} = \dfrac{1}{(\sqrt[4]{16})^3} = \dfrac{1}{2^3} = \dfrac{1}{8}$

b. $(-8)^{-4/3} = \dfrac{1}{(-8)^{4/3}} = \dfrac{1}{(\sqrt[3]{-8})^4} = \dfrac{1}{(-2)^4} = \dfrac{1}{16}$

c. $125^{-2/3} = \dfrac{1}{125^{2/3}} = \dfrac{1}{(\sqrt[3]{125})^2} = \dfrac{1}{5^2} = \dfrac{1}{25}$ ■

Operations with Rational Exponents

The properties of exponents that we studied in Chapter 1 can be extended to rational exponents. If this is done, we have the following results.

LAWS OF EXPONENTS

> Let r, s, and t be rational numbers. If a and b are real numbers, the indicated expressions exist:
>
> I. $a^r \cdot a^s = a^{r+s}$ II. $\dfrac{a^r}{a^s} = a^{r-s}$ III. $(a^r)^s = a^{r \cdot s}$
>
> IV. $(a^r b^s)^t = a^{rt} b^{st}$ V. $\left(\dfrac{a^r}{b^s}\right)^t = \dfrac{a^{rt}}{b^{st}}$

EXAMPLE 5   Using the laws of exponents

If x and y are positive, simplify:

a. $x^{1/3} \cdot x^{1/4}$ **b.** $\dfrac{x^{-2/3}}{x^{1/5}}$ **c.** $(y^{3/5})^{-1/6}$

SOLUTION

a. $x^{1/3} \cdot x^{1/4} = x^{1/3+1/4}$ Law I

$\qquad\qquad\quad = x^{4/12+3/12}$ The LCD is 12.

$\qquad\qquad\quad = x^{7/12}$ Add exponents.

b. $\dfrac{x^{-2/3}}{x^{1/5}} = x^{-2/3-1/5}$ Law II

$\qquad\quad = x^{-10/15-3/15}$ The LCD is 15.

$\qquad\quad = x^{-13/15}$

$\qquad\quad = \dfrac{1}{x^{13/15}}$ $a^{-n} = \dfrac{1}{a^n}$

c. $(y^{3/5})^{-1/6} = y^{(3/5) \cdot (-1/6)}$ Law III

$\qquad\qquad\quad = y^{-1/10}$ $\frac{3}{5} \cdot \left(-\frac{1}{6}\right) = -\frac{1}{10}$

$\qquad\qquad\quad = \dfrac{1}{y^{1/10}}$ ■

EXAMPLE 6   More practice using the laws of exponents

If x and y are positive, simplify:

a. $(x^{1/5} y^{3/4})^{-20}$ **b.** $\dfrac{x^{1/3} y^{3/4}}{x^{2/3} y^{1/4}}$ **c.** $x^{2/3}(x^{-1/3} + y^{1/5})$

SOLUTION

a. $(x^{1/5} y^{3/4})^{-20} = (x^{1/5})^{-20} \cdot (y^{3/4})^{-20}$ Law IV

$\qquad\qquad\qquad\quad = x^{(1/5)(-20)} \cdot y^{(3/4)(-20)}$ Law III

$\qquad\qquad\qquad\quad = x^{-4} \cdot y^{-15}$

$\qquad\qquad\qquad\quad = \dfrac{1}{x^4 y^{15}}$ Definition of negative exponent

b. $\dfrac{x^{1/3}y^{3/4}}{x^{2/3}y^{1/4}} = x^{1/3-2/3} \cdot y^{3/4-1/4}$ Law II

$$= x^{-1/3} \cdot y^{1/2}$$

$$= \frac{y^{1/2}}{x^{1/3}}$$

c. We first use the distributive property and then simplify.

$$x^{2/3}(x^{-1/3} + y^{1/5}) = x^{2/3} \cdot x^{-1/3} + x^{2/3} \cdot y^{1/5}$$

$$= x^{2/3+(-1/3)} + x^{2/3}y^{1/5}$$

$$= x^{1/3} + x^{2/3}y^{1/5} \qquad \blacksquare$$

The laws of exponents for rational numbers provide a good way of simplifying some of the expression we have studied. In Example 3, we evaluated $(8^{2/3})$ and $(-27)^{2/3}$. If we write 8 as 2^3, we can write

$$8^{2/3} = (2^3)^{2/3}$$

$$= 2^{(3)\cdot(2/3)} \qquad \text{Law III}$$

$$= 2^2$$

$$= 4$$

Similarly, since $-27 = (-3)^3$,

$$(-27)^{2/3} = [(-3)^3]^{2/3} \qquad \text{Law III}$$

$$= (-3)^2$$

$$= 9$$

Use these ideas when working the exercises! One word of caution, however. We cannot evaluate $[(-2)^2]^{1/2}$ using law III. If we do, we obtain

$$[(-2)^2]^{1/2} = (-2)^{(2)(1/2)}$$

$$= (-2)^1$$

$$= -2$$

If we use the order of operations we have studied and square -2 first, we have

$$[(-2)^2]^{1/2} = [4]^{1/2}$$

$$= 2$$

The problem is that $a^{m/n}$ was defined provided m and n have no common factors. In this case, $m = 2$ and $n = 2$, so they have a common factor. To remedy this situation, we make the following definition.

DEFINITION OF $(a^m)^{1/n}$

> If m and n are positive even integers, then
>
> $$(a^m)^{1/n} = |a|^{m/n}$$

Thus $[(-2)^2]^{1/2} = |2|^{2/2} = |2|^1 = 2$. Note that $[(-2)^{1/2}]^2$ is not defined since $(-2)^{1/2} = \sqrt{-2}$, which is not a real number. We shall discuss this further in Section 5.2.

GRAPH IT

You can approximate the roots in Examples 1, 2, 3, and 4 in different ways, but be aware that different graphers have different procedures for this. To find $\sqrt[3]{-64}$ in Example 1a with a TI-82, press MATH 4 −64 ENTER and you get −4. You can also use the $\sqrt[x]{\ }$ feature to find roots. (To find $\sqrt[4]{16}$ with a TI-82, press 4 MATH 5 16 ENTER and you get 2.) Note that the first entry is the type of root you want (square, cube, fourth, fifth, and so on). Of course, trying to find $(-25)^{3/2}$ in Example 3c gets you an error message! Do you see why? On the other hand, if in Example 3b you press $(-27)^{2/3}$, you get an error message. Instead, enter $((-27)^2)^{(1/3)}$ or $((-27)^{(1/3)})^2$ and you get 9, as shown in Window 1. As you can see, you have to know some algebra to use your grapher!

```
((-27)^2)^(1/3)
                9
((-27)^(1/3))^2
                9
```

WINDOW 1

A few years ago you had to find square and cube roots using a long table. No more! You can make your own table with a TI-82. For example, to construct a table of square roots, press 2nd TblSet and set TblMin = 1 and ΔTbl = 1. Now to tell the grapher you want a square root table, enter $Y_1 = X^{(1\div2)}$. Remember that to enter the exponent $\frac{1}{2}$, you press ∧ (1 ÷ 2) or the $\sqrt{\ }$ key. Now press 2nd TABLE. You see two columns. The X column shows the number whose square root you want and the Y1 column shows the corresponding square root. For example (Window 2), the second line shows a 2 for X and 1.4142 for Y1, meaning that the square root of 2 is 1.4142. You can make tables of cube, fourth, or nth roots by entering $Y_1 = x^{1/3}$ or $x^{1/4}$ or $x^{1/n}$.

X	Y1
1	1
2	1.4142
3	1.7321
4	2
5	2.2361
6	2.4495
7	2.6458
X=2	

WINDOW 2

There is at least one more way to approximate roots with a grapher. If you want the root of a specific number such as $\sqrt[5]{243}$, go to the home screen and enter 243^(1/5), then press ENTER. (Don't forget the parentheses when entering the exponents!) The answer is 3.

Finally, you can verify Examples 5a and b by graphing the original problem as Y_1, the answer as Y_2, and verifying that the graphs coincide.

EXERCISE 5.1

A **In Problems 1–12, evaluate if possible.**

1. $\sqrt{4}$

2. $\sqrt{25}$

3. $\sqrt[3]{8}$

4. $\sqrt[3]{125}$

5. $\sqrt[3]{-8}$

6. $\sqrt[3]{-125}$

7. $\sqrt[3]{\dfrac{-1}{64}}$

8. $\sqrt[3]{\dfrac{-1}{27}}$

9. $\sqrt[4]{16}$

10. $\sqrt[4]{625}$ b

11. $\sqrt[5]{32}$

12. $\sqrt[5]{\dfrac{-1}{243}}$

B **In Problems 13–40, evaluate if possible.**

13. $9^{1/2}$

14. $16^{1/2}$

15. $(-4)^{1/2}$

16. $-4^{1/2}$

17. $27^{1/3}$

18. $125^{1/3}$

19. $81^{1/4}$

20. $16^{1/4}$

21. $\left(\dfrac{-1}{8}\right)^{1/3}$

22. $\left(\dfrac{-1}{27}\right)^{1/3}$

23. $\left(\dfrac{-1}{256}\right)^{1/4}$

24. $\left(\dfrac{1}{256}\right)^{1/4}$

25. $27^{2/3}$

26. $(-27)^{2/3}$

27. $125^{2/3}$

28. $216^{2/3}$

29. $\left(\dfrac{1}{8}\right)^{2/3}$

30. $\left(\dfrac{1}{81}\right)^{3/4}$

31. $(-8)^{4/3}$

32. $(-27)^{4/3}$

33. $(32)^{4/5}$

34. $(-32)^{4/5}$

35. $-32^{4/5}$

36. $(-64)^{5/3}$

37. $64^{-2/3}$

38. $27^{-2/3}$

39. $[(-7)^4]^{1/4}$

40. $[(-11)^6]^{1/6}$

C **In Problems 41–70, simplify and write the expression as a product or quotient with positive exponents. All letters represent positive numbers.**

41. $x^{1/7} \cdot x^{2/7}$

42. $y^{1/6} \cdot y^{1/6}$

43. $x^{-1/9} \cdot x^{-4/9}$

44. $y^{-5/2} \cdot y^{-3/2}$

45. $\dfrac{x^{4/5}}{x^{2/5}}$

46. $\dfrac{y^{5/7}}{y^{2/7}}$

47. $\dfrac{z^{2/3}}{z^{-1/3}}$

48. $\dfrac{a^{4/5}}{a^{-3/5}}$

49. $(x^{1/5})^{10}$

50. $(y^{1/3})^{12}$

51. $(z^{1/3})^{-6}$

52. $(a^{1/4})^{-8}$

53. $(b^{2/3})^{-6/5}$

54. $(c^{2/7})^{-7/8}$

55. $(a^{2/3}b^{3/4})^{-12}$

56. $(x^{1/8}y^{2/3})^{-24}$ **57.** $\left(\dfrac{a^{2/3}}{b^{3/5}}\right)^{-15}$ **58.** $\left(\dfrac{x^{1/2}}{y^{3/5}}\right)^{-20}$

59. $\left(\dfrac{x^{-2/5}}{y^{3/4}}\right)^{-40}$ **60.** $\left(\dfrac{x^{-1/3}}{y^{3/8}}\right)^{-48}$

61. $x^{1/3}(x^{2/3}+y^{1/2})$ **62.** $x^{-4/5}(y^{1/3}+x^{-1/5})$

63. $y^{3/4}(x^{1/2}-y^{1/2})$ **64.** $y^{2/3}(y^{1/2}-x^{2/3})$

65. $\dfrac{x^{1/6}\cdot x^{-5/6}}{x^{1/3}}$ **66.** $\dfrac{(x^{1/3}\cdot x^{1/2})^2}{x^{1/2}}$

67. $\dfrac{(x^{1/3}\cdot y^{-1/2})^6}{(y^{1/2})^{-4}}$ **68.** $\left(\dfrac{x^{4/3}\cdot y^{1/2}}{x^{1/3}}\right)^{-1/2}$

69. $\dfrac{(x^{1/4}\cdot y^2)^4}{(x^{2/3}\cdot y)^{-3}}$ **70.** $\left(\dfrac{-8a^{-3}b^{12}}{c^{15}}\right)^{-1/3}$

APPLICATIONS

If air resistance is neglected, the terminal velocity v of a falling body in meters per second is given by

$$v = (20h + v_0)^{1/2}$$

71. Find v if $h = 10$ and $v_0 = 25$ m/sec.

72. Find v if a body is dropped ($v_0 = 0$) from a height of 45 m.

73. If the velocity as measured in feet per second is

$$v = (64h + v_0)^{1/2}$$

find v if $h = 12$ ft and $v_0 = 16$ ft/sec.

74. Find v if a body is dropped ($v_0 = 0$) from a height of 25 ft.

In Problems 75–80, evaluate $\sqrt{b^2 - 4ac}$.

75. $a = 1, b = 5, c = 4$ **76.** $a = 1, b = 3, c = 2$

77. $a = 2, b = -3, c = -20$ **78.** $a = \dfrac{1}{2}, b = -\dfrac{1}{12}, c = -1$

79. $a = \dfrac{1}{12}, b = \dfrac{1}{3}, c = -1$ **80.** $a = \dfrac{1}{12}, b = \dfrac{1}{2}, c = \dfrac{2}{3}$

SKILL CHECKER

Write:

81. $\dfrac{2x}{xy^2}$ with a denominator of $8x^3y^3$

82. $\dfrac{1}{5xy}$ with a denominator of $125x^3y^2$

83. $\dfrac{3xy}{x^2y^3}$ with a denominator of $16x^4y^4$

84. $\dfrac{2}{27x^5}$ with a denominator of $81x^8$

85. $\dfrac{5}{8x^4}$ with a denominator of $32x^5$

USING YOUR KNOWLEDGE
Using a Calculator to Find Rational Roots and Radicals

86. You already know that $\sqrt{x} = x^{1/2}$. Use your knowledge to write

$$\sqrt{\sqrt{x}}$$

using exponents.

87. If you have a calculator with a $\boxed{\sqrt{\ }}$ key, you can find $\sqrt{9}$ by simply pressing $\boxed{9}\,\boxed{\sqrt{\ }}$. You can also find $\sqrt[4]{16}$ using the $\boxed{\sqrt{\ }}$ key and the results of Problem 86.
   a. Find $\sqrt[4]{16}$.
   b. Find $\sqrt[4]{4096}$.

88. a. Write $\sqrt[3]{x}$ using exponents.
   b. Write $\sqrt[3]{\sqrt{x}}$ using a single exponent.
   c. If you have a calculator with a $\boxed{\sqrt{\ }}$ and a $\boxed{\sqrt[3]{\ }}$ key, find $\sqrt[6]{729}$.

CALCULATOR CORNER

Many of the numerical evaluations in this section can be done with the $\boxed{y^x}$ key in your calculator. This key will raise the number y to the x power. Thus to find 2^3, enter $\boxed{2}\,\boxed{y^x}\,\boxed{3}\,\boxed{=}$. The answer is 8. To find $\sqrt[3]{-64}$, first recall that $\sqrt[3]{-64} = (-64)^{1/3}$. Thus we enter

$$\boxed{64}\,\boxed{+/-}\,\boxed{y^x}\,\boxed{(}\,\boxed{1}\,\boxed{\div}\,\boxed{3}\,\boxed{)}\,\boxed{=}$$

Note that $\frac{1}{3}$ has to be written in parentheses.

If your instructor permits, use the $\boxed{y^x}$ key on your calculator to find the following values.

89. $\sqrt[3]{-\dfrac{1}{8}}$ **90.** $(-125)^{1/3}$ **91.** $\sqrt[4]{\dfrac{1}{16}}$

WRITE ON . . .

92. Explain why a square root of 4 is -2 but $\sqrt{4} = 2$.

93. How many square roots does every positive real number a have? Name them when $a = 36$.

94. Explain why $\sqrt[n]{a} = a^{1/n}$ is not a real number when a is negative and n is even.

95. Explain what is meant by "the nth root of a real number a."

96. Explain why the even root of a negative number is not a real number. (For example, $\sqrt{-4}$ is not a real number.)

If you know how to do these problems, you have learned your lesson!

Simplify (x and y positive):

97. $(x^{1/4}y^{3/5})^{-20}$

98. $(x^{1/7}y^{3/14})^{-21}$

99. $\dfrac{x^{1/4}y^{2/3}}{x^{3/4}y^{1/3}}$

100. $\dfrac{x^{1/5}y^{3/7}}{x^{2/5}y^{2/7}}$

101. $x^{2/5}(x^{-1/5} + y^{1/4})$

102. $x^{3/7}(x^{-3/7} + y^{3/4})$

103. $x^{1/3} \cdot x^{1/5}$

104. $y^{1/5} \cdot y^{1/4}$

105. $\dfrac{x^{-2/3}}{x^{1/4}}$

106. $\dfrac{y^{-3/4}}{y^{1/5}}$

107. $(y^{3/4})^{-1/5}$

108. $(x^{2/3})^{-2/5}$

Evaluate if possible:

109. $81^{-3/4}$

110. $(-27)^{-4/3}$

111. $216^{-2/3}$

112. $27^{2/3}$

113. $(-64)^{-2/3}$

114. $(-36)^{3/2}$

115. $49^{1/2}$

116. $(-216)^{1/3}$

117. $\left(\dfrac{1}{81}\right)^{1/4}$

118. $\sqrt{-25}$

119. $\sqrt[3]{-125}$

120. $\sqrt[3]{\dfrac{-1}{27}}$

5.2

SIMPLIFYING RADICALS

To succeed, review how to:

1. Use the laws of exponents (pp. 31–37).
2. Factor perfect square trinomials (pp. 165–170).

Objectives:

A Simplify radical expressions.

B Rationalize the denominator of a fraction.

C Reduce the order of a radical expression.

getting started Radical Speeding

Have you seen your local police measuring skid marks at the scene of an accident? The speed s (in miles per hour) a car was traveling if it skidded d feet after the brakes were applied on a dry concrete road is given by $s = \sqrt{24d}$. A car leaving a 50-ft skid mark was traveling at a speed $s = \sqrt{24 \cdot 50} = \sqrt{1200}$. How can we simplify this? By factoring 1200 into factors that are perfect squares. Thus $\sqrt{1200} = \sqrt{100 \cdot 4 \cdot 3} = 20\sqrt{3}$ or about 35 mi/hr. Note that we assumed $\sqrt{100 \cdot 4 \cdot 3} = \sqrt{100} \cdot \sqrt{4} \cdot \sqrt{3}$. Is this true?

In this section, we shall study how to simplify radical expressions by using properties analogous to the properties of rational exponents we studied in Section 5.1. First, let's recall the relationship between rational exponents and radicals. In general, the nth root of a number a has the following definition.

nTH ROOT

$$a^{1/n} = \sqrt[n]{a} \qquad (a \geq 0)$$

where n is a positive integer.

From this definition, we can derive three important relationships involving radicals that can be proved using the properties of exponents discussed previously. *In the discussion that follows, we shall assume that when the index of a radical is even, the radicand is nonnegative.*

 Properties of Radicals

LAWS FOR SIMPLIFYING RADICAL EXPRESSIONS

$$\text{I. } \sqrt[n]{a^n} = a \qquad (a \geq 0)$$

$$\text{II. } \sqrt[n]{ab} = \sqrt[n]{a}\,\sqrt[n]{b} \qquad (a, b \geq 0) \qquad \text{Product rule}$$

$$\text{III. } \sqrt[n]{\frac{a}{b}} = \frac{\sqrt[n]{a}}{\sqrt[n]{b}} \qquad (a \geq 0, b > 0) \qquad \text{Quotient rule}$$

The first of these laws is equivalent to the definition of the principal nth root of a. Thus $\sqrt[n]{a^n} = (a^n)^{1/n} = a^{n \cdot 1/n} = a$. The other two laws are obtained as follows:

$$\text{II. } \sqrt[n]{ab} = (ab)^{1/n} = a^{1/n} \cdot b^{1/n} = \sqrt[n]{a} \cdot \sqrt[n]{b}$$

$$\text{III. } \sqrt[n]{\frac{a}{b}} = \left(\frac{a}{b}\right)^{1/n} = \frac{a^{1/n}}{b^{1/n}} = \frac{\sqrt[n]{a}}{\sqrt[n]{b}}$$

We have already mentioned that when m and n are even, $(a^m)^{1/n} = |a|^{m/n}$. For $m = n$, we have the following definition.

DEFINITION OF $\sqrt[n]{a^n}$

$$\sqrt[n]{a^n} = |a| \qquad n \text{ is an \textbf{even} positive integer.}$$

$$\sqrt[n]{a^n} = a \qquad n \text{ is an \textbf{odd} positive integer.}$$

Thus for even indices, $\sqrt{3^2} = |3| = 3$, $\sqrt{(-3)^2} = |-3| = 3$, and $\sqrt[4]{(-x)^4} = |-x|$. When the index is odd, absolute value is not necessary. Thus, $\sqrt[3]{-8} = -2$ and $\sqrt[5]{-x^5} = -x$.

EXAMPLE 1   Simplifying expressions containing radicals

Simplify:

a. $\sqrt[4]{(-2)^4}$ **b.** $\sqrt[8]{(-x)^8}$

c. $\sqrt[9]{(-x)^9}$ **d.** $\sqrt{x^2 + 8x + 16}$

SOLUTION

a. Since the index 4 is even, $\sqrt[4]{(-2)^4} = |-2| = 2$.

b. The index is even. Thus $\sqrt[8]{(-x)^8} = |-x|$. No further simplification is possible, since we don't know if x is positive or negative.

c. $\sqrt[9]{(-x)^9} = -x$. (Absolute value is not necessary since the index 9 is not even.)

d. We start by factoring: $x^2 + 8x + 16 = (x + 4)^2$. Thus

$$\sqrt{x^2 + 8x + 16} = \sqrt{(x + 4)^2}$$

$$= |x + 4|$$

(Absolute value is needed, since the index 2 is even.)   ■

If an expression does not have a perfect root, we can sometimes simplify it by factoring out any perfect roots from the radicand. To help you, we list the first few square, cube, and fourth roots in Table 2.

TABLE 2 Partial List of Square, Cube, and Fourth Roots

Square Roots	Cube Roots	Fourth Roots
$\sqrt{0} = 0$	$\sqrt[3]{0} = 0$	$\sqrt[4]{0} = 0$
$\sqrt{1} = 1$	$\sqrt[3]{1} = 1$	$\sqrt[4]{1} = 1$
$\sqrt{4} = 2$	$\sqrt[3]{8} = 2$	$\sqrt[4]{16} = 2$
$\sqrt{9} = 3$	$\sqrt[3]{27} = 3$	$\sqrt[4]{81} = 3$
$\sqrt{16} = 4$	$\sqrt[3]{64} = 4$	$\sqrt[4]{256} = 4$
$\sqrt{25} = 5$	$\sqrt[3]{125} = 5$	$\sqrt[4]{625} = 5$

To simplify $\sqrt[3]{40}$, we find a factor of 40 that has a perfect cube root. This factor is $8 = 2^3$. Thus $40 = 2^3 \cdot 5$ and

$$\sqrt[3]{40} = \sqrt[3]{2^3 \cdot 5}$$
$$= \sqrt[3]{2^3} \cdot \sqrt[3]{5} \quad \text{Law II}$$
$$= 2\sqrt[3]{5} \quad \text{Law I}$$

EXAMPLE 2   Simplifying square and cube roots

Simplify:

a. $\sqrt{48}$ **b.** $\sqrt[3]{54}$ **c.** $\sqrt[3]{128a^4 b^6}$

SOLUTION

a. Since $48 = 16 \cdot 3 = 4^2 \cdot 3$,

$$\sqrt{48} = \sqrt{4^2 \cdot 3}$$
$$= \sqrt{4^2} \cdot \sqrt{3} \quad \text{Note that } \sqrt{4^2} = 4.$$
$$= 4\sqrt{3}$$

b. Since $54 = 27 \cdot 2 = 3^3 \cdot 2$,

$$\sqrt[3]{54} = \sqrt[3]{3^3 \cdot 2}$$
$$= \sqrt[3]{3^3} \cdot \sqrt[3]{2} \quad \text{Note that } \sqrt[3]{3^3} = 3.$$
$$= 3\sqrt[3]{2}$$

c. We factor 128 into factors that have perfect cube roots. Since 64 is a perfect cube and $128 = 64 \cdot 2$,

$$\sqrt[3]{128a^4 b^6} = \sqrt[3]{4^3 \cdot 2 \cdot a^3 \cdot a \cdot (b^2)^3} \quad \text{Factor.}$$
$$= \sqrt[3]{4^3 \cdot 2} \cdot \sqrt[3]{a^3} \cdot \sqrt[3]{a} \cdot \sqrt[3]{(b^2)^3} \quad \text{Law II}$$
$$= 4 \cdot \sqrt[3]{2} \cdot a \cdot \sqrt[3]{a} \cdot b^2 = 4ab^2 \sqrt[3]{2a} \quad \text{Law I}$$

You can save time by writing all perfect cube factors together like this:

$$\sqrt[3]{128a^4 b^6} = \sqrt[3]{[4^3 a^3 (b^2)^3] \cdot 2a} = 4ab^2 \sqrt[3]{2a} \quad \blacksquare$$

The third law mentioned in this section can be used to change a radical into a form in which the radicand contains no fractions. For example,

$$\sqrt{\frac{3}{16}} = \frac{\sqrt{3}}{\sqrt{16}} = \frac{\sqrt{3}}{4}$$

$$\sqrt[3]{\frac{7}{8}} = \frac{\sqrt[3]{7}}{\sqrt[3]{8}} = \frac{\sqrt[3]{7}}{2}$$

If the denominator does not have a perfect root, multiply the numerator and denominator by a factor that will yield a perfect root. To simplify

$$\sqrt{\dfrac{3}{8}}$$

multiply the denominator by 2 to obtain 16, which has a perfect root. Of course, you must also multiply the numerator by 2 so that you get an equivalent fraction. Thus

$$\sqrt{\dfrac{3}{8}} = \sqrt{\dfrac{3 \cdot 2}{8 \cdot 2}} = \sqrt{\dfrac{6}{16}} = \dfrac{\sqrt{6}}{4}$$

EXAMPLE 3   Simplifying using the quotient rule and perfect roots

Simplify:

a. $\sqrt{\dfrac{7}{32}}$ **b.** $\sqrt[3]{\dfrac{9}{x^3}}$ **c.** $\sqrt[4]{\dfrac{2}{27x^5}}$

SOLUTION

a. We multiply the denominator and numerator by 2 because $32 \cdot 2 = 64$ is a perfect square:

$$\sqrt{\dfrac{7}{32}} = \sqrt{\dfrac{7 \cdot 2}{32 \cdot 2}} = \sqrt{\dfrac{14}{64}} = \dfrac{\sqrt{14}}{\sqrt{64}} = \dfrac{\sqrt{14}}{8}$$

b. Since $\sqrt[3]{x^3} = x$, we just simplify the denominator.

$$\sqrt[3]{\dfrac{9}{x^3}} = \dfrac{\sqrt[3]{9}}{\sqrt[3]{x^3}} = \dfrac{\sqrt[3]{9}}{x}$$

c. To obtain a perfect fourth power, multiply 27 by 3 to obtain $81 = 3^4$ and x^5 by x^3 to obtain x^8.

$$\sqrt[4]{\dfrac{2}{27x^5}} = \sqrt[4]{\dfrac{2 \cdot 3x^3}{27x^5 \cdot 3x^3}} = \sqrt[4]{\dfrac{6x^3}{81x^8}} = \dfrac{\sqrt[4]{6x^3}}{\sqrt[4]{81x^8}} = \dfrac{\sqrt[4]{6x^3}}{3x^2}$$ ∎

Rationalizing the Denominator

As we saw in Example 3, in some cases the denominator of a fraction contains radical expressions. For example,

$$\dfrac{\sqrt{3}}{\sqrt{5}}$$

has $\sqrt{5}$ in the denominator. To simplify

$$\dfrac{\sqrt{3}}{\sqrt{5}}$$

we use the fundamental property of fractions to multiply numerator and denominator by $\sqrt{5}$, which causes the denominator to be rational. Thus

$$\frac{\sqrt{3}}{\sqrt{5}} = \frac{\sqrt{3} \cdot \sqrt{5}}{\sqrt{5} \cdot \sqrt{5}} \qquad \text{Fundamental property of fractions}$$

$$= \frac{\sqrt{15}}{\sqrt{5^2}} \qquad \text{Law II}$$

$$= \frac{\sqrt{15}}{5} \qquad \text{Law I}$$

This process is called **rationalizing the denominator.** To rationalize the denominator in the expression

$$\frac{\sqrt{7}}{\sqrt{3x}}$$

$(x > 0)$, we have to change the fraction to an equivalent one without a radical in the denominator. As before, we multiply the denominator and numerator by $\sqrt{3x}$ to obtain

$$\frac{\sqrt{7}}{\sqrt{3x}} = \frac{\sqrt{7} \cdot \sqrt{3x}}{\sqrt{3x} \cdot \sqrt{3x}} \qquad \text{Fundamental property of fractions}$$

$$= \frac{\sqrt{21x}}{\sqrt{3^2 x^2}} \qquad \text{Law II}$$

$$= \frac{\sqrt{21x}}{3x} \qquad \text{Law I}$$

EXAMPLE 4   Rationalizing the denominator

Rationalize the denominators:

a. $\dfrac{\sqrt{11}}{\sqrt{6}}$ **b.** $\dfrac{\sqrt{3}}{\sqrt{5x}}$ $(x > 0)$ **c.** $\dfrac{\sqrt{5}}{\sqrt{18x^2}}$

SOLUTION   To obtain a perfect square in the denominator, multiply by $\sqrt{6}$.

a. $\dfrac{\sqrt{11}}{\sqrt{6}} = \dfrac{\sqrt{11} \cdot \sqrt{6}}{\sqrt{6} \cdot \sqrt{6}} = \dfrac{\sqrt{66}}{6}$

b. To obtain a perfect square in the denominator, multiply by $\sqrt{5x}$.

$$\frac{\sqrt{3}}{\sqrt{5x}} = \frac{\sqrt{3} \cdot \sqrt{5x}}{\sqrt{5x} \cdot \sqrt{5x}} = \frac{\sqrt{15x}}{5x}$$

c. To convert $\sqrt{18}$ to a perfect square, multiply by $\sqrt{2}$.

$$\frac{\sqrt{5}}{\sqrt{18x^2}} = \frac{\sqrt{5} \cdot \sqrt{2}}{\sqrt{18x^2} \cdot \sqrt{2}} = \frac{\sqrt{10}}{\sqrt{36x^2}} = \frac{\sqrt{10}}{6x^2}$$

Note that x^2 is *not* under the radical, so x^2 does not change. ∎

When the radical in the denominator is of index n, we must make the radicand an exact nth power. For example, to rationalize

$$\frac{\sqrt[3]{5}}{\sqrt[3]{3x}}$$

we convert $\sqrt[3]{3}$ to a perfect cube root by multiplying by $\sqrt[3]{3^2}$, and we convert $\sqrt[3]{x}$ to a perfect cube root by multiplying by $\sqrt[3]{x^2}$. We can combine these two steps and multiply by $\sqrt[3]{3^2 x^2}$ to obtain

$$\frac{\sqrt[3]{5}}{\sqrt[3]{3x}} = \frac{\sqrt[3]{5} \cdot \sqrt[3]{3^2 \cdot x^2}}{\sqrt[3]{3x} \cdot \sqrt[3]{3^2 \cdot x^2}}$$ 　Multiply the numerator and denominator by $\sqrt[3]{3^2 \cdot x^2}$ to make the denominator a perfect cube root.

$$= \frac{\sqrt[3]{5 \cdot 9 \cdot x^2}}{\sqrt[3]{3^3 x^3}}$$ 　Law II

$$= \frac{\sqrt[3]{45x^2}}{3x}$$ 　Law I

EXAMPLE 5　　Making the radicand an exact nth power

Rationalize the denominator:

a. $\dfrac{1}{\sqrt[3]{5x}}$

b. $\dfrac{\sqrt[5]{5}}{\sqrt[5]{8x^4}}$

SOLUTION

a. To convert $\sqrt[3]{5x}$ to a perfect cube root, multiply by $\sqrt[3]{5^2 x^2}$.

$$\frac{1}{\sqrt[3]{5x}} = \frac{1 \cdot \sqrt[3]{5^2 x^2}}{\sqrt[3]{5x} \cdot \sqrt[3]{5^2 x^2}}$$

$$= \frac{\sqrt[3]{25x^2}}{\sqrt[3]{5^3 x^3}}$$

$$= \frac{\sqrt[3]{25x^2}}{5x}$$

b. $\dfrac{\sqrt[5]{5}}{\sqrt[5]{8x^4}} = \dfrac{\sqrt[5]{5}}{\sqrt[5]{2^3 x^4}}$ 　Write $8x^4$ as $2^3 x^4$.

$$= \frac{\sqrt[5]{5} \cdot \sqrt[5]{2^2 \cdot x}}{\sqrt[5]{2^3 \cdot x^4} \cdot \sqrt[5]{2^2 \cdot x}}$$ 　Use the fundamental property of fractions.

$$= \frac{\sqrt[5]{20x}}{\sqrt[5]{2^5 \cdot x^5}}$$ 　Law II

$$= \frac{\sqrt[5]{20x}}{2x}$$ 　Law I 　■

Reducing the Index of a Radical Expression

The index of a radical expression can sometimes be reduced by writing the radical as a power with a rational exponent and then reducing the exponent. For example, if $x \geq 0$,

$$\sqrt[6]{x^3} = (x^3)^{1/6}$$
$$= x^{3 \cdot 1/6}$$
$$= x^{1/2}$$
$$= \sqrt{x}$$

Similarly, for $x \geq 0$ and $y \geq 0$,

$$\sqrt[4]{64x^2y^2} = \sqrt[4]{(8xy)^2}$$
$$= [(8xy)^2]^{1/4}$$
$$= [8xy]^{2 \cdot 1/4}$$
$$= (8xy)^{1/2} = \sqrt{8xy} = 2\sqrt{2xy}$$

EXAMPLE 6   Reducing the index

Reduce the index:

a. $\sqrt[4]{\dfrac{16}{81}}$

b. $\sqrt[6]{27c^3d^3}$ $(c \geq 0, d \geq 0)$

SOLUTION

a. $\sqrt[4]{\dfrac{16}{81}} = \left[\left(\dfrac{2}{3}\right)^4\right]^{1/4} = \dfrac{2}{3}$

b. $\sqrt[6]{27c^3d^3} = \sqrt[6]{(3cd)^3}$
$$= [(3cd)^3]^{1/6}$$
$$= [3cd]^{1/2}$$
$$= \sqrt{3cd}$$ ■

We have used different techniques to *simplify* radical expressions. To make sure that the resulting radicals are simplified, use these rules.

RULES FOR SIMPLIFYING RADICAL EXPRESSIONS

A radical expression is in **simplified** form if

1. All exponents in the radicand (the expression under the radical) are less than the index.

2. There are no fractions under the radical sign.

3. There are no radicals in the denominator.

4. The index is as low as possible.

EXAMPLE 7   Simplifying radical expressions

Simplify:

$$\sqrt[6]{\dfrac{a^2}{16c^{10}}}$$

SOLUTION To make the denominator a perfect sixth root, note that $16 = 2^4$ and then multiply numerator and denominator under the radical by $2^2 c^2$. We have

$$\sqrt[6]{\frac{a^2}{16c^{10}} \cdot \frac{2^2 c^2}{2^2 c^2}} = \sqrt[6]{\frac{2^2 a^2 c^2}{2^6 c^{12}}}$$

$$= \frac{\sqrt[6]{2^2 a^2 c^2}}{2c^2}$$

$$= \frac{(2^2 a^2 c^2)^{1/6}}{2c^2}$$

$$= \frac{[(2ac)^2]^{1/6}}{2c^2}$$

$$= \frac{(2ac)^{1/3}}{2c^2}$$

$$= \frac{\sqrt[3]{2ac}}{2c^2} \qquad ■$$

GRAPH IT

Most of the numerical problems in this section can be verified using a grapher. To verify Example 1a, enter the information shown in Window 1 and press ENTER. The answer 2 appears. Note how parentheses were used to enter the expression. Moreover, you have to *know* that taking the fourth root is the same as raising to the 1/4 power. In this regard, the grapher is no help; you still have to know your algebra.

You can verify Example 3a in a similar way: Enter the original problem,

$$\sqrt{\frac{7}{32}}$$

and the simplified version,

$$\frac{\sqrt{14}}{8}$$

You should get the approximation 0.4677071733 in both cases, as shown in Window 2. Try verifying the results in Examples 4a and 6a using this idea.

If the examples to be verified are not numerical, you can still graph the original problem and answer to make sure the graphs are the same. For example, to verify the results of Example 3c, let

$$Y_1 = \left(\frac{2}{27x^5}\right)^{1/4} \qquad \text{and} \qquad Y_2 = \frac{(6x^3)^{1/4}}{3x^2}$$

Graph both of these using a standard window first; then regraph using a $[-1, 10]$ by $[-1, 10]$ window for a better view. The graph in both cases is shown in Window 3.

You can verify the rest of the problems involving one variable using this technique. However, you won't be able to verify the algebraic results of Example 6b with your grapher! Do you see why? Our favorite moral: Some problems have to be done algebraically.

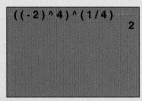

WINDOW 1

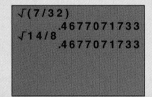

WINDOW 2

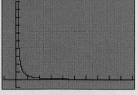

WINDOW 3

EXERCISE 5.2

A **In Problems 1–24, simplify the expression given. (*Hint:* Some answers require absolute values.)**

1. $\sqrt{(-5)^2}$

2. $\sqrt{(5)^2}$

3. $\sqrt[3]{-64}$

4. $\sqrt[3]{-125}$

5. $\sqrt[6]{(-x)^6}$

6. $\sqrt[5]{(-x)^5}$

7. $\sqrt{x^2 + 12x + 36}$

8. $\sqrt{4x^2 + 12x + 9}$

9. $\sqrt{9x^2 - 12x + 4}$

10. $\sqrt{16x^2 + 8x + 1}$

11. $\sqrt{16x^3y^3}$

12. $\sqrt{81x^3y^4}$

13. $\sqrt[3]{40x^4y}$

14. $\sqrt[3]{81x^3y^6}$

15. $\sqrt[4]{x^5y^7}$

16. $\sqrt[4]{162x^4y^7}$

17. $\sqrt[5]{-243a^{10}b^{17}}$

18. $\sqrt[5]{-32a^{15}b^{20}}$

19. $\sqrt{\dfrac{13}{49}}$

20. $\sqrt{\dfrac{17}{64}}$

21. $\sqrt{\dfrac{17}{4x^2}}$

22. $\sqrt{\dfrac{19}{64x^4}}$

23. $\sqrt[3]{\dfrac{3}{64x^3}}$

24. $\sqrt[3]{\dfrac{-7}{27x^6}}$

B **In Problems 25–40, rationalize the denominator. (Assume all variables represent positive real numbers.)**

25. $\sqrt{\dfrac{2}{3}}$

26. $\sqrt{\dfrac{4}{5}}$

27. $\dfrac{-\sqrt{2}}{\sqrt{7}}$

28. $\dfrac{-\sqrt{3}}{\sqrt{11}}$

29. $\sqrt{\dfrac{5}{2a}}$

30. $\sqrt{\dfrac{7}{36}}$

31. $\sqrt{\dfrac{5}{32ab}}$

32. $\sqrt{\dfrac{5}{8ab}}$

33. $-\sqrt{\dfrac{3}{2a^3b^3}}$

34. $-\sqrt{\dfrac{3}{8ab^3}}$

35. $\dfrac{\sqrt{x}\,\sqrt{xy^3}}{\sqrt{y}}$

36. $\dfrac{\sqrt{xy}\,\sqrt{xy^4}}{\sqrt{y}}$

37. $-\sqrt[3]{\dfrac{7}{9}}$

38. $-\sqrt[3]{\dfrac{3}{32}}$

39. $\sqrt[3]{\dfrac{3}{16x^2}}$

40. $\sqrt[3]{\dfrac{5}{16x}}$

C **In Problems 41–50, reduce the index (order) of the given radical and simplify. (Assume the variables represent positive real numbers.)**

41. $\sqrt[6]{9}$

42. $\sqrt[6]{4}$

43. $\sqrt[4]{4a^2}$

44. $\sqrt[4]{9a^2}$

45. $\sqrt[4]{25x^6y^2}$

46. $\sqrt[4]{36x^2y^6}$

47. $\sqrt[4]{49x^{10}y^6}$

48. $\sqrt[4]{100x^{10}y^{10}}$

49. $\sqrt[6]{8a^3b^3}$

50. $\sqrt[6]{27a^3b^9}$

In Problems 51–55, simplify the radical expression. (Assume all variables represent positive real numbers.)

51. $\sqrt[6]{\dfrac{a^4}{b^8}}$

52. $\sqrt[4]{\dfrac{c^6}{4b^2}}$

53. $\sqrt[4]{\dfrac{64a^2}{9b^6}}$

54. $\sqrt[4]{\dfrac{4c^2y^6}{9b^4}}$

55. $\sqrt[6]{\dfrac{b^3a^3}{8x^3}}$

APPLICATIONS

56. A body starting at rest takes t seconds to fall a distance of d feet, where

$$t = \sqrt{\dfrac{d}{16}}$$

  a. Simplify this expression.

  b. How long would it take an object starting at rest to fall 100 ft?

57. The radius of a sphere is given by

$$r = \sqrt[3]{\dfrac{3V}{4\pi}}$$

where V is the volume of the sphere and π is about $\frac{22}{7}$.

  a. Simplify $\sqrt[3]{\dfrac{3V}{4\pi}}$.

  b. If the volume of a sphere is 36π ft^3, what is its radius?

58. The root-mean-square velocity, $\bar{v}$, of a gas particle is given by the formula

$$\bar{v} = \dfrac{\sqrt{3kT}}{\sqrt{m}}$$

where k is a constant, T is the temperature (in degrees Kelvin), and m is the mass of the particle. Rationalize the denominator of the expression on the right-hand side.

59. The mass m of an object depends on its speed v and the speed of light c. The relationship is given by the formula

$$m = \frac{m_0}{\sqrt{1 - \dfrac{v^2}{c^2}}}$$

where m_0 is the *rest mass*, the mass when $v = 0$. Simplify the expression on the right-hand side and rationalize the denominator.

60. Have you heard of supersonic airplanes with a speed of Mach 2? Mach 2 means that the speed of the plane is *twice* the speed of sound. The Mach number M can be found from the formula

$$M = \sqrt{\frac{2}{\gamma}} \ \sqrt{\frac{P_2 - P_1}{P_1}}$$

where P_1 and P_2 are air pressures. Simplify this expression and write the answer with a rationalized denominator.

SKILL CHECKER

61. Multiply $(a + b)(a - b)$.

62. Use the result of Problem 61 to multiply $(\sqrt{x} + \sqrt{y})(\sqrt{x} - \sqrt{y})$.

63. Use the result of Problem 61 to multiply $(x^{3/2} + y^{3/2})(x^{3/2} - y^{3/2})$.

Divide:

64. $\dfrac{6 + 3y}{3}$ 65. $\dfrac{3xy + 6x^2 y}{3xy}$ 66. $\dfrac{12x^2 y^3 + 18xy^3}{6xy}$

Evaluate and simplify $\sqrt{b^2 - 4ac}$ in each case:

67. $a = 2$, $b = -1$, and $c = -6$

68. $a = 2$, $b = -5$, and $c = -12$

69. $a = 6$, $b = -4$, and $c = -2$

70. $a = -1$, $b = 1$, and $c = 12$

USING YOUR KNOWLEDGE   **From Radicals to Rational and Back**

We've studied rational exponents and radical expressions. We will now use this knowledge to translate one notation to the other.

71. In Problem 59,

$$m = \frac{m_0}{\sqrt{1 - \dfrac{v^2}{c^2}}}$$

Rationalize the denominator and write the result using rational exponents.

72. In Problem 60, simplify the expression defining M and write the result using rational exponents.

73. The period T of a pendulum is

$$T = \sqrt{\frac{2\pi L}{g}}$$

where L is the length and g is the gravity constant. Simplify this expression and write the result using rational exponents.

74. The pressure P of a gas is related to its volume V by the formula $P = kV^{-7/5}$, where k is the proportionality constant. Write this formula using radical notation.

75. The average speed v of oxygen molecules is given by the formula $v = (3kT)^{1/2} m^{-1/2}$. Write this formula in simplified form using radicals.

CALCULATOR CORNER

Some of the simplifications we have made can be checked using a calculator with a $\boxed{\sqrt[x]{y}}$ key. To access this feature, you usually have to press the $\boxed{\text{2nd}}$ or $\boxed{\text{2ndF}}$ key first and then the $\boxed{\sqrt[x]{y}}$ key. The calculator will then find the xth root of y.

In Example 2, we learned that $\sqrt[3]{54} = 3\sqrt[3]{2}$. To check this, enter $\boxed{54}\ \boxed{\text{2nd}}\ \boxed{\sqrt[x]{y}}\ \boxed{3}\ \boxed{=}$. The display shows 3.77976315. Now enter $\boxed{3}\ \boxed{\times}\ \boxed{2}\ \boxed{\sqrt[x]{y}}\ \boxed{3}\ \boxed{=}$. The same result appears, so our answer is correct. If your instructor agrees, check the numerical problems in this section (Problems 25–28, for example) with a calculator.

WRITE ON . . .

76. Law I states that $\sqrt[n]{a^n} = a$ for $a \geq 0$. What happens if $a < 0$? Explain and give examples.

77. Write the procedure you use to rationalize a radical denominator in a quotient.

78. State the conditions under which $\sqrt[n]{a^n} = (\sqrt[n]{a})^n = a$.

79. Use some counterexamples to show that $(a^2 + b^2)^{1/2} \neq a + b$.

80. Use some counterexamples to show that $(a^{1/2} + b^{1/2})^2 \neq a + b$.

MASTERY TEST   If you know how to do these problems, you have learned your lesson!

Simplify:

81. $\sqrt{\dfrac{11}{12}}$

82. $\sqrt[3]{\dfrac{6}{x^3}}$

83. $\sqrt[4]{\dfrac{3}{8x^6}}$

84. $\sqrt{32}$

85. $\sqrt[3]{32}$

86. $\sqrt[3]{81a^6b^4}$

87. $\sqrt[6]{\dfrac{a^3}{8x^3}}$

88. $\sqrt[8]{\dfrac{x^4}{16a^4}}$

89. $\sqrt[4]{(-10)^4}$

90. $\sqrt[6]{(-x)^6}$

91. $\sqrt[7]{(-x)^7}$

92. $\sqrt{x^2 + 10x + 25}$

Reduce the index:

93. $\sqrt[8]{\dfrac{81}{256}}$

94. $\sqrt[6]{4c^2d^2}$

Rationalize the denominator:

95. $\dfrac{\sqrt{7}}{\sqrt{5}}$

96. $\dfrac{\sqrt[3]{5}}{\sqrt[3]{4}}$

97. $\dfrac{\sqrt{5}}{\sqrt{6x}}$ $(x > 0)$

98. $\dfrac{\sqrt{11}}{\sqrt{32x^3}}$

5.3

OPERATIONS WITH RADICALS

To succeed, review how to:

1. Combine like terms (pp. 47–48).
2. Remove parentheses using the distributive property (pp. 149–151).
3. Write a fraction with a specified denominator (p. 205).
4. Reduce fractions (p. 208).

Objectives:

A Add and subtract radical expressions that are similar.

B Multiply and divide radical expressions.

C Rationalize the denominators of radical expressions involving sums or differences.

getting started

Radical Flight

How fast can this plane travel? The answer is classified information, but it is known to exceed twice the speed of sound (747 mi/hr). It is also said that the plane's speed is more than Mach 2. As noted in Problem 60, page 298, the formula for calculating the Mach number M is

$$M = \sqrt{\frac{2}{\gamma}}\,\sqrt{\frac{P_2 - P_1}{P_1}}$$

where P_1 and P_2 are air pressures. This expression can be simplified by multiplying both radical expressions and rationalizing the denominator. In this section we will add, subtract, multiply, and divide radical expressions.

A   Adding and Subtracting Similar Radical Expressions

In Section 1.4 we combined like terms using the distributive property. Thus

$$3x + 5x = (3 + 5)x = 8x$$

$$7x - 4x = (7 - 4)x = 3x$$

Similarly,

$$3\sqrt{2} + 5\sqrt{2} = (3 + 5)\sqrt{2} = 8\sqrt{2}$$

$$7\sqrt[3]{7} - 4\sqrt[3]{7} = (7 - 4)\sqrt[3]{7} = 3\sqrt[3]{7}$$

Thus we can combine *like* (*similar*) radical expressions. Here is the definition.

LIKE (SIMILAR)
RADICAL EXPRESSIONS

> Radical expressions with the same index and the same radicand are **like (similar) expressions.**

If the expressions don't appear to be similar or like, we must try to simplify them first. Thus to add $\sqrt{75} + \sqrt{27}$, we proceed as follows:

$$\sqrt{75} + \sqrt{27} = \sqrt{25 \cdot 3} + \sqrt{9 \cdot 3}$$
$$= \sqrt{25} \cdot \sqrt{3} + \sqrt{9} \cdot \sqrt{3} \qquad \sqrt{ab} = \sqrt{a}\sqrt{b}$$
$$= 5\sqrt{3} + 3\sqrt{3}$$
$$= (5 + 3)\sqrt{3} = 8\sqrt{3} \qquad \text{Add like radicals.}$$

The subtraction of similar (like) radicals is done in the same way. Thus

$$\sqrt{80} - \sqrt{20} = \sqrt{16 \cdot 5} - \sqrt{4 \cdot 5}$$
$$= \sqrt{16} \cdot \sqrt{5} - \sqrt{4} \cdot \sqrt{5}$$
$$= 4\sqrt{5} - 2\sqrt{5}$$
$$= (4 - 2) \cdot \sqrt{5} = 2\sqrt{5} \qquad \text{Subtract like radicals.}$$

EXAMPLE 1   Adding and subtracting radical expressions

Perform the indicated operations:

a. $\sqrt{175} + \sqrt{28}$ **b.** $\sqrt{98} - \sqrt{32}$

c. $3\sqrt{18x} - 5\sqrt{8x}$ **d.** $5\sqrt[3]{80x} - 3\sqrt[3]{270x}$

SOLUTION

a. $\sqrt{175} + \sqrt{28} = \sqrt{25 \cdot 7} + \sqrt{4 \cdot 7}$
$$= \sqrt{25} \cdot \sqrt{7} + \sqrt{4} \cdot \sqrt{7}$$
$$= 5\sqrt{7} + 2\sqrt{7} = 7\sqrt{7}$$

b. $\sqrt{98} - \sqrt{32} = \sqrt{49 \cdot 2} - \sqrt{16 \cdot 2}$
$$= \sqrt{49} \cdot \sqrt{2} - \sqrt{16} \cdot \sqrt{2}$$
$$= 7\sqrt{2} - 4\sqrt{2} = 3\sqrt{2}$$

c. $3\sqrt{18x} - 5\sqrt{8x} = 3\sqrt{9 \cdot 2x} - 5\sqrt{4 \cdot 2x}$
$$= 3\sqrt{9} \cdot \sqrt{2x} - 5\sqrt{4} \cdot \sqrt{2x}$$
$$= 3 \cdot 3 \cdot \sqrt{2x} - 5 \cdot 2 \cdot \sqrt{2x}$$
$$= 9\sqrt{2x} - 10\sqrt{2x}$$
$$= -\sqrt{2x}$$

d. This time we must find factors of 80 and 270 that are perfect cubes: $80 = 8 \cdot 10 = 2^3 \cdot 10$ and $270 = 27 \cdot 10 = 3^3 \cdot 10$. Thus

$$5\sqrt[3]{80x} - 3\sqrt[3]{270x} = 5\sqrt[3]{2^3 \cdot 10x} - 3\sqrt[3]{3^3 \cdot 10x}$$
$$= 5\sqrt[3]{2^3} \cdot \sqrt[3]{10x} - 3\sqrt[3]{3^3} \cdot \sqrt[3]{10x}$$
$$= 5 \cdot 2 \cdot \sqrt[3]{10x} - 3 \cdot 3\sqrt[3]{10x}$$
$$= 10\sqrt[3]{10x} - 9\sqrt[3]{10x}$$
$$= \sqrt[3]{10x} \qquad ■$$

We now show you how to combine like radicals that are fractions.

EXAMPLE 2   Subtracting radical expressions containing fractions

Perform the indicated operations:

a. $3\sqrt{\dfrac{1}{2}} - 5\sqrt{\dfrac{1}{8}}$

b. $3\sqrt[3]{\dfrac{3x}{4x^2}} - \sqrt[3]{\dfrac{3}{32x}}$

SOLUTION

a. We first simplify each of the radicals by making the denominator a perfect square.

$$\sqrt{\dfrac{1}{2}} = \sqrt{\dfrac{1 \cdot 2}{2 \cdot 2}} = \dfrac{\sqrt{2}}{2} \qquad \text{and} \qquad \sqrt{\dfrac{1}{8}} = \sqrt{\dfrac{1 \cdot 2}{8 \cdot 2}} = \dfrac{\sqrt{2}}{4}$$

Thus

$$3\sqrt{\dfrac{1}{2}} - 5\sqrt{\dfrac{1}{8}} = 3 \cdot \dfrac{\sqrt{2}}{2} - 5 \cdot \dfrac{\sqrt{2}}{4} \qquad \text{Substitute } \dfrac{\sqrt{2}}{2} \text{ for } \sqrt{\dfrac{1}{2}} \text{ and } \dfrac{\sqrt{2}}{4} \text{ for } \sqrt{\dfrac{1}{8}}.$$

$$= \dfrac{3\sqrt{2}}{2} - \dfrac{5\sqrt{2}}{4} \qquad \text{Multiply.}$$

$$= \dfrac{2 \cdot 3\sqrt{2}}{2 \cdot 2} - \dfrac{5\sqrt{2}}{4} \qquad \text{Since the LCD of 2 and 4 is 4, write the first fraction with 4 as the denominator.}$$

$$= \dfrac{6\sqrt{2} - 5\sqrt{2}}{4} \qquad \text{Use 4 as the denominator.}$$

$$= \dfrac{\sqrt{2}}{4} \qquad \text{Subtract.}$$

b. We first ① make the denominators perfect cubes and find the cube root, ② find the LCD of the resulting fractions, and then ③ subtract.

$$\sqrt[3]{\dfrac{3x}{4x^2}} = \sqrt[3]{\dfrac{3x \cdot 2x}{4x^2 \cdot 2x}} = \dfrac{\sqrt[3]{6x^2}}{2x} \qquad \text{① Make the denominators perfect cubes and take the cube root.}$$

$$\sqrt[3]{\dfrac{3}{32x}} = \sqrt[3]{\dfrac{3 \cdot 2x^2}{32x \cdot 2x^2}} = \dfrac{\sqrt[3]{6x^2}}{4x}$$

Thus

$$3\sqrt[3]{\frac{3x}{4x^2}} - \sqrt[3]{\frac{3}{32x}} = 3 \cdot \frac{\sqrt[3]{6x^2}}{2x} - \frac{\sqrt[3]{6x^2}}{4x}$$

$$= \frac{3\sqrt[3]{6x^2}}{2x} - \frac{\sqrt[3]{6x^2}}{4x}$$

$$= \frac{2 \cdot 3\sqrt[3]{6x^2}}{2 \cdot 2x} - \frac{\sqrt[3]{6x^2}}{4x}$$ ② Since the LCD is $4x$, multiply numerator and denominator of the first fraction by 2.

$$= \frac{6\sqrt[3]{6x^2} - \sqrt[3]{6x^2}}{4x}$$

$$= \frac{5\sqrt[3]{6x^2}}{4x}$$ ③ Subtract.
$x \neq 0$ ∎

Multiplying and Dividing Radical Expressions

The distributive property, in conjunction with the fact that $\sqrt{a} \cdot \sqrt{b} = \sqrt{ab}$, can be used to simplify radical expressions that contain parentheses. For example,

$$\sqrt{2} \cdot (\sqrt{3} + \sqrt{5}) = \sqrt{2} \cdot \sqrt{3} + \sqrt{2} \cdot \sqrt{5}$$ Distributive property

$$= \sqrt{6} + \sqrt{10}$$

Similarly, if $x \geq 0$, then

$$\sqrt{2x} \cdot (\sqrt{x} + \sqrt{3}) = \sqrt{2x} \cdot \sqrt{x} + \sqrt{2x} \cdot \sqrt{3}$$

$$= \sqrt{2x^2} + \sqrt{6x}$$ Since $\sqrt{ab} = \sqrt{a} \cdot \sqrt{b}$, $\sqrt{2x^2} = \sqrt{2} \cdot \sqrt{x^2}$.

$$= \sqrt{2}\sqrt{x^2} + \sqrt{6x}$$ $\sqrt{x^2} = x$ if $x \geq 0$.

$$= x\sqrt{2} + \sqrt{6x}$$

EXAMPLE 3   Multiplying radical expressions

Perform the indicated operations:

a. $\sqrt{3}(\sqrt{5} + \sqrt{12})$ **b.** $\sqrt{3x}(\sqrt{x} - \sqrt{5}), x \geq 0$

c. $\sqrt[3]{3x}(\sqrt[3]{9x^2} - \sqrt[3]{18x})$

SOLUTION

a. $\sqrt{3}(\sqrt{5} + \sqrt{12}) = \sqrt{3} \cdot \sqrt{5} + \sqrt{3} \cdot \sqrt{12}$

$$= \sqrt{15} + \sqrt{36}$$

$$= \sqrt{15} + 6$$

b. $\sqrt{3x}(\sqrt{x} - \sqrt{5}) = \sqrt{3x}\sqrt{x} - \sqrt{3x}\sqrt{5}$

$$= \sqrt{3x^2} - \sqrt{15x}$$

$$= x\sqrt{3} - \sqrt{15x}$$

c. $\sqrt[3]{3x}(\sqrt[3]{9x^2} - \sqrt[3]{18x}) = \sqrt[3]{3x}\sqrt[3]{9x^2} - \sqrt[3]{3x}\sqrt[3]{18x}$

$$= \sqrt[3]{27x^3} - \sqrt[3]{54x^2}$$

$$= \sqrt[3]{3^3 \cdot x^3} - \sqrt[3]{3^3 \cdot 2 \cdot x^2}$$

$$= 3x - 3\sqrt[3]{2x^2}$$ ∎

If we wish to obtain the product of two binomials that contain radicals, we first simplify the radicals involved (if possible) and then use FOIL. For example, to find the product $(\sqrt{98} + \sqrt{27})(\sqrt{72} + \sqrt{75})$, we proceed as follows.

$(\sqrt{98} + \sqrt{27})(\sqrt{72} + \sqrt{75})$

$\quad = (\sqrt{49 \cdot 2} + \sqrt{9 \cdot 3})(\sqrt{36 \cdot 2} + \sqrt{25 \cdot 3})$ Factor under each radical.

$\quad = (7\sqrt{2} + 3\sqrt{3})(6\sqrt{2} + 5\sqrt{3})$ Simplify.

$\quad \overset{F}{} \quad\quad \overset{O}{} \quad\quad \overset{I}{} \quad\quad \overset{L}{}$

$\quad = 7 \cdot 6 \cdot \sqrt{2} \cdot \sqrt{2} + 7 \cdot 5 \cdot \sqrt{2} \cdot \sqrt{3} + 3 \cdot 6 \cdot \sqrt{3} \cdot \sqrt{2} + 3 \cdot 5 \cdot \sqrt{3} \cdot \sqrt{3}$

     Use FOIL.

$\quad = 42\sqrt{2^2} + 35\sqrt{2} \cdot \sqrt{3} + 18\sqrt{3} \cdot \sqrt{2} + 15\sqrt{3^2}$ Simplify.

$\quad = 42 \cdot 2 + 35\sqrt{6} + 18\sqrt{6} + 15 \cdot 3$ Since $\sqrt{2^2} = 2$, $\sqrt{3}\sqrt{2} = \sqrt{6}$ and $\sqrt{3^2} = 3$.

$\quad = 84 + 53\sqrt{6} + 45$ Multiply.

$\quad = 129 + 53\sqrt{6}$ Combine like terms.

EXAMPLE 4   Using FOIL to multiply binomials containing radicals

Find the product: $(\sqrt{63} + \sqrt{75})(\sqrt{28} - \sqrt{27})$

SOLUTION   We first simplify the radicals and then use FOIL.

$(\sqrt{63} + \sqrt{75})(\sqrt{28} - \sqrt{27}) = (\sqrt{9 \cdot 7} + \sqrt{25 \cdot 3})(\sqrt{4 \cdot 7} - \sqrt{9 \cdot 3})$

$\quad\quad = (3\sqrt{7} + 5\sqrt{3})(2\sqrt{7} - 3\sqrt{3})$

$\quad\quad \overset{F}{} \quad \overset{O}{} \quad \overset{I}{} \quad \overset{L}{}$

$\quad\quad = 6\sqrt{7^2} - 9\sqrt{21} + 10\sqrt{21} - 15\sqrt{3^2}$

$\quad\quad = 6 \cdot 7 + \sqrt{21} - 15 \cdot 3$

$\quad\quad = 42 + \sqrt{21} - 45$

$\quad\quad = -3 + \sqrt{21}$ ∎

EXAMPLE 5   Multiplying radical expressions

Multiply:

a. $(\sqrt{3} + 2)^2$ **b.** $(3 - \sqrt{2})^2$ **c.** $(\sqrt{3} + \sqrt{2})(\sqrt{3} - \sqrt{2})$

SOLUTION

a. Since $(x + a)^2 = x^2 + 2ax + a^2$,

$\quad (\sqrt{3} + 2)^2 = (\sqrt{3})^2 + 2 \cdot 2 \cdot \sqrt{3} + 2^2$

$\quad\quad\quad\quad = 3 + 4\sqrt{3} + 4$

$\quad\quad\quad\quad = 7 + 4\sqrt{3}$

b. Since $(x - a)^2 = x^2 - 2ax + a^2$,

$\quad (3 - \sqrt{2})^2 = 3^2 - 2 \cdot \sqrt{2} \cdot 3 + (\sqrt{2})^2$

$\quad\quad\quad\quad = 9 - 6\sqrt{2} + 2$

$\quad\quad\quad\quad = 11 - 6\sqrt{2}$

c. Since $(x + a)(x - a) = x^2 - a^2$,

$\quad (\sqrt{3} + \sqrt{2})(\sqrt{3} - \sqrt{2}) = (\sqrt{3})^2 - (\sqrt{2})^2$

$\quad\quad\quad\quad\quad\quad = 3 - 2$

$\quad\quad\quad\quad\quad\quad = 1$ ∎

In Chapter 6 some of the answers will be of the form

$$\frac{6 + \sqrt{8}}{2}$$

We can simplify this expression in two ways.

METHOD 1   Write

$$\frac{6 + \sqrt{8}}{2}$$

in lowest terms by ① writing $\sqrt{8}$ as $\sqrt{4 \cdot 2} = 2\sqrt{2}$, ② factoring, and then ③ reducing. The procedure looks like this:

$$\frac{6 + \sqrt{8}}{2} = \overset{①}{\frac{6 + 2\sqrt{2}}{2}} = \overset{②}{\frac{2(3 + \sqrt{2})}{2}} = \overset{③}{3 + \sqrt{2}}$$

METHOD 2   If we view

$$\frac{6 + \sqrt{8}}{2}$$

as a division of a binomial by a monomial, we can solve the problem by ① writing $\sqrt{8}$ as $2\sqrt{2}$; then we ② write each term in the numerator over the common denominator and ③ reduce. The procedure looks like this:

$$\frac{6 + \sqrt{8}}{2} = \overset{①}{\frac{6 + 2\sqrt{2}}{2}} = \overset{②}{\frac{6}{2} + \frac{2\sqrt{2}}{2}} = \overset{③}{2} + \sqrt{2} \qquad \text{As before}$$

EXAMPLE 6   Simplifying radical expressions

Simplify:

$$\frac{6 + \sqrt{18}}{3}$$

SOLUTION
METHOD 1   Since $\sqrt{18} = \sqrt{9 \cdot 2} = 3\sqrt{2}$, we have

$$\frac{6 + \sqrt{18}}{3} = \overset{①}{\frac{6 + 3\sqrt{2}}{3}} = \overset{②}{\frac{3(2 + \sqrt{2})}{3}} = \overset{③}{2 + \sqrt{2}}$$

METHOD 2   We can also do this problem by dividing individual terms. Thus

$$\frac{6 + \sqrt{18}}{3} = \overset{①}{\frac{6 + 3\sqrt{2}}{3}} = \overset{②}{\frac{6}{3} + \frac{3\sqrt{2}}{3}} = \overset{③}{2 + \sqrt{2}} \qquad ∎$$

Rationalizing Denominators

We know how to rationalize the denominator in expressions of the form

$$\frac{a}{\sqrt{b}}$$

We now show you how to rationalize radical expressions that contain sums or differences involving radicals in the denominator. The procedure involves the concept of *conjugate expressions*.

CONJUGATE

> The numbers $a + b$ and $a - b$ are **conjugates** of each other.

Here are some numbers and their conjugates.

Number	Conjugate
$3 + \sqrt{2}$	$3 - \sqrt{2}$
$-4 + \sqrt{5}$	$-4 - \sqrt{5}$
$7 - \sqrt{3}$	$7 + \sqrt{3}$
$-8 - \sqrt{6}$	$-8 + \sqrt{6}$

Since $(a + b)(a - b) = a^2 - b^2$, the product of a number and its conjugate is $a^2 - b^2$. Now suppose we want to rationalize the denominator in

$$\frac{3}{3 + \sqrt{3}}$$

We use the fundamental property of fractions and multiply the numerator and denominator of

$$\frac{3}{3 + \sqrt{3}}$$

by the conjugate of $3 + \sqrt{3}$, that is, by $3 - \sqrt{3}$, to obtain

$$\frac{3}{3 + \sqrt{3}} = \frac{3 \cdot (3 - \sqrt{3})}{(3 + \sqrt{3})(3 - \sqrt{3})}$$

$$= \frac{3 \cdot (3 - \sqrt{3})}{3^2 - (\sqrt{3})^2} \qquad \text{Since } (a + b)(a - b) = a^2 - b^2,$$
$$\qquad\qquad\qquad\qquad (3 + \sqrt{3})(3 - \sqrt{3}) = 3^2 - (\sqrt{3})^2.$$

$$= \frac{3 \cdot (3 - \sqrt{3})}{9 - 3} \qquad \text{Note that } (\sqrt{3})^2 = \sqrt{3} \cdot \sqrt{3}$$
$$\qquad\qquad\qquad\qquad = \sqrt{9}$$
$$\qquad\qquad\qquad\qquad = 3$$

$$= \frac{\overset{1}{\cancel{3}} \cdot (3 - \sqrt{3})}{\underset{2}{\cancel{6}}}$$

$$= \frac{3 - \sqrt{3}}{2}$$

> **NOTE** To rationalize radical expressions that contain sums or differences and that involve square root radicals in the denominator, multiply numerator and denominator by the conjugate of the denominator.

EXAMPLE 7   Rationalizing denominators

Rationalize the denominator:

$$\frac{\sqrt{x}}{\sqrt{x} - \sqrt{y}}$$

where x and y represent positive numbers

SOLUTION   We first multiply numerator and denominator by $\sqrt{x} + \sqrt{y}$, the conjugate of $\sqrt{x} - \sqrt{y}$.

$$\frac{\sqrt{x}}{\sqrt{x} - \sqrt{y}} = \frac{\sqrt{x}(\sqrt{x} + \sqrt{y})}{(\sqrt{x} - \sqrt{y})(\sqrt{x} + \sqrt{y})} \qquad \text{Fundamental property of fractions}$$

$$= \frac{\sqrt{x}(\sqrt{x} + \sqrt{y})}{(\sqrt{x})^2 - (\sqrt{y})^2} \qquad (\sqrt{x} + \sqrt{y})(\sqrt{x} - \sqrt{y}) = (\sqrt{x})^2 - (\sqrt{y})^2$$

$$= \frac{\sqrt{x}(\sqrt{x} + \sqrt{y})}{x - y} \qquad (\sqrt{x})^2 = x \text{ and } (\sqrt{y})^2 = y$$

$$= \frac{\sqrt{x^2} + \sqrt{xy}}{x - y} \qquad \text{Use the distributive law.}$$

$$= \frac{x + \sqrt{xy}}{x - y} \qquad \sqrt{x^2} = x \text{ for } x > 0 \qquad ■$$

You can use a similar procedure to rationalize the *numerator* of a radical expresion. We discuss how to do this in the *Using Your Knowledge* section.

GRAPH IT

As we did in Section 5.2, we can verify numerical problems (Examples 1a, b, 2a, 3a, 4, 5, and 6) using a grapher. Let's verify the results of Example 4. Enter $(\sqrt{63} + \sqrt{75})(\sqrt{28} - \sqrt{27})$ ENTER . It helps to know that $\sqrt{n} = n^{1/2}$, so you can enter $\sqrt{63}$ as $63^{1/2}$ (or 2nd √ 63); $\sqrt{75}$ as $75^{1/2}$ (or 2nd √ 75); and so on as you wish. Now enter the answer $-3 + \sqrt{21}$ ENTER . In both cases, the result is 1.582575695 as shown in Window 1.

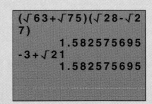

WINDOW 1

Example 2b can be verified graphically by entering

$$Y_1 = 3\sqrt[3]{\frac{3x}{4x^2}} - \sqrt[3]{\frac{3}{32x}} \qquad \text{and} \qquad Y_2 = \frac{5\sqrt[3]{6x^2}}{4x}$$

As you can see in Window 2, the two graphs are identical. When you enter Y_2, make sure it is entered as $(5(6x^2)^{(1/3)})/4x$. (You can also use MATH 4 to find cube roots.) What does the vertical line segment at $x = 0$ mean? Is that line segment really part of the graph?

Now verify the results in Examples 3b and 3c.

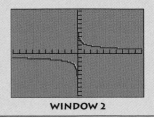

WINDOW 2

EXERCISE 5.3

A **In Problems 1–30, perform the indicated operations. (Where the index is even, assume all variables are positive.)**

1. $12\sqrt{2} + 3\sqrt{2}$

2. $15\sqrt{3} + 2\sqrt{3}$

3. $\sqrt{80a} + \sqrt{125a}$

4. $\sqrt{98a} + \sqrt{32a}$

5. $\sqrt{50} - \sqrt{32}$

6. $\sqrt{75} + \sqrt{12}$

7. $\sqrt{50a^2} - \sqrt{200a^2}$

8. $\sqrt{48a^2} - \sqrt{363a^2}$

9. $2\sqrt{300} - 9\sqrt{12} - 7\sqrt{48}$

10. $\sqrt{175} + \sqrt{567} - \sqrt{63}$

11. $\sqrt[3]{40} + \sqrt[3]{625}$

12. $\sqrt[3]{54} + \sqrt[3]{16}$

13. $\sqrt[3]{81} - 3\sqrt[3]{375}$

14. $\sqrt[3]{24} - \sqrt[3]{81}$

15. $2\sqrt[3]{-24} - 4\sqrt[3]{-81} - \sqrt[3]{375}$

16. $10\sqrt[3]{-40} - 2\sqrt[3]{-135} + 4\sqrt[3]{-320}$

17. $\sqrt[3]{3a} - \sqrt[3]{24a} + \sqrt[3]{375a}$

18. $\sqrt[3]{r^5} - \sqrt[3]{8r^5} - r\sqrt[3]{64r^2}$

19. $\dfrac{3\sqrt[3]{3}}{2} - \dfrac{\sqrt[3]{3}}{3}$

20. $\dfrac{4}{5} - \dfrac{\sqrt[3]{2}}{2}$

21. $\sqrt{\dfrac{1}{2}} + \sqrt{\dfrac{1}{3}} + \sqrt{\dfrac{1}{6}}$

22. $\sqrt{\dfrac{25}{3}} - 2\sqrt{\dfrac{16}{3}} + 2\sqrt{\dfrac{4}{3}}$

23. $3\sqrt{\dfrac{1}{12}} - \sqrt{\dfrac{1}{15}} + 5\sqrt{\dfrac{3}{5}}$

24. $\sqrt{\dfrac{2}{3}} - \sqrt{\dfrac{1}{6}} + \sqrt{\dfrac{1}{2}}$

25. $\sqrt{\dfrac{x}{y}} + \sqrt{\dfrac{y}{x}} - \sqrt{\dfrac{1}{xy}}$

26. $2\sqrt[3]{\dfrac{1}{5}} + 6\sqrt[3]{\dfrac{1}{40}}$

27. $6\sqrt[3]{\dfrac{3}{5}} + 6\sqrt[3]{\dfrac{81}{40}}$

28. $2a\sqrt[3]{\dfrac{a}{5}} + 6\sqrt[3]{\dfrac{a^4}{40}}$

29. $\sqrt[3]{\dfrac{2}{3x}} - \sqrt[3]{\dfrac{3}{32x}} - \sqrt[3]{\dfrac{-2}{9x}}$

30. $3\sqrt[3]{\dfrac{x^5}{4}} - 3x\sqrt[3]{\dfrac{x^2}{108}}$

B **In Problems 31–64, perform the indicated operations. (Where the index is even, assume all variables are positive.)**

31. $3(5 - \sqrt{2})$

32. $-2(\sqrt{2} - 3)$

33. $\sqrt[3]{2}(\sqrt[3]{4} + 3)$

34. $\sqrt[3]{3}(\sqrt[3]{9} + 2)$

35. $2\sqrt{3}(7\sqrt{5} + 5\sqrt{3})$

36. $2\sqrt{5}(5\sqrt{2} + 3\sqrt{5})$

37. $3\sqrt[3]{5}(2\sqrt[3]{3} - \sqrt[3]{25})$

38. $4\sqrt[3]{2}(3\sqrt[3]{4} - 3\sqrt[3]{2})$

39. $-4\sqrt{7}(2\sqrt{3} - 5\sqrt{2})$

40. $-3\sqrt{2}(5\sqrt{7} - 2\sqrt{3})$

41. $(5\sqrt{3} + \sqrt{5})(3\sqrt{3} + 2\sqrt{5})$

42. $(2\sqrt{2} + 5\sqrt{3})(3\sqrt{2} + \sqrt{3})$

43. $(3\sqrt{6} - 2\sqrt{3})(4\sqrt{6} + 5\sqrt{3})$

44. $(3\sqrt{5} - 2\sqrt{3})(2\sqrt{5} + 3\sqrt{3})$

45. $(7\sqrt{5} - 11\sqrt{7})(5\sqrt{5} + 8\sqrt{7})$

46. $(2\sqrt{3} - 5\sqrt{2})(3\sqrt{3} + 2\sqrt{2})$

47. $(1 + \sqrt{2})(1 - \sqrt{2})$

48. $(2 + \sqrt{3})(2 - \sqrt{3})$

49. $(2 + 3\sqrt{3})(2 - 3\sqrt{3})$

50. $(5 + 5\sqrt{2})(5 - 5\sqrt{2})$

51. $(\sqrt{3} + \sqrt{2})^2$

52. $(\sqrt{2} + \sqrt{3})^2$

53. $(a + \sqrt{b})^2$

54. $(\sqrt{a} + b)^2$

55. $(\sqrt{3} - \sqrt{2})^2$

56. $(\sqrt{2} - \sqrt{3})^2$

57. $(a - \sqrt{b})^2$

58. $(\sqrt{b} - a)^2$

59. $(\sqrt{a} - \sqrt{b})^2$

60. $(\sqrt{b} - \sqrt{a})^2$

61. $\dfrac{3 + \sqrt{18}}{3}$

62. $\dfrac{5 + \sqrt{50}}{5}$

63. $\dfrac{6 - \sqrt{27}}{12}$

64. $\dfrac{8 - \sqrt{32}}{4}$

C **In Problems 65–75, rationalize the denominator. (Assume all variables represent positive numbers.)**

65. $\dfrac{3 + \sqrt{3}}{\sqrt{2}}$

66. $\dfrac{2 + \sqrt{5}}{\sqrt{3}}$

67. $\dfrac{2}{3 - \sqrt{2}}$

68. $\dfrac{6}{2 - \sqrt{2}}$

69. $\dfrac{4a}{3 - \sqrt{5}}$

70. $\dfrac{3a}{4 - \sqrt{3}}$

71. $\dfrac{3a + 2b}{3 + \sqrt{2}}$

72. $\dfrac{5a + b}{2 + \sqrt{3}}$

73. $\dfrac{\sqrt{a} + b}{\sqrt{a} - b}$

74. $\dfrac{a + \sqrt{b}}{a - \sqrt{b}}$

75. $\dfrac{\sqrt{a} + \sqrt{2b}}{\sqrt{a} - \sqrt{2b}}$

SKILL CHECKER

In Problems 76–80, solve the equation.

76. $x + 5 = 9$ **77.** $2x + 3 = 25$

78. $x^2 - 15x + 50 = 0$ **79.** $x^2 - 3x + 2 = 0$

80. $x^2 + 6x + 5 = 0$

USING YOUR KNOWLEDGE — Rationalizing Numerators in Calculus

In this section we learned how to rationalize the denominator of a fraction involving the sum or difference of radical expressions. In calculus, we sometimes have to rationalize the *numerator* of a fraction that involves the sum or difference of radical expressions. The idea is the same: Multiply numerator and denominator by the *conjugate* of the numerator. To rationalize the numerator in

$$\frac{\sqrt{3} + \sqrt{2}}{5}$$

we proceed as follows.

$$\frac{\sqrt{3} + \sqrt{2}}{5} = \frac{(\sqrt{3} + \sqrt{2})(\sqrt{3} - \sqrt{2})}{5(\sqrt{3} - \sqrt{2})}$$

Multiply numerator and denominator by the conjugate of the numerator.

$$(\sqrt{3} + \sqrt{2})(\sqrt{3} - \sqrt{2}) = (\sqrt{3})^2 - (\sqrt{2})^2$$

$$= \frac{3 - 2}{5(\sqrt{3} - \sqrt{2})}$$

$$= \frac{1}{5(\sqrt{3} - \sqrt{2})}$$

In Problems 81–90, rationalize the numerator.

81. $\dfrac{\sqrt{5} + \sqrt{2}}{3}$ **82.** $\dfrac{\sqrt{5} + \sqrt{3}}{4}$

83. $\dfrac{\sqrt{x} - \sqrt{2}}{5}$ **84.** $\dfrac{\sqrt{5} - \sqrt{x}}{5}$

85. $\dfrac{\sqrt{x} + \sqrt{y}}{x}$ **86.** $\dfrac{\sqrt{x} - \sqrt{y}}{x}$

87. $\dfrac{\sqrt{x} + \sqrt{y}}{\sqrt{x}}$ **88.** $\dfrac{\sqrt{x} + \sqrt{y}}{\sqrt{y}}$

89. $\dfrac{\sqrt{x} - \sqrt{y}}{\sqrt{x}}$ **90.** $\dfrac{\sqrt{x} - \sqrt{y}}{\sqrt{y}}$

WRITE ON . . .

91. Why is it impossible to combine $\sqrt{3x} + \sqrt[3]{3x}$ into a single term?

92. Explain why $\sqrt{a + b} \neq \sqrt{a} + \sqrt{b}$ and give examples.

93. Explain why $\sqrt[3]{x} \cdot \sqrt[3]{x} \neq x$. What factor do you need in the box, $\sqrt[3]{x} \cdot \sqrt[3]{x} \cdot \square = x$, to make the statement true?

94. What does it mean when we say "rationalize the denominator"?

95. State what conditions have to be met for a radical expression to be simplified.

MASTERY TEST

If you know how to do these problems, you have learned your lesson!

Rationalize the denominator:

96. $\dfrac{\sqrt{y}}{\sqrt{y} + \sqrt{x}}$ **97.** $\dfrac{\sqrt{y}}{\sqrt{y} - \sqrt{x}}$

Reduce to lowest terms:

98. $\dfrac{10 + \sqrt{50}}{5}$ **99.** $\dfrac{20 + \sqrt{32}}{4}$

Perform the indicated operations:

100. $(\sqrt{27} + \sqrt{28})(\sqrt{75} - \sqrt{112})$

101. $(\sqrt{28} - \sqrt{27})(\sqrt{112} + \sqrt{75})$

102. $\sqrt{2}(\sqrt{3} + \sqrt{10})$

103. $\sqrt{5x}(\sqrt{x} - \sqrt{3}), \quad x \geq 0$

104. $\sqrt[3]{2x}(\sqrt[3]{4x^2} - \sqrt[3]{16x})$ **105.** $5\sqrt{\dfrac{1}{2}} - 7\sqrt{\dfrac{1}{8}}$

106. $4\sqrt[3]{\dfrac{5x}{4x^2}} - \sqrt[3]{\dfrac{5}{32x}}$ **107.** $2\sqrt[3]{250x} - 4\sqrt[3]{16x}$

108. $3\sqrt{20x} - 5\sqrt{45x}$ **109.** $\sqrt{98} - \sqrt{50}$

110. $\sqrt{44} + \sqrt{99}$

5.4

SOLVING EQUATIONS CONTAINING RADICALS

To succeed, review how to:

1. Solve linear and quadratic equations (pp. 69, 186–190).

2. Square a radical expression (p. 303).

Objectives:

A. Solve equations involving radicals.

B. Solve applications requiring the solution of radical equations.

getting started

Radical Curves

If a traffic engineer wants the speed limit v on this curve to be 45 mi/hr, what radius should the curve have? The speed v (in miles per hour) a car can travel on a concrete highway curve without skidding is $v = \sqrt{9r}$, where r is the radius of the curve in feet. Since $v = 45$, to find the answer we must find r in the equation

$$45 = \sqrt{9r}$$
$$(45)^2 = (\sqrt{9r})^2 \qquad \text{Square both sides.}$$
$$2025 = 9r \qquad (\sqrt{9r})^2 = 9r$$
$$\frac{2025}{9} = r \qquad \text{Divide by 9.}$$
$$225 = r$$

Thus the curve must have a radius r of 225 ft or more. We can check this by substituting 225 for r in the equation

$$45 = \sqrt{9r} \quad \text{to obtain} \quad 45 = \sqrt{9 \cdot 225} = \sqrt{9} \cdot \sqrt{225} = 3 \cdot 15 \qquad \text{A true statement}$$

Thus the curve must have at least a 225-ft radius.

Solving Equations Containing Radicals

In algebra, the equation $45 = \sqrt{9r}$ is called a **radical equation** and can be solved by squaring both sides of the equation. Sometimes, however, squaring both sides introduces **extraneous solutions**—that is, solutions that do not satisfy the original equation. For example, the equation $x = 2$ has one solution, 2. Squaring both sides gives $x^2 = 4$. The equation $x^2 = 4$ has two solutions, 2 and -2.

We introduced the extraneous solution -2 when we squared both sides. However, all solutions of the equation $x = 2$ are solutions of $x^2 = 4$.

POWER RULE FOR EQUATIONS

All solutions of the equation $P = Q$ are solutions of the equation $P^n = Q^n$, where n is a natural number.

This rule tells us that when we raise both sides of an equation to a power, the solutions of the *original* equation are *always* solutions of the new equation. However, the new equation may have extraneous solutions that have to be discarded. Because of this, *the solutions of the new equations must be checked in the original equation and extraneous solutions discarded.* Here is the procedure we need to solve equations containing radicals.

PROCEDURE

> **To Solve Equations Containing Radicals**
>
> 1. **Isolate** one radical that contains variables on one side of the equation.
> 2. **Raise** each side of the equation to a power that is the same as the index of the radical.
> 3. **Simplify.**
> 4. **Repeat** steps 1–3 *if* the equation still contains a radical term.
> 5. **Solve** the resulting linear or quadratic equation using the appropriate methods.
> 6. **Check** all proposed (trial) solutions in the original equation.

EXAMPLE 1   Solving equations containing one radical

Solve:

a. $\sqrt{x + 5} = 3$ **b.** $\sqrt{x + 1} = x - 1$ **c.** $\sqrt{x - 1} - x = -1$

SOLUTION

a. We use the six-step procedure. In this case, the radical containing the variable is already isolated.

1. $\sqrt{x + 5} = 3$ Given.

2. $(\sqrt{x + 5})^2 = 3^2$ Square each side.

3. $x + 5 = 9$ $(\sqrt{x + 5})^2 = x + 5$

4. There are no radical terms left.

5. $x = 4$ Subtract 5 on both sides.

6. Substituting the proposed solution 4 for x in the original equation gives

$$\sqrt{4 + 5} \overset{?}{=} 3$$

$$\sqrt{9} = 3$$

a true statement. Thus the solution of $\sqrt{x + 5} = 3$ is 4.

b. Using the six-step procedure, we note that the radical is already isolated, so we square both sides to eliminate the radical.

1. $\sqrt{x + 1} = x - 1$ Given.

2. $(\sqrt{x + 1})^2 = (x - 1)^2$ Square each side.

3. $x + 1 = x^2 - 2x + 1$ Expand $(x - 1)^2$.

4. There are no radical terms left.

5. $0 = x^2 - 3x$ Subtract x and 1 to write the resulting quadratic equation in standard form.

$0 = x(x - 3)$ Factor.

$x = 0$ or $x - 3 = 0$ Use the zero-factor property.

$x = 0$ or $x = 3$ Solve $x - 3 = 0$.

Thus the proposed solutions are 0 and 3.

6. Substituting 0 for x in the original equation, we have

$$\sqrt{0 + 1} \overset{?}{=} 0 - 1$$

$$1 \overset{?}{=} -1$$ A false statement

Thus zero is not a solution. Substituting 3 for x in $\sqrt{x + 1} = x - 1$, we have

$$\sqrt{3 + 1} \stackrel{?}{=} 3 - 1$$

$$\sqrt{4} = 2 \qquad \text{A true statement}$$

The solution of $\sqrt{x + 1} = x - 1$ is 3. (Discard the extraneous solution, 0!)

c. We first have to isolate the radical on one side (step 1), so we start by adding x on both sides. Then we proceed as before.

	$\sqrt{x - 1} - x = -1$	Given.
1.	$\sqrt{x - 1} = x - 1$	Add x on both sides. (Now $\sqrt{x - 1}$ is isolated.)
2.	$(\sqrt{x - 1})^2 = (x - 1)^2$	Square each side.
3.	$x - 1 = x^2 - 2x + 1$	Expand on the right.

4. There are no radicals left.

5.	$0 = x^2 - 3x + 2$	Subtract x and add 1.
	$0 = (x - 2)(x - 1)$	Factor.
	$x - 2 = 0 \quad \text{or} \quad x - 1 = 0$	Use the zero-factor property.
	$x = 2 \quad \text{or} \quad x = 1$	Solve.

6. The proposed solutions are 2 and 1. Substituting 2 for x, we have

$$\sqrt{x - 1} - x = -1$$

$$\sqrt{2 - 1} - 2 \stackrel{?}{=} -1$$

$$\sqrt{1} - 2 = -1 \qquad \text{A true statement}$$

Thus 2 is a solution. Now we check the proposed solution 1 by substitution.

$$\sqrt{x - 1} - x = -1$$

$$\sqrt{1 - 1} - 1 \stackrel{?}{=} -1$$

$$\sqrt{0} - 1 = -1 \qquad \text{A true statement}$$

Thus 1 is also a solution. The solutions are 1 and 2. ∎

Sometimes we have radicals on both sides of the equation. In such cases, we must isolate one of the radicals and raise both sides of the equation to the appropriate power. Since we still have radicals on one side of the equation, we isolate them and square again (step 4 in the procedure). Thus to solve $\sqrt{x - 11} = \sqrt{x} - 1$, we first square both sides of the equation. Note that the right-hand side contains an expression of the form $(x - a)^2 = x^2 - 2ax + a^2$. Using our six-step procedure, we have

1.	$\sqrt{x - 11} = \sqrt{x} - 1$	$\sqrt{x - 11}$ is isolated.
2.	$(\sqrt{x - 11})^2 = (\sqrt{x} - 1)^2$	Square each side.
3.	$x - 11 = x - 2 \cdot 1 \cdot \sqrt{x} + 1$	Simplify.
	$x - 11 = x - 2\sqrt{x} + 1$	
4.	$-11 = -2\sqrt{x} + 1$	Subtract x.
	$-12 = -2\sqrt{x}$	Subtract 1.
	$6 = \sqrt{x}$	Divide by -2.
5.	$36 = x$	Square each side.

Since we still have a radical term ($\sqrt{x}$), we isolate $\sqrt{x}$. This is step 4 in the procedure.

6. The solution is 36. You can verify this by substituting in the original equation.

EXAMPLE 2   Solving equations containing two radicals

Solve: $\sqrt{x-5} - \sqrt{x} = -1$

SOLUTION   In step 1, we add $\sqrt{x}$ to both sides of the equation so that $\sqrt{x-5}$ is isolated.

	$\sqrt{x-5} - \sqrt{x} = -1$	Given.
1.	$\sqrt{x-5} = \sqrt{x} - 1$	Add $\sqrt{x}$ to isolate $\sqrt{x-5}$.
2.	$x - 5 = (\sqrt{x} - 1)^2$	Square each side.
3.	$x - 5 = x - 2\sqrt{x} + 1$	$(\sqrt{x} - 1)^2 = x - 2\sqrt{x} + 1$
4.	$-5 = -2\sqrt{x} + 1$	Subtract x.
	$-6 = -2\sqrt{x}$	Subtract 1. We have to isolate $\sqrt{x}$.
	$3 = \sqrt{x}$	Divide both sides by -2.
5.	$9 = x$	Square each side.

6. The proposed solution is 9. Since $\sqrt{9-5} - \sqrt{9} = \sqrt{4} - \sqrt{9} = 2 - 3 = -1$, 9 is the correct solution.   ■

The sides of an equation can be raised to powers greater than 2. For example, to solve the equation $\sqrt[4]{x} = 2$, we raise both sides of the equation to the fourth power. (Step 2 says that we have to *raise* both sides of the equation to a power that is the same as the index of the radical, which is 4.) Thus

$$(\sqrt[4]{x})^4 = 2^4$$

$$x = 16$$

Here is another example.

EXAMPLE 3   Solving equations containing a cube root

Solve: $\sqrt[3]{x-2} = 3$

SOLUTION   The radical is isolated, so we go to step 2.

2. $(\sqrt[3]{x-2})^3 = 3^3$ Cube each side.

3. $x - 2 = 27$ Simplify.

4. There are no radicals left.

5. $x = 29$ Add 2.

6. Substituting 29 for x in the original equation gives

$$\sqrt[3]{29 - 2} \stackrel{?}{=} 3$$

$$\sqrt[3]{27} = 3$$

Since this statement is true, 29 is the solution of $\sqrt[3]{x-2} = 3$.   ■

EXAMPLE 4   Solving equations containing a fourth root

Solve: $\sqrt[4]{x-1} + 3 = 0$

SOLUTION   We use our six-step procedure.

$$\sqrt[4]{x-1} + 3 = 0 \qquad \text{Given.}$$

1. $\sqrt[4]{x-1} = -3$ Subtract 3 to isolate $\sqrt[4]{x-1}$.

2. $(\sqrt[4]{x-1})^4 = (-3)^4$ Raise to the fourth power.

3. $x - 1 = 81$ Simplify.

4. There are no radicals left.

5. $x = 82$ Add 1 to solve $x - 1 = 81$.

6. Substitute 82 for x in $\sqrt[4]{x-1} + 3 = 0$ to obtain

$$\sqrt[4]{82-1} + 3 = 0$$

$$\sqrt[4]{81} + 3 = 0$$

$$3 + 3 = 0 \qquad \text{(False!)}$$

Thus 82 is an extraneous solution. There is no real-number solution for this equation. The solution set is $\varnothing$.

Now look at step 1. Can you tell that there is no solution to the equation? Why?

■

B   Solving Applications Involving Radicals

A common use of radicals occurs in problems involving free-falling objects. (See Problems 43 and 44.) Here is an example.

EXAMPLE 5   Free-falling object

If an object is dropped from a height of h feet, the relationship between its velocity v (in feet per second) when it hits the ground and the height h is $v^2 = 2gh$, where $g = 32$ feet per second per second.
a. Solve for v.
b. If an object is dropped from a height of 81 ft, what is its velocity when it hits the ground?

SOLUTION
a. We have to solve the equation $v^2 = 2gh$ for v. Taking the square root of both sides of $v^2 = 2gh$, we have $v = \sqrt{2gh}$.
b. Here we have to find v when $h = 81$ and $g = 32$. Substituting 81 for h and 32 for g in $v = \sqrt{2gh}$, we obtain

$$v = \sqrt{2gh} = \sqrt{2 \cdot 32 \cdot 81} = \sqrt{64 \cdot 81} = 8 \cdot 9 = 72$$

Thus the velocity v of the object when it hits the ground is 72 ft/sec. (Remember that v is in feet per second.)

■

EXAMPLE 6   Call lengths in cell phones

Do you have a "cell" phone? How long are your calls? According to the Cellular Telecommunications Industry Association, the average length of a call for 1990, 1991, and 1992 was 2.20, 2.38, and 2.58 min, respectively. The average length can be approximated by $L(t) = \sqrt{t} + 5$, where $L(t)$ represents the length of the call in minutes t years after 1990.

a. Use the formula to approximate the average length of a call for 1990. (The actual length was 2.20 min).

b. In what year would you expect the average length of a call to be 3 min?

SOLUTION

a. Since t is the number of years after 1990, in 1990 $t = 0$ and $L(0) = \sqrt{0 + 5} = 2.24$ min.

b. To predict when the call length will be 3 min, we have to find t when $L(t) = 3$. Thus we have to solve the equation

$$\sqrt{t + 5} = 3$$

$$t + 5 = 9 \qquad \text{Square both sides.}$$

$$t = 4 \qquad \text{Subtract 5 from both sides.}$$

Thus 4 years after 1990, that is, in 1994, the average length of a cellular call was expected to be 3 min. ∎

GRAPH IT

In this section we shall use the **solve** feature of your grapher. On a TI-82, this feature is found in your [MATH] menu, under number 0. But even with a "solver" you have to be careful!

Let's try Example 1c. To solve an equation with the solve feature, you must write the equation in the form $A(x) = 0$. Since the given equation is $\sqrt{x - 1} - x = -1$, add 1 to both sides to obtain $\sqrt{x - 1} - x + 1 = 0$. From the home screen, start by pressing [MATH] 0; then enter ($\sqrt{x - 1} - x + 1$, X, 3). This means that the left-hand side of the equation is $\sqrt{x - 1} - x + 1$, the variable is X, and your guess for the solution is 3. Press [ENTER] and the answer 2 appears (see Window 1). But we need a new guess to get other possible answers. Press [2nd] [ENTER] and place the cursor over the 3. Enter your new guess, say, 0, and press [ENTER]. What did you get? An error message! Why? Let's look at the graph of $Y_1 = \sqrt{x - 1} - x + 1$ using a window $[-1, 3]$ by $[-1, 2]$, as shown in Window 2. The graph is defined for values of x that are greater than or equal to 1. You can see that by looking at the term $\sqrt{x - 1}$, which is defined only when $x - 1$ is nonnegative—that is, when x is greater than or equal to 1! This means that our guess of 0 is not an acceptable value. (Note that when $x = 0$, $\sqrt{x - 1} = \sqrt{0 - 1} = \sqrt{-1}$, which is not a real number). Thus when solving an equation using the solve feature, it's a good idea to graph the equation and make sure that your guesses for x are based on the graph.

Now let's pick a number near 1, but less than 2, say, 1.1. You get the answer 2 again. Since the graph indicates that one of the answers may be 1, pick 1 for your guess. Finally, you get the second answer, 1. If this method is not to your liking, there's another possibility. First, isolate the radical in Example 1c, to obtain $\sqrt{x - 1} = x - 1$. Now graph $Y_1 = \sqrt{x - 1}$ and $Y_2 = x - 1$. The value of x where the two curves intersect is a solution of the equation. Can you see from Window 3 that the two points are $x = 1$ and $x = 2$? (With a TI-82, you can find the point of intersection by pressing [2nd] [CALC] 5 and following the prompts. Try it!)

Our moral here is: It's probably easier to do these problems algebraically!

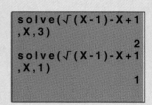

WINDOW 1

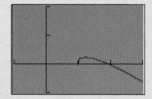

WINDOW 2

$Y_1 = \sqrt{x - 1}$

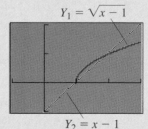

$Y_2 = x - 1$

WINDOW 3

EXERCISE 5.4

A In Problems 1–20, solve the given equation.

1. $\sqrt{x} = 4$

2. $\sqrt{3x} = 6$

3. $\sqrt{x + 6} = 7$

4. $\sqrt{x - 3} = 10$

5. $\sqrt{\dfrac{x}{2}} = 3$

6. $\sqrt{\dfrac{3x}{2}} = 3$

7. $\sqrt[4]{x + 1} + 2 = 0$

8. $\sqrt[4]{x + 3} + 5 = 0$

9. $\sqrt[3]{3x - 1} = \sqrt[3]{5x - 7}$ 10. $\sqrt[3]{5x - 3} = \sqrt[3]{7x - 5}$

11. $\sqrt{x + 4} = x + 2$ 12. $\sqrt{x + 3} = x + 1$

13. $\sqrt{x + 3} = x - 3$ 14. $\sqrt{x + 9} = x - 3$

15. $\sqrt[3]{y + 8} = -2$ 16. $\sqrt[3]{y + 4} = -1$

17. $\sqrt{x + 5} - x = -7$ 18. $\sqrt{x + 5} - x = -1$

19. $\sqrt{x - 5} - x = -7$ 20. $\sqrt{x - 1} - x = -3$

In Problems 21–30, you are required to square twice to eliminate all radicals.

21. $\sqrt{y + 1} = \sqrt{y} + 1$ 22. $\sqrt{y - 4} = 2 + \sqrt{y}$

23. $\sqrt{y + 8} - \sqrt{y} = 2$ 24. $\sqrt{y + 5} - \sqrt{y} = 1$

25. $\sqrt{x + 3} = \sqrt{x} + \sqrt{3}$ 26. $\sqrt{x + 5} = \sqrt{x} + \sqrt{5}$

27. $\sqrt{5x - 1} + \sqrt{x + 3} = 4$ 28. $\sqrt{2x - 1} + \sqrt{x + 3} = 3$

29. $\sqrt{x - 3} + \sqrt{2x + 1} = 2\sqrt{x}$

30. $\sqrt{x + 4} + \sqrt{3x + 9} = \sqrt{x + 25}$

In Problems 31–40, solve for x or y.

31. $\sqrt{x - a} = b$ 32. $\sqrt{x + a} = b$

33. $\sqrt[3]{a - by} = c$ 34. $\sqrt[3]{a - by} = a b^3 - y$

35. $\sqrt{\dfrac{x}{a}} = b$ 36. $\sqrt{\dfrac{x}{b}} = \dfrac{a}{b}$

37. $\sqrt{\dfrac{a}{b - x}} = \sqrt{b}$ 38. $\sqrt{\dfrac{b}{a - x}} = \sqrt{a}$

39. $\sqrt[3]{3x - a} = \sqrt[3]{b - a}$ 40. $\sqrt[3]{2x - b} = \sqrt[3]{b - 2a}$

B Solve these applications.

41. The radius r of a sphere is given by

$$r = \sqrt{\dfrac{S}{4\pi}}$$

where S is the surface area. If the surface area of a sphere is 942 ft^2, find its radius. (Use $\pi = 3.14$.)

42. The radius r of a cone is given by

$$r = \sqrt{\dfrac{3V}{\pi h}}$$

where V is the volume and h is the height. If a 10-cm-high cone contains 94.26 cm^3 of ice cream, what is its radius? (Use $\pi = 3.142$.)

43. The time t (in seconds) it takes a body to fall d feet is given by

$$t = \sqrt{\dfrac{2d}{g}}$$

where g is the gravitational acceleration.
a. Solve for d.
b. How far would a body fall in 3 sec? (Use $g = 32.2$ ft/sec^2.)

44. After traveling d feet, the velocity v (in feet per second) of a falling body starting from rest is given by $v = \sqrt{2gd}$.
a. Solve for d.
b. If a body that started from rest is traveling at 44 ft/sec, how far has it fallen? (Use $g = 32$ ft/sec^2.)

45. A pendulum of length L (in feet) takes

$$t = 2\pi \sqrt{\dfrac{L}{g}}$$

seconds to go through a complete cycle.
a. Solve for L.
b. If a pendulum takes 2 sec to go through one complete cycle, how long is the pendulum? $\left(\text{Let } g = 32 \text{ ft/sec}^2 \text{ and } \pi = \frac{22}{7}.\right)$

SKILL CHECKER

Simplify by removing parentheses and collecting like terms:

46. $(5 + 4x) + (7 - 2x)$ 47. $(3 + 4x) + (8 + 2x)$

48. $(9 + 2x) - (2 + 4x)$ 49. $(6 + 5x) - (7 - 3x)$

50. $(8 + 3x) - (5 - 4x)$

Rationalize the denominator:

51. $\dfrac{2 + 3\sqrt{2}}{4 + \sqrt{2}}$ 52. $\dfrac{4 + 4\sqrt{3}}{3 + 2\sqrt{3}}$

53. $\dfrac{2 - \sqrt{2}}{5 - 3\sqrt{2}}$ 54. $\dfrac{3 - \sqrt{3}}{5 - \sqrt{3}}$

55. $\dfrac{\sqrt{x} - \sqrt{y}}{\sqrt{x} + \sqrt{y}}$

USING YOUR KNOWLEDGE   **The Roads with Radicals**

Suppose you are the engineer designing several roads. We mentioned at the beginning of this section that the velocity v (miles per hour) that a car can travel on a concrete highway curve without skidding is $v = \sqrt{9r}$, where r (in feet) is the radius of the curve.

Use your knowledge to determine the radius of the curve on a highway exit in which you want the speed to be as follows.

56. 25 mi/hr   **57.** 30 mi/hr

58. 35 mi/hr   **59.** 40 mi/hr

WRITE ON . . .

60. Consider the equation $\sqrt{x+1} + 2 = 0$.
   a. What should the first step be in solving this equation?
   b. List reasons that show that this equation has no real-number solutions.

61. Consider the equation $\sqrt{x+3} = -\sqrt{2x-3}$. List reasons that show that this equation has no real-number solutions.

62. What is your definition of a "proposed" or "trial" solution when you solve equations involving radicals?

63. Why is it necessary to check proposed solutions in the original equation when you solve equations involving radicals?

MASTERY TEST

If you know how to do these problems, you have learned your lesson!

Solve, if possible:

64. $\sqrt[3]{x-5} = 2$ **65.** $\sqrt[4]{x+3} + 16 = 0$

66. $\sqrt{x-3} - \sqrt{x} = -1$ **67.** $\sqrt{x+1} - x = 1$

68. $\sqrt{x+1} = x - 5$ **69.** $\sqrt{x+2} + 3 = 0$

70. $\sqrt{x-2} - 2 = 0$ **71.** $\sqrt{x} + \sqrt{2x+1} = 1$

72. The power used by an appliance is given by

$$I = \sqrt{\dfrac{P}{R}}$$

where I is the current (in amps), R is the resistance (in ohms), and P is the power (in watts).
   a. Solve for R.
   b. Find the resistance R for an electric oven rated at 1500 watts and drawing $I = 10$ amps of current.

5.5

COMPLEX NUMBERS

To succeed, review how to:

1. Remove parentheses and collect like terms in an expression (pp. 45–48).

2. Rationalize the denominator of an expression (pp. 18–20).

Objectives:

A Write the square root of a negative integer in terms of i.

B Add and subtract complex numbers.

C Multiply and divide complex numbers.

D Find powers of i.

---

getting started **Italian Sums and Products**

Can you find two numbers whose sum is 10 and whose product is 40? Girolamo Cardan (1501–1576), an Italian mathematician, claimed the answer is $5 + \sqrt{-15}$ and $5 - \sqrt{-15}$. Obviously, the sum is 10, but what about the product? Said Cardan, "Putting aside the mental tortures involved, multiply $5 + \sqrt{-15}$ by $5 - \sqrt{-15}$, making $25 - (-15)$, whence the product is 40." To see this better, let $a = 5$ and $b = \sqrt{-15}$. Now $(a + b)(a - b) = a^2 - b^2 = 5^2 - (\sqrt{-15})^2 = 25 - (-15)$.

What is the problem? Well, $\sqrt{-15}$ is not a real number! To solve this, Carl Friedrich Gauss (1777–1855) (at right) developed a new set of numbers containing elements that are square roots of negative numbers. One of these numbers is i, and it is defined as follows.

i is a number such that $i^2 = -1$; that is, $i = \sqrt{-1}$.

Writing Square Roots of Negative Numbers in Terms of i

With the preceding definition of i, the square root of any negative real number can be written as the product of a real number and i. Thus

$$\sqrt{-4} = \sqrt{-1} \cdot \sqrt{4} = i2 \quad \text{or} \quad 2i$$
$$\sqrt{-3} = \sqrt{-1 \cdot 3} = \sqrt{-1} \cdot \sqrt{3} = i\sqrt{3}$$

Since it is easy to confuse $\sqrt{3}i$ and $\sqrt{3i}$, when possible, we write products involving radicals and i as factors with the i in front; that is, we may write $i\sqrt{5}$ instead of $\sqrt{5}i$. Of course, both notations are acceptable.

EXAMPLE 1   Writing expressions in terms of i

Write the given expression in terms of i.

a. $\sqrt{-9}$ **b.** $\sqrt{-18}$

SOLUTION

a. $\sqrt{-9} = \sqrt{-1 \cdot 9} = \sqrt{-1}\sqrt{9} = 3i$
b. $\sqrt{-18} = \sqrt{-1 \cdot 18} = \sqrt{-1}\sqrt{18} = i\sqrt{18} = i\sqrt{9 \cdot 2} = i3\sqrt{2} = 3\sqrt{2}i$ ∎

The numbers $3i$ and $i3\sqrt{3} = 3\sqrt{3}i$ are called **pure imaginary numbers**. We can form a new set of numbers by adding these imaginary numbers to real numbers as follows.

COMPLEX NUMBER

> If a and b are real numbers, then any number of the form
>
> $$\underset{\underset{\text{Real part}}{\uparrow} \quad \underset{\text{Imaginary part}}{\uparrow}}{a + bi}$$
>
> is called a **complex number**.

In the complex number $a + bi$, a is called the *real* part and bi the *imaginary* part. Thus the number $-3 + 4i$ is a complex number whose real part is -3 and whose imaginary part is $4i$. Similarly, $2 - 3i$ is a complex number with 2 as its real part and $-3i$ as its imaginary part.

> **NOTE** Real numbers and pure imaginary numbers are also complex numbers. For example, the real numbers $\sqrt{2}$, 0, $-0.\overline{3}$, and $\frac{1}{5}$ are complex numbers. The pure imaginary numbers $\sqrt{2}i$, $-3i$, and $-\frac{4}{5}i$ are also complex numbers.

Adding and Subtracting Complex Numbers

To add (or subtract) complex numbers, we add (or subtract) the real parts and the imaginary parts separately. The rules for these operations are as follows.

RULES FOR ADDING AND SUBTRACTING COMPLEX NUMBERS

> For a, b, c, and d real numbers,
>
> $$(a + bi) + (c + di) = (a + c) + (b + d)i$$
> $$(a + bi) - (c + di) = (a - c) + (b - d)i$$

You will find that these operations are similar to combining like terms in a polynomial. For example, $(3 + 4i) + (8 + 2i) = (3 + 8) + (4 + 2)i = 11 + 6i$, and $(9 + 2i) - (2 + 4i) = (9 - 2) + (2 - 4)i = 7 - 2i$.

 The sum or difference of two complex numbers is always a complex number and should be written in the form $a + bi$.

EXAMPLE 2   Adding and subtracting complex numbers

Find:

a. $(5 + 4i) + (7 - 2i)$ **b.** $(6 + 5i) - (7 - 3i)$

SOLUTION

a. $(5 + 4i) + (7 - 2i) = (5 + 7) + [4 + (-2)]i$
$$= 12 + 2i$$

b. $(6 + 5i) - (7 - 3i) = (6 - 7) + [5 - (-3)]i$
$$= -1 + 8i$$ ■

Multiplying and Dividing Complex Numbers

The commutative, associative, and distributive properties of real numbers also apply to complex numbers. These properties can be used to find the product and quotient of complex numbers. In practice, we multiply complex numbers using the rule to multiply binomials (FOIL) and replacing i^2 by -1. Thus

$$\overset{\text{F}\quad\text{O}\quad\text{I}\quad\text{L}}{(3 + 4i)(2 + 3i) = 6 + 9i + 8i + 12i^2}$$

$$= 6 + 9i + 8i - 12 \qquad \text{Since } i^2 = -1, 12i^2 = -12.$$

$$= -6 + 17i$$

Note that the answer is written in the form $a + bi$ since the product of two complex numbers is always a complex number.

EXAMPLE 3   Multiplying complex numbers

Find the product:

a. $(2 - 5i)(3 + 7i)$ **b.** $-3i(4 - 7i)$

SOLUTION

a. $\overset{\text{F}\quad\text{O}\quad\text{I}\quad\text{L}}{(2 - 5i)(3 + 7i) = 6 + 14i - 15i - 35i^2}$

$$= 6 - i + 35 \qquad \text{Since } i^2 = -1, -35i^2 = 35.$$

$$= 41 - i$$

b. $-3i(4 - 7i) = -12i + 21i^2 \qquad \text{By the distributive property}$

$$= -12i - 21 \qquad \text{Since } i^2 = -1, 21i^2 = -21.$$

$$= -21 - 12i$$ ■

 Expressions such as $\sqrt{-9}$ and $\sqrt{-18}$ should be written in the form bi *before* any other operations are carried out.

For example, to multiply $\sqrt{-9} \cdot \sqrt{-4}$, we must first write $\sqrt{-9} \cdot \sqrt{-4} = 3i \cdot 2i$. Then, $3i \cdot 2i = 6i^2 = 6(-1) = -6$. If we were to use the product rule for radicals, we would get $\sqrt{-9} \cdot \sqrt{-4} = \sqrt{(-9) \cdot (-4)} = \sqrt{36} = 6$. This is *not correct!* Remember to write $\sqrt{-9}$ as $3i$ and $\sqrt{-4}$ as $2i$ *before* multiplying.

EXAMPLE 4   Multiplying square roots of negative numbers

Find:

a. $\sqrt{-16}(3 + \sqrt{-8})$ **b.** $\sqrt{-36}(\sqrt{-3} - \sqrt{-18})$

SOLUTION

a. We first write the square roots of negative numbers in terms of i and then proceed as usual. Since $\sqrt{-16} = 4i$ and $\sqrt{-8} = i2\sqrt{2} = 2i\sqrt{2}$, we write

$$\sqrt{-16}(3 + \sqrt{-8}) = 4i(3 + 2i\sqrt{2}) \qquad \sqrt{-16} = 4i \text{ and } \sqrt{-8} = 2i\sqrt{2} \text{ or } 2\sqrt{2}i$$

$$= 12i + 8i^2\sqrt{2} \qquad \text{Use the distributive property.}$$

$$= 12i - 8\sqrt{2} \qquad \text{Since } i^2 = -1,\ 8\sqrt{2}i^2 = -8\sqrt{2}.$$

$$= -8\sqrt{2} + 12i \qquad \text{Write in the form } a + bi.$$

b. Since $\sqrt{-36} = 6i$ and $\sqrt{-18} = i3\sqrt{2} = 3i\sqrt{2}$, we write

$$\sqrt{-36}(\sqrt{-3} - \sqrt{-18}) = 6i(i\sqrt{3} - 3i\sqrt{2})$$

$$= 6i^2\sqrt{3} - 18i^2\sqrt{2}$$

$$= -6\sqrt{3} + 18\sqrt{2} \qquad \blacksquare$$

To find the *quotient* of two complex numbers, we use the rationalizing process developed in Section 5.2 and the assumption that

$$\frac{a + bi}{c} = \frac{a}{c} + \frac{bi}{c}$$

For example, to find

$$\frac{2 + 3i}{4 - i}$$

we proceed as follows.

$$\frac{2 + 3i}{4 - i} = \frac{(2 + 3i)(4 + i)}{(4 - i)(4 + i)} \qquad \text{Multiply the numerator and denominator by the conjugate of } 4 - i, \text{ that is, } 4 + i.$$

$$= \frac{8 + 2i + 12i + 3i^2}{16 + 4i - 4i - i^2}$$

$$= \frac{8 + 2i + 12i - 3}{16 + 4i - 4i + 1}$$

$$= \frac{5 + 14i}{17}$$

$$= \frac{5}{17} + \frac{14}{17}i$$

PROCEDURE

Dividing One Complex Number by Another

To divide one complex number by another, multiply the numerator and denominator by the conjugate of the denominator. Note that the conjugate of $a + bi$ is $a - bi$ and $(a + bi)(a - bi) = a^2 - (bi)^2 = a^2 + b^2$.

EXAMPLE 5   Dividing complex numbers

Find:

a. $\dfrac{5 + 4i}{3 + 2i}$ **b.** $\dfrac{2 - 4i}{5 - 3i}$ **c.** $\dfrac{3 - 2i}{i}$

SOLUTION

a. $\dfrac{5 + 4i}{3 + 2i} = \dfrac{(5 + 4i)(3 - 2i)}{(3 + 2i)(3 - 2i)}$ Multiply by the conjugate of $3 + 2i$.

$\phantom{\dfrac{5 + 4i}{3 + 2i}} = \dfrac{15 - 10i + 12i - 8i^2}{3^2 + 2^2}$ Since $(3 + 2i)(3 - 2i) = 3^2 + 2^2$

$\phantom{\dfrac{5 + 4i}{3 + 2i}} = \dfrac{15 - 10i + 12i + 8}{13}$

$\phantom{\dfrac{5 + 4i}{3 + 2i}} = \dfrac{23 + 2i}{13}$

$\phantom{\dfrac{5 + 4i}{3 + 2i}} = \dfrac{23}{13} + \dfrac{2}{13}i$

b. $\dfrac{2 - 4i}{5 - 3i} = \dfrac{(2 - 4i)(5 + 3i)}{(5 - 3i)(5 + 3i)}$

$\phantom{\dfrac{2 - 4i}{5 - 3i}} = \dfrac{10 + 6i - 20i - 12i^2}{5^2 + 3^2}$

$\phantom{\dfrac{2 - 4i}{5 - 3i}} = \dfrac{10 + 6i - 20i + 12}{34}$

$\phantom{\dfrac{2 - 4i}{5 - 3i}} = \dfrac{22 - 14i}{34}$

$\phantom{\dfrac{2 - 4i}{5 - 3i}} = \dfrac{22}{34} - \dfrac{14}{34}i$

$\phantom{\dfrac{2 - 4i}{5 - 3i}} = \dfrac{11}{17} - \dfrac{7}{17}i$

c. Since the conjugate of $a + bi$ is $a - bi$, the conjugate of $0 + 1i$ is $0 - 1i = -i$. Multiplying numerator and denominator of the fraction by $-i$, we have

$\dfrac{3 - 2i}{i} = \dfrac{(3 - 2i)(-i)}{i \cdot (-i)}$

$\phantom{\dfrac{3 - 2i}{i}} = \dfrac{-3i + 2i^2}{-i^2}$

$\phantom{\dfrac{3 - 2i}{i}} = \dfrac{-3i - 2}{1}$

$\phantom{\dfrac{3 - 2i}{i}} = -2 - 3i$ ∎

D | Finding Powers of i

We already know that, by definition, $i^2 = -1$. If we assume that the laws of exponents hold, we can write any power of i as one of the numbers 1, -1, i, or $-i$. Thus

$$i^1 = i \qquad\qquad i^5 = i \cdot i^4 = i \cdot (1) = i$$

$$i^2 = -1 \qquad\qquad i^6 = i \cdot i^5 = i \cdot i = -1$$

$$i^3 = i \cdot i^2 = i(-1) = -i \qquad i^7 = i \cdot i^6 = i(-1) = -i$$

$$i^4 = i^2 \cdot i^2 = (-1)(-1) = 1 \qquad i^8 = i \cdot i^7 = i \cdot (-i) = 1$$

Since $i^4 = 1$, the easiest way to simplify higher powers of i is to write them in terms of i^4. Thus to find i^{20}, we write

$$i^{20} = (i^4)^5 = (1)^5 = 1$$

Similarly,

$$i^{21} = (i^4)^5 \cdot i = 1 \cdot i = i$$

$$i^{22} = (i^4)^5 \cdot i^2 = 1 \cdot (-1) = -1$$

$$i^{23} = (i^4)^5 \cdot i^3 = 1 \cdot i^3 = -i$$

Note that dividing the exponent, 20 in this case, by 4 will give you the answer!

If the remainder is 0, the answer is 1 (as in i^{20}).

If the remainder is 1, the answer is i (as in i^{21}).

If the remainder is 2, the answer is -1 (as in i^{22}).

If the remainder is 3, the answer is $-i$ (as in i^{23}).

After that, the answers repeat.

EXAMPLE 6 | Finding powers of i

Find:

a. i^{53} **b.** i^{47} **c.** i^{-3} **d.** i^{-1} **e.** i^{-34}

SOLUTION

a. Dividing 53 by 4, we obtain 13 with a remainder of 1. Thus the answer is i. To show this, write

$$i^{53} = (i^4)^{13} \cdot i$$

$$= 1 \cdot i = i$$

b. If we divide 47 by 4, the remainder is 3. Thus the answer is $-i$. Note that

$$i^{47} = (i^4)^{11} \cdot i^3$$

$$= 1 \cdot i^3$$

$$= -i$$

c. By definition of negative exponents,

$$i^{-3} = \frac{1}{i^3}$$

$$i^{-3} = \frac{1 \cdot i}{i^3 \cdot i} \qquad \text{Write with a denominator of } i^4 = 1.$$

$$= \frac{i}{1} = i$$

d. $i^{-1} = \dfrac{1}{i} = \dfrac{1 \cdot i^3}{i \cdot i^3} = \dfrac{i^3}{i^4} = i^3 = -i$

e. $i^{-34} = \dfrac{1}{i^{34}} = \dfrac{1 \cdot i^2}{i^{34} \cdot i^2} = \dfrac{i^2}{i^{36}} = \dfrac{i^2}{1} = -1$ ∎

The introduction of the complex numbers completes the development of our number system. We started with the natural numbers (N) and whole numbers (W), studied the integers (I) and the rational numbers (Q), and then concentrated on the real numbers (R). Now we have developed the complex numbers (C), which include all the other numbers we have discussed. In set language, the relationship is

$$N \subseteq W \subseteq I \subseteq Q \subseteq R \subseteq C$$

as shown in Figure 1.

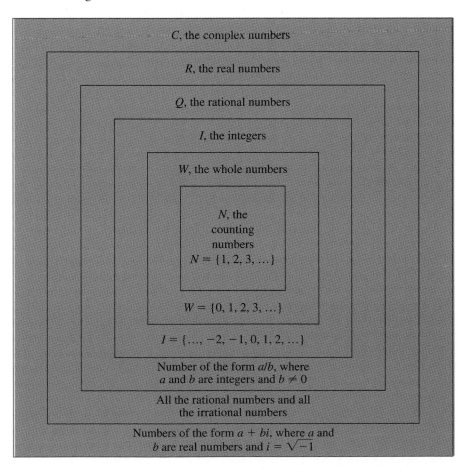

C, the complex numbers

R, the real numbers

Q, the rational numbers

I, the integers

W, the whole numbers

N, the counting numbers
$N = \{1, 2, 3, \ldots\}$

$W = \{0, 1, 2, 3, \ldots\}$

$I = \{\ldots, -2, -1, 0, 1, 2, \ldots\}$

Number of the form a/b, where a and b are integers and $b \neq 0$

All the rational numbers and all the irrational numbers

Numbers of the form $a + bi$, where a and b are real numbers and $i = \sqrt{-1}$

FIGURE 1
The Number System

EXERCISE 5.5

A **In Problems 1–10, write the given expression in terms of *i*.**

1. $\sqrt{-25}$

2. $\sqrt{-81}$

3. $\sqrt{-50}$

4. $\sqrt{-98}$

5. $4\sqrt{-72}$

6. $3\sqrt{-200}$

7. $-3\sqrt{-32}$

8. $-5\sqrt{-64}$

9. $4\sqrt{-28} + 3$

10. $7\sqrt{-18} + 5$

B **In Problems 11–30, perform the indicated operations.
(Write the answer in the form *a* + *bi*.)**

11. $(4 + i) + (2 + 3i)$

12. $(7 + 3i) + (2 + i)$

13. $(3 - 2i) - (5 + 4i)$

14. $(4 - 5i) - (2 + 3i)$

15. $(-3 - 5i) + (-2 - i)$

16. $(-7 - 3i) + (-2 - i)$

17. $(3 + \sqrt{-4}) - (5 - \sqrt{-9})$

18. $(-2 - \sqrt{-16}) - (3 - \sqrt{-25})$

19. $(-5 + \sqrt{-1}) + (-2 + 3\sqrt{-1})$

20. $(-3 + 2\sqrt{-1}) + (-4 + 5\sqrt{-1})$

21. $(3 - 4i) + (5 + 3i)$

22. $(3 - 7i) + (3 + 4i)$

23. $(4 + \sqrt{-9}) + (6 + \sqrt{-4})$

24. $(-3 - \sqrt{-25}) + (5 - \sqrt{-16})$

25. $(2 - \sqrt{-2}) - (5 + \sqrt{-2})$

26. $(3 + \sqrt{-50}) - (7 + \sqrt{-2})$

27. $(-5 - \sqrt{-2}) - (-4 - \sqrt{-18})$

28. $(-8 - \sqrt{-125}) - (-2 - \sqrt{-5})$

29. $(-4 + \sqrt{-20}) + (-3 + \sqrt{-5})$

30. $(-7 + \sqrt{-24}) + (-3 + \sqrt{-6})$

C **In Problems 31–70, perform the indicated operations.
(Write the answer in the form *a* + *bi*.)**

31. $3(4 + 2i)$

32. $5(4 + 3i)$

33. $-4(3 - 5i)$

34. $-3(7 - 4i)$

35. $\sqrt{-4}(3 + 2i)$

36. $\sqrt{-9}(2 + 5i)$

37. $\sqrt{-3}(3 + \sqrt{-3})$

38. $\sqrt{-5}(2 - \sqrt{-5})$

39. $3i(3 + 2i)$

40. $7i(4 + 3i)$

41. $4i(3 - 7i)$

42. $-5i(2 - 3i)$

43. $-\sqrt{-16}(-5 - \sqrt{-25})$

44. $-\sqrt{-25}(-3 - \sqrt{-9})$

45. $(3 + i)(2 + 3i)$

46. $(2 + 3i)(4 + 5i)$

47. $(3 - 2i)(3 + 2i)$

48. $(4 - 3i)(4 + 3i)$

49. $(3 + 2\sqrt{-4})(4 - \sqrt{-9})$

50. $(-3 + 3\sqrt{-9})(-2 + 5\sqrt{-4})$

51. $(2 + 3\sqrt{-3})(2 - 3\sqrt{-3})$

52. $(4 + 2\sqrt{-5})(4 - 2\sqrt{-5})$

53. $\dfrac{3}{i}$

54. $\dfrac{5}{i}$

55. $\dfrac{6}{-i}$

56. $\dfrac{3}{-2i}$

57. $\dfrac{i}{1 + 2i}$

58. $\dfrac{2i}{1 + 3i}$

59. $\dfrac{3i}{1 - 2i}$

60. $\dfrac{4i}{2 - 3i}$

61. $\dfrac{3 + 4i}{1 - 2i}$

62. $\dfrac{3 + 5i}{1 - 3i}$

63. $\dfrac{4 + 3i}{2 + 3i}$

64. $\dfrac{5 + 4i}{3 + 2i}$

65. $\dfrac{3}{\sqrt{-4}}$

66. $\dfrac{-4}{\sqrt{-9}}$

67. $\dfrac{3 + \sqrt{-5}}{4 + \sqrt{-2}}$

68. $\dfrac{2 + \sqrt{-2}}{1 + \sqrt{-3}}$

69. $\dfrac{-1 - \sqrt{-2}}{-3 - \sqrt{-3}}$

70. $\dfrac{-1 - \sqrt{-3}}{-2 - \sqrt{-2}}$

D **In Problems 71–80, write the answer as 1, −1, *i*, or −*i*.**

71. i^{40}

72. i^{28}

73. i^{19}

74. i^{38}

75. i^{21}

76. i^{-44}

77. i^{-32}

78. i^{53}

79. i^{65}

80. i^{16}

APPLICATIONS

81. The impedance in a circuit is the measure of how much a circuit impedes (hinders) the flow of current through it. If the impedance of a resistor is $Z_1 = 5 + 3i$ ohms and the impedance of another resistor is $Z_2 = 3 - 2i$ ohms, what is the total impedance (sum) of the two resistors when placed in series?

82. Repeat Problem 81 where the impedance of the first resistor is $3 + 7i$ and the impedance of the second is $4 - 5i$.

83. If two resistors Z_1 and Z_2 are connected in parallel, their total impedance is given by

$$Z_T = \frac{Z_1 \cdot Z_2}{Z_1 + Z_2}$$

Find the total impedance of the resistors of Problem 81.

84. Use the formula in Problem 83 and find the total impedance of the resistors of Problem 82.

SKILL CHECKER

Solve:

85. $x^2 - 4 = 0$

86. $x^2 - 16 = 0$

87. $x^2 = 25$

88. $x^2 = 36$

USING YOUR KNOWLEDGE — Absolute Values of Complex Numbers

If x is a real number, the absolute value of x is defined as follows.

$$|x| = \begin{cases} x, & \text{if } x > 0 \\ 0, & \text{if } x = 0 \\ -x, & \text{if } x < 0 \end{cases}$$

Thus

$$|5| = 5 \qquad \text{Because } 5 > 0$$

$$|0| = 0$$

$$|-8| = -(-8) = 8 \qquad \text{Because } -8 < 0$$

How can we define the absolute value of a complex number? The definition is

$$|a + bi| = \sqrt{a^2 + b^2}$$

In Problems 89–92, use this definition to find each value.

89. $|3 + 4i|$

90. $|12 + 5i|$

91. $|2 - 3i|$

92. $|5 - 7i|$

WRITE ON . . .

93. Explain why it is incorrect to use the product rule for radicals to multiply $\sqrt{-4} \cdot \sqrt{-9}$. What answer do you get if you do? What should the answer be? Under what conditions is $\sqrt{a} \cdot \sqrt{b} = \sqrt{a \cdot b}$?

94. Write a short paragraph explaining the relationship among the real numbers, the imaginary numbers, and the complex numbers.

95. Write in your own words the procedure you use to
   a. Find the conjugate of a complex number.
   b. Add (or subtract) two complex numbers.
   c. Find the quotient of two complex numbers.

MASTERY TEST
If you know how to do these problems, you have learned your lesson!

Find:

96. i^{42}

97. i^{23}

98. i^{-8}

99. i^{-2}

100. $\dfrac{3 + 5i}{2 + 3i}$

101. $\dfrac{3 - 5i}{4 - 3i}$

102. $\dfrac{2 - 3i}{i}$

103. $\sqrt{-25}(6 + \sqrt{-8})$

104. $\sqrt{-49}(\sqrt{-3} - \sqrt{-27})$

105. $(2 - 4i)(2 + 6i)$

106. $-4i(5 - 8i)$

107. $(7 + 3i) - (4 + 5i)$

108. $(9 - 2i) - (12 - 3i)$

109. $(2 + 3i) + (-3 + 4i)$

110. $(-3 - 4i) + (-5 - 7i)$

Write in terms of i:

111. $\sqrt{-50}$

112. $\sqrt{-25}$

research questions	Sources of information for these questions can be found in the *Bibliography* at the end of this text.

1. Who invented the square root sign $\sqrt{}$ (called a radix)?

2. What mathematician coined the word *imaginary,* and in what book did the term first appear?

3. Who was the first mathematician to use the notation i for $\sqrt{-1}$, and in what book did this symbol first appear?

4. The author's *Algebra* (1770) remarks that "such numbers, which by their nature are impossible, are ordinarily called imaginary or fanciful numbers because they exist only in the imagination." Write a short paragraph about this author and his other contributions to mathematics.

5. Who was the first person to attempt (not successfully) to represent the complex numbers graphically, and what was the name of the book in which he tried?

6. The graphical representation of complex numbers occurred at nearly the same time to three mathematicians with no connection or knowledge of each other. Who were these three mathematicians?

7. Write a short paragraph about the mathematical contributions of each of the three people named in Question 6.

SUMMARY

SECTION	ITEM	MEANING	EXAMPLE
5.1	nth root	If a and x are real numbers and n is a positive integer, x is an nth root of a if $x^2 = a$.	The square root of 25 is 5, the cube root of -8 is -2, and the fourth root of -16 is not a real number.
5.1A	$\sqrt{}$ Radicand Index (order) $\sqrt[n]{a}$, (a radical expression)	A radical sign In $\sqrt[n]{a}$, a is the radicand. In $\sqrt[n]{a}$, the index is n. The nth root of a	In $\sqrt[3]{22}$, 22 is the radicand. In $\sqrt[3]{22}$, the index is 3. $\sqrt[3]{64} = 4$, $\sqrt[5]{-32} = -2$, and $\sqrt{-25}$ is not a real number.
5.1B	$a^{1/n}$ $a^{m/n}$ $a^{-m/n}$	$a^{1/n} = \sqrt[n]{a}$, if it exists. $(\sqrt[n]{a})^m = \sqrt[n]{a^m}$ $\dfrac{1}{(a^{m/n})}$	$8^{1/3} = \sqrt[3]{8} = 2$ and $32^{1/5} = \sqrt[5]{32} = 2$ $16^{3/2} = (\sqrt{16})^3 = 4^3 = 64$ $4^{-3/2} = \dfrac{1}{4^{3/2}} = \dfrac{1}{8}$
5.1C	$(a^m)^{1/n}$ (m and n positive, even integers)	$|a|^{m/n}$ If r, s, and t are rational numbers and a is a real number, the following properties apply: I. $a^r \cdot a^s = a^{r+s}$ II. $\dfrac{a^r}{a^s} = a^{r-s}$ $(a \neq 0)$ III. $(a^r)^s = a^{rs}$ IV. $(a^r b^s)^t = a^{rt} b^{st}$ V. $\left(\dfrac{a^r}{b^s}\right)^t = \dfrac{a^{rt}}{b^{st}}$ $(a \neq 0, b \neq 0)$	$[(-4)^2]^{1/2} = |-4| = 4$ I. $x^4 \cdot x^{-1} = x^{4+(-1)} = x^3$ II. $\dfrac{x^7}{x^5} = x^{7-5} = x^2$ III. $(x^2)^3 = x^{2 \cdot 3} = x^6$ IV. $(x^3 y^4)^5 = x^{3 \cdot 5} \cdot y^{4 \cdot 5} = x^{15} y^{20}$ V. $\left(\dfrac{x^3}{y^4}\right)^5 = \dfrac{x^{3 \cdot 5}}{y^{4 \cdot 5}} = \dfrac{x^{15}}{y^{20}}$

SECTION	ITEM	MEANING	EXAMPLE
5.2A	Law I Law II Law III	$\sqrt[n]{a^n} = a$, for $a \geq 0$ $\sqrt[n]{ab} = \sqrt[n]{a} \cdot \sqrt[n]{b}$ $\sqrt[n]{\dfrac{a}{b}} = \dfrac{\sqrt[n]{a}}{\sqrt[n]{b}}$	$\sqrt[4]{2^4} = 2$ $\sqrt{18} = \sqrt{9} \cdot \sqrt{2} = 3\sqrt{2}$ $\sqrt[3]{\dfrac{8}{27}} = \dfrac{\sqrt[3]{8}}{\sqrt[3]{27}} = \dfrac{2}{3}$
5.2C	Simplified form of a radical	A radical is in simplified form if the radicand has no factors with exponents greater than or equal to the index, there are no fractions under the radical sign or radicals in the denominator, and the index is as low as possible.	$\dfrac{\sqrt[3]{2ac}}{2c^2}$ is in simplified form.
5.3A	Like radicals	Radicals with the same index and the same radicand	$\sqrt[3]{3x^2}$ and $-7\sqrt[3]{3x^2}$ are like radicals.
5.3C	Conjugates	$a + b$ and $a - b$ are conjugates.	$3 + \sqrt{2}$ and $3 - \sqrt{2}$ are conjugates.
5.4A	Extraneous solution	A proposed solution that does not satisfy the original equation	0 is an extraneous solution of $\sqrt{x + 1} = x - 1$.
5.5A	i Complex number	$\sqrt{-1}$ A number that can be written in the form $a + bi$, where a and b are real	 $3 + 7i$ and $-4 - 8i$ are complex numbers.
5.5B	Addition and subtraction of complex numbers	$(a + bi) \pm (c + di) = (a \pm c) + (b \pm d)i$	$(3 + 2i) + (5 + 3i) = 8 + 5i$ $(4 - 2i) - (5 + 3i) = -1 - 5i$
5.5C	Multiplication of complex numbers	$(a + bi)(c + di) = (ac - bd) + (ad + bc)i$	$(2 + 3i)(4 - 5i) =$ $[2 \cdot 4 - 3 \cdot (-5)] + [2 \cdot (-5) + 3 \cdot 4]i =$ $23 + 2i$

REVIEW EXERCISES

(If you need help with these exercises, look in the section indicated in brackets.)

In Problems 1–4, find the real roots.

1. [5.1A]
   a. $\sqrt{-8}$
   b. $\sqrt[3]{-64}$

2. [5.1A]
   a. $\sqrt{-9}$
   b. $\sqrt[3]{-125}$

3. [5.1B]
   a. $(-27)^{1/3}$
   b. $(-64)^{1/3}$

4. [5.1B]
   a. $\left(\dfrac{1}{16}\right)^{1/4}$
   b. $\left(\dfrac{1}{256}\right)^{1/4}$

In Problems 5–8, evaluate if possible.

5. [5.1B]
   a. $125^{2/3}$
   b. $64^{2/3}$

6. [5.1B]
   a. $(-25)^{3/2}$
   b. $(-36)^{3/2}$

7. [5.1B]
   a. $(-8)^{-2/3}$
   b. $(-64)^{-2/3}$

8. [5.1B]
   a. $8^{-2/3}$
   b. $64^{-2/3}$

In Problems 9–18, simplify (x and y are positive).

9. [5.1C]
   a. $x^{1/5} \cdot x^{1/3}$
   b. $x^{1/5} \cdot x^{1/4}$

10. [5.1C]
a. $\dfrac{x^{-1/4}}{x^{1/5}}$
b. $\dfrac{x^{-1/3}}{x^{1/5}}$

11. [5.1C]
a. $(x^{1/3}y^{2/5})^{-15}$
b. $(x^{1/6}y^{2/5})^{-30}$

12. [5.1C]
a. $x^{3/5}(x^{-1/5}+y^{3/5})$
b. $x^{4/5}(x^{-1/5}+y^{3/5})$

13. [5.2A]
a. $\sqrt[4]{(-7)^4}$
b. $\sqrt[4]{(-6)^4}$

14. [5.2A]
a. $\sqrt[8]{(-x)^8}$
b. $\sqrt[4]{(-x)^4}$

15. [5.2A]
a. $\sqrt[3]{48}$
b. $\sqrt[3]{56}$

16. [5.2A]
a. $\sqrt[3]{16x^4y^6}$
b. $\sqrt[3]{16x^8y^{15}}$

17. [5.2A]
a. $\sqrt{\dfrac{5}{243}}$
b. $\sqrt{\dfrac{5}{1024}}$

18. [5.2A]
a. $\sqrt[3]{\dfrac{1}{x^3}}$
b. $\sqrt[3]{\dfrac{5}{x^3}}$

In Problems 19–22, rationalize the denominator.

19. [5.2B]
a. $\dfrac{\sqrt5}{\sqrt{11}}$
b. $\dfrac{\sqrt5}{\sqrt{13}}$

20. [5.2B]
a. $\dfrac{\sqrt2}{\sqrt{5x}},\,x>0$
b. $\dfrac{\sqrt3}{\sqrt{5x}},\,x>0$

21. [5.2B]
a. $\dfrac{1}{\sqrt[3]{5x}}$
b. $\dfrac{1}{\sqrt[3]{7x}}$

22. [5.2B]
a. $\dfrac{\sqrt[5]{1}}{\sqrt[5]{16x^3}}$
b. $\dfrac{\sqrt[5]{5}}{\sqrt[5]{16x^3}}$

In Problems 23–24, reduce the index of the expression.

23. [5.2C]
a. $\sqrt[4]{\dfrac{256}{81}}$
b. $\sqrt[4]{\dfrac{625}{81}}$

24. [5.2C]
a. $\sqrt[6]{81c^4d^4}$
b. $\sqrt[6]{625c^4d^4}$

25. [5.2C] Simplify.
a. $\sqrt[6]{\dfrac{3a^2}{243c^{16}}}$
b. $\sqrt[6]{\dfrac{9a^2}{81c^{16}}}$

In Problems 26–31, perform the indicated operations.

26. [5.3A]
a. $\sqrt8+\sqrt{32}$
b. $\sqrt{18}+\sqrt{32}$

27. [5.3A]
a. $\sqrt{63}-\sqrt{28}$
b. $\sqrt{112}-\sqrt{63}$

28. [5.3A]
a. $3\sqrt{\dfrac12}-3\sqrt{\dfrac18}$
b. $4\sqrt{\dfrac12}-3\sqrt{\dfrac18}$

29. [5.3A]
a. $5\sqrt[3]{\dfrac{3}{4x}}-\sqrt[3]{\dfrac{3}{32x}}$
b. $6\sqrt[3]{\dfrac{3}{4x}}-\sqrt[3]{\dfrac{3}{32x}}$

30. [5.3B]
a. $\sqrt2(\sqrt{18}+\sqrt3)$
b. $\sqrt2(\sqrt{32}+\sqrt3)$

31. [5.3B]
a. $\sqrt[3]{2x}(\sqrt[3]{24x^2}-\sqrt[3]{81x})$
b. $\sqrt[3]{3x}(\sqrt[3]{16x^2}-\sqrt[3]{54x})$

In Problems 32–35, find the product.

32. [5.3B]
a. $(\sqrt{27}+\sqrt{18})(\sqrt{12}+\sqrt8)$
b. $(\sqrt{12}+\sqrt{18})(\sqrt{12}+\sqrt8)$

33. [5.3B]
a. $(\sqrt3+4)(\sqrt3+4)$
b. $(\sqrt3+5)(\sqrt3+5)$

34. [5.3B]
a. $(3-\sqrt3)^2$
b. $(4-\sqrt3)^2$

35. [5.3B]
a. $(\sqrt6+\sqrt3)(\sqrt6-\sqrt3)$
b. $(\sqrt7+\sqrt3)(\sqrt7-\sqrt3)$

36. [5.3B] Reduce.
a. $\dfrac{20-\sqrt{50}}{5}$
b. $\dfrac{30-\sqrt{50}}{5}$

37. [5.3C] Rationalize the denominator.
a. $\dfrac{\sqrt y}{\sqrt x-\sqrt y}$
b. $\dfrac{\sqrt x}{\sqrt x-\sqrt y}$

38. [5.4A] Solve.
a. $\sqrt{x-2}=-2$
b. $\sqrt{x-2}=-3$

39. [5.4A] Solve.
a. $\sqrt{x+5}=x-1$
b. $\sqrt{x+6}=x-6$

40. [5.4A] Solve.
a. $\sqrt{x-3}-x=-3$
b. $\sqrt{x-4}-x=-4$

41. [5.4A] Solve.

a. $\sqrt{x-3} - \sqrt{x} = -1$ b. $\sqrt{x-5} - \sqrt{x} = -1$

42. [5.4A] Solve.

a. $\sqrt[3]{x-3} = 3$ b. $\sqrt[3]{x-3} = 4$

43. [5.4A] Solve.

a. $\sqrt[4]{x+1} + 1 = 0$ b. $\sqrt[4]{x-2} + 16 = 0$

44. [5.4B] The distance d (in feet) from a light source of intensity I (in foot-candles) is

$$d = \sqrt{\frac{k}{I}}$$

where k is a constant.

a. Solve for I.   b. Solve for k.

45. [5.5A] Write in terms of i.

a. $\sqrt{-100}$ b. $\sqrt{-121}$

46. [5.5A] Write in terms of i.

a. $\sqrt{-72}$ b. $\sqrt{-50}$

47. [5.5B] Find.

a. $(3+5i) + (7-2i)$ b. $(4+7i) + (2-4i)$

48. [5.5B] Find.

a. $(3+5i) - (7-2i)$ b. $(4+7i) - (2-4i)$

49. [5.5C] Multiply.

a. $(3+2i)(5-3i)$ b. $(4+5i)(2-3i)$

50. [5.5C] Multiply.

a. $\sqrt{-16}(4 - \sqrt{-72})$ b. $\sqrt{-36}(4 - \sqrt{-72})$

51. [5.5C] Divide.

a. $\dfrac{2+3i}{4+3i}$ b. $\dfrac{3-5i}{4-3i}$

52. [5.5D] Write the answer as 1, -1, i, or $-i$.

a. i^{38} b. i^{75}

53. [5.5D] Write the answer as 1, -1, i, or $-i$.

a. i^{-14} b. i^{-27}

PRACTICE TEST

(Answers on pages 329–331.)

1. Find, if possible.

a. $\sqrt[3]{-64}$ b. $\sqrt{-36}$

2. Find, if possible.

a. $(-27)^{1/3}$ b. $\left(\dfrac{1}{16}\right)^{1/4}$

3. Evaluate.

a. $8^{2/3}$ b. $(-25)^{3/2}$

4. Evaluate.

a. $(-27)^{-2/3}$ b. $8^{-2/3}$

5. If x is positive, simplify.

a. $x^{1/2} \cdot x^{2/3}$ b. $\dfrac{x^{-1/3}}{x^{2/5}}$

6. If x is positive, simplify.

a. $(x^{1/5}y^{3/4})^{-20}$ b. $x^{3/5}(x^{-1/5} + y^{3/5})$

7. Simplify.

a. $\sqrt[4]{(-5)^4}$ b. $\sqrt[4]{(-x)^4}$

8. Simplify.

a. $\sqrt[3]{40}$ b. $\sqrt[3]{54a^4b^{12}}$

9. Simplify.

a. $\sqrt{\dfrac{5}{243}}$ b. $\sqrt[3]{\dfrac{5}{x^6}}$

10. Rationalize the denominator.

a. $\dfrac{\sqrt{5}}{\sqrt{6}}$ b. $\dfrac{\sqrt{2}}{\sqrt{5x}}, x > 0$

11. Rationalize the denominator.

a. $\dfrac{1}{\sqrt[3]{3x}}$ b. $\dfrac{\sqrt[5]{7}}{\sqrt[5]{16x^3}}$

12. Reduce the index.

a. $\sqrt[4]{\dfrac{16}{81}}$ b. $\sqrt[6]{81c^4d^4}$

13. Simplify $\sqrt[6]{\dfrac{a^2}{16c^{16}}}$.

14. Perform the indicated operations.

a. $\sqrt{32} + \sqrt{98}$ b. $\sqrt{112} - \sqrt{28}$

15. Perform the indicated operations.

a. $4\sqrt{\dfrac{1}{2}} - 3\sqrt{\dfrac{1}{8}}$ b. $5\sqrt[3]{\dfrac{3}{4x}} - \sqrt[3]{\dfrac{3}{32x}}$

16. Perform the indicated operations.
    a. $\sqrt{2}(\sqrt{8} + \sqrt{5})$ b. $\sqrt[3]{3x}(\sqrt[3]{9x^2} - \sqrt[3]{16x})$

17. Find the product.
    a. $(\sqrt{27} + \sqrt{50})(\sqrt{12} + \sqrt{18})$
    b. $(\sqrt{3} + 3)(\sqrt{3} + 3)$

18. Find the product.
    a. $(3 - \sqrt{3})^2$ b. $(\sqrt{6} + \sqrt{3})(\sqrt{6} - \sqrt{3})$

19. Reduce $\dfrac{10 - \sqrt{75}}{5}$.

20. Rationalize the denominator in $\dfrac{\sqrt{x}}{\sqrt{x} + \sqrt{y}}$.

21. a. Solve $\sqrt{x + 2} = -2$.   b. Solve $\sqrt{x + 2} = x - 10$.

22. Solve $\sqrt{x - 3} - x = -3$.

23. Solve $\sqrt{x - 9} - \sqrt{x} = -1$.

24. Solve $\sqrt[3]{x - 3} - 3 = 0$.

25. The time t (in seconds) it takes a free-falling object to fall d feet is given by

$$t = \sqrt{\dfrac{d}{16}}$$

    a. Solve for d.

    b. An object dropped from the top of a building hits the ground after 4 sec. How tall is the building?

26. Write in terms of i.
    a. $\sqrt{-64}$ b. $\sqrt{-98}$

27. Find.
    a. $(4 + 5i) + (6 - 14i)$ b. $(5 + 2i) - (7 - 8i)$

28. Multiply.
    a. $(2 + 2i)(4 - 3i)$ b. $\sqrt{-9}(3 - \sqrt{-8})$

29. Find.
    a. $\dfrac{4 + 3i}{2 + 3i}$ b. $\dfrac{4 - 3i}{3 - 5i}$

30. Write the answer as 1, -1, i, or $-i$.
    a. i^{56} b. i^{-9}

ANSWERS TO PRACTICE TEST

Answer	If you missed: Question	Review: Section	Examples	Page
1a. -4	1a	5.1A	1	281–282
1b. Not a real number	1b	5.1A	1	281–282
2a. -3	2a	5.1B	2	282–283
2b. $\dfrac{1}{2}$	2b	5.1B	2	282–283
3a. 4	3a	5.1B	3	283–284
3b. Not a real number	3b	5.1B	3	283–284
4a. $\dfrac{1}{9}$	4a	5.1B	4	284
4b. $\dfrac{1}{4}$	4b	5.1B	4	284
5a. $x^{7/6}$	5a	5.1C	5	285
5b. $\dfrac{1}{x^{11/15}}$	5b	5.1C	5	285

Answer	If you missed:		Review:	
	Question	Section	Examples	Page
6a. $\dfrac{1}{x^4 y^{15}}$	6a	5.1C	6	285–286
6b. $x^{2/5} + x^{3/5} y^{3/5}$	6b	5.1C	6	285–286
7a. 5	7a	5.2A	1	290
7b. $\lvert -x \rvert = \lvert x \rvert$	7b	5.2A	1	290
8a. $2\sqrt[3]{5}$	8a	5.2A	2	291
8b. $3ab^4\sqrt[3]{2a}$	8b	5.2A	2	291
9a. $\dfrac{\sqrt{15}}{27}$	9a	5.2A	3	292
9b. $\dfrac{\sqrt[3]{5}}{x^2}$	9b	5.2A	3	292
10a. $\dfrac{\sqrt{30}}{6}$	10a	5.2B	4	293
10b. $\dfrac{\sqrt{10x}}{5x}$	10b	5.2B	4	293
11a. $\dfrac{\sqrt[3]{9x^2}}{3x}$	11a	5.2B	5	294
11b. $\dfrac{\sqrt[5]{14x^2}}{2x}$	11b	5.2B	5	294
12a. $\dfrac{2}{3}$	12a	5.2C	6	295
12b. $\sqrt[3]{9c^2 d^2}$	12b	5.2C	6	295
13. $\dfrac{\sqrt[3]{2ac}}{2c^3}$	13	5.2C	7	295–296
14a. $11\sqrt{2}$	14a	5.3A	1	300–301
14b. $2\sqrt{7}$	14b	5.3A	1	300–301
15a. $\dfrac{5\sqrt{2}}{4}$	15a	5.3A	2	301–302
15b. $\dfrac{9\sqrt[3]{6x^2}}{4x}$	15b	5.3A	2	301–302
16a. $4 + \sqrt{10}$	16a	5.3B	3	302
16b. $3x - 2\sqrt[3]{6x^2}$	16b	5.3B	3	302
17a. $48 + 19\sqrt{6}$	17a	5.3B	4, 5	303
17b. $12 + 6\sqrt{3}$	17b	5.3B	4, 5	303

Answer	If you missed:	Review:		
	Question	Section	Examples	Page
18a. $12 - 6\sqrt{3}$	18a	5.3B	5	303
18b. 3	18b	5.3B	5	303
19. $2 - \sqrt{3}$	19	5.3B	6	304
20. $\dfrac{x - \sqrt{xy}}{x - y}$	20	5.3C	7	306
21a. No solution	21a	5.4A	1	310–311
21b. 14	21b	5.4A	1	310–311
22. 3 or 4	22	5.4A	1	310–311
23. 25	23	5.4A	2	312
24. 30	24	5.4A	3, 4	312–313
25a. $d = 16t^2$	25a	5.4B	5	313
25b. 256 ft	25b	5.4B	5	313
26a. $8i$	26a	5.5A	1	317
26b. $7\sqrt{2}i$	26b	5.5A	1	317
27a. $10 - 9i$	27a	5.5B	2	318
27b. $-2 + 10i$	27b	5.5B	2	318
28a. $14 + 2i$	28a	5.5C	3	318
28b. $6\sqrt{2} + 9i$	28b	5.5C	4	319
29a. $\dfrac{17}{13} - \dfrac{6}{13}i$	29a	5.5C	5	320
29b. $\dfrac{27}{34} + \dfrac{11}{34}i$	29b	5.5C	5	320
30a. 1	30a	5.5D	6	321–322
30b. $-i$	30b	5.5D	6	321–322

6

Quadratic Equations and Inequalities

In Section 6.1, we solve quadratic equations by extracting roots and completing the square. Completing the square is then used to develop the quadratic formula (Section 6.2). Next, we use the discriminant to classify and verify the solutions of quadratic equations and even to create some quadratic equations in Section 6.3 when we are given their solution set. In Section 6.4, we generalize the techniques used in solving quadratic equations to solve higher-degree equations that can be written in quadratic form, by using substitutions. We end the chapter by applying our knowledge to the solution of quadratic, polynomial, and rational inequalities.

Many civilizations, starting with the Babylonians in 2000 B.C., were familiar with quadratic equations, whose solutions were found by following verbal instructions given in a text. We believe that the Arabian mathematician al-Khowarizmi (ca. 820) was the first to classify quadratic equations into three types: $x^2 + ax = b$, $x^2 + b = ax$, and $x^2 = ax + b$, where a and b are positive. These three types of equations were solved by using a few general rules, and their correctness was demonstrated geometrically. In general, the quadratic equation $x^2 + px = q$ was solved by the method of "completion of squares."

Although the Arabs recognized the existence of two solutions for quadratic equations, they listed only the positive ones. In the twelfth century A.D., the Hindu mathematician Bhaskara affirmed the existence and validity of negative as well as positive solutions. What types of quadratic equations were being solved? A Babylonian clay tablet from 2000 B.C. states, "I have added the area and two-thirds of my square and it is 0;36 $\left[\frac{36}{60}\right]$. What is the side of my square?" Can you translate this into an equation? The problem itself was solved by using a quadratic formula equivalent to the one we discuss in this chapter!

6.1

SOLVING QUADRATICS BY COMPLETING THE SQUARE

To succeed, review how to:

1. Take the square root of a number (p. 291).
2. Rationalize the denominator of a fraction (pp. 292–294).
3. Add, subtract, multiply, and divide complex numbers (pp. 317–321).
4. Expand $(x \pm a)^2$ (pp. 153–154).

Objectives:

A Solve quadratic equations of the form $ax^2 + b = 0$.

B Solve equations of the form $a(x + b)^2 = c$.

C Solve quadratic equations by completing the square.

getting started

Antennas and Quadratic Equations

Look at the square "grids" on the antenna. The area of each one is 49 ft^2. Can we find the dimensions of each grid? If the length of one side is x feet, the area of the square is x^2 square feet. Since the area is also 49 ft^2, we have $x^2 = 49$.

We solve this equation by taking (**extracting**) the square roots of both sides. Because a nonzero number has two square roots, we have

$$x = \sqrt{49} = 7 \qquad \text{or} \qquad x = -\sqrt{49} = -7$$

The solutions (or roots) are 7 and −7. This procedure is usually shortened to

$$x^2 = 49$$
$$x = \pm\sqrt{49} = \pm 7$$

where the notation ± 7 (read "plus or minus 7") means that $x = 7$ or $x = -7$. If x represents the *length* of a side, the answer −7 must be discarded because the length of an object cannot be negative.

In Chapter 2, we studied methods of solving *linear equations*—that is, equations in which the variable involved has an exponent of 1 (the first power). We are now ready to discuss equations containing the *second* (but no higher) power of the unknown.

Such equations are called *second-degree*, or *quadratic*, equations; these can be written in *standard form*. Here is the definition.

QUADRATIC EQUATION IN STANDARD FORM

$$ax^2 + bx + c = 0 \qquad (a, b, \text{ and } c \text{ are real numbers, } a \neq 0)$$

is a **quadratic equation in standard form**.

The procedure used to solve these equations is similar to that employed in Chapter 2 and consists of applying certain transformations to obtain equivalent equations whose solution set is evident.

Thus the solutions of $x^2 = 25$ are found by taking the square root of both sides of the equation to get

$$x = \pm\sqrt{25} = \pm 5$$

In general, we have the following.

SOLUTIONS OF $x^2 = a$

The solutions of $x^2 = a$, where a is a real number, are $\sqrt{a}$ and $-\sqrt{a}$, abbreviated as $\pm\sqrt{a}$.

Solving Quadratic Equations of the Form $ax^2 + b = 0$

To solve the equation $9x^2 - 5 = 0$, we need to isolate the x^2 and take the square roots of both sides. Here's how we do it:

$$9x^2 - 5 = 0 \qquad \text{Given.}$$

$$9x^2 = 5 \qquad \text{Add 5 to both sides.}$$

$$x^2 = \frac{5}{9} \qquad \text{Divide both sides by 9 (now } x^2 \text{ is by itself).}$$

$$x = \pm\sqrt{\frac{5}{9}} = \pm\frac{\sqrt{5}}{3} \qquad \text{Take square roots of both sides.}$$

The solutions are

$$\frac{\sqrt{5}}{3} \qquad \text{and} \qquad -\frac{\sqrt{5}}{3}$$

and the solution set is

$$\left\{ \frac{\sqrt{5}}{3}, -\frac{\sqrt{5}}{3} \right\}$$

EXAMPLE 1   Solving $ax^2 + b = 0$

Solve:

a. $4x^2 - 7 = 0$ **b.** $3x^2 + 24 = 0$ **c.** $5x^2 - 4 = 0$

SOLUTION

a. The idea is to isolate x^2 and then extract square roots.

$$4x^2 - 7 = 0 \qquad \text{Given.}$$

$$4x^2 = 7 \qquad \text{Add 7 to both sides.}$$

$$x^2 = \frac{7}{4} \qquad \text{Divide both sides by 4 (now } x^2 \text{ is isolated).}$$

$$x = \pm\sqrt{\frac{7}{4}} = \pm\frac{\sqrt{7}}{2} \qquad \text{Extract square roots.}$$

The solutions are

$$\frac{\sqrt{7}}{2} \qquad \text{and} \qquad -\frac{\sqrt{7}}{2}$$

and the solution set is

$$\left\{ \frac{\sqrt{7}}{2}, -\frac{\sqrt{7}}{2} \right\}$$

b.

$$3x^2 + 24 = 0 \qquad \text{Given.}$$

$$3x^2 = -24 \qquad \text{Add } -24 \text{ to both sides.}$$

$$x^2 = -8 \qquad \text{Divide both sides by 3.}$$

$$x = \pm\sqrt{-8} \qquad \text{Extract square roots.}$$

$$x = \pm 2\sqrt{2}\,i \qquad \text{Simplify: } \sqrt{-8} = \sqrt{4}\sqrt{-2} = 2\sqrt{2}\,i.$$

The solutions are $2\sqrt{2}\,i$ and $-2\sqrt{2}\,i$. Note that the answers are **complex numbers** given in the *simplified* form, $\pm 2\sqrt{2}\,i$ instead of $\pm\sqrt{8}\,i$. (Make sure you write $\pm 2\sqrt{2}\,i$, *not* $\pm 2\sqrt{2i}$.)

c.

$$5x^2 - 4 = 0 \qquad \text{Given.}$$

$$5x^2 = 4 \qquad \text{Add 4 to both sides.}$$

$$x^2 = \frac{4}{5} \qquad \text{Divide both sides by 5.}$$

$$x = \pm\sqrt{\frac{4}{5}} = \pm\frac{2}{\sqrt{5}} \qquad \text{Extract square roots.}$$

$$= \pm\frac{2 \cdot \sqrt{5}}{\sqrt{5} \cdot \sqrt{5}} \qquad \text{Rationalize the denominator.}$$

$$= \pm\frac{2\sqrt{5}}{5}$$

The solutions are

$$\frac{2\sqrt{5}}{5} \qquad \text{and} \qquad -\frac{2\sqrt{5}}{5}$$

and the solution set is

$$\left\{ \frac{2\sqrt{5}}{5}, \; -\frac{2\sqrt{5}}{5} \right\}$$

Note that the answer should be written with a *rationalized* denominator.   ■

B   Solving Equations of the Form $a(x + b)^2 = c$

The method used to solve equations of the form $ax^2 + c = 0$ can also be used to solve equations of the form $(x + a)^2 = b$. Thus to solve the equation $(x + 3)^2 = 16$, we proceed as follows:

$$(x + 3)^2 = 16 \qquad \text{Given.}$$

$$x + 3 = \pm\sqrt{16} = \pm 4 \qquad \text{Extract square roots.}$$

$$x + 3 = 4 \quad \text{or} \quad x + 3 = -4$$

$$x = 1 \quad \text{or} \quad x = -7 \qquad \text{Solve } x + 3 = 4 \text{ and } x + 3 = -4.$$

The solutions are 1 and -7 and the solution set is $\{1, -7\}$.

To solve the equation $(x - 2)^2 - 9 = 0$, we add 9 to both sides and then extract square roots to obtain

$$(x - 2)^2 - 9 = 0$$
$$(x - 2)^2 = 9$$
$$x - 2 = \pm\sqrt{9} = \pm 3$$
$$x - 2 = 3 \quad \text{or} \quad x - 2 = -3$$
$$x = 5 \quad \text{or} \quad x = -1$$

The solutions for $(x - 2)^2 - 9 = 0$ are 5 and -1, and the solution set is $\{5, -1\}$.

EXAMPLE 2   Solving $a(x + b)^2 = c$

Solve:

a. $(x - 1)^2 = 27$

b. $3(x - 2)^2 + 25 = 0$

SOLUTION

a. $(x - 1)^2 = 27$ Given.

$\quad\quad x - 1 = \pm\sqrt{27} = \pm 3\sqrt{3}$ Extract square roots.

$\quad x - 1 = 3\sqrt{3} \quad\quad \text{or} \quad\quad x - 1 = -3\sqrt{3}$

$\quad\quad x = 1 + 3\sqrt{3} \quad\quad \text{or} \quad\quad x = 1 - 3\sqrt{3}$ Solve for x.

The solutions are $1 + 3\sqrt{3}$ and $1 - 3\sqrt{3}$, and the solution set is $\{1 + 3\sqrt{3},$ $1 - 3\sqrt{3}\}$. This can be verified by substituting $1 + 3\sqrt{3}$ and $1 - 3\sqrt{3}$ in the equation $(x - 1)^2 = 27$. Since $[(1 + 3\sqrt{3}) - 1]^2 = (3\sqrt{3})^2 = 3^2 \cdot 3 = 27$ and $[(1 - 3\sqrt{3}) - 1]^2 = (-3\sqrt{3})^2 = 3^2 \cdot 3 = 27$, the result is correct.

b. $\quad 3(x - 2)^2 + 25 = 0$ Given.

$\quad\quad 3(x - 2)^2 = -25$ Subtract 25 from both sides.

$\quad\quad (x - 2)^2 = \dfrac{-25}{3}$ Divide both sides by 3 [now the $(x - 2)^2$ is isolated].

$\quad\quad x - 2 = \pm\sqrt{\dfrac{-25}{3}} = \pm\dfrac{5i}{\sqrt{3}}$ Extract square roots.

$\quad\quad\quad = \pm\dfrac{5i \cdot \sqrt{3}}{\sqrt{3} \cdot \sqrt{3}}$ Rationalize the denominator.

$\quad\quad x = 2 \pm \dfrac{5\sqrt{3}}{3} i$ Solve for x by adding 2.

The solutions are

$$2 + \frac{5\sqrt{3}}{3} i \quad \text{and} \quad 2 - \frac{5\sqrt{3}}{3} i$$

and the solution set is

$$\left\{ 2 + \frac{5\sqrt{3}}{3} i, \; 2 - \frac{5\sqrt{3}}{3} i \right\}$$

We leave the verification to you.   ∎

EXAMPLE 3   Writing as $a(x + b)^2 = c$ and solving

Solve: $3x^2 + 12x + 12 = 20$

SOLUTION   If we subtract 20 from both sides, we obtain $3x^2 + 12x - 8 = 0$, which is *not* factorable using integer coefficients, so we try to factor the left side first.

$$3x^2 + 12x + 12 = 20 \qquad \text{Given.}$$

$$3(x^2 + 4x + 4) = 20 \qquad \text{Factor out the 3.}$$

$$3(x + 2)^2 = 20 \qquad \text{Factor } x^2 + 4x + 4 \text{ as } (x + 2)^2.$$

$$(x + 2)^2 = \frac{20}{3} \qquad \text{Divide by 3.}$$

$$x + 2 = \pm\sqrt{\frac{20}{3}} = \pm\frac{\sqrt{20}}{\sqrt{3}} \qquad \text{Extract square roots.}$$

$$= \pm\frac{\sqrt{20} \cdot \sqrt{3}}{\sqrt{3} \cdot \sqrt{3}} \qquad \text{Rationalize the denominator.}$$

$$= \pm\frac{\sqrt{60}}{3}$$

$$= \pm\frac{2\sqrt{15}}{3} \qquad \text{Simplify: } \sqrt{60} = \sqrt{4 \cdot 15} = 2\sqrt{15}.$$

$$x = -2 \pm \frac{2\sqrt{15}}{3} \qquad \text{Solve for } x \text{ by subtracting 2.}$$

The solutions are

$$-2 + \frac{2\sqrt{15}}{3} \qquad \text{and} \qquad -2 - \frac{2\sqrt{15}}{3}$$

and the solution set is

$$\left\{ -2 + \frac{2\sqrt{15}}{3}, \; -2 - \frac{2\sqrt{15}}{3} \right\}$$

We leave the verification to you.   ■

Solving by Completing the Square

The solutions of Examples 2 and 3 were obtained by writing the equation in the form $a(x + b)^2 = c$. Suppose we have an equation that is not of this form. We can make it of this form if we learn a technique called **completing the square**. As you recall,

$$(x + a)^2 = x^2 + 2ax + a^2 \qquad \text{and} \qquad (x - a)^2 = x^2 - 2ax + a^2$$

In both cases, the last term is the square of one-half the coefficient of x. How can we make $x^2 + 10x$ a perfect square trinomial? Since the coefficient of x is 10, and in a perfect square trinomial, the coefficient of x is $2a$, $10 = 2a$ and a must be 5. Thus we make $x^2 + 10x$ a perfect square trinomial by adding 5^2. We then have

$$x^2 + 10x + 5^2 = (x + 5)^2$$

Now consider the equation $x^2 + 8x = 12$. To make $x^2 + 8x$ a perfect square trinomial, we note that the coefficient of x is 8, so the number to be added is the square of one-half of 8—that is, 4^2. Adding 4^2 to both sides, we have

$$x^2 + 8x + 4^2 = 12 + 4^2$$

$$(x + 4)^2 = 12 + 4^2 \qquad \text{Factor } x^2 + 8x + 4^2 \text{ as } (x+4)^2.$$

And, $(x + 4)^2 = 28$, the form we want! Now take the square roots of both sides:

$$x + 4 = \pm\sqrt{28} = \pm 2\sqrt{7} \qquad \text{Simplify the radical.}$$

$$x = -4 \pm 2\sqrt{7} \qquad \text{Subtract 4 from both sides.}$$

The solutions are $-4 + 2\sqrt{7}$ and $-4 - 2\sqrt{7}$, and the solution set is $\{-4 + 2\sqrt{7}, -4 - 2\sqrt{7}\}$. Note that in the equation $x^2 + 8x = 12$, the variables x and x^2 are on one side of the equation.

EXAMPLE 4 Solving by completing the square

Solve by completing the square: $x^2 + 10x - 2 = 0$

SOLUTION

$$x^2 + 10x - 2 = 0 \qquad \text{Given.}$$

$$x^2 + 10x = 2 \qquad \text{Add 2 to both sides so that only the variable terms are on one side.}$$

$$x^2 + 10x + 5^2 = 2 + 5^2 \qquad \text{Add } 5^2, \text{ the square of one-half of the coefficient of } x\text{—that is, add } \left[\frac{1}{2} \cdot 10\right]^2\text{—to both sides.}$$

$$(x + 5)^2 = 27 \qquad \text{Factor on the left.}$$

$$x + 5 = \pm\sqrt{27} = \pm 3\sqrt{3} \qquad \text{Extract square roots and simplify.}$$

$$x = -5 \pm 3\sqrt{3} \qquad \text{Subtract 5.}$$

The solutions are $-5 + 3\sqrt{3}$ and $-5 - 3\sqrt{3}$, and the solution set is $\{-5 + 3\sqrt{3}, -5 - 3\sqrt{3}\}$. ■

When the coefficient of x^2 is not 1 and the left-hand side of the equation is not factorable using integer coefficients, we first isolate the terms containing the variable on one side of the equation and then divide each term by the coefficient of x^2. For example, to solve $3x^2 + 2x - 6 = 0$, we add 6 to both sides (so all the terms containing the variable are isolated on one side), and then divide by 3, the coefficient of x^2. Here are the steps.

$$3x^2 + 2x - 6 = 0 \qquad \text{Given.}$$

STEP 1 Add 6 to isolate the variables on the left-hand side.

$$3x^2 + 2x = 6 \qquad \text{Add 6 to both sides.}$$

STEP 2 Divide each term by 3, the coefficient of x^2.

$$x^2 + \frac{2x}{3} = 2 \qquad \text{Divide each term by 3.}$$

STEP 3 Add the square of one-half the first-degree term's coefficient to both sides; that is, add

$$\left[\frac{1}{2}\left(\frac{2}{3}\right)\right]^2 = \left(\frac{1}{3}\right)^2$$

to both sides

$$x^2 + \frac{2x}{3} + \left(\frac{1}{3}\right)^2 = 2 + \left(\frac{1}{3}\right)^2$$

The first-degree term's coefficient is $\frac{2}{3}$, so add
$\left(\frac{1}{2} \cdot \frac{2}{3}\right)^2 = \left(\frac{1}{3}\right)^2$
to both sides.

STEP 4 Factor the left-hand side, and simplify the right-hand side.

$$\left(x + \frac{1}{3}\right)^2 = \frac{19}{9}$$

STEP 5 Extract square roots.

$$x + \frac{1}{3} = \pm\sqrt{\frac{19}{9}} = \pm\frac{\sqrt{19}}{3}$$

STEP 6 Solve the resulting equation by subtracting $\frac{1}{3}$ from both sides and simplifying.

$$x = \pm\frac{\sqrt{19}}{3} - \frac{1}{3} \qquad \text{Subtract } \frac{1}{3}.$$

$$x = \frac{-1 \pm \sqrt{19}}{3}$$

The solution set is

$$\left\{\frac{-1 + \sqrt{19}}{3}, \frac{-1 - \sqrt{19}}{3}\right\}$$

Here is a summary of this procedure.

PROCEDURE

> **Solving a Quadratic Equation by Completing the Square**
>
> 1. Write the equation with the variables in descending order on the left and the numbers on the right.
> 2. If the coefficient of the square term is not 1, divide each term by this coefficient.
> 3. Add the square of one-half of the coefficient of the first-degree term to both sides.
> 4. Rewrite the left-hand side as a perfect square in factored form and simplify the right-hand side.
> 5. Extract the square root of both sides and rationalize the denominator if necessary.
> 6. Solve the resulting equation.

EXAMPLE 5 Completing the square when the coefficient of x^2 is not 1

Solve by completing the square: $3x^2 - 3x - 1 = 0$

SOLUTION We use the six-step procedure.

$$3x^2 - 3x - 1 = 0$$ Given.

STEP 1 $3x^2 - 3x = 1$ Add 1 so the variables are isolated.

STEP 2 $x^2 - x = \dfrac{1}{3}$ Divide each term by 3.

STEP 3 $x^2 - x + \left(\dfrac{1}{2}\right)^2 = \dfrac{1}{3} + \left(\dfrac{1}{2}\right)^2$ The first-degree term's coefficient is -1, so add $\left[\dfrac{1}{2}(-1)\right]^2 = \left(-\dfrac{1}{2}\right)^2 = \left(\dfrac{1}{2}\right)^2$ to each side.

STEP 4 $\left(x - \dfrac{1}{2}\right)^2 = \dfrac{7}{12}$ Factor the left-hand side, simplify the right-hand side.

STEP 5 $x - \dfrac{1}{2} = \pm\sqrt{\dfrac{7}{12}} = \pm\dfrac{\sqrt{7}\cdot\sqrt{3}}{\sqrt{12}\cdot\sqrt{3}}$ Extract square roots and rationalize the denominator.

$$= \pm\dfrac{\sqrt{21}}{6}$$ Simplify.

STEP 6 $x - \dfrac{1}{2} = \dfrac{\sqrt{21}}{6}$ or $x - \dfrac{1}{2} = -\dfrac{\sqrt{21}}{6}$ Solve.

$$x = \dfrac{1}{2} + \dfrac{\sqrt{21}}{6} \quad \text{or} \quad x = \dfrac{1}{2} - \dfrac{\sqrt{21}}{6}$$ Check this.

The solutions are

$$\dfrac{1}{2} + \dfrac{\sqrt{21}}{6} \quad \text{and} \quad \dfrac{1}{2} - \dfrac{\sqrt{21}}{6}, \quad \text{or} \quad \dfrac{3 + \sqrt{21}}{6} \quad \text{and} \quad \dfrac{3 - \sqrt{21}}{6}$$

The solution set is thus

$$\left\{ \dfrac{3 + \sqrt{21}}{6}, \dfrac{3 - \sqrt{21}}{6} \right\}$$ ∎

GRAPH IT

We've already mentioned that there are several methods that can be used to solve equations:

1. Graphing (and using the TRACE and ZOOM keys to approximate the solutions).

2. Finding roots (press 2nd CALC 2).

3. Finding intersections (press 2nd CALC 5).

4. Using the solve feature (press MATH 0).

In this section we shall use the third technique. (If your grapher doesn't have this feature, use the first technique.) Keep in mind that the answers (solutions) you will get are decimal approximations. (If the algebraic solution is

$$\dfrac{\sqrt{2}}{2}$$

the grapher will give it as .7071067812.) We start every problem by placing all variables on the left (Y_1) and all numbers on the right (Y_2) of the equation. You have to do this algebraically! We then find the intersection of Y_1 and Y_2. The value of x thus obtained is the solution of the equation.

(*continued on next page*)

Let's try Example 5, which we rewrite as $3x^2 - 3x = 1$. Graph $Y_1 = 3x^2 - 3x$ and $Y_2 = 1$, using a $[-1, 3]$ by $[-2, 3]$ window. Now press $\boxed{\text{2nd}}$ $\boxed{\text{CALC}}$ 5. Move the cursor to a point on Y_1 and to the left of the intersection point. The grapher prompt is "First curve?" Press $\boxed{\text{ENTER}}$. Press $\boxed{\text{ENTER}}$ again when the grapher asks "Second Curve?" and when it asks for a "Guess?" The intersection is given as X $= -.2637626$ and Y $= 1$. Thus the solution (not shown) is $x = -0.2637626$. Check to see that this decimal approximation corresponds to the solution

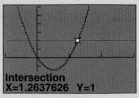

Intersection
X=1.2637626   Y=1

WINDOW 1
Using the intersection
feature

$$x = \frac{3 - \sqrt{21}}{6}$$

Now press $\boxed{\text{2nd}}$ $\boxed{\text{CALC}}$ 5 again and move the cursor to a point to the left of the second intersection. Press $\boxed{\text{ENTER}}$ three times. The intersection occurs when X $=1.2637626$, which corresponds to

$$\frac{3 + \sqrt{21}}{6}$$

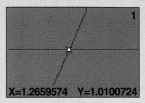

X=1.2659574   Y=1.0100724

WINDOW 2
Using the trace and zoom
features

(see Window 1). In case you don't have the intersect feature, Window 2 shows the result obtained by graphing and approximating the points of intersection (X$=1.2659574$).

Try Example 2b, with $Y_1 = 3(x - 2)^2$, $Y_2 = -25$, and a $[-10, 10]$ by $[-30, 5]$ window. Do the curves intersect? Can you find real-number solutions?

EXERCISE 6.1

A In Problems 1–20, solve the equation.

1. $x^2 = 64$

2. $x^2 = 81$

3. $x^2 = -121$

4. $x^2 = -144$

5. $x^2 - 169 = 0$

6. $x^2 - 100 = 0$

7. $x^2 + 4 = 0$

8. $x^2 + 25 = 0$

9. $36x^2 - 49 = 0$

10. $36x^2 - 81 = 0$

11. $4x^2 + 81 = 0$

12. $9x^2 + 64 = 0$

13. $3x^2 - 25 = 0$

14. $5x^2 - 16 = 0$

15. $5x^2 + 36 = 0$

16. $11x^2 + 49 = 0$

17. $3x^2 - 100 = 0$

18. $4x^2 - 13 = 0$

19. $13x^2 + 81 = 0$

20. $11x^2 + 4 = 0$

B In Problems 21–40, solve the equation.

21. $(x + 5)^2 = 4$

22. $(x + 3)^2 = 9$

23. $x^2 + 4x + 4 = -25$

24. $x^2 + 2x + 1 = -16$

25. $(x - 6)^2 = 18$

26. $(x - 2)^2 = 50$

27. $x^2 - 2x + 1 = -28$

28. $x^2 - 6x + 9 = -32$

29. $(x - 1)^2 - 50 = 0$

30. $(x - 2)^2 - 18 = 0$

31. $(x - 5)^2 - 32 = 0$

32. $(x - 2)^2 - 4 = 0$

33. $(x - 9)^2 + 64 = 0$

34. $(x - 2)^2 + 25 = 0$

35. $3x^2 + 6x + 3 = 96$

36. $3x^2 + 30x + 75 = 72$

37. $7(x - 2)^2 - 350 = 0$

38. $3(x - 1)^2 - 54 = 0$

39. $7(x - 5)^2 + 189 = 0$

40. $5(x - 3)^2 + 250 = 0$

C In Problems 41–70, solve by completing the square.

41. $x^2 + 6x + 5 = 0$

42. $x^2 + 4x + 3 = 0$

43. $x^2 + 8x + 15 = 0$

44. $x^2 + 8x + 7 = 0$

45. $x^2 + 6x + 10 = 0$

46. $x^2 + 12x + 37 = 0$

47. $x^2 - 10x + 24 = 0$

48. $x^2 + 12x - 28 = 0$

49. $x^2 - 10x + 21 = 0$

50. $x^2 - 2x - 143 = 0$

51. $x^2 - 8x + 17 = 0$

52. $x^2 - 14x + 58 = 0$

53. $2x^2 + 4x + 3 = 0$

54. $2x^2 + 7x + 6 = 0$

55. $3x^2 + 6x + 78 = 0$

56. $9x^2 + 6x + 2 = 0$

57. $25y^2 - 25y + 6 = 0$

58. $4y^2 - 16y + 15 = 0$

59. $4y^2 - 4y + 5 = 0$

60. $9x^2 - 12x + 13 = 0$

61. $4x^2 - 7 = 4x$

62. $2x^2 - 18 = -9x$

63. $2x^2 + 1 = 4x$

64. $2x^2 + 3 = 6x$

65. $(x + 3)(x - 2) = -4$

66. $(x + 4)(x - 1) = -6$

67. $2x(x + 5) - 1 = 0$

68. $2x(x - 4) = 2(9 - 8x) - x$

69. $2x(x + 3) - 10 = 0$

70. $4x(x + 1) - 5 = 0$

APPLICATIONS

71. The distance traveled in t seconds by an object dropped from a height h is given by $h = 16t^2$. How long would it take an object dropped from a height of 64 ft to hit the ground?

72. Use the formula given in Problem 71 to find the time it takes an object dropped from a height of 32 ft to hit the ground.

73. The amount of money A received at the end of 2 yr when P dollars are invested at a compound rate r is $A = P(1 + r)^2$. Find the rate of interest r (written as a percent) if a person invested $100 and received $121 at the end of 2 yr.

74. Use the formula of Problem 73 to find the rate of interest r (written as a percent) if a person invested $100 and received $144 at the end of 2 yr.

SKILL CHECKER

Evaluate

$$\frac{-b \pm \sqrt{b^2 - 4ac}}{2a}$$

for the given values of a, b, and c:

75. $a = 1, b = -9, c = 0$

76. $a = 1, b = -6, c = 0$

77. $a = 1, b = -2, c = -2$

78. $a = 1, b = -4, c = -4$

79. $a = 8, b = 7, c = -1$

80. $a = 3, b = -2, c = -5$

81. $a = 3, b = -8, c = 7$

82. $a = 4, b = -3, c = 5$

83. $a = 1, b = 2, c = 6$

84. $a = 1, b = 2, c = 5$

USING YOUR KNOWLEDGE   Average, Demand, and Bacteria

Many applications of mathematics require finding the maximum or the minimum of certain algebra expressions. Thus a certain business may wish to find the price at which a product will bring *maximum* profits, while a team of engineers may be interested in *minimizing* the amount of carbon monoxide produced by auto-

mobiles. Now suppose you are the manufacturer of a product whose average manufacturing cost $\bar{C}$ (in dollars), based on producing x (thousand) units, is given by the expression

$$\bar{C} = x^2 - 8x + 18$$

How many units should be produced to minimize the cost per unit? If we consider the right-hand side of the equation, we can complete the square and leave the equation unchanged by adding and subtracting the appropriate number. Thus

$$\bar{C} = x^2 - 8x + 18$$
$$= (x^2 - 8x +) + 18$$
$$= (x^2 - 8x + 4^2) + 18 - 4^2$$

Then

$$\bar{C} = (x - 4)^2 + 2$$

Now for $\bar{C}$ to be as small as possible (minimizing the cost), we make $(x - 4)^2$ zero by letting $x = 4$; then $\bar{C} = 2$. This tells us that when 4 (thousand) units are produced, the minimum cost is 2. That is, the minimum average cost per unit is $2.

Use your knowledge about completing the square to solve the following problems.

85. A manufacturer's average cost $\bar{C}$ (in dollars), based on manufacturing x (thousand) items, is given by

$$\bar{C} = x^2 - 4x + 6$$

  a. How many units should be produced to minimize the cost per unit?

  b. What is the minimum average cost per unit?

86. The demand D for a certain product depends on x (thousand) units produced and is given by

$$D = x^2 - 2x + 3$$

For what number of units is the demand at its lowest?

87. Have you seen people adding chlorine to their swimming pools? This is done to reduce the number of bacteria present in the water. Suppose that after t days, the number of bacteria per cubic centimeter is given by the expression

$$B = 20t^2 - 120t + 200$$

In how many days will the number of bacteria be at its lowest?

WRITE ON . . .

88. Explain why $\sqrt{49}$ has only one answer, 7, but $x^2 = 49$ has two solutions, 7 and -7.

89. We have solved $x^2 - 49 = 0$ by adding 49 to both sides and then extracting roots. Describe another procedure you can use

to solve $x^2 - 49 = 0$. Does your method work when solving $x^2 - 2 = 0$?

90. What is the first step in solving a quadratic equation by completing the square?

 If you know how to do these problems, you have learned your lesson!

Solve by completing the square:

91. $5x^2 - 5x = 1$

92. $2x^2 - 2x - 1 = 0$

93. $x^2 + 12x = 8$

94. $x^2 + 6x - 1 = 0$

Solve:

95. $2x^2 + 12x + 18 = 27$

96. $3x^2 + 6x + 3 = 20$

97. $5(x - 3)^2 + 36 = 0$

98. $2(x - 1)^2 + 49 = 0$

99. $(x - 2)^2 = 24$

100. $(x + 1)^2 = 18$

101. $3x^2 - 16 = 0$

102. $3x^2 + 54 = 0$

103. $5x^2 + 60 = 0$

104. $9x^2 - 4 = 0$

6.2

THE QUADRATIC FORMULA: APPLICATIONS

To succeed, review how to:

1. Find the square root of a number (p. 291).
2. Simplify square roots (p. 291).
3. Write fractions in lowest terms (p. 208).
4. Solve a quadratic equation by completing the square (pp. 338–341).

Objectives:

 Solve equations using the quadratic formula.

 Solve factorable cubic equations.

 Solve applications involving quadratic equations.

getting started **Rockets and Quadratics**

The man standing by the first liquid-fueled rocket is Dr. Robert Goddard. His rocket went up 41 ft! Do you know how long it took it to go up that high? The height (in feet) of the rocket is given by

$$h = -16t^2 + v_0 t$$

where v_0 is the initial velocity (51.225 ft/sec). We can substitute 41 for h, the height, and 51.225 for v_0, the initial velocity, and solve for t in the equation

$$-16t^2 + 51.225t = 41$$

Unfortunately, this equation is not factorable, and to complete the square, we would have to divide by -16 and add

$$\left(-\frac{51.225}{32}\right)^2$$

to both sides. There must be a better way to solve this equation! In this section we derive a formula that will give us a more efficient method for solving this type of problem.

Let's solve the general quadratic equation $ax^2 + bx + c = 0$ $(a \neq 0)$ using the procedure we studied for completing the square and then using the resulting formula, which solves *any* given quadratic equation by substituting specific values for a, b, and c. Using the six-step procedure for completing the square, we have

$$ax^2 + bx + c = 0 \quad (a \neq 0) \qquad \text{Given.}$$

$$ax^2 + bx = -c \qquad \text{Add } -c \text{ to both sides.}$$

$$x^2 + \frac{b}{a}x = -\frac{c}{a} \qquad \text{Divide each term by } a.$$

$$x^2 + \frac{b}{a}x + \left(\frac{b}{2a}\right)^2 = \left(\frac{b}{2a}\right)^2 - \frac{c}{a} \qquad \text{Add the square of one-half the coefficient of } x, \text{ that is, } \left(\frac{1}{2} \cdot \frac{b}{a}\right)^2 \text{ to both sides.}$$

$$\left(x + \frac{b}{2a}\right)^2 = \frac{b^2}{4a^2} - \frac{c}{a} \qquad \text{Factor the left-hand side, simplify the right-hand side. Since the LCD on the right is } 4a^2; \text{ write}$$

$$\left(x + \frac{b}{2a}\right)^2 = \frac{b^2}{4a^2} - \frac{4ac}{4a^2} \qquad -\frac{c}{a} \text{ as } -\frac{4ac}{4a^2}$$

$$\left(x + \frac{b}{2a}\right)^2 = \frac{b^2 - 4ac}{4a^2} \qquad \text{Then combine } \frac{b^2}{4a^2} \text{ and } -\frac{4ac}{4a^2}.$$

$$x + \frac{b}{2a} = \frac{\pm\sqrt{b^2 - 4ac}}{2a} \qquad \text{Extract square roots.}$$

$$x = -\frac{b}{2a} \pm \frac{\sqrt{b^2 - 4ac}}{2a} \qquad \text{Add } -\frac{b}{2a}.$$

$$x = \frac{-b \pm \sqrt{b^2 - 4ac}}{2a} \qquad \text{Combine.}$$

Remember, we only have to do this once. Now we have the **quadratic formula** to use!

Solving Equations Using the Quadratic Formula

Any time we have a quadratic equation in the standard form $ax^2 + bx + c = 0$, we can find the solutions by simply substituting the values for a, b, and c in the quadratic formula.

SOLUTIONS OF A QUADRATIC EQUATION IN STANDARD FORM

The solutions of $ax^2 + bx + c = 0$ are

$$x = \frac{-b \pm \sqrt{b^2 - 4ac}}{2a}$$

EXAMPLE 1   Using the quadratic formula to solve equations

Solve: $8x^2 + 7x + 1 = 0$

SOLUTION   The equation is written in standard form:

$$\underbrace{8x^2}_{a\,=\,8} + \underbrace{7x}_{b\,=\,7} + \underbrace{1}_{c\,=\,1} = 0$$

where it's clear that $a = 8$, $b = 7$, and $c = 1$. Substituting the values of a, b, and c in the formula, we obtain

$$x = \frac{-7 \pm \sqrt{(7)^2 - 4(8)(1)}}{2(8)}$$ Let $a = 8$, $b = 7$, and $c = 1$.

$$= \frac{-7 \pm \sqrt{49 - 32}}{16}$$ Since $(7)^2 = 49$ and $-4(8)(1) = -32$

$$= \frac{-7 \pm \sqrt{17}}{16}$$ $\sqrt{49 - 32} = \sqrt{17}$

$$= \frac{-7 \pm \sqrt{17}}{16}$$

Thus

$$x = \frac{-7 + \sqrt{17}}{16} \quad \text{or} \quad x = \frac{-7 - \sqrt{17}}{16}$$

The solution set is

$$\left\{ \frac{-7 + \sqrt{17}}{16}, \frac{-7 - \sqrt{17}}{16} \right\}$$ ∎

EXAMPLE 2   More practice using the quadratic formula

Solve: $2x^2 = 2x + 1$

SOLUTION   We proceed by steps as before. We write the equation in standard form by subtracting $2x$ and 1 from both sides of $2x^2 = 2x + 1$ to obtain

$$\overbrace{2x^2}^{a = 2} - \overbrace{2x}^{b = -2} - \overbrace{1}^{c = -1} = 0$$ In standard form

where $a = 2$, $b = -2$, and $c = -1$. Substituting these values in the quadratic formula, we have

$$x = \frac{-(-2) \pm \sqrt{(-2)^2 - 4(2)(-1)}}{2(2)}$$ Let $a = 2$, $b = -2$, and $c = -1$.

$$= \frac{2 \pm \sqrt{4 + 8}}{4}$$ Since $(-2)^2 = 4$ and $-4(2)(-1) = 8$

$$= \frac{2 \pm \sqrt{12}}{4}$$ $\sqrt{4 + 8} = \sqrt{12}$

$$= \frac{2 \pm \sqrt{4 \cdot 3}}{4}$$ $\sqrt{12} = \sqrt{4 \cdot 3}$

$$= \frac{2 \pm 2\sqrt{3}}{4}$$ $\sqrt{4 \cdot 3} = \sqrt{4} \cdot \sqrt{3} = 2\sqrt{3}$

$$= \frac{\overset{1}{2}(1 \pm \sqrt{3})}{\underset{2}{\cancel{4}}}$$

Thus

$$x = \frac{1 + \sqrt{3}}{2} \quad \text{or} \quad x = \frac{1 - \sqrt{3}}{2}$$

The solution set is

$$\left\{ \frac{1 + \sqrt{3}}{2}, \frac{1 - \sqrt{3}}{2} \right\} \qquad \blacksquare$$

EXAMPLE 3   Using the quadratic formula when there is no constant term

Solve: $9x = x^2$

SOLUTION   Subtracting $9x$ from both sides of $9x = x^2$, we have

$$0 = x^2 - 9x$$

or

$$\underbrace{x^2}_{a = 1} - \underbrace{9x}_{b = -9} + \underbrace{0}_{c = 0} = 0 \qquad \text{In standard form}$$

where $a = 1$, $b = -9$, and $c = 0$ (because the c term is missing). Substituting these values in the formula, we obtain

$$x = \frac{-(-9) \pm \sqrt{(-9^2 - 4(1)(0)}}{2(1)} \qquad \text{Let } a = 1, b = -9, \text{ and } c = 0.$$

$$= \frac{9 \pm \sqrt{81 - 0}}{2} \qquad \text{Since } (-9)^2 = 81 \text{ and } -4(1)(0) = 0$$

$$= \frac{9 \pm \sqrt{81}}{2} \qquad \sqrt{81 - 0} = \sqrt{81}$$

$$= \frac{9 \pm 9}{2} \qquad \pm\sqrt{81} = \pm 9$$

Thus

$$x = \frac{9 + 9}{2} = \frac{18}{2} = 9 \quad \text{or} \quad x = \frac{9 - 9}{2} = \frac{0}{2} = 0$$

The solutions are 9 and 0, and the solution set is $\{9, 0\}$. $\blacksquare$

 The expression $x^2 - 9x = x(x - 9) = 0$ could have been solved by factoring. Try factoring the equation *before* you use the quadratic formula.

EXAMPLE 4   Clearing fractions before using the quadratic formula

Solve:

$$\frac{x^2}{4} + \frac{2}{3}x = -\frac{1}{3}$$

SOLUTION   We have to write the equation in standard form, but first we clear fractions by multiplying each term by the LCM of 4 and 3—that is, by 12:

$$12 \cdot \frac{x^2}{4} + 12 \cdot \frac{2}{3}x = -\frac{1}{3} \cdot 12$$

$$3x^2 + 8x = -4$$

We then add 4 to obtain the equivalent equation

$$\underbrace{3x^2}_{a\,=\,3} + \underbrace{8x}_{b\,=\,8} + \underbrace{4}_{c\,=\,4} = 0 \qquad \text{In standard form}$$

where $a = 3$, $b = 8$, and $c = 4$. Substituting in the formula

$$x = \frac{-8 \pm \sqrt{64 - 4(3)(4)}}{2(3)} \qquad \text{Let } a = 3, b = 8, \text{ and } c = 4.$$

$$= \frac{-8 \pm \sqrt{64 - 48}}{6} \qquad \text{Since } -4(3)(4) = -48$$

$$= \frac{-8 \pm \sqrt{16}}{6} \qquad \sqrt{64 - 48} = \sqrt{16}$$

$$= \frac{-8 \pm 4}{6} \qquad \pm\sqrt{16} = \pm 4$$

Thus

$$x = \frac{-8 + 4}{6} = \frac{-4}{6} = -\frac{2}{3} \qquad \text{or} \qquad x = \frac{-8 - 4}{6} = \frac{-12}{6} = -2$$

The solutions are $-\frac{2}{3}$ and -2, and the solution set is $\left\{-\frac{2}{3}, -2\right\}$. ■

CAUTION Now, a word of warning. As you recall, some quadratic equations have imaginary-number solutions. Such solutions can be obtained by using the quadratic formula, as shown in Example 5.

EXAMPLE 5   Solving quadratic equations with imaginary-number solutions

Solve: $3x^2 + 3x = -2$

SOLUTION   We add 2 to both sides of $3x^2 + 3x = -2$ so the equation is in standard form. We then have

$$\underbrace{3x^2}_{a\,=\,3} + \underbrace{3x}_{b\,=\,3} + \underbrace{2}_{c\,=\,2} = 0$$

where $a = 3$, $b = 3$, and $c = 2$. Now we have

$$x = \frac{-3 \pm \sqrt{(3)^2 - 4(3)(2)}}{2(3)}$$

$$= \frac{-3 \pm \sqrt{9 - 24}}{6}$$

$$= \frac{-3 \pm \sqrt{-15}}{6}$$

$$= \frac{-3 \pm \sqrt{15}i}{6} \qquad \text{Note that } \sqrt{-15} = \sqrt{15}i. \text{ Make sure } i \text{ is outside the radical sign!}$$

The solutions are thus

$$\frac{-3 + \sqrt{15}\,i}{6} \quad \text{and} \quad \frac{-3 - \sqrt{15}\,i}{6}$$

and the solution set is

$$\left\{\frac{-3 + \sqrt{15}\,i}{6}, \frac{-3 - \sqrt{15}\,i}{6}\right\}$$

∎

B   Solving Factorable Cubic Equations

Now let's consider $x^3 - 27 = 0$, which is *not* a quadratic equation. We will solve it by factoring. Here's how we do it.

$$(x - 3)(x^2 + 3x + 9) = 0 \qquad \text{Factor.}$$

$$x - 3 = 0 \quad \text{or} \quad x^2 + 3x + 9 = 0 \qquad \text{Zero-factor property}$$

$$x = 3 \quad \text{or} \quad x^2 + 3x + 9 = 0$$

Since the second equation does not factor, we use the quadratic formula to solve it. For this equation, $a = 1$, $b = 3$, and $c = 9$. Substituting in the quadratic formula, we have

$$x = \frac{-3 \pm \sqrt{3^2 - 4(1)(9)}}{2 \cdot 1}$$

$$= \frac{-3 \pm \sqrt{9 - 36}}{2}$$

$$= \frac{-3 \pm \sqrt{-27}}{2}$$

$$= \frac{-3 \pm \sqrt{9 \cdot 3 \cdot (-1)}}{2}$$

$$= \frac{-3 \pm 3\sqrt{3}\,i}{2}$$

The solutions of $x^3 - 27 = 0$ are thus

$$3, \quad \frac{-3 + 3\sqrt{3}\,i}{2}, \quad \text{and} \quad \frac{-3 - 3\sqrt{3}\,i}{2}$$

and the solution set is

$$\left\{3, \frac{-3 + 3\sqrt{3}\,i}{2}, \frac{-3 - 3\sqrt{3}\,i}{2}\right\}$$

Note that $x^3 - 27 = 0$ *cannot* be completely solved by extracting roots—that is, by writing

$$x^3 - 27 = 0$$

$$x^3 = 27 \qquad \text{Add 27.}$$

$$x = \sqrt[3]{27} \qquad \text{Extract roots.}$$

$$x = 3$$

As you can see, this method yields only one real-number solution when there are actually three solutions: one real-number and two imaginary-number solutions.

EXAMPLE 6　Solving a factorable cubic equation

Solve: $8x^3 - 27 = 0$

SOLUTION　We factor the equation, use the zero-factor property, and then use the quadratic formula.

$$8x^3 - 27 = 0 \qquad \text{Given.}$$

$$(2x - 3)(4x^2 + 6x + 9) = 0 \qquad \text{Factor.}$$

$$2x - 3 = 0 \quad \text{or} \quad 4x^2 + 6x + 9 = 0$$

$$x = \frac{3}{2} \quad \text{or} \quad 4x^2 + 6x + 9 = 0$$

The second equation is a quadratic equation with $a = 4$, $b = 6$, and $c = 9$. We solve it with the quadratic formula:

$$x = \frac{-6 \pm \sqrt{6^2 - 4(4)(9)}}{2 \cdot 4}$$

$$= \frac{-6 \pm \sqrt{36 - 144}}{8}$$

$$= \frac{-6 \pm \sqrt{-108}}{8}$$

$$= \frac{-6 \pm \sqrt{36 \cdot (3) \cdot (-1)}}{8}$$

$$= \frac{-6 \pm 6\sqrt{3}\,i}{8}$$

$$= \frac{2(-3 \pm 3\sqrt{3}\,i)}{2 \cdot 4}$$

$$= \frac{-3 \pm 3\sqrt{3}\,i}{4}$$

The solutions of $8x^3 - 27 = 0$ are

$$\frac{3}{2}, \qquad \frac{-3 + 3\sqrt{3}\,i}{4}, \qquad \text{and} \qquad \frac{-3 - 3\sqrt{3}\,i}{4}$$

and the solution set is

$$\left\{ \frac{3}{2}, \frac{-3 + 3\sqrt{3}\,i}{4}, \frac{-3 - 3\sqrt{3}\,i}{4} \right\} \qquad \blacksquare$$

By the way, we left Dr. Goddard's rocket up in the air! How long *did* it fly? The equation was

$$-16t^2 + 51.225t = 41$$

or, in standard form,

$$-16t^2 + 51.225t - 41 = 0$$

Here $a = -16$, $b = 51.225$, and $c = -41$, so the quadratic formula gives

$$t = \frac{-51.225 \pm \sqrt{51.225^2 - 4(-16)(-41)}}{2 \cdot (-16)}$$

With a calculator, we get $t \approx 1.6$. Thus it took the rocket about 1.6 sec to reach 41 ft.

Applications Involving Quadratic Equations

Quadratic equations are used in many fields: rockets, economics, and marketing, to name a few. Here are some examples.

EXAMPLE 7   What's the price?

In business, when the price p (in dollars) of a product increases, the demand d decreases and is given by $d = 300/p$. On the other hand, when the price p increases, the supply s producers are willing to sell increases and is given by $s = 100p - 50$. In economic theory, the point at which the supply equals the demand, $s = d$, is called the **equilibrium point**. Find the price p at the equilibrium point.

SOLUTION   Since $s = d$ at equilibrium, we have

$$100p - 50 = \frac{300}{p}$$

$$p(100p - 50) = p \cdot \frac{300}{p} \qquad \text{Multiply by } p.$$

$$100p^2 - 50p = 300$$

$$2p^2 - p = 6 \qquad \text{Divide by 50.}$$

$$2p^2 - p - 6 = 0 \qquad \text{Subtract 6.}$$

Since $a = 2$, $b = -1$, and c -6, we have

$$p = \frac{-(-1) \pm \sqrt{(-1)^2 - 4(2)(-6)}}{2 \cdot 2}$$

$$= \frac{1 \pm \sqrt{1 + 48}}{4}$$

$$= \frac{1 \pm \sqrt{49}}{4}$$

$$= \frac{1 \pm 7}{4}$$

Thus

$$p = \frac{1 + 7}{4} = 2 \qquad \text{or} \qquad p = \frac{1 - 7}{4} = -\frac{3}{2}$$

Since the price must be positive, we use $p = \$2$. Note that in this case, we obtain two rational roots, which means that the original equation was factorable. Before you use the quadratic formula, always try to factor. You will often save time! ■

EXAMPLE 8   Where's the beef?

According to the U.S. Department of Agriculture, per capita consumption of beef declined from 1985 to 1992, while poultry consumption rose in the same period. If $B(t)$ and $P(t)$ represent per capita consumption of beef and poultry (in pounds), respectively, and t represents the number of years after 1985, beef and poultry consumption can be described by these equations.

$$\text{Beef:} \qquad B(t) = 0.2t^2 - 3.23t + 75$$

$$\text{Poultry:} \qquad P(t) = 2.12t + 45$$

a. Was the consumption of beef and poultry ever the same?
b. In what year would this happen?

SOLUTION

a. To find out whether the consumption of beef and poultry would ever be the same, we have to determine whether $B(t) = P(t)$ has a real-number solution. So we assume $B(t) = P(t)$, or

$$0.2t^2 - 3.23t + 75 = 2.12t + 45$$

$$0.2t^2 - 5.35t + 30 = 0 \qquad \text{Subtract } 2.12t \text{ and } 45.$$

Here $a = 0.2$, $b = -5.35$, and $c = 30$. Substituting in the quadratic formula, we have

$$t = \frac{-(-5.35) \pm \sqrt{(-5.35)^2 - 4 \cdot (0.2)(30)}}{2 \cdot (0.2)}$$

$$= \frac{5.35 \pm \sqrt{28.6225 - 24}}{0.4}$$

$$= \frac{5.35 \pm \sqrt{4.6225}}{0.4}$$

$$= \frac{5.35 \pm 2.15}{0.4}$$

Thus

$$t = \frac{7.5}{0.4} = 18.75 \qquad \text{or} \qquad t = \frac{3.2}{0.4} = 8$$

which means that the consumption of beef and poultry would be the same for the specified values of t.

b. The consumption of beef and poultry would be the same $t = 18.75$ years from 1985, that is, in 2003.75 (2004) or $t = 8$ yr from 1985, that is, in 1993. ■

So far we've used graphers to verify our work. They are even more efficient in predicting the *type* of solutions we will get when we solve quadratic equations. Thus if you graph the equations in Examples 1–4, you will notice that the graphs cross the horizontal axis at two points and the equations have two solutions that the grapher gives as decimal approximations.

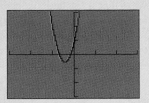

WINDOW 1
$Y_1 = 8x^2 + 7x + 1$
$Y_2 = 0$

In Example 1, we enter $Y_1 = 8x^2 + 7x + 1$ using a $[-3, 3]$ by $[-3, 3]$ window (see Window 1). To find where Y_1 intersects the horizontal axis, we can enter $Y_2 = 0$ (which is the horizontal axis) and use [2nd] [CALC] 5, the intersection feature of your grapher. (If your grapher doesn't have this feature, you can use the [TRACE] and [ZOOM] keys or the root feature [2nd] [CALC] 2.) This will confirm that

$$-0.6951941 \approx \frac{-7 - \sqrt{17}}{16}$$

is indeed one of the solutions. The other solution is given as

$$-0.1798059 \approx \frac{-7 + \sqrt{17}}{16}$$

as shown in the example.

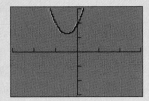

WINDOW 2
$Y_1 = 3x^2 + 3x + 2$

What happens if you try to solve $3x^2 + 3x = -2$ from Example 5? First, write the equation as $3x^2 + 3x + 2 = 0$ and graph $Y_1 = 3x^2 + 3x + 2$. As you can see, the curve does not cut the horizontal axis in Window 2. This means that there are no real-number solutions for this equation. The solutions are *imaginary* numbers, and you have to find them algebraically. (Even with the best grapher available, you have to know your algebra!)

Finally, try to find the solutions of $8x^3 - 27 = 0$. The graph is shown in Window 3 and, as before, only one solution, $x = 1.5 = \frac{3}{2}$, is a real number. You have to use the quadratic equation to find the other two solutions! (By the way, if you want to see a more complete graph, try a $[-2, 2]$ by $[-50, 50]$ window.)

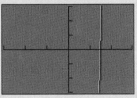

WINDOW 3
$Y_1 = 8x^3 - 27$

EXERCISE 6.2

A In Problems 1–34, solve the equation.

1. $x^2 + x - 2 = 0$

2. $x^2 + 4x - 1 = 0$

3. $x^2 + 4x = -1$

4. $x^2 + 6x = -5$

5. $x^2 - 3x = 2$

6. $x^2 - 4x = 12$

7. $7y^2 = 12y - 5$

8. $7x^2 = 6x - 1$

9. $5y^2 + 8y = -5$

10. $5y^2 + 6y = -5$

11. $7y + 6 = -2y^2$

12. $7y + 3 = -2y^2$

13. $\dfrac{x^2}{5} - \dfrac{x}{2} = \dfrac{-3}{10}$

14. $\dfrac{x^2}{4} - \dfrac{x}{2} = -\dfrac{1}{8}$

15. $\dfrac{x^2}{7} + \dfrac{x}{2} = \dfrac{-3}{14}$

16. $\dfrac{x^2}{8} + \dfrac{x}{2} = -\dfrac{1}{8}$

17. $\dfrac{x^2}{2} - \dfrac{3x}{4} = \dfrac{-1}{8}$

18. $\dfrac{x^2}{10} - \dfrac{x}{5} = \dfrac{3}{2}$

19. $\dfrac{x^2}{8} = -\dfrac{x}{4} - \dfrac{1}{8}$

20. $\dfrac{x^2}{12} = -\dfrac{x}{4} - \dfrac{1}{3}$

21. $6x = 4x^2 + 1$

22. $6x = 9x^2 - 4$

23. $3x = 1 - 3x^2$

24. $3x = 2x^2 - 5$

25. $x(x + 2) = 2x(x + 1) - 4$

26. $x(4x - 7) - 10 = 6x^2 - 7x$

27. $6x(x + 5) = (x + 15)^2$

28. $6x(x + 1) = (x + 3)^2$

29. $(x - 2)^2 = 4x(x - 1)$

30. $(x - 4)^2 = 4x(x - 2)$

B In Problems 31–34, solve the equation.

31. $x^3 - 8 = 0$

32. $x^3 - 1 = 0$

33. $8x^3 - 1 = 0$

34. $27x^3 - 1 = 0$

C Applications

35. According to the U.S. Department of Agriculture, annual per capita consumption of whole milk declined from 1970 to 1989 while that of lowfat milk increased in the same period. If $W(t)$ and $L(t)$ represent the annual per capita consumption of whole milk and lowfat milk (in gallons), respectively, and t represents the number of years after 1970, milk consumption can be described by the following equations:

Whole: ↓ $W(t) = 0.013t^2 - 0.96t + 25.4$

Lowfat: ↑ $L(t) = 0.4t + 6.03$

a. Can the consumption of whole and lowfat milk ever be the same?

b. In what year could this happen?

36. Find the price p (dollars) at the equilibrium point if the supply is $s = 30p - 50$ and the demand is $d = \frac{20}{p}$.

37. Find the price p (dollars) at the equilibrium point if the supply is $s = 30p - 50$ and the demand is $d = \frac{10}{p}$.

38. Find the price p (dollars) at the equilibrium point if the supply is $s = 20p - 60$ and the demand is $d = \frac{30}{p}$.

39. The bending moment M of a simple beam is given by $M = 20x - x^2$. For what values of x is $M = 40$?

40. Use the formula given in Problem 39 to find the values of x for which $M = 60$.

The maximum safe length L for which a beam will support a load d is given by $aL^2 + bL + c = d$, where a, b, c, and d depend on the materials and structures used. In Problems 41 and 42, find:

41. L when $a = 400$, $b = 200$, $c = 200$, and $d = 800$

42. L when $a = 5$, $b = 0$, $c = 100$, and $d = 180$

SKILL CHECKER

Simplify $\sqrt{b^2 - 4ac}$ using:

43. $a = 3$, $b = -2$, $c = -1$ 44. $a = 2$, $b = 3$, $c = -1$

45. $a = 3$, $b = -5$, $c = 4$ 46. $a = 3$, $b = -1$, $c = 1$

Find the product:

47. $(2x + 1)(3x - 4)$ 48. $(3x + 1)(2x - 5)$

49. $(3x - 7)(4x + 3)$ 50. $(4x - 8)(3x + 5)$

USING YOUR KNOWLEDGE

A Different Way of Finding the Quadratic Formula

In this section we derived the quadratic formula by completing the square. The procedure depends on making the x^2 coefficient 1. But there is another way to derive the quadratic formula. See if you can state the reasons for each of the

steps, which are given consecutively in Problems 51–57. (Note that $ax^2 + bx + c = 0$ is given.)

51. $4a^2x^2 + 4abx + 4ac = 0$

52. $4a^2x^2 + 4abx = -4ac$

53. $4a^2x^2 + 4abx + b^2 = b^2 - 4ac$

54. $(2ax + b)^2 = b^2 - 4ac$

55. $2ax + b = \pm\sqrt{b^2 - 4ac}$

56. $2ax = -b \pm \sqrt{b^2 - 4ac}$

57. $x = \dfrac{-b \pm \sqrt{b^2 - 4ac}}{2a}$

CALCULATOR CORNER

Using a Calculator with the Quadratic Formula

Any calculator that has a store ($\boxed{\text{STO}}$) and recall ($\boxed{\text{RCL}}$) keys can be extremely helpful in finding the solutions of a quadratic equation with the quadratic formula. Of course, the solutions you obtain are being approximated by decimals. It's especially convenient to start with the radical part in the solution of the quadratic equation and then store this value so you can evaluate both solutions without having to backtrack or write down any intermediate steps. Let's look at the equation of Example 1:

$$8x^2 + 7x + 1 = 0$$

Using the quadratic formula, one solution is obtained by entering

$7\ \boxed{x^2}\ \boxed{-}\ 4\ \boxed{\times}\ 8\ \boxed{\times}\ 1\ \boxed{=}$

$\boxed{\sqrt{x}}\ \boxed{\text{STO}}\ 7\ \boxed{+/-}\ \boxed{+}\ \boxed{\text{RCL}}\ \boxed{=}\ \boxed{\div}\ 2\ \boxed{\div}\ 8\ \boxed{=}$

The display shows -0.1798059 (which was given as

$$\frac{-7 + \sqrt{17}}{16}$$

in the example). To obtain the other solution, enter

$7\ \boxed{+/-}\ \boxed{-}\ \boxed{\text{RCL}}\ \boxed{=}\ \boxed{\div}\ 2\ \boxed{\div}\ 8\ \boxed{=}$

which yields -0.6951194. In general, to solve the equation $ax^2 + bx + c = 0$ using your calculator, enter the following:

$\boxed{b}\ \boxed{x^2}\ \boxed{-}\ 4\ \boxed{\times}\ \boxed{a}\ \boxed{\times}\ \boxed{c}\ \boxed{=}\ \boxed{\sqrt{x}}\ \boxed{\text{STO}}\ \boxed{b}\ \boxed{+/-}\ \boxed{+}\ \boxed{\text{RCL}}\ \boxed{=}$

$\boxed{\div}\ 2\ \boxed{\div}\ \boxed{a}\ \boxed{=}\ \boxed{b}\ \boxed{+/-}\ \boxed{-}\ \boxed{\text{RCL}}\ \boxed{=}\ \boxed{\div}\ 2\ \boxed{\div}\ \boxed{a}\ \boxed{=}$

Note that if $b^2 - 4ac < 0$, the calculator will give you an error message when you press $\boxed{\sqrt{x}}$. In such cases, you will have to change the sign before pressing $\boxed{\sqrt{x}}$ and supply the i in the final answer. (Try this in Example 6.)

WRITE ON . . .

58. Why do we have the restriction $a \neq 0$ when solving the equation $ax^2 + bx + c = 0$?

59. Explain the difference between linear, quadratic, and cubic equations.

60. The term $\sqrt{a}$ is a real number only when a is nonnegative. Use this fact to write a procedure that enables you to determine whether a quadratic equation with real coefficients has real-number solutions or imaginary-number solutions.

MASTERY TEST

If you know how to do these problems, you have learned your lesson!

Solve:

61. $27x^3 - 8 = 0$

62. $64x^3 - 1 = 0$

63. $3x^2 + 2x = -1$

64. $2x^2 + 3x = -2$

65. $\dfrac{x^2}{4} - \dfrac{3}{8}x = \dfrac{1}{4}$

66. $\dfrac{x^2}{2} - \dfrac{1}{4}x = \dfrac{3}{2}$

67. $6x = x^2$

68. $7x - x^2 = 0$

69. $x^2 = 4x + 4$

70. $x^2 = 2 + 2x$

71. $3x^2 + 2x - 5 = 0$

72. $2x^2 - 3x = 4$

73. The supply s of a certain product that producers are willing to sell is given by $s = 100p - 50$, where p is the price in dollars. If the demand d for this product is given by $d = \frac{50}{p}$, find the price p when $s = d$, the equilibrium point.

6.3

THE DISCRIMINANT AND ITS APPLICATIONS

To succeed, review how to:

1. Evaluate and simplify expressions that contain radicals (pp. 291–292).

2. Multiply two binomials (pp. 151–153).

Objectives:

A Use the discriminant to determine the number and type of solutions of a quadratic equation.

B Use the discriminant to determine whether a quadratic expression is factorable and then factor it.

C Find a quadratic equation with specified solutions.

D Verify the solutions of a quadratic equation.

getting started

Breaking Even Through Discriminants

A merchant in this mall wants to know if she is going to "break even." First, her daily costs, C, are represented by

$$C = 0.001x^2 + 10x + 100$$

where x is the number of items sold, and the corresponding revenue is $R = 20x - 0.01x^2$. The break-even point occurs when the cost C equals the revenue R—that is, when

$$C = R$$
$$0.001x^2 + 10x + 100 = 20x - 0.01x^2$$
$$0.011x^2 - 10x + 100 = 0 \qquad \text{In standard form}$$

She can break even only *if* this equation has real-number solutions. How do we ascertain that? By using the $b^2 - 4ac$ under the radical in the quadratic formula! Keep reading to see how.

The quadratic formula

$$x = \frac{-b \pm \sqrt{b^2 - 4ac}}{2a}$$

gives the solutions to any quadratic equation in standard form, so we can find out what type of solutions the equation has by looking at the expression under the radical, $b^2 - 4ac$. To do this, we need the following definition.

DISCRIMINANT

The expression $b^2 - 4ac$ under the radical is called the **discriminant** D; that is, $D = b^2 - 4ac$.

To find the discriminant of $0.011x^2 - 10x + 100 = 0$, note that $a = 0.011, b = -10$, and $c = 100$. Thus

$$b^2 - 4ac = (-10)^2 - 4(0.011)(100)$$
$$= 100 - 4.4$$
$$= 95.6$$

This means that the solutions of the equation are

$$\frac{-(-10) + \sqrt{95.6}}{2(0.011)} \quad \text{and} \quad \frac{-(-10) - \sqrt{95.6}}{2(0.011)}$$

That is, there are two real-number positive solutions to the equation, and our merchant can break even.

Using the Discriminant to Classify Solutions

If we calculate the discriminant of a quadratic equation, we can predict whether the solutions will be rational, irrational, or imaginary as well as the number of solutions we will have. This is important in applied problems where irrational or imaginary solutions are not always acceptable. The different possibilities are listed in Table 1.

TABLE 1 Solutions to $ax^2 + bx + c = 0$ Based on the Discriminant

The discriminant of $ax^2 + bx + c = 0$ is $b^2 - 4ac$. If a, b, and c are integers ($a \neq 0$), the type and number of solutions are as follows:	
Discriminant, $b^2 - 4ac$	**Solutions, $\dfrac{-b \pm \sqrt{b^2 - 4ac}}{2a}$**
Positive, not the square of an integer	*Two* different *irrational* solutions
Positive, and the square of an integer	*Two* different *rational* solutions
Negative	*Two* different *imaginary* solutions
Zero	*One rational* solution

Here are some examples.

Equation	$b^2 - 4ac$	Solutions (Roots)
$4x^2 - 3x - 5 = 0$	$(-3)^2 - 4(4)(-5) = 89$	Two irrational numbers
$x^2 - 2x - 3 = 0$	$(-2)^2 - 4(1)(-3) = 16$	Two rational numbers
$4x^2 - 3x + 5 = 0$	$(-3)^2 - 4(4)(5) = -71$	Two imaginary numbers
$4x^2 - 4x + 1 = 0$	$(-4)^2 - 4(4)(1) = 0$	One rational number

EXAMPLE 1   Using the discriminant to classify the solution

Given: $4x^2 - kx = -1$

a. Find the discriminant.

b. Find k so that the equation has exactly one rational solution.

SOLUTION

a. We first write the equation in standard form by adding 1 to both sides to obtain $4x^2 - kx + 1 = 0$. Thus $a = 4$, $b = -k$, $c = 1$, and

$$b^2 - 4ac = (-k)^2 - 4(4)(1)$$
$$= k^2 - 16$$

b. For this equation to have one rational solution, the discriminant $k^2 - 16$ must be zero:

$$0 = k^2 - 16$$
$$16 = k^2$$
$$\pm\sqrt{16} = k$$
$$\pm 4 = k$$

Thus when k is 4 or -4, the discriminant is zero, and the equation has one rational number as its solution.   ■

Determining Whether a Quadratic Expression Is Factorable

We have now learned that if the discriminant D is a perfect square, $ax^2 + bx + c = 0$ has two rational solutions, say r and s. Thus $a(x - r)(x - s) = 0$, which means that $ax^2 + bx + c$ is factorable into factors with *integer* coefficients. Here is our result.

FACTORABLE QUADRATIC EXPRESSIONS

> If $b^2 - 4ac$ is a perfect square, then $ax^2 + bx + c$ is factorable using integer coefficients.

To find out whether $12x^2 + 20x - 25$ is factorable, we must find $b^2 - 4ac = (20)^2 - 4(12)(-25) = 1600$. Since 1600 is a perfect square $[(40)^2 = 1600]$, $12x^2 + 20x - 25$ is factorable. (Contrast this technique with the ac test we learned in Section 3.1 or with trial and error!)

EXAMPLE 2   Using the discriminant to determine whether the expression is factorable

Use the discriminant to determine whether $20x^2 + 10x - 32$ is factorable.

SOLUTION   Here $a = 20$, $b = 10$, and $c = -32$.

$$b^2 - 4ac = (10)^2 - 4(20)(-32)$$
$$= 100 + 2560$$
$$= 2660$$

Since 2660 is not a perfect square, $20x^2 + 10x - 32$ is not factorable.   ■

Now that we know how to use the discriminant to determine whether a quadratic is factorable, we shall learn how to factor it. To factor $12x^2 + 20x - 25$, we first solve the corresponding quadratic equation $12x^2 + 20x - 25 = 0$. Recall that $b^2 - 4ac = 1600$; thus

$$x = \frac{-20 \pm \sqrt{1600}}{2(12)} = \frac{-20 \pm 40}{24}$$

The solutions are

$$\frac{-20 + 40}{24} = \frac{5}{6} \quad \text{and} \quad \frac{-20 - 40}{24} = -\frac{5}{2}$$

We now *reverse* the steps for solving a quadratic equation by factoring. Since

$$x = \frac{5}{6} \quad \text{or} \quad x = -\frac{5}{2}$$

$$6x = 5 \quad \text{or} \quad 2x = -5$$

$$6x - 5 = 0 \quad \text{or} \quad 2x + 5 = 0$$

That is, $(6x - 5)(2x + 5) = 0$. (You can check this by multiplying.) Thus, $12x^2 + 20x - 25 = (6x - 5)(2x + 5)$.

EXAMPLE 3   Factoring a quadratic expression

Factor if possible: $12x^2 + x - 35$

SOLUTION

$$b^2 - 4ac = (1)^2 - 4(12)(-35) = 1 + 1680 = 1681$$

Since $1681 = 41^2$, the expression is factorable. We use the quadratic formula to solve the related equation $12x^2 + x - 35 = 0$.

$$x = \frac{-1 + \sqrt{1681}}{2(12)} = \frac{-1 \pm 41}{24}$$

$$x = \frac{5}{3} \quad \text{and} \quad x = -\frac{7}{4}$$

Now we reverse the steps for solving a quadratic equation by factoring.

$$x = \frac{5}{3} \quad \text{or} \quad x = -\frac{7}{4}$$

$$3x = 5 \quad \text{or} \quad 4x = -7$$

$$3x - 5 = 0 \quad \text{or} \quad 4x + 7 = 0$$

Multiplying, $(3x - 5)(4x + 7) = 0$. Thus $12x^2 + x - 35 = (3x - 5)(4x + 7)$. You can check that the factorization is correct by multiplying the factors to obtain $12x^2 + x - 35$. ■

Finding Quadratic Equations with Specified Solutions

The process just discussed can be used to find a quadratic equation with specified solutions as shown in Example 4.

EXAMPLE 4   Testing, testing . . .

A professor wants to create a test question involving a quadratic equation whose solutions are $\left\{\frac{4}{5}, -\frac{2}{3}\right\}$. What should the quadratic equation be?

SOLUTION   In essence, you have to work backward! Since the solutions are $\frac{4}{5}$ and $-\frac{2}{3}$, the professor wants:

$$x = \frac{4}{5} \qquad \text{or} \qquad x = -\frac{2}{3}$$

$$x - \frac{4}{5} = 0 \qquad \text{or} \qquad x + \frac{2}{3} = 0 \qquad \text{Subtract } \tfrac{4}{5}, \text{ add } \tfrac{2}{3}.$$

$$\left(x - \frac{4}{5}\right)\left(x + \frac{2}{3}\right) = 0 \qquad \text{By the zero-factor property}$$

$$x^2 + \frac{2}{3}x - \frac{4}{5}x - \frac{8}{15} = 0 \qquad \text{Use FOIL.}$$

$$15x^2 + 10x - 12x - 8 = 0 \qquad \text{Multiply each term by the LCM of 3, 5, and 15, which is 15.}$$

$$15x^2 - 2x - 8 = 0 \qquad \text{Simplify.}$$

Thus $15x^2 - 2x - 8 = 0$ is a quadratic equation whose solution set is $\left\{\frac{4}{5}, -\frac{2}{3}\right\}$. ■

Verifying the Solutions of a Quadratic Equation

In the process of solving Example 3, we found out that the solutions of the equation $12x^2 + x - 35 = 0$ are $\frac{5}{3}$ and $-\frac{7}{4}$. To verify this, we can substitute these values in the original equation. But there is another way. If the equation is

$$ax^2 + bx + c = 0$$

then dividing by a, we can rewrite it as

$$x^2 + \frac{b}{a}x + \frac{c}{a} = 0 \qquad\qquad\qquad \textbf{(1)}$$

If the solutions are r_1 and r_2, we can also write

$$(x - r_1)(x - r_2) = 0$$

or

$$x^2 - (r_1 + r_2)x + r_1 r_2 = 0 \qquad\qquad\qquad \textbf{(2)}$$

Comparing equations (1) and (2), we see that

$$\frac{b}{a} = -(r_1 + r_2) \quad \text{and} \quad \frac{c}{a} = r_1 r_2$$

This discussion can be summarized as follows.

SUM AND PRODUCT OF THE SOLUTIONS OF A QUADRATIC EQUATION

If r_1 and r_2 are the solutions of the equation $ax^2 + bx + c = 0$, then

$$r_1 + r_2 = -\frac{b}{a} \quad \text{and} \quad r_1 r_2 = \frac{c}{a}$$

That is, the sum of the solutions of a quadratic equation is $-\frac{b}{a}$, and the product of the solution is $\frac{c}{a}$.

We can now verify that $\frac{5}{3}$ and $-\frac{7}{4}$ are solutions of $12x^2 + x - 35 = 0$. The sum of the solutions is

$$\frac{5}{3} + \left(-\frac{7}{4}\right) = \frac{20}{12} + \left(-\frac{21}{12}\right) = -\frac{1}{12} = -\frac{b}{a}$$

The product is

$$\frac{5}{3} \cdot \left(-\frac{7}{4}\right) = -\frac{35}{12} = \frac{c}{a}$$

so our results are correct.

EXAMPLE 5   Using the sum and product properties to verify solutions

Use the sum and product properties to see whether the solutions of $3x^2 + 5x - 2 = 0$ are

a. $-\frac{1}{3}$ and 2   **b.** $\frac{1}{3}$ and -2

SOLUTION

a. In the equation $3x^2 + 5x - 2 = 0$, $a = 3$, $b = 5$, and $c = -2$. The sum of the solutions is

$$-\frac{b}{a} = -\frac{5}{3}$$

Since $-\frac{1}{3} + 2 = -\frac{1}{3} + \frac{6}{3} = \frac{5}{3}$, $-\frac{1}{3}$ and 2 cannot be the solutions.

b. The sum of the proposed solutions is $\frac{1}{3} + \left(-\frac{6}{3}\right) = -\frac{5}{3} = -\frac{b}{a}$. The product of the solutions of $3x^2 + 5x - 2 = 0$ must be

$$\frac{c}{a} = -\frac{2}{3}$$

The product of the proposed solutions is $\frac{1}{3} \cdot (-2) = -\frac{2}{3}$. Thus $\frac{1}{3}$ and -2 are the correct solutions. ■

GRAPH IT

Can you use your grapher to determine what type of solutions a quadratic equation will have *without* calculating the discriminant? Almost! If the graph of a quadratic expression crosses the horizontal axis at *two* points, you know that the corresponding quadratic equation has *two* solutions, but in general, you will *not* know if they are rational or irrational. For example, $x^2 - 4$ cuts the horizontal axis twice: at 2 and at -2. In this case, it's obvious that there are two rational solutions for $x^2 - 4 = 0$.

Now try graphing $x^2 - 3 = 0$ using a $[-3, 3]$ by $[-3, 3]$ window. The curve crosses the horizontal axis twice, so we have two solutions, but we don't know if they are rational or irrational (Window 1). If you let $Y_1 = x^2 - 3$ and $Y_2 = 0$ and press $\boxed{\text{2nd}}$ $\boxed{\text{CALC}}$ 5, one of the points of intersection is given as $x = 1.7320508$, which seems like an irrational number since it does not repeat and does not terminate. As a matter of fact, 1.7320508 is an approximation for $\sqrt{3}$, which is irrational. But looks are deceiving. Try graphing

$$x^2 - \frac{1}{169} = 0$$

using the same technique (Window 2). This time the intersection is given as $x = 0.07692308$, which seems to be irrational (because it does not appear to repeat or terminate), but in fact, 0.07692308 is an approximation for $\frac{1}{13}$! Here our moral is, If the curve cuts the horizontal axis in two places, the corresponding quadratic equation has *two* solutions, but you *still* have to use the discriminant to determine whether the solutions are rational or irrational. If your calculator has a "Root" feature, press $\boxed{\text{2nd}}$ $\boxed{\text{CALC}}$ 2 to find the roots.

The rest of the story is simple. If the graph cuts the horizontal axis in *one* place, there is *one* solution, but we don't know if it is rational or irrational. [Try $x^2 - 4x + 4 = 0$ whose solution is $x = 2$, which is rational, and $(x - \sqrt{2})^2 = 0$ whose solution is $x = \sqrt{2}$, which is irrational.] If the graph does not touch the horizontal axis, there are no real-number solutions. The *two* resulting solutions are imaginary. (Try $x^2 + 4 = 0$.)

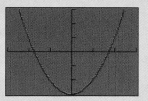
WINDOW 1
$Y_1 = x^2 - 3$
$Y_2 = 0$

WINDOW 2
$Y_1 = x^2 - \dfrac{1}{169}$
$Y_2 = 0$

EXERCISE 6.3

A In Problems 1–10, find the discriminant and determine the number and type of solutions.

1. $3x^2 + 5x - 2 = 0$
2. $3x^2 - 2x + 5 = 0$
3. $4x^2 = 4x - 1$
4. $2x^2 = 2x + 5$
5. $x^2 - 10x = -25$
6. $x^2 - 5x = 5$
7. $4x^2 - 5x + 3 = 0$
8. $5x^2 - 7x + 8 = 0$
9. $x^2 - 2 = \dfrac{5}{2}x$
10. $x^2 + \dfrac{1}{5} = \dfrac{2}{5}x$

In Problems 11–20, determine the value of k that will make the given equation have exactly one rational solution.

11. $x^2 - 4kx + 64 = 0$
12. $3x^2 + kx + 3 = 0$
13. $kx^2 - 10x = 5$
14. $2kx^2 - 12x = -9$
15. $2x^2 = kx - 8$
16. $3x^2 = kx - 3$
17. $25x^2 - kx = -4$
18. $4x^2 + 9kx = -1$
19. $x^2 + 8x = k$
20. $2x^2 - 4x = k$

B In Problems 21–30, use the discriminant to determine whether the given polynomial is factorable into factors with integer coefficients. If it is, use the technique of Example 3 to factor it.

21. $10x^2 - 7x + 8$
22. $10x^2 - 7x + 1$
23. $12x^2 - 17x + 6$
24. $12x^2 - 17x + 2$
25. $27x^2 + 51x - 56$
26. $15x^2 + 52x - 83$
27. $15x^2 + 52x - 84$
28. $27x^2 - 57x - 40$
29. $12x^2 - 61x + 60$
30. $30x^2 - 19x - 140$

C In Problems 31–40, find a quadratic equation with integer coefficients and the given solution set.

31. $\{3, 4\}$

32. $\{-1, 3\}$

33. $\{-5, -7\}$

34. $\{-3, -4\}$

35. $\left\{3, -\frac{2}{3}\right\}$

36. $\left\{-5, -\frac{2}{7}\right\}$

37. $\left\{\frac{1}{2}, -\frac{1}{2}\right\}$

38. $\left\{\frac{1}{3}, -\frac{1}{3}\right\}$

39. $\left\{0, -\frac{1}{5}\right\}$

40. $\left\{-\frac{3}{4}, 0\right\}$

D In Problems 41–45, find (a) the sum of the solutions, (b) the product of the solutions, and (c) determine if the two given values are the solutions of the given equation.

41. $4x^2 - 6x + 5 = 0$; the proposed solutions are $\frac{1}{2}$ and $\frac{5}{2}$.

42. $2x^2 + 9x = 35$; the proposed solutions are $-\frac{7}{2}$ and -1.

43. $5x^2 + 13x = 6$; the proposed solutions are $\frac{2}{5}$ and -3.

44. $4 - 3x = 7x^2$; the proposed solutions are $\frac{7}{4}$ and 1.

45. $-2 - 5x = 2x^2$; the proposed solutions are $\frac{1}{2}$ and 2.

46. If d is a constant and 3 is one solution of the equation $2x^2 - dx + 5 = 0$, use the product property to find the other solution.

47. If k is a constant and -5 is one solution of $3x^2 + kx = 40$, use the product property to find the other solution.

48. If the sum of the solutions of the equation $2x^2 - kx = 4$ is 3, find the value of k.

49. If the sum of the solutions of the equation $10x^2 + (k - 2)x = 3$ is $-\frac{13}{10}$, find k.

50. If the sum of the solutions of $3x^2 + (2k - 5)x + 8 = 0$ is 4, find k.

SKILL CHECKER

Solve:

51. $x + 2\sqrt{x} - 3 = 0$

52. $x + 4\sqrt{x} - 12 = 0$

53. $x^2 + 6x + 5 = 0$

54. $x^2 - 14x - 15 = 0$

USING YOUR KNOWLEDGE   Take a Dive

The highest regularly performed head-first dives are made at La Quebrada in Acapulco, Mexico. The height h (in meters) above the water of the diver when he is x meters away from the cliff is given by $h = -x^2 + 2x + 27$. Use the discriminant to find out whether:

55. The diver will ever be 27.5 m above the water. How many times will this occur?

56. The diver will ever be 28 m above the water. How many times will this occur?

57. The diver will ever be 29 m above the water.

WRITE ON . . .

58. Why do you think the expression $b^2 - 4ac$ is called the *discriminant*?

59. Can a quadratic equation with integer coefficients have exactly one imaginary solution? Explain why or why not.

60. Can a quadratic equation with integer coefficients have exactly one irrational solution? Explain why or why not.

61. Can a quadratic equation with integer coefficients have exactly one rational and one irrational solution? Explain why or why not.

MASTERY TEST   If you know how to do these problems, you have learned your lesson!

Use the sum and product properties to see whether the solutions of $4x^2 - 12x + 5 = 0$ are

62. $\frac{1}{2}$ and $-\frac{5}{2}$

63. $\frac{1}{2}$ and $\frac{5}{2}$

Use the discriminant to determine whether the expression is factorable. If it is, factor it.

64. $12x^2 + 23x + 10$

65. $12x^2 + x - 35$

Find a quadratic equation with integer coefficients whose solution set is

66. $\{-2, 4\}$

67. $\left\{-1, \frac{2}{3}\right\}$

68. Consider $4x^2 + kx = -4$.
    a. Find the discriminant.
    b. Find k so that the equation has one rational solution.

6.4

SOLVING EQUATIONS IN QUADRATIC FORM

To succeed, review how to:

1. Find the LCD of two or more rational expressions (p. 225).

2. Solve quadratic equations by factoring or by using the quadratic formula (pp. 186–190, 345–349).

Objectives:

 A Solve equations involving rational expressions by converting them to quadratic equations.

B Solve equations that are quadratic in form by substitution.

getting started

A Work Project

This team can finish laying the blocks in 2 days. If each works alone, the woman takes 3 days more than the man. How long would it take each of them working alone to finish the job? If we assume that the man can finish in d days, he will do $\frac{1}{d}$ of the work each day. The woman will finish in $d + 3$ days and do

$$\frac{1}{d + 3}$$

of the work each day. The work done each day is

Work done by man	+	work done by woman	=	work done together
$\frac{1}{d}$	+	$\frac{1}{d + 3}$	=	$\frac{1}{2}$

To solve this equation, we multiply each term by the LCD, $2d(d + 3)$, to obtain

$$2d(d + 3) \cdot \frac{1}{d} + 2d(d + 3) \cdot \frac{1}{d + 3} = 2d(d + 3) \cdot \frac{1}{2}$$

$2(d + 3) + 2d = d(d + 3)$	Simplify.
$2d + 6 + 2d = d^2 + 3d$	Remove parentheses.
$0 = d^2 - d - 6$	Write in standard form.
$0 = (d - 3)(d + 2)$	Factor.
$d - 3 = 0$ or $d + 2 = 0$	By the zero-factor property
$d = 3$ or $d = -2$	Solve.

Since d is the number of days, $d = -2$ has to be discarded. Thus it takes the man working alone 3 days to finish and the woman working alone 3 days more—that is, 6 days—to finish.

 Solving Equations That Contain Rational Expressions

As we have seen, equations involving rational expressions can lead to quadratic equations that can be solved by factoring or by using the quadratic formula. When solving such equations, make sure that the proposed solutions are checked in the original equations to avoid zero denominators. If a zero denominator occurs, discard the proposed solution as an *extraneous* solution.

EXAMPLE 1   Solving equations that contain rational expressions

Solve:

a. $\dfrac{4}{x^2 - 4} - \dfrac{1}{x - 2} = 1$ **b.** $\dfrac{-12}{x^2 - 9} + \dfrac{1}{x - 3} = 1$

SOLUTION

a. Since $x^2 - 4 = (x + 2)(x - 2)$, the LCD is $(x + 2)(x - 2)$. Multiplying each term by the LCD $(x + 2)(x - 2)$ of the fractions, we have

$$(x + 2)(x - 2) \cdot \frac{4}{(x^2 - 4)} - (x + 2)(x - 2) \cdot \frac{1}{x - 2} = (x + 2)(x - 2) \cdot 1$$

$$4 - (x + 2) = x^2 - 4 \qquad \text{Simplify.}$$

$$x^2 + x - 6 = 0 \qquad \text{Write in standard form.}$$

$$(x + 3)(x - 2) = 0 \qquad \text{Factor.}$$

$$x = -3 \quad \text{or} \quad x = 2 \qquad \begin{array}{l}\text{Solve} \\ x + 3 = 0 \text{ and} \\ x - 2 = 0.\end{array}$$

The proposed solutions are -3 and 2. However, $x = 2$ is not a solution, since

$$\frac{1}{x - 2}$$

is not defined for $x = 2$. Discard $x = 2$ as an extraneous solution. Thus the only solution is -3, which you can check in the original equation.

b. Since $x^2 - 9 = (x + 3)(x - 3)$, the LCD is $(x + 3)(x - 3)$. Multiplying each term by the LCD, we have

$$(x + 3)(x - 3) \cdot \frac{-12}{x^2 - 9} + (x + 3)(x - 3) \cdot \frac{1}{x - 3} = (x + 3)(x - 3) \cdot 1$$

$$-12 + x + 3 = x^2 - 9 \qquad \text{Simplify.}$$

$$x^2 - x = 0 \qquad \text{Write in standard form.}$$

$$x(x - 1) = 0 \qquad \text{Factor.}$$

$$x = 0 \quad \text{or} \quad x = 1 \qquad \begin{array}{l}\text{Solve } x = 0 \text{ and} \\ x - 1 = 0.\end{array}$$

This time both solutions satisfy the original equation. Thus the solutions are 0 and 1. Check this! ∎

B   Solving Equations by Substitution

Some equations that are not quadratic equations can be written in quadratic form and solved by an appropriate substitution. We illustrate several of these substitutions in the next examples.

EXAMPLE 2   Solving equations by substitution

Solve: $x^4 - 10x^2 + 9 = 0$

SOLUTION   This equation can be written as

$$(x^2)^2 - 10(x^2) + 9 = 0 \qquad \text{Given.}$$

$$u^2 - 10u + 9 = 0 \qquad \text{Let } u = x^2.$$

$$(u - 9)(u - 1) = 0 \qquad \text{Factor.}$$

$$u = 9 \quad \text{or} \quad u = 1 \qquad \text{Solve } u - 9 = 0 \text{ and } u - 1 = 0.$$

$$x^2 = 9 \quad \text{or} \quad x^2 = 1 \qquad \text{Substitute } x^2 = u.$$

$$x = \pm 3 \quad \text{or} \quad x = \pm 1 \qquad \text{Extract roots.}$$

The solutions of $x^4 - 10x^2 + 9 = 0$ are 3, 1, -3, and -1. ■

EXAMPLE 3   Solving equations that are quadratic in form

Solve: $(x^2 - x)^2 - (x^2 - x) - 30 = 0$

SOLUTION   This equation is already quadratic in form. If we let $u = x^2 - x$, we can write

$$u^2 - u - 30 = 0$$

$$(u - 6)(u + 5) = 0 \qquad \text{Factor.}$$

$$u = 6 \quad \text{or} \quad u = -5 \qquad \text{Solve } u - 6 = 0 \text{ and } u + 5 = 0.$$

$$x^2 - x = 6 \quad \text{or} \quad x^2 - x = -5 \qquad \text{Substitute } u = x^2 - x.$$

$$x^2 - x - 6 = 0 \quad \text{or} \quad x^2 - x + 5 = 0 \qquad \text{In standard form}$$

The first equation can be solved by factoring:

$$x^2 - x - 6 = (x - 3)(x + 2) = 0$$

Thus $x = 3$ or $x = -2$. We use the quadratic formula for $x^2 - x + 5 = 0$. Here $a = 1$, $b = -1$, and $c = 5$. Thus

$$x = \frac{-(-1) \pm \sqrt{(-1)^2 - 4(1)(5)}}{2(1)} = \frac{1 \pm \sqrt{-19}}{2}$$

$$= \frac{1 \pm \sqrt{19}\,i}{2}$$

The solutions of $(x^2 - x)^2 - (x^2 - x) - 30 = 0$ are

$$3, \quad -2, \quad \frac{1 + \sqrt{19}\,i}{2}, \quad \text{and} \quad \frac{1 - \sqrt{19}\,i}{2} \qquad ■$$

EXAMPLE 4   Solving equations containing rational exponents

Solve: $x^{1/2} - 8x^{1/4} + 15 = 0$

SOLUTION   To write the equation in quadratic form, we let $u = x^{1/4}$, which makes $u^2 = x^{1/2}$.

$$x^{1/2} - 8x^{1/4} + 15 = 0$$

$$u^2 - 8u + 15 = 0 \qquad \text{Substitute.}$$

$$(u - 3)(u - 5) = 0 \qquad \text{Factor.}$$

$$u = 3 \quad \text{or} \quad u = 5 \qquad \text{Solve } u - 3 = 0 \text{ and } u - 5 = 0.$$

$$x^{1/4} = 3 \quad \text{or} \quad x^{1/4} = 5 \qquad \text{Substitute } u = x^{1/4}.$$

$$x = 3^4 \quad \text{or} \quad x = 5^4 \qquad \text{Raise each side to the fourth power.}$$

$$x = 81 \quad \text{or} \quad x = 625$$

You can verify that the answers are correct by substituting in the original equation. ■

Finally, we solve some equations involving radicals. For example, the equation $x + 2\sqrt{x} - 3 = 0$ can be written as a quadratic equation if we let $u = \sqrt{x}$. This makes $u^2 = x$ so that we can write

$$x + 2\sqrt{x} - 3 = 0$$

$$u^2 + 2u - 3 = 0$$

$$(u + 3)(u - 1) = 0 \qquad \text{Factor.}$$

$$u = -3 \quad \text{or} \quad u = 1 \qquad \text{Solve } u + 3 = 0 \text{ and } u - 1 = 0.$$

$$\sqrt{x} = -3 \quad \text{or} \quad \sqrt{x} = 1 \qquad \text{Substitute } u = \sqrt{x}.$$

But the square root of x is never negative, so the equation $\sqrt{x} = -3$ has no solution. The solution of $\sqrt{x} = 1$ is 1. Thus the equation $x + 2\sqrt{x} - 3 = 0$ has only one solution, 1. You can verify that this solution is correct by direct substitution into the equation.

EXAMPLE 5   Using substitution to solve equations containing radicals

Solve: $x - 4\sqrt{x} + 3 = 0$

SOLUTION   We let $u = \sqrt{x}$, which makes $u^2 = x$.

$$x - 4\sqrt{x} + 3 = 0$$

$$u^2 - 4u + 3 = 0$$

$$(u - 3)(u - 1) = 0 \qquad \text{Factor.}$$

$$u = 3 \quad \text{or} \quad u = 1 \qquad \text{Solve } u - 3 = 0 \text{ and } u - 1 = 0.$$

$$\sqrt{x} = 3 \quad \text{or} \quad \sqrt{x} = 1 \qquad \text{Substitute } u = \sqrt{x}.$$

$$x = 9 \quad \text{or} \quad x = 1 \qquad \text{Square both sides.}$$

We leave it to you to verify that both solutions satisfy the original equation. ■

EXAMPLE 6   Solving equations containing negative exponents

Solve: $5x^{-4} - 4x^{-2} - 1 = 0$

SOLUTION   We can solve this equation by making the substitution $u = x^{-2}$, but this time we first convert the equation to one with positive exponents. Since

$$x^{-4} = \frac{1}{x^4} \qquad \text{and} \qquad x^{-2} = \frac{1}{x^2}$$

we have

$$5x^{-4} - 4x^{-2} - 1 = \frac{5}{x^4} - \frac{4}{x^2} - 1 = 0$$

Multiplying each term by the LCD, x^4, we obtain

$$x^4 \cdot \frac{5}{x^4} - x^4 \cdot \frac{4}{x^2} - x^4 \cdot 1 = 0$$

$5 - 4x^2 - x^4 = 0$	Simplify.
$-x^4 - 4x^2 + 5 = 0$	Write in descending order.
$x^4 + 4x^2 - 5 = 0$	Multiply by -1.
$u^2 + 4u - 5 = 0$	Let $u = x^2$.
$(u + 5)(u - 1) = 0$	Factor.
$u = -5$ or $u = 1$	Solve $u + 5 = 0$ and $u - 1 = 0$.
$x^2 = -5$ or $x^2 = 1$	Substitute x^2 for u.
$x = \pm\sqrt{5}i$ or $x = \pm 1$	Extract roots.

We leave it to you to verify that the four solutions satisfy the original equation. ∎

GRAPH IT

As we have mentioned, your grapher can be used to solve the equations we have studied in several ways. One way, of course, is to use the solve feature. When doing so, it's usually easy to obtain one solution, but in many cases, the second solution is hard to obtain. Try Example 1b if you don't believe this! Moreover, there may be cases in which the graph does not cross the horizontal axis. This means that there are no real-number solutions.

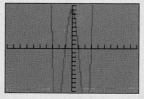

WINDOW 1

Now let's try to solve Example 2 by graphing. If your grapher has the intersection feature (look under the [MATH] menu), it will be easier to graph $Y_1 = x^4 - 10x^2 + 9$ and $Y_2 = 0$ using a standard window and then pressing [2nd] [CALC] 5 to find the intersection. (If you don't have this feature, just graph Y_1 and find the points at which the graph crosses the horizontal axis.) If you have a root feature, press [2nd] [CALC] 2 to find the solutions. In either case, the solutions are -3, -1, 1, and 3, as shown in Window 1.

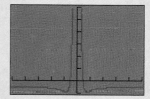

WINDOW 2

What about Example 6? Since some of the solutions are real and some imaginary, what does this graph look like? Enter $Y_1 = 5x^{-4} - 4x^{-2} - 1$ and graph. Of course, you can see only the real solutions, -1 and 1, as shown in Window 2. How do you know there are more solutions? You need more information! The original equation is equivalent to $x^4 + 4x^2 - 5 = 0$, which has degree 4, so the equation must have four solutions (some may repeat). Since we found only two solutions, we suspect that the other two solutions may be imaginary. (Note that an nth-degree equation has n solutions but some of them may repeat. At this time, the only way to check for repeated solutions is to solve the problem algebraically!)

EXERCISE 6.4

In Problems 1–10, solve the equation.

1. $\dfrac{x}{x + 4} + \dfrac{x}{x + 1} = 0$

2. $\dfrac{x}{x + 2} + \dfrac{x}{x + 3} = 0$

3. $\dfrac{x - 1}{x + 11} - \dfrac{2}{x - 1} = 0$

4. $\dfrac{x + 1}{x - 2} - \dfrac{8}{x - 1} = 0$

5. $\dfrac{x}{x - 1} - \dfrac{x}{x + 1} = 0$

6. $\dfrac{x}{x + 4} + \dfrac{x}{x - 2} = \dfrac{-1}{2}$

7. $\dfrac{x}{x + 2} - \dfrac{x}{x + 1} = \dfrac{-1}{6}$

8. $\dfrac{x}{x + 1} - \dfrac{x}{x - 1} = \dfrac{-3}{4}$

9. $\dfrac{x}{x + 4} + \dfrac{x}{x + 2} = \dfrac{-4}{3}$

10. $\dfrac{2x}{x - 2} + \dfrac{x}{x - 1} = \dfrac{7}{6}$

B In Problems 11–39, solve by substitution.

11. $x^4 - 13x^2 + 36 = 0$ **12.** $x^4 - 5x^2 + 4 = 0$

13. $4x^4 + 35x^2 = 9$ **14.** $3x^4 + 2x^2 = 8$

15. $3y^4 = 5y^2 + 2$ **16.** $6y^4 = 7y^2 - 2$

17. $x^6 + 7x^3 - 8 = 0$ **18.** $x^6 - 26x^3 - 27 = 0$

19. $(x + 1)^2 - 3(x + 1) = 40$ **20.** $(x + 2)^2 - 2(x + 2) = 8$

21. $(y^2 - y)^2 - 8(y^2 - y) = 9$ **22.** $(y^2 - y)^2 - 4(y^2 - y) = 12$

23. $x^{1/2} + 3x^{1/4} - 10 = 0$ **24.** $x^{1/2} + 4x^{1/4} - 12 = 0$

25. $y^{2/3} - 5y^{1/3} = -6$ **26.** $y^{2/3} + 5y^{1/3} = -6$

27. $x + \sqrt{x} - 6 = 0$ **28.** $x - \sqrt{x} - 30 = 0$

29. $(x^2 - 4x) - 8\sqrt{x^2 - 4x} + 15 = 0$

30. $(x^2 + 3x) + 5\sqrt{x^2 + 3x} - 14 = 0$

31. $z + 3 - \sqrt{z + 3} - 6 = 0$ **32.** $z + 4 - \sqrt{z + 4} - 12 = 0$

33. $3\sqrt{x} - 5\sqrt[4]{x} + 2 = 0$ **34.** $x^{-2} + 2x^{-1} - 3 = 0$

35. $x^{-2} + 2x^{-1} - 8 = 0$ **36.** $8x^{-4} - 9x^{-2} + 1 = 0$

37. $3x^{-4} - 5x^{-2} - 2 = 0$ **38.** $6x^{-4} + x^{-2} - 1 = 0$

39. $6x^{-4} + 5x^{-2} - 4 = 0$

APPLICATIONS

40. Working together, two workers can complete a job in 6 hr. If they work alone, one of them takes 9 hr more than the other to finish. How long does it take for each of them working alone to finish the job?

41. Working together, Jack and Jill can shovel the snow in the driveway in 6 hr. It takes Jack 5 hr more than Jill to do the job by himself. How long does it take each of them working alone to finish the job?

SKILL CHECKER

Solve:

42. $\dfrac{x}{5} - \dfrac{x}{3} \le \dfrac{x - 5}{5}$ **43.** $\dfrac{7x + 2}{6} \ge \dfrac{3x - 2}{4}$

44. $\dfrac{8x - 23}{6} + \dfrac{1}{3} \ge \dfrac{5x}{2}$

USING YOUR KNOWLEDGE   A Rental Problem

45. A group of students rented a cabin for $1600. When two of the group failed to pay their shares, the cost to each of the

remaining students was $40 more. Use your knowledge of the RSTUV procedure to find how many students were in the group.

$n = $ number of students in the group

$n - 2 = $ number of students who paid

$\dfrac{1600}{n} = $ amount each student should have paid in dollars

46. A group of students rented a bus for $720. If there had been six more students, the price per student would have been $6 less. How many students were in the group?

WRITE ON . . .

47. Find the number of solutions for $x - 1 = 0$, $x^2 - 1 = 0$, and $x^3 - 1 = 0$. Make a conjecture regarding the number of solutions for $x^4 - 1 = 0$. (If you want to solve the equation $x^4 - 1 = 0$ first, do it by factoring.)

48. Which method would you use to solve $x + 2\sqrt{x} - 3 = 0$, the substitution method or isolating the radical and squaring both sides? Explain. Which method has fewer steps?

49. In Example 6, we solved the problem by transforming the given equation into an equivalent one with positive exponents. Now solve Example 6 using the substitution $u = x^{-2}$. Which method do you prefer? Explain why.

50. The equation $x^4 - 1 = 0$ has four solutions, two real and two imaginary. (Solve $x^4 - 1 = 0$ by factoring if you don't believe this!) How many solutions do you think the equation $x^6 - 1 = 0$ has? How many do you think are real? How many do you think are imaginary? Make a conjecture about the number and nature of the solutions (real or imaginary) of $x^{2n} - 1 = 0$.

MASTERY TEST   If you know how to do these problems, you have learned your lesson!

Solve:

51. $x - 5\sqrt{x} + 4 = 0$ **52.** $2x - 3\sqrt{x} = -1$

53. $x^{1/2} - 5x^{1/4} + 6 = 0$ **54.** $x^{1/4} - 5x^{1/2} = -4$

55. $x^{-4} - 9x^{-2} + 14 = 0$ **56.** $x^{-4} - 7x^{-2} = -12$

57. $(x^2 - x)^2 - (x^2 - x) - 2 = 0$

58. $4(x^2 + 1)^2 - 7(x^2 + 1) = 2$

59. $\dfrac{6}{x^2 - 9} - \dfrac{1}{x - 3} = 1$ **60.** $\dfrac{6}{x^2 - 4} + \dfrac{1}{x - 2} = 1$

61. $x^4 - 5x^2 + 4 = 0$ **62.** $x^4 - 6x^2 + 5 = 0$

6.5

NONLINEAR INEQUALITIES

To succeed, review how to:

1. Solve a quadratic equation (pp. 186–190, 345–349).

2. Solve and graph linear inequalities (pp. 108–112).

Objectives:

A Solve quadratic inequalities.

B Solve inequalities of degree 3 or higher.

C Solve rational inequalities.

D Solve an application involving inequalities.

getting started

Skidding to Quadratics

Have you seen an officer measuring skid marks at the scene of an accident? The distance d (in feet) in which a car traveling v miles per hour can be stopped is given by

$$d = 0.05v^2 + v$$

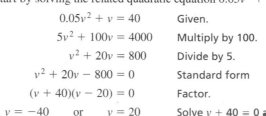

The skid marks of one accident were more than 40 ft long. The accident occurred in a 20 mi/hr zone. Was the driver going over the speed limit? To answer this question, we must solve the quadratic inequality $0.05v^2 + v > 40$. Let's start by solving the related quadratic equation $0.05v^2 + v = 40$.

$0.05v^2 + v = 40$	Given.
$5v^2 + 100v = 4000$	Multiply by 100.
$v^2 + 20v = 800$	Divide by 5.
$v^2 + 20v - 800 = 0$	Standard form
$(v + 40)(v - 20) = 0$	Factor.
$v = -40 \quad$ or $\quad v = 20$	Solve $v + 40 = 0$ and $v - 20 = 0$.

Now divide the number line into three regions, A, B, and C, using the solutions (*critical values*) -40 and 20 as boundaries (see the following table). To find the regions where $0.05v^2 + v > 40$, choose a test point in each of the regions, and test whether $0.05v^2 + v > 40$ for each of the points. Let's select the points -50, 0, and 30 from A, B, and C, respectively.

For $v = -50$	For $v = 0$	For $v = 30$
$0.05v^2 + v$	$0.05v^2 + v$	$0.05v^2 + v$
$= 0.05(-50)^2 + (-50)$	$= 0.05(0)^2 + 0$	$= 0.05(30)^2 + 30$
$= 125 - 50$	$= 0 < 40$	$= 45 + 30$
$= 75 > 40$		$= 75 > 40$
Thus $0.05v^2 + v > 40$.	Thus $0.05v^2 + v < 40$.	Thus $0.05v^2 + v > 40$.
Use this interval: $(-\infty, -40)$.	Discard this interval.	Use this interval: $(20, \infty)$.

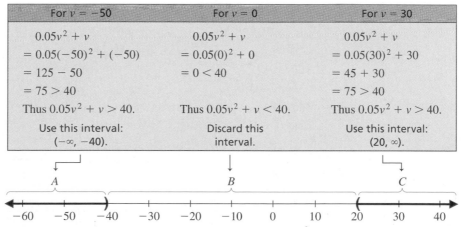

The solution set is the union of the two intervals where the inequality $0.05v^2 + v > 40$ is true, that is, $(-\infty, -40) \cup (20, \infty)$. Note that the end points -40 and 20 are *not* included in the solution set. Also, since the velocity must be positive, we discard the interval $(-\infty, -40)$ and conclude that the car was going more than 20 mi/h.

Solving Quadratic Inequalities

The procedure used in the *Getting Started* is based on the fact that, if you select a test point in one of the chosen intervals and the inequality is satisfied by the test point, then *every* point in the interval satisfies the inequality. The procedure we just used to solve the speed limit problem can be generalized to solve any polynomial inequality. Here are the steps.

PROCEDURE

Procedure to Solve Polynomial Inequalities

1. Write the inequality in standard form.

2. Find the *critical values* by solving the related equation.

3. Separate the number line into intervals using the critical values found in step 2 as boundaries.

4. Choose a test point for each interval and determine if the point satisfies the **original inequality**.

5. State the solution set, which consists of all the intervals in which the test point satisfies the original inequality.

EXAMPLE 1   Using the five-step procedure to solve a quadratic inequality

Solve: $(x - 1)(x + 3) < 0$

SOLUTION   We use the five-step procedure.

1. The inequality $(x - 1)(x + 3) < 0$ is already in standard form.

2. The related equation is $(x - 1)(x + 3) = 0$, and the critical values are $x = 1$ and $x = -3$.

3. Separate the number line into three regions using -3 and 1 as boundaries.

4. Select the test points: -4 from A, 0 from B, and 2 from C.

For $x = -4$	For $x = 0$	For $x = 2$
$(x - 1)(x + 3)$	$(x - 1)(x + 3)$	$(x - 1)(x + 3)$
$= (-4 - 1)(-4 + 3)$	$= (0 - 1)(0 + 3)$	$= (2 - 1)(2 + 3)$
$= (-5)(-1)$	$= (-1)(3)$	$= (1)(5)$
$= 5 > 0$	$= -3 < 0$	$= 5 > 0$
Thus $(x - 1)(x + 3) > 0$.	Thus $(x - 1)(x + 3) < 0$	Thus $(x - 1)(x + 3) > 0$.
Discard this interval.	Use this interval: $(-3, 1)$.	Discard this interval.

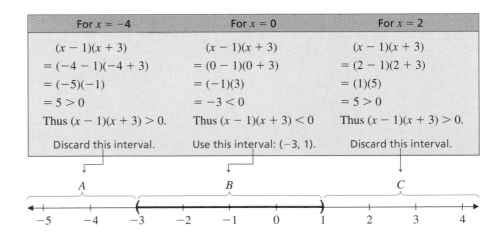

5. The solution set is the interval $(-3, 1)$.

EXAMPLE 2   More practice solving quadratic inequalities

Solve: $x^2 - 6x \geq -7$

SOLUTION   Again, we use the five-step procedure.

1. We first add 7 to both sides so that the inequality is in the standard form, $x^2 - 6x + 7 \geq 0$.

2. The related equation is $x^2 - 6x + 7 = 0$. Since $x^2 - 6x + 7$ is *not* factorable, we use the quadratic formula with $a = 1$, $b = -6$, and $c = 7$, and we obtain

$$x = \frac{-(-6) \pm \sqrt{(-6)^2 - 4(1)(7)}}{2(1)} = \frac{6 \pm \sqrt{8}}{2} = \frac{6 \pm 2\sqrt{2}}{2} = 3 \pm \sqrt{2}$$

The critical values are thus $x = 3 + \sqrt{2}$ and $x = 3 - \sqrt{2}$.

3. Separate the number line into three regions using $3 + \sqrt{2}$ and $3 - \sqrt{2}$ as boundaries. Approximating $\sqrt{2}$, as 1.4, the boundaries are $3 + 1.4 = 4.4$ and $3 - 1.4 = 1.6$.

4. Select the test points: 0 from *A*, 3 from *B*, and 6 from *C*.

For $x = 0$, we want to know if $x^2 - 6x \geq -7$.	For $x = 3$, we want to know if $x^2 - 6x \geq -7$.	For $x = 6$, we want to know if $x^2 - 6x \geq -7$.
$x^2 - 6x$ $= (0)^2 - 6(0)$ $= 0 - 0$ $= 0$ Thus $x^2 - 6x = 0 \geq -7$.	$x^2 - 6x$ $= (3)^2 - 6(3)$ $= 9 - 18$ $= -9$ Thus $x^2 - 6x = -9 < -7$.	$x^2 - 6x$ $= (6)^2 - 6(6)$ $= 36 - 36$ $= 0$ Thus $x^2 - 6x = 0 \geq -7$.
Use this interval: $(-\infty, 3 - \sqrt{2}]$.	Discard this interval.	Use this interval: $[3 + \sqrt{2}, \infty)$.

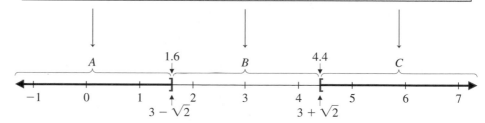

5. The solution set is $(-\infty, 3 - \sqrt{2}] \cup [3 + \sqrt{2}, \infty)$. Note that the critical values $3 - \sqrt{2}$ and $3 + \sqrt{2}$ are part of the solution set since they satisfy the inequality $x^2 - 6x \geq -7$. (Check this, but first read the caution note for a good rule of thumb.)   ■

CAUTION The end points of an interval are included in the solution set when the original inequality sign is $\leq$ or $\geq$; they are omitted when the original inequality sign is $>$ or $<$.

Solving Polynomial Inequalities

Suppose we wish to solve the inequality

$$(x + 3)(x - 2)(x - 4) \geq 0$$

The critical values are $x = -3$, $x = 2$, and $x = 4$, so we separate the number line into *four* regions, this time using -3, 2, and 4 as boundaries. Now select the points -4 from A, 0 from B, 3 from C, and 5 from D, and determine which of these points satisfy the inequality

$$(x + 3)(x - 2)(x - 4) \geq 0$$

for $x = -4$	For $x = 0$	For $x = 3$	For $x = 5$
$(x + 3)(x - 2)(x - 4)$	$(x + 3)(x - 2)(x - 4)$	$(x + 3)(x - 2)(x - 4)$	$(x + 3)(x - 2)(x - 4)$
$= (-4 + 3)(-4 - 2)(-4 - 4)$	$= (0 + 3)(0 - 2)(0 - 4)$	$= (3 + 3)(3 - 2)(3 - 4)$	$= (5 + 3)(5 - 2)(5 - 4)$
$= (-1)(-6)(-8)$	$= (3)(-2)(-4)$	$= (6)(1)(-1)$	$= (8)(3)(1)$
$= -48 < 0$	$= 24 > 0$	$= -6 < 0$	$= 24 > 0$
Thus $(x + 3)(x - 2)(x - 4)$	Thus $(x + 3)(x - 2)(x - 4)$	Thus $(x + 3)(x - 2)(x - 4)$	Thus $(x + 3)(x - 2)(x - 4)$
$= -48 < 0$.	$= 24 \geq 0$.	$= -6 < 0$.	$= 24 \geq 0$.
Discard this interval.	Use this interval: $[-3, 2]$.	Discard this interval.	Use this interval: $[4, \infty)$.

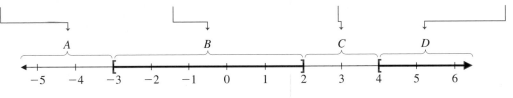

The solution set is the union of the two intervals selected, $[-3, 2] \cup [4, \infty)$. Note that -3, 2, and 4 are part of the solution set since they satisfy the inequality $(x + 3)(x - 2)(x - 4) \geq 0$.

EXAMPLE 3   Using the five-step procedure to solve polynomial inequalities

Solve: $(x + 1)(x - 3)(x + 4) \leq 0$

SOLUTION   We follow our five-step procedure.

1. The inequality is in standard form.
2. The related equation is $(x + 1)(x - 3)(x + 4) = 0$, and the critical values are -1, 3, and -4.
3. Separate the number line into four regions using -1, 3, and -4 as boundaries.
4. Select the test points: -5 from A, -3 from B, 0 from C, and 4 from D.

For $x = -5$	For $x = -3$	For $x = 0$	For $x = 4$
$(x + 1)(x - 3)(x + 4)$	$(x + 1)(x - 3)(x + 4)$	$(x + 1)(x - 3)(x + 4)$	$(x + 1)(x - 3)(x + 4)$
$= (-5 + 1)(-5 - 3)(-5 + 4)$	$= (-3 + 1)(-3 - 3)(-3 + 4)$	$= (0 + 1)(0 - 3)(0 + 4)$	$= (4 + 3)(4 - 3)(4 + 4)$
$= (-4)(-8)(-1)$	$= (-2)(-6)(1)$	$= (1)(-3)(4)$	$= (7)(1)(8)$
$= -32 < 0$	$= 12 > 0$	$= -12 < 0$	$= 56 > 0$
Thus $(x + 1)(x - 3)(x + 4)$	Thus $(x + 1)(x - 3)(x + 4)$	Thus $(x + 1)(x - 3)(x + 4)$	Thus $(x + 1)(x - 3)(x + 4)$
$= -32 \le 0$.	$= 12 > 0$.	$= -12 \le 0$.	$= 56 > 0$.
Use this interval: $(-\infty, -4]$.	Discard this interval.	Use this interval: $[-1, 3]$.	Discard this interval.

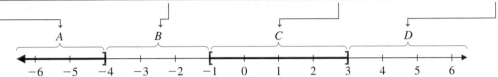

5. The solution set is the union of the two intervals selected,
$(-\infty, -4] \cup [-1, 3]$.

Solving Rational Inequalities

In Chapter 4, we studied quotients of polynomials called *rational expressions*. We are now ready to consider *rational inequalities* such as

$$\frac{x - 3}{x + 2} \le 0$$

This time, the **critical values** are the numbers that make the *numerator* $(x - 3)$ and *denominator* $(x + 2)$ zero, that is, 3 and -2. Of course, the critical values that make the denominator zero are excluded from the solution set. As before, we separate the number line into three regions, A, B, and C using -2 and 3 as boundaries, and then select test points from each of these regions to determine whether they satisfy our inequality. Let's select -3 from A, 0 from B, and 4 from C.

For $x = -3$, we want to know if	For $x = 0$, we want to know if	For $x = 4$, we want to know if
$\dfrac{x - 3}{x + 2} \le 0.$	$\dfrac{x - 3}{x + 2} \le 0.$	$\dfrac{x - 3}{x + 2} \le 0.$
$\dfrac{-3 - 3}{-3 + 2} = \dfrac{-6}{-1} = 6 > 0$	$\dfrac{0 - 3}{0 + 2} = \dfrac{-3}{2} < 0$	$\dfrac{4 - 3}{4 + 2} = \dfrac{1}{6} > 0$
Thus $\dfrac{x - 3}{x + 2} = 6 > 0.$	Thus $\dfrac{x - 3}{x + 2} = \dfrac{-3}{2} \le 0.$	Thus $\dfrac{x - 3}{x + 2} = \dfrac{1}{6} > 0.$
Discard this interval.	Use this interval: $(-2, 3]$.	Discard this interval.

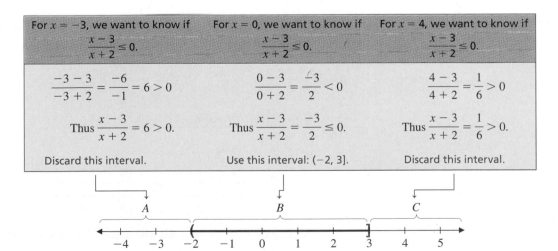

The solution set is the interval $(-2, 3]$.

EXAMPLE 4   Using the five-step procedure to solve rational inequalities

Solve:

$$\frac{x}{x-2} \geq 2$$

SOLUTION   We still follow our five-step procedure.

1. To write the inequality in standard form, we first subtract 2 from both sides and get a common denominator to simplify.

$$\frac{x}{x-2} \geq 2 \qquad \text{Given.}$$

$$\frac{x}{x-2} - 2 \geq 0 \qquad \text{Subtract 2.}$$

$$\frac{x}{x-2} - \frac{2(x-2)}{x-2} \geq 0 \qquad 2 = \frac{2(x-2)}{x-2}$$

$$\frac{x - 2x + 4}{x-2} \geq 0 \qquad \text{Remove parentheses.}$$

$$\frac{4-x}{x-2} \geq 0 \qquad \text{Simplify.}$$

2. The critical points are 4 and 2. The 2 will cause the denominator to be zero, so it is not included in the solution set. The 4 is included since it will make the rational expression zero.

3. Separate the number line into three regions using 2 and 4 as boundaries.

4. Select the test points: 0 from *A*, 3 from *B*, and 5 from *C*.

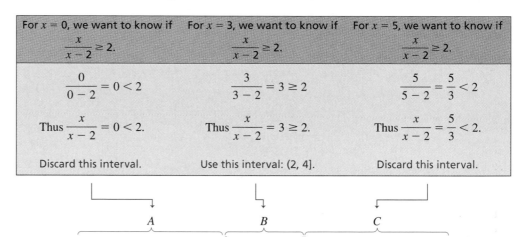

5. The solution set is (2, 4], which does not include 2 but does include 4, since 4 satisfies the inequality

$$\frac{x}{x-2} \geq 2$$

You have probably noticed that all the solution sets we've obtained are either intervals or unions of intervals. There are other possibilities, and they are based on the fact that for any real number a, $a^2 \geq 0$. Table 2 gives you some of these unusual solution sets, where a and x represent real numbers.

TABLE 2 Unusual Solution Sets

Explanation	Example	Solution Set
$a^2 \geq 0$ for every real number x.	$(x - 1)^2 \geq 0$ for every real number x.	The solution set of $(x - 1)^2 \geq 0$ is the set of all real numbers.
$a^2 \leq 0$ only when $a = 0$.	$(x - 1)^2 \leq 0$ only when $x - 1 = 0$, that is, when $x = 1$.	The solution set of $(x - 1)^2 \leq 0$ is $\{1\}$.
a^2 is never negative, so $a < 0$ is false.	$(x - 1)^2$ is never negative, so $(x - 1)^2 < 0$ is false.	The solution set of $(x - 1)^2 < 0$ is the empty set $\varnothing$.

You can check all these examples by using the same five-step procedure we've been using.

D Solving an Application Involving Inequalities

Example 5 shows how to apply the material we've been studying to a projectile motion problem. Remember our RSTUV method? We will use it again here.

problem solving

EXAMPLE 5   Problem for a rocket scientist

The height h (in feet) of a rocket is given by $h = 64t - 16t^2$, where t is the time in seconds. During what time interval will the rocket be higher than 48 ft above the ground?

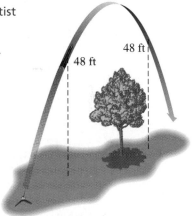

48 ft

48 ft

SOLUTION   We'll use the RSTUV method. The problem asks for a certain time interval in which $h > 48$.

1. Read the problem.

2. Select the unknown.

The unknown is the time t.

3. Think of a plan.

We have to find values of t for which $h = 64t - 16t^2 > 48$.

4. Use the five-step procedure to solve the inequality.

   1. Subtract 48 from both sides and rewrite in descending order. $-16t^2 + 64t - 48 > 0$

   2. The related equation is $-16t^2 + 64t - 48 = 0$

$$t^2 - 4t + 3 = 0 \qquad \text{Divide by } -16.$$

$$(t - 1)(t - 3) = 0 \qquad \text{Factor.}$$

   The critical values are 1 and 3.

   3. Separate the number line into three regions, A, B, and C with 1 and 3 as boundaries.

4. Select the test points: 0 from *A*, 2 from *B*, and 4 from *C*.

For $t = 0$, $64(0) - 16(0)^2 = 0 < 48$.   Discard interval *A*.

For $t = 2$, $64(2) - 16(2)^2 = 128 - 64 = 64 > 48$.   Use interval *B*: (1, 3).

For $t = 4$, $64(4) - 16(4)^2 = 256 - 256 = 0 < 48$.   Discard interval *C*.

5. The solution set is (1, 3), which means that the rocket is more than 48 ft high when the time *t* is between 1 and 3 sec.

5. Verify the solution.

When $t = 1$, the height of the rocket is $h = 64(1) - 16(1)^2 = 48$, that is, exactly 48 ft. If *t* is more than 1 and less than 3, the height is more than 48. Now on the way down, when $t = 3$, the height is $h = 64(3) - 16(3)^2 = 48$ ft again. ■

GRAPH IT

Your grapher simplifies the work in this section more than in any other work we've encountered. To solve a polynomial inequality with a grapher,

1. Write the inequality in standard form.

2. Graph the related equation.

3. The solution set consists of the intervals on the **horizontal** axis, the ***x*-axis**, for which the corresponding portions of the graph satisfy the inequality (above the *x*-axis if the inequality is > or ≥ below the *x*-axis if the inequality is < or ≤).

Thus to do Example 1, graph the related equation $Y_1 = (x - 1)(x + 3)$ in a standard window. The portion of the graph *below* the horizontal axis (less than zero) has corresponding *x*-values between -3 and 1. Hence the solution set is as before, the interval $(-3, 1)$ (see Window 1).

In Example 2, graph $Y_1 = x^2 - 6x + 7$. Pressing the [TRACE] or [ZOOM] 1, or a decimal window the points at which the graph cuts the *x*-axis can be approximated by $(-\infty, 1.59) \cup (4.41, \infty)$. Since we want the points for which the graph is ≥ 0, the solution set consists of the points in the interval $[1.59, 4.41]$ (see Window 2). Now you try Example 3!

For Example 4, graph

$$Y_1 = \frac{(4 - x)}{(x - 2)}$$

using a $[-5, 5]$ by $[-5, 5]$ window or, if you want to avoid some of the algebra, graph

$$Y_1 = \left(\frac{x}{(x - 2)}\right) - 2$$

an equivalent form of

$$\frac{(4 - x)}{(x - 2)}$$

In either case, be very careful with the placement of parentheses! Clearly, the solution set consists of the *x*-values whose corresponding *y*-values (vertical) are above the *x*-axis— that is, the values between 2 and 4. Of course, the 2 is excluded since it will yield a zero denominator, so the solution set is as before, (2, 4] (see Window 3).

You can also graph

$$Y_1 = \frac{x}{(x - 2)}$$

and $Y_2 = 2$ and find the interval where Y_1 is above Y_2 as shown in Window 4.

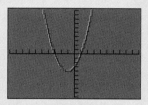

WINDOW 1
$Y_1 = (x - 1)(x + 3)$

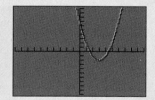

WINDOW 2
$Y_1 = x^2 - 6x + 7$

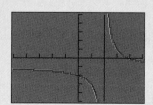

WINDOW 3
$Y_1 = \dfrac{(4 - x)}{x - 2}$

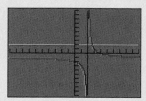

WINDOW 4

EXERCISE 6.5

A **In Problems 1–16, solve and graph the given inequality.**

1. $(x + 1)(x - 3) > 0$

2. $(x - 1)(x + 2) < 0$

3. $x(x + 4) \leq 0$

4. $(x - 1)x \geq 0$

5. $x^2 - x - 2 \leq 0$

6. $x^2 - x - 6 \leq 0$

7. $x^2 - 3x \geq 0$

8. $x^2 + 2x \leq 0$

9. $x^2 - 3x + 2 < 0$

10. $x^2 - 2x - 3 > 0$

11. $x^2 + 2x - 3 < 0$

12. $x^2 + x - 2 < 0$

13. $x^2 + 10x \leq -25$

14. $x^2 + 8x \leq -16$

15. $x^2 - 8x \geq -16$

16. $x^2 - 6x \geq -9$

In Problems 17–20, use the quadratic formula to find the critical values.

17. $x^2 - x \geq 1$ $(\sqrt{5} \approx 2.2)$ **18.** $x^2 - x \geq 3$ $(\sqrt{13} \approx 3.6)$

19. $x^2 - x \leq 4$ $(\sqrt{17} \approx 4.1)$ **20.** $x^2 - x \leq 5$ $(\sqrt{21} \approx 4.6)$

B **In Problems 21–24, solve and graph the inequality.**

21. $(x + 1)(x - 2)(x + 3) \geq 0$

22. $(x - 1)(x + 2)(x - 3) \geq 0$

23. $(x - 1)(x - 2)(x - 3) \leq 0$

24. $(x - 2)(x - 3)(x - 4) \leq 0$

C **In Problems 25–34, solve and graph the inequality.**

25. $\dfrac{2}{x - 2} \geq 0$

26. $\dfrac{3}{x - 1} \leq 0$

27. $\dfrac{x + 5}{x - 1} > 2$

28. $\dfrac{2x - 3}{x + 3} < 1$

29. $\dfrac{3x - 4}{2x - 1} < 1$

30. $\dfrac{x - 1}{x + 5} > 1$

31. $\dfrac{1}{x - 1} < \dfrac{1}{x + 2}$

32. $\dfrac{1}{x + 1} > \dfrac{1}{x - 2}$

33. $\dfrac{4}{x} + 6 > \dfrac{2}{x} + 2$

34. $\dfrac{3}{x} + 1 < \dfrac{1}{x} - 2$

In Problems 35–38, find all values of x for which the given expression is a real number. (*Hint:* $\sqrt{a}$ is a real number if $a \geq 0$.)

35. $\sqrt{x^2 - 9}$ **36.** $\sqrt{x^2 - 4x + 4}$

37. $\sqrt{x^2 - 6x + 5}$ **38.** $\sqrt{3x - 8}$

D **Applications**

39. The equivalent resistance of two electric circuits is given by $R^2 - 3R + 1$. When is this resistance more than 5 ohms?

40. The bending moment of a beam is $M = 20 - x^2$, where x is the distance, in feet, from one end of the beam. At what distances will the bending moment be more than 4?

41. The number N of water mites in a water sample depends on the temperature T in degrees Fahrenheit and is given by $N = 110T - T^2$. At what temperatures will the number of mites exceed 1000?

42. The profit P in a business varies in an 8-hr day according to the formula $P = 15t - 5t^2$, where t is the time in hours. During what hours is there a profit; that is, during what hours is $P > 0$?

43. The height h (in feet) of a projectile is $h = 48t - 16t^2$, where t is the time in seconds. During what time interval will the projectile be higher than 32 ft above the ground?

44. The distance d (in feet) in which a car traveling v miles per hour can be stopped is given by

$$d = 0.05v^2 + v$$

At what speed will it take more than 120 ft to stop the car?

Find y for the given value of x:

45. $y = 3x + 6$, $x = 2$ **46.** $y = 3x + 6$, $x = -2$

47. $y = -\dfrac{2}{3}x + 4$, $x = 3$ **48.** $y = -\dfrac{2}{3}x + 4$, $x = -3$

Find x for the given value of y:

49. $y = 3x + 6$, $y = 0$ **50.** $y = 2x + 8$, $y = 0$

USING YOUR KNOWLEDGE   Stopping Jaguars

In Problems 51–53, give your answers to the nearest tenth of a unit. We already know that the distance d (in feet) in which a car traveling v miles per hour can be stopped is given by

$$d = 0.05v^2 + v$$

51. For what values of v is $50 \le d \le 60$? (*Hint:* $\sqrt{11} \approx 3.32$ and $\sqrt{13} \approx 3.61$.)

52. For what values of v is $80 \le d \le 90$? (*Hint:* $\sqrt{17} \approx 4.12$ and $\sqrt{19} \approx 4.36$.)

53. According to the *Guinness Book of Records*, the longest skid mark on a public road was made by a Jaguar automobile involved in an accident in England. The skid mark was 950 ft long. If you assume that the car actually stopped after the brakes were applied and that it traveled for 950 ft, how fast was the car traveling? (Use a calculator to solve this problem!)

WRITE ON . . .

Explain in your own words:

54. Why the solution set of $(x - 1)^2 \ge 0$ is the set of all real numbers.

55. Why the solution set of $(x - 1)^2 \le 0$ is the single point $\{1\}$.

56. Why the solution set of $(x - 1)^2 < 0$ is the empty set $\varnothing$.

57. Why the solution set of the rational inequality

$$\frac{ax + b}{cx + d} \le 0$$

cannot include more than one end point (where a, b, and c are integers).

MASTERY TEST   If you know how to do these problems, you have learned your lesson!

Solve and graph the solution set:

58. $\dfrac{x}{x - 1} \le 1$

59. $\dfrac{x + 2}{x} > 1$

60. $(x + 3)(x - 2)(x + 1) \ge 0$

61. $(x - 4)(x + 3)(x - 1) \le 0$

62. $x^2 + x \ge 2$

63. $x^2 + x \le 6$

64. $(x + 3)(x - 2) < 0$

65. $(x + 1)(x - 3) \ge 0$

66. The profit P for a restaurant varies in an 8-hr day according to the formula $P = 3t - t^2$, where t is the time in hours. During what hours is there a profit; that is, during what hours is $P > 0$?

research questions   Sources of information for these questions can be found in the *Bibliography* at the end of this book.

1. Write a paper explaining how Euclid used theorems on areas to solve the quadratic equation $x^2 + b^2 = ax$.

2. Write a paper detailing Al-Khowarizmi's geometric demonstration of his rules to solve the quadratic equation $x^2 + 10x = 39$.

3. Why did they call Al-Khowarizmi's method of solving quadratics "completing the square"?

4. We have mentioned that Al-Khowarizmi was probably the foremost Arabic algebraist. The other one was dubbed "The Reckoner from Egypt." Who was this person, what did he write, and on what subjects?

SUMMARY

SECTION	ITEM	MEANING	EXAMPLE
6.1	Quadratic equation in standard form $\pm\sqrt{a}$	An equation of the form $ax^2 + bx + c = 0 \ (a \neq 0)$ The solutions of $x^2 = a$	$2x^2 - 3x + 7 = 0$ is a quadratic equation. $\pm\sqrt{3}$ are the solutions of $x^2 = 3$.
6.1C	Completing the square	1. Write the equation with the variables in descending order on the left and numbers on the right. 2. Divide each term by the coefficient of $x^2 \ (\neq 1)$. 3. Add the square of one-half the coefficient of x to both sides. 4. Factor the perfect square trinomial on the left. 5. Solve the resulting equation.	$5x^2 - 5x - 1 = 0$ Given. 1. $5x^2 - 5x = 1$ 2. $x^2 - x = \dfrac{1}{5}$ 3. $x^2 - x + \left(-\dfrac{1}{2}\right)^2$ $ = \dfrac{1}{5} + \left(-\dfrac{1}{2}\right)^2$ 4. $\left(x - \dfrac{1}{2}\right)^2 = \dfrac{9}{20}$ 5. $x - \dfrac{1}{2} = \pm\sqrt{\dfrac{9}{20}}$ $ = \pm\dfrac{3\sqrt{5}}{10}$ $ x = \dfrac{1}{2} \pm \dfrac{3\sqrt{5}}{10}$
6.2A	Quadratic formula	The solutions of $ax^2 + bx + c = 0$ are $x = \dfrac{-b \pm \sqrt{b^2 - 4ac}}{2a}.$	The solutions of $3x^2 + 2x - 5 = 0$ are $\dfrac{-2 \pm \sqrt{4 + 60}}{6} = 1$ and $-\dfrac{5}{3}.$
6.3	Discriminant	The discriminant of $ax^2 + bx + c = 0$ is $D = b^2 - 4ac.$	The discriminant of $3x^2 + 2x - 5 = 0$ is 64.
6.3A	Types of solutions for $ax^2 + bx + c = 0$ where a, b, and c are rational numbers	1. If $D > 0$ and D is not a perfect square, two different irrational solutions 2. If $D > 0$ and $D = N^2$, two different rational solutions 3. If $D < 0$, two different imaginary solutions 4. If $D = 0$, one rational solution	1. $4x^2 - 3x - 5 = 0 \ (D = 89)$ has two different irrational solutions. 2. $x^2 - 2x - 3 = 0 \ (D = 16)$ has two different rational solutions. 3. $4x^2 - 3x + 5 = 0 \ (D = -71)$ has two different imaginary solutions. 4. $4x^2 - 4x + 1 = 0 \ (D = 0)$ has one rational solution.
6.3B	Factorable quadratic expressions	If D is a perfect square, $ax^2 + bx + c$ is factorable.	$20x^2 + 10x - 30$ is factorable $(D = 2500)$. $20x^2 + 10x - 30 = 10(2x + 3)(x - 1)$

SECTION	ITEM	MEANING	EXAMPLE
6.3D	Sum and product of the solutions	The sum and product of the solutions of $ax^2 + bx + c = 0$ are $-\dfrac{b}{a}$ and $\dfrac{c}{a}$, respectively.	The sum and product of the solutions of $4x^2 - 12x + 5 = 0$ are 3 and $\dfrac{5}{4}$, respectively.
6.4A	Equations that are quadratic in form	Equations that can be written as quadratics by use of appropriate substitutions	$x^4 - 5x^2 + 4 = 0$, $(x^2 - x)^2 - (x^2 - x) - 2 = 0$, and $x - 5\sqrt{x} + 4 = 0$
6.5A	Quadratic inequality	An inequality that can be written as $ax^2 + bx + c < 0$. (The symbol $<$ can be replaced by $>$, $\le$, or $\ge$.)	$x^2 + x - 6 < 0$ and $x^2 + x - 2 \ge 0$

REVIEW EXERCISES

(If you need help with these exercises, look in the section indicated in brackets.)

1. [6.1A] Solve.
    a. $16x^2 - 49 = 0$ b. $25x^2 - 16 = 0$

2. [6.1A] Solve.
    a. $5x^2 + 30 = 0$ b. $6x^2 + 42 = 0$

3. [6.1B] Solve.
    a. $(x - 3)^2 = 32$ b. $(x - 5)^2 = 50$

4. [6.1B] Solve.
    a. $2(x - 2)^2 + 25 = 0$ b. $3(x - 3)^2 + 64 = 0$

5. [6.1B] Solve.
    a. $5x^2 - 10x + 5 = 12$ b. $12x^2 + 12x + 3 = 16$

6. [6.1C] Solve by completing the square.
    a. $x^2 - 8x - 9 = 0$ b. $x^2 + 12x + 32 = 0$

7. [6.1C] Solve by completing the square.
    a. $4x^2 + 4x - 3 = 0$ b. $16x^2 - 24x + 7 = 0$

8. [6.2A] Solve by the quadratic formula.
    a. $3x^2 + 5x - 2 = 0$ b. $5x^2 - 9x - 2 = 0$

9. [6.2A] Solve by the quadratic formula.
    a. $3x^2 = 2x + 4$ b. $4x^2 = 6x + 3$

10. [6.2A] Solve by the quadratic formula.
    a. $16x = x^2$ b. $12x = x^2$

11. [6.2A] Solve by the quadratic formula.
    a. $x^2 + \dfrac{x}{15} = \dfrac{1}{3}$ b. $\dfrac{x^2}{2} + \dfrac{9x}{10} = \dfrac{1}{5}$

12. [6.2A] Solve by the quadratic formula.
    a. $3x^2 - 2x = -1$ b. $5x^2 - 2x = -4$

13. [6.2B] Solve.
    a. $8x^3 - 125 = 0$ b. $125x^3 - 8 = 0$

14. [6.2C] The demand d is given by
$$d = \frac{450}{p}$$
and the supply s is given by $s = 100p - 150$.
    a. Find the equilibrium point (at which $d = s$).
    b. Replace the values of d and s in part a by
$$d = \frac{50}{p}$$
and $s = 150p - 100$ and find the equilibrium point.

15. [6.3A] For the given equation, find the discriminant and determine the value of k so that the equation has exactly one rational solution.
    a. $16x^2 - kx = -1$ b. $8x^2 - kx = -2$

16. [6.3B] Use the discriminant to determine whether the given quadratic is factorable into factors with integer coefficients. If so, factor it.
    a. $3x^2 - 11x - 6$ b. $18x^2 + 13x + 2$

17. [6.3B] Factor into factors with integer coefficients if possible.
    a. $18x^2 - 9x - 5$ b. $18x^2 + 13x + 1$

18. [6.3C] Find a quadratic equation with integer coefficients whose solution set is
    a. $\{-2, 3\}$ b. $\left\{\dfrac{1}{4}, -\dfrac{2}{3}\right\}$

19. **[6.3D]** Without solving the equation, find the sum and the product of the solutions of
    a. $15x^2 + 4x - 3 = 0$ **b.** $9x^2 - 12x - 5 = 0$

20. **[6.3D]** Use the sum and product properties to check whether
    a. $\frac{1}{3}$ and $-\frac{3}{5}$ are the solutions of the equation in Problem 19, part a.
    b. $\frac{1}{3}$ and $-\frac{5}{3}$ are the solutions of the equation in Problem 19, part b.

21. **[6.4A]** Solve.
    a. $\dfrac{8}{x^2 - 16} - \dfrac{1}{x - 4} = 1$ **b.** $\dfrac{-24}{x^2 - 36} + \dfrac{2}{x + 6} = 1$

22. **[6.4B]** Solve.
    a. $(x^2 + x)^2 + 2(x^2 + x) - 8 = 0$
    b. $(x^2 - 3x)^2 - 4(x^2 - 3x) - 12 = 0$

23. **[6.4B]** Solve.
    a. $x^{1/2} - 4x^{1/4} = -3$ **b.** $x^{1/2} + x^{1/4} = 6$

24. **[6.4B]** Solve.
    a. $x^{2/3} + x^{1/3} = 12$ **b.** $x^{2/3} - 5x^{1/3} = 6$

25. **[6.4B]** Solve.
    a. $x - 2\sqrt{x} = 3$ **b.** $x - 4\sqrt{x} = 5$

26. **[6.4B]** Solve.
    a. $3x^{-4} - 4x^{-2} + 1 = 0$ **b.** $3x^{-4} - 2x^{-2} - 1 = 0$

27. **[6.5A]** Solve and graph.
    a. $(x - 2)(x + 3) < 0$
    b. $(x + 2)(x - 3) < 0$

28. **[6.5A]** Solve and graph.
    a. $x^2 + 4x \geq 0$
    b. $x^2 - 3x \geq 0$

29. **[6.5A]** Solve and graph.
    a. $x^2 + 4x \geq 8$
    b. $x^2 + 6x \geq 18$

30. **[6.5B]** Solve and graph.
    a. $(x - 1)(x - 2)(x - 3) \leq 0$
    b. $(x + 1)(x + 2)(x - 3) \leq 0$

31. **[6.5B]** Solve and graph.
    a. $(x + 1)(x + 2)(x - 3) \geq 0$
    b. $(x - 1)(x + 2)(x + 3) \geq 0$

32. **[6.5B]** Solve and graph.
    a. $\dfrac{x + 2}{x - 2} \leq 2$
    b. $\dfrac{x - 2}{x + 2} \geq 3$

PRACTICE TEST

(Answers on pages 382–383.)

1. Solve.
   a. $25x^2 - 4 = 0$ **b.** $18x^2 + 3 = 0$

2. Solve.
   a. $(x - 1)^2 = 45$ **b.** $2(x - 2)^2 = 49$

3. Solve $3x^2 + 6x + 3 = 5$.

4. Solve by completing the square: $x^2 - 6x + 5 = 0$.

5. Solve by completing the square: $9x^2 - 6x - 1 = 0$.

6. Solve by the quadratic formula: $7x^2 + 5x - 2 = 0$.

7. Solve by the quadratic formula: $3x^2 = 3x + 2$.

8. Solve by the quadratic formula: $32x = x^2$.

9. Solve by the quadratic formula:
   $$\frac{x^2}{16} + \frac{5x}{4} = \frac{11}{4}$$

10. Solve by the quadratic formula: $4x^2 + 3x = -2$.

11. Solve $64x^3 - 27 = 0$.

12. If the demand d is given by
    $$d = \frac{500}{p}$$
    and the supply s is given by $s = 200p - 150$, find the equilibrium point (at which $s = d$).

13. For the equation $4x^2 - kx = -9$
    a. Find the discriminant.
    b. Find k so that the equation has exactly one rational solution.

14. Use the discriminant to determine whether $12x^2 - 4x - 21$ is factorable into factors with integer coefficients.

15. Factor $35x^2 - 32x - 12$ into factors with integer coefficients if possible.

16. Find a quadratic equation with integer coefficients whose solution set is $\left\{\frac{2}{3}, -\frac{1}{4}\right\}$.

17. a. Find the sum and the product of the solutions of the equation $6x^2 + 7x - 5 = 0$.

   b. Use the sum and product properties to check whether the solutions are $-\frac{5}{3}$ and $\frac{1}{2}$.

18. Solve.

   a. $\dfrac{6}{x^2 - 9} - \dfrac{1}{x - 3} = 1$

   b. $\dfrac{2}{x^2 - 1} - \dfrac{3}{x - 1} = 1$

19. Solve.

   a. $x^4 - 7x^2 + 6 = 0$

   b. $(x^2 - 2x)^2 - (x^2 - 2x) - 6 = 0$

20. Solve.

   a. $x^{1/2} - 3x^{1/4} + 2 = 0$ **b.** $x - 5\sqrt{x} + 6 = 0$

21. Solve $3x^{-4} - 2x^{-2} - 1 = 0$.

22. Solve and graph $(x + 1)(x - 2) < 0$.

23. Solve and graph $x^2 + 2x \geq 15$.

24. Solve and graph $(x + 1)(x - 2)(x - 3) \leq 0$.

25. Solve and graph

$$\frac{x + 1}{x - 1} \leq 3$$

ANSWERS TO PRACTICE TEST

Answer	If you missed: Question	Review: Section	Examples	Page
1a. $\dfrac{2}{5}, -\dfrac{2}{5}$	1a	6.1A	1a	335
1b. $\dfrac{\sqrt{6}}{6}i, -\dfrac{\sqrt{6}}{6}i$	1b	6.1A	1b	336
2a. $1 \pm 3\sqrt{5}$	2a	6.1B	2a	337
2b. $2 \pm \dfrac{7\sqrt{2}}{2}$ or $\dfrac{4 \pm 7\sqrt{2}}{2}$	2b	6.1B	2b	337
3. $-1 \pm \dfrac{\sqrt{15}}{3}$ or $\dfrac{-3 \pm \sqrt{15}}{3}$	3	6.1B	3	338
4. $1, 5$	4	6.1C	4	339
5. $\dfrac{1 \pm \sqrt{2}}{3}$	5	6.1C	5	340–341
6. $\dfrac{2}{7}, -1$	6	6.2A	1	345–346
7. $\dfrac{3 \pm \sqrt{33}}{6}$	7	6.2A	2	346–347
8. $0, 32$	8	6.2A	3	347
9. $2, -22$	9	6.2A	4	347–348
10. $\dfrac{-3 \pm \sqrt{23}i}{8}$ or $\dfrac{-3}{8} \pm \dfrac{\sqrt{23}}{8}i$	10	6.2A	5	348–349

Answer	If you missed:		Review:	
	Question	Section	Examples	Page
11. $\dfrac{3}{4}, -\dfrac{3}{8} \pm \dfrac{3\sqrt{3}}{8}i$ or $\dfrac{-3 \pm 3\sqrt{3}i}{8}$	11	6.2A	6	350
12. $p = 2$	12	6.2B	7	351–352
13a. $k^2 - 144$	13a	6.3A	1a	357
13b. $k = \pm 12$	13b	6.3A	1b	357
14. Yes, $D = 32^2$.	14	6.3B	2	357
15. $(5x - 6)(7x + 2)$	15	6.3B	3	358
16. $12x^2 - 5x - 2 = 0$	16	6.3C	4	359
17a. Sum: $-\dfrac{7}{6}$; product: $-\dfrac{5}{6}$	17a	6.3D	5a	360
17b. Yes	17b	6.3D	5b	360
18a. -4	18a	6.4A	1a	364
18b. $0, -3$	18b	6.4A	1b	364
19a. $\pm 1, \pm\sqrt{6}$	19a	6.4B	2	365
19b. $3, -1, 1 \pm i$	19b	6.4B	3	365
20a. $1, 16$	20a	6.4B	4	365–366
20b. $4, 9$	20b	6.4B	5	366
21. $\pm 1, \pm\sqrt{3}i$	21	6.4C	6	366–367
22. $(-1, 2)$ (number line from -3 to 3, open interval from -1 to 2)	22	6.5A	1	370
23. $(-\infty, -5] \cup [3, \infty)$ (number line from -6 to 4)	23	6.5A	2	371
24. $(-\infty, -1] \cup [2, 3]$ (number line from -2 to 4)	24	6.5B	3	372–373
25. $(-\infty, 1) \cup [2, \infty)$ (number line from -1 to 5)	25	6.5B	4	374

Graphs and Functions

In this chapter we generalize our work with equations to include equations in *two* variables and their graphs. To do this, we introduce the Cartesian coordinate system and learn how to graph points. We then discuss the distance between two points, the slope of a line, and the different ways in which the equation for a line can be written. As before, after discussing equations we study inequalities. We then discuss applications to three types of variation: direct, inverse, and joint. We end the chapter by studying functions and their graphs.

The inventor of the Cartesian coordinate system, Rene Descartes, was born March 31, 1596, near Tours, France. Trained as a gentleman and educated in Latin, Greek, and rhetoric, he maintained a healthy skepticism toward all he was taught. Disenchanted with these studies, he took an interlude of pleasure in Paris, later moving to the suburb of St. Germain, where he worked on mathematics for 2 years. In 1637, at age 41, his friends persuaded him to print his masterpiece, known as the *Discourse on Method,* which included an essay on geometry that is probably the most important thing he ever did. His *Analytic Geometry,* a combination of algebra and geometry, revolutionized the study of geometry and made much of modern mathematics possible.

In 1649, Queen Christine of Sweden chose Descartes as her private philosophy teacher. Unfortunately, she insisted that her lessons begin promptly at five o'clock in the morning in the ice-cold library of her palace. This proved to be too much for the frail Descartes, who soon caught "inflammation of the lungs," from which he died on February 11, 1650, at the age of 54. What price for fame!

7.1

THE RECTANGULAR COORDINATE SYSTEM

To succeed, review how to:

1. Evaluate an expression (pp. 42–49).
2. Solve linear equations (p. 69).

Objectives:

A Given an ordered pair of numbers, find its graph, and vice versa.

B Graph lines by finding two or more points satisfying the equation of the line.

C Graph lines by finding the *x*- and *y*-intercepts.

D Graph horizontal and vertical lines.

getting started   Hurricane Preparedness

The map shows the **coordinates** of the hurricane to be near the vertical line indicating 90° of longitude and the horizontal line indicating 25° of latitude. If we agree to list the longitude first, we can identify this point by using the ordered pair

$$(90, \quad 25)$$
This is the longitude. ⟶↑   ↑⟶ This is the latitude.

So if we know the *coordinates* of the hurricane, we can find it on the map and if we know the *point* where the hurricane is located on the map, we can find its coordinates. This is an example of a rectangular coordinate system, which we shall study in this section.

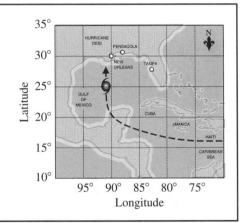

A   Graphing and Finding Ordered Pairs

In algebra we use a similar system to locate points by using **ordered pairs** of numbers. We draw two perpendicular number lines called the *x*-axis and the *y*-axis, intersecting at a point *O* called the **origin**. (See Figure 1.) On the *x*-axis, *positive* is to the right, whereas on the *y*-axis, *positive* is up. The two axes divide the plane into four regions called **quadrants**. These quadrants are numbered counterclockwise using Roman numerals and starting in the upper right-hand region, as shown. The whole arrangement is called a **Cartesian coordinate system**, a **rectangular coordinate system**, or simply a **coordinate plane**.

Every point *P* in the plane can be associated with an ordered pair of real numbers (*x*, *y*), and every ordered pair (*x*, *y*) can be associated with a point *P* in the plane. For example, in Figure 2 the point *A* can be associated with the ordered pair (1, 3), the point *B* with the ordered pair (−3, 2), and the point *C* with the ordered pair (0, −2). The **graphs** of the points *A*(1, 3), *B*(−3, 2), and *C*(0, −2) are indicated by the blue dots.

FIGURE 1

FIGURE 2

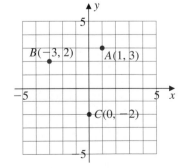

A NOTE ABOUT COORDINATES

If the x-coordinate is *positive*, the point is to the *right* of the vertical axis.
If the x-coordinate is *negative*, the point is to the *left* of the vertical axis.
If the x-coordinate is *zero*, the point is *on* the vertical axis.
If the y-coordinate is *positive*, the point is *above* the horizontal axis.
If the y-coordinate is *negative*, the point is *below* the horizontal axis.
If the y-coordinate is *zero*, the point is *on* the horizontal axis.

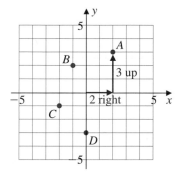

FIGURE 3

EXAMPLE 1   Graphing ordered pairs

Graph: The ordered pairs $A(2, 3)$, $B(-1, 2)$, $C(-2, -1)$, and $D(0, -3)$

SOLUTION   To graph the ordered pair $(2, 3)$, we start at the origin and move 2 units to the *right* and then 3 units *up*, reaching the point whose coordinates are $(2, 3)$. The other three pairs are graphed in a similar manner and are shown in Figure 3. ■

Since every point P in the plane is associated with an ordered pair (x, y), we should be able to find the coordinates of any point, as shown in Example 2.

EXAMPLE 2   Finding coordinates

Determine: The coordinates of each of the points shown in Figure 4

SOLUTION   Point A is 4 units to the *right* of the origin, and 1 unit *above* the horizontal axis. Thus the ordered pair corresponding to A is $(4, 1)$. The coordinates of the other four points can be found in a similar manner. Here is the summary.

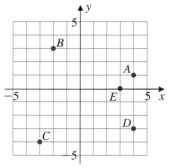

FIGURE 4

Point	Start at the origin, move:	Coordinates
A	4 units *right*, 1 unit *up*	$(4, 1)$
B	2 units *left*, 3 units *up*	$(-2, 3)$
C	3 units *left*, 4 units *down*	$(-3, -4)$
D	4 units *right*, 3 units *down*	$(4, -3)$
E	3 units *right*, 0 units *up*	$(3, 0)$

■

 Graphing Lines

The equation $2x + 3y = 12$ is an equation in two variables, x and y. If in this equation x is replaced by 3 and y by 2, we obtain the *true* statement $2 \cdot 3 + 3 \cdot 2 = 12$. Thus we say that the *ordered pair* $(3, 2)$ is a **solution** of the equation $2x + 3y = 12$ or that it **satisfies** the equation. On the other hand, $(2, 3)$ is *not* a solution of $2x + 3y = 12$ because $2 \cdot 2 + 3 \cdot 3 = 4 + 9 \neq 12$.

The **solution set** of an equation in two variables can be written using set notation. For example, the solution set of $2x + 3y = 12$ can be written as $\{(x, y) | 2x + 3y = 12\}$ or, if we decide to solve for y, as

$$\left\{(x, y) \,\middle|\, y = \frac{-2}{3}x + 4\right\}$$

The solution set of the equation $2x + 3y = 12$ consists of *infinitely* many points, so it would be impossible to list all these points. However, we can find some of these points by substituting values for one of the variables and then computing the corresponding values for the other variable. For example, if we replace x by -3 in the equation

$$y = \frac{-2}{3}x + 4$$

we have

$$y = \frac{-2}{3} \cdot (-3) + 4 = 6$$

$$\text{If } x = 0, \quad y = \frac{-2}{3} \cdot (0) + 4 = 4$$

$$\text{If } x = 3, \quad y = \frac{-2}{3} \cdot (3) + 4 = 2$$

$$\text{If } x = 6, \quad y = \frac{-2}{3} \cdot (6) + 4 = 0$$

$$\text{If } x = 9, \quad y = \frac{-2}{3} \cdot (9) + 4 = -2$$

x	y	Ordered Pair
-3	6	$(-3, 6)$
0	4	$(0, 4)$
3	2	$(3, 2)$
6	0	$(6, 0)$
9	-2	$(9, -2)$

These ordered pairs can be entered in a table (as shown on the left) and then graphed as shown in Figure 5.

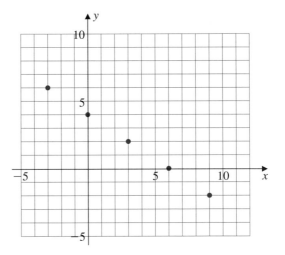

FIGURE 5

The points appear to be on a straight line. In fact, it can be proved that every *solution* of

$$y = \frac{-2}{3}x + 4 \quad \text{(or } 2x + 3y = 12\text{)}$$

corresponds to a point on the line, and vice versa. Thus the line shown in Figure 6, obtained by joining the points shown in Figure 5, is the *graph* of

$$y = \frac{-2}{3}x + 4 \quad \text{(or } 2x + 3y = 12\text{)}$$

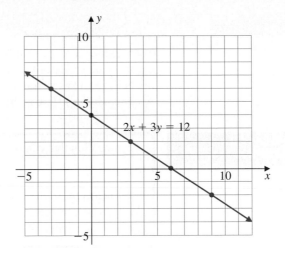

FIGURE 6

The line in Figure 6 represents only a *part* of the complete graph, which continues without end in both directions, as indicated by the arrows in the figure. Although it takes only *two* points to determine a straight line, we shall include a third point here in our solutions to make sure that we have drawn the correct line.

EXAMPLE 3 Graphing lines

Graph: $x + 2y = 8$

SOLUTION

For $x = -2$,	For $x = 0$,	For $x = 2$,
$-2 + 2y = 8$	$0 + 2y = 8$	$2 + 2y = 8$
$2y = 10$	$2y = 8$	$2y = 6$
$y = 5$	$y = 4$	$y = 3$
We have $(-2, 5)$.	We have $(0, 4)$.	We have $(2, 3)$.

Thus we graph the points $(-2, 5)$, $(0, 4)$, and $(2, 3)$ and then draw a line through them, as shown in Figure 7. ∎

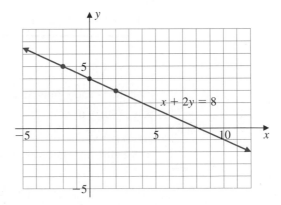

FIGURE 7

Graphing Lines Using Intercepts

The equation $x + 2y = 8$ has a *straight* line as its graph. It can be shown that any equation in two variables x and y that is of the form $Ax + By = C$ (where A and B are not both zero) has a straight line for its graph. (This is why $Ax + By = C$ is called a *linear equation*.) Here is the definition.

STANDARD FORM OF LINEAR EQUATIONS

> Any equation that can be written in the form
>
> $$Ax + By = C$$
>
> is a **linear equation** in two variables, and the graph is a *straight line*. The form $Ax + By = C$ is called the **standard form** of the equation, and A, B, and C are real numbers where A and B are not both 0.

> **NOTE** We prefer to write $3x + 2y = 6$ instead of $\frac{1}{2}x + \frac{1}{3}y = 1$; that is, we use the standard form with integer coefficients for A, B, and C.

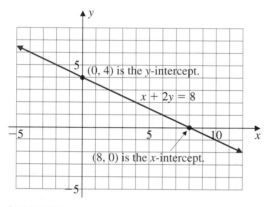

FIGURE 8

Since two points determine a straight line, it is sufficient to locate two points in order to graph a linear equation. The easiest points to compute are those involving zeros—that is, points of the form $(x, 0)$ and $(0, y)$. For example, if we let $x = 0$ in the equation $x + 2y = 8$, we obtain $2y = 8$, or $y = 4$. Thus $(0, 4)$ is a point on the graph. Similarly, if we let $y = 0$ in the equation $x + 2y = 8$, we have $x = 8$. Thus $(8, 0)$ is also on the graph. Since the points $(8, 0)$ and $(0, 4)$ are the points at which the line crosses the x- and y-axes, respectively, they are called the ***x*- and *y*-intercepts**. These two intercepts, as well as the line determined by them (the graph of $x + 2y = 8$), are shown in Figure 8. Here is the procedure we used to find the intercepts.

PROCEDURE

> **Finding the Intercepts**
>
> To find the x-intercept, let $y = 0$ and find x: $(x, 0)$ is the x-intercept.
>
> To find the y-intercept, let $x = 0$ and find y: $(0, y)$ is the y-intercept.

EXAMPLE 4   Finding intercepts

Find the x- and y-intercepts and then graph the lines:

a. $y = 3x + 6$ **b.** $2x + 3y = 0$

SOLUTION

a. We first find the x- and y-intercepts. For $x = 0$, $y = 3x + 6$ becomes

$$y = 3(0) + 6 = 6$$

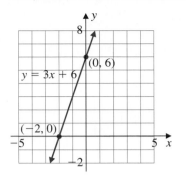

FIGURE 9

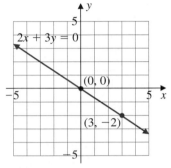

FIGURE 10

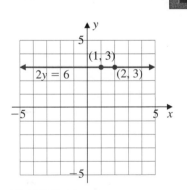

FIGURE 11

Thus $(0, 6)$ is the y-intercept. For $y = 0$, $y = 3x + 6$ becomes

$$0 = 3x + 6$$

$$-6 = 3x \qquad \text{Subtract 6.}$$

$$x = -2 \qquad \text{Divide by 3.}$$

Thus $(-2, 0)$ is the x-intercept. We join the points $(-2, 0)$ and $(0, 6)$ with a line and obtain the graph of $y = 3x + 6$ shown in Figure 9.

b. As before, we try to find the x- and y-intercepts. For $x = 0$, $2x + 3y = 0$ becomes

$$2(0) + 3y = 0$$

or $\qquad\qquad\qquad y = 0$

Thus $(0, 0)$ is the y-intercept. Note that if we now let $y = 0$, we get the ordered pair $(0, 0)$ again. This is because any line of the form $Ax + By = 0$ (where A and B are not both zero) goes through the origin. In these cases, we select a different value for x, say $x = 3$. Then $2x + 3y = 0$ becomes

$$2(3) + 3y = 0$$

$$3y = -6 \qquad \text{Subtract 6.}$$

$$y = -2 \qquad \text{Divide by 3.}$$

Thus $(3, -2)$ is another point on the line. We join the points $(0, 0)$ and $(3, -2)$ with a line and obtain the graph of $2x + 3y = 0$ shown in Figure 10. ∎

D   Graphing Horizontal and Vertical Lines

It's worth noting that the procedure we have discussed works *only* for equations that can be written in the form $Ax + By = C$, where A and B are not zero. Thus since the equation $2y = 6$ can be written in the form $0 \cdot x + 2y = 6$, we *cannot* graph $2y = 6$ by finding its intercepts. The equation $2y = 6$ assigns to every value of x a y-value of 3. As a matter of fact, if we solve for $2y = 6$, we get $y = 3$, which has no specific x-coordinate. Thus for $x = 1$, $y = 3$, and for $x = 2$, $y = 3$ (y is always 3). If we graph the points $(1, 3)$ and $(2, 3)$ and connect them with a straight line, we see that the result is a horizontal line, as shown in Figure 11. Similarly, the equation $2x = 6$ assigns an x-value of 3 to every y. Thus for $y = 1$, $x = 3$, and for $y = 5$, $x = 3$. If we graph the points $(3, 1)$ and $(3, 5)$ and draw a straight line through them, we see that the result is a vertical line, as shown in Figure 12. In general, we have the following.

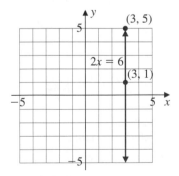

FIGURE 12

HORIZONTAL AND VERTICAL LINES

The graph of $y = C$ is a *horizontal* line.
The graph of $x = C$ is a *vertical* line.

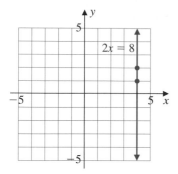

FIGURE 13

EXAMPLE 5   Graphing horizontal and vertical lines

Graph:

a. $2x = 8$ **b.** $5y = -10$

SOLUTION

a. Since $2x = 8$ is equivalent to $x = 4$, the graph of $2x = 8$ is a vertical line for which $x = 4$. If we choose the solutions $(4, 1)$ and $(4, 2)$ and draw a straight line through them, we obtain the graph of $2x = 8$ shown in Figure 13.

b. Since $5y = -10$ is equivalent to $y = -2$, $5y = -10$ is a horizontal line for which $y = -2$. If we choose the solutions $(1, -2)$ and $(2, -2)$ and draw a straight line through them, we obtain the graph of $5y = -10$ shown in Figure 14. ■

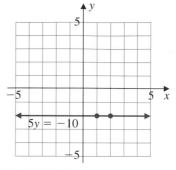

FIGURE 14

GRAPH IT

The examples in this section can easily be done with a grapher. In Example 1, we are going to plot (graph) data. We explore a feature of the TI-82 not used until now, the [STAT] key. Let's graph the points in the example. Start by clearing any old [LIST] by pressing [STAT] 4. When "**ClrList**" appears, press [2nd] [L1] [,] [2nd] [L2] [,] [2nd] [L3] [,] up to [2nd] [L6] and [ENTER] to clear all the lists. Now enter [STAT] 1 and enter the x-coordinates of points A, B, C, and D $(2, -1, -2, 0)$ under L1 and the y-coordinates $(3, 2, -1, -3)$ under L2. Press [2nd] [STAT PLOT] 1 [ENTER], go to "TYPE" and select the first type of graph, called a *scattergram*. Finally, press [ZOOM] 9, and the graph of the points appears. (See Window 1.) Note that pressing [ZOOM] 9 selects the correct window for you! You should be able to do Example 3 on your own, provided you remember how to solve for y!

Example 4 is done by *first* graphing the line $Y_1 = 3x + 6$ in a standard window and *then* finding the intercepts by simply examining the graph shown in Window 2. (By the way, you can turn **off** the graph of the points in Example 1 by pressing [2nd] [STAT PLOT] 1 then selecting "OFF" and pressing [ENTER].) Example 5b can be done by first solving for y and then graphing $Y_1 = -2$. (Can you figure out how to graph Example 5a?) The moral of our story continues to be: You still have to know some algebra even if you have a grapher!

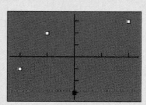

WINDOW 1

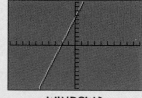

WINDOW 2

EXERCISE 7.1

A **In Problems 1–10, graph the ordered pair and state in which quadrant (if any) each lies.**

1. $(3, 4)$ **2.** $(4, 3)$ **3.** $(-4, 3)$ **4.** $(-3, 4)$ **5.** $(-3, -2)$

6. $(-2, -3)$ **7.** $(0, -2)$ **8.** $(2, 0)$ **9.** $\left(\frac{1}{2}, -3\right)$ **10.** $\left(3, -\frac{1}{2}\right)$

In Problems 11–20, determine the coordinates of each of the points in Figure 15.

11. A **12.** B

13. C **14.** D

15. E **16.** F

17. G **18.** H

19. I **20.** J

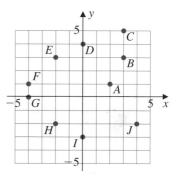

FIGURE 15

B In Problems 21–26, complete the ordered pairs to satisfy the equation and then graph it.

21. $y = x + 3$
$(-2, \ ), (-1, \ ), (0, \ ), (1, \ ), (2, \ )$

22. $y = 2x + 1$
$(-1, \ ), (0, \ ), (1, \ )$

23. $x - y = 4$
$(-1, \ ), (0, \ ), (1, \ )$

24. $x - 3y = 6$
$(0, \ ), (3, \ ), (-3, \ )$

25. $2x - y - 3 = 0$
$(-1, \ ), (0, \ ), (1, \ )$

26. $2x + y + 2 = 0$
$(-1, \ ), (0, \ ), (1, \ )$

37. $2x - 5y = -10$

38. $2x - 3y = -6$

39. $x - y - 5 = 0$

40. $2x - 3y - 12 = 0$

D In Problems 41–50, determine whether the given line is horizontal or vertical and then graph the line.

41. $-\dfrac{7}{2}x = 14$

42. $-\dfrac{1}{2}x = 2$

43. $\dfrac{3}{2}x = 6$

C In Problems 27–40, find the x- and y-intercepts and then graph the equation.

27. $y = x - 5$

28. $2y = 4x - 2$

44. $-\dfrac{5}{2}y = 10$

29. $2x + 3y = 6$

30. $3x + 2y = 6$

45. $-\dfrac{3}{4}x = 3$

31. $2x - y = 4$

32. $3x - y = 3$

46. $-\dfrac{3}{7}y = -\dfrac{6}{7}$

33. $2x + y - 4 = 0$

34. $3x + y - 3 = 0$

47. $-\dfrac{1}{3} + y = \dfrac{2}{3}$

35. $y + 4x = 0$

36. $y + 3x = 0$

48. $-\dfrac{1}{5} + y = \dfrac{4}{5}$

49. $\dfrac{2}{3} = x - \dfrac{4}{3}$

50. $\dfrac{3}{4} = x - \dfrac{5}{4}$

APPLICATIONS

51. The number N of chirps a cricket makes per minute is given by

$$N = 4(T - 40)$$

where T is the temperature in degrees Fahrenheit. The ordered pairs corresponding to this equation are of the form (T, N).

a. Does the ordered pair (60, 80) satisfy the equation?

b. Based on your answer to part a, how many chirps does a cricket make when the temperature is 60°F?

c. How many chirps does a cricket make when the temperature is 80°F?

52. Based on the formula of Problem 51, crickets stop chirping when $N = 0$. The corresponding ordered pair will be $(T, 0)$.

a. Find T.

b. At what temperature will crickets stop chirping?

Are you exercising too hard? Your target zone can tell you. It works like this. Take your pulse after exercising, find your age on the x-axis in Figure 16, and follow a vertical line up to the lower edge of the shaded area; then go across to the number on the y-axis at the left. That pulse rate is the lower limit for your target zone. To find the upper limit, continue vertically on your age line to the top of the shaded area and then go across

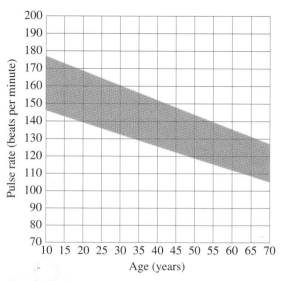

FIGURE 16

to the y-axis to locate your pulse rate. Use this information and Figure 16 to solve Problems 53–56.

53. What is the lower limit pulse rate for a 20-year-old person? Write the answer as an ordered pair (age, limit).

54. What is the upper limit pulse rate for a 20-year-old person? Write the answer as an ordered pair.

55. What is the upper limit pulse rate for a 45-year-old person? Write the answer as an ordered pair.

56. What is the lower limit pulse rate for a 50-year-old person? Write the answer as an ordered pair.

SKILL CHECKER

Find the quotients:

57. $\dfrac{3 - 6}{3 - 1}$ **58.** $\dfrac{6 - 3}{1 - 3}$

59. $\dfrac{3 - (-6)}{1 - 4}$ **60.** $\dfrac{4 - (-2)}{2 - 5}$

61. $\dfrac{4 - (-2)}{3 - (-6)}$ **62.** $\dfrac{2 - (-6)}{4 - (-8)}$

USING YOUR KNOWLEDGE   The Long, Hot Summer Exercise

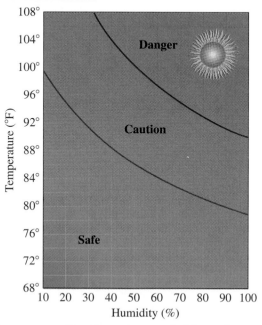

The ideas presented in this section are vital for understanding graphs. For example, do you exercise in the summer? To determine the risk of exercising in the heat, you must know how

to read the graph. To do this, first find the temperature on the *y*-axis, then read across from it to the right, stopping at the vertical line representing the relative humidity. Thus on a 90°F day, if the humidity is less than 30%, the weather is in the safe zone.

63. If the humidity is 50%, how high can the temperature rise and still be in the safe zone for exercising? (Answer to the nearest degree.)

64. If the humidity is 70%, at what temperature will the danger zone start?

65. If the temperature is 100°F, what does the humidity have to be so that it is safe to exercise?

66. Between what temperatures should you use caution when exercising if the humidity is 80%?

67. If you start jogging at 1 P.M. when the temperature is 86°F and the humidity is 60%, how many degrees can the temperature rise before you get to the danger zone?

WRITE ON . . .

68. What does the graph of an equation represent to you? Can you ever draw the *complete* graph of a line?

69. Why are *linear equations* named that way?

70. What happens when $A = 0$ in the equation $Ax + By = C$?

71. What happens when $B = 0$ in the equation $Ax + By = C$?

72. Why does setting $x = 0$ give the *y*-intercept for a line?

73. Why does setting $y = 0$ give the *x*-intercept for a line?

74. If two points determine a line, why did we use three points when graphing lines?

75. Explain in your own words the procedure you use to graph lines by (a) graphing points, and (b) using the intercepts.

MASTERY TEST If you know how to do these problems, you have learned your lesson!

Graph:

76. $3x = -9$

77. $2y = -6$

78. $2x - y = 6$

79. $-3x - y = -6$

80. Find the *x*- and *y*-intercepts of $2x - 3y = 6$ and then graph the line.

81. Find the *x*- and *y*-intercepts of $3x - 2y = -6$ and then graph the line.

82. Graph the points $A(3, 2)$, $B(-1, 4)$, $C(0, -2)$, $D(-1, -2)$, and $E(2, -1)$.

83. What are the coordinates of the points A, B, and C?

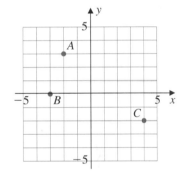

7.2

THE DISTANCE FORMULA AND THE SLOPE OF A LINE

To succeed, review how to:

1. Use the Pythagorean Theorem (p. 191).
2. Add, subtract, multiply, and divide integers (pp. 15–19).
3. Find the reciprocal of a number (p. 22).

Objectives:

A. Find the distance between two points in the Cartesian plane.

B. Find the slope of a line passing through two given points.

C. Use the definition of slope to decide whether two lines are perpendicular, parallel, or neither.

D. Graph a line given its slope and a point on the line.

getting started

Performance and Slopes

In the graph, the distance along the *x*-axis shows the number of weeks elapsed, and the distance on the *y*-axis shows the amount of deposits (in milligrams) in a car engine. The solid line shows that the deposits *increase* when using ordinary fuels but *decrease* after adding Texaco System³ gasoline.

In mathematics we have a formula for finding the distance between any two points in the plane. We also measure how fast lines rise (increase) or fall (decrease) by measuring *slopes*. We discuss both of these ideas in this section.

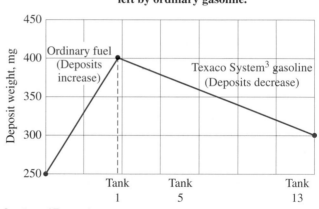

In the BMW test, Texaco's new System³ gasoline removed performance-robbing deposits left by ordinary gasoline.

Courtesy of Texaco, Inc.

A   Using the Distance Formula

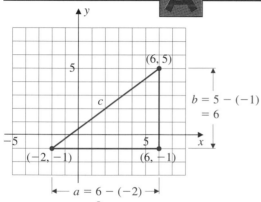

FIGURE 17

In the triangle of Figure 17, the distance a between $(-2, -1)$ and $(6, -1)$ is $6 - (-2) = 8$, and the distance b between $(6, 5)$ and $(6, -1)$ is $5 - (-1) = 6$. To find the distance c, we need to use the Pythagorean Theorem studied in Chapter 3. According to that theorem, if the legs of a right triangle are a and b and the hypotenuse is c, then

$$c^2 = a^2 + b^2$$

Thus

$$c^2 = 8^2 + 6^2 = 64 + 36$$

$$c^2 = 100$$

$$c = \pm\sqrt{100} = \pm 10$$

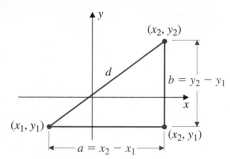

FIGURE 18

Since c represents a distance, we discard the negative answer and conclude that $c = 10$. We can repeat a similar argument to find the distance d between any two points (x_1, y_1) and (x_2, y_2), as shown in Figure 18. As before, the distance a is $|x_2 - x_1|$ and the distance b is $|y_2 - y_1|$. By the Pythagorean Theorem, we have

$$d^2 = |x_2 - x_1|^2 + |y_2 - y_1|^2$$
$$d = \sqrt{(x_2 - x_1)^2 + (y_2 - y_2)^2}$$

Note that since the square of a number is the same as the square of its opposite, we don't need the absolute-value signs when we take the square root. Here is a summary of what we have done.

THE DISTANCE FORMULA

The distance between the points (x_1, y_1) and (x_2, y_2) is

$$d = \sqrt{(x_2 - x_1)^2 + (y_2 - y_1)^2}$$

EXAMPLE 1   Using the distance formula

Find the distance between the given points:

a. $A(1, 1)$ and $B(5, 4)$ **b.** $C(2, 3)$ and $D(-2, 5)$

c. $E(-2, 1)$ and $F(-2, 3)$

SOLUTION

a. If we let $x_1 = 1$, $y_1 = 1$, $x_2 = 5$, and $y_2 = 4$, then

$$d = \sqrt{(5 - 1)^2 + (4 - 1)^2} = \sqrt{(4)^2 + (3)^2} = \sqrt{25} = 5$$

b. Here $x_1 = 2$, $y_1 = 3$, $x_2 = -2$, and $y_2 = 5$. Thus

$$d = \sqrt{(-2 - 2)^2 + (5 - 3)^2} = \sqrt{(-4)^2 + (2)^2} = \sqrt{20} = 2\sqrt{5}$$

c. Now $x_1 = -2$, $y_1 = 1$, $x_2 = -2$, and $y_2 = 3$. Hence

$$d = \sqrt{[-2 - (-2)]^2 + (3 - 1)^2} = \sqrt{[0]^2 + (2)^2} = \sqrt{4} = 2$$

Note that EF is a vertical line, so that $d = |3 - 1| = 2$. ■

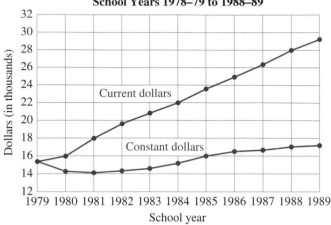

FIGURE 19

Source: © National Education Association. 1989. *Estimates of School Statistics 1988–89*. Washington, DC: NEA. Reprinted by permission.

B Finding the Slope of a Line

A second property of a line segment joining two points in a plane is its "steepness" (inclination). For example, suppose you want to find the average yearly raise in teachers' average salaries, as shown in Figure 19. According to Figure 19, the salary in 1979 was about 16 (thousand); in 1989, it was about 30 (thousand). The change is

$$30 - 16 = 14 \text{ (thousand)}$$

during the 10-yr period, and the average yearly raise is given by the ratio

$$\frac{\text{increase in ten years}}{\text{number of years}} = \frac{30 - 16}{10} = \frac{14}{10} = 1.4 \text{ (thousand)}$$

Thus teachers' average salaries went up about \$1400 (1.4 thousand) each year. In mathematics the ratio of the change (difference) in y to the change in x is called the *slope* of the line segment and is denoted by the letter m. Thus

$$m = \frac{\text{change in } y}{\text{change in } x} = \frac{\text{difference in } y\text{'s}}{\text{difference in } x\text{'s}} = \frac{\text{rise } (\uparrow)}{\text{run } (\rightarrow)}$$

We summarize this discussion with the following definition and formula.

DEFINITION OF SLOPE

If $A(x_1, y_1)$ and $B(x_2, y_2)$ are any two distinct points on a line L (that is not parallel to the y-axis), then the **slope** of L, denoted by m, is

$$m = \frac{y_2 - y_1}{x_2 - x_1}$$

NOTE The slope is simply a number that measures the "steepness" of a line. For positive numbers, the larger the number, the steeper the line.

When using this definition, it doesn't matter which point is taken for A and which for B. For example, the slope of the line passing through the points $A(0, -6)$ and $B(3, 3)$ is

$$m = \frac{3 - (-6)}{3 - 0} = \frac{9}{3} = 3$$

If we choose A to be $(3, 3)$ and B to be $(0, -6)$, the slope is the same,

$$m = \frac{-6 - 3}{0 - 3} = \frac{-9}{-3} = 3$$

Thus A and B can be interchanged without changing the value of the resulting slope.

On the left side (x_2, y_2) is the first point and (x_1, y_1) is the second point. | On the right side (x_1, y_1) is the first point and (x_2, y_2) is the second point.

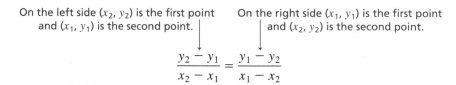

$$\frac{y_2 - y_1}{x_2 - x_1} = \frac{y_1 - y_2}{x_1 - x_2}$$

EXAMPLE 2   Finding slopes

Find the slope of the line passing through the given points:

a. $A(-3, 1), B(-1, -2)$ **b.** $A(-1, 5), B(-1, 6)$

c. $A(1, 1), B(2, 3)$ **d.** $A(3, 4), B(1, 4)$

SOLUTION

a. The slope, as shown in Figure 20, is

$$m = \frac{-2 - 1}{-1 - (-3)} = \frac{-3}{2} \quad \begin{matrix} \longleftarrow \text{rise} \\ \longleftarrow \text{run} \end{matrix}$$

Note that $\frac{-3}{2}$ is in standard form, so if we start at A, we move 3 units *down* (the change in y) and 2 units *right* (the change in x), ending at B.

b. The slope is

$$m = \frac{6 - 5}{-1 - (-1)} = \frac{1}{0}$$

which is undefined. Notice that the line passing through $A(-1, 5)$ and $B(-1, 6)$ in Figure 21 has as its equation $x = -1$ and is a line *parallel* to the y-axis. The fact that the change in x is zero in lines parallel to the y-axis (that is, *vertical* lines) requires that they be excluded from the definition; their slope is *undefined*.

c. The slope, as shown in Figure 22, is

$$m = \frac{3 - 1}{2 - 1} = \frac{2}{1} = 2$$

The change in y (rise) is 2 units up, the change in x (run) is 1 unit right.

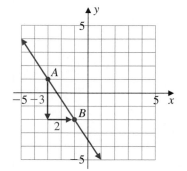

FIGURE 20

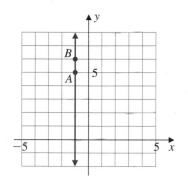

FIGURE 21

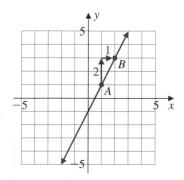

FIGURE 22

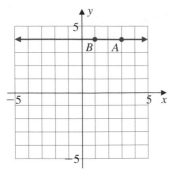

FIGURE 23

d. The slope, as shown in Figure 23, is

$$m = \frac{4 - 4}{1 - 3} = \frac{0}{-2}$$

Since

$$\frac{0}{-2} = 0$$

the slope is zero. Note that the line is parallel to the x-axis and has equation $y = 4$, a horizontal line. Horizontal lines have zero slope. ∎

We can see from Example 2 that the following in Table 1 hold true.

TABLE 1 Slope Summary

A line that *falls* from left to right has a *negative* slope.	The slope of a *vertical* line is *undefined*. Since $x_2 - x_1 = 0$, so m is *undefined*.	A line that *rises* from left to right has a *positive* slope.	A *horizontal* line has zero slope. Since $y_2 - y_1 = 0$,
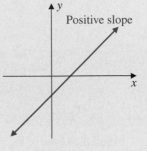	$$m = \frac{y_2 - y_1}{x_2 - x_1} = \frac{y_2 - y_1}{0}$$ 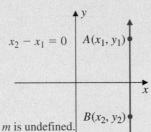	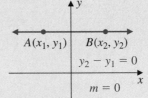	$$m = \frac{y_2 - y_1}{x_2 - x_1} = \frac{0}{x_2 - x_1}$$ $$m = 0$$

Parallel and Perpendicular Lines

The definition of slope can be used to determine when two line segments are parallel. Since two parallel lines have the same steepness (inclination) and thus the same slope, we have the following.

SLOPES OF PARALLEL LINES

Two nonvertical lines L_1 and L_2 with slopes m_1 and m_2 are parallel if and only if $m_1 = m_2$.

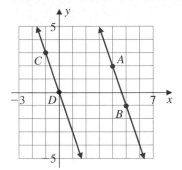

FIGURE 24

NOTE If L_1 and L_2 are vertical lines, they are parallel.

For example, consider the two lines, one passing through points $A(4, 2)$ and $B(5, -1)$ and the other through points $C(-1, 3)$ and $D(0, 0)$, as shown in Figure 24. The slope of AB is

$$m_1 = \frac{-1 - 2}{5 - 4} = \frac{-3}{1} = -3$$

and the slope of CD is

$$m_2 = \frac{0 - 3}{0 - (-1)} = \frac{-3}{1} = -3$$

Thus $m_1 = m_2$, so both lines have the same slope and are parallel.

It is shown in more advanced courses that two lines with slopes m_1 and m_2 are perpendicular (meet at a 90° angle) if

$$m_1 = \frac{-1}{m_2}$$

that is, if their slopes are *negative reciprocals*. This means that

$$\text{if} \quad m_1 = \frac{-1}{m_2}, \quad \text{then} \quad m_1 \cdot m_2 = -\frac{1}{m_2} \cdot m_2 = -1$$

SLOPES OF PERPENDICULAR LINES

The lines L_1 and L_2 with slopes m_1 and m_2, respectively, are perpendicular if and only if the slopes are **negative reciprocals**; that is, $m_1 \cdot m_2 = -1$.

We can show that the line passing through $A(4, 2)$ and $B(5, -1)$ is perpendicular to the line passing through $C(2, 1)$ and $D(5, 2)$. Since the slope of AB is

$$m_1 = \frac{-1 - 2}{5 - 4} = \frac{-3}{1} = -3$$

and that of CD is

$$m_2 = \frac{2 - 1}{5 - 2} = \frac{1}{3}$$

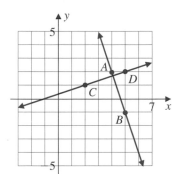

FIGURE 25

we have $m_1 \cdot m_2 = -3 \cdot \frac{1}{3} = -1$. These perpendicular lines can be graphed as shown in Figure 25.

EXAMPLE 3   Finding whether lines are parallel or perpendicular

A line L_1 has slope $\frac{2}{3}$; find:
a. Whether another line passing through $A(5, 1)$ and $B(8, 3)$ is parallel or perpendicular to L_1
b. Whether the line passing through $C(-1, -4)$ and $D(-3, -1)$ is parallel or perpendicular to L_1

SOLUTION

a. The slope of AB is

$$m = \frac{3 - 1}{8 - 5} = \frac{2}{3}$$

Since the slope of L_1 is also $\frac{2}{3}$, line AB is parallel to L_1.

b. The slope of CD is

$$m = \frac{-1 - (-4)}{-3 - (-1)} = \frac{3}{-2} = -\frac{3}{2}$$

Since the slope of L_1 is $\frac{2}{3}$ and

$$\frac{2}{3} \cdot \left(-\frac{3}{2}\right) = -1$$

line CD is perpendicular to L_1. ■

EXAMPLE 4   Finding y when lines are perpendicular

The line through $A(4, y)$ and $B(-2, -5)$ is perpendicular to a line whose slope is $-\frac{2}{3}$; find y.

SOLUTION   The slope of the line through points A and B is

$$\frac{y - (-5)}{4 - (-2)} = \frac{y + 5}{6}$$

If the line through A and B is perpendicular to a line whose slope is $-\frac{2}{3}$, the slope of the line through A and B must be $\frac{3}{2}$ (the negative reciprocal of $-\frac{2}{3}$). Thus

$$\frac{y + 5}{6} = \frac{3}{2}$$

$$\frac{6 \cdot (y + 5)}{6} = \frac{6 \cdot 3}{2} \qquad \text{Multiply both sides by 6.}$$

$$y + 5 = 9 \qquad \text{Simplify.}$$

$$y = 4 \qquad \text{Subtract 5 from both sides.}$$

Thus the line through $A(4, 4)$ and $B(-2, -5)$ has slope

$$\frac{4 - (-5)}{4 - (-2)} = \frac{9}{6} = \frac{3}{2}$$

which makes it perpendicular to a line whose slope is $-\frac{2}{3}$, and our result is correct. ■

D   **Graphing Lines Using the Slope and a Point**

We can graph a line if we know its slope and a point on the line. For example, suppose a line goes through the point $(1, -2)$ and has slope $\frac{2}{3}$. To graph the line, we recall that the slope of a line is the ratio

$$m = \frac{\text{change in } y}{\text{change in } x} = \frac{2}{3}$$

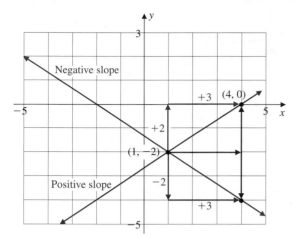

FIGURE 26

We start at the point $(1, -2)$ and go 2 units up (change in y is 2) and 3 units right (change in x is 3), ending at $(4, 0)$. We then draw a line through the points $(1, -2)$ and $(4, 0)$ to obtain the graph shown in blue in Figure 26. If the slope of the line had been $\frac{-2}{3}$, we would move 2 units down (change in y) and then go 3 units right (change in x), ending at $(4, -4)$. This second line with the negative slope is shown in red.

> **NOTE** Recall from Section 4.1 that the standard form of $-\frac{2}{3}$ is $\frac{-2}{3}$. This explains why we move 2 units *down* (because the change in y, the rise, is negative) and 3 units *right* (the change in x is positive).

EXAMPLE 5   Graphing lines using the slope and a point

Graph: A line that goes through the point $(1, 2)$ and has slope $-\frac{3}{4}$

SOLUTION   First, note that the standard form of $-\frac{3}{4}$ is $\frac{-3}{4}$. We start at the point $(1, 2)$ and go 3 units down (the change in y). We then go right 4 units (the change in x), ending at $(5, -1)$, as shown in Figure 27. The graph is obtained by drawing a line through the points $(1, 2)$ and $(5, -1)$. ■

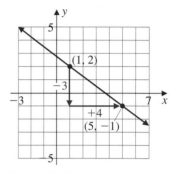

FIGURE 27

GRAPH IT

The slope of a line is a number that measures its steepness. Let's see what this means graphically. Graph $Y_1 = x$, $Y_2 = 4x$, $Y_3 = 8x$, and $Y_4 = 50x$ (see Window 1). What happens to the steepness of the line as the coefficient of x *increases* from 1 to 50? Now graph $Y_1 = -x$, $Y_2 = -4x$, $Y_3 = -8x$, and $Y_4 = -50x$ (see Window 2). What happens to the steepness of the line as the coefficient of x *decreases* from -1 to -50?

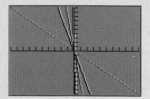

WINDOW 1

You can use a grapher to find the slope of a line. To do this, you need two points and the fact that if you have an equation written in the form $y = ax + b$, a is the slope of the line. Let's do Example 2a. The two given points are $A(-3, 1)$ and $B(-1, -2)$. If your grapher has a $\boxed{\text{LIST}}$ key, make two lists: L1 for the x-coordinates and L2 for the y-coordinates. Press $\boxed{\text{STAT}}$ 1 and enter -3 and -1 under L1 and 1 and -2 under L2. The strokes needed to give you the equation of a line of the form $y = ax + b$ using a technique called "linear regression" are $\boxed{\text{STAT}}$ $\boxed{\blacktriangleright}$ 5 $\boxed{\text{ENTER}}$. The grapher will tell you that the slope $a = -1.5 = -\frac{3}{2}$, as shown in Window 3. Try Example 2b and see what happens!

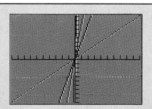

WINDOW 2

Now let's try Example 3a. Since we want to know if a line passing through $A(5, 1)$ and $B(8, 3)$ is parallel or perpendicular to a line with slope $\frac{2}{3}$, we first find the slope of the line going through A and B. As before, we make two lists, L1 and L2 by pressing $\boxed{\text{STAT}}$ 1 and entering 5 and 8 under L1 and 1 and 3 under L2 and then pressing $\boxed{\text{STAT}}$ $\boxed{\blacktriangleright}$ 5 $\boxed{\text{ENTER}}$. The slope a shown in Window 4 is given as .6666666667, which you should recognize as $\frac{2}{3}$. Since the two lines have the same slope, they are parallel. Now you try Example 3b.

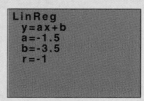

WINDOW 3

```
LinReg
y=ax+b
a=-1.5
b=-3.5
r=-1
```

WINDOW 4

```
LinReg
y=ax+b
a=.6666666667
b=-2.333333333
r=1
```

EXERCISE 7.2

A B In Problems 1–10, find the distance between the points and the slope of the line passing through the points.

1. $A(2, 4)$, $B(-1, 0)$

2. $A(3, -2)$, $B(8, 10)$

3. $C(-4, -5)$, $D(-1, 3)$

4. $C(5, 7)$, $D(-2, 3)$

5. $E(4, 8)$, $G(1, -1)$

6. $H(-2, -2)$, $I(6, -4)$

7. $A(3, -1)$, $B(-2, -1)$

8. $C(-2, 3)$, $D(4, 3)$

9. $E(-1, 2)$, $F(-1, -4)$

10. $G(-3, 2)$, $H(-3, 5)$

C In Problems 11–20, determine whether lines AB and CD are parallel, perpendicular, or neither.

11. $A(1, 6)$, $B(-1, 4)$ and $C\left(1, \dfrac{-7}{2}\right)$, $D\left(\dfrac{7}{2}, -1\right)$

12. $A(0, 4)$, $B(1, -1)$ and $C(0, 1)$, $D\left(\dfrac{7}{2}, -1\right)$

13. $A(2, 0)$, $B(4, 5)$ and $D\left(\dfrac{7}{2}, 0\right)$, $E(1, 1)$

14. $A(1, 1)$, $B(-1, 2)$ and $C(1, -1)$, $D(0, -3)$

15. $A(-1, 1)$, $B(1, 2)$ and $C(1, -1)$, $D(0, -1)$

16. $A(1, 1)$, $B(3, 3)$ and $C(1, -1)$, $D(0, 2)$

17. $A(1, -1)$, $B\left(2, \dfrac{-1}{2}\right)$ and $C(2, -2)$, $D(1, 0)$

18. $A(1, 1)$, $B\left(\dfrac{1}{5}, 0\right)$ and $C(1, 1)$, $D\left(0, \dfrac{9}{5}\right)$

19. $A(0, 1)$, $B(14, -1)$ and $C\left(0, \dfrac{3}{2}\right)$, $D\left(\dfrac{7}{2}, 1\right)$

20. $A(2, -2)$, $B(1, -7)$ and $C(1, -3)$, $D(0, -8)$

In Problems 21–30, find x or y.

21. The line through $A(x, 4)$ and $B(6, 8)$ is parallel to a line whose slope is 1.

22. The line through $A(x, 5)$ and $B(-2, 3)$ is parallel to a line whose slope is $\frac{2}{5}$.

23. The line through $A(x, 2)$ and $B(2, 6)$ is perpendicular to a line whose slope is $\frac{1}{2}$.

24. The line through $A(x, 4)$ and $B(-3, -4)$ is perpendicular to a line whose slope is $\frac{2}{5}$.

25. The line through $A(x, -6)$ and $B(-2, -1)$ is perpendicular to a line whose slope is $-\frac{2}{3}$.

26. The line through $A(2, y)$ and $B(3, 4)$ is parallel to a line whose slope is 2.

27. The line through $A(3, y)$ and $B(1, -2)$ is parallel to a line whose slope is -3.

28. The line through $A(2, y)$ and $B(1, -4)$ is perpendicular to a line whose slope is $\frac{1}{3}$.

29. The line through $A(x, 4)$ and $(3, 5)$ is perpendicular to the horizontal line $y = 5$.

30. The line through $A(-4, 2)$ and $(x, 7)$ is perpendicular to the horizontal line $y = -3$.

D In Problems 31–40, graph the line with the indicated slope and passing through the given point.

31. Slope 2 through $(1, 1)$

32. Slope $\frac{2}{3}$ through $(1, 2)$

33. Slope $-\frac{2}{3}$ through $(1, 1)$

34. Slope $-\frac{3}{4}$ through $(-1, -1)$

35. Slope 0 through $(2, 3)$

36. Slope 0 through $(3, 2)$

37. Slope 0 through $(0, 0)$

38. Slope undefined through $(-1, 2)$

39. Slope undefined through $(2, -1)$

40. Slope undefined through $(0, 0)$

The midpoint of the line segment AB joining the points $A(x_1, y_1)$ and $B(x_2, y_2)$ is the point (x_m, y_m), where

$$x_m = \frac{x_1 + x_2}{2} \quad \text{and} \quad y_m = \frac{y_1 + y_2}{2}$$

In Problems 41–48, find the midpoint of the line segment AB.

41. $A(3, 4)$ and $B(7, 2)$

42. $A(2, 5)$ and $B(-6, 3)$

43. $A(0, -8)$ and $B(-8, 0)$ 44. $A(-3, -5)$ and $B(1, 3)$

45. $A(-5, -2)$ and $B(-6, 2)$ 46. $A(0, -4)$ and $B(-5, -2)$

47. $A(0, -4)$ and $B(-8, 0)$ 48. $A(0, 0)$ and $B(-10, -5)$

APPLICATIONS

49. A pilot is flying from New Orleans, with coordinates $(90, 30)$, to Philadelphia, with coordinates $(76, 40)$. What are the coordinates of the point exactly halfway between the two cities?

50. A pilot is flying from Buenos Aires, with coordinates $(58, 36)$, to Sao Paulo, with coordinates $(46, 24)$. What are the coordinates of the point exactly halfway between the two cities?

In Problems 51–56, use the Pythagorean Theorem and the distance formula to determine whether the three points form the vertices (corners) of a right triangle. (*Hint:* If a and b are the shorter sides of a triangle and c is the longest side and if $c^2 = a^2 + b^2$, then the triangle is a right triangle.)

51. $A(2, 2)$, $B(0, 5)$, $C(-20, 12)$

52. $A(0, 6)$, $B(-3, 0)$, $C(9, -6)$

53. $A(2, 2)$, $B(0, 5)$, $C(-19, -12)$

54. $A(0, 0)$, $B(6, 0)$, $C(3, 3)$

55. $A(2, 2)$, $B(-4, -14)$, $C(-20, -8)$

56. $A(3, 2)$, $B(0, -4)$, $C(12, -10)$

57. The U.S. population has been growing according to the equation

$$y = 2.2x + 180$$

where y is the population (in millions) and x is the number of years after 1960. For example, in 1970, $x = 10$ and $y = 2.2 \cdot 10 + 180 = 202$. This means that the U.S. population was about 202 million in 1970. Use the equation $y = 2.2x + 180$ to find the population in each of the following years:

a. 1980   b. 1990   c. 2000

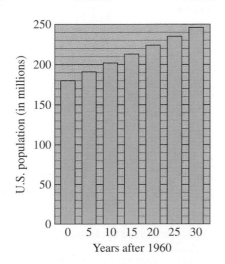

d. In what year will the U.S. population reach 312 million?

e. Graph the equation $y = 2.2x + 180$.

58. The municipal waste y (in millions of tons) generated in the United States is $y = 3.4x + 88$, where x is the number of years after 1960. Use the equation $y = 3.4x + 88$ to find the amount of waste generated in each of the following years:

a. 1980

b. 1990

c. 2000

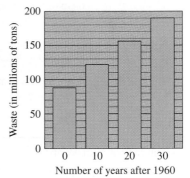

d. In what year will the amount of waste generated reach 241 million tons?

e. Graph the equation $y = 3.4x + 88$.

SKILL CHECKER

Solve for y:

59. $6x + 3y = 12$ 60. $2x + 3y = 6$

61. $3y - 2x = 12$ 62. $5y - 2x = 10$

USING YOUR KNOWLEDGE
Distances and Slopes

The following problems require using your knowledge about a topic previously discussed.

63. A line has x-intercept 2 and y-intercept -3. What is the slope of the line?

64. A line has x-intercept -3 and y-intercept 2. What is the slope of the line?

65. Find the slope of a line parallel to the line through the points $(3, \sqrt{3})$ and $(2, 3\sqrt{3})$.

66. Find the slope of a line parallel to the line through the points $(5, \sqrt{8})$ and $(6, \sqrt{2})$.

67. Find the slope of a line perpendicular to a second line passing through the points $(3, \sqrt{8})$ and $(2, \sqrt{32})$.

68. Find the slope of a line perpendicular to a second line passing through the points $(5, \sqrt{27})$ and $(4, -\sqrt{3})$.

69. The line through the points $(1, c)$ and $(2, c^2)$ is parallel to another line whose slope is 2. What are the possible values for c?

70. The line through the points $(3, 2c)$ and $(4, c^2)$ is parallel to another line whose slope is 3. What is the value of c?

71. The line through the points $(4, c)$ and $(5, c^2)$ is perpendicular to another line whose slope is $-\frac{1}{2}$. What is the value of c?

WRITE ON . . .

Explain in your own words:

72. Why the slope of a horizontal line is zero.

73. Why the slope of a vertical line is undefined.

74. Why two parallel lines have the same slope.

What does it mean graphically:

75. If the slope of a line is positive?

76. If the slope of a line is negative?

MASTERY TEST   If you know how to do these problems, you have learned your lesson!

Graph:

77. The line through the point $(2, 1)$ with slope $-\frac{1}{3}$.

78. The line through the point $(-2, 1)$ with slope $\frac{1}{4}$.

Find x or y:

79. The line through $A(5, y)$ and $B(-1, -4)$ is perpendicular to a line whose slope is $-\frac{3}{4}$.

80. The line through $A(x, 2)$ and $B(3, 4)$ is parallel to a line whose slope is 1.

81. Line L_1 has slope $\frac{3}{2}$. Find whether the line passing
   a. Through $A(6, 3)$ and $B(3, 5)$ is parallel or perpendicular to L_1.
   b. Through $A(-6, -8)$ and $B(-4, -5)$ is parallel or perpendicular to L_1.

82. Find the slope of the line passing through the points
   a. $A(-4, 2)$ and $B(-2, -3)$ **b.** $A(-2, 4)$ and $B(-2, 7)$
   c. $A(2, 1)$ and $B(4, 5)$ **d.** $A(2, 3)$ and $B(1, 3)$

83. Find the distance between the points
   a. $A(2, 2)$ and $B(8, 10)$ **b.** $C(-3, 2)$ and $D(5, 4)$
   c. $E(-4, 3)$ and $F(-4, 7)$ **d.** $G(-3, 2)$ and $H(1, 2)$

7.3

EQUATIONS OF LINES

To succeed, review how to:

1. Graph lines when two points are given (pp. 387–389).

2. Write a linear equation in standard form (p. 390).

3. Solve an equation for a specified variable (p. 77).

Objectives:

Find the equation and the graph of a line given

A Two points.

B One point and the slope.

C The slope and the y-intercept.

D One point and the fact that the line is parallel or perpendicular to a given line.

getting started   **How's the Weather?**

This thermometer has Celsius (centigrade) and Fahrenheit scales. Do you know how to convert temperatures in the Celsius scale to Fahrenheit? One way is to graph ordered pairs of numbers in which the first coordinate represents the Celsius temperature and the second, the corresponding Fahrenheit temperature. Since at 0°C, the corresponding Fahrenheit temperature is 32°, one such point is (0, 32). Another one is (100, 212). Thus the graph of the equation can be drawn as shown in Figure 28. As you know, the graph of a line can be obtained by using *any* two

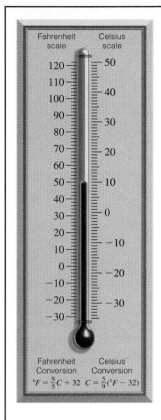

given points on the line. To obtain the equation of the line—that is, to find F in terms of C—we select a point on the line and assign it coordinates (C, F). The slope of the line going through $(0, 32)$ and $(100, 212)$ is

$$m = \frac{212 - 32}{100 - 0} = \frac{180}{100} = \frac{9}{5}$$

The points $(0, 32)$ and $(100, 212)$ correspond to the freezing and boiling points of water, respectively.

The slope of the line going through $(0, 32)$ and (C, F) is

$$m = \frac{F - 32}{C - 0} = \frac{F - 32}{C}$$

Since $(0, 32)$, (C, F), and $(100, 212)$ are on the same line, these slopes are equal. Thus

$$\frac{F - 32}{C} = \frac{9}{5}$$

$$F - 32 = \frac{9}{5}C \qquad \text{Multiply both sides by } C.$$

$$F = \frac{9}{5}C + 32 \quad \text{Add 32 to both terms.}$$

So to convert Celsius temperatures (C) to Fahrenheit (F), substitute the Celsius temperature C in $F = \frac{9}{5}C + 32$. For example, when the temperature is 10°C, the corresponding Fahrenheit temperature is $F = \frac{9}{5}(10) + 32 = 50°$. What we did here was find the equation of a line. There are several ways to do this, and we discuss each of them in this section.

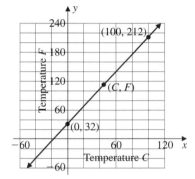

FIGURE 28

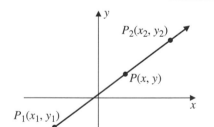

FIGURE 29

A Finding Equations Given Two Points

We are now ready to work with the graph of a line to find its equation. (In Section 7.1, we worked from an equation of a line to draw its graph.) In general, if a line goes through two points $P_1(x_1, y_1)$ and $P_2(x_2, y_2)$, as shown in Figure 29, an equation for this line can be found as follows:

1. Select a general point $P(x, y)$ on the line.

2. The slope of the line P_1P_2 is

$$m = \frac{y_2 - y_1}{x_2 - x_1} \qquad (x_2 \neq x_1)$$

3. The slope of the line P_1P is

$$m = \frac{y - y_1}{x - x_1} \qquad (x \neq x_1)$$

4. Since the slopes are equal,

$$\frac{y - y_1}{x - x_1} = \frac{y_2 - y_1}{x_2 - x_1}$$

or $\quad y - y_1 = \dfrac{y_2 - y_1}{x_2 - x_1} \cdot (x - x_1) \qquad$ Multiply both sides by $(x - x_1)$.

Since

$$\frac{y_2 - y_1}{x_2 - x_1} = m$$

it's easier to remember this equation as $y - y_1 = m(x - x_1)$.

We summarize this discussion as follows.

THE TWO-POINT FORM OF A LINE

An equation of a line going through the points (x_1, y_1) and (x_2, y_2) is given by

$$y - y_1 = \frac{y_2 - y_1}{x_2 - x_1} \cdot (x - x_1) \qquad (x_2 \neq x_1) \tag{1}$$

EXAMPLE 1   Finding an equation given two points

Find, write in standard form, and graph: An equation of the line going through the points $(5, 2)$ and $(6, 4)$.

SOLUTION   Letting $(x_1, y_1) = (5, 2)$ and $(x_2, y_2) = (6, 4)$, and substituting in equation (1), we have

$$y - 2 = \frac{4 - 2}{6 - 5} \cdot (x - 5) \qquad \text{Two-point form}$$

$$y - 2 = 2(x - 5)$$

$$y - 2 = 2x - 10 \qquad \text{Distributive property}$$

$$8 = 2x - y \qquad \text{Subtract } y \text{ and add 10.}$$

In standard form, an equation of this line is $2x - y = 8$. The graph is shown in Figure 30. ■

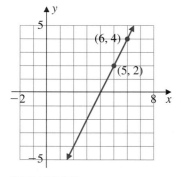

FIGURE 30

Finding Equations Given a Point and the Slope

If we replace

$$\frac{y_2 - y_1}{x_2 - x_1}$$

by m in equation (1) we obtain the following.

THE POINT-SLOPE FORM OF A LINE

$$y - y_1 = m(x - x_1) \tag{2}$$

The point-slope form enables us to find an equation of a line when a point $P(x_1, y_1)$ and the slope m are given. We use this equation in Example 2.

EXAMPLE 2 Finding an equation given a point and the slope

Find, write in standard form, and graph: An equation of a line with slope $m = -2$ and point $(3, 5)$

SOLUTION Here $m = -2, (x_1, y_1) = (3, 5)$. Substituting in equation (2), we get

$$y - 5 = -2(x - 3)$$

$$y - 5 = -2x + 6 \qquad \text{Distributive property}$$

$$2x + y = 11 \qquad \text{Add } 2x \text{ and } 5 \text{ to both sides.}$$

To graph the equation, we start at $(3, 5)$. Since

$$m = -2 = \frac{-2}{1}$$

go 2 units down (the change in y) and 1 unit right (the change in x), ending at $(4, 3)$. The graph is the line through the points $(3, 5)$ and $(4, 3)$, as shown in Figure 31. ■

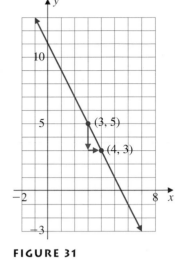

FIGURE 31

Finding Equations Given the Slope and the y-Intercept

If, in $y - y_1 = m(x - x_1)$ the point $P(x_1, y_1)$ is on the y-axis, then $x_1 = 0$. If we let $y_1 = b$, then $P(x_1, y_1) = P(0, b)$. Thus the point-slope form of the equation is $y - b = m(x - 0) = mx$. Solving for y by adding b to both sides, we have the following.

THE SLOPE-INTERCEPT FORM OF A LINE

$$y = mx + b \qquad\qquad (3)$$

slope ⌐ ⌐ y-intercept

EXAMPLE 3 Finding an equation given the slope and the y-intercept

Find the slope-intercept form and graph: An equation of a line with slope 5 and y-intercept 3

SOLUTION Using equation (3) with $m = 5$ and $b = 3$, we find the required equation to be $y = 5x + 3$. To graph this line, we start at the y-intercept, $(0, 3)$. Since the slope is $5 = \frac{5}{1}$, we go 5 units up and 1 unit right, ending up at $(1, 8)$. The graph is the line drawn through the points $(0, 3)$ and $(1, 8)$, as shown in Figure 32. ■

Equation (3) is especially convenient because we can find the slope and y-intercept of a given line by just writing the equation of the line in the form $y = mx + b$. For example, to find the slope and y-intercept of the line $2x + 5y = 7$, we solve for y to obtain

$$y = -\frac{2}{5}x + \frac{7}{5}$$

Thus comparing with equation (3), we see that m (the slope) is $-\frac{2}{5}$, and b (the y-intercept) is $\frac{7}{5}$.

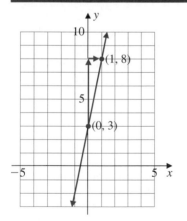

FIGURE 32

EXAMPLE 4   Finding the slope and the y-intercept of a line

Find the slope and y-intercept of: The line $6x + 3x = 12$

SOLUTION   We have to solve for y. Thus

$$3y = -6x + 12 \qquad \text{Subtract } 6x.$$

$$y = -2x + 4 \qquad \text{Divide by 3.}$$

$$\underset{\text{slope}}{\uparrow} \qquad \underset{\text{intercept}}{\uparrow}$$

The slope is -2 and the y-intercept is 4.   ■

D   Finding the Equation of a Line Through a Given Point and Parallel or Perpendicular to a Given Line

We are already familiar with particular aspects of the slopes of parallel lines and perpendicular lines. For example, we can use the fact that two parallel lines have identical slopes to find an equation of a line that is parallel to a given line. Likewise, the fact that two perpendicular lines have slopes that are negative reciprocals of each other can be used to find an equation of a line that is perpendicular to a given line. We illustrate these ideas next.

EXAMPLE 5   Finding an equation of a line through a given point and parallel or perpendicular to a given line

Find: An equation of the line passing through the point (2, 1) and
a. Parallel to the line $y - x = 1$
b. Perpendicular to the line $y - x = 1$

SOLUTION
a. We first write the equation $y - x = 1$ in the slope-intercept form: $y = x + 1$. The slope of this line is 1. If we wish to construct another line parallel to $y = x + 1$ passing through (2, 1), we simply use the point-slope form, with $(x_1, y_1) = (2, 1)$ and $m = 1$, the same slope as that of $y = x + 1$. Thus

$$y - 1 = 1(x - 2)$$

$$y = x - 1$$

b. The line $y - x = 1$ has slope 1. A line perpendicular to this line must have a slope $\frac{-1}{1} = -1$. Using the point-slope form again, we find that an equation of the line perpendicular to $y - x = 1$ and passing through (2, 1) is

$$y - 1 = -1(x - 2)$$

$$y = -x + 3$$

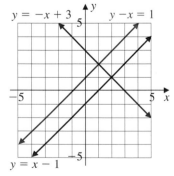

FIGURE 33

All the lines involved are shown in Figure 33.   ■

Before you go on to the exercises, Table 2 gives you a summary of most of the formulas dealing with linear equations that you can refer to.

TABLE 2 Linear Equations

If you want an *equation*	Write
In standard form	$Ax + By = C$
For a vertical line	$x = C$ (C a constant)
For a horizontal line	$y = C$ (C a constant)
For a line going through (x_1, y_1) and (x_2, y_2)	$y - y_1 = \dfrac{y_2 - y_1}{x_2 - x_1}(x - x_1)$
For a line with slope m going through (x_1, y_1)	$y - y_1 = m(x - x_1)$
For a line with slope m and y-intercept b	$y = mx + b$
If you want the *formula*	**Write**
For the distance d between (x_1, y_1) and (x_2, y_2)	$d = \sqrt{(x_2 - x_1)^2 + (y_2 - y_1)^2}$
For the slope of a line passing through (x_1, y_1) and (x_2, y_2)	$m = \dfrac{y_2 - y_1}{x_2 - x_1}$
For the slope of a vertical line	Undefined
For the slope of a horizontal line	$m = 0$
For the slope of a line parallel to the line $y = mx + b$	m
For the slope of a line perpendicular to the line $y = mx + b$	$-\dfrac{1}{m}$

As you recall from Section 7.2, we can use a grapher to find the equation of a line given two points. To do Example 1, press [STAT] 1 and enter 5 and 6 under L_1 and 2 and 4 under L_2, then press [STAT] [▶] 5 [ENTER] to get the equation $y = ax + b$, with $a = 2$ and $b = -8$, that is, the line $y = 2x - 8$, which is equivalent to $2x - y = 8$. Now graph the two points we entered in the table by using [2nd] [STAT PLOT]. Does the line pass through the two points? (See Window 1.)

WINDOW 1

In Example 2, we are given the point $(3, 5)$. Since the slope is $-2 = \frac{-2}{1}$, move 1 unit right and 2 units down to reach the point $(4, 3)$. We now have two points, so we use the same procedure and find the equation of the line $y = ax + b$, with $a = -2$ and $b = 11$—that is, the line $y = -2x + 11$, which is equivalent to $2x + y = 11$. Examples 3 and 4 are best done algebraically, so let's try Example 5. We are given the point $(2, 1)$, and we want a line parallel to $y - x = 1$—that is, $y = x + 1$, which has slope $1 = \frac{1}{1}$. This means that the change in x is 1 and the change in y is 1. Starting at the point $(2, 1)$, go 1 unit right and 1 unit up, ending at $(3, 2)$. We now have two points, $(2, 1)$ and $(3, 2)$, so we can follow our previous steps to make two lists and get an equation of the form $y = ax + b$. For $(2, 1)$ and $(3, 2)$, $a = 1$ and $b = -1$, that is, $y = x - 1$ as before. Now check Example 5b.

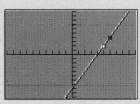

WINDOW 2
$Y_1 = x - 1$
$Y_2 = -x + 3$

Finally, a word of warning about your grapher. Look at Figure 33 in Example 5. The lines $y = x - 1$ and $y = -x + 3$ are *perpendicular*. Graph these two lines using your grapher and a standard window (see Window 2). *They don't look perpendicular!* Why? Because the standard window is *not* square (the units on the x-axis are larger than the units on the y-axis). To remedy this, try a *square* window by pressing [ZOOM] 5 (see Window 3). Your faith in your grapher should now be restored.

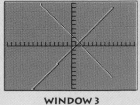

WINDOW 3

EXERCISE 7.3

A In Problems 1–4, graph the line that passes through the given points. Then find the equation of the line in standard form.

1. $(1, -1)$ and $(2, 2)$

2. $(-3, -4)$ and $(-2, 0)$

3. $(3, 2)$ and $(2, 3)$

4. $(3, 0)$ and $(0, 5)$

B In Problems 5–8, find the point-slope form of the equation of the line with the given slope and passing through the given point, then graph the line.

5. Slope 2, point $(-3, 5)$

6. Slope $\dfrac{1}{2}$, point $(2, 3)$

7. Slope -3, point $(-1, -2)$

8. Slope $\dfrac{-1}{3}$, point $(2, -4)$

C In Problems 9–12, find the slope-intercept form of the equation of the line with the given slope and intercept. Then write your answer in standard form.

9. Slope 5, y-intercept 2

10. Slope $\dfrac{1}{4}$, y-intercept 2

11. Slope $\dfrac{-1}{5}$, y-intercept $\dfrac{-1}{3}$

12. Slope $\dfrac{-1}{7}$, y-intercept $\dfrac{-1}{9}$

D In Problems 13–20, find the slope-intercept form of the equation of the line passing

13. Through $(1, -2)$ and parallel to $y = 2x + 1$

14. Through $(-1, -2)$ and parallel to $2y = -4x + 5$

15. Through $(-5, 3)$ and parallel to $2y + 6x = 8$

16. Through $(-3, -5)$ and parallel to $3y - 6x = 12$

17. Through $(1, 1)$ and perpendicular to $2y = x + 6$

18. Through $(2, 3)$ and perpendicular to $3y = -x + 5$

19. Through $(-2, -4)$ and perpendicular to $y - x = 3$

20. Through $(-3, 5)$ and perpendicular to $2y - x = 5$

In Problems 21–26, find the slope-intercept equation of the given line.

21. $x - 3y = 5$

22. $4x + 5y = 20$

23. $2y = 6 - 5x$

24. $2y = -3x + 6$

25. $x = 4 - 8y$

26. $2x = 6 - 4y$

In Problems 27–38, find an equation of the line described.

27. A line passing through the point $(3, 4)$ and with slope 0.

28. A line passing through the point $(-2, -4)$ and with slope 0.

29. The slope of the line is undefined, and it passes through the point $(-2, 4)$.

30. The slope of the line is undefined, and it passes through the point $(-4, -5)$.

31. A vertical line passing through $(-2, 3)$.

32. A vertical line passing through $(-3, -1)$.

33. A horizontal line passing through $(3, 2)$.

34. A horizontal line passing through $(-3, -4)$.

35. The line with x-intercept 2 and y-intercept 4.

36. The line with x-intercept -3 and y-intercept -1.

37. The line with x-intercept -3 and y-intercept 2.

38. The line with x-intercept -4 and y-intercept 3.

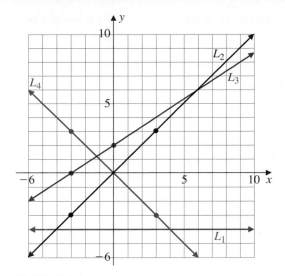

FIGURE 34

In Problems 39–42, find an equation for the specified line shown in Figure 34.

39. L_1 **40.** L_2

41. L_3 **42.** L_4

In Problems 43–48, graph the line using the given information. Then write the equation of the line in slope-intercept form.

43. y-intercept 2, slope $\dfrac{2}{3}$ **44.** y-intercept 3, slope $-\dfrac{2}{3}$

45. y-intercept $\dfrac{1}{2}$, slope -2 **46.** y-intercept $\dfrac{5}{2}$, slope -1

47. y-intercept -2, slope 0   **48.** y-intercept 3, slope 0

APPLICATIONS

49. When sunglasses are sold at the regular price of $6, a student purchases two pairs. If they are sold for $4 at the flea market, the student purchases six pairs. Let the ordered pair (d, p) represent the demand and price for sunglasses.
 a. Find the demand equation.

 b. How many pairs would the student buy if sunglasses are selling for $2?

50. When skateboards are sold for $80, 5 of them are sold each day. When they are on sale for $40, 13 of them are sold daily. Let (d, p) represent the demand and price for skateboards.
 a. Find the demand equation.

 b. For 10 skateboards to be sold on a given day, what should the price be?

51. A wholesaler will supply 50 sets of video games at $35. If the price drops to $20, she will provide only 20. Let (s, p) represent the supply and the price for video games.
 a. Find the supply equation.

 b. At what price would the wholesaler stop selling the video games; in other words, when is the supply zero?

 c. If the price went up to $40 per game, how many games would she be willing to supply?

52. When T-shirts are selling for $6, a retailer is willing to supply 6 of them each day. If the price increases to $12, he's willing to supply 12 T-shirts.
 a. Find the supply equation.

 b. If the T-shirts were free, how many would he supply?

53. The supply s of a product is given by $s = 3p - 6$, while the demand d is $d = -2p + 14$. What will the price p be when the supply s equals the demand d? (This price is called the *equilibrium price*.)

54. The supply and demand for a product are given by $s = 3p - 10$ and $d = -2p + 40$, respectively. At what price p will the supply equal the demand?

SKILL CHECKER

Find the x- and y-intercepts of the line:

55. $2x - 4y = 8$ **56.** $3x - 2y = 6$

57. $y = 2x + 6$ **58.** $y = -4x + 8$

Solve:

59. $3(x - 1) = 6x + 3$ **60.** $4(x - 1) = 8x + 4$

USING YOUR KNOWLEDGE   Business by Candle Lights

In economics and business, the slope m and the y-intercept of an equation play an important role. Let's see how.

Suppose you wish to go into the business of manufacturing fancy candles. First, you have to buy some ingredients such as wax, paint, and so on. Assume all these ingredients cost you $100. This is the *fixed cost*. Now suppose it costs $2 to manufacture each candle. This is the *marginal cost*. What would be the total cost y if the marginal cost is $2, x units are produced, and the fixed cost is $100? The answer is

$$\underbrace{y}_{\substack{\text{Total} \\ \text{cost}}} = \underbrace{2x}_{\substack{\text{Cost for} \\ x\ \text{units}}} + \underbrace{100}_{\substack{\text{Fixed} \\ \text{cost}}}$$

In general, an equation of the form

$$y = mx + b$$

gives the total cost y of producing x units, where m is the cost of producing 1 unit and b is the fixed cost.

61. Find the total cost y of producing x units of a product costing $2 per unit if the fixed cost is $50.

62. Find the total cost y of producing x units of a product whose production cost is $7 per unit if the fixed cost is $300.

63. The total cost y of producing x units of a certain product is given by

$$y = 2x + 75$$

a. What is the production cost for each unit?

b. What is the fixed cost?

WRITE ON . . .

64. How do you decide what formula to use when you're asked to find an equation of a line?

Write the procedure you use to draw the graph of a line:

65. When a point and the slope are given.

66. When the y-intercept and the slope are given.

67. When the x- and y-intercepts are given.

MASTERY TEST

If you know how to do these problems, you have learned your lesson!

68. Find an equation of the line passing through the point $(1, 1)$ and
a. Parallel to $2y - 6x = 5$.
b. Perpendicular to $2y - 6x = 5$.

69. Find the slope and the y-intercept of the line $8x + 4y = 16$.

70. Find the slope and the y-intercept of the line $y = 3$.

71. A line has slope 3 and y-intercept 2. Find the slope-intercept form of the equation of the line and then graph it.

72. The slope of a line is undefined and it passes through the point $(-2, 0)$. What is the equation of this line?

73. Find an equation of the line with slope -3 and passing through the point $(1, 2)$ and then graph it.

74. Find an equation of the line passing through points $(3, 1)$ and $(4, 3)$, write it in standard form, and then graph it.

75. Find an equation of the line passing through points $(3, 2)$ and $(4, 2)$ and then graph it.

7.4

LINEAR INEQUALITIES IN TWO VARIABLES

To succeed, review how to:

1. Find the *x*- and *y*-intercepts of a line (pp. 390–391).
2. Solve linear equations (p. 69).
3. Graph lines (pp. 387–391).
4. Solve inequalities involving absolute values (pp. 124–126).

Objectives:

A Graph linear inequalities.

B Graph inequalities involving absolute values.

C Graph systems of linear inequalities.

---

getting started

Renting Cars

Suppose you want to rent a car for a few days. Here are some prices for an intermediate car rented in Florida in a recent year:

Rental A: $36 per day, $0.15 per mile
Rental B: $49 per day, $0.33 per mile

The total cost *C* for the Rental A car is

$$C = \underbrace{36d}_{\substack{\text{Cost for} \\ d \text{ days}}} + \underbrace{0.15m}_{\substack{\text{Cost for} \\ m \text{ miles}}}$$

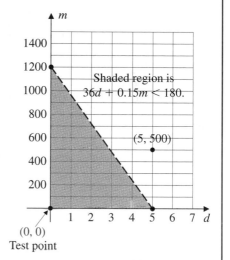

Now suppose you want the cost *C* to be $180. Then $180 = 36d + 0.15m$. We graph this equation by finding the intercepts. When $d = 0$, $180 = 0.15m$, or

$$m = \frac{180}{0.15} = 1200$$

When $m = 0$, $180 = 36d$, or

$$d = \frac{180}{36} = 5$$

We join (0, 1200) and (5, 0) with a line and then graph the discrete points corresponding to 1, 2, 3, or 4 days.

But what if you want the cost to be less than $180? We then have

$$36d + 0.15m < 180$$

We have graphed the points on the line $36d + 0.15m = 180$. Where are the points for which $36d + 0.15m < 180$? As the graph shows, the line $36d + 0.15m = 180$ divides the plane into three parts:

1. The points *below* the line

2. The points *on* the line

3. The points *above* the line

It can be shown that if any point on one side of the line $Ax + By = C$ satisfies the inequality $Ax + By < C$, then all points on that side satisfy the inequality and no point on the other side of the line does. Let's select (0, 0) as a test point. Since

$$36 \cdot 0 + 0.15 \cdot 0 < 180$$

$$0 < 180$$

is true, all points below the line (shown shaded) satisfy the inequality. [As a check, note that (5, 500), which is above the line, doesn't satisfy the inequality, since $36 \cdot 5 + 0.15 \cdot 500 < 180$ is a false statement.] The line $36d + 0.15m = 180$ is *not* part of the graph and is shown dashed. To apply this result to our rental-car problem, note that *d* must be an integer. The solution to our problem consists of the points on the heavy-line segments at $d = 1, 2, 3,$ and 4. The graph shows, for instance, that you can rent a car for 2 days and go about 700 mi at a cost less than $180. What we have just set up is called a *linear inequality,* which we shall now examine in more detail.

Graphing Linear Inequalities

**LINEAR INEQUALITY
IN TWO VARIABLES**

A **linear inequality** is a statement that can be written in the form

$$Ax + By \leq C \qquad \text{or} \qquad Ax + By \geq C$$

where A and B are not both zero.

Here is a summary of the procedure we used to graph the linear inequality in the *Getting Started*.

PROCEDURE

Graphing a Linear Inequality

1. Graph the line associated with the inequality. If the inequality involves $\leq$ or $\geq$, draw a solid line; this means the line is included in the solution. If the inequality involves $<$ or $>$, draw the line dashed, which means the line is not part of the solution.
2. Choose a test point [(0, 0) if possible] not on the line.
3. If the test point satisfies the inequality, shade the region containing the test point; otherwise, shade the region on the other side of the line.

EXAMPLE 1   Graphing a linear inequality

Graph: $x - 2y < -4$

SOLUTION   We follow the three-step procedure.

1. We first graph the line $x - 2y = -4$ by finding the intercepts.

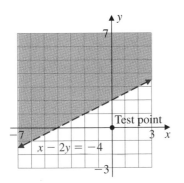

FIGURE 35

x	y	
0	2	When $x = 0$, $-2y = -4$ and $y = 2$
-4	0	When $y = 0$, $x - 2 \cdot 0 = -4$ and $x = -4$

Note that the line itself is shown dashed to indicate that it isn't part of the solution (Figure 35).

2. We select an easy test point and see if it satisfies the inequality. If it does, the solution lies on the same side of the line as the test point; otherwise, the solution is on the other side of the line. An easy point is (0, 0), which is *below* the line. If we substitute $x = 0$ and $y = 0$ in the inequality $x - 2y < -4$, we obtain

$$0 - 2 \cdot 0 < -4$$
$$0 < -4$$

which is false.

3. Thus the point (0, 0) is not part of the solution. Because of this, the solution consists of the points *above* (on the other side of) the line $x - 2y = -4$ and is shown shaded. ∎

In Example 1, the test point was not part of the solution for the given inequality. Now we give an example in which the test point is part of the solution for the inequality.

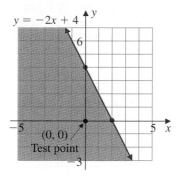

$y = -2x + 4$

(0, 0)
Test point

FIGURE 36

EXAMPLE 2 Graphing an inequality where the test point is part of the solution

Graph: $y \leq -2x + 4$

SOLUTION We use our three-step procedure.

1. As usual, we first graph the line $y = -2x + 4$.

x	y	
0	4	When $x = 0$, $y = 4$.
2	0	When $y = 0$, $0 = -2x + 4$, or $x = 2$.

The graph of the line is shown in Figure 36.

2. Now we select the point (0, 0) as a test point. When $x = 0$ and $y = 0$, we have

$$y \leq -2x + 4$$
$$0 \leq -2 \cdot 0 + 4$$
$$0 \leq 4$$

which is true.

3. Thus all the points on the same side of the line as (0, 0)—that is, the points *below* the line—are solutions of $y \leq -2x + 4$. These solutions are shown shaded in Figure 36. This time, the line is *solid* because it's part of the solution. ∎

> If you solve an inequality for *y* obtaining:
> (1) $y > ax + b$, the solution set consists of all the points *above* $y = ax + b$
> (2) $y < ax + b$, the solution set consists of all the points *below* $y = ax + b$

EXAMPLE 3 Graphing an inequality without using a test point

Graph: $x \geq -1$

SOLUTION We first graph the vertical line $x = -1$. This time, we don't even need a test point! All points to the *right* of this line have x-coordinates greater than -1 (points to the left have x-coordinates less than -1). The graph, which includes the line $x = -1$, is shown in Figure 37. ∎

$x = -1$

FIGURE 37

Graphing Absolute-Value Inequalities

As you recall from Section 2.7,

$$|x| \leq a \quad \text{is equivalent to} \quad -a \leq x \leq a$$

Thus if we graph $|x| \leq 1$, we must graph all the points satisfying the inequality $-1 \leq x \leq 1$ (that is, the points between -1 and 1) as well as the lines $x = -1$ and $x = 1$. These points are shown in Figure 38.

EXAMPLE 4 Graphing an absolute-value inequality

Graph: $|y| \leq 2$

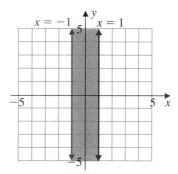

$x = -1$ $x = 1$

FIGURE 38

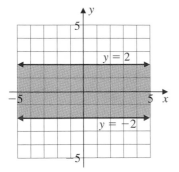

FIGURE 39

SOLUTION   Since $|y| \leq 2$ is equivalent to $-2 \leq y \leq 2$, the graph, shown in Figure 39, consists of all points bounded by the horizontal lines $y = -2$ and $y = 2$, as well as these two lines; these lines are therefore shown as solid lines.   ■

EXAMPLE 5   Graphing an absolute-value inequality

Graph: $|x + 1| > 2$

SOLUTION   The inequality $|x + 1| > 2$ is equivalent to

$$x + 1 > 2 \qquad \text{or} \qquad x + 1 < -2$$
$$x > 1 \qquad \text{or} \qquad x < -3$$

Thus the graph of $|x + 1| > 2$ consists of all points to the *right* of the vertical line $x = 1$ and all points to the *left* of the line $x = -3$. Note that the boundary lines $x = 1$ and $x = -3$ are *not* part of the graph. They are therefore shown as dashed lines in Figure 40.   ■

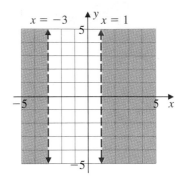

FIGURE 40

Graphing Systems of Linear Inequalities

If we graph two or more linear inequalities **on the same coordinate system**, we have a *system of linear inequalities.* The solution set of the system is the set of points that satisfies *all* the inequalities in the system—that is, the region that is *common* to every graph in the system. We illustrate how to find this solution set in Examples 6 and 7.

EXAMPLE 6   Graphing systems of linear inequalities

Graph the solution set of the system of inequalities:

$$x \leq 0 \qquad \text{and} \qquad y \geq 2$$

SOLUTION   Since $x = 0$ is a vertical line corresponding to the y-axis, $x \leq 0$ consists of the points to the *left* of or *on* the line $x = 0$, as shown in Figure 41. The condition $y \geq 2$ defines all points *above* or *on* the line $y = 2$, as shown in Figure 42. The solution set is the set satisfying both conditions $x \leq 0$ **and** $y \geq 2$; this is the **intersection** of the two solution sets and is the region common to both graphs, as shown by the darker region in Figure 43.   ■

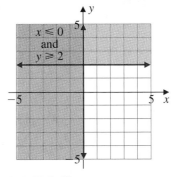

FIGURE 43

FIGURE 41

FIGURE 42

EXAMPLE 7   More practice graphing systems of linear inequalities

Graph the solution set of the system of inequalities:

$$y + x \geq 2 \qquad \text{and} \qquad y - x \leq 2$$

SOLUTION   First, we graph the line $y + x = 2$. When $x = 0$, $y = 2$, so we graph $(0, 2)$. When $y = 0$, $x = 2$, and we graph $(2, 0)$. Joining the points $(0, 2)$ and $(2, 0)$ with a line, the graph of $y + x = 2$. Use $(0, 0)$ as a test point. Does $(0, 0)$ satisfy

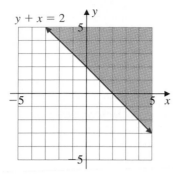

FIGURE 44

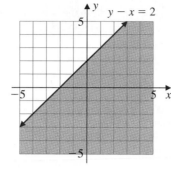

FIGURE 45

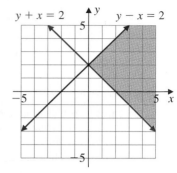

FIGURE 46

$y + x \geq 2$? Letting $x = 0$ and $y = 0$ we obtain $0 + 0 \geq 2$, which is not true. Thus the solution set consists of all points *above* or *on* the line $y + x = 2$, as shown in Figure 44.

Now graph the line $y - x = 2$ (Figure 45). When $x = 0$, $y = 2$, so we graph $(0, 2)$. When $y = 0$, $x = -2$, so we graph $(-2, 0)$ and join it with $(0, 2)$ with a line, the graph of $y - x = 2$. Using $(0, 0)$ as a test point for $y - x \leq 2$, we let $x = 0$ and $y = 0$ to obtain $0 - 0 \leq 2$, which is true, so we shade all the points *below* or *on* the line $y - x = 2$.

The solution set of the system is the **intersection** of the solution sets of $y + x \geq 2$ and $y - x \leq 2$, as shown in Figure 46. Note that we have shown the solution sets of $y + x \geq 2$ and $y - x \leq 2$ separately to illustrate the procedure step by step. When you do the exercises, try to graph the solution set of *all* inequalities involved on the same coordinate axes and use different colors to distinguish between the different regions. ■

GRAPH IT

Your grapher will be a tremendous help when you graph linear inequalities, but before you can use it properly, you need to do some work on your own. In Example 1, you simply graphed $x - 2y = -4$ and used a test point to find the region that satisfied $x - 2y < -4$. If you want to use your grapher to do this problem, you *first* have to solve for y in $x - 2y < -4$ (add $2y + 4$ to both sides and then divide each term by 2) to obtain

$$\frac{x}{2} + 2 < y \qquad \text{or equivalently,} \qquad y > \frac{x}{2} + 2$$

Now graph the associated equation,

$$y = \frac{x}{2} + 2$$

The solution set of

$$y > \frac{x}{2} + 2$$

(which is equivalent to the original inequality $x - 2y < -4$) consists of the points *above* the graph of the line, as shown in Window 1. (Since we want the y's that are strictly greater than

$$\frac{x}{2} + 2$$

$x - 2y < -4 \ \text{ or } \ y > \dfrac{x}{2} + 2$

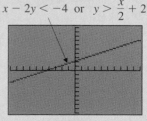

WINDOW 1

(continued on next page)

we do *not* include the points on the line itself in the solution set.) If you have a draw feature, you can press $\boxed{\text{DRAW}}$ to tell the grapher to "draw" the points above the line

$$y = \frac{x}{2} + 2$$

as shown in Window 2.

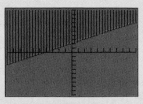

WINDOW 2

Example 2 is much easier to do because the inequality is already solved for *y*. We graph $y = -2x + 4$ and find the points *below* or *on* the line since we are graphing the inequality $y \le -2x + 4$; this is shown in Window 3. (Remember, if the inequality sign is $<$ or $>$, do *not* include the line in the solution set. If the inequality sign is $\le$ or $\ge$, *do* include the line.)

Examples 3, 4, and 5 are done easier algebraically, but where the grapher will save you some time is in the graphing of systems of linear inequalities. Let's try Example 7. Start by solving each inequality for *y* to obtain the equivalent system $y \ge 2 - x$ and $y \le x + 2$. Then graph the corresponding lines $y = 2 - x$ and $y = x + 2$. The solution set consists of all points *above* or *on* $y = 2 - x$ (since we want $y \ge 2 - x$) and *below* or *on* $y = x + 2$ (since we want $y \le x + 2$), as shown in Window 4.

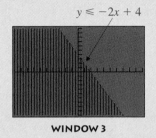

$y \le -2x + 4$

WINDOW 3

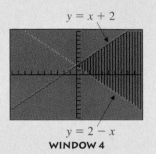

$y = x + 2$

$y = 2 - x$

WINDOW 4

EXERCISE 7.4

A **In Problems 1–20, graph the inequalities.**

1. $x + 2y > 4$

2. $x + 3y > 3$

3. $-2x - 5y \le -10$

4. $-3x - 2y \le -6$

5. $y \ge 2x - 2$

6. $y \ge -2x + 4$

7. $6 < 3x - 2y$

8. $6 < 2x - 3y$

9. $4x + 3y \ge 12$

10. $-3y \ge 6x + 6$

11. $10 < -5x + 2y$

12. $4 < -2x - 4y$

13. $2x \ge 2y - 4$

14. $2x \ge 4y + 2$

15. $2y < -4x + 8$

16. $3y < -6x + 9$

17. $x \ge -3$

18. $x \ge -4$

19. $y < 3$

20. $y < -2$

B In Problems 21–37, graph the inequalities.

21. $|x| < 1$

22. $|x| < 3$

23. $|y| < 4$

24. $|y| < 3$

25. $|x| \geq 1$

26. $|x| \geq 2$

27. $|y| \geq 2$

28. $|y| \geq 4$

29. $|x + 2| < 1$

30. $|x + 3| < 1$

31. $|y + 2| < 1$

32. $|y + 2| < 2$

33. $|x + 1| \geq 3$

34. $|x + 2| \geq 1$

35. $|x - 1| \leq 2$

36. $|x - 2| \leq 1$

37. $|y - 2| < 1$

38. $|y - 3| < 1$

C In Problems 39–48, graph the solution set of the given system of inequalities.

39. $x - y \geq 2$ and $x + y \leq 6$

40. $x + 2y \leq 3$ and $x \leq y$

41. $2x - 3y \leq 6$ and $4x - 3y \geq 12$

42. $2x - 5y \leq 10$ and $3x + 2y \leq 6$

43. $2x - 3y \leq 5$ and $x \geq y$

44. $x \leq 2y$ and $x + y < 4$

45. $x + 3y \leq 6$ and $x \geq y$

46. $2x - y \leq 2$ and $x \leq y$

47. $x - y \leq 1$ and $3x - y < 3$

48. $x - y \geq -2$ and $x + y \leq 6$

SKILL CHECKER

Solve:

49. $56 = k \cdot 14$

50. $39 = k \cdot 13$

51. $60 = \dfrac{k}{3}$

52. $40 = \dfrac{k}{5}$

**The More You Drive,
the Costlier It Gets.**

**You can use what you know about linear inequalities to save
money when you rent a car. Here are the rental prices for an
intermediate car obtained from a telephone survey conducted
in Tampa, Florida, in a recent year:**

>   Rental A: **$41 a day, unlimited mileage**
>
>   Rental B: **$36 a day, $0.15 per mile**

53. If you compare the Rental A price ($41) with the Rental B
    price, you see that Rental B appears to be cheaper.
    a. How far can you drive a Rental B car in one day if you wish
       to spend exactly $41? (Answer to the nearest mile.)
    b. If you are planning on driving 100 mi in one day, which car
       would you rent, Rental B or Rental A?

54. How far can you drive a Rental B car in 1 day if you wish to
    spend exactly $42? (Answer to the nearest mile.)

55. Based on your answers to Problems 53 and 54, which is the
    cheaper rental price? (*Hint:* It has to do with the miles you
    drive.)

WRITE ON . . .

56. Write the procedure you would use to graph the solution set of
    the inequality $ax + by > c$.

57. How would you describe the graph of the inequality $x \geq k$
    (where k is a constant)?

58. Explain what it means when a dashed line is used as the
    boundary in the graph of the solution set of a linear inequality.

59. Explain what it means when a solid line is used as the bound-
    ary in the graph of the solution set of a linear inequality.

60. Is it possible for a system of linear inequalities to have *no*
    solution? If so, find an example of such a system.

**If you know how to do these prob-
lems, you have learned your lesson!**

Graph:

61. $|x + 2| > 3$ 62. $|x - 1| < 4$

63. $|y| \leq 1$ 64. $|y| > 3$

65. $x \geq -2$ 66. $x < -3$

67. $3x - 2y < -6$ 68. $2x - 3y > -6$

69. $x > 2$ and $y < -1$ 70. $x \leq -3$ and $y > 2$

71. $x - 2y \leq -2$ and $2y - x < 4$

72. $2x - 3y > 6$ and $3x - 2y < -6$

7.5

VARIATION

To succeed, review how to:

1. Evaluate an expression (pp. 42–49).

2. Solve linear equations (p. 69).

Objectives:

Write an equation expressing:

A Direct variation.

B Inverse variation.

C Joint variation.

D Solve applications involving direct, inverse, and joint variation.

getting started

Pendulums and Gas Mileage

As the length L of the string increases, the time T it takes the pendulum to make a full back and forth swing increases. What is the formula relating the length L and the time T? Galileo Galilei discovered that the time T (in seconds) it takes for one swing of the pendulum varies *directly* as the square root of the length L of the pendulum. In the same manner, the number m of miles you drive a car is *proportional to,* or *varies directly as,* the number g of gallons of gas used. This means that the ratio

$$\frac{m}{g} \text{ is a constant:} \qquad \frac{m}{g} = k \qquad \text{or} \qquad m = kg$$

In 1992, an Audi 100 TD1 diesel car drove 1338.1 mi on a single tank of fuel, a world record for an unmodified production car. If the fuel tank capacity was 17.62 gal, and $\frac{m}{g} = k$ as before, what is k? Can you explain what k means? By the way, the drivers, Stuart Bladon and Robert Procter, drove the length of Great Britain and then returned to Scotland on this one tank of fuel!

 Direct Variation

Do you get higher grades when you study more hours? If this is the case, your grades vary directly or are directly proportional to the number of hours you study. Here is the definition for direct variation.

DIRECT VARIATION

y **varies directly as** x if there is a constant k such that

$$y = kx$$

(k is usually called the constant of variation.)

Of course, other words can be used to indicate direct variation. Here's a list of some of these words and how they translate into an equation.

English Phrase	Translation
y varies with x	$y = kx$
y varies directly as t	$y = kt$
y is proportional to v	$y = kv$
v varies as the square of t	$v = kt^2$
p varies as the cube of r	$p = kr^3$
T varies as the square root of L	$T = k\sqrt{L}$

EXAMPLE 1   Mustaches and variations

The length L of a mustache varies directly as the time t that it takes to grow.
a. Write an equation of variation.
b. The longest mustache on record was grown by Masuriya Din. His mustache grew 56 in. (on each side) over a 14-yr period. Find k and explain what it represents.

SOLUTION
a. Since the length L varies directly as the time t,

$$L = kt$$

b. We know that when $L = 56$, $t = 14$. Thus

$$56 = k \cdot 14$$
$$4 = k$$

This means that Mr. Din's mustache grew 4 in. each year.   ∎

Inverse Variation

Sometimes, as one quantity increases, a related quantity decreases proportionately. For example, the *more* time you spend practicing a task, the *less* time it will take you to do the task. In such cases, we say that the quantities *vary inversely as* each other.

INVERSE VARIATION

y **varies inversely as** x if there is a constant k such that

$$y = \frac{k}{x}$$

Here are some other words that also mean "vary inversely."

English Phrase	Translation
y varies inversely with x	$y = \dfrac{k}{x}$
y is inversely proportional to x	$y = \dfrac{k}{x}$
v varies inversely as the square of t	$v = \dfrac{k}{t^2}$
p varies inversely as the cube of r	$p = \dfrac{k}{r^3}$
T varies inversely as the square root of L	$T = \dfrac{k}{\sqrt{L}}$

EXAMPLE 2   Speeds and distances

The speed s that a car travels is inversely proportional to the time t it takes to travel a given distance.

a. Write the equation of variation.
b. If a car travels at 60 mi/hr for 3 hr, what is k, and what does it represent?

SOLUTION
a. The equation is

$$s = \frac{k}{t}$$

b. We know that $s = 60$ when $t = 3$. Substituting 60 for s and 3 for t,

$$60 = \frac{k}{3}$$

$$k = 180$$

In this case, k represents the total distance traveled, 180 mi.   ■

EXAMPLE 3   Deafening sound

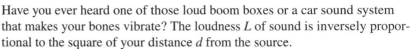

Have you ever heard one of those loud boom boxes or a car sound system that makes your bones vibrate? The loudness L of sound is inversely proportional to the square of your distance d from the source.

a. Write an equation of variation.
b. The loudness of rock music coming from a boom box 5 ft away is 100 dB (decibels). Find k.
c. If you move to 10 ft away from the boom box, how loud is the sound?

SOLUTION
a. The equation is

$$L = \frac{k}{d^2}$$

b. We know that $L = 100$ for $d = 5$ so that

$$100 = \frac{k}{5^2} = \frac{k}{25}$$

Multiplying both sides by 25, we find that $k = 2500$.
c. Since $k = 2500$,

$$L = \frac{2500}{d^2} \qquad \text{Substitute 2500 for } k.$$

When $d = 10$,

$$L = \frac{2500}{10^2} = 25 \text{ dB}$$

By the way, 100 dB is only 20 dB from the threshold of pain, which causes immediate and permanent hearing loss!   ■

Expressing Joint Variation

Besides the direct and inverse variations we have discussed so far, there can be variation involving a third variable. A variable z can vary *jointly* with the variables x and y. For example, labor costs c vary jointly with the number of workers w used and the number of hours h that they work. The formal expression of joint variation is given here.

JOINT VARIATION

z **varies jointly** with x and y if there is a constant k such that

$$z = kxy$$

The statement z *is proportional to x and y* is sometimes used to mean z *varies jointly with variables x and y*. Thus the fact that labor costs c vary jointly with the number w of workers used and the number h of hours worked can be expressed as $c = kwh$, where k is a constant.

EXAMPLE 4　　The lifting force

The lifting force P exerted by the atmosphere on the wings of an airplane varies jointly with the wing area A in square feet and the square of the plane's speed V in miles per hour. Suppose the lift is 1200 lb for a wing area of 100 ft^2 and a speed of 75 mi/hr.

a. Find an equation of variation.
b. Find k.
c. Find the lifting force on a wing area of 60 ft^2 when $V = 125$.

SOLUTION

a. Since P varies jointly with the area A and the square of the velocity V, we have $P = kAV^2$.

b. When the lift $P = 1200$, we know that $A = 100$ and $V = 75$. Substituting these values in the equation $P = kAV^2$, we obtain

$$1200 = k \cdot 100 \cdot (75)^2$$

Dividing both sides by $100 \cdot 75^2$, we find

$$k = \frac{1200}{100 \cdot 75^2} = \frac{12}{75^2} = \frac{4}{1875}$$

Thus $P = kAV^2$ becomes

$$P = \frac{4}{1875} AV^2$$

c. $P = \dfrac{4}{1875} AV^2$, where $A = 60$ and $V = 125$.

$$= \frac{4}{1875}(60)(125^2)$$

$$= 2000 \text{ lb}$$

The lifting force is 2000 lb. ■

Solving an Application

EXAMPLE 5 An awful lot of snow

Figure 47 shows the number of gallons of water, g (in millions), produced by an inch of snow in different cities. Note that the larger the area of the city, the more gallons of water are produced, so g is directly proportional to A, the area of the city (in square miles).

a. Write an equation of variation.

b. If the area of St. Louis is about 62 mi^2, what is k?

c. Find the amount of water produced by 1 in. of snow falling in Anchorage, Alaska, with an area of 1700 mi^2.

SOLUTION

a. Since g is directly proportional to A, $g = kA$.

b. From Figure 47, we can see that $g = 100$ (million) is the number of gallons of water produced by 1 in. of snow in St. Louis. Since it is given that $A = 62$, $g = kA$ becomes

$$100 = k \cdot 62 \qquad \text{or} \qquad k = \frac{100}{62} = \frac{50}{31}$$

c. For Anchorage, $A = 1700$, thus $g = \frac{50}{31} \cdot 1700 = 2742$ million gallons of water. ∎

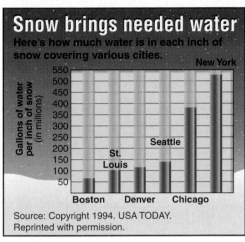

Snow brings needed water

Here's how much water is in each inch of snow covering various cities.

Source: Copyright 1994. USA TODAY. Reprinted with permission.

FIGURE 47

GRAPH IT

How do you recognize different types of variation with your grapher? If you have several points, you can graph them and examine the result. In Example 1, we concluded that Mr. Din's mustache grew 4 in. each year. This means that at the end of the first year, his mustache was 4 in. long; at the end of the second year, it was 8 in. long; and at the end of 14 yr, it was 56 in. long. You now have three ordered pairs: (1, 4), (2, 8), and (14, 56). You can make a list by pressing [STAT] 1 and then entering 1, 2, and 14 under L_1 and 4, 8, and 56 under L_2. Press [2nd] [STAT PLOT] 1 to turn plot on; select the first type of graph, L_1, L_2, and ■. So that we don't have to contend with deciding what type of window to use, press [ZOOM] 9. As you can see in Window 1, the points are on a line. (Remember that the length varies directly as the time.)

If you want to find the equation of the line passing through the three points in the graph, press [STAT] [▶] 5 [ENTER]. The grapher will tell you that $y = ax + b$, where $a = 4$ and $b = 0$—that is, $y = 4x$ as before.

To get a feel for inverse variation, in Example 2, select a $[-1, 200]$ by $[-1, 5]$ window with Xscl = 10 and Yscl = 10. Using Y_1 instead of x and x instead of t, graph

$$Y_1 = \frac{180}{x}$$

The result is shown in Window 2. Can you see that as x increases, Y_1 decreases? Try looking at the graph

$$L = \frac{2500}{d^2}$$

What window do you need? What does the graph look like?

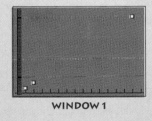

WINDOW 1

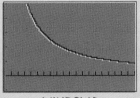

WINDOW 2

EXERCISE 7.5

A In Problems 1–5, write an equation of variation using k as the constant.

1. The tension T on a spring varies directly with the distance s it is stretched. (This fact is usually called Hooke's law.)

2. The distance s a body falls in t seconds is directly proportional to the square of t.

3. The weight W of a dam varies directly with the cube of its height h.

4. The kinetic energy KE of a moving body is proportional to the square of its velocity v.

5. The weight W of a human brain is directly proportional to the body weight B.

B In Problems 6–8, write an equation of variation using k as the constant.

6. In a circuit with constant voltage, the current I varies inversely with the resistance R of the circuit.

7. For a wire of fixed length, the resistance R varies inversely with the square of its diameter D.

8. The intensity of illumination I from a source of light varies inversely with the square of the distance d from the source.

C In Problems 9–20, write an equation of variation using k as the constant.

9. The annual interest I received on a savings account varies jointly with the principal P (the amount in the account) and the interest rate r paid by the bank.

10. The cost C of a building varies jointly as the number w of workers used to build it and the cost of materials m.

11. The amount of oil A used by a ship traveling at a uniform speed varies jointly with the distance s and the square of the speed v.

12. The power P in an electric circuit varies jointly with the resistance R and the square of the current I.

13. The volume V of a rectangular container of fixed length varies jointly with its depth d and width w.

14. The force of attraction F between two spheres of mass m_1 and m_2 varies directly as the product of the masses and inversely as the square of the distance d between their centers.

15. The illumination I in foot-candles upon a wall varies directly with the intensity i in candlepower of the source of light and inversely with the square of the distance d from the light.

16. The strength S of a horizontal beam of rectangular cross section and of length L varies jointly as the breadth b and the square of the depth d and inversely as the length L.

17. The electrical resistance R of a wire of uniform cross section varies directly as its length L and inversely as its cross-sectional area A.

18. The electrical resistance R of a wire varies directly as the length L and inversely as the square of its diameter d.

19. The weight W of a body varies inversely as the square of its distance d from the center of the Earth.

20. z varies directly as the cube of x and inversely as the square of y.

D Applications

21. The amount of annual interest I you receive on a savings account is directly proportional to the amount of money m you have in the account.
   a. Write an equation of variation.
   b. If $480 produces $26.40 in interest, what is k?
   c. How much annual interest would you receive if the account had $750?

22. The number of revolutions, R (rev), a record makes as it is being played varies directly as the time t that it is on the turntable.
   a. Write an equation of variation.
   b. A record that lasted $2\frac{1}{2}$ min made 112.5 rev. What is k?
   c. If a record makes 108 rev, how long does it take to play it?

23. The distance d an automobile travels after the brakes have been applied varies directly as the square of its speed s.
   a. Write an equation of variation.
   b. If the stopping distance for a car going 30 mi/hr is 54 ft, what is k?
   c. What is the stopping distance for a car going 60 mi/hr?

24. The weight of a person varies directly as the cube of the person's height h (in inches). The **threshold weight** T (in pounds) for a person is defined as "the crucial weight, above which the mortality (risk) for the patient rises astronomically."
   a. Write an equation of variation relating T and h.
   b. If $T = 196$ when $h = 70$, find k written as a fraction.
   c. To the nearest pound, what is the threshold weight T for a person 75 in. tall?

25. The number S of new songs a rock band needs to stay on top each year is inversely proportional to the number of years y the band has been in the business.
a. Write an equation of variation.

b. If, after 3 yr in the business, the band needs 50 new songs, how many songs will it need after 5 yr?

26. When the distance is set at infinity, the *f*-number on a camera lens varies inversely as the diameter d of the aperture (opening).
a. Write an equation of variation.

b. If the *f*-number on a camera is 8 when the aperture is $\frac{1}{2}$ in., what is k?

c. Find the *f*-number when the aperture is $\frac{1}{4}$ in.

27. The weight W of an object varies inversely as the square of its distance d from the center of the Earth.
a. Write an equation of variation.

b. An astronaut weighs 121 lb on the surface of the Earth. If the radius of the Earth is 3960 mi, find the value of k for this astronaut. (Do not multiply out your answer.)

c. What will this astronaut weigh when she is 880 mi above the surface of the Earth?

28. The number of miles m you can drive in your car is directly proportional to the amount of fuel g in your gas tank.
a. Write an equation of variation.

b. The greatest distance yet driven without refueling on a single fill in a standard vehicle is 1691.6 miles. If the twin tanks used to do this carried a total of 38.2 gal of fuel, what is k?

c. How many miles per gallon is this?

29. The distance d (in miles) traveled by a car is directly proportional to the average speed s (in mi/hr) of the car, even when driving in reverse!
a. Write an equation of variation.

b. The highest average speed attained in any nonstop reverse drive of more than 500 mi is 28.41 mi/hr. If the distance traveled was 501 mi, find k.

c. What does k represent?

30. Have you called in on a radio contest lately? According to Don Burley, a radio talk-show host in Kansas City, the listener response to a radio call-in contest is directly proportional to the size of the prize.
a. If 40 listeners call when the prize is $100, write an equation of variation using N for the number of listeners and P for the prize in dollars.

b. How many calls would you expect for a $5000 prize?

31. The number of chirps C a cricket makes each minute is directly proportional to 37 less than the temperature F in degrees Fahrenheit.

a. If a cricket chirps 80 times when the temperature is $57°F$, what is the equation of variation?

b. How many chirps per minute would the cricket make when the temperature is $90°F$?

32. According to George Flick, the ship's surgeon of the SS *Constitution*, the number of hours H your life is shortened by smoking cigarettes varies jointly as N and $t + 10$, where N is the number of cigarettes you smoke and t is the time in minutes it takes you to smoke each cigarette. If it takes 5 min to smoke a cigarette and smoking 100 of them shortens your lifespan by 25 hr, how long would smoking 2 packs a day for a year (360 days) shorten your lifespan? (*Note:* There are 20 cigarettes in a pack.)

33. The concentration of carbon dioxide (CO_2) in the atmosphere has been increasing due to human activities such as automobile emissions, electricity generation, and deforestation. In 1965, CO_2 concentration was 319.9 parts per million (ppm), and 23 years later, it increased to 351.3 ppm. The *increase* of carbon dioxide concentration, I, in the atmosphere is directly proportional to number of years n elapsed since 1965.
a. Write an equation of variation for I.

b. Find k.

c. What would you predict the CO_2 concentration to be in the year 2000?

34. The *increase, I,* in the *percent* of college graduates in the United States among persons 25 years and older between 1930 and 1990 is proportional to the square of the number of years after 1930. In 1940, the increase was about 5%.
a. Write an equation of variation for I if n is the number of years elapsed since 1930.

b. Find k.

c. What would you predict the percent increase to be in the year 2000?

35. The *simple interest I* in an account varies jointly as the time t and the principal P. After one quarter (3 months), an $8000 principal earned $100 in interest. How much would a $10,000 principal earn in 5 months?

36. At depths of more than 1000 m (a kilometer), water temperature T (in degrees Celsius) in the Pacific Ocean varies inversely as the water depth d (in meters). If the water temperature at 4000 m is $1°C$, what would it be at 8000 m?

37. Anthropologists use the cephalic index C in the study of human races and groupings. This index is directly proportional to the width w and inversely proportional to the length L of the head. The width of the head in a skull found in 1921 and named Rhodesian man was 15 cm, and its length was 21 cm. If the cephalic index of Rhodesian man was 98, what would the cephalic index of Cro-Magnon man be, whose head was 20 cm long and 15 cm wide?

SKILL CHECKER

Graph:

38. $x + y = 3$

39. $2x - y = 2$

40. $2x + \dfrac{1}{2}y = 2$

41. $y = -x - 3$

42. $y = -4x + 4$

USING YOUR KNOWLEDGE

The Pressure of Diving

The equation for direct variation between x and y ($y = kx$) and the equation of a line of slope m passing through the origin ($y = mx$) are very similar. Look at the following table, which gives the water pressure in pounds per square inch exerted on a diver.

Depth of diver (ft)	10	25	40	55
Pressure on diver (lb/in.²)		4.2	10.5	16.8	23.1

43. Graph these points using x as the depth and y as the pressure.

44. What is the slope of the resulting line?

45. As it turns out, the pressure p on a diver is directly proportional to the depth d. Write an equation of variation.

46. Use one of the points in the table to find k.

47. What is the relationship between k and the slope found in Problem 44?

48. Predict the pressure on a diver at a depth of 125 ft.

WRITE ON . . .

49. Explain the difference between direct variation and inverse variation.

50. Find two different ways of expressing the idea that "y is directly proportional to x."

51. Find two different ways of expressing the idea that "y is inversely proportional to x."

52. Find two different ways of expressing the idea that y and z are directly proportional to x.

MASTERY TEST

If you know how to do these problems, you have learned your lesson!

53. The wind force F on a vertical surface varies jointly with the area A of the surface and the square of the wind velocity V. Suppose the wind force on 1 ft² of surface is 1.8 lb when $V = 20$ mi/hr.
  a. Find an equation of variation.
  b. Find k.
  c. Find the force on a 2-ft² vertical surface when $V = 60$ mi/hr.

54. The f-number on a camera varies inversely as the diameter a of the aperture when the distance is set at infinity.
  a. Write an equation of variation.
  b. Find k when the f-number is 8 and $a = \frac{1}{2}$.
  c. Find a if the f-number is 16.

55. The principal P invested is inversely proportional to the annual rate of interest r.
  a. Write an equation of variation.
  b. Find k when $r = 10\%$ and $P = \$100$.

56. Hair length L is proportional to time t.
  a. Write an equation of variation.
  b. Find k if your hair grew 0.6 in. in 2 months.

7.6

INTRODUCTION TO FUNCTIONS: LINEAR FUNCTIONS

To succeed, review how to:

1. Graph a line (pp. 387–391).

2. Evaluate an expression (pp. 42–49).

Objectives:

A Find the domain and range of a relation.

B Use the vertical line test to determine if a relation is a function.

C Find the domain of a function defined by an equation.

D Find the value of a function.

getting started

Traveling Ants

Did you know that there is a linear relationship between the temperature and the rate of travel (speed) of certain ants? If y is the speed in centimeters per second and x is the temperature in degrees Celsius, the relationship is

$$y = \frac{1}{6}(x - 4)$$

Thus if the temperature is 10°C, the speed is

$$y = \frac{1}{6}(10 - 4) = \frac{1}{6} \cdot 6 = 1 \text{ cm/sec}$$

If the temperature is 16°C, the speed is

$$y = \frac{1}{6}(16 - 4) = \frac{1}{6} \cdot 12 = 2 \text{ cm/sec}$$

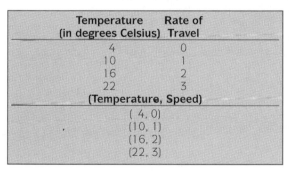

Temperature (in degrees Celsius)	Rate of Travel
4	0
10	1
16	2
22	3
(Temperature, Speed)	
(4, 0)	
(10, 1)	
(16, 2)	
(22, 3)	

We can make a table showing two related sets of numbers, one for the temperature x and the other for the speed y, as shown. (By the way, x has to be greater than 4°C and less than 35°C. Why?)

The numbers in the first column are called values of the **independent variable** because they are chosen *independently* of the second number. The numbers in the second column are called values of the **dependent variable** because they *depend* on the values of the numbers in the first column. The numbers in our table are written as ordered pairs, and these ordered pairs can also be written as the set

$$\{(4, 0), (10, 1), (16, 2), (22, 3)\}$$

The concept of ordered pairs begins our investigation into an important idea in algebra: the concept of a function.

A **Finding the Domain and Range**

Any set of ordered pairs is a relation, which we define as follows.

RELATION, DOMAIN, AND RANGE

A **relation** is a set of ordered pairs. The set of all first coordinates is the **domain** of the relation, and the set of all second coordinates is the **range** of the relation.

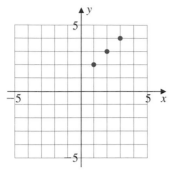

FIGURE 48

Now suppose $S = \{(1, 5), (2, 7), (3, 9)\}$. The domain D is the set of all first coordinates—that is, $D = \{1, 2, 3\}$—and the range R is the set of all second coordinates—that is, $R = \{5, 7, 9\}$.

The relation S can also be written by giving the rule used to obtain the ordered pairs. Thus

$$S = \{(x, y)\,|\,y = 2x + 3\}$$

Here $x = 1, 2,$ or 3.

EXAMPLE 1   Finding the domain and range of a relation

Find the domain and range of: The relation $A = \{(1, 2), (2, 3), (3, 4)\}$

SOLUTION   The domain of A is the set of first coordinates, so $D = \{1, 2, 3\}$. The range of A is the set of second coordinates, so $R = \{2, 3, 4\}$. The graph of the relation A is shown in Figure 48.   ∎

Since a relation is a set of ordered pairs, relations can be graphed in the Cartesian plane. We can then identify the domain and range by examining the graph of the relation.

> **NOTE**   When no domain is specified, the domain is assumed to be the set of all real numbers for which the relation is defined.

EXAMPLE 2   Finding the graph and the domain and range of a relation

Find the graph, the domain, and the range of: The relation $\{(x, y)\,|\,y = 2x - 4\}$

SOLUTION   The graph of the relation is the graph of the equation $y = 2x - 4$, shown in Figure 49. The domain of this relation is the set of all real numbers since any real number x can be used as the first coordinate. Similarly, the range of y is the set of all real numbers. Some of the ordered pairs in the relation are $(0, -4), (1, -2),$ and $(2, 0)$. Note that the graph of the relation $\{(x, y)\,|\,y = 2x - 4\}$ is the graph of its ordered pairs and the graph is a "picture" of the relation.   ∎

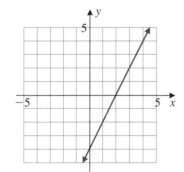

FIGURE 49

EXAMPLE 3   More practice finding the domain and range

Find the domain and range of:

a.

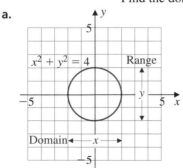

b.

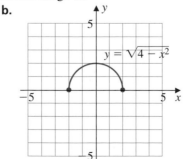

c.

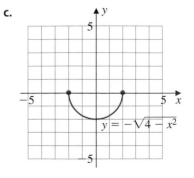

SOLUTION

a. The graph is a circle of radius 2 centered at the origin. From the graph, it is clear that x and y can be any real numbers between -2 and 2, inclusive. Thus the domain is $D = \{x \mid -2 \leq x \leq 2\}$, and the range is $R = \{y \mid -2 \leq y \leq 2\}$.

b. The graph is the top half of the circle from part a. The domain is $D = \{x \mid -2 \leq x \leq 2\}$, and the range is $R = \{y \mid 0 \leq y \leq 2\}$.

c. The graph is the bottom half of the circle from part a. The domain is $D = \{x \mid -2 \leq x \leq 2\}$, and the range is $R = \{y \mid -2 \leq y \leq 0\}$.   ■

Functions and the Vertical Line Test

The relation in Example 3, part a, allows *two* values of y for the same value of x. For instance, if $x = 0$, then $y = \pm 2$. On the other hand, the relations in parts b and c allow only *one* value of y for each value of x. These two relations are *functions*. Here is the definition.

FUNCTION

> A **function** is a relation in which no two different ordered pairs have the same first coordinate.

Thus the relation $\{(1, 2), (2, 3), (3, 4)\}$ is a function since no two ordered pairs have the same first coordinate. On the other hand, the relation $\{(1, 2), (2, 3), (1, 3)\}$ is not a function since $(1, 2)$ and $(1, 3)$ have the same first coordinate.

The graph of a relation can be used to determine whether the relation is a function. Since any two points with the same first coordinate will be on a vertical line parallel to the y-axis, if any vertical line intersects the graph more than once, the relation is *not* a function. Testing to see if a relation is a function by determining whether a vertical line crosses the graph more than once is called the **vertical line test**.

VERTICAL LINE TEST

> If a vertical line parallel to the y-axis intersects the graph of a relation more than once, the relation is *not* a function.

Using this test, we can see that the relation in Example 3, part a, is *not* a function. (A vertical line crosses the graph in more than one place). On the other hand, the graphs in parts b and c are functions.

EXAMPLE 4   Using the vertical line test

Use the vertical line test to determine whether the graph of the relation defines a function:

a.

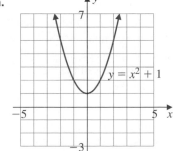

b.

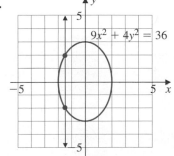

This vertical line crosses the graph in *two* points.

SOLUTION
a. The graph is a **parabola**. The relation is a function since no vertical line crosses the graph more than once.
b. The graph is an **ellipse**. Since we can draw a vertical line that crosses the graph in more than one point, the relation is not a function. ■

Finding the Domain of a Function

When relations are defined by means of an equation, the domain is the set of all possible replacements for the variable x that result in real numbers for y. Thus we cannot replace x by values that will produce zero in the denominator or the square root of a negative number. For example, the domain of $y = \frac{1}{x}$ is the set of all real numbers *except* 0 and the domain of $y = \sqrt{x}$ is the set of all real numbers $x \geq 0$.

EXAMPLE 5   Finding the domain of a function

Find the domain of the function defined by:

a. $y = \dfrac{1}{(x - 2)(x + 3)}$ b. $y = \sqrt{x - 3}$

SOLUTION
a. Since we cannot replace x by values that will produce zero in the denominator, we must avoid the case in which

$$x - 2 = 0 \qquad \text{or} \qquad x + 3 = 0$$
$$x = 2 \qquad\qquad\qquad x = -3$$

Thus the domain of

$$y = \dfrac{1}{(x - 2)(x + 3)}$$

is the set of all real numbers except 2 and -3.
b. The square root of a negative number is not a real number, thus we must make the expression under the radical, $x - 3$, nonnegative, that is,

$$x - 3 \geq 0$$

or $x \geq 3$

The domain is $\{x \mid x \geq 3\}$. ■

Finding the Value of a Function

Water pressure is a *function* of the depth, as you can see in the photo. The higher pressure at the lower holes of the can makes water squirt out in a flat trajectory, whereas the lower pressure at the upper holes produces only a weak stream. It is known that the pressure y of the water (in pounds per square foot) at a depth of x feet (x is a positive number) is

$$y = 62.5x$$

If we wish to emphasize that the pressure y is a function of the depth x, we can use the **function notation** $f(x)$ (read "f of x") and write

$$f(x) = 62.5x$$

To find the pressure at 2 ft below the surface, we find the value of y when $x = 2$. Thus if $x = 2$, then $y = (62.5)(2) = 125$. Using functional notation, we would write $f(2) = (62.5)(2) = 125$. This table illustrates both notations:

y in terms of x $y = 62.5x$	Function Notation $f(x) = 62.5x$
If $x = 2$, then $y = (62.5)(2) = 125$.	$f(2) = (62.5)(2) = 125$
If $x = 4$, then $y = (62.5)(4) = 250$.	$f(4) = (62.5)(4) = 250$
If $x = 5$, then $y = (62.5)(5) = 312.5$.	$f(5) = (62.5)(5) = 312.5$

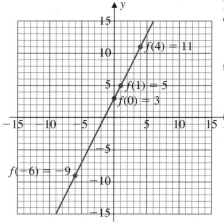

If $y = f(x)$, the symbols $f(x)$ and y are interchangeable; they both represent the value of the function for the given value of x. Thus if

$$y = f(x) = 2x + 3$$

then

$$f(1) = 2(1) + 3 = 5$$
$$f(0) = 2(0) + 3 = 3$$
$$f(-6) = 2(-6) + 3 = -9$$
$$f(4) = 2(4) + 3 = 11$$
$$f(a) = 2(a) + 3 = 2a + 3$$
$$f(w + 2) = 2(w + 2) + 3 = 2w + 7$$

and so on. Whatever appears between the parentheses in $f(\ )$ is to be substituted for x in the rule that defines $f(x)$.

Instead of describing a function in set notation, we frequently say "the function defined by $f(x) = \ldots$," where the ellipsis dots are to be replaced by the expression for the value of the function. For instance, "the function defined by $f(x) = 2x + 3$" has the same meaning as "the function $f = \{(x, y) \mid y = 2x + 3\}$." The function $f(x) = 2x + 3$ is an example of a linear function because its graph is a straight line. In general, we have the following definition.

LINEAR FUNCTION

A **linear function** is a function that can be written in the form

$$f(x) = mx + b$$

where m and b are real numbers.

EXAMPLE 6   Finding the values of a function

Let $f(x) = 3x + 5$ and find:

a. $f(4)$ **b.** $f(2)$ **c.** $f(2) + f(4)$ **d.** $f(x + 1)$

SOLUTION

a. Since $f(x) = 3x + 5$,

$$f(4) = 3 \cdot 4 + 5 = 12 + 5 = 17$$

b. $f(2) = 3 \cdot 2 + 5 = 6 + 5 = 11$

c. Since $f(2) = 11$ and $f(4) = 17$,

$$f(2) + f(4) = 11 + 17 = 28$$

d. $f(x + 1) = 3(x + 1) + 5 = 3x + 8$ ■

EXAMPLE 7   Finding the values of a function

Let $f = \{(3, 2), (4, -2), (5, 0)\}$ and find:

a. $f(3)$ **b.** $f(4)$ **c.** $f(3) + f(4)$

SOLUTION

a. In the ordered pair $(3, 2)$, $x = 3$ and $y = 2$. Thus $f(3) = 2$.

b. In the ordered pair $(4, -2)$, $x = 4$ and $y = -2$. Thus $f(4) = -2$.

c. $f(3) + f(4) = 2 + (-2) = 0$ ■

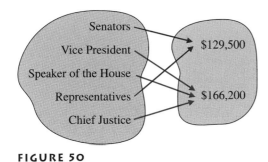

FIGURE 50

The rule or correspondence that assigns a range value to each domain value can also be shown by a method called **mapping**. For example, see Figure 50 where the mapping shown is a function that assigns to each federal government title its corresponding salary. The value of the function can be found by inspecting the "map." Note that in this case,

$$f(\text{Senator}) = \$129,500$$

$$f(\text{Vice President}) = \$166,200$$

$$f(\text{Speaker}) = \$166,200$$

$$f(\text{Representative}) = \$129,500$$

$$f(\text{Chief Justice}) = \$166,200$$

The idea of the function is of great importance in algebra, so make sure you understand this concept and its notation. Here are three ways in which a function can be viewed.

EXPRESSIONS OF A FUNCTION

1. A **function** is a rule or a correspondence that assigns exactly **one** *range* value to each *domain* value (as in Example 6).

2. A **function** is a relation in which **no two** ordered pairs have the same first coordinate (as in Example 7).

3. A **function** is a *mapping* that assigns exactly **one** range element to each domain element.

GRAPH IT

Your grapher is the perfect tool to determine whether a set of ordered pairs is a function. Using the vertical line test or the definition of a function, make sure that no two ordered pairs have the same first component. In Example 1, make two lists. (See *Graph It* on page 427 for how to do this.) Enter 1, 2, and 3 on the first and 2, 3, and 4 on the second. To graph the points on the list, press [2nd] [STAT PLOT] 1 [ENTER]; then press [ZOOM] 9 and the graph of the corresponding ordered pairs appears. Since no two ordered pairs have the same first component, the relation is a function. You can also ascertain the domain and range by using the [TRACE] feature, which gives you the *x*- and *y*-values for each of the points. (If you want to see both axes, adjust the window so that Xmin and Ymin are zero as shown in Window 1.)

(continued on facing page)

For Example 2, simply graph $Y_1 = 2x - 4$ in a standard window. Clearly, the domain and range consists of all real numbers since the line extends indefinitely right and left, as shown in Window 2 (all real numbers x), and up and down (all real numbers y).

Now, let's try Example 5 with

$$Y_1 = \frac{1}{(x - 2)(x + 3)}$$

using a decimal window $[-4.7, 4.7]$ by $[-4.7, 4.7]$ on a TI-82 and using dots instead of a connected graph. Do you see why the domain doesn't include 2 or -3? If you don't see it yet, use the $\boxed{\text{TRACE}}$ and $\boxed{\text{ZOOM}}$ features. Do you ever get a y-value for $x = 2$ or $x = -3$? (See Window 3.) Try part b and see what you get for the domain! To enter $y = \sqrt{x - 3}$, you have to press $\boxed{\text{Y=}}$ and enter $(x - 3)\wedge(1/2)$ or use the $\boxed{\sqrt{\ }}$ key.

Your grapher can also evaluate functions, as long as you tell it what the function is. To do Example 6, enter $Y_1 = 3x + 5$. Now go to the home screen and give x a value, say, 4, by pressing 4 $\boxed{\text{STO}}$ $\boxed{\text{X,T,}\theta}$ $\boxed{\text{ENTER}}$. Next, tell the grapher that you want to evaluate the function Y_1. Press $\boxed{\text{2nd}}$ $\boxed{\text{Y-VARS}}$ 1 1 $\boxed{\text{ENTER}}$. (See Window 4.) You can also do the evaluation on your home screen. To do this on a TI-82, press 4 $\boxed{\text{STO}}$ $\boxed{\text{X,T,}\theta}$ $\boxed{\text{ENTER}}$; then press 3 $\boxed{\text{X,T,}\theta}$ $\boxed{+}$ 5 $\boxed{\text{ENTER}}$ and the value 17 appears.

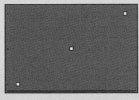

WINDOW 1

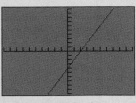

WINDOW 2

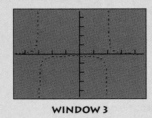

WINDOW 3

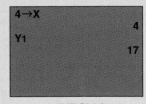

WINDOW 4

EXERCISE 7.6

A In Problems 1–10, find the domain and the range, and determine whether the relation is a function.

1. $\{(-3, 0), (-2, 1), (-1, 2)\}$

2. $\{(-1, -2), (0, -1), (1, 0)\}$

3. $\{(3, 0), (4, 0), (5, 0)\}$

4. $\{(0, 1), (0, 2), (0, 3)\}$

5. $\{(1, 2), (1, 3), (2, 2), (2, 3)\}$

6. $\{(2, 1), (1, 2), (3, 4), (4, 3)\}$

7.
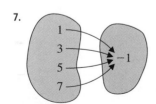

8.
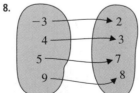

9.
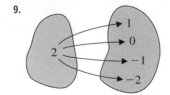

10.
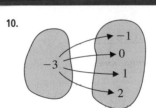

A **B** In Problems 11–30, give the domain and the range, and use the vertical line test to determine whether the relation is a function.

11.
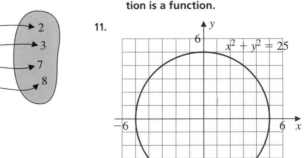

12.

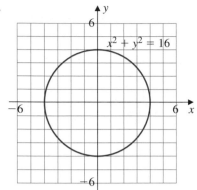

$x^2 + y^2 = 16$

13.

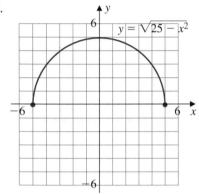

$y = \sqrt{25 - x^2}$

14.

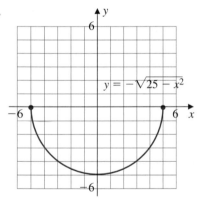

$y = -\sqrt{25 - x^2}$

15.

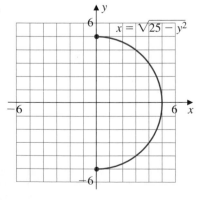

$x = \sqrt{25 - y^2}$

16.

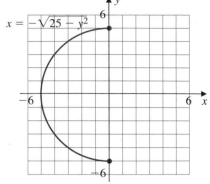

$x = -\sqrt{25 - y^2}$

17.

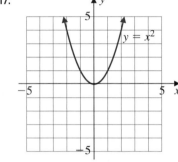

$y = x^2$

18.

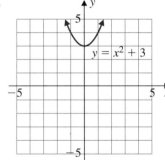

$y = x^2 + 3$

19.

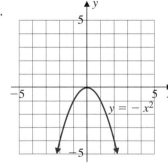

$y = -x^2$

20.

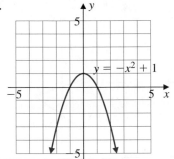

$y = -x^2 + 1$

24.

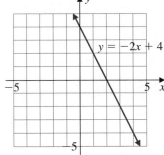

$y = -2x + 4$

21.

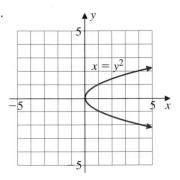

$x = y^2$

25.

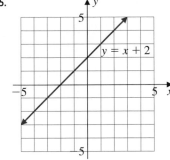

$y = x + 2$

22.

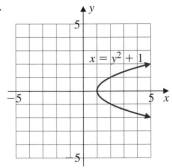

$x = y^2 + 1$

26.

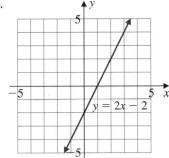

$y = 2x - 2$

23.

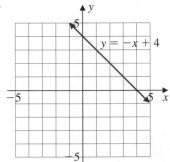

$y = -x + 4$

27.

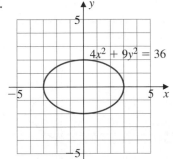

$4x^2 + 9y^2 = 36$

28.

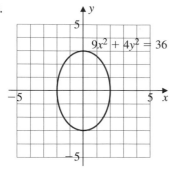

$$9x^2 + 4y^2 = 36$$

29.

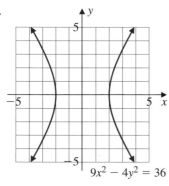

$$9x^2 - 4y^2 = 36$$

30.

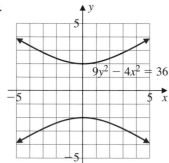

$$9y^2 - 4x^2 = 36$$

In Problems 31–36, use the vertical line test to determine whether the graphs represent functions.

31.

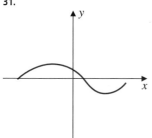

32.

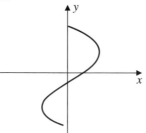

33.

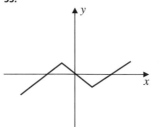

34.

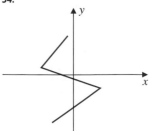

35.

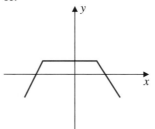

36.

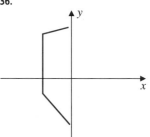

In Problems 37–50, find the domain of the function defined by the given equation.

37. $y = \sqrt{x - 5}$

38. $y = \sqrt{x + 5}$

39. $y = \sqrt{4 - 2x}$

40. $y = \sqrt{6 + 3x}$

41. $y = \sqrt{x^2 + 1}$

42. $y = \sqrt{x^2 + 2}$

43. $y = \dfrac{1}{x - 5}$

44. $y = \dfrac{1}{x + 5}$

45. $y = \dfrac{x + 2}{x + 5}$

46. $y = \dfrac{x + 3}{x - 6}$

47. $y = \dfrac{x}{x^2 + 3x + 2}$

48. $y = \dfrac{x}{x^2 + 5x + 6}$

49. $y = \dfrac{x - 1}{x^2 - 16}$

50. $y = \dfrac{x + 2}{x^2 - 9}$

In Problems 51–56, the definition of a function is given.

51. Given: $f(x) = 3x + 1$. Find:
  a. $f(0)$ **b.** $f(2)$ **c.** $f(-2)$

52. Given: $g(x) = -2x + 1$. Find:
  a. $g(0)$ **b.** $g(1)$ **c.** $g(-1)$

53. Given: $F(x) = \sqrt{x - 1}$. Find:
  a. $F(1)$ **b.** $F(5)$ **c.** $F(26)$

54. Given: $G(x) = x^2 + 2x - 1$. Find:
   a. $G(0)$ **b.** $G(2)$ **c.** $G(-2)$

55. Given: $f(x) = \dfrac{1}{3x + 1}$. Find:

   a. $f(1)$ **b.** $f(1) - f(2)$ **c.** $\dfrac{f(1) - f(2)}{3}$

56. Given: $f(x) = \dfrac{x - 2}{x + 3}$. Find:
   a. $f(2)$ **b.** $f(3)$ **c.** $f(3) - f(2)$

The functions defined by $f(x) = 3x - 4$ and $g(x) = x^2 + 2x + 4$ are used in Problems 57–60.

57. Find:
   a. $f(3)$

   b. $g(3)$

   c. $f(3) + g(3)$

58. Find:
   a. $f(4)$

   b. $g(4)$

   c. $f(4) - g(4)$

59. Find:
   a. $f(-2)$

   b. $g(-3)$

   c. $f(-2) \cdot g(-3)$

60. Find:
   a. $f(-1)$

   b. $g(-2)$

   c. $\dfrac{f(-1)}{g(-2)}$

The functions $f = \{(1, 3), (-1, 5), (-3, 7), (-5, 9)\}$ and $g = \{(-2, 4), (0, 6), (2, 8), (4, 10)\}$ are used in Problems 61–64.

61. Find:
   a. $f(1)$ **b.** $g(-2)$ **c.** $f(1) + g(-2)$

62. Find:
   a. $f(-1)$ **b.** $g(0)$ **c.** $f(-1) - g(0)$

63. Find:
   a. $f(-3)$ **b.** $g(2)$ **c.** $f(-3) \cdot g(2)$

64. Find:
   a. $f(-5)$ **b.** $g(4)$ **c.** $\dfrac{f(-5)}{g(4)}$

APPLICATIONS

65. The revenue obtained from selling x textbooks is given by $R(x) = 30x - 0.0005x^2$. The cost of producing the books is $C(x) = 100{,}000 + 6x$.
   a. Find the profit function $P(x) = R(x) - C(x)$.
   b. Find the profit when 10,000 books are sold.

66. The Fahrenheit temperature reading F is a function of the Celsius temperature reading C. This function is given by
$$F(C) = \frac{9}{5}C + 32$$
   a. If the temperature is 15°C, what is the Fahrenheit temperature?

   b. Water boils at 100°C. What is the corresponding Fahrenheit temperature?

   c. The freezing point of water is 0°C or 32°F. How many Fahrenheit degrees below freezing is a temperature of -10°C?

   d. The lowest temperature attainable is -273°C; this is the zero point on the absolute temperature scale. What is the corresponding Fahrenheit temperature?

67. When you exercise, your pulse rate should be within a certain *target zone*. The *upper limit U* of your target zone when exercising is a function of age a (in years) and is given by
$$U(a) = -a + 190 \qquad \text{(your pulse or heart rate)}$$
Find the highest safe heart rate for a person who is
   a. 50 years old
   b. 60 years old

68. The lower limit L of your target zone when exercising is a function of a (in years) and is given by
$$L(a) = -\frac{2}{3}a + 150$$
The target zone for a person a years old consists of all the heart rates between $L(a)$ and $U(a)$, inclusive. Thus if a person's heart rate is R, that person's target zone is described by $L(a) \le R \le U(a)$. Find the target zone for a person who is
   a. 30 years old
   b. 45 years old

69. The ideal weight w (in pounds) of a man is a function of his height h (in inches). This function is defined by
$$w(h) = 5h - 190$$
   a. If a man is 70 in. tall, what should his weight be?
   b. If a man weighs 200 lb, what should his height be?

70. The cost C in dollars of renting a car for 1 day is a function of the number of miles traveled, m. For a car renting for $20 per day and 20¢ per mile, this function is given by
$$C(m) = 0.20m + 20$$
   a. Find the cost of renting a car for 1 day and driving 290 mi.
   b. If an executive paid $60.60 after renting a car for 1 day, how many miles did she drive?

71. The pressure P (in pounds per square foot) at a depth of d feet below the surface of the ocean is a function of the depth. This function is given by
$$P(d) = 63.9d$$
What is the pressure on a submarine at a depth of
   a. 10 ft?
   b. 100 ft?

72. If a ball is dropped from a point above the surface of the Earth, the distance s (in meters) that the ball falls in t seconds is a function of t. This function is given by

$$s(t) = 4.9t^2$$

Find the distance that the ball falls in

a. 2 sec   **b.** 5 sec

73. Your shoe size S is a *linear* function of the length L (in inches) of your foot and is given by

$$S = m(L) = 3L - 22 \quad \text{(for men)}$$
$$S = f(L) = 3L - 21 \quad \text{(for women)}$$

a. What is the independent variable?

b. What is the dependent variable?

c. If the length of a man's foot is 11 in., what is his shoe size?

d. If the length of a woman's foot is 11 in., what is her shoe size?

74. The speed S (in centimeters per second) of an ant is a *linear* function of the temperature C (in degrees Celsius) and is given by

$$S = f(C) = \frac{1}{6}(C - 4)$$

a. What is the independent variable?

b. What is the dependent variable?

c. What is the speed of an ant on a hot day when the temperature is 28°C?

d. What is the speed of an ant on a cold day when the temperature is 10°C?

75. According to FBI data, the number of robberies (per 100,000 population) is a *quadratic* function of t, the number of years after 1980 and is given by

$$R(t) = 1.85t^2 - 19.14t + 262$$

a. What was the number of robberies (per 100,000) in 1980?

b. What do you predict that the number of robberies would be in the year 2000?

76. Do you recycle bottles, newspaper, and plastic? The waste recovered (in millions of tons) is a

quadratic function of t, the number of years after 1960 and is given by

$$G(t) = 0.04t^2 - 0.59t + 7.42$$

a. How many million tons of waste was recovered in 1960?

b. How many million tons of waste would you expect to be recovered in the year 2000?

Blood alcohol levels

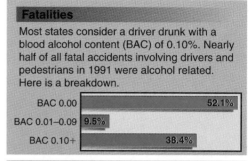

Fatalities

Most states consider a driver drunk with a blood alcohol content (BAC) of 0.10%. Nearly half of all fatal accidents involving drivers and pedestrians in 1991 were alcohol related. Here is a breakdown.

BAC 0.00 — 52.1%
BAC 0.01–0.09 — 9.5%
BAC 0.10+ — 38.4%

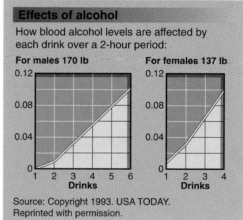

Effects of alcohol

How blood alcohol levels are affected by each drink over a 2-hour period:

For males 170 lb   For females 137 lb

Source: Copyright 1993. USA TODAY. Reprinted with permission.

77. The graph shows that your blood alcohol level L is related to the number of drinks D you consume and, of course, your weight, sex, and the time period in which the drinks were consumed.

a. Do the two graphs represent functions?

b. What are the domain and range for the graph representing males?

c. What are the domain and range for the graph representing females?

d. Use the points $(2, 0.01)$ and $(6, 0.10)$ to find the slope in the first graph.

e. Use the point-slope form to find the equation for $C = f(D)$ in part (d).

78. The graphs show the worldwide emissions (in parts per trillion, ppt) of two ozone-destroying chemicals: CFC-12 and CFC-11.

a. Do both graphs represent functions?

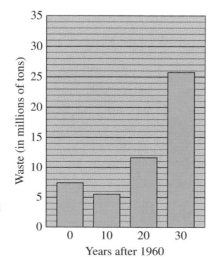

Waste (in millions of tons) vs. Years after 1960

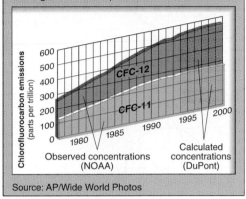

Decline in ozone-destroying chemicals seen

Worldwide emissions of two ozone-destroying chemicals are slowing sooner than expected. The gradual repair of the Earth's ozone layer should begin by 2000. Ozone molecules absorb some of the Sun's harmful ultraviolet radiation before it reaches the Earth's surface, making life on Earth possible.

Source: AP/Wide World Photos

b. What is the domain of the functions?

c. What is the range for CFC-11?

d. What is the range for CFC-12?

e. From 1980 to 1992, CFC-12 emissions can be modeled by the linear function $f(t) = 20t + 300$, where t is the number of years elapsed after 1980. What is $f(0)$, and is it close to the value for 1980 shown on the graph?

f. What is $f(10)$, and is it close to the corresponding value on the graph?

g. If you use this same linear model, what would be the CFC-12 emissions in the year 2000? Is this prediction consistent with the predictions in the graph? What is the difference in parts per trillion between the linear model and the observed values?

WRITE ON . . .

What is the difference between:

79. A function and a relation when they are written as a set of ordered pairs?

80. The graph of a function and the graph of a relation?

Write your own definition of:

81. The domain of a function. **82.** The range of a function.

True or false? Explain your answer.

83. Every relation is a function.

84. Every function is a relation.

Everything You Always Wanted to Know about Relations (in Mathematics)

A special kind of relation that's important in mathematics is called an **equivalence relation**. A relation R is an equivalence relation if it has the following three properties:

a. Reflexive property. If a is an element of the domain of R, then (a, a) is an element of R.

b. Symmetric property. If (a, b) is an element of R, then (b, a) is an element of R.

c. Transitive property. If (a, b) and (b, c) are both elements of R, then (a, c) is an element of R.

A very simple example of an equivalence relation is

$$R = \{(x, y) \mid y = x, x \text{ an integer}\}$$

To show that R is an equivalence relation, we check the above three properties:

a. Reflexive: If a is an integer, then $a = a$, so (a, a) is an element of R.

b. Symmetric: Suppose that (a, b) belongs to R. Then, by the definition of R, a and b are integers and $b = a$. But if $b = a$, then $a = b$, so (b, a) also belongs to R.

c. Transitive: Suppose that (a, b) and (b, c) both belong to R. Then $b = a$ and $c = b$, so $c = a$. Hence (a, c) also belongs to R.

Because R has all three properties, it is an equivalence relation.

The pairs in a relation don't have to be numbers, and some interesting relations occur outside the field of numbers. For example,

$$R = \{(x, y) \mid y \text{ is a member of the same family as } x, x \text{ is a person}\}$$

is a relation. Is R an equivalence relation?

We check the three properties as before:

a. Reflexive: Given a person A, is (A, A) an element of R? Yes, A is obviously a member of the same family as A.

b. Symmetric: Suppose that (A, B) is an element of R. Then B is a member of the same family as A. But then A is a member of the same family as B, so (B, A) is an element of R.

c. Transitive: Suppose that (A, B) and (B, C) both belong to R. Then A, B, and C are all members of the same family. Thus (A, C) is an element of R.

Again, we see that R has all three properties, so it is an equivalence relation.

Can you determine which of the following are equivalence relations?

85. $R = \{(x, y) \mid x \text{ and } y \text{ are triangles and } y \text{ is similar to } x\}$
(*Note:* Here, *is similar to* means *has the same shape as*.)

86. $R = \{(x, y) \mid x \text{ and } y \text{ are integers and } y > x\}$

87. $R = \{(x, y) \mid x$ and y are positive integers and y has the same parity as $x\}$; that is, y is odd if x is odd and y is even if x is even

88. $R = \{(x, y) \mid x$ and y are boys and y is the brother of $x\}$

89. $R = \{(x, y) \mid x$ and y are positive integers and when x and y are divided by 3, y leaves the same remainder as $x\}$

90. $R = \{(x, y) \mid x$ is a fraction

$$\frac{a}{b}$$

where a and b are integers ($b \neq 0$), and y is an equivalent fraction

$$\frac{ma}{mb}$$

where m is a nonzero integer$\}$

MASTERY TEST   If you know how to do these problems, you have learned your lesson!

91. If $f = \{(4, 3), (5, -1), (6, 0)\}$, find:
  a. $f(4)$ **b.** $f(5)$ **c.** $f(4) - f(5)$

92. If $f(x) = 2x - 3$, find:
  a. $f(4)$ **b.** $f(2)$ **c.** $f(x + 1)$

Find the domain of:

93. $f(x) = \dfrac{1}{x - 2}$ **94.** $y = \sqrt{x - 3}$

Use the vertical line test to determine whether the relation shown is a function:

95.

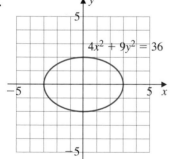

$4x^2 + 9y^2 = 36$

96.

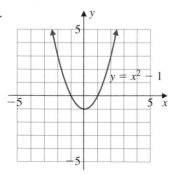

$y = x^2 - 1$

Find the domain and range of the relations:

97.

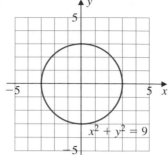

$x^2 + y^2 = 9$

98.

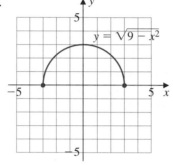

$y = \sqrt{9 - x^2}$

99.

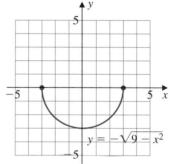

$y = -\sqrt{9 - x^2}$

100. Find the graph, the domain, and the range of the relation $\{(x, y) \mid y = 2x + 4\}$.

101. Find the domain and range of the relation $B = \{(7, 8), (8, 9), (9, 10)\}$.

research questions	**Sources of information for most of these questions can be found in the *Bibliography* at the end of this book.**

1. Some historians claim that the official birthday of analytic geometry is November 10, 1619. Investigate and write a report on why this is so and the events that led Descartes to the discovery of analytic geometry.

2. Find out what led Descartes to make his famous pronouncement, "*Je pense, donc je suis*" (I think, therefore I am), and write a report about the contents of one of his works, *La Geometrie*.

3. She was 19, a capable ruler, a good classicist, a remarkable athlete, and an expert hunter and horsewoman. Find out who this queen was and what connections she had with Descartes.

4. Upon her arrival at the University of Stockholm, one newspaper reporter wrote "Today we do not herald the arrival of some vulgar, insignificant prince of noble blood. No, the Princess of Science has honored our city with her arrival." Write a report identifying this woman and discussing the circumstances leading to her arrival in Sweden.

5. When she was 6 years old, the "princess's" room was decorated with a unique type of wallpaper. Write a paragraph about this wallpaper and its influence on her career.

6. In 1888, the "princess" won the Prix Bordin offered by the French Academy of Sciences. Write a report about the contents of her prize-winning essay and the motto that accompanied it.

SUMMARY

SECTION	ITEM	MEANING	EXAMPLE
7.1C	Linear equation in standard form	Any equation that can be written in the form $Ax + By = C$ (A, B, C real numbers, A and B not both zero)	$7x + 8y = 3$ and $2x = 3y - 2$ are linear equations.
	x-intercept	The x-coordinate of the point at which the graph crosses the x-axis	The x-intercept of $2x + y = 6$ is $x = 3$.
	y-intercept	The y-coordinate of the point at which the graph crosses the y-axis	The y-intercept of $2x + y = 6$ is $y = 6$.
7.1D	Horizontal line	A line whose equation can be written in the form $y = C$, C a constant	$y = 4$, $y = -3$, $y = 0.25$, and $2y = 6$ are equations of horizontal lines.
	Vertical line	A line whose equation can be written in the form $x = C$, C a constant	$x = 4$, $x = -3$, $x = 0.25$, and $2x = 6$ are equations of vertical lines.
7.2A	Distance formula	The distance between the points (x_1, y_1) and (x_2, y_2) is $\sqrt{(x_2 - x_1)^2 + (y_2 - y_1)^2}$.	The distance between $(3, 4)$ and $(6, 8)$ is $\sqrt{3^2 + 4^2} = \sqrt{9 + 16} = 5$.
7.2B	Slope	The slope of the line through (x_1, y_1) and (x_2, y_2) is $m = \dfrac{y_2 - y_1}{x_2 - x_1}$.	The slope of the line through $(3, 5)$ and $(9, 7)$ is $m = \dfrac{7 - 5}{9 - 3} = \dfrac{1}{3}$.
7.2C	Slopes of parallel lines	Two different lines L_1 and L_2 with slopes m_1 and m_2 are parallel if and only if $m_1 = m_2$.	The lines $y = -2x + 3$ and $y = -2x + 9$ are parallel.
	Slopes of perpendicular lines	Two lines L_1 and L_2 with slopes m_1 and m_2 are perpendicular if and only if $m_1 = -\dfrac{1}{m_2}$.	A line perpendicular to the line $y = -2x + 3$ will have a slope of $\frac{1}{2}$, since the slope of $y = -2x + 3$ is -2.

SECTION	ITEM	MEANING	EXAMPLE
7.3A	Two-point form of a line	An equation of the line going through the points (x_1, y_1) and (x_2, y_2) is $$y - y_1 = \frac{y_2 - y_1}{x_2 - x_1}(x - x_1).$$	An equation of the line through the points $(3, 1)$ and $(4, 3)$ is $$y - 1 = \frac{3 - 1}{4 - 3}(x - 3).$$
7.3B	Point-slope form of a line	An equation of the line going through the point (x_1, y_1) and with slope m is $y - y_1 = m(x - x_1)$.	An equation of the line going through the point $(2, 5)$ and with slope -3 is $y - 5 = -3(x - 2)$.
7.3C	Slope-intercept form of a line	An equation of the line with slope m and y-intercept b is $y = mx + b$.	An equation of the line with slope 3 and y-intercept -4 is $y = 3x - 4$.
7.4A	Linear inequality in two variables	A linear inequality is a statement that can be written in the form $Ax + By \le C$ or $Ax + By \ge C$.	$3x + 5y \le 10$ is a linear inequality.
7.4B	Absolute-value inequality	An inequality of the form $|x| \le a$ or $|x| \ge a$	$|x| < 3$ and $|x - 1| \ge 5$ are absolute-value inequalities.
7.5A	Direct variation	y varies directly as x if there is a constant k such that $y = kx$.	If you are paid an hourly rate, the salary s you receive varies directly as the number of hours h you work, so $s = kh$.
7.5B	Inverse variation	y varies inversely as x if there is a constant k such that $y = \dfrac{k}{x}$.	The intensity I of the light you get from a reflector varies inversely with the square of the distance d you are from the reflector, so $I = \dfrac{k}{d^2}$.
7.5C	Joint variation	z varies jointly with x and y if there is a constant k such that $z = kxy$.	The interest I received varies jointly with the principal P and the interest rate r, so $I = kPr$.
7.6A	Relation	A set of ordered pairs	$\{(2, -1), (4, 6), (5, 2)\}$
	Domain	The set of first coordinates of a relation	$\{2, 4, 5\}$ is the domain of the preceding relation.
	Range	The set of second coordinates of a relation	$\{-1, 2, 6\}$ is the range of the preceding relation.
7.6B	Function	A relation in which no two different ordered pairs have the same first coordinate	$\{(1, 2), (2, 4), (3, 6)\}$ is a function.
	Vertical line test	If any vertical line intersects the graph of a relation more than once, the relation is *not* a function.	$\{(1, 3), (1, 4)\}$ is *not* a function.
7.6D	Function notation	Use of a letter such as f to denote a function and $f(x)$ to mean the value of the function for the given value of x.	$f = \{(x, y) \mid y = x^2\}$. For this function, $f(x) = x^2$, $f(2) = 4$, and $f(-3) = 9$.
	Linear function	A function that can be written in the form $f(x) = mx + b$ where m and b are real numbers.	$f(x) = 3x + 8$ is a linear function. $f(x) = 3x^2 + 8$ is *not* a linear function.

REVIEW EXERCISES

(If you need help with these exercises, look in the section indicated in brackets.)

1. [7.1A] Graph.
   a. $A(1, 4)$, $B(-3, 1)$, $C(-3, -2)$, and $D(3, -1)$
   b. $A(4, 1)$, $B(-1, 3)$, $C(-2, -3)$, and $D(1, -3)$

2. [7.1B] Graph.
   a. $x + 2y = 4$ **b.** $2x - y = 2$

3. [7.1C] Find the x- and y-intercepts and graph the line.
   a. $y = 3x + 3$ **b.** $y = 2x - 4$

4. [7.1D] Graph on the same coordinate system.
   a. $2x = 6$ **b.** $3y = 6$

5. [7.1D] Graph on the same coordinate system.
   a. $2x = -6$ **b.** $3y = -6$

6. [7.2A] Find the distance between the two given points.
   a. $A(1, 3)$ and $B(6, 15)$
   b. $A(2, 3)$ and $B(-2, 5)$
   c. $A(2, -4)$ and $B(6, -4)$

7. [7.2B] Find the slope of the line through the given points.
   a. $A(-3, 2)$ and $B(1, 0)$
   b. $A(4, -2)$ and $B(4, -7)$

8. [7.2B] Find the slope of the line through the given points.
   a. $A(3, 4)$ and $B(4, 3)$
   b. $A(1, -1)$ and $B(-3, 5)$

9. [7.2C] A line L has slope $\frac{3}{4}$. Determine whether the line through the two given points is parallel or perpendicular to L.
   a. $A(1, 3)$ and $B(-2, 7)$
   b. $A(1, 3)$ and $B(5, 6)$

10. [7.2C] A line L has slope -2. Determine whether the line through the two given points is parallel or perpendicular to L.
   a. $A(2, -1)$ and $B(1, 1)$
   b. $A(3, -1)$ and $B(2, -3)$

11. [7.2C] The line through $(2, 4)$ and $(5, y)$ is perpendicular to a line with the given slope. Find y.
   a. $m = \dfrac{3}{2}$ **b.** $m = -2$

12. [7.2D] A line passes through the point $(-1, 2)$ and has the given slope. Graph the line.
   a. $m = -\dfrac{1}{2}$ **b.** $m = \dfrac{2}{3}$

13. [7.3A] Find an equation of the line through the two given points, and write the equation in standard form.
   a. $A(2, 5)$ and $B(-1, 2)$
   b. $A(-4, 3)$ and $B(-2, -2)$

14. [7.3B] Find an equation of the line with slope 2 and passing through the given point, and write the equation in standard form.
   a. $A(-1, -3)$ **b.** $A(5, 0)$

15. [7.3C]
   a. A line has slope 3 and y-intercept 2. Find the slope-intercept equation of this line.
   b. Repeat part a if the slope is -3 and the y-intercept is 4.

16. [7.3C] Find the slope and the y-intercept of the line.
   a. $4x - 2y = 8$ **b.** $3x + 6y = 12$

17. [7.3C] Find the slope and the y-intercept of the line.
   a. $\dfrac{x}{2} + \dfrac{y}{4} = 1$ **b.** $\dfrac{x}{3} - \dfrac{y}{4} = -1$

18. [7.3D] Find an equation of the line through the point $(2, 1)$ and that is parallel to the line
   a. $2x + y = 7$ **b.** $3x - y = 4$

19. [7.3D] Find an equation of the line through the point $(2, 1)$ and that is perpendicular to the line
   a. $2x + 3y = 7$ **b.** $3x - 2y = 4$

20. [7.4A] Graph.
   a. $2x + y < -4$ **b.** $x - 2y < 2$

21. **[7.4A]** Graph.

   a. $y \leq 2x + 2$

   b. $y \leq -x + 3$

22. **[7.4A]** Graph.

   a. $x \geq -3$

   b. $x \geq 2$

23. **[7.4B]** Graph.

   a. $|y| \leq 2$

   b. $|y| \geq 2$

24. **[7.4B]** Graph.

   a. $|x - 2| > 1$

   b. $|x - 1| \leq 1$

25. **[7.5A]** The gas in a closed container exerts a pressure P on the walls of the container. This pressure varies directly as the temperature T of the gas.

   a. Write an equation of variation using k for the constant of variation.

   b. If the pressure is 3 lb/in.^{2} when the temperature is 360°F, find k.

26. **[7.5B]** If the temperature of a gas is held constant, the pressure P varies inversely as the volume V.

   a. Write an equation of variation using k for the constant of variation.

   b. A pressure of 1600 lb/in.^{2} is exerted by 2 ft^3 of air in a cylinder fitted with a piston. Find k.

27. **[7.5B]** The force F with which the Earth attracts an object above the Earth's surface varies inversely as the square of the distance d from the center of the Earth.

   a. Write an equation of variation using k for the constant of variation.

   b. An astronaut weighs 120 lb on the Earth's surface, and the radius of Earth is about 4000 mi. Find the value of k.

28. **[7.5B]** What would the astronaut of Problem 27 weigh if she is on a [space voyage] 1000 mi above the Earth's surface?

29. **[7.5C]** The horsepower h that a rotating shaft can safely transmit varies jointly as the cube of its diameter d and the number of revolutions r it makes per minute.

   a. Write an equation of variation using k for the constant of variation.

   b. A 2-in. shaft at a speed of 1000 rev/min can safely transmit 400 hp (horsepower). Find k for this shaft.

30. **[7.5C]** Find the horsepower that the shaft of Problem 29, part b, can safely transmit at a speed of 1500 rev/min.

31. **[7.6A]** Find the domain and the range.

   a. $\{(0, 5), (2, 9), (3, 10), (5, 8)\}$

   b. $\{(0, 6), (2, 10), (3, 11), (5, 9)\}$

32. **[7.6A]** Find the domain and the range and graph the relation.

   a. $\{(x, y) \,|\, y = 1 - x\}$

   b. $\{(x, y) \,|\, y = 2 - x\}$

   c. $\{(x, y) \,|\, y = 3 - x\}$

33. **[7.6B]** Find the domain and the range.

   a.

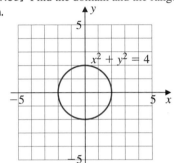

   b.

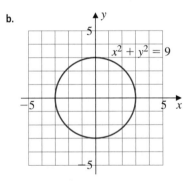

34. **[7.6B]** Find the domain and the range.

   a.

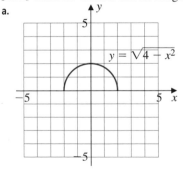

b.

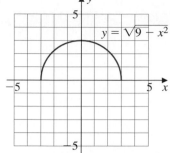

35. [7.6B] Find the domain and the range.

a.

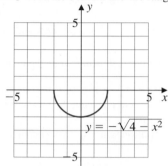

b.

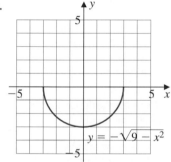

36. [7.6B] Use the vertical line test to determine whether the relation defines a function.

a.

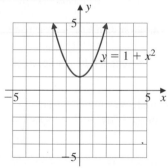

b.

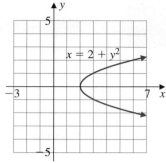

37. [7.6C] Find the domain of the function.

a. $y = \dfrac{2}{x - 1}$

b. $y = \dfrac{2}{x - 2}$

38. [7.6C] Find the domain of the function.
a. $y = \sqrt{x - 3}$
b. $y = \sqrt{x - 4}$

39. [7.6D] Let $f(x) = x - 4$. Find:
a. $f(2)$
b. $f(1)$
c. $f(2) - f(1)$

40. [7.6D] Let $f(x) = x^2 - 3$. Find:
a. $f(2)$
b. $f(1)$
c. $f(2) - f(1)$

41. [7.6D] Let $f = \{(2, 0), (3, 3), (1, -1)\}$. Find:
a. $f(2)$
b. $f(1)$
c. $f(2) - f(1)$

42. [7.6D] Let $f = \{(2, 1), (3, 4), (1, 0)\}$. Find:
a. $f(2)$
b. $f(1)$
c. $f(2) - f(1)$

PRACTICE TEST

(Answers on pages 452–457.)

1. Graph $A(3, 2)$, $B(4, -2)$, $C(-1, -2)$, and $D(-1, 3)$.

2. Graph $3x - y = 3$.

3. Find the x- and y-intercepts of $y = 3x + 2$ and then graph the line.

4. Graph.
   a. $3x = -9$ b. $2y = -4$

5. Find the distance between the two given points.
   a. $A(4, 3)$ and $B(7, 7)$
   b. $A(3, 4)$ and $B(5, -2)$
   c. $A(-3, 2)$ and $B(-3, 5)$

6. Find the slope of the line through the two given points.
   a. $A(-2, 2)$ and $B(1, 1)$
   b. $A(-2, 4)$ and $B(-2, 8)$
   c. $A(3, 4)$ and $B(4, 3)$
   d. $A(4, 4)$ and $B(5, 4)$

7. A line L_1 has slope $\frac{3}{2}$. Determine whether the line through the two given points is parallel or perpendicular to L_1.
   a. $A(4, 2)$ and $B(1, 4)$
   b. $A(-1, -4)$ and $B(-4, -6)$

8. The line through $A(1, -2)$ and $B(-1, y)$ is perpendicular to a line with slope $-\frac{2}{3}$. Find y.

9. A line goes through the point $(3, -1)$ and has slope $-\frac{1}{2}$. Graph this line.

10. Find an equation of the line through $(4, 3)$ and $(2, 4)$. Then write the equation in standard form and graph it.

11. Find an equation of the line with slope -2 and passing through the point $(2, -3)$. Then graph the line.

12. A line has slope 3 and y-intercept 2. Find the slope-intercept equation of this line and graph the line.

13. Find the slope and the y-intercept of the line $6x + 3y = 12$.

14. Find an equation of the line through the point $(1, 2)$ and
   a. Parallel to the line $2x - 3y = 5$.
   b. Perpendicular to the line $2x - 3y = 5$.

15. Graph $2x - y < -2$.

16. Graph $y \leq 2x - 2$.

17. Graph $x \geq -3$.

18. Graph $|y| \leq 2$.

19. Graph $|x - 1| > 2$.

20. An enclosed gas exerts a pressure P on the walls of the container. This pressure is directly proportional to the temperature T of the gas.
   a. Write an equation of variation with k as the constant of variation.
   b. The pressure of a certain gas is 4 lb/in.^{2} when the temperature is 460°F. Find k.

21. If the temperature of a gas is held constant, the pressure P varies inversely as the volume V.
   a. Write an equation of variation with k as the constant of variation.
   b. A pressure of 1760 lb/in.^{2} is exerted by 4 ft^3 of air in a cylinder fitted with a piston. Find k.

22. The force F with which the Earth attracts an object above the Earth's surface varies inversely as the square of the distance d from the center of the Earth.
   a. Write an equation of variation with k as the constant of variation.
   b. A meteorite weighs 40 lb on the Earth's surface, and the radius of Earth is about 4000 mi. Find k.

23. What would the weight of the meteorite in Problem 22 be at a distance of 1000 mi above the Earth's surface?

24. The horsepower that a rotating shaft can safely transmit varies jointly as the cube of its diameter d and the number of revolutions, n, it makes per minute.
   a. Write an equation of variation with k as the constant of variation.
   b. A 2-in. shaft can safely transmit 400 hp at a speed of 1400 rev/min. Find k.

25. Use the value of k from Problem 24 to find the horsepower that a 4-in. shaft can safely transmit at a speed of 2100 rev/min.

26. Find the domain and range of the relation $\{(1, 3), (2, 5), (3, 7), (4, 9)\}$.

27. Find the domain and the range of the relation.

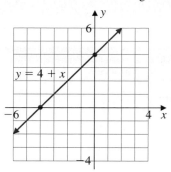

$y = 4 + x$

28. Find the domain and the range of the relation.

a.

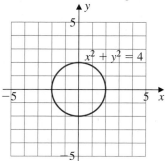

$x^2 + y^2 = 4$

b.

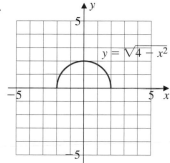

$y = \sqrt{4 - x^2}$

c.

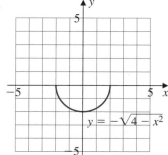

$y = -\sqrt{4 - x^2}$

29. a. Find the domain and the range of the relation.

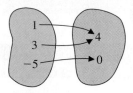

b. Find $f(-5)$.

c. Find $f(1) + f(3)$.

d. Is this relation a function?

30. Use the vertical line test to determine whether the graph of the given relation defines a function.

a.

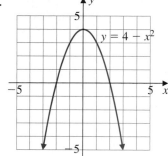

$y = 4 - x^2$

b.

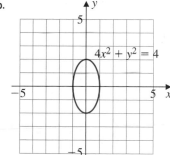

$4x^2 + y^2 = 4$

31. Find the domain of the function.

a. $y = \dfrac{2}{x + 3}$ **b.** $y = \sqrt{x + 2}$

32. Let $f(x) = 4x - 3$. Find:

a. $f(2)$ **b.** $f(1)$ **c.** $f(2) - f(1)$

33. Let $f = \{(1, 4), (2, -1), (3, 2)\}$. Find:

a. $f(2)$ **b.** $f(1)$ **c.** $f(2) - f(1)$

ANSWERS TO PRACTICE TEST

Answer	If you missed: Question	Review: Section	Examples	Page
1.	1	7.1A	1	387
2. $3x - y = 3$	2	7.1B	3	389
3. $y = 3x + 2$	3	7.1C	4	390–391
4a. $3x = -9$	4a	7.1D	5	392

Answer	If you missed:		Review:	
	Question	Section	Examples	Page
4b.	4b	7.1D	5	392
5a. 5	5a	7.2A	1	397
5b. $2\sqrt{10}$	5b	7.2A	1	397
5c. 3	5c	7.2A	1	397
6a. $-\dfrac{1}{3}$	6a	7.2B	2	399–400
6b. Undefined	6b	7.2B	2	399–400
6c. -1	6c	7.2B	2	399–400
6d. 0	6d	7.2B	2	399–400
7a. Perpendicular	7a	7.2C	3	401–402
7b. Neither	7b	7.2C	3	401–402
8. $y = -5$	8	7.2C	4	402
9.	9	7.2D	5	403

Answer	If you missed:		Review:	
	Question	Section	Examples	Page
10. $x + 2y = 10$	10	7.3A	1	408
11. $2x + y = 1$	11	7.3B	2	409
12. $y = 3x + 2$	12	7.3C	3	409
13. $m = -2$, $b = 4$	13	7.3C	4	410
14a. $2x - 3y = -4$	14a	7.3D	5	410
14b. $3x + 2y = 7$	14b	7.3D	5	410

Answer	If you missed:		Review:	
	Question	Section	Examples	Page
15.	15	7.4A	1	416

$2x - y < -2$

| 16. | 16 | 7.4A | 2 | 417 |

$y \leqslant 2x - 2$

| 17. | 17 | 7.4A | 3 | 417 |

$x \geqslant -3$

| 18. | 18 | 7.4B | 4 | 417–418 |

$|y| \leqslant 2$

Answer	If you missed:		Review:	
	Question	Section	Examples	Page
19. $|x - 1| > 2$	19	7.4B	5	418
20a. $P = kT$	20a	7.5A	1	424
20b. $k = \dfrac{1}{115}$	20b	7.5A	1	424
21a. $P = \dfrac{k}{V}$	21a	7.5B	2	425
21b. $k = 7040$	21b	7.5B	2	425
22a. $F = \dfrac{k}{d^2}$	22a	7.5B	3	425
22b. $k = 640{,}000{,}000$	22b	7.5B	3	425
23. 25.6 lb	23	7.5B	3	425
24a. $H = kd^3 n$	24a	7.5C	4	426
24b. $k = \dfrac{1}{28}$	24b	7.5C	4	426
25. 4800 hp	25	7.5C	4	426
26. $D = \{1, 2, 3, 4\}$; $R = \{3, 5, 7, 9\}$	26	7.6A	1	432
27. The domain and range are the set of real numbers.	27	7.6A	2	432
28a. $D = \{x \mid -2 \le x \le 2\}$; $R = \{y \mid -2 \le y \le 2\}$	28a	7.6A	3	432–433
28b. $D = \{x \mid -2 \le x \le 2\}$; $R = \{y \mid 0 \le y \le 2\}$	28b	7.6A	3	432–433
28c. $D = \{x \mid -2 \le x \le 2\}$; $R = \{y \mid -2 \le y \le 0\}$	28c	7.6A	3	432–433
29a. $D = \{1, 3, -5\}$, $R = \{4, 0\}$	29a	7.6A	3	432–433
29b. 0	29b	7.6A	3	432–433
29c. 8	29c	7.6A	3	432–433
29d. Yes	29d	7.6D	7	436
30a. A function	30a	7.6B	4	433–434

Answer	If you missed:		Review:	
	Question	Section	Examples	Page
30b. Not a function	30b	7.6B	4	433–434
31a. All real numbers except -3	31a	7.6C	5	434
31b. All real numbers greater than or equal to -2	31b	7.6C	5	434
32a. 5	32a	7.6D	6	435–436
32b. 1	32b	7.6D	6	435–436
32c. 4	32c	7.6D	6	435–436
33a. -1	33a	7.6D	7	436
33b. 4	33b	7.6D	7	436
33c. -5	33c	7.6D	7	436

8

Solving Systems of Linear Equations

This chapter is devoted to solving systems of linear equations using five methods: *graphical, substitution, elimination, matrices,* and *determinants.* We start by solving *graphically* systems of two variables with integer solutions. When the solutions are not integers and one of the variables is isolated, it's easier to use the *substitution* method. In general, the *elimination* method is the preferred method for systems with two or three unknowns (Section 8.2). We then apply these techniques to solve applications in Section 8.3.

The accessibility of computers and graphers, which reduce tedious computations, suggest that we use a more mechanical (algorithmic) way of solving systems of equations. Two methods widely used involve *matrices* (arrays of numbers showing the coefficients of the variables and the constants) and *determinants* (numerical arrays corresponding to matrices) that can be easily evaluated. We use these two methods in Sections 8.4 and 8.5.

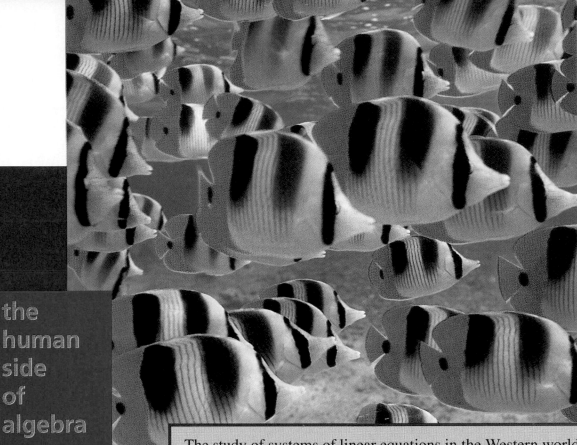

The study of systems of linear equations in the Western world was initiated by Gottfried Wilhelm Leibniz. In 1693, Leibniz solved a system of three equations by eliminating two of the unknowns and using a determinant to obtain the solution. It was Colin Maclaurin, however, who used determinants to solve simultaneous linear equations in two, three, and four unknowns. Maclaurin was born in Scotland and educated at the University of Glasgow, which he entered at the incredible age of 11! He became professor of mathematics at Aberdeen at 19 and taught at the prestigious University of Edinburgh at 25.

Ironically, Maclaurin's name is associated with a portion of analysis (the Maclaurin series) that was really discovered by Brook Taylor. The general Taylor series, in turn, had been known long before to James Gregory and Jean Bernoulli. If Maclaurin's name is recalled in connection with a series he did not first discover, this is compensated by the fact that a contribution he did make bears the name of someone else who discovered and printed it later: Cramer's rule, published in 1750, was probably known to Maclaurin as early as 1729!

8.1

SYSTEMS WITH TWO VARIABLES

To succeed, review how to:

1. Find the x- and y-intercepts of a line (pp. 390–391).

2. Graph a line (pp. 387–389).

3. Find the slope of a line given its equation (pp. 409–410).

Objectives:

Find the solution of a system of two linear equations using:

 The graphical method

 The substitution method

 The elimination method

 Solve applications involving systems of equations

getting started

Radio and Color Television Sales

Have you bought a radio or a color television lately? From 1990 through 1995, the number of radios (in millions) sold and the number of color televisions (in millions) sold in the United States can be represented by

$$\text{Color televisions sold:} \quad y = 0.9x + 20.4$$

$$\text{Radios sold:} \quad y = -0.3x + 21.6$$

where x represents the year, with $x = 0$ corresponding to 1990. This *system* of *two* equations can be graphed as shown. In which year were as many radios sold as televisions? As you can see, this happened during the first year—that is, when $x = 1$. In this year, the number of units (in millions) sold was $y = 0.9(1) + 20.4 = 21.3$. If we denote the year by x and the number of units sold by y, the point $(x, y) = (1, 21.3)$ represents the point at which the curves *intersect*. This point is the *solution* of the system of equations. Note that the point $(1, 21.3)$ *satisfies* both equations.

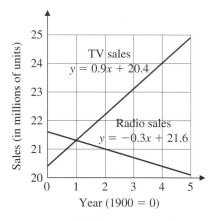

We can also solve this system of equations by *substitution*. Since the variable y is the same in both equations, we substitute $0.9x + 20.4$ for y in the second equation to obtain

$$0.9x + 20.4 = -0.3x + 21.6$$
$$0.9x = -0.3x + 1.2 \qquad \text{Subtract 20.4 from both sides.}$$
$$1.2x = 1.2 \qquad \text{Add } 0.3x \text{ to both sides.}$$
$$x = 1 \qquad \text{Divide both sides by 1.2.}$$

If you substitute 1 for x in the first equation (as we did before), you obtain 21.3 for the value of y.

These are the graphical and substitution methods for solving a system of equations; we investigate these and other methods in this section.

 Finding Solutions Using the Graphical Method

As we have seen, the solution set of a system of linear equations can be estimated by graphing the equations on the same axes and determining the coordinates of any points of intersection. This is called the **graphical method**, and we use it in

Example 1. Remember, the *solution* of the system is a point (an ordered pair) that satisfies both equations.

EXAMPLE 1 Using the graphical method to solve a system with one solution

Use the graphical method to find the solution of the system:

$$2x - y = 2$$
$$y = x - 1$$

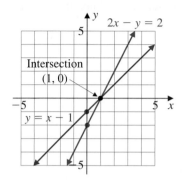

FIGURE 1

SOLUTION We first graph the equation $2x - y = 2$ using the x- and y-intercepts shown in the table.

When $x = 0$, $2x - y = 2$ becomes $2(0) - y = 2$, or $y = -2$.

When $y = 0$, $2x - y = 2$ becomes $2x - 0 = 2$, or $x = 1$.

x	y
0	-2
1	0

Thus the points $(0, -2)$ and $(1, 0)$ are on the graph of the equation. We draw a line through these two points (the blue line in Figure 1). The two points and the complete graph are shown in color. We then graph $y = x - 1$.

When $x = 0$, $y = x - 1$ becomes $y = 0 - 1$, or $y = -1$.

When $y = 0$, $y = x - 1$ becomes $0 = x - 1$, or $x = 1$.

x	y
0	-1
1	0

We then join the points $(0, -1)$ and $(1, 0)$ with a line. The graph for $y = x - 1$ is shown in red in Figure 1. Since the lines intersect at $(1, 0)$, the point $(1, 0)$ is the solution of the system of equations, and the system is said to be **consistent** because it has *one* solution.

CHECK: For $x = 1$, $y = 0$, $2x - y = 2$ becomes $2(1) - 0 = 2$ (true).
For $x = 1$, $y = 0$, $y = x - 1$ becomes $0 = 1 - 1$ (true).
Thus $(1, 0)$ is the correct solution for the system; the solution set is $\{(1, 0)\}$. ∎

EXAMPLE 2 Using the graphical method to solve a system with no solution

Use the graphical method to find the solution of the system:

$$y - 2x = 4$$
$$3y - 6x = 18$$

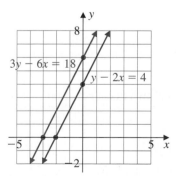

FIGURE 2

SOLUTION We first graph the equation $y - 2x = 4$ by using the x- and y-intercepts shown in the table.

$y - 2(0) = 4$, so $y = 4$

$0 - 2x = 4$, so $x = -2$

x	y
0	4
-2	0

The two points, as well as the completed graph, are shown in red in Figure 2. We then graph $3y - 6x = 18$ using the accompanying table.

$3y - 6(0) = 18$, so $y = 6$

$3(0) - 6x = 18$, so $x = -3$

x	y
0	6
-3	0

The graph of $3y - 6x = 18$ is shown in blue in Figure 2. The two lines appear to be parallel and do not intersect. If we examine the equations more carefully, we see that by dividing the second equation ($3y - 6x = 18$) by 3, we get $y - 2x = 6$. Thus one equation says $y - 2x = 4$ and the other says that $y - 2x = 6$. Obviously, both equations cannot be true at the same time. The slope of $y - 2x = 4$ is 2 and the slope of $3y - 6x = 18$ is also 2. The lines are parallel, so their graphs cannot intersect. Thus there is *no solution* for this system. The solution set is $\varnothing$ and the system is said to be **inconsistent**. ■

EXAMPLE 3   Using the graphical method to solve a system with infinitely many solutions

Use the graphical method to solve the system:

$$2x + \frac{1}{2}y = 2$$

$$y = -4x + 4$$

SOLUTION   We use the *x*- and *y*-intercepts shown in the table to graph $2x + \frac{1}{2}y = 2$.

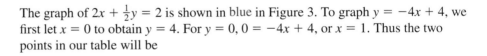

$2(0) + \dfrac{1}{2}y = 2$, so $y = 4$

$2x + \dfrac{1}{2}(0) = 2$, so $x = 1$

x	y
0	4
1	0

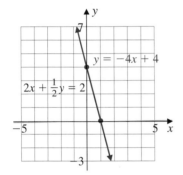

$2x + \frac{1}{2}y = 2$

$y = -4x + 4$

FIGURE 3

The graph of $2x + \frac{1}{2}y = 2$ is shown in blue in Figure 3. To graph $y = -4x + 4$, we first let $x = 0$ to obtain $y = 4$. For $y = 0$, $0 = -4x + 4$, or $x = 1$. Thus the two points in our table will be

x	y
0	4
1	0

But these points are exactly the same as those obtained in the preceding table. What does this mean? It means that the graphs of the lines $2x + \frac{1}{2}y = 2$ and $y = -4x + 4$ **coincide** (are the same). Thus a solution of one equation is automatically a solution of the other. This solution set is $\{(x, y)|y = -4x + 4\}$. In fact, there are *infinitely* many solutions. Such a system is called **dependent**. ■

As you can see from the preceding examples, a system of two linear equations in two variables can have

1. One solution (the lines intersect as in Example 1)

2. No solution (the lines are parallel as in Example 2)

3. Infinitely many solutions (the lines coincide as in Example 3)

The three possibilities are shown in Figure 4. Remember, a system of linear equations is called *consistent* if it has at least one solution, *inconsistent* if it has no solution, and *dependent* if it has infinitely many solutions.

Consistent system

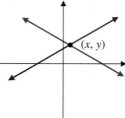

Inconsistent system

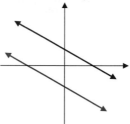

Dependent system

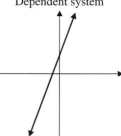

(x, y)

Two lines that *intersect* at the point (x, y); the solution is (x, y).

Two lines that are *parallel* and thus do not *intersect;* there is no solution, and the solution set is $\varnothing$.

Two lines that *coincide;* there are infinitely many solutions.

FIGURE 4

B **Finding Solutions Using the Substitution Method**

Sometimes the graphical method isn't accurate enough to determine the exact solutions of a system of equations. For example, to solve the system

$$y = 2.5x + 72$$
$$y = 3x + 70$$

we can start by letting $x = 0$ in the first equation to obtain $y = 2.5(0) + 72$, or $y = 72$. To graph this point, we need a piece of graph paper with 72 units, or else we have to make each division on the graph paper 10 units. But there is a way out. Keep in mind that we are looking for a point that satisfies *both* equations and notice that the second equation tells us that y is $3x + 70$. Thus we can *substitute* $3x + 70$ for y in the equation $y = 2.5x + 72$ to obtain

$$3x + 70 = 2.5x + 72$$
$$3x = 2.5x + 2 \qquad \text{Subtract 70.}$$
$$0.5x = 2 \qquad \text{Subtract 2.5}x.$$
$$x = \frac{2}{0.5} \qquad \text{Divide by 0.5.}$$
$$x = 4$$

Now substitute 4 for x in the equation $y = 3x + 70$ to obtain $y = 3(4) + 70 = 82$.

The solution of the system is $(4, 82)$, and the method we used is called the **substitution method**.

> **NOTE** The substitution method is most useful when both equations are solved for one of the variables in terms of the other (as in the system we just solved) or when at least one of the equations is in this form, as in Example 1.

To solve the system of Example 1 by substitution, we write

$$2x - y = 2$$
$$y = x - 1$$

Since $y = x - 1$, we substitute $x - 1$ for y in the first equation, $2x - y = 2$, which becomes

$$2x - (x - 1) = 2$$

$$2x - x + 1 = 2$$

$$x + 1 = 2$$

$$x = 1$$

Substituting 1 for x in the second equation, $y = x - 1$, gives

$$y = 1 - 1 = 0$$

Thus the solution is $(1, 0)$, and the solution set is $\{(1, 0)\}$ as was shown in Example 1.

EXAMPLE 4   Using the substitution method to solve a system with one solution

Use the substitution method to solve the system:

$$3y + x = 9 \qquad \textbf{(1)}$$

$$x - 3y = 0 \qquad \textbf{(2)}$$

SOLUTION   We solve equation (2) for x to obtain $x = 3y$. Substituting $3y$ for x in equation (1)

$$x = 3y$$

$$3y + x = 9 \quad \text{becomes} \quad 3y + 3y = 9$$

$$6y = 9$$

$$y = \frac{3}{2}$$

Since $x = 3y$ and $y = \frac{3}{2}$,

$$x = 3\left(\frac{3}{2}\right) = \frac{9}{2}$$

Thus the solution of $3y + x = 9$ and $x - 3y = 0$ is $\left(\frac{9}{2}, \frac{3}{2}\right)$, and the solution set is $\left\{\left(\frac{9}{2}, \frac{3}{2}\right)\right\}$.

CHECK $3\left(\frac{3}{2}\right) + \frac{9}{2} = 9 \qquad \text{and} \qquad \frac{9}{2} - 3 \cdot \left(\frac{3}{2}\right) = 0$

The system is consistent. You can verify this by graphing $3x + y = 9$ and $x - 3y = 0$ as shown in Figure 5. ∎

How do we recognize that a system such as $x + y = 3$ and $y = -x - 3$ is inconsistent? We can do this with the substitution method as follows. The system is

$$x + y = 3 \qquad \textbf{(1)}$$

$$y = -x - 3 \qquad \textbf{(2)}$$

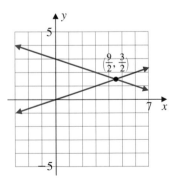

FIGURE 5

The system $3y + x = 9$ and $x - 3y = 0$ is **consistent**; its solution is $\left(\frac{9}{2}, \frac{3}{2}\right)$.

We substitute $-x - 3$ for y in equation (1), and

$$y = -x - 3$$

$$x + y = 3 \quad \text{becomes} \quad x + (-x - 3) = 3$$

$$-3 = 3$$

Since $-3 = 3$ is a contradiction, the system $x + y = 3$ and $y = -x - 3$ is inconsistent; it has *no* solution (Figure 6). Graphically, the lines are parallel, so the solution set is $\varnothing$.

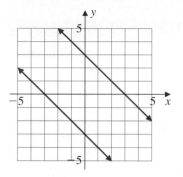

FIGURE 6

The system $x + y = 3$ and $y = -x - 3$ is **inconsistent**; the lines are parallel.

EXAMPLE 5 Using the substitution method to solve a system with no solution

Solve the system:

$$x + y = 5 \qquad \textbf{(1)}$$

$$y = -x \qquad \textbf{(2)}$$

SOLUTION Substituting $y = -x$ in equation (1), we get

$$x + (-x) = 5$$

$$0 = 5$$

Since this is a contradiction, the system $x + y = 5$ and $y = -x$ has *no* solution (Figure 7). The system is inconsistent and the solution set is $\varnothing$. ∎

EXAMPLE 6 Using the substitution method to solve a system with infinitely many solutions

Solve the system:

$$x + 2y = 4 \qquad \textbf{(1)}$$

$$2x = 8 - 4y \qquad \textbf{(2)}$$

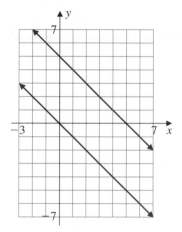

FIGURE 7

The system $x + y = 5$ and $y = -x$ is **inconsistent**; the lines are parallel.

SOLUTION Solving equation (1) for x, we obtain $x = 4 - 2y$. We now substitute $4 - 2y$ for x in equation (2).

$$2x = 8 - 4y$$

$$2(4 - 2y) = 8 - 4y$$

$$8 - 4y = 8 - 4y$$

Since this equation is an identity, any value of y will make it true. Thus there are infinitely many solutions (Figure 8). The system is dependent. For example, if $x = 0$ in equation (1),

$$2y = 4 \qquad \text{and} \qquad y = 2$$

Thus $(0, 2)$ is a solution of the system. If we let $y = 0$ in equation (2),

$$2x = 8 - 4(0) \qquad \text{and} \qquad x = 4$$

Thus $(4, 0)$ is another solution. You can continue to obtain solutions by assigning numbers to one of the variables in either equation and solving for the other variable. The solution set is obtained by solving $x + 2y = 4$ for y and writing

$$\left\{ (x, y) \mid y = -\frac{x}{2} + 2 \right\}$$

∎

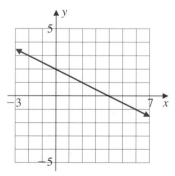

FIGURE 8

The system $x + 2y = 4$ and $2x = 8 - 4y$ is **dependent**; it has infinitely many solutions, and the lines coincide.

Solving by Elimination

A third method for solving a system of linear equations is called the **elimination method**. The key step in this method is *to obtain coefficients that differ only in sign* for one of the variables so that this variable is eliminated by adding the two equations. The procedure is given next.

PROCEDURE

> **Procedure for Solving a System of Two Equations in Two Unknowns by Elimination**
>
> 1. Multiply both sides of the equations (as needed) by numbers that make the coefficients of one of the variables opposites.
> 2. Add the two equations.
> 3. Solve for the remaining variable.
> 4. Substitute this solution into one of the given equations and solve for the second variable.
> 5. Check the solution.

We illustrate this method in Example 7.

EXAMPLE 7   Using the elimination method to solve a system with one solution

Solve the given system by the elimination method:

$$2x + 3y = 8 \tag{1}$$
$$x - y = -1 \tag{2}$$

SOLUTION   To eliminate y, multiply both sides of equation (2) by 3. Here are the steps.

1. Write the system.

$$2x + 3y = 8 \tag{1}$$
$$x - y = -1 \tag{2}$$

2. Multiply both sides of equation (2) by 3 so that the y's are eliminated when the result is added to the first equation.

3.

$2x + 3y =$	8	Same
$3x - 3y =$	-3	Multiply by 3.
$5x \quad\quad =$	5	Add.
$x \quad\quad =$	1	Solve.

4. Substituting $x = 1$ in equation (2) (which is simpler), we get

$$1 - y = -1$$

so

$$2 = y$$

5. The solution is $(1, 2)$, as can easily be checked since

$$2(1) + 3(2) = 8 \tag{1}$$
$$1 - 2 = -1 \tag{2}$$

The system is *consistent* (see Figure 9).   ■

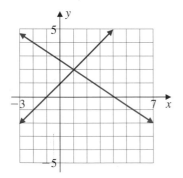

FIGURE 9

The system $2x + 3y = 8$ and $x - y = -1$ is **consistent**; the solution is $(1, 2)$.

In many cases, it is not enough to multiply one of the equations by a number to eliminate one of the variables. For example, to find the common solution of

$$2x + 3y = -1 \qquad \textbf{(1)}$$
$$3x + 2y = -4 \qquad \textbf{(2)}$$

we multiply *both* equations by numbers chosen so that the coefficients of one of the variables become opposites in the resulting equations. Thus we have the following steps.

1. Write the system.

$$2x + 3y = -1 \qquad \textbf{(1)}$$
$$3x + 2y = -4 \qquad \textbf{(2)}$$

2. $4x + 6y = -2$ Multiply both sides of equation (1) by 2.

$\underline{-9x - 6y = 12}$ Multiply both sides of equation (2) by -3.

3. $-5x \quad\;\; = 10$ Add.

$\quad\;\; x \quad\;\; = -2$ Solve.

4. Substituting $x = -2$ in equation (1), we get

$$2(-2) + 3y = -1$$
$$-4 + 3y = -1$$
$$3y = 3$$
$$y = 1$$

5. The solution of the system is $(-2, 1)$. The system is *consistent* (see Figure 10).

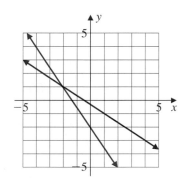

FIGURE 10

The system $2x + 3y = -1$ and $3x + 2y = -4$ is **consistent**; the solution is $(-2, 1)$.

EXAMPLE 8   Using the elimination method to solve a system with no solutions

Solve the system:

$$3x + 2y = 6 \qquad \textbf{(1)}$$
$$6x + 4y = 10 \qquad \textbf{(2)}$$

SOLUTION   We proceed by steps as before.

1. Write the system.

$$3x + 2y = 6 \qquad \textbf{(1)}$$
$$6x + 4y = 10 \qquad \textbf{(2)}$$

2. $-6x - 4y = -12$ Multiply both sides of equation (1) by -2.

$\underline{6x + 4y = \;\;\; 10}$ Same

3. $0 = -2$ Add.

4. Since we obtained the contradiction $0 = -2$, this system is *inconsistent* and has no solution.

5. There is no ordered pair (x, y) that is a common solution of both equations. Note that the slope of $3x + 2y = 6$ is $-\frac{3}{2}$, the same as that of the line $6x + 4y = 10$. Thus the lines are parallel (see Figure 11). ■

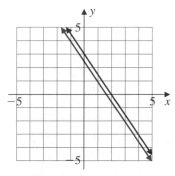

FIGURE 11

The system $3x + 2y = 6$ and $6x + 4y = 10$ is **inconsistent**; the lines are parallel.

EXAMPLE 9   Using the elimination method to solve a system
with infinitely many solutions

Solve the system:

$$\frac{x}{6} + \frac{y}{2} = 1 \tag{1}$$

$$\frac{x}{9} + \frac{y}{3} = \frac{2}{3} \tag{2}$$

SOLUTION   Since it's easier to work with integer coefficients, we multiply the first equation by 6, the LCD of $\frac{x}{6}$ and $\frac{y}{2}$, and the second equation by 9, the LCD of $\frac{x}{9}, \frac{y}{3},$ and $\frac{2}{3}$.

1. Write the system.

$$\frac{x}{6} + \frac{y}{2} = 1 \tag{1}$$

$$\frac{x}{9} + \frac{y}{3} = \frac{2}{3} \tag{2}$$

2. $x + 3y = 6$ Multiply both sides of equation (1) by 6.

                $x + 3y = 6$ Multiply both sides of equation (2) by 9.

3. $0 = 0$ Subtract.

4. Since we obtain the true statement $0 = 0$, the equations are *dependent,* and the system has infinitely many solutions (see Figure 12). For example, if we substitute 6 for x in equation (1) we have

$$\frac{6}{6} + \frac{y}{2} = 1$$

$$1 + \frac{y}{2} = 1$$

$$\frac{y}{2} = 0$$

$$y = 0$$

Thus $(6, 0)$ is a solution. We can substitute 12 for x,

$$\frac{12}{6} + \frac{y}{2} = 1$$

$$2 + \frac{y}{2} = 1$$

$$\frac{y}{2} = -1$$

$$y = -2$$

and we have another solution, $(12, -2)$. The solution set is obtained by solving $x + 3y = 6$ for y and then is written as $\left\{(x, y) \,\middle|\, y = -\frac{x}{3} + 2\right\}$.

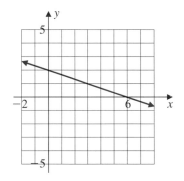

FIGURE 12

The system $\frac{x}{6} + \frac{y}{2} = 1$ and $\frac{x}{9} + \frac{y}{3} = \frac{2}{3}$ is **dependent**; the lines coincide, and there are infinitely many solutions.

5. Note that in step 2, we came up with identical equations. Any time we have two equivalent (or identical) equations, the system is dependent, and we write the solution set as $\{(x, y)\,|\,y = mx + b\}$. ■

Which Method Shall I Use?

We have discussed three methods for solving systems of equations: graphical, substitution, and elimination. If you are wondering how to decide which method to use, Table 1 gives you some guidelines.

TABLE 1 Solving Systems of Equations: A Summary

Method	Suggested Use	Disadvantages
Graphical	When the coefficients of the variables and the solutions are integers. You get a picture of the situation.	If the solutions are not integers, they are hard to read on the graph.
Substitution	When one of the variables is isolated (alone) on one side of the equation.	If fractions are involved, you may have much computation.
Elimination	When fractions, decimals, or variables with coefficients that are the same or negative inverses of each other ($2x$ and $-2x$, for example) are present.	You may have lots of computations involving signed numbers.

Solving Applications Involving Systems of Equations

Have you bought any good books lately? The price for art books and the number purchased in 1980, 1985, and 1990 are shown in the table (these are according to the *Statistical Abstract of the United States*). Note that as the price of the book goes up, the number of books purchased by consumers (that is, the *demand*) goes down; see Figure 13. (Consumers aren't willing to pay that much for a book!)

Demand Function, *D*

Price (in dollars)	Quantity (in hundreds)
28	17
35	15
42	13

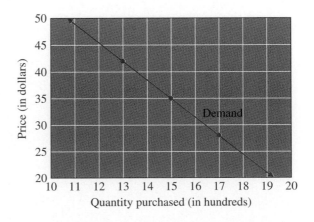

FIGURE 13

On the other hand, booksellers *supply* fewer books when prices go down and more books when prices go up (because the margin of profit is larger with the higher price); this is shown in Figure 14.

Supply Function, S

Price (in dollars)	Quantity (in hundreds)
26	12
35	15
44	18

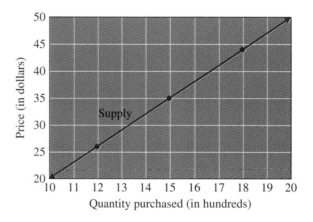

FIGURE 14

Now let's graph both functions on the same axes. As price increases, demand decreases. As price increases, supply increases. The point at which the graphs intersect is called the **equilibrium point**; see Figure 15. At this point, the number of books that booksellers are willing to supply is the same number that consumers are willing to buy. By definition, if D is the demand function, S is the supply function, and p is the price, the equilibrium point is where demand equals supply; that is,

$$D(p) = S(p)$$

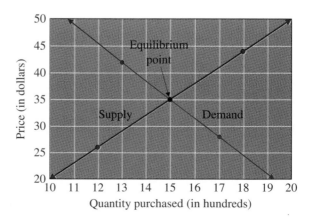

FIGURE 15

EXAMPLE 10   A supply and demand problem

Find the equilibrium point if the supply and demand functions are given by:

$$S(p) = 20p + 200 \quad \text{and} \quad D(p) = 740 - 100p$$

SOLUTION   We use the RSTUV method. The equilibrium point is the point at which

$$D(p) = S(p)$$

$740 - 100p = 20p + 200$	Substitute $740 - 100p$ for $D(p)$ and $20p + 200$ for $S(p)$.
$540 - 100p = 20p$	Subtract 200 from both sides.
$540 = 120p$	Add $100p$ to both sides.
$\dfrac{540}{120} = p$	Divide both sides by 120.
$\$4.50 = p$	Divide 540 by 120.

Thus the price at which the supply equals the demand is $4.50. At this point, the number of books suppliers are willing to offer is

$$S(4.5) = 20(4.5) + 200$$
$$= 90 + 200$$
$$= 290 \text{ books}$$ ■

problem solving   **EXAMPLE 11**   Geometry problems

Two angles are complementary. If the measure of one of the angles is 20° more than the other, what are the measures of these angles?

SOLUTION   We use the RSTUV method.

1. Read the problem.

We are asked to find the measures of these angles.

2. Select the unknown.

There are two angles involved. Let x be the measure of one of the angles and y the measure of the other angle.

3. Think of a plan. Do you know what complementary angles are? They are angles whose sum is 90°.

We have to translate the problem, but to do so, we must know that *complementary* angles are angles whose sum is 90°. With this information, we have

$x + y = 90$	The sum of the angles is 90°.
$y = x + 20$	One of the angles is 20° more than the other.

4. Use the substitution method to solve the system.

Substituting $y = x + 20$ in $x + y = 90$, we obtain

$x + (x + 20) = 90$	
$2x + 20 = 90$	Add like terms.
$2x = 70$	Subtract 20.
$x = 35$	Divide by 2.

Thus one angle is 35° and the other 20° more, or 55°.

5. Verify the answer.

Since the sum of the measures of the two angles is $35° + 55° = 90°$, the angles are complementary and our result is correct.   ■

GRAPH IT

If you are using a grapher, you have to solve for y to graph each of the equations in a system. In doing so, you will know, even before graphing, whether the sys-tem is inconsistent (the lines will be parallel), dependent (the lines will coincide), or consistent [the lines will intersect at one point (x, y)]. Thus to use the grapher, write the equation as

$$y = m_1x + b_1$$
$$y = m_2x + b_2$$

1. The system is *inconsistent* when $m_1 = m_2$ and $b_1 \neq b_2$.

2. The system is *dependent* when $m_1 = m_2$ and $b_1 = b_2$.

3. The system is *consistent* otherwise.

Thus in Example 1, we have $y = 2x - 2$ and $y = x - 1$, so the system is consistent. The graph is shown in Window 1 and the solution is clearly $(1, 0)$. You can verify this:

WINDOW 1

(a) Numerically. After entering $Y_1 = 2x - 2$ and $Y_2 = x - 1$, go to the home screen of a TI-82 and store the value $x = 1$ by pressing 1 $\boxed{\text{STO}}$ $\boxed{\text{X,T,}\theta}$ $\boxed{\text{ENTER}}$. Now let's evaluate Y_1. Press $\boxed{\text{2nd}}$ $\boxed{\text{Y-VARS}}$ 1 $\boxed{\text{ENTER}}$ (the variable Y_1 should now be showing). Press $\boxed{\text{ENTER}}$ again. The answer is zero. [You can also evaluate $Y_1(1)$ by entering $\boxed{\text{2nd}}$ $\boxed{\text{Y-VARS}}$ 1 $\boxed{\text{ENTER}}$ $\boxed{(}$ 1 $\boxed{)}$.] Do the same for Y_2 and you should get zero again. Thus $(1, 0)$ satisfies both equations.

(b) Graphically. By repeatedly using the $\boxed{\text{TRACE}}$ and $\boxed{\text{ZOOM}}$ keys of your grapher, you can find the intersection of two curves. (With a TI-82, press $\boxed{\text{2nd}}$ $\boxed{\text{CALC}}$ 5 and $\boxed{\text{ENTER}}$ three times to find the coordinates of the intersection of two functions.)

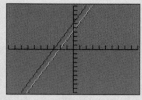

WINDOW 2

To do Example 2, solve each of the equations for y and graph $y = 2x + 4$ and $y = 2x + 6$. Since both lines have the same slope, their graphs are parallel and the system is inconsistent as before. The graph is shown in Window 2.

In Example 3, we solve for y in $2x + \frac{1}{2}y = 2$ to obtain $y = -4x + 4$. Since the other equation is also $y = -4x + 4$, the system is dependent. The graph is shown in Window 3.

You can now verify the results of Examples 4, 5, and 6, but remember that you can tell, even before you graph, the type of system you have by simply looking at the resulting equivalent equations

$$y = m_1x + b_1$$
$$y = m_2x + b_2$$

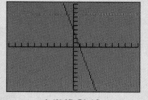

WINDOW 3

EXERCISE 8.1

A **In Problems 1–10, solve by the graphical method. Label each system as consistent (one solution), inconsistent (no solution), or dependent (many solutions).**

1. $x - 2y = 6$
   $\quad y = 2x$

2. $2x + y = 4$
   $\quad y = 2x$

3. $y = x - 3$
   $\quad y = 2x - 4$

4. $y = x - 1$
   $\quad y = 3x - 3$

5. $2y = -x + 4$
   $\quad y = -2x + 4$

6. $2y = -x + 2$
   $\quad y = -2x + 2$

7. $2x - y = -2$
   $\quad y = 2x + 4$

8. $2x + y = -2$
   $\quad y = -2x + 4$

9. $3x + 4y = 12$
   $\quad 8y = 24 - 6x$

10. $2x - 3y = 6$
    $\quad 6x = 18 + 9y$

In Problems 11–26, solve by the substitution method. Label each system as consistent (one solution), inconsistent (no solution), or dependent (many solutions).

11. $y = 2x - 4$
$-2x = y - 4$

12. $y = 2x + 2$
$-x = y + 1$

13. $x + y = \dfrac{5}{2}$
$3x + y = \dfrac{9}{2}$

14. $x + y = \dfrac{5}{2}$
$3x + y = \dfrac{3}{2}$

15. $y - 4 = 2x$
$y = 2x + 2$

16. $y + 5 = 4x$
$y = 4x + 7$

17. $x = 8 - 2y$
$x + 2y = 4$

18. $x = 4 - 2y$
$x - 2y = 0$

19. $x + 2y = 4$
$x = -2y + 4$

20. $x + 3y = 6$
$x = -3y + 6$

21. $x = 2y + 1$
$y = 2x + 1$

22. $y = 3x + 2$
$x = 3y + 2$

23. $2x - y = -4$
$4x = 4 + 2y$

24. $5x + y = 5$
$5x = 15 - 3y$

25. $x = 5 - y$
$0 = x - 4y$

26. $x = 3 - y$
$0 = 2x - y$

In Problems 27–56, solve each system by elimination. Label the system consistent (one solution), inconsistent (no solution), or dependent (infinitely many solutions).

27. $x + y = 8$
$x - y = 2$

28. $x + y = 3$
$x - y = 1$

29. $x + 4y = 2$
$x - 4y = -2$

30. $x - 5y = 15$
$x + 5y = -5$

31. $-x - 2y = -2$
$x - 2y = -2$

32. $x + 3y = -7$
$-x + 2y = -3$

33. $2x + y = 7$
$3x - 2y = 0$

34. $2x + y = 4$
$3x - 2y = -1$

35. $2x - 2y = 6$
$x + y = 2$

36. $3x - 3y = -15$
$x - y = -5$

37. $3x + 5y = 1$
$-6x - 10y = 2$

38. $5x - 2y = 4$
$-10x + 4y = 1$

39. $2x + y = 8$
$3x - y = 7$

40. $x - 3y = -2$
$x + 3y = 4$

41. $2x - 5y = 9$
$4x - 10y = 18$

42. $3x + 5y = 26$
$5x + 3y = 22$

43. $6x + 5y = 12$
$9x - 4y = -5$

44. $5x + 4y = 6$
$4x - 3y = 11$

45. $2x - 3y = 16$
$x - y = 7$

46. $3x - 2y = 35$
$x - 5y = 42$

47. $18x - 15y = 1$
$10x - 12y = 3$

48. $6x - 9y = -2$
$3x - 5y = -6$

49. $\dfrac{x}{3} + \dfrac{y}{6} = \dfrac{2}{3}$
$\dfrac{2}{5}x + \dfrac{y}{4} = \dfrac{1}{5}$

50. $\dfrac{x}{6} + \dfrac{y}{3} = \dfrac{1}{2}$
$\dfrac{3}{5}x + \dfrac{y}{4} = \dfrac{17}{20}$

51. $\dfrac{5}{6}x + \dfrac{y}{4} = 7$
$\dfrac{2}{3}x - \dfrac{y}{8} = 3$

52. $\dfrac{1}{5}x + \dfrac{2}{5}y = 1$
$\dfrac{1}{4}x - \dfrac{1}{3}y = \dfrac{-5}{12}$

53. $\dfrac{2}{x} + \dfrac{3}{y} = \dfrac{-1}{2}$
$\dfrac{3}{x} - \dfrac{2}{y} = \dfrac{17}{12}$

Hint: Multiply the first equation by 2, and the second by 3 and add, or let $a = \frac{1}{x}$, $b = \frac{1}{y}$ and solve the system

$$2a + 3b = -\dfrac{1}{2}$$

$$3a - 2b = \dfrac{17}{12}$$

54. $\dfrac{4}{x} + \dfrac{2}{y} = \dfrac{26}{21}$
$\dfrac{2}{x} - \dfrac{1}{y} = \dfrac{-1}{21}$

Hint: Multiply the second equation by 2 and add, or let $a = \frac{1}{x}$, $b = \frac{1}{y}$ and solve the resulting system.

55. $\dfrac{2}{x} - \dfrac{1}{y} = 0$
$\dfrac{3}{x} + \dfrac{5}{y} = \dfrac{13}{4}$

Hint: Multiply the first equation by 5 and add, or let $a = \frac{1}{x}$, $b = \frac{1}{y}$ and solve the resulting system.

56. $\dfrac{1}{x} - \dfrac{3}{y} = \dfrac{-13}{10}$
$\dfrac{5}{x} + \dfrac{2}{y} = 2$

Hint: Multiply the first equation by -5 and add.

APPLICATIONS

In Problems 57–61, find the price p at the equilibrium point for each pair of demand and supply functions.

57. $D(p) = 620 - 10p$
$S(p) = 20p + 200$

58. $D(p) = 200 - 2p$
$S(p) = 50 + p$

59. $D(p) = 1500 - 5p$
$S(p) = 300 + p$

60. $D(p) = 1100 - 8p$
$S(p) = 200 + 17p$

61. $D(p) = 450 - 5p$
$S(p) = 2p + 170$

62. The demand function for a certain product is $D(p) = 500 - 5p$, and the supply function is $S(p) = 8p + 110$. What will the price of this product be at the equilibrium point? What will the demand for the product be at the equilibrium point?

63. The demand function for skate boards is $D(p) = 960 - 16p$, where p is the price in dollars. If the manufacturer is willing to supply $S(p) = 14p + 120$ boards when the price is p, what is the price of each skate board when the equilibrium point is reached? What is the demand at that price; that is, how many skate boards sell at that price?

64. The demand function for sunglasses is $D(p) = 24 - 0.4p$, where p is the price in dollars. If the supply function is $S(p) = p + 10$, what will the price of sunglasses be when the equilibrium point is reached? How many sunglasses would suppliers be willing to offer at that price?

65. Find the price p (in dollars) and the supply and the demand at the equilibrium point when the demand function is given by $D(p) = 480 - 3p$ and the supply function is $S(p) = 25p + 60$.

66. A store will buy 80 digital cellular phones if the price is \$350 and 120 if the price is \$300. The wholesaler is willing to supply 60 phones if the price is \$280 and 140 if the price is \$370. If the supply and demand functions are linear, make a graph of the given information and find the equilibrium point.

In Problems 67–70, (a) write a system of two equations describing the situation, and (b) find the measure of each angle.

67. Two angles are complementary, and the measure of one of the angles is 15° more than the other.

Note that the sum of the measures of the two angles is 90°.

68. Two angles are complementary, and the measure of one of the angles is twice the measure of the other angle.

Note that the sum of the measures of the two angles is 90°.

69. Two angles are supplementary, and the measure of one of the angles is four times the measure of the other.

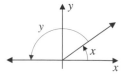

Note that the sum of the measures of the two angles is 180°.

70. Two angles are supplementary, and the measure of one of the angles is 30° more than the other.

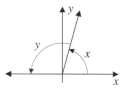

Note that the sum of the measures of the two angles is 180°.

71. The heaviest pair of African elephant tusks on record weighed a total of 465 lbs. One of the tusks weighed 15 lb more than the other.
    a. Write a system of two equations representing this situation.
    b. What does each tusk weigh?

72. Do you know who invented basketball? It was Dr. James Naismith who in 1921 created a game that could be played indoors in Massachusetts. Although the height of the players has changed dramatically, the size of the court hasn't. The perimeter (distance around) of the maximum-size basketball court is 288 ft, and the length L of the court is 44 ft more than the width W.
    a. Write a system of two equations representing this situation.
    b. What are the dimensions of the maximum-size basketball court?

73. Do you smoke? How many cigarettes a year? Together, smokers in the United States and Japan consume 4637 cigarettes per person each year. If the average Japanese smoker consumes 437 cigarettes a year more than the average U.S. smoker, what is the annual consumption of cigarettes per person in each country?

74. The average American drinks 29 more Cokes per year than the average Mexican. If together they consume 555 Cokes per year, how many Cokes are consumed each year by the average American and how many by the average Mexican?

75. Find the solution to this ancient Babylonian problem:

There are two silver rings; $\frac{1}{7}$ of the first ring and $\frac{1}{11}$ of the second ring are broken off so that what is broken off weighs 1 shekel. The first, diminished by $\frac{1}{7}$, weighs as much as the second diminished by $\frac{1}{11}$. What did the silver rings originally weigh?

(*Hint:* Consider the system of equations

$$\frac{x}{7} + \frac{y}{11} = 1 \quad \text{and} \quad \frac{6x}{7} = \frac{10y}{11}$$

Clear denominators by multiplying by the LCD, then use the substitution method to solve the system.)

76. Is there a nuclear reactor near your city? The total number of reactors in the United States is 160, but not all of them are in operation. The number of operable reactors is four more than double the number of inoperable reactors. How many U.S. reactors are operable and how many are inoperable?

77. According to the *Guinness Book of World Records,* the sum of the ages of the two oldest cats is 70 yr. The difference of their ages is 2 yr. What are their ages?

78. According to the *Guinness Book of World Records,* the sum of the heights of the shortest and tallest persons on record is 130 in. If the difference in their heights was 84 in., how tall was each?

79. The height of the Empire State Building and its antenna is 1472 ft. The difference in height between the building and the antenna is 1028 ft. How tall is the antenna and how tall is the building?

80. The height of the Eiffel Tower and its antenna is 1052 ft. The difference in height between the tower and the antenna is 920 ft. How tall is the antenna and how tall is the tower?

81. At one time, the combined weight of the McCreary brothers was 1300 lb. Their weight difference was 20 lb. What was the weight of each brother?

SKILL CHECKER

Find the x- and y-intercept of each line:

82. $2x + y = 6$

83. $3x + 2y = 12$

84. $-2x + 3y = 9$

85. $-3x + 2y = 4$

86. $-2x - 3y = 7$

USING YOUR KNOWLEDGE **Tweedledee and Tweedledum**

Have you read *Alice in Wonderland?* Do you know who the author of this book is? The answer, of course, is Lewis Carroll. Did you also know that he was an accomplished mathematician and logician? These interests show up occasionally in his children's books. Take, for example, *Through the Looking Glass,* where one of the characters, Tweedledee, is talking to Tweedledum:

Tweedledee: The sum of your weight and twice mine is 361 pounds.

Tweedledum: Contrariwise, the sum of your weight and twice mine is 360 pounds.

87. If Tweedledee weighs x pounds and Tweedledum weighs y pounds, can you use the knowledge gained in this section to find their weights?

WRITE ON ...

88. How would you know if a system of linear equations is consistent when you are using:
  a. the graphical method?   **b.** the substitution method?
  c. the elimination method?

89. How would you know if a system of linear equations is inconsistent when you are using:
  a. the graphical method?   **b.** the substitution method?
  c. the elimination method?

90. How would you know if a system of linear equations is dependent when you are using:
  a. the graphical method?   **b.** the substitution method?
  c. the elimination method?

91. Write the possible advantages and disadvantages of each of the methods we have studied: graphical, substitution, and elimination.

MASTERY TEST If you know how to do these problems, you have learned your lesson!

Solve the system by elimination:

92. $5x = 6 - 4y$
  $3y = 4x - 11$

93. $3x = 6 - 3y$
  $2y = 5x + 11$

94. $\dfrac{x}{3} + \dfrac{y}{9} = \dfrac{2}{3}$
  $\dfrac{x}{2} + \dfrac{y}{6} = 1$

95. $\dfrac{x}{3} + \dfrac{y}{4} = \dfrac{11}{6}$
  $\dfrac{x}{6} + \dfrac{y}{8} = \dfrac{7}{12}$

96. $3x + 2y = 8$
  $x - y = 1$

97. $2x + 3y = 9$
  $x - y = 2$

Solve the system by substitution:

98. $x - 3y = 4$
  $2x = 8 + 6y$

99. $x - y = 2$
  $y = x$

100. $2y + x = 8$
  $x - 3y = 0$

Use the graphical method to solve the system and classify the system as consistent, inconsistent, or dependent:

101. $2x + 2y = 8$
  $2x - y = 2$

102. $2x + y = 4$
  $2y + 4x = 6$

103. $2x + y = 5$
  $2x + 4y = 8$

104. $x + \dfrac{1}{2}y = -2$
  $y = -2x - 4$

105. $y - 3x = 3$
  $2y - 6x = 12$

106. $x - y = -1$
  $y = -x - 1$

8.2

SYSTEMS WITH THREE VARIABLES

To succeed, review how to:

1. Solve linear equations (pp. 69–73).

2. Evaluate an expression (pp. 42–44).

3. Solve a system of two equations in two unknowns (pp. 463–469).

Objectives:

A Solve a system of three equations and three unknowns by the elimination method.

B Determine whether a system of three equations in three unknowns is consistent, inconsistent, or dependent.

getting started

Coffee Mixing: An Application

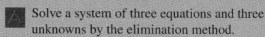
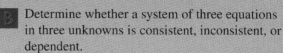

The man has three coffees that sell for \$13.00, \$14.00, and \$15.00 per pound, respectively. If he calls these coffees A, B, and C and decides to make 50 lb of a mixture containing x pounds of A, y pounds of B, and z pounds of C, then

$$x + y + z = 50 \qquad \textbf{(1)}$$

In this mixture he will have twice as much of brand B as of brand C. Thus

$$y = 2z \qquad \textbf{(2)}$$

Finally, if he sells the 50 lb at \$14.20 per pound, the total price will be \$710.00 so that

$$13x + 14y + 15z = 710 \qquad \textbf{(3)}$$

Now we rewrite equations (1), (2), and (3) in standard form:

$$x + \quad y + \quad z = \ 50 \qquad\qquad\qquad\qquad\qquad\qquad \textbf{(4)}$$
$$y - \quad 2z = \quad 0 \qquad \text{Subtract } 2z \text{ from both terms of (2).} \qquad \textbf{(5)}$$
$$13x + 14y + 15z = 710 \qquad\qquad\qquad\qquad\qquad\qquad \textbf{(6)}$$

We have obtained a system of linear equations in three unknowns. There are many ways to solve this system. If we use the elimination method, we take two *different* pairs of equations and eliminate the same variable from each pair. We first note that equation (5) *does not* contain x. Because of this fact, we select equations (4) and (6) and eliminate x by multiplying equation (4) by -13 and adding equation (6), as follows:

$$-13x - 13y - 13z = -650 \qquad \textbf{(7)}$$
$$\underline{13x + 14y + 15z = \ \ 710} \qquad \textbf{(6)}$$
$$y + \quad 2z = \quad 60 \qquad \text{Add.} \qquad \textbf{(8)}$$

The new system, consisting of equations (5) and (8), is a system of two equations in two unknowns. We solve it using the techniques of Section 8.1. To eliminate z from this system, we add equations (5) and (8), obtaining

$$y - 2z = \ 0 \qquad \textbf{(5)}$$
$$\underline{y + 2z = 60} \qquad \textbf{(8)}$$
$$2y \quad\ \ = 60 \qquad \text{Add.} \qquad \textbf{(9)}$$
$$y = 30$$

Substituting $y = 30$ in the equation gives

$$y - 2z = 0$$
$$30 - 2z = 0$$
$$z = 15$$

Using $y = 30$ and $z = 15$ in equation (4) gives

$$x + 30 + 15 = 50$$
$$x + 45 = 50$$
$$x = 5$$

Thus the solution for the system is the ordered triple (5, 30, 15). You can check this by substituting 5 for x, 30 for y, and 15 for z in each of the original equations. This means the man mixed 5 lb of the $13.00 coffee, 30 lb of the $14.00 coffee, and 15 lb of the $15.00 coffee.

Solving Equations by Elimination

In Section 8.1, we used the graphical, substitution, and elimination methods to solve systems of equations involving two variables in two equations. When we have three variables, the graphical method isn't used because a three-dimensional coordinate system is required, and the substitution method isn't used when one or more of the equations has only two variables. However, we can use the elimination method to solve a system of three linear equations in three unknowns, as we illustrated in the *Getting Started*. Now let's examine this method in more detail. To solve a system of three linear equations in three unknowns by elimination, we use the following procedure.

PROCEDURE

Solving a Three-Equation System with Three Unknowns by Elimination

1. Select a pair of equations and eliminate one variable from this pair.
2. Select a *different* pair of equations and eliminate the same variable as in step 1.
3. Solve the pair of equations resulting from steps 1 and 2. (Use the procedure outlined in Section 8.1.)
4. Substitute the values found in step 3 in the simplest of the original equations, and then solve for the third variable.
5. Check by substituting the values in each of the original equations.

EXAMPLE 1   Using the elimination method to solve a three-equation system with one solution

Solve the system:

$$x + y + z = 12 \quad \text{(1)}$$
$$2x - y + z = 7 \quad \text{(2)}$$
$$x + 2y - z = 6 \quad \text{(3)}$$

If we add these two, z is eliminated. (from (1) and (2))
If we add these two, z is eliminated. (from (2) and (3))

SOLUTION   It's easiest to eliminate z from the two pairs of equations, (1) and (3), and (2) and (3).

1. Adding (1) and (3), we obtain

$$2x + 3y = 18 \quad \text{(4)}$$

2. Adding (2) and (3), we have

$$3x + y = 13 \tag{5}$$

3. We now have the system

$$2x + 3y = 18 \tag{4}$$

$$3x + y = 13 \tag{5}$$

Multiplying equation (5) by -3, we have

$$2x + 3y = 18 \tag{4}$$

$$\underline{-9x - 3y = -39} \tag{6}$$

$$-7x = -21 \qquad \text{Add.}$$

$$x = 3$$

Substituting 3 for x in equation (5), we get $9 + y = 13$, or $y = 4$.

4. Since we know now that x is 3 and y is 4, we can substitute these values in equation (1) to obtain $3 + 4 + z = 12$. Solving, we find $z = 5$.

5. The solution of the system is (3, 4, 5), as can easily be verified:

$$3 + 4 + 5 = 12 \tag{1}$$

$$2(3) - 4 + 5 = 7 \tag{2}$$

$$3 + 2(4) - 5 = 6 \tag{3}$$

The system is *consistent*. ∎

EXAMPLE 2   Using the elimination method to solve a system of three equations with no solution

Solve the system:

$$x + 3y - z = 1 \tag{1}$$
$$x - y + z = 4 \tag{2}$$
$$3x + y + z = 3 \tag{3}$$

If we add these two, z is eliminated.
If we add these two, z is eliminated.

SOLUTION   Adding first (1) and (2) and then (1) and (3), we obtain

$$2x + 2y = 5 \qquad \text{The sum of (1) and (2)} \tag{4}$$

$$4x + 4y = 4 \qquad \text{The sum of (1) and (3)} \tag{5}$$

Multiplying equation (4) by -2 to eliminate x, we have

$$-4x - 4y = -10 \qquad \text{This is } -2(2x + 2y) = -2 \cdot 5. \tag{6}$$

$$\underline{4x + 4y = 4} \tag{5}$$

$$0 = -6 \qquad \text{Add.} \tag{7}$$

Since it's impossible for 0 to be equal to -6, this system has *no solution;* it is an *inconsistent system.* ∎

EXAMPLE 3 Using the elimination method to solve a system of three equations with infinitely many solutions

Solve the system:

$$x - 2y + 3z = 4 \qquad (1)$$
$$2x - y + z = 1 \qquad (2)$$
$$x + y - 2z = -3 \qquad (3)$$

SOLUTION We use the five-step procedure.

1. Multiplying equation (2) by -2 and adding to equation (1) (to eliminate y), we get

$$x - 2y + 3z = 4 \qquad (1)$$
$$\underline{-4x + 2y - 2z = -2} \qquad (4)$$
$$-3x + z = 2 \qquad (5)$$

2. Adding equations (2) and (3) to eliminate y, we obtain

$$2x - y + z = 1 \qquad (2)$$
$$\underline{x + y - 2z = -3} \qquad (3)$$
$$3x - z = -2 \qquad (6)$$

3. We now have the system

$$-3x + z = 2 \qquad (5)$$
$$\underline{3x - z = -2} \qquad (6)$$
$$0 = 0 \quad \text{Add.} \qquad (7)$$

Thus the system is *dependent* and has infinitely many solutions of the form (x, y, z). One such solution is obtained if we let $x = 0$ in equation (6), which gives $z = 2$.

4. Substituting $x = 0$ and $z = 2$ in equation (2), we have

$$2 \cdot 0 - y + 2 = 1 \qquad (2)$$
$$-y + 2 = 1$$
$$y = 1$$

5. So $(0, 1, 2)$ is *one* of the solutions for the system, as can be easily checked by substituting $x = 0$, $y = 1$, and $z = 2$ in the original system. Here are some more solutions you can check: $(1, 6, 5)$ and $(-1, -4, -1)$. ∎

B · Consistent, Inconsistent, and Dependent Systems

As was the case with the solution of two equations in two unknowns, the solution of three equations in three unknowns always produces one of three different possibilities.

POSSIBLE SOLUTIONS OF THREE EQUATIONS IN THREE UNKNOWNS

The system is *consistent* and *independent,* as in Example 1; it has one solution consisting of an ordered triple (*x, y, z*).

The system is *inconsistent,* as in Example 2. It has no solution.

The system is *consistent* and *dependent,* as in Example 3. It has infinitely many solutions.

In the case of two unknowns, the fact that a system is *consistent* tells us that if we graph the lines associated with the system, the lines *intersect.* If we graph a linear equation in three unknowns, the graph is a plane. Thus if three linear equations in three unknowns have a solution, it means that the three planes corresponding to their equations intersect at a point, as shown in Figure 16.

If the equations are *inconsistent,* the planes do not intersect at a common point, as shown in Figures 17 and 18. Finally, if the equations are *dependent,* the three planes can intersect in a line, as in Figure 19 (or the three planes can be coincident). Any point on the intersection is a solution; consequently, there are infinitely many solutions.

FIGURE 16

One solution: Planes intersect in exactly one point.

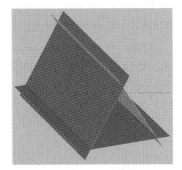

FIGURE 17

No solution: Planes intersect two at a time but have no common point of intersection. The equations are *inconsistent,* and the lines of intersection are parallel.

FIGURE 18

No solution: Three parallel planes with no point of inter-section. The equations are *inconsistent.*

FIGURE 19

Infinitely many solutions: The three planes intersect along a common line. The equations are *dependent.*

GRAPH IT

The graphical method is not practical when solving a system of three equations in three unknowns because the result-ing graphs would have to be in three dimensions; that is, you would need a coordinate system with three axes, *x, y,* and *z.* Graphers with three-dimensional capabilities are expected to be available in the near future but, at present, the best you can do with a grapher when solving systems of equations in three unknowns is to follow steps 1 and 2 of the procedure given in the text. At that point, you will have a linear system with two equations and two unknowns. To finish, you then solve each of the two equations for the same variable, graph the resulting equations, and find the point of intersection, if any. Since you will then have the values (answer) for two of the three variables, substitute these two values in any of the original equations and solve that equation for the third value.

Thus to solve Example 1, follow steps 1 and 2 to obtain the system given by equations (4) and (5)

$$2x + 3y = 18$$
$$3x + y = 13$$

(continued on facing page)

Or, solving for y in each equation,

$$y = -\frac{2}{3}x + 6$$

$$y = -3x + 13$$

Now graph these two lines (see Window 1). The intersection is at $x = 3$ and $y = 4$. (If your grapher has the intersection feature, press 2nd CALC 5 and follow the instructions). Now substitute $x = 3$ and $y = 4$ in

$$x + y + z = 12$$

$$3 + 4 + z = 12$$

$$7 + z = 12$$

$$z = 5$$

Thus we have $x = 3$, $y = 4$, and $z = 5$; that is, the solution set is the point $(3, 4, 5)$ as before.

To do Example 2, proceed with steps 1 and 2, as in the text, to obtain the system

$$2x + 2y = 5$$

$$4x + 4y = 4$$

Solving each equation for y,

$$y = -x + \frac{5}{2}$$

$$y = -x + 1$$

Clearly, these two equations are inconsistent and have no solution. You can confirm this by graphing them; you will obtain the parallel lines shown in Window 2. Now try Example 3.

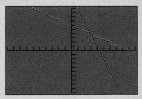

WINDOW 1

$y = -\frac{2}{3}x + 6$

$y = -3x + 13$

WINDOW 2

$y = -x + \frac{5}{2}$

$y = -x + 1$

EXERCISE 8.2

A B In Problems 1–20, solve the system. Label the system as consistent, inconsistent, or dependent.

1. $x + y + z = 12$
$x - y + z = 6$
$x + 2y - z = 7$

2. $x + y + z = 13$
$x - 2y + 4z = 10$
$3x + y - 3z = 5$

3. $x + y + z = 4$
$2x + 2y - z = -4$
$x - y + z = 2$

4. $2x - y + z = 3$
$x + 4y - z = 6$
$3x + 2y + 3z = 16$

5. $2x - y + z = 3$
$x + 2y + z = 12$
$4x - 3y + z = 1$

6. $x - 3y - 2z = -12$
$2x + y - 3z = -1$
$3x - 2y - z = -5$

7. $x - 2y - 3z = 2$
$x - 4y - 13z = 14$
$-3x + 5y + 4z = 2$

8. $2x + 2y + z = 3$
$-x + y - z = 5$
$3x + 5y + z = 8$

9. $2x + 4y + 3z = 3$
$10x - 8y - 9z = 0$
$4x + 4y - 3z = 2$

10. $9x + 4y - 10z = 6$
$6x - 8y + 5z = -1$
$12x + 12y - 15z = 10$

11. $x - 2y - z = 3$
$2x - 5y + z = -1$
$x - 2y - z = -3$

12. $x - 3y + 6z = -8$
$3x - 2y - 10z = 11$
$5x - 6y - 2z = 7$

13. $2x + y + z = 5$
$-x + 2y - z = 3$
$3x + 4y + z = 10$

14. $x - 3y + z = 2$
$x + 2y - z = 1$
$-7x + y + z = -10$

15. $x + y = 5$
$y + z = 3$
$x + z = 7$

16. $x + 2y = -1$
$2y + z = 0$
$x + 2z = 11$

17. $x - 2y = 0$
$y - 2z = 5$
$x + y + z = 8$

18. $2y + z = 9$
$z - 2y = 1$
$x + y + z = 1$

19.
$$5x - 3z = 2$$
$$2z - y = -5$$
$$x + 2y - 4z = 8$$

20.
$$5x - 2z = 1$$
$$3z - y = 6$$
$$x + 2y - z = -1$$

21. Find a, b, and c so that the ordered pairs $(1, 0)$, $(-1, -4)$, and $(2, 5)$ satisfy the equation $y = ax^2 + bx + c$.

[*Hint:* For $(1, 0)$ $x = 1$ and $y = 0$, and

$$y = ax^2 + bx + c \qquad \textbf{(1)}$$
$$0 = a + b + c$$

Do the same for $(-1, -4)$ and $(2, 5)$ and then solve the resulting system for a, b, and c.]

22. Show that the system

$$2x + \qquad 4z = 6 \qquad \textbf{(1)}$$
$$3x + y + z = -1 \qquad \textbf{(2)}$$
$$2y - z = -2 \qquad \textbf{(3)}$$
$$x - y + z = -5 \qquad \textbf{(4)}$$

does not have a solution. [*Hint:* Solve the system consisting of equations (1), (2), and (4) and then show that the solution does not satisfy equation (3).]

23. Find the solution set of

$$2x + \qquad 4z = 6$$
$$3x + y + z = -1$$
$$2y - z = -2$$
$$x - y - 2z = -5$$

24. Find a value of k so that the system

$$5x - y + 2z = 2 \qquad \textbf{(1)}$$
$$3x + y - 3z = 7 \qquad \textbf{(2)}$$
$$x + 5y + z = 5 \qquad \textbf{(3)}$$
$$x + ky - z = 9 \qquad \textbf{(4)}$$

has a solution. [*Hint:* Solve the system consisting of equations (1), (2), and (3). Then substitute the values of x, y, and z in equation (4) and solve for k.]

25. Find a value of k so that the system

$$2x + \qquad 4z = 6 \qquad \textbf{(1)}$$
$$3x + y + z = -1 \qquad \textbf{(2)}$$
$$2y - z = -2 \qquad \textbf{(3)}$$
$$x - y + kz = -5 \qquad \textbf{(4)}$$

has a solution. (See Problem 24.)

| **APPLICATIONS** |

In Problems 26–36, write a system of three equations in three unknowns and solve the problem.

26. The sum of three numbers is 48. The second number is double the first, and the third number is triple the first. What are the numbers?

27. The sum of three numbers is 49. The second number is 8 more than the first, and the third is 9 more than the second. Find the numbers.

28. The sum of the measures of the three angles in a triangle is 180°. The measure of one of the angles is twice that of the smallest angle, and the measure of the largest angle is three times that of the smallest. Find the measure of each angle.

29. The sum of the measures of three angles in a triangle is 180°. The sum of the measures of the first and the second angle is 20° less than that of the third angle, and the measure of the second angle is 20° more than that of the first. Find the measure of each angle.

30. The sum of the salaries of the President, Vice President, and Chief Justice of the United States is $521,200. The President makes $39,400 more than the Vice President. If the Vice President and the Chief Justice make the same salary, how much does each of these people make?

31. In a survey of 1000 people, the number of respondents complaining about corns, heel pain, or ingrown toenails was 150. Two more people suffered heel pain than ingrown toenails, and the number of the respondents with either heel pain or ingrown toenails exceeded the number of those with corns by 10. Write a system of three equations and three unknowns, solve it, and find the number of people complaining about corns only, heel pain only, and ingrown toenails only.

32. Do you take vitamins? The annual amount spent on vitamins C, E, and B in the United States reached $885 million! Vitamin C is the most popular of the three; $90 million more was spent on it than on vitamin B, and $15 million more was spent on vitamin E than on B. Find the amount of money spent on each of these vitamins.

33. Who makes all the dough? Pizza Hut, Domino's, and Little Caesar's account for 48% of the total pizza market. Pizza Hut sells 11% more than Domino's, while Domino's sells 2% more than Little Caesar's. Find each company's market share (percent).

34. Do you play a musical instrument? More than 57 million adults do! The number of piano, guitar, and organ players combined is 46 million. The number of guitar and organ players exceeds by 4 million the number of piano players, but the number of guitar players is only 2 million less than those playing piano. Find out how many adults play each of these instruments.

35. Who are the richest Americans? In a recent year, Bill Gates (computer software), John Kluge (media), and the Walton family (retailing) had combined fortunes totaling $16.9 billion. Kluge's fortune was $0.4 billion more than Walton and $0.8 billion less than Gates. Find out just how rich these people were during this particular year.

36. If you went to school in Japan, then moved to Israel, and finally ended up in the United States the third year, you would have attended school for a total of 639 days. In Israel, you would have attended 36 more days than in the United States. In Japan, you would have attended even longer: 27 more days than in Israel. How long is the school year in each of these countries?

SKILL CHECKER

Solve:

37. A person bought some bonds yielding 5% annually and some certificates yielding 7%. If the total investment amounts to $20,000 and the interest received is $1160, how much is invested in bonds and how much in certificates?

38. A car leaves a town going north at 30 mi/hr. Two hours later, another car leaves the same town traveling on the same road in the same direction at 40 mi/hr. How far from the town does the second car overtake the first one?

39. How many gallons of a 30% solution must be mixed with 40 gal of a 12% solution to obtain a 20% solution?

40. A person can complete a job in 3 hr. Another person does it in 2 hr. How long would it take to complete the job if both people work together?

41. A plane traveled 840 mi with a 30-mi/hr tail wind in the same time it took to travel 660 mi against the wind. What is the plane's speed in still air?

USING YOUR KNOWLEDGE
Crickets, Ants, and the Weather

Can you tell how hot it is by listening to the crickets? You can if you know the formula! The number of chirps n a cricket makes per minute is related to the temperature F, in degrees Fahrenheit, by

$$F = \frac{n}{4} + 40 \qquad \textbf{(1)}$$

Ants are also affected by temperature changes. As a matter of fact, the crawling speed d (in centimeters per second) of a certain ant is related to the temperature C, in degrees Celsius, by

$$C = 6d + 4 \qquad \textbf{(2)}$$

The relationship between C and F is given by

$$C = \frac{5}{9}(F - 32) \qquad \textbf{(3)}$$

42. Use the substitution method with equations (2) and (3) to solve for F in terms of d.

43. Substitute the expression obtained for F in Problem 42 in equation (1) to find the relationship between d and n.

44. Now use your answer from Problem 43 to solve this: If the cricket is chirping 112 times a minute, how fast is the ant crawling?

WRITE ON . . .

45. Why is the graphical method difficult to use when solving a system of equations in three unknowns?

46. When would you use the substitution method when solving a system of three equations in three unknowns?

47. If you graph the linear equation $2x + y = 4$, what is the resulting graph? What do you think the graph will be if you graph the equation $2x + y + z = 4$?

48. How do you know if a system of three equations in three unknowns is inconsistent?

49. How do you know if a system of three equations in three unknowns is dependent?

MASTERY TEST
If you know how to do these problems, you have learned your lesson!

Solve:

50. $\begin{aligned} x + 2y + z &= -10 \\ x + y - z &= -3 \\ 5x + 7y - z &= -29 \end{aligned}$

51. $\begin{aligned} x + y + z &= 4 \\ x - 2y - z &= 1 \\ 2x - y - 2z &= -1 \end{aligned}$

52. $\begin{aligned} x + 8y - z &= 8 \\ -x + 2y - z &= 4 \\ 2x + y + z &= 2 \end{aligned}$

53. $\begin{aligned} 2x + 2y - 6z &= 5 \\ -x - y + 3z &= 4 \\ 3x - y + z &= 2 \end{aligned}$

54. $\begin{aligned} x + y + z &= 4 \\ x - y + z &= 2 \\ 2x + 2y - z &= -4 \end{aligned}$

55. $\begin{aligned} x + 2y + z &= 4 \\ -3x + 4y - z &= -4 \\ -2x - 4y - 2z &= -8 \end{aligned}$

8.3

COIN, DISTANCE-RATE-TIME, INVESTMENT, AND GEOMETRY PROBLEMS

To succeed, review how to:

1. Use the RSTUV method (p. 89).
2. Solve a system of linear equations in two or three unknowns (pp. 463–469, 477–479).

Objectives:

A — Solve coin problems with two or more unknowns.

B — Solve general problems with two or more unknowns.

C — Solve rate, time, and distance (motion) problems with two or more unknowns.

D — Solve investment problems with two or more unknowns.

E — Solve geometry problems with two or more unknowns.

getting started

Money Problems!

The pile of money contains $3.25 in nickels and dimes. There are five more nickels than dimes. How many nickels and how many dimes are in the pile?

As usual, we use the RSTUV method to solve this problem.

1. *Read the problem:* We are asked to find the number of nickels and dimes.

2. *Select the unknown:* Let n be the number of nickels and d the number of dimes.

3. *Think of a plan:* We have to translate two statements, yielding two equations and two unknowns. First note that:

> If you have 1 nickel, you have $0.05(1)
>
> If you have 2 nickels, you have $0.05(2)
>
> If you have n nickels, you have $0.05(n)

Similarly,

> If you have d dimes, you have $0.10(d)

Now we can translate the statement.

$$\begin{array}{ccccc} \text{The pile} & & \text{in nickels} & \text{and} & \text{dimes.} \\ \text{contains \$3.25} & & & & \\ \left\{\begin{array}{c}\text{Total} \\ \text{amount}\end{array}\right\} & = & \left\{\begin{array}{c}\text{amount} \\ \text{in nickels}\end{array}\right\} & + & \left\{\begin{array}{c}\text{amount} \\ \text{in dimes}\end{array}\right\} \\ 3.25 & = & 0.05n & + & 0.10d \\ 325 & = & 5n & + & 10d \qquad \text{Multiply by 100.} \\ 65 & = & n & + & 2d \qquad \text{Divide by 5.} \end{array}$$

The statement "there are five more nickels than dimes" means that

$$\begin{array}{ccccc} \left\{\begin{array}{c}\text{The number} \\ \text{of nickels}\end{array}\right\} & = & \left\{\begin{array}{c}\text{5 more} \\ \text{than}\end{array}\right\} & + & \left\{\begin{array}{c}\text{the number} \\ \text{of dimes}\end{array}\right\} \\ n & = & 5 & + & d \end{array}$$

Thus the complete problem can be reduced to the system of equations

$$65 = n + 2d$$

$$n = 5 + d$$

4. *Use the substitution method to solve this system:* Substituting $5 + d$ for n in the first equation gives

$$65 = (5 + d) + 2d$$

$= 5 + d + 2d$	Remove parentheses.
$= 5 + 3d$	Collect like terms.

$$65 = 5 + 3d$$

$60 = 3d$	Subtract 5.
$20 = d$	Divide by 3.

Since $n = 5 + d$, we substitute 20 for d:

$$n = 5 + 20 = 25$$

Hence we have 25 nickels and 20 dimes.

5. *Verify the answer:* We do have 5 more nickels (25 in total) than dimes (20 in total) and \$3.25 (\$1.25 in nickels and \$2.00 in dimes).

Solving Coin Problems

Coin problems are a classic way to practice setting up systems of equations.

problem solving

EXAMPLE 1   Nickels and dimes

Jack has \$3 in nickels and dimes. He has twice as many nickels as he has dimes. How many nickels and how many dimes does he have?

SOLUTION   As usual, we use the RSTUV method.

1. Read the problem.

We are asked to find the number of nickels and dimes, so we need two variables.

2. Select the unknown.

Let n be the number of nickels and d the number of dimes.

3. Think of a plan.

Translate the problem. Jack has \$3 (300 cents) in nickels and dimes:

$$300 = 5n + 10d$$

He has twice as many nickels as he has dimes:

$$n = 2d$$

We then have the system

$$5n + 10d = 300$$

$$n = 2d$$

4. Use the substitution method to solve the system.

This time it's easy to use the substitution method.

$$5n + 10d = 300 \rightarrow 5(2d) + 10d = 300 \qquad \text{Let } n = 2d.$$

$$10d + 10d = 300 \qquad \text{Simplify.}$$

$$20d = 300 \qquad \text{Combine like terms.}$$

$$d = 15 \qquad \text{Divide by 20.}$$

$$n = 2(15) = 30 \qquad \text{Substitute } d = 15 \text{ in } n = 2d.$$

Thus Jack has 15 dimes ($1.50) and 30 nickels ($1.50).

5. Verify the answer.

It's easy to verify this answer because we know Jack has $3 and twice as many nickels as dimes. ∎

 Solving General Problems

We can also use systems of equations to solve other problems. Here is an interesting one.

 problem solving

EXAMPLE 2   A weighty matter

The greatest weight differential ever recorded for a married couple is 1300 lb—the case of Jon Brower Minnoch and his wife Jeannette. Their combined weight is 1520 lb. What is the weight of each of the Minnochs? (He is the heavier one.)

SOLUTION

1. Read the problem.

We are asked to find the weights of the Minnochs, so we need two variables.

2. Select the unknown.

Let h be the weight of Jon and w be the weight of Jeannette.

3. Think of a plan.

Translate the problem. The weight differential is 1300 lb:

$$h - w = 1300$$

Their combined weight is 1520 lb:

$$h + w = 1520$$

We then have the system

$$h - w = 1300$$
$$h + w = 1520$$

4. Use the elimination method to solve the system.

We use the elimination method.

$$h - w = 1300$$
$$\underline{h + w = 1520}$$
$$2h = 2820 \qquad \text{Add.}$$
$$h = 1410 \qquad \text{Divide by 2.}$$

Substitute $h = 1410$ in $h + w = 1520$:

$$1410 + w = 1520$$
$$w = 110$$

Thus Jeannette weighs 110 lb and Jon weighs 1410 lb.

5. Verify the answer.

You can verify this by going to the *Guinness Book of Records*! ∎

Motion Problems

Remember the motion problems we solved in Chapter 2? They can also be solved using two variables. The procedure is about the same! We write the given information in a chart labeled $R \times T = D$, and as usual, use the RSTUV method.

problem solving

EXAMPLE 3   The current in Norway

The world's strongest current is the Salstraumen in Norway. The current is so strong that a boat, which can go 48 mi downstream (with the current) in 1 hr, takes 4 hr to go the same 48 mi upstream (against the current). How fast is the current flowing?

SOLUTION

1. Read the problem.

We are asked to find the speed of the current, so we need two variables: the speed of the boat and the speed of the current.

2. Select the unknown.

Let x be the speed of the boat in still water and y be the speed of the current. Then $(x + y)$ is the speed of the boat downstream; $(x - y)$ is the speed of the boat upstream.

3. Think of a plan.

We enter this information in a chart:

	R	$\times$	T	$=$	D		
Downstream	$x + y$		1		48	$\rightarrow$	$x + y = 48$
Upstream	$x - y$		4		48	$\rightarrow$	$(x - y)4 = 48$

4. Use the elimination method to solve the system.

Our system of equations is simplified as follows:

$$x + y = 48 \quad \xrightarrow{\text{leave as is}} \quad x + y = 48$$

$$(x - y)4 = 48 \quad \xrightarrow{\text{divide by 4}} \quad \underline{x - y = 12}$$

$$2x = 60 \qquad \text{Add.}$$

$$x = 30 \qquad \text{Divide by 2.}$$

$$30 + y = 48 \qquad \text{Substitute } x = 30 \text{ in } x + y = 48.$$

$$y = 18 \qquad \text{Subtract 30.}$$

Thus the speed of the boat in still water is $x = 30$ mi/hr and the speed of the current is 18 mi/hr.

5. Verify the answer.

The verification is left to you. You may wish to compare how we solved this type of problem using one variable in Section 2.4 with the method used here. ■

Investment Problems

The investment problems we studied in Section 2.4 are easier to solve if we use three variables. We illustrate their solution next.

problem solving

EXAMPLE 4   Tracking investments

An investor divides $20,000 among three investments at 6%, 8%, and 10%. If the total annual income is $1700 and the income of the 10% investment exceeds the income from the 6% and 8% investments by $300, how much was invested at each rate?

1. Read the problem.

2. Select the unknown.

3. Think of a plan.

SOLUTION

We are asked to find how much was invested at each rate, so we need three variables.

Let x be the amount invested at 6%, y the amount invested at 8%, and z the amount invested at 10%.

Let's use a chart with the heading:

$$\text{Principal} \times \text{Rate} = \text{Interest}$$

We enter the information in the chart:

	Principal	×	Rate	=	Interest
1st investment	x		6%		$0.06x$
2nd investment	y		8%		$0.08y$
3rd investment	z		10%		$0.10z$
Total	$20,000				$1700

Looking at the column labeled *principal,* we see that

$$x + y + z = 20{,}000 \tag{1}$$

From the column labeled *interest,* we can see that the total interest earned is $1700; thus

$$0.06x + 0.08y + 0.10z = 1700$$

Multiplying by 100 gives

$$6x + 8y + 10z = 170{,}000 \tag{2}$$

We also know that

$$\left\{\begin{array}{c}\text{The income} \\ \text{from the 10\%} \\ \text{investment}\end{array}\right\} \left\{\begin{array}{c}\text{exceeds the income} \\ \text{from the 6\% and} \\ \text{8\% investments}\end{array}\right\} \quad \text{(by \$300)}$$

$$0.10z \quad = \quad 0.06x + 0.08y \quad + \quad 300$$

$$10z \quad = \quad 6x + 8y \quad + \quad 30{,}000 \qquad \text{Multiply by 100.}$$

$$-6x - 8y + 10z = 30{,}000 \tag{3}$$

4. Use the elimination method to solve the system.

We now have the system

$$x + y + z = 20{,}000 \tag{1}$$

$$6x + 8y + 10z = 170{,}000 \tag{2}$$

$$-6x - 8y + 10z = 30{,}000 \tag{3}$$

Adding equations (2) and (3), we have

$$20z = 200{,}000 \tag{4}$$

$$z = 10{,}000$$

Now multiply equation (1) by 6 and add it to equation (3):

$$6x + 6y + 6z = 120{,}000 \tag{5}$$

$$\underline{-6x - 8y + 10z = 30{,}000} \tag{3}$$

$$-2y + 16z = 150{,}000 \tag{6}$$

Substitute 10,000 for z in equation (6):

$$-2y + 16(10,000) = 150,000 \qquad \textbf{(7)}$$
$$-2y + \quad 160,000 = 150,000$$
$$-2y \qquad\qquad\quad = -10,000$$
$$y = 5000$$

Finally, substitute 10,000 for z and 5000 for y in equation (1):

$$x + 5000 + 10,000 = 20,000$$
$$x + \qquad\quad 15,000 = 20,000$$
$$x = 5000$$

Thus $x = 5000$, $y = 5000$, and $z = 10,000$.

5. Verify the answer.

You can verify that if we invest \$5000 at 6%, \$5000 at 8%, and \$10,000 at 10%, the conditions of the problem are satisfied. ∎

Solving Geometry Problems

Problems involving the perimeter of a rectangle also involve two unknowns. Example 5 shows how to set up a system of two equations to solve such problems.

problem solving   **EXAMPLE 5**   A "check" using geometry and perimeter

The check is in the mail? Certainly not the physically largest check ever written, which had a perimeter of 202 ft and was 39 ft longer than wide! Can you find the dimensions of this check given by InterMortgage of Leeds, England, to a Yorkshire TV telephone appeal?

SOLUTION   We continue to use the RSTUV procedure.

1. Read the problem.

We are asked to find the dimensions of the check. Even though the problem doesn't specify it, we assume that the check is rectangular in shape.

2. Select the unknown.

Since we want the dimensions of the check, we let W be its width and L its length.

3. Think of a plan.

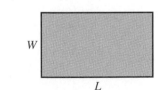

First, translate the problem. A picture helps.

Since the perimeter P is 202 ft and the perimeter is also $2W + 2L$, we have the equation

$$2W + 2L = 202 \qquad \textbf{(1)}$$

Also, "the check was 39 ft longer than wide" means

$$L = W + 39 \qquad \textbf{(2)}$$

Note that equation (1) can be simplified if we divide each of its terms by 2. Since $L = W + 39$, it will be easier to use the substitution method.

We have the system

$$2W + 2L = 202 \qquad \textbf{(1)}$$
$$L = W + 39 \qquad \textbf{(2)}$$

First, divide each term in equation (1) by 2 to obtain

$$W + L = 101 \qquad \textbf{(3)}$$
$$L = W + 39 \qquad \textbf{(2)}$$

4. Use the substitution method to solve the system.

Substituting $W + 39$ for L in equation (3), we have

$$W + (W + 39) = 101$$
$$2W + 39 = 101 \quad \text{Simplify.}$$
$$2W = 62 \quad \text{Subtract 39 from both sides.}$$
$$W = 31F \quad \text{Divide both sides by 2.}$$

From equation (2), $L = W + 39$ and we know that $W = 31$, so

$$L = 31 + 39 = 70$$

Thus the dimensions of the check are 31 ft wide by 70 ft long.

5. Verify the answer.

The verification that the perimeter is 202 ft and that the check is 39 ft longer than wide is left to you. ■

GRAPH IT

You can use your grapher as a tool to assist you in solving the problems in this section. There are two important considerations:

1. Which variable will you designate as the independent variable (the one that is easier to solve for)?

2. What size window will let you see the part of the graph at which the given lines intersect? (Use a window that contains the x- and y-intercepts of the lines involved.)

For instance, in Example 1, go through the first three steps of the RSTUV method (your grapher can't do this part for you!) to obtain the system

$$65 = n + 2d \quad \text{and} \quad n = 5 + d$$

Since it's easier to solve for n, we have the equivalent system

$$n = 65 - 2d \quad \text{and} \quad n = 5 + d$$

Graph

$$Y_1 = 65 - 2X \quad \text{and} \quad Y_2 = 5 + X$$

In the first equation, when $X = 0$, $Y_1 = 65$ and when $Y_1 = 0$, $X = 32.5$, so we use a [0, 35] by [0, 65] window, with a scale of 5 for X and Y (Xscl = 5, Yscl = 5). Using your intersection feature (2nd CALC 5 and ENTER three times; see Window 1), we get $X = 20$ and $Y = 25$. Thus, $d = 20$ and $n = 25$ as before; that is, we have 20 dimes and 25 nickels. Now you can verify that these two numbers satisfy the conditions of the original problem.

In Example 2, we have the system $h - w = 1300$ and $h + w = 1520$, or

$$h = 1300 + w \quad \text{and} \quad h = 1520 - w$$

Graph

$$Y_1 = 1300 + X \quad \text{and} \quad Y_2 = 1520 - X$$

using a [−1300, 1600] by [0, 1600] window with a scale of 100 for both the X and Y. Using your intersection feature, we obtain $X = 110$ and $Y = 1410$ (Window 2). Thus $w = 110$ and $h = 1410$ as before. Check this out with the conditions of the problem.

For Example, 3, we graph the system

$$Y_1 = 48 - X \quad \text{and} \quad Y_2 = X - 12$$

with a [0, 50] by [0, 50] window and an X- and Y-scale of 5, which yields $X = 30$ and $Y = 18$ as the point of intersection (Window 3).

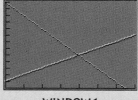
WINDOW 1

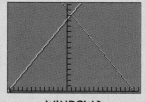
WINDOW 2

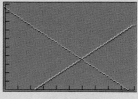
WINDOW 3

Now it's your turn to do Example 4. But wait, we have three variables there. It's actually easier to do that problem algebraically! In other problems with three variables, reduce the system to two equations and two unknowns and graph. Then substitute your answers for x and y into any of the equations and solve for z. Now try Example 5!

EXERCISE 8.3

A **In Problems 1–10, use two unknowns to solve these coin problems.**

1. Natasha has $6.25 in nickels and dimes. If she has twice as many dimes as she has nickels, how many dimes and how many nickels does she have?

2. Mida has $2.25 in nickels and dimes. She has four times as many dimes as nickels. How many dimes and how many nickels does she have?

3. Dora has $5.50 in nickels and quarters. She has twice as many quarters as she has nickels. How many of each coin does she have?

4. Mongo has 20 coins consisting of nickels and dimes. If the nickels were dimes and the dimes were nickels, he would have 50¢ more than he now has. How many nickels and how many dimes does he have?

5. Desi has 10 coins consisting of pennies and nickels. Strangely enough, if the nickels were pennies and the pennies were nickels, she would have the same amount of money as she now has. How many pennies and nickels does she have?

6. Don has $26 in his pocket. If he has only 1-dollar bills and 5-dollar bills, and he has a total of 10 bills, how many of each bill does he have?

7. A person went to the bank to deposit $300. The money was in 10- and 20-dollar bills, 25 bills in all. How many of each did the person have?

8. A woman has $5.95 in nickels and dimes. If she has a total of 75 coins, how many nickels and how many dimes does she have?

9. A man has $7.05 in nickels, dimes, and quarters. The quarters are worth $4.60 more than the dimes and the dimes are worth 25¢ more than the nickels. How many nickels, dimes, and quarters does the man have?

10. Amy has $2.50 consisting of nickels, dimes, and quarters in her piggy bank. She has the same amount in nickels and dimes, and twice as much in nickels as she has in quarters. How many nickels, dimes, and quarters does Amy have?

B **In Problems 11–20, use two unknowns to solve these general problems.**

11. The sum of two numbers is 102. Their difference is 16. What are the numbers?

12. The difference between two numbers is 28. Their sum is 82. What are the numbers?

13. The sum of two integers is 126. If one of the integers is 5 times the other, what are the integers?

14. The difference between two integers is 245. If one of the integers is 8 times the other, find the integers.

15. The difference between two numbers is 16. One of the numbers is 5 times the other. What are the numbers?

16. The sum of two numbers is 116. One of the numbers is 50 less than the other. What are the numbers?

17. Longs Peak is 145 ft higher than Pikes Peak. If you were to put these two peaks one on top of the other, you would still be 637 ft short of reaching the elevation of Mt. Everest, 29,002 ft. Find the elevations of Longs Peak and Pikes Peak.

18. The height of the Empire State building and its antenna is 1472 ft. The difference in height between the building and the antenna is 1028 ft. How tall is the antenna and how tall is the building?

19. The largest sundae ever made contained about 6700 lb of topping. The topping flavors were chocolate, butterscotch, and caramel. There was the same amount of butterscotch as caramel but 600 lb more of chocolate than butterscotch. How many pounds of each were included in the topping?

20. The largest pancake ever made used buckwheat flour, Puritan mix, and 15 gal of syrup. The flour and mix weighed 100 lb more than the 15 gal of syrup. What was the weight of the syrup if the whole pancake weighed 4100 lb? By the way, 68 lb of butter were added before it was consumed!

C **In Problems 21–25, use two unknowns to solve these motion problems.**

21. A plane flies 540 mi with a tail wind in $2\frac{1}{4}$ hr. The plane makes the return trip against the same wind and takes 3 hr. Find the speed of the plane in still air and the speed of the wind.

22. A motor boat runs 45 mi downstream in $2\frac{1}{2}$ hr and 39 mi upstream in $3\frac{1}{4}$ hr. Find the speed of the boat in still water and the speed of the current.

23. A motorboat can travel 15 mi/hr downstream and 9 mi/hr upstream on a certain river. Find the rate of the current and the rate at which the boat can travel in still water.

24. It takes a motorboat $1\frac{1}{3}$ hr to go 20 mi downstream and $2\frac{2}{9}$ hr to return. Find the rate of the current and the rate at which the boat can travel in still water.

25. A plane flying with the wind took 2 hr for a 1000-mi flight and $2\frac{1}{2}$ hr for the return flight. Find the wind velocity and the speed of the plane in still air.

In Problems 26–34, use two or more unknowns to solve these investment problems.

26. Two sums of money totaling $20,000 earn 8% and 10% annual interest, respectively. If the interest from both investments amounts to $1900, how much is invested at each rate?

27. An investor invested $10,000, part at 6% and the rest at 8%. Find the amount invested at each rate if the annual income from the two investments is $720.

28. Andy Cabazos has $20,000 in three investments paying 6%, 8%, and 10%. The total interest on the 6% and 8% investments is $300 less than that obtained from the 10% investment. If his annual income from these investments is $1700, how much does he have invested at each rate?

29. Marlene McGuire invested $25,000 in municipal bonds. The first investment paid 6%, the second 8%, and the third, 10%. If her annual income from these bonds was $2000 and the interest she received on the combined 6% and 8% investments equaled the interest on the 10% investment, how much money did she have in each category?

30. Marc Goldstein divided $20,000 into three parts. One part yielded 4%, another 8%, and the third one, 6%. If his total return was $1080 and he made $40 less on his 8% investment than on his 4% investment, what amount did he invest in each category?

In Problems 31–34, use two unknowns to solve these geometry problems.

31. The perimeter of the largest flag is 1520 ft. If the length of the flag exceeds the width by 250 ft, what are the dimensions of the flag?

32. The largest flag *actually flown* from a flagpole was 98 ft longer than wide. If its perimeter was 1016 ft, what were the dimensions of the flag? (It was a Brazilian flag, flown in Brazil.)

33. The world's largest quilt boasted a 438-ft perimeter with its length being 49 ft more than its width. What were the dimensions of the quilt? By the way, it took 7000 North Dakotans to make it!

34. If you walked around the largest rectangular swimming pool in the world in Morocco, you would end up walking 3640 ft. If the pool is 1328 ft longer than it is wide, what are the dimensions of the pool?

SKILL CHECKER

Solve the system:

35. $2x - y + z = 3$
    $x + y = -1$
    $3x - y - 2z = 7$

36. $2x - y + 2z = 3$
    $2x + 2y - z = 0$
    $-x + 2y + 2z = -12$

USING YOUR KNOWLEDGE   The A, B, Cs of Vitamins

In this section we solved coin, general, distance, investment, and geometry problems. What kind of problem is left out? Mixture problems! Use your knowledge to solve these mixture problems.

A dietician wants to arrange a diet composed of three basic foods A, B, and C. The diet must include 170 units of calcium, 90 units of iron, and 110 units of vitamin B. The table gives the number of units per ounce of each of the needed ingredients contained in each of the basic foods.

Nutrient	Units per Ounce		
	Food A	Food B	Food C
Calcium	15	5	20
Iron	5	5	10
Vitamin B	10	15	10

If *a*, *b*, and *c* are the number of ounces of basic foods A, B, and C taken by an individual, write an equation indicating:

37. The amount of calcium needed

38. The amount of iron needed

39. The amount of vitamin B needed

We shall come back to this problem and show you a new way to solve it in the next section!

WRITE ON . . .

40. State the advantages and disadvantages of solving word problems using systems of equations instead of the techniques we studied in Chapter 2.

41. Write a word problem that uses the system

$$x + y = 10$$
$$x - y = 2$$

to obtain its solution.

42. Write the method you use to solve word problems.

   If you know how to do these problems, you have learned your lesson!

43. The perimeter of a rectangle is 170 cm. If its length is 15 cm more than its width, what are the dimensions of the rectangle?

44. An investor divides $20,000 among three investments at 6%, 8%, and 10%. If her total income is $1740 and the income of

the 10% investment exceeds the income from the 6% and 8% investments by $260, how much did she invest at each rate?

45. A plane flies 1200 mi with a tail wind in 3 hr. It takes 4 hr to fly the same distance against the wind. Find the wind velocity and the velocity of the plane in still air.

46. A change machine gave Jill $2 in nickels and dimes. She had twice as many nickels as dimes. How many nickels and how many dimes did the machine give Jill?

47. The McGuire twins weighed a total of 1466 lb. If their weight differential was 20 lb, what was the weight of each of the twins?

8.4

MATRICES

To succeed, review how to:

1. Solve a system of three equations in three unknowns (pp. 477–479).

2. Recognize whether a system is inconsistent or dependent (pp. 479–480).

Objectives:

A Perform elementary operations on systems of equations.

B Solve systems of linear equations using matrices.

C Solve applications using matrices.

---

getting started

Tweedledee and Tweedledum

Do you remember Tweedledee and Tweedledum from Problem 87 in Exercise 8.1? Here's what they said:

Tweedledee: The sum of your weight and twice mine is 361 pounds.

Tweedledum: Contrariwise, the sum of your weight and twice mine is 360 pounds.

If Tweedledee weighs x pounds and Tweedledum weighs y pounds, the two sentences can be translated as

$$2x + y = 361$$
$$x + 2y = 360$$

Alice meets Tweedledee and Tweedledum in *Through the Looking Glass.*

We are going to solve this system again, but this time we shall also write the equivalent operations using *matrices*.

MATRIX

A **matrix** (plural matrices) is a rectangular array of numbers enclosed in brackets.

For example,

$$\begin{bmatrix} 2 & 1 \\ 1 & 1 \end{bmatrix} \quad \text{and} \quad \begin{bmatrix} -5 & 3 & 2 \\ 4 & 0 & 1 \end{bmatrix}$$

are matrices. The first matrix has two rows and two columns (2×2), while the second one has two rows and three columns (2×3). A matrix derived from a system of linear equations (each written in standard form with the constant terms on the right) is called the **augmented matrix** of the system. For example, the augmented matrix of the system

$$\begin{aligned} 2x + \ y &= 361 \\ x + 2y &= 360 \end{aligned} \quad \text{is} \quad \begin{bmatrix} 2 & 1 & | & 361 \\ 1 & 2 & | & 360 \end{bmatrix}$$

Now compare the solution of the two-equation system with its solution using matrices:

$$
\begin{aligned}
2x + y &= 361 \\
x + 2y &= 360
\end{aligned}
\qquad
\left[
\begin{array}{cc|c}
2 & 1 & 361 \\
1 & 2 & 360
\end{array}
\right]
$$

$$
\begin{aligned}
2x + y &= 361 \\
-2x - 4y &= -720
\end{aligned}
\qquad
\left[
\begin{array}{cc|c}
2 & 1 & 361 \\
-2 & -4 & -720
\end{array}
\right]
\qquad \text{Multiply the second equation by } -2.
$$

$$
\begin{aligned}
2x + y &= 361 \\
-3y &= -359
\end{aligned}
\qquad
\left[
\begin{array}{cc|c}
2 & 1 & 361 \\
0 & -3 & -359
\end{array}
\right]
\qquad \text{Add the two equations.}
$$

$$
\begin{aligned}
2x + y &= 361 \\
y &= 119\tfrac{2}{3}
\end{aligned}
\qquad
\left[
\begin{array}{cc|c}
2 & 1 & 361 \\
0 & 1 & 119\tfrac{2}{3}
\end{array}
\right]
\qquad \text{Divide the second equation by } -3.
$$

Now you can substitute $119\tfrac{2}{3}$ for y in $2x + y = 361$ and solve for x. Did you note that the operations performed on the equations were identical to those performed on the matrices? In this section we shall solve systems of three equations using matrices.

Solving Systems of Linear Equations Using Matrices

Suppose we wish to solve a system of three linear equations in three unknowns. This means that we wish to find all sets of values of (x, y, z) that make all three equations true. We first need to consider what changes can be made in the system to yield an **equivalent system**, a system that has exactly the same solutions as the original system. We need to consider only three simple operations.

PROCEDURE

> **Elementary Operations on Systems of Equations**
>
> 1. **The order of the equations may be changed.** This clearly cannot affect the solutions.
> 2. **Any of the equations may be multiplied by any nonzero real number.** If (m, n, p) is a solution of $ax + by + cz = d$, then, for $k \neq 0$, it is also a solution of $kax + kby + kcz = kd$, and conversely. (Why?)
> 3. **Any equation of the system may be replaced by the sum (term by term) of itself and any other equation of the system.** (You can show this by doing Problem 11 of Exercise 8.5.)

These three **elementary operations** are used to simplify systems of equations and to find their solutions. Let's use them to solve the system

$$
\begin{aligned}
2x - y + z &= 3 \\
x + y &= -1 \\
3x - y - 2z &= 7
\end{aligned}
$$

STEP 1 In the second equation of the system, x and y have unit coefficients, so we interchange the first two equations to get the following more convenient arrangement:

$$
\begin{aligned}
x + y &= -1 \tag{1}
\end{aligned}
$$

System I
$$
\begin{aligned}
2x - y + z &= 3 \tag{2}
\end{aligned}
$$
$$
\begin{aligned}
3x - y - 2z &= 7 \tag{3}
\end{aligned}
$$

STEP 2 To make the coefficients of x in equations (1) and (2) the same in absolute value but opposite in sign, multiply both sides of equation (1) by -2:

$$-2x - 2y = +2$$

STEP 3 To eliminate x between equations (1) and (2), add the equation obtained in step 2 to equation (2) to get

$$-3y + z = 5 \qquad\qquad \textbf{(4)}$$

and restore equation (1) by dividing out the -2. The system now reads

System II

$$
\begin{aligned}
x + y &= -1 & \textbf{(1)} \\
-3y + z &= 5 & \textbf{(4)} \\
3x - y - 2z &= 7 & \textbf{(3)}
\end{aligned}
$$

Now we proceed in a similar way to eliminate x between equations (1) and (3).

STEP 4 Multiply both sides of equation (1) by -3:

$$-3x - 3y = 3$$

STEP 5 Add this last equation to equation (3) to get

$$-4y - 2z = 10 \qquad\qquad \textbf{(5)}$$

and restore equation (1) by dividing out the -3. The system now reads

System III

$$
\begin{aligned}
x + y &= -1 & \textbf{(1)} \\
-3y + z &= 5 & \textbf{(2)} \\
-4y - 2z &= 10 & \textbf{(5)}
\end{aligned}
$$

We eliminate y between equations (2) and (5) in the following way:

STEP 6 Multiply both sides of equation (2) by -4 and both sides of equation (5) by 3 to get

$$
\begin{aligned}
12y - 4z &= -20 \\
-12y - 6z &= 30
\end{aligned}
$$

STEP 7 Add the last two equations to get

$$-10z = 10 \qquad\qquad \textbf{(6)}$$

and restore the second equation by dividing out the -4. The system is now

System IV

$$
\begin{aligned}
x + y &= -1 & \textbf{(1)} \\
-3y + z &= 5 & \textbf{(2)} \\
-10z &= 10 & \textbf{(6)}
\end{aligned}
$$

STEP 8 It's easy to solve system IV: Equation (6) immediately gives $z = -1$. Then, by substitution into equation (2), we get

$$
\begin{aligned}
-3y - 1 &= 5 & \textbf{(2)} \\
-3y &= 6 \\
y &= -2
\end{aligned}
$$

By substituting $y = -2$ into equation (1) we find

$$x - 2 = -1 \qquad\qquad \textbf{(1)}$$

$$x = 1$$

Thus the solution of the system is $x = 1$, $y = -2$, $z = -1$. This is easily checked in the given set of equations.

A general system of the same form as system IV may be written

$$ax + by + cz = d$$

System V
$$ey + fz = g$$

$$hz = k$$

A system such as system V, in which the first unknown, x, is missing from the second and third equations and the second unknown, y, is missing from the third equation, is said to be in **echelon form**. A system in echelon form is quite easy to solve. The third equation immediately yields the value of z. Back-substitution of this value into the second equation yields the value of y. Finally, back-substitution of the values of y and z into the first equation yields the value of x. Briefly, we say that we solve the system by **back-substitution**.

Every system of linear equations can be brought into echelon form by the use of the elementary operations. It remains for us only to organize and make the procedure more efficient by employing the augmented matrix.

Let's compare the augmented matrices of system I and system IV:

System I
(original system)

Main → diagonal
$$\left[\begin{array}{ccc|c} 2 & -1 & 1 & 3 \\ 1 & 1 & 0 & -1 \\ 3 & -1 & -2 & 7 \end{array}\right]$$

System IV

Main → diagonal
$$\left[\begin{array}{ccc|c} 1 & 1 & 0 & -1 \\ 0 & -3 & 1 & 5 \\ 0 & 0 & -10 & 10 \end{array}\right]$$

The augmented matrix for system IV shows that the system is in echelon form because it has only 0's below the main diagonal (the diagonal left to right of the coefficient matrix). We should be able to obtain the second matrix from the first by performing operations corresponding to the elementary operations on the equations. These operations are called **elementary row operations**, and they always yield matrices of equivalent systems. Such matrices are called **row-equivalent**. If two matrices A and B are row-equivalent, we write $A \sim B$. The elementary row operations are as follows.

PROCEDURE

Elementary Row Operations on Matrices
1. Change the order of the rows.
2. Multiply all the elements of a row by any nonzero number.
3. Replace any row by the element-by-element sum of itself and any other row.

Compare the elementary row operations with the elementary operations on a system. They are analogous!

We illustrate the procedure by showing the transition from the matrix of the original system I to that of system IV. To explain what is happening at each step, we use the notation R_1, R_2, and R_3 for the respective rows of the matrix, along with the following typical abbreviations:

Step	Notation	Meaning
1	$R_1 \longleftrightarrow R_2$	Interchange R_1 and R_2.
2	$2 \times R_1$	Multiply each element of R_1 by 2.
3	$2 \times R_1 + R_2 \rightarrow R_2$	Replace R_2 by $2 \times R_1 + R_2$.

Thus we write step 1 like this:

$$\begin{bmatrix} 2 & -1 & 1 & | & 3 \\ 1 & 1 & 0 & | & -1 \\ 3 & -1 & -2 & | & 7 \end{bmatrix} \sim \begin{bmatrix} 1 & 1 & 0 & | & -1 \\ 2 & -1 & 1 & | & 3 \\ 3 & -1 & -2 & | & 7 \end{bmatrix}$$
$$R_1 \longleftrightarrow R_2$$

Next, we proceed to get 0's in the second and third rows of the first column:

$$\begin{bmatrix} 1 & 1 & 0 & | & -1 \\ 2 & -1 & 1 & | & 3 \\ 3 & -1 & -2 & | & 7 \end{bmatrix} \sim \begin{bmatrix} 1 & 1 & 0 & | & -1 \\ 0 & -3 & 1 & | & 5 \\ 0 & -4 & -2 & | & 10 \end{bmatrix}$$
$$-2 \times R_1 + R_2 \rightarrow R_2$$
$$-3 \times R_1 + R_3 \rightarrow R_3$$

To complete the procedure, we get a 0 in the third row of the second column:

$$\begin{bmatrix} 1 & 1 & 0 & | & -1 \\ 0 & -3 & 1 & | & 5 \\ 0 & -4 & -2 & | & 10 \end{bmatrix} \sim \begin{bmatrix} 1 & 1 & 0 & | & -1 \\ 0 & -3 & 1 & | & 5 \\ 0 & 0 & -10 & | & 10 \end{bmatrix}$$
$$-4 \times R_2 + 3 \times R_3 \rightarrow R_3$$

Now we need to verify the result, which is the augmented matrix of system IV. Of course, once we obtain this last matrix, we can solve the system by back-substitution, as before.

We shall now provide additional illustrations of this procedure, but these will help you only if you take a pencil and paper and carry out the detailed row operations as they are indicated.

EXAMPLE 1   Using matrices to solve a system of equations with one solution

Solve the system:

$$\begin{aligned} 2x - y + 2z &= 3 \\ 2x + 2y - z &= 0 \\ -x + 2y + 2z &= -12 \end{aligned}$$

SOLUTION   The augmented matrix is

$$\begin{bmatrix} 2 & -1 & 2 & | & 3 \\ 2 & 2 & -1 & | & 0 \\ -1 & 2 & 2 & | & -12 \end{bmatrix} \sim \begin{bmatrix} 2 & -1 & 2 & | & 3 \\ 0 & 3 & -3 & | & -3 \\ 0 & 3 & 6 & | & -21 \end{bmatrix}$$

$$-R_1 + R_2 \rightarrow R_2$$
$$R_1 + 2 \times R_3 \rightarrow R_3$$

$$\sim \begin{bmatrix} 2 & -1 & 2 & | & 3 \\ 0 & 1 & -1 & | & -1 \\ 0 & 0 & 9 & | & -18 \end{bmatrix}$$

$$-R_2 + R_3 \rightarrow R_3$$
$$\frac{1}{3} \times R_2 \rightarrow R_2$$

This last matrix is in echelon form and corresponds to the system

$$2x - y + 2z = 3 \qquad\qquad \textbf{(1)}$$
$$y - z = -1 \qquad\qquad \textbf{(2)}$$
$$9z = -18 \qquad\qquad \textbf{(3)}$$

We solve this system by back-substitution. Equation (3) immediately yields $z = -2$. Equation (2) then becomes

$$y + 2 = -1$$

so that

$$y = -3$$

Equation (1) then becomes

$$2x + 3 - 4 = 3$$

so that

$$x = 2$$

The final answer, $x = 2$, $y = -3$, $z = -2$, can be checked in the given system.   ∎

EXAMPLE 2   Using matrices to solve a system of equations with no solution

Solve the system:

$$2x - y + 2z = 3$$
$$2x + 2y - z = 0$$
$$4x + y + z = 5$$

SOLUTION

$$\begin{bmatrix} 2 & -1 & 2 & | & 3 \\ 2 & 2 & -1 & | & 0 \\ 4 & 1 & 1 & | & 5 \end{bmatrix} \sim \begin{bmatrix} 2 & -1 & 2 & | & 3 \\ 0 & 3 & -3 & | & -3 \\ 0 & 3 & -3 & | & -1 \end{bmatrix}$$

$$-R_1 + R_2 \rightarrow R_2$$
$$-2 \times R_1 + R_3 \rightarrow R_3$$

$$\sim \begin{bmatrix} 2 & -1 & 2 & | & 3 \\ 0 & 3 & -3 & | & -3 \\ 0 & 0 & 0 & | & 2 \end{bmatrix}$$

$$-R_2 + R_3 \rightarrow R_3$$

The final matrix is in echelon form, and the last line corresponds to the equation

$$0x + 0y + 0z = 2$$

which is false for all values of x, y, z. Hence the given system has *no solution*. ∎

NOTE It is important to notice that if reduction to echelon form introduces any row with all 0's to the left and a nonzero number to the right of the vertical line, then the system has no solution.

EXAMPLE 3 Using matrices to solve a system of equations with infinitely many solutions

Solve the system:

$$\begin{aligned}
2x - y + 2z &= 3 \\
2x + 2y - z &= 0 \\
4x + y + z &= 3
\end{aligned}$$

SOLUTION

$$\begin{bmatrix} 2 & -1 & 2 & | & 3 \\ 2 & 2 & -1 & | & 0 \\ 4 & 1 & 1 & | & 3 \end{bmatrix} \sim \begin{bmatrix} 2 & -1 & 2 & | & 3 \\ 0 & 3 & -3 & | & -3 \\ 0 & 3 & -3 & | & -3 \end{bmatrix}$$

$$\begin{aligned} -R_1 + R_2 &\to R_2 \\ -2 \times R_1 + R_3 &\to R_3 \end{aligned}$$

$$\sim \begin{bmatrix} 2 & -1 & 2 & | & 3 \\ 0 & 3 & -3 & | & -3 \\ 0 & 0 & 0 & | & 0 \end{bmatrix}$$

$$-R_2 + R_3 \to R_3$$

The last matrix is in echelon form, and the last line corresponds to the equation

$$0x + 0y + 0z = 0$$

which is true for all values of x, y, and z. Thus any solution of the first two equations will be a solution of the system. The first two equations are

$$2x - y + 2z = 3 \qquad \textbf{(1)}$$
$$3y - 3z = -3 \qquad \textbf{(2)}$$

This system is equivalent to

$$2x - y = 3 - 2z \qquad \textbf{(3)}$$
$$y = -1 + z \qquad \text{Solve equation (2) for } y. \qquad \textbf{(4)}$$

Suppose we let $z = k$, where k is any real number. Then equation (4) gives $y = -1 + k$. Substitution into equation (3) results in

$$2x + 1 - k = 3 - 2k$$

so that

$$2x = 2 - k$$

$$x = 1 - \frac{1}{2}k$$

Thus if k is any real number, then $x = 1 - \frac{1}{2}k$, $y = k - 1$, $z = k$ is a solution of the system. You can verify this by substitution in the original system. We see that the system in this example has infinitely many solutions, because the value of k may be quite arbitrarily chosen. For instance, if $k = 2$, then the solution is $x = 0$, $y = 1$, $z = 2$; if $k = 5$, then $x = -\frac{3}{2}$, $y = 4$, $z = 5$; if $k = -4$, then $x = 3$, $y = -5$, $z = -4$; and so on. ∎

In the final echelon form, the system has infinitely many solutions if there is no row with all 0's to the left and a nonzero number to the right of the vertical line but there *is* a row with all 0's to both the left and the right.

Examples 1–3 illustrate the three possibilities for three linear equations in three unknowns. The system may have **one unique solution** as in Example 1; the system may have **no solution** as in Example 2; or the system may have **infinitely many solutions** as in Example 3. The final echelon form of the matrix always shows which case is at hand.

Solving Applications Using Matrices

problem solving

MATRICES AND NUTRITION

Here is a problem from *Using Your Knowledge,* Section 8.3. Now let's solve this problem using matrices.

EXAMPLE 4   Where are the nutrients?

A dietitian wants to arrange a diet composed of three basic foods A, B, and C. The diet must include 170 units of calcium, 90 units of iron, and 110 units of vitamin B. The table gives the number of units per ounce of each of the needed ingredients contained in each of the basic foods.

Nutrient	Units per Ounce		
	Food A	Food B	Food C
Calcium	15	.5	20
Iron	5	5	10
Vitamin B	10	15	10

SOLUTION

1. Read the problem.

If a, b, and c are the number of ounces of basic foods A, B, and C taken by an individual, find the number of ounces of each of the basic foods needed to meet the diet requirements.

2. Select the unknown.

We want to find the values of a, b, and c, the number of ounces of basic foods A, B, and C taken by an individual.

3. Think of a plan.

We write a system of equations: What is the amount of calcium needed? Since the individual gets 15 units of calcium from A, 5 from B, and 20 from C, the amount of calcium is

$$15a + 5b + 20c = 170$$

What is the amount of iron needed?

$$5a + 5b + 10c = 90$$

What is the amount of vitamin B needed?

$$10a + 15b + 10c = 110$$

4. Use matrices to solve the system.

Write the equations obtained using matrices. The simplified system of three equations and three unknowns, obtained by dividing each term in each of the equations by 5, is written as

$$3a + b + 4c = 34$$
$$a + b + 2c = 18$$
$$2a + 3b + 2c = 22$$

The corresponding augmented matrix is thus

$$\left[\begin{array}{ccc|c} 3 & 1 & 4 & 34 \\ 1 & 1 & 2 & 18 \\ 2 & 3 & 2 & 22 \end{array}\right]$$

$$\left[\begin{array}{ccc|c} 3 & 1 & 4 & 34 \\ 1 & 1 & 2 & 18 \\ 2 & 3 & 2 & 22 \end{array}\right] \sim \left[\begin{array}{ccc|c} 1 & 1 & 2 & 18 \\ 3 & 1 & 4 & 34 \\ 2 & 3 & 2 & 22 \end{array}\right] \sim \left[\begin{array}{ccc|c} 1 & 1 & 2 & 18 \\ 0 & -2 & -2 & -20 \\ 0 & 1 & -2 & -14 \end{array}\right]$$

$$R_1 \longleftrightarrow R_2 \qquad \begin{array}{c} R_2 - 3R_1 \rightarrow R_2 \\ R_3 - 2R_1 \rightarrow R_3 \end{array}$$

$$\sim \left[\begin{array}{ccc|c} 1 & 1 & 2 & 18 \\ 0 & 1 & 1 & 10 \\ 0 & 1 & -2 & -14 \end{array}\right] \sim \left[\begin{array}{ccc|c} 1 & 1 & 2 & 18 \\ 0 & 1 & 1 & 10 \\ 0 & 0 & -3 & -24 \end{array}\right]$$

$$\frac{R_2}{-2} \rightarrow R_2 \qquad\qquad R_3 - R_2 \rightarrow R_3$$

From the third row, $-3c = -24$, or $c = 8$. Substituting in the second row, we have $b + 8 = 10$, or $b = 2$. Finally, substituting in the first row, we have $a + 2 + 2(8) = 18$, or $a = 0$.

5. Verify the solution.

Substitute $a = 0$, $b = 2$, and $c = 8$ into the first equation to obtain $15 \cdot 0 + 5 \cdot 2 + 20 \cdot 8 = 10 + 160 = 170$. Then use the same procedure to check the second and third equations. Thus the required answer is 0 oz of food A, 2 oz of food B, and 8 oz of food C. ■

EXAMPLE 5 Nailing down the problem

Tom Jones, who was building a workshop, went to the hardware store and bought 1 lb each of three kinds of nails: small, medium, and large. After completing part of the work, Tom found that he had underestimated the number of small and large nails he needed. So he bought another pound of the small nails and 2 lb more of the large nails. After some more work, he again ran short of nails and had to buy another pound of each of the small and the medium nails. Upon looking over his bills, he found that the hardware store charged him $2.10 for nails the first time, $2.30 the second time, and $1.20 the third time. The prices for the various sizes of nails were not listed. Find these prices.

SOLUTION

1. Read the problem.

Note that there are three types of nails costing $2.10, $2.30, and $1.20.

2. Select the unknown.

We let x, y, and z be the prices in cents per pound for the small, medium, and large nails, respectively.

3. Think of a plan.

Then we know that

$$x + y + z = 210$$
$$x \quad\quad + 2z = 230$$
$$x + y \quad\quad = 120$$

4. Use matrices to solve the system.

We solve this system as follows:

$$\begin{bmatrix} 1 & 1 & 1 & | & 210 \\ 1 & 0 & 2 & | & 230 \\ 1 & 1 & 0 & | & 120 \end{bmatrix} \sim \begin{bmatrix} 1 & 1 & 1 & | & 210 \\ 0 & 1 & -1 & | & -20 \\ 0 & 0 & 1 & | & 90 \end{bmatrix}$$

$$R_1 - R_2 \rightarrow R_2$$
$$R_1 - R_3 \rightarrow R_3$$

The second matrix is in echelon form, and the solution of the system is easily found by back-substitution to be $x = 50$, $y = 70$, $z = 90$, giving the schedule of prices shown.

Tom's Schedule of Nail Prices

Nail Size	Price Per Pound
Small	50¢
Medium	70¢
Large	90¢

5. Verify the solution.

Substitute $x = 50$, $y = 70$, and $z = 90$ in $x + y + z = 210$ to obtain $50 + 70 + 90 = 210$, which is true. ∎

GRAPH IT

Graphers use matrix operations (too advanced to discuss in detail here) to solve systems of equations. Let's illustrate the procedure to solve Example 1. Let A be the coefficient matrix and B the constant matrix defined by

$$A = \begin{bmatrix} 2 & -1 & 2 \\ 2 & 2 & -1 \\ -1 & 2 & 2 \end{bmatrix} \quad B = \begin{bmatrix} 3 \\ 0 \\ -12 \end{bmatrix}$$

```
[A]⁻¹[B]
              [[2 ]
               [-3]
               [-2]]
```

To enter A, press $\boxed{\text{MATRX}}$ $\boxed{\blacktriangleleft}$ 1 3 $\boxed{\text{ENTER}}$ 3 $\boxed{\text{ENTER}}$. This tells the grapher you are about to enter a 3 × 3 matrix. Now enter the values for A by pressing

2 $\boxed{\text{ENTER}}$ $\boxed{(-)}$ 1 $\boxed{\text{ENTER}}$ 2 $\boxed{\text{ENTER}}$ 2 $\boxed{\text{ENTER}}$ 2 $\boxed{\text{ENTER}}$ $\boxed{(-)}$ 1 $\boxed{\text{ENTER}}$ $\boxed{(-)}$ 1 $\boxed{\text{ENTER}}$ 2 $\boxed{\text{ENTER}}$ 2 $\boxed{\text{ENTER}}$

To enter B, press $\boxed{\text{MATRX}}$ $\boxed{\blacktriangleleft}$ 2 3 $\boxed{\text{ENTER}}$ 1 $\boxed{\text{ENTER}}$. Next, enter the values for B by pressing 3 $\boxed{\text{ENTER}}$ 0 $\boxed{\text{ENTER}}$ $\boxed{(-)}$ 12 $\boxed{\text{ENTER}}$. Press $\boxed{\text{2nd}}$ $\boxed{\text{QUIT}}$ to go to the home screen. Finally, press $\boxed{\text{MATRX}}$ 1 $\boxed{x^{-1}}$ $\boxed{\text{MATRX}}$ 2 $\boxed{\text{ENTER}}$. The solution is as shown in the window. This means that $x = 2$, $y = -3$, and $z = -2$, as obtained in Example 1.

See if you can follow the same procedure to solve Example 2, but don't get alarmed if you get an error message. (There is no solution to the system in Example 2.)

EXERCISE 8.4

A B In Problems 1–10, find all the solutions (if there are any).

1.
$x + y - z = 3$
$x - 2y + z = -3$
$2x + y + z = 4$

2.
$x + 2y - z = 5$
$2x + y + z = 1$
$x - y + z = -1$

3.
$2x - y + 2z = 5$
$2x + y - z = -6$
$3x + 2z = 3$

4.
$x + 2y - z = 0$
$2x + 3y = 3$
$2y + z = -1$

5.
$3x + 2y + z = -5$
$2x - y - z = -6$
$2x + y + 3z = 4$

6.
$4x + 3y - z = 12$
$2x - 3y - z = -10$
$x + y - 2z = -5$

7.
$x + y + z = 3$
$x - 2y + z = -3$
$3x + 3z = 5$

8.
$x + y + z = 3$
$x - 2y + 3z = 5$
$5x - 4y + 11z = 20$

9.
$x + y + z = 3$
$x - 2y + z = -3$
$x + z = 1$

10.
$x - y - 2z = -1$
$x + 2y + z = 5$
$5x + 4y - z = 13$

11. Show that elementary row operation 3 (the replacement operation) yields an equivalent system for

$$2x - y + z = 3$$
$$x + y = -1$$
$$3x - y - 2z = 7$$

[*Hint:* Consider the first two equations of the system. Show that if (m, n, p) satisfies both these equations, then it satisfies the system consisting of the first equation and the sum of the first two equations, and conversely.]

APPLICATIONS

In Problems 12–18, use matrices to solve the system.

12. The sum of $8.50 is made up of nickels, dimes, and quarters. The number of dimes is equal to the number of quarters plus twice the number of nickels. The value of the dimes exceeds the combined value of the nickels and the quarters by $1.50. How many of each coin are there?

13. The Mechano Distributing Company has three types of vending machines, which dispense snacks as listed in the table. Mechano fills all the machines once a day and finds them all sold out before the next day. The total daily sales are candy, 760; peanuts, 380; sandwiches, 660. How many of each type of machine does Mechano have?

Mechano Distributing Company Data

Snack	Vending Machine Type		
	I	II	III
Candy	20	24	30
Peanuts	10	18	10
Sandwiches	0	30	30

14. Suppose the total daily income from the various types of machines in Problem 13 is as follows: type I, $32.00; type II, $159.00; type III, $192.00. What is the selling price for each type of snack? (Use your answers from Problem 13.)

15. Gro-Kwik Garden Supply has three types of fertilizer, which contain chemicals A, B, and C in the percentages shown in the table. In what proportions must Gro-Kwik mix these three types to get an 8-8-8 fertilizer (one that has 8% of each of the three chemicals)?

Gro-Kwik Garden Supply Data

Chemical	Amount of Fertilizer		
	I	II	III
A	6%	8%	12%
B	6%	12%	8%
C	8%	4%	12%

16. Three water supply valves, A, B, and C, are connected to a tank. If all three valves are opened, the tank is filled in 8 hr. The tank can also be filled by opening A for 8 hr and B for 12 hr, while keeping C closed, or by opening B for 10 hr and C for 28 hr, while keeping A closed. Find the time needed by each valve to fill the tank by itself. (*Hint:* Let x, y, and z, respectively, be the fractions of the tank that valves A, B, and C can fill alone in 1 hr.)

17. A 2 × 2 matrix (2 rows and 2 columns)

$$\begin{bmatrix} a & b \\ c & d \end{bmatrix}$$

is said to be **singular** if $ad - bc = 0$. Otherwise, it is nonsingular. Determine whether the following matrices are singular or nonsingular.

a. $\begin{bmatrix} 1 & 2 \\ 2 & 4 \end{bmatrix}$ **b.** $\begin{bmatrix} 2 & -3 \\ 3 & 5 \end{bmatrix}$ **c.** $\begin{bmatrix} 0 & 2 \\ 2 & 4 \end{bmatrix}$

18. The product of two matrices,

$$\begin{bmatrix} a & b \\ c & d \end{bmatrix} \quad \text{and} \quad \begin{bmatrix} x & y \\ z & w \end{bmatrix}$$

is found by a row-column multiplication defined as follows:

$$\begin{bmatrix} a & b \\ c & d \end{bmatrix} \begin{bmatrix} x & y \\ z & w \end{bmatrix} \quad \begin{bmatrix} ax + bz & ay + bw \\ cx + dz & cy + dw \end{bmatrix}$$

If this product equals

$$\begin{bmatrix} 1 & 0 \\ 0 & 1 \end{bmatrix}$$

each of the two matrices is said to be the **multiplicative inverse** of the other. To find the inverse of

$$\begin{bmatrix} a & b \\ c & d \end{bmatrix}$$

you have to find a matrix

$$\begin{bmatrix} x & y \\ z & w \end{bmatrix}$$

such that

$$\begin{bmatrix} a & b \\ c & d \end{bmatrix} \begin{bmatrix} x & y \\ z & w \end{bmatrix} = \begin{bmatrix} 1 & 0 \\ 0 & 1 \end{bmatrix}$$

This means that you must solve the systems

$$\begin{array}{ccc} ax + bz = 1 & & ay + bw = 0 \\ & \text{and} & \\ cx + dz = 0 & & cy + dw = 1 \end{array}$$

To solve the first system, we can write

$$\begin{bmatrix} a & b & | & 1 \\ c & d & | & 0 \end{bmatrix} \sim \begin{bmatrix} ac & bc & | & c \\ ac & ad & | & 0 \end{bmatrix} \sim \begin{bmatrix} ac & bc & | & c \\ 0 & ad - bc & | & -c \end{bmatrix}$$

Now we see that the second equation has a unique solution if and only if $ad - bc \neq 0$.

Use these ideas to find the inverse of

a. $\begin{bmatrix} 3 & 2 \\ 2 & 1 \end{bmatrix}$
b. $\begin{bmatrix} 1 & -2 \\ 2 & 1 \end{bmatrix}$

SKILL CHECKER

Evaluate:

19. $(8)(-5) - (22)(5)$

20. $(13)(-4) - (21)(3)$

21. $(2)(-4) - (5)(93)$

22. $\dfrac{(5)(-1) - (3)(1)}{(1)(-1) - (3)(1)}$

23. $\dfrac{(1)(3) - (3)(5)}{(1)(-3) - (1)(93)}$

24. $\dfrac{(-3)(1) - (-2)(5)}{(2)(1) - (-1)(-3)}$

USING YOUR KNOWLEDGE · **Finding Your Identity**

Instead of stopping with the echelon form of the matrix and then using back-substitution to solve a system of equations, many people prefer to transform the matrix of coefficients into the **identity matrix** (a square matrix with 1's along the diagonal and 0's everywhere else) and then read off the solution by inspection. If the final augmented matrix reads

$$\begin{bmatrix} 1 & 0 & 0 & | & a \\ 0 & 1 & 0 & | & b \\ 0 & 0 & 1 & | & c \end{bmatrix}$$

then the solution of the system is $x = a$, $y = b$, $z = c$. You need only read the column to the right of the vertical line.

Suppose, for example, that we have reduced the augmented matrix to the form

$$\begin{bmatrix} 2 & -1 & 2 & | & 3 \\ 0 & 3 & -3 & | & -3 \\ 0 & 0 & 2 & | & 5 \end{bmatrix}$$

We can now divide R_2 by 3 and R_3 by 2 to obtain

$$\begin{bmatrix} 2 & -1 & 2 & | & 3 \\ 0 & 1 & -1 & | & -1 \\ 0 & 0 & 1 & | & \frac{5}{2} \end{bmatrix}$$

To get 0's in the off-diagonal places of the matrix to the left of the vertical line, we first add R_2 to R_1 to get the new R_1:

$$\begin{bmatrix} 2 & 0 & 1 & | & 2 \\ 0 & 1 & -1 & | & -1 \\ 0 & 0 & 1 & | & \frac{5}{2} \end{bmatrix}$$

Next, we subtract R_3 from R_1 and add R_3 to R_2, with the result

$$\begin{bmatrix} 2 & 0 & 0 & | & -\frac{1}{2} \\ 0 & 1 & 0 & | & \frac{3}{2} \\ 0 & 0 & 1 & | & \frac{5}{2} \end{bmatrix}$$

Finally, we divide R_1 by 2 to obtain

$$\begin{bmatrix} 1 & 0 & 0 & | & -\frac{1}{4} \\ 0 & 1 & 0 & | & \frac{3}{2} \\ 0 & 0 & 1 & | & \frac{5}{2} \end{bmatrix}$$

from which we can see by inspection that the solution of the system is $x = -\frac{1}{4}$, $y = \frac{3}{2}$, $z = \frac{5}{2}$.

Keep in mind that after you bring the augmented matrix into echelon form, you perform additional elementary row operations to get 0's in all the off-diagonal places of the coefficient matrix and 1's on the diagonal. Of course, we assume that the system has a unique solution. If this is not the case, you will see what the situation is when you obtain the echelon form and you will also see that it's impossible to transform the coefficient matrix into the identity matrix if the system doesn't have a unique solution. Use these ideas to solve Problems 1–10 in Exercise 8.4.

If you know how to do these problems, you have learned your lesson!

WRITE ON . . .

When solving systems of equations using matrices, explain how you recognize:

25. A consistent system

26. An inconsistent system

27. A dependent system

28. Describe the relationship between elementary row operations on matrices and elementary operations on systems of equations.

Solve using matrices:

29. $2x + 3y - z = -1$
$3x + 4y + 2z = 14$
$x - 6y - 5z = 4$

30. $5x + 2y + 4z = -5$
$7x + 8y - 2z = 13$
$2x - 5y + 3z = 4$

31. $5x + 6y - 30z = 13$
$2x + 4y - 12z = 6$
$x + 2y - 6z = 8$

32. $x + y + z = 4$
$x - 2y + z = 7$
$2x - y + 2z = 11$

33. $3x - y + z = 3$
$x - 2y + 2z = 1$
$2x + y - z = 2$

34. $x - y + z = 3$
$x + 2y - z = 3$
$2x - 2y + 2z = 6$

8.5

DETERMINANTS AND CRAMER'S RULE

To succeed, review how to:

1. Evaluate expressions involving integers (pp. 42–44).

Objectives:

A Evaluate a 2 × 2 determinant.

B Use Cramer's rule to solve a system of two equations in two unknowns.

C Use minors to evaluate 3 × 3 determinants.

D Use Cramer's rule to solve a system of three equations.

getting started **Leibniz and Matrices**

Gottfried Wilhelm Leibniz developed the theory of determinants. In 1693, Leibniz studied and used *determinants* to solve systems of simultaneous equations. A **determinant** is a square array of numbers of the form

$$\begin{vmatrix} a_1 & b_1 \\ a_2 & b_2 \end{vmatrix}$$

The numbers $a_1, a_2, b_1,$ and b_2 are called the *elements* of the determinant. As you can see, this determinant has *two* rows and *two* columns. For this reason,

$$\det A = \begin{vmatrix} a_1 & b_1 \\ a_2 & b_2 \end{vmatrix}$$

is called a two-by-two (2 × 2) determinant. Every **square matrix** (a matrix with the same number of rows and columns) has a determinant associated with it.

Gottfried Wilhelm Leibniz
(1646–1716)

Evaluating 2 × 2 Determinants

The value of

$$\det A = \begin{vmatrix} a_1 & b_1 \\ a_2 & b_2 \end{vmatrix}$$

is defined to be

$$a_1 b_2 - a_2 b_1$$

which can be obtained by multiplying along the diagonals, as indicated in the following definition.

DETERMINANT

> The determinant of the matrix $\begin{bmatrix} a_1 & b_1 \\ a_2 & b_2 \end{bmatrix}$ is denoted by $\begin{vmatrix} a_1 & b_1 \\ a_2 & b_2 \end{vmatrix}$ and is defined as
>
> $$\begin{vmatrix} a_1 & b_1 \\ a_2 & b_2 \end{vmatrix} = a_1 b_2 - a_2 b_1$$

For example,

$$\begin{vmatrix} 2 & 5 \\ -3 & -9 \end{vmatrix} = (2)(-9) - (-3)(5) = -18 + 15 = -3$$

EXAMPLE 1   Evaluating determinants

Evaluate:

a. $\begin{vmatrix} -3 & 7 \\ -5 & 4 \end{vmatrix}$ **b.** $\begin{vmatrix} -3 & 6 \\ -5 & 10 \end{vmatrix}$

SOLUTION

a. $\begin{vmatrix} -3 & 7 \\ -5 & 4 \end{vmatrix} = (-3)(4) - (-5)(7) = -12 - (-35) = 23$

b. $\begin{vmatrix} -3 & 6 \\ -5 & 10 \end{vmatrix} = (-3)(10) - (-5)(6) = -30 - (-30) = 0$ ∎

Notice in Example 1(b) that the second-column elements of the determinant are both the same multiple of the corresponding first-column elements:

$$6 = (-2)(-3) \qquad \text{and} \qquad 10 = (-2)(-5)$$

In general, if the elements of one row are just some constant k times the elements of the other row (or if this is true of the columns), then the value of the determinant is zero. This is very easy to see:

$$\begin{vmatrix} a & b \\ ka & kb \end{vmatrix} = akb - bka = 0$$

This result will be of use to us in *Using Your Knowledge* in Exercise 8.5.

Using Cramer's Rule to Solve a System of Two Equations

One of the important applications of determinants is in the solution of a system of linear equations. Let's look at the simplest case, two equations in two unknowns. Such a system can be written

$$a_1x + b_1y = d_1 \qquad \textbf{(1)}$$

$$a_2x + b_2y = d_2 \qquad \textbf{(2)}$$

We can eliminate y from this system by multiplying equation (1) by b_2 and equation (2) by b_1 and then subtracting, as follows.

$$a_1b_2x + b_1b_2y = d_1b_2 \qquad \text{This is } b_2 \text{ times equation (1).}$$

$$\underline{(-)\, a_2b_1x + b_1b_2y = d_2b_1} \qquad \text{This is } b_1 \text{ times equation (2).}$$

$$a_1b_2x - a_2b_1x = d_1b_2 - d_2b_1 \qquad \text{Subtract the second equation from the first.}$$

$$(a_1b_2 - a_2b_1)x = d_1b_2 - d_2b_1 \qquad \text{Factor out } x.$$

Now if the quantity $a_1b_2 - a_2b_1 \neq 0$, we can divide by this quantity to get

$$x = \frac{d_1b_2 - d_2b_1}{a_1b_2 - a_2b_1} \qquad \text{Solve for } x.$$

Notice that the denominator, $a_1b_2 - a_2b_1$, which we shall denote by D, can be written as the determinant

$$D = \begin{vmatrix} a_1 & b_1 \\ a_2 & b_2 \end{vmatrix} = a_1b_2 - a_2b_1$$

This determinant is naturally called the **determinant of the coefficients**. Notice also that the numerator, $d_1b_2 - d_2b_1$, can be obtained from the denominator by replacing the coefficients of x (the a's) by the corresponding constant terms, the d's. We denote the numerator of x by D_x. Thus

Constant terms replace the a's. $\qquad D_x = \begin{vmatrix} d_1 & b_1 \\ d_2 & b_2 \end{vmatrix} \qquad$ y coefficients unchanged

We can now write the solution for x in the form

$$x = \frac{D_x}{D}$$

A similar procedure shows that the solution for y is

$$y = \frac{a_1d_2 - a_2d_1}{a_1b_2 - a_2b_1}$$

Notice that the denominator, $a_1b_2 - a_2b_1$, is again the determinant D. The numerator, $a_1d_2 - a_2d_1$, which we shall denote by D_y, can be formed from D by replacing the coefficients of y (the b's) by the corresponding d's. Hence

x coefficients unchanged $\qquad D_y = \begin{vmatrix} a_1 & d_1 \\ a_2 & d_2 \end{vmatrix} \qquad$ Constant terms replace the y coefficients.

and we can write the solution for y in the form

$$y = \frac{D_y}{D}$$

We summarize these results as follows.

CRAMER'S RULE FOR SOLVING A SYSTEM OF TWO EQUATIONS IN TWO UNKNOWNS

The system

$$a_1x + b_1y = d_1$$
$$a_2x + b_2y = d_2$$

has as its solution

$$x = \frac{D_x}{D} \quad \text{and} \quad y = \frac{D_y}{D}$$

where $D \neq 0$ and

$$D_x = \begin{vmatrix} d_1 & b_1 \\ d_2 & b_2 \end{vmatrix}, \quad D_y = \begin{vmatrix} a_1 & d_1 \\ a_2 & d_2 \end{vmatrix}, \quad D = \begin{vmatrix} a_1 & b_1 \\ a_2 & b_2 \end{vmatrix}$$

EXAMPLE 2   Using Cramer's rule to solve a system of two equations

Use Cramer's rule to solve the system:

$$2x + 3y = 7$$
$$5x + 9y = 11$$

SOLUTION

$$D = \begin{vmatrix} 2 & 3 \\ 5 & 9 \end{vmatrix} = 18 - 15 = 3 \qquad \text{Use the coefficients of the variables.}$$

$$D_x = \begin{vmatrix} 7 & 3 \\ 11 & 9 \end{vmatrix} = 63 - 33 = 30 \qquad \begin{array}{l} \text{These are the constant terms.} \\ \text{These are the coefficients of } y. \end{array}$$

$$D_y = \begin{vmatrix} 2 & 7 \\ 5 & 11 \end{vmatrix} = 22 - 35 = -13 \qquad \begin{array}{l} \text{These are the coefficients of } x. \\ \text{These are the constant terms.} \end{array}$$

Therefore,

$$x = \frac{D_x}{D} = \frac{30}{3} = 10 \qquad y = \frac{D_y}{D} = \frac{-13}{3} = -\frac{13}{3}$$

CHECK   Substituting $x = 10$, $y = -\frac{13}{3}$ in the equations, we obtain

$$(2)(10) + (3)\left(-\frac{13}{3}\right) = 20 - 13 = 7$$

$$(5)(10) + (9)\left(-\frac{13}{3}\right) = 50 - 39 = 11$$

Thus the answers do satisfy the given equations, and the solution is $x = 10$, $y = -\frac{13}{3}$, or $\left(10, -\frac{13}{3}\right)$.   ■

Using Minors to Evaluate 3 × 3 Determinants

We can also solve a system of three linear equations in three unknowns by using determinants. A 3 × 3 determinant—that is, a determinant with three rows and three columns—is written as

$$\det A = \begin{vmatrix} a_1 & b_1 & c_1 \\ a_2 & b_2 & c_2 \\ a_3 & b_3 & c_3 \end{vmatrix}$$

The value of a 3 × 3 determinant can be defined by using the idea of a *minor* of an element in a determinant. Here is the definition of minor.

MINOR

In the determinant

$$\begin{vmatrix} a_1 & b_1 & c_1 \\ a_2 & b_2 & c_2 \\ a_3 & b_3 & c_3 \end{vmatrix}$$

the **minor** of an element is the determinant that remains after deleting the row and column in which the element appears. Thus the minor of a_1 is

$$\begin{vmatrix} b_2 & c_2 \\ b_3 & c_3 \end{vmatrix}$$

the minor of b_1 is

$$\begin{vmatrix} a_2 & c_2 \\ a_3 & c_3 \end{vmatrix}$$

and the minor of c_1 is

$$\begin{vmatrix} a_2 & b_2 \\ a_3 & b_3 \end{vmatrix}$$

The value of a 3 × 3 determinant is defined as follows:

$$\begin{vmatrix} a_1 & b_1 & c_1 \\ a_2 & b_2 & c_2 \\ a_3 & b_3 & c_3 \end{vmatrix} = a_1 \begin{vmatrix} b_2 & c_2 \\ b_3 & c_3 \end{vmatrix} - b_1 \begin{vmatrix} a_2 & c_2 \\ a_3 & c_3 \end{vmatrix} + c_1 \begin{vmatrix} a_2 & b_2 \\ a_3 & b_3 \end{vmatrix}$$

The 2 × 2 determinants in this equation are clearly minors of the elements in the first row of the 3 × 3 determinant on the left; the right-hand side is called the **expansion** of the 3 × 3 determinant by minors along the first row. An expansion with numbers is carried out in Example 3.

EXAMPLE 3   Using minors to evaluate a 3 × 3 determinant

Expand by minors along the first row:

$$\begin{vmatrix} 1 & 1 & 1 \\ 1 & 2 & 1 \\ 1 & 1 & 2 \end{vmatrix}$$

SOLUTION

$$\begin{vmatrix} 1 & 1 & 1 \\ 1 & 2 & 1 \\ 1 & 1 & 2 \end{vmatrix} = (1)\begin{vmatrix} 2 & 1 \\ 1 & 2 \end{vmatrix} - (1)\begin{vmatrix} 1 & 1 \\ 1 & 2 \end{vmatrix} + (1)\begin{vmatrix} 1 & 2 \\ 1 & 1 \end{vmatrix}$$

$$= 3 - 1 - 1 = 1 \qquad \blacksquare$$

It's also possible to expand a determinant by the minors of *any row* or *any column*. To do this, it's necessary to define the *sign array* of a determinant.

SIGN ARRAY

> For a 3×3 determinant, the **sign array** is the following arrangement of alternating signs:
>
> $$\begin{array}{ccc} + & - & + \\ - & + & - \\ + & - & + \end{array}$$

To obtain the expansion of

$$\begin{vmatrix} a_1 & b_1 & c_1 \\ a_2 & b_2 & c_2 \\ a_3 & b_3 & c_3 \end{vmatrix}$$

along a particular row or column, we simply write the corresponding sign from the array in front of each term in the expansion. For example, to expand

$$\begin{vmatrix} 1 & 1 & 1 \\ 1 & 2 & 1 \\ 1 & 1 & 2 \end{vmatrix}$$

along the *second row,* we write

$$\begin{vmatrix} 1 & 1 & 1 \\ 1 & 2 & 1 \\ 1 & 1 & 2 \end{vmatrix} = -(1)\begin{vmatrix} 1 & 1 \\ 1 & 2 \end{vmatrix} + (2)\begin{vmatrix} 1 & 1 \\ 1 & 2 \end{vmatrix} - (1)\begin{vmatrix} 1 & 1 \\ 1 & 1 \end{vmatrix}$$

$$= -1(1) + 2(1) - 1(0) = 1$$

Note that we used the signs of the *second row* of the sign array, $- + -$, for the first, second, and third terms, respectively.

EXAMPLE 4   Expanding determinants by minors

Expand the given determinant along the third column:

a. $\begin{vmatrix} 0 & 1 & 1 \\ 1 & 2 & -1 \\ 1 & -1 & 3 \end{vmatrix}$

b. $\begin{vmatrix} 1 & 1 & 0 \\ 0 & -1 & 1 \\ 2 & -1 & -3 \end{vmatrix}$

SOLUTION

a. $\begin{vmatrix} 0 & 1 & 1 \\ 1 & 2 & -1 \\ 1 & -1 & 3 \end{vmatrix} = +(1)\begin{vmatrix} 1 & 2 \\ 1 & -1 \end{vmatrix} - (-1)\begin{vmatrix} 0 & 1 \\ 1 & -1 \end{vmatrix} + (3)\begin{vmatrix} 0 & 1 \\ 1 & 2 \end{vmatrix}$

$$= (1)(-3) + (1)(-1) + (3)(-1) = -3 - 1 - 3 = -7$$

b. $\begin{vmatrix} 1 & 1 & 0 \\ 0 & -1 & 1 \\ 2 & -1 & -3 \end{vmatrix} = +(0)\begin{vmatrix} 0 & -1 \\ 2 & -1 \end{vmatrix} - (1)\begin{vmatrix} 1 & 1 \\ 2 & -1 \end{vmatrix} + (-3)\begin{vmatrix} 1 & 1 \\ 0 & -1 \end{vmatrix}$

$$= 0 - (1)(-3) + (-3)(-1) = 6 \qquad \blacksquare$$

NOTE A last word before you do the problems. When expanding a 3×3 determinant, it's easier to expand along the row or column containing the most zeros, since the coefficient of the resulting minors would then be zero. Do it this way—trust us, it will save you time.

Cramer's Rule for Solving a System of Three Equations

We can solve the system

$$a_1x + b_1y + c_1z = d_1$$
$$a_2x + b_2y + c_2z = d_2$$
$$a_3x + b_3y + c_3z = d_3$$

in exactly the same manner as we solved the system of equations with two unknowns. The details are similar to those for a system of two equations. We state these results here.

CRAMER'S RULE FOR SOLVING A SYSTEM OF THREE EQUATIONS IN THREE UNKNOWNS

The system

$$a_1x + b_1y + c_1z = d_1$$
$$a_2x + b_2y + c_2z = d_2$$
$$a_3x + b_3y + c_3z = d_3$$

1. Has the unique solution

$$x = \frac{D_x}{D}, \qquad y = \frac{D_y}{D}, \qquad z = \frac{D_z}{D}$$

where

$$D = \begin{vmatrix} a_1 & b_1 & c_1 \\ a_2 & b_2 & c_2 \\ a_3 & b_3 & c_3 \end{vmatrix}, \qquad D_x = \begin{vmatrix} d_1 & b_1 & c_1 \\ d_2 & b_2 & c_2 \\ d_3 & b_3 & c_3 \end{vmatrix},$$

$$D_y = \begin{vmatrix} a_1 & d_1 & c_1 \\ a_2 & d_2 & c_2 \\ a_3 & d_3 & c_3 \end{vmatrix}, \qquad \text{and} \qquad D_z = \begin{vmatrix} a_1 & b_1 & d_1 \\ a_2 & b_2 & d_2 \\ a_3 & b_3 & d_3 \end{vmatrix}, \qquad D \neq 0$$

2. Is *inconsistent* and has no solution if $D = 0$ and any one of D_x, D_y, D_z is different from zero.

3. Has *no unique* solution if $D = 0$ and $D_x = D_y = D_z = 0$. (In this case, the system either has no solution or infinitely many solutions. This situation is studied more fully in advanced algebra.)

Note carefully that D is the determinant of the coefficients; D_x is formed from D by replacing the coefficients of x (the a's) by the corresponding d's; D_y is formed from D by replacing the coefficients of y (the b's) by the corresponding d's; and D_z is formed similarly.

EXAMPLE 5   Using Cramer's rule to solve a system of three equations in three unknowns

Use Cramer's rule to solve the system:

$$x + y + 2z = 7$$
$$x - y - 3z = -6$$
$$2x + 3y + z = 4$$

SOLUTION   To use Cramer's rule, we first have to evaluate D, the determinant of the coefficients. If D is not zero, then we calculate the other three determinants, D_x, D_y, and D_z. For the given system,

$$D = \begin{vmatrix} 1 & 1 & 2 \\ 1 & -1 & -3 \\ 2 & 3 & 1 \end{vmatrix} = (1)\begin{vmatrix} -1 & -3 \\ 3 & 1 \end{vmatrix} - (1)\begin{vmatrix} 1 & -3 \\ 2 & 1 \end{vmatrix} + (2)\begin{vmatrix} 1 & -1 \\ 2 & 3 \end{vmatrix}$$

$$= (1)(-1 + 9) - (1)(1 + 6) + (2)(3 + 2)$$

$$= 8 - 7 + 10 = 11$$

$$D_x = \begin{vmatrix} 7 & 1 & 2 \\ -6 & -1 & -3 \\ 4 & 3 & 1 \end{vmatrix} = (7)\begin{vmatrix} -1 & -3 \\ 3 & 1 \end{vmatrix} - (1)\begin{vmatrix} -6 & -3 \\ 4 & 1 \end{vmatrix} + (2)\begin{vmatrix} -6 & -1 \\ 4 & 3 \end{vmatrix}$$

$$= (7)(-1 + 9) - (1)(-6 + 12) + (2)(-18 + 4)$$

$$= 56 - 6 - 28 = 22$$

$$D_y = \begin{vmatrix} 1 & 7 & 2 \\ 1 & -6 & -3 \\ 2 & 4 & 1 \end{vmatrix} = (1)\begin{vmatrix} -6 & -3 \\ 4 & 1 \end{vmatrix} - 7\begin{vmatrix} 1 & -3 \\ 2 & 1 \end{vmatrix} + (2)\begin{vmatrix} 1 & -6 \\ 2 & 4 \end{vmatrix}$$

$$= (1)(-6 + 12) - (7)(1 + 6) + (2)(4 + 12)$$

$$= 6 - 49 + 32 = -11$$

$$D_z = \begin{vmatrix} 1 & 1 & 7 \\ 1 & -1 & -6 \\ 2 & 3 & 4 \end{vmatrix} = (1)\begin{vmatrix} -1 & -6 \\ 3 & 4 \end{vmatrix} - (1)\begin{vmatrix} 1 & -6 \\ 2 & 4 \end{vmatrix} + (7)\begin{vmatrix} 1 & -1 \\ 2 & 3 \end{vmatrix}$$

$$= (1)(-4 + 18) - (1)(4 + 12) + (7)(3 + 2)$$

$$= 14 - 16 + 35 = 33$$

Thus by Cramer's rule,

$$x = \frac{D_x}{D} = \frac{22}{11} = 2, \qquad y = \frac{D_y}{D} = \frac{-11}{11} = -1, \qquad z = \frac{D_z}{D} = \frac{33}{11} = 3$$

You can check the solution $(2, -1, 3)$ by substituting in the given equations.   ∎

EXAMPLE 6　　More practice using Cramer's rule to solve a system of three equations in three unknowns

Use Cramer's rule to solve the system:

$$x + y - z = 2$$
$$2x + y + z = 4$$
$$-x - y + z = 3$$

SOLUTION　By Cramer's rule, if there is a unique solution, it is given by

$$x = \frac{D_x}{D}, \qquad y = \frac{D_y}{D}, \qquad \text{and} \qquad z = \frac{D_z}{D}$$

However,

$$D = \begin{vmatrix} 1 & 1 & -1 \\ 2 & 1 & 1 \\ -1 & -1 & 1 \end{vmatrix} = (1)\begin{vmatrix} 1 & 1 \\ -1 & 1 \end{vmatrix} - (1)\begin{vmatrix} 2 & 1 \\ -1 & 1 \end{vmatrix} + (-1)\begin{vmatrix} 2 & 1 \\ -1 & -1 \end{vmatrix}$$

$$= 2 - 3 + 1 = 0$$

and

$$D_x = \begin{vmatrix} 2 & 1 & -1 \\ 4 & 1 & 1 \\ 3 & -1 & 1 \end{vmatrix}$$

$$= (2)\begin{vmatrix} 1 & 1 \\ -1 & 1 \end{vmatrix} - (1)\begin{vmatrix} 4 & 1 \\ 3 & 1 \end{vmatrix} + (-1)\begin{vmatrix} 4 & 1 \\ 3 & -1 \end{vmatrix}$$

$$= 4 - 1 + 7 = 10 \neq 0$$

Hence the system is inconsistent and has no solution. ∎

GRAPH IT

Most graphers have a feature that calculates the determinant of a matrix. Let's start by erasing any matrices you may have in memory by pressing [2nd] [MEM] 2 4 [ENTER]. (Some graphers let you clear a matrix by storing zero to that matrix. Press 0 [STO] [2nd] [A] and you get the 0 matrix.) In any event, let's enter the matrix A in Example 1a and find its determinant. As you recall,

$$A = \begin{vmatrix} -3 & 7 \\ -5 & 4 \end{vmatrix}$$

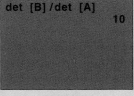

WINDOW 1
Det [A] from Example 1

Press [MATRX] [▶] [▶] [ENTER] to edit matrix A. Now enter 2 × 2 for the number of rows and columns and then the values of the matrix A. Leave the home screen by pressing [2nd] [QUIT]. To find the determinant on the TI-82, press [MATRX] [▶] 1 [MATRX] 1 [ENTER] and you get the answer 23 (Window 1). On the TI-81, press [MATRX] 5 [2nd] [A] and you get the same result.

Now let's do Example 2. Since the grapher uses the matrices A, B, C, D, and so on, let $A = D$, $B = D_x$, and $C = D_y$. Enter the values for A, B, and C as defined in Example 2. In your home screen, matrices A and B should look like those shown in Window 2. Since $x = \frac{B}{A}$, press [MATRX] [▶] 1 [MATRX] 2 (this will enter det [B] on the home screen). Then press [÷] [MATRX] [▶] 1 [MATRX] 1 [ENTER] (this will divide by det [A]). The result is 10, as shown in Window 3. You can use the same procedure to find y. Now that you know how to find determinants, you can do the rest of the examples!

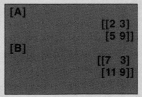

WINDOW 2
Matrices A and B from Example 2

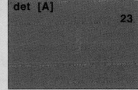

WINDOW 3
Finding the solution for x in Example 2

EXERCISE 8.5

A In Problems 1–10, evaluate the determinant.

1. $\begin{vmatrix} 1 & 1 \\ 0 & 2 \end{vmatrix}$

2. $\begin{vmatrix} 2 & -1 \\ 4 & 3 \end{vmatrix}$

3. $\begin{vmatrix} -3 & -2 \\ 5 & 1 \end{vmatrix}$

4. $\begin{vmatrix} 2 & -1 \\ -3 & 1 \end{vmatrix}$

5. $\begin{vmatrix} -2 & 0 \\ 5 & -3 \end{vmatrix}$

6. $\begin{vmatrix} 5 & 2 \\ -10 & -4 \end{vmatrix}$

7. $\begin{vmatrix} \dfrac{1}{2} & \dfrac{-1}{4} \\ \dfrac{1}{2} & \dfrac{3}{4} \end{vmatrix}$

8. $\begin{vmatrix} \dfrac{1}{5} & \dfrac{1}{10} \\ \dfrac{1}{2} & \dfrac{1}{4} \end{vmatrix}$

9. $\begin{vmatrix} \dfrac{3}{5} & \dfrac{1}{2} \\ \dfrac{-1}{4} & \dfrac{-1}{2} \end{vmatrix}$

10. $\begin{vmatrix} \dfrac{4}{5} & \dfrac{-1}{3} \\ \dfrac{-1}{2} & \dfrac{1}{2} \end{vmatrix}$

B In Problems 11–30, solve the system using Cramer's rule. If the system is dependent or inconsistent, state that fact.

11. $x + y = 5$
    $3x - y = 3$

12. $x + y = 9$
    $x - y = 3$

13. $x + y = 9$
    $x - y = -1$

14. $2x + y = -1$
    $x - 2y = -13$

15. $4x + 9y = 3$
    $3x + 7y = 2$

16. $5x + 2y = 32$
    $3x + y = 18$

17. $x - y = -1$
    $x - 2y = -6$

18. $x - 2y = -13$
    $3x - 2y = -19$

19. $2x + 3y = -13$
    $6x + 9y = -39$

20. $4x + 5y = -2$
    $12x + 15y = -6$

21. $x - y = 1$
    $x - 2y = 4$

22. $x - 2y = 4$
    $4x - 5y = 7$

23. $x + 3y = 6$
    $2x + 6y = 5$

24. $x - 2y = 3$
    $-x + 2y = 6$

25. $x = 7y + 3$
    $2x + 3y = 23$

26. $x = 3y + 1$
    $2x + 3y = 20$

27. $y = -3x + 17$
    $2x - y = 8$

28. $y = -2x + 14$
    $3x - y = 11$

29. $\dfrac{x}{2} - \dfrac{y}{3} = \dfrac{-1}{6}$
    $\dfrac{x}{3} + \dfrac{y}{4} = \dfrac{-7}{12}$

30. $\dfrac{x}{3} - \dfrac{y}{5} = \dfrac{4}{3}$
    $\dfrac{x}{4} - \dfrac{y}{3} = \dfrac{1}{12}$

C In Problems 31–40, evaluate the determinant.

31. $\begin{vmatrix} 1 & 3 & 2 \\ 2 & 4 & 1 \\ 3 & 6 & 5 \end{vmatrix}$

32. $\begin{vmatrix} 1 & 3 & 5 \\ 2 & 0 & 10 \\ -3 & 1 & -15 \end{vmatrix}$

33. $\begin{vmatrix} 1 & 2 & 3 \\ 4 & 5 & 6 \\ 7 & 8 & 9 \end{vmatrix}$

34. $\begin{vmatrix} 1 & 1 & 1 \\ 2 & 3 & 1 \\ 2 & 4 & 1 \end{vmatrix}$

35. $\begin{vmatrix} 2 & 1 & 3 \\ 1 & 2 & -1 \\ 3 & 1 & 5 \end{vmatrix}$

36. $\begin{vmatrix} -1 & 1 & -1 \\ -2 & 2 & -6 \\ 3 & -3 & 4 \end{vmatrix}$

37. $\begin{vmatrix} 1 & 1 & 6 \\ 1 & 1 & 4 \\ 1 & -1 & 2 \end{vmatrix}$

38. $\begin{vmatrix} 1 & 4 & 0 \\ 1 & -3 & 1 \\ 0 & 8 & -1 \end{vmatrix}$

39. $\begin{vmatrix} 0 & -1 & 2 \\ 2 & 1 & -3 \\ 1 & -3 & 1 \end{vmatrix}$

40. $\begin{vmatrix} -3 & 2 & -4 \\ 1 & -1 & 3 \\ 1 & 2 & 10 \end{vmatrix}$

D In Problems 41–50, solve the system using Cramer's rule. If the system is inconsistent or has no unique solution, state that fact. You can use minors to expand the resulting determinants.

41. $x + y + z = 6$
    $2x - 3y + 3z = 5$
    $3x - 2y - z = -4$

42. $x + y + z = 13$
    $3x + y - 3z = 5$
    $x - 2y + 4z = 10$

43. $6x + 5y + 4z = 5$
    $5x + 4y + 3z = 5$
    $4x + 3y + z = 7$

44. $3x + 2y + z = 4$
    $4x + 3y + z = 5$
    $5x + y + z = 9$

45. $x - 2y + 3z = 15$
    $5x + 7y - 11z = -29$
    $-13x + 17y + 19z = 37$

46. $2x - y + z = 3$
    $x + 2y + z = 12$
    $4x - 3y + z = 1$

47. $5x + 3y + 5z = 3$
    $3x + 5y + z = -5$
    $2x + 2y + 3z = 7$

48. $x + y = 5$
    $y + z = 3$
    $x + z = 7$

49. $2y + z = 9$
    $-2y + z = 1$
    $x + y + z = 1$

50. $x - y = 3$
    $y - z = 3$
    $x + z = 9$

In Problems 51–60, show that the statement is true.

51. $\begin{vmatrix} a & b & 0 \\ c & d & 0 \\ e & f & 0 \end{vmatrix} = 0$

52. $\begin{vmatrix} a & b & c \\ d & e & f \\ 0 & 0 & 0 \end{vmatrix} = 0$

53. $\begin{vmatrix} a & b & c \\ 1 & 2 & 3 \\ a & b & c \end{vmatrix} = 0$ **54.** $\begin{vmatrix} 1 & a & a \\ 2 & b & b \\ 3 & c & c \end{vmatrix} = 0$

55. $\begin{vmatrix} 1 & 2 & 3 \\ 3 & 1 & 2 \\ k & 2k & 3k \end{vmatrix} = k\begin{vmatrix} 1 & 2 & 3 \\ 3 & 1 & 2 \\ 1 & 2 & 3 \end{vmatrix}$

56. $\begin{vmatrix} 1 & 2 & 3k \\ 3 & 2 & k \\ 0 & 1 & 2k \end{vmatrix} = k\begin{vmatrix} 1 & 2 & 3 \\ 3 & 2 & 1 \\ 0 & 1 & 2 \end{vmatrix}$

57. $\begin{vmatrix} kb_1 & b_1 & 1 \\ kb_2 & b_2 & 2 \\ kb_3 & b_3 & 3 \end{vmatrix} = 0$ **58.** $\begin{vmatrix} b_1 & b_2 & b_3 \\ kb_1 & kb_2 & kb_3 \\ 1 & 2 & 3 \end{vmatrix} = 0$

59. $\begin{vmatrix} 1 & 1 & 1 \\ 2 & a & a \\ 3 & b & b \end{vmatrix} = 0$ **60.** $\begin{vmatrix} 0 & 0 & 0 \\ a & b & c \\ d & e & f \end{vmatrix} = 0$

SKILL CHECKER

Factor:

61. $x^2 + 4x + 3$ **62.** $x^2 - 4x + 4$

63. $x^2 + 2x - 3$ **64.** $-x^2 - 4x - 3$

65. $-x^2 + 4x - 3$ **66.** $-x^2 + 2x + 15$

USING YOUR KNOWLEDGE **Determining the Equations of Lines**

Determinants provide a very convenient and neat way of writing the equation of a line through two points. This is one of the problems that we studied in Section 7.3. Suppose the line is to pass through the points (x_1, y_1) and (x_2, y_2); then an equation of the line can be written in the form

$$\begin{vmatrix} x & y & 1 \\ x_1 & y_1 & 1 \\ x_2 & y_2 & 1 \end{vmatrix} = 0$$

It's quite easy to verify this fact. First, think of expanding the determinant by minors along the first row. The coefficients of x and y will be constants, so the equation is linear. Next, you can see that (x_1, y_1) and (x_2, y_2) both satisfy the equation, because if you substitute either of these pairs for (x, y) in the first row of the determinant, the result is a determinant with two identical rows and this you know is zero. See Problem 53. Thus the equation is that of the desired line.

As an illustration, let's find an equation of the line through $(1, 3)$ and $(-5, -2)$. The determinant form of this equation is

$$\begin{vmatrix} x & y & 1 \\ 1 & 3 & 1 \\ -5 & -2 & 1 \end{vmatrix} = 0$$

or

$$[3 - (-2)]x - [1 - (-5)]y + [-2 - (-15)](1) = 0$$

where the quantities in brackets are the expanded minors of the first-row elements. The final equation is

$$5x - 6y + 13 = 0$$

In Problems 67–72, use the determinant method to find the equation of the line through the two given points.

67. $(2, 7)$ and $(0, 3)$

68. $(10, 12)$ and $(-7, 1)$

69. $(-1, 4)$ and $(8, 2)$

70. $(5, 0)$ and $(0, -3)$

71. $(a, 0)$ and $(0, b)$, $ab \neq 0$

72. (a, b) and $(0, 0)$, $(a, b) \neq (0, 0)$

WRITE ON . . .

How can you tell:

73. If a system of equations is consistent when using determinants to solve it?

74. If a system of equations is inconsistent when using determinants to solve it?

75. If a system of equations is dependent when using determinants to solve it?

76. If a system of linear equations has a determinant $D = 0$, what do you know about the system?

77. Explain why a determinant that has a column or row of zeros has a value of zero. (See Problems 51 and 52.)

MASTERY TEST

If you know how to do these problems, you have learned your lesson!

Expand the determinant by minors:

78. Along the third row

$$\begin{vmatrix} 0 & 1 & 1 \\ 1 & 2 & -1 \\ 1 & -1 & 3 \end{vmatrix}$$

79. Along the third row

$$\begin{vmatrix} 1 & 1 & 0 \\ 0 & -1 & 1 \\ 2 & -1 & -3 \end{vmatrix}$$

80. Along the first column

$$\begin{vmatrix} 1 & 1 & 1 \\ 1 & 2 & 1 \\ 1 & 1 & 2 \end{vmatrix}$$

Use Cramer's rule to solve the system:

81. $\begin{aligned} x - y - z &= 2 \\ x + 2y + z &= 6 \\ -x + y + z &= 4 \end{aligned}$

82. $\begin{aligned} x + y + z &= 6 \\ x - y + z &= 2 \\ x + y - 2z &= 4 \end{aligned}$

83. Evaluate the determinant

$$\begin{vmatrix} 2 & -3 & 4 \\ -1 & 2 & 1 \\ 1 & 1 & -2 \end{vmatrix}$$

84. Use Cramer's rule to solve the system:

$$\begin{aligned} 2x + 3y &= 13 \\ 5x - 4y &= 21 \end{aligned}$$

Evaluate:

85. $\begin{vmatrix} -2 & 6 \\ -3 & 5 \end{vmatrix}$

86. $\begin{vmatrix} -2 & 8 \\ -3 & 12 \end{vmatrix}$

research questions

Sources of information for these questions can be found in the *Bibliography* at the end of this book.

1. Leibniz was probably the first person to use elimination and determinants to solve a system of three equations in three unknowns. But the evidence of a systematic method of solving a system of three equations appeared in China probably in about 250 B.C. Here is the problem:

There are three grades of corn. After threshing, three bundles of top grade, two bundles of medium grade, and one bundle of low grade make 39 dou (a measure of volume). Two bundles of top grade, three bundles of medium grade, and one bundle of low grade will produce 34 dou. The yield of one bundle of top grade, two bundles of medium grade, and three bundles of low grade is 26 dou. How many dou are contained in each bundle of each grade?

a. Write a system of three equations in three unknowns representing this situation.

b. Find out the name of the book where this problem originated and write a short paragraph detailing the contents of the book.

c. Write a short paragraph explaining how the Chinese solved this problem.

2. Who introduced the term *matrix* into mathematical literature?

3. The mathematician viewed as the "creator of the theory of matrices" wrote a book called *A Memoir on the Theory of Matrices*. Who was this mathematician, what were his contributions to the study of matrices, and what was the name of the only theorem appearing in the book?

4. Who really invented Cramer's rule? Write a paragraph supporting your findings.

5. Write a report detailing the uses of matrices in science with particular emphasis on the work of Werner Heisenberg, W. J. Duncan, and A. R. Collar.

6. Who received the 1973 Nobel prize for economics, and how did he use matrices?

SUMMARY

SECTION	ITEM	MEANING	EXAMPLE
8.1A	Consistent system	Graphs intersect at one point. There is one solution.	$2x - y = 2$ and $y = x - 1$ form a consistent system intersecting at $(1, 0)$.
	Inconsistent system	Graphs are parallel lines. There is no solution.	$y - 2x = 4$ and $3y - 6x = 18$ form an inconsistent system.
	Dependent system	Graphs coincide. There are infinitely many solutions.	$2x + \frac{1}{2}y = 2$ and $y = -4x + 4$ form a dependent system.
8.1B	Substitution method (two variables)	A method where one equation is solved for a variable that is substituted into the other equation.	$y = 2x$ and $2x + y = 4$ can be solved by substituting $y = 2x$ into $2x + y = 4$ to obtain $$2x + 2x = 4 \text{ or}$$ $$4x = 4$$ $$x = 1$$ Thus $y = 2 \cdot 1 = 2$.
8.1C	Elimination method (two variables)	A method where equations are multiplied by suitable numbers so that addition eliminates one of the variables	For the system $$x - 2y = 4$$ $$x + y = 6$$ multiplying the second equation by 2 yields $$x - 2y = 4$$ $$2x + 2y = 12$$ so that addition eliminates the y, leaving $3x = 16$ or $x = \frac{16}{3}$.
8.2A	Elimination method (three variables)	A method in which two equations are selected and one variable is eliminated. Then a different pair of equations is selected, and the same variable is eliminated. The system is then solved as in 8.1C.	Consider the system $$2x - y + z = 4 \quad (1)$$ $$-x - y - z = 0 \quad (2)$$ $$-x + 2y - z = 2 \quad (3)$$ Add (1) and (2), then add (1) and (3). We get $$x - 2y = 4$$ $$x + y = 6$$ Solve this system as in 8.2A.

SECTION	ITEM	MEANING	EXAMPLE
8.2B	Consistent system	A system with one solution consisting of an ordered triple of the form (x, y, z)	The system $x + y + z = 6$ $x - y - z = -4$ $x + y - z = 0$ is consistent. The solution is $(1, 2, 3)$.
	Inconsistent system	A system with no solution	The system $x + y + z = 4$ $-x - y - z = 3$ $x + 2y + z = 5$ is inconsistent.
	Dependent system	A system with infinitely many solutions	The system $x + y + z = 4$ $-x - y - z = -4$ $x + 2y - z = 5$ is dependent.
8.4	Matrix	A rectangular array of numbers	$\begin{bmatrix} 2 & -3 \\ -1 & 0 \end{bmatrix}$ is a matrix.
	Augmented matrix	A matrix consisting of the coefficients of the variables and the constants in a system of equations	$\begin{bmatrix} 1 & 2 & 3 & -1 \\ 2 & 3 & 4 & 0 \\ -1 & -2 & 3 & 5 \end{bmatrix}$ is the augmented matrix for the system $x + 2y + 3z = -1$ $2x + 3y + 4z = 0$ $-x - 2y + 3z = 5$
8.5	Determinant	$\det A = \begin{vmatrix} a_1 & b_1 \\ a_2 & b_2 \end{vmatrix}$	$\begin{vmatrix} 2 & -4 \\ -3 & 5 \end{vmatrix}$ is a 2 × 2 determinant.
8.5A	The value of det A	$a_1b_2 - a_2b_1$	The value of $\begin{vmatrix} 2 & -4 \\ -3 & 5 \end{vmatrix}$ is $(2)(5) - (-3)(-4) = -2$.
8.5B	Cramer's rule for a 2 × 2 system	The solution to the system $a_1x + b_1y = d_1$ $a_2x + b_2y = d_2$ is given by $x = \dfrac{D_x}{D}$ and $y = \dfrac{D_y}{D}$ $(D \neq 0)$ where $D = \begin{vmatrix} a_1 & b_1 \\ a_2 & b_2 \end{vmatrix}$, $D_x = \begin{vmatrix} d_1 & b_1 \\ d_2 & b_2 \end{vmatrix}$, and $D_y = \begin{vmatrix} a_1 & d_1 \\ a_2 & d_2 \end{vmatrix}$	The solution of $2x + 3y = 7$ $5x + 9y = 11$ is $x = \dfrac{D_x}{D} = \dfrac{\begin{vmatrix} 7 & 3 \\ 11 & 9 \end{vmatrix}}{\begin{vmatrix} 2 & 3 \\ 5 & 9 \end{vmatrix}} = \dfrac{30}{3} = 10$ $y = \dfrac{D_y}{D} = \dfrac{\begin{vmatrix} 2 & 7 \\ 5 & 11 \end{vmatrix}}{\begin{vmatrix} 2 & 3 \\ 5 & 9 \end{vmatrix}} = \dfrac{-13}{3}$

SECTION	ITEM	MEANING	EXAMPLE
8.5C	Minor	The minor of an element of a determinant is the determinant that remains after deleting the row and column in which the element appears.	The minor of 6 in $\begin{vmatrix} 3 & 4 & 6 \\ 1 & 2 & 3 \\ 0 & 1 & 4 \end{vmatrix}$ is $\begin{vmatrix} 1 & 2 \\ 0 & 1 \end{vmatrix}$.
	The value of $\begin{vmatrix} a_1 & b_1 & c_1 \\ a_2 & b_2 & c_2 \\ a_3 & b_3 & c_3 \end{vmatrix}$	$a_1 \begin{vmatrix} b_2 & c_2 \\ b_3 & c_3 \end{vmatrix} - b_1 \begin{vmatrix} a_2 & c_2 \\ a_3 & c_3 \end{vmatrix} + c_1 \begin{vmatrix} a_2 & b_2 \\ a_3 & b_3 \end{vmatrix}$	

REVIEW EXERCISES

(If you need help with these exercises, look in the section indicated in brackets.)

1. **[8.1A]** Use the graphical method to solve the system.
   a. $2x - y = 2$
   $y = 3x - 4$
   b. $x - 2y = 0$
   $y = x - 2$

2. **[8.1A]** Use the graphical method to solve the system.
   a. $2y - x = 3$
   $4y = 2x + 7$
   b. $3y + x = 5$
   $2x = 8 - 6y$

3. **[8.1A]** Use the graphical method to solve the system.
   a. $3x + 2y = 6$
   $y = 3 - \dfrac{3}{2}x$
   b. $x + 2y = 4$
   $2x = 8 - 4y$

4. **[8.1B]** Solve by the substitution method.
   a. $2x - y = 4$
   $x + y = 5$
   b. $2x + 3y = 10$
   $x - y = -1$

5. **[8.1B]** Solve by the substitution method.
   a. $2x + 4y = 7$
   $x = -2y - 1$
   b. $2y + x = 5$
   $3x = 10 - 6y$

6. **[8.1C]** Solve by the substitution method.
   a. $2y - x = 5$
   $2x = 4y - 10$
   b. $x + 5y = 5$
   $y = 1 - \dfrac{x}{5}$

7. **[8.1C]** Solve by the elimination method.
   a. $x - 3y = 7$
   $2x - y = 9$
   b. $2x + 3y = 4$
   $x + y = 1$

8. **[8.1C]** Solve by the elimination method.
   a. $2x + 3y = 7$
   $6x + 9y = 14$
   b. $3x - 4y = 5$
   $6x - 8y = 15$

9. **[8.1C]** Solve by the elimination method.
   a. $\dfrac{x}{5} + \dfrac{y}{2} = \dfrac{1}{5}$
   $2x + 5y = 2$
   b. $\dfrac{x}{3} - \dfrac{y}{4} = 2$
   $\dfrac{x}{6} - \dfrac{y}{8} = 1$

10. **[8.1C]** Solve the system.
   a. $4y = -5x - 2$
   $4x = -3y - 1$
   b. $2x = 3y - 1$
   $2y = 3x - 1$

11. **[8.2A]** Solve the system.
   a. $x - y + z = 4$
   $x + y - z = 0$
   $2x - y + z = 6$
   b. $2x - 3y + z = 3$
   $2x + 3y + z = -3$
   $2x - 9y + z = 9$

12. **[8.3A]** Solve the system.
   a. $2x + y - 2z = 4$
   $-x + y + 2z = 2$
   $3y + 2z = 12$
   b. $2x + 2y + z = 4$
   $-2x - y + z = 2$
   $-2x + 3z = 9$

13. **[8.2A]** Solve the system.

    a. $x + 2y + 3z = 6$ **b.** $x + 2y \quad\ = 4$
       $x - 2y - z = -2$ $y + 2z = 6$
       $x \quad\ + z = 2$ $2x + 2y - 4z = -4$

14. **[8.3A]** How many of each coin do Joey and Alice have?

    a. Joey has $4 in nickels and dimes, and he has five more nickels than dimes.

    b. Alice has $2 in nickels and dimes, and she has five fewer nickels than dimes.

15. **[8.3B]** How tall are these buildings?

    a. The total height of a building and a flagpole on the roof is 200 ft. The building is nine times as high as the flagpole.

    b. The total height of a building and a flagpole on the roof is 180 ft. The building is eight times as high as the flagpole.

16. **[8.3C]** Find the speed of each current.

    a. A motorboat can go 12 mi downstream on a river in 20 min. It takes this boat 30 min to go upstream the same 12 mi.

    b. A motorboat can go 6 mi downstream on a river in 15 min. It takes this boat 20 min to go upstream the same 6 mi.

17. **[8.3D]** Find out how much Bill and Betty have invested at each rate.

    a. Bill has three investments totaling $40,000. These investments earn interest at 4%, 6%, and 8%. Bill's annual income from these investments is $2600. The income from the 8% investment exceeds the total income from the other two investments by $600.

    b. Betty has three investments totaling $45,000. These investments earn interest at 4%, 6%, and 8%. Betty's annual income from these investments is $2900. The income from the 8% investment exceeds the total income from the other two investments by $300.

18. **[8.3E]** What are the dimensions of each rectangle?

    a. The perimeter of a rectangle is 100 in. and the length is 30 in. more than the width.

    b. The perimeter of a rectangle is 80 in. and the length is three times the width.

19. **[8.4B]** Use matrices to solve the system.

    a. $2x - y - z = 3$ **b.** $2x - 6y + 2z = 4$
       $x + y + z = 6$ $2x + y - 4z = 6$
       $3x + 2y + z = 10$ $-x + 3y - z = -2$

20. **[8.4B]** Use matrices to solve the system.

    a. $3x + y - 2z = 1$ **b.** $x + y + z = 2$
       $9x + 3y - 6z = 6$ $2x - y + z = -1$
       $-2x - y + 3z = -1$ $x - y - z = 0$

21. **[8.5A]** Evaluate.

    a. $\begin{vmatrix} 3 & 5 \\ 2 & -4 \end{vmatrix}$ **b.** $\begin{vmatrix} -4 & 5 \\ -6 & 4 \end{vmatrix}$

22. **[8.5B]** Solve by Cramer's rule.

    a. $2x + 5y = -8$ **b.** $4x + 2y = 1$
       $3x - 4y = 11$ $2x - 6y = 4$

23. **[8.5C]** Evaluate.

    a. $\begin{vmatrix} 1 & -2 & -2 \\ 3 & 0 & -1 \\ 4 & 1 & 2 \end{vmatrix}$ **b.** $\begin{vmatrix} 0 & 2 & 4 \\ 1 & 2 & 0 \\ 2 & 1 & 3 \end{vmatrix}$

24. **[8.5C]** Expand by minors along the first row.

    a. $\begin{vmatrix} 1 & -1 & 1 \\ 2 & 3 & 1 \\ 1 & 3 & 2 \end{vmatrix}$ **b.** $\begin{vmatrix} 4 & -2 & -1 \\ 2 & 5 & -2 \\ 1 & -2 & 2 \end{vmatrix}$

25. **[8.5C]** Expand by minors along the second column.

    a. $\begin{vmatrix} 1 & 0 & 5 \\ 3 & 2 & 1 \\ 5 & 3 & -1 \end{vmatrix}$ **b.** $\begin{vmatrix} 1 & 2 & 1 \\ 0 & 4 & -2 \\ 3 & 6 & -2 \end{vmatrix}$

26. **[8.5C]** Expand by minors along the third column.

    a. $\begin{vmatrix} 1 & 3 & 0 \\ 0 & 1 & -2 \\ 2 & 4 & 3 \end{vmatrix}$ **b.** $\begin{vmatrix} 3 & 1 & 5 \\ 1 & 0 & -2 \\ 6 & 1 & 3 \end{vmatrix}$

27. **[8.5D]** Solve by Cramer's rule.

    a. $x + 2y + z = 6$ **b.** $x + y + 2z = -3$
       $x + y - z = 7$ $x - y + 2z = 1$
       $2x - y + 2z = -3$ $x + 2y - z = -2$

28. **[8.5D]** Solve by Cramer's rule.

    a. $2x + y + z = 6$ **b.** $x + 2y \quad\ = 0$
       $x - y + z = 7$ $y - z = 2$
       $5x + y + 3z = 8$ $2x + y + 3z = 5$

PRACTICE TEST

(Answers are on pages 522–523.)

1. Use the graphical method to solve the system.
$$x - 3y = 3$$
$$y = x - 1$$

2. Use the graphical method to solve the system.
$$y - 3x = -3$$
$$3y = 9x + 9$$

3. Use the graphical method to solve the system.
$$3x + 2y = 6$$
$$x = 2 - \frac{2}{3}y$$

4. Use the substitution method to solve the system.
$$x - 2y = 4$$
$$x = 1 + y$$

5. Use the substitution method to solve the system.
$$2x - 3y = 6$$
$$4x = 6y + 7$$

6. Use the substitution method to solve the system.
$$2x - y = 6$$
$$2y = 4x - 12$$

7. Solve by the elimination method.
$$3x + 4y = 5$$
$$x + y = 1$$

8. Solve by the elimination method.
$$3x + 4y = 5$$
$$6x + 8y = 9$$

9. Solve the system.
$$\frac{x}{2} + \frac{y}{3} = 2$$
$$\frac{x}{4} + \frac{y}{6} = 1$$

10. Solve the system.
$$2x = 3y - 10$$
$$2y = 3x + 10$$

11. Solve the system.
$$x + y + z = 2$$
$$2x + y - z = 5$$
$$x + y - z = 4$$

12. Solve the system.
$$2x + y - 2z = 4$$
$$-x + y + 2z = 2$$
$$3y + 2z = 12$$

13. Solve the system.
$$2x + y + z = 4$$
$$-x + 2y + z = 3$$
$$-2x + 9y + 5z = 16$$

14. Pedro has $3.50 in nickels and dimes. He has 10 more nickels than dimes. How many of each coin does he have?

15. The total height of a building and a flagpole on the roof is 240 ft. The building is nine times as high as the flagpole. How high is the building?

16. A motorboat can go 10 mi downstream on a river in 20 min. It takes 30 min for this boat to go back upstream the same 10 mi. Find the speed of the current.

17. Annie has three investments totaling $60,000. These investments earn interest at 4%, 6%, and 8%. Annie's total annual income from these investments is $4000. The income from the 8% investment exceeds the total income from the other two investments by $800. Find how much she has invested at each rate.

18. Use matrices to solve the system.
$$x + y + z = -2$$
$$2x + y - z = 3$$
$$-x + y + z = 0$$

19. Evaluate.

a. $\begin{vmatrix} 4 & -2 \\ 5 & -6 \end{vmatrix}$

b. $\begin{vmatrix} 4 & -2 \\ 6 & -3 \end{vmatrix}$

20. Solve by Cramer's rule.
$$2x - 3y = 5$$
$$6x - 4y = 7$$

21. Evaluate.
$$\begin{vmatrix} -2 & 4 & -2 \\ 1 & 0 & 3 \\ 4 & 5 & 2 \end{vmatrix}$$

22. Expand by minors along the first row.
$$\begin{vmatrix} 3 & 2 & -1 \\ 1 & 3 & 2 \\ 1 & 1 & -1 \end{vmatrix}$$

23. Expand by minors along the third column.
$$\begin{vmatrix} 1 & 1 & -2 \\ 2 & -2 & 1 \\ 0 & 2 & 1 \end{vmatrix}$$

24. Solve by Cramer's rule.

$$x + 2y + z = 6$$
$$x + y - z = 7$$
$$2x - y + 2z = -3$$

25. Solve by Cramer's rule.

$$x + 2y + z = 6$$
$$x + y - z = 7$$
$$3x + 5y + z = 8$$

ANSWERS TO PRACTICE TEST

Answer	If you missed:	Review:		
	Question	Section	Examples	Pages
1. The solution is $(0, -1)$.	1	8.1A	1	461
2. Inconsistent; no solution	2	8.1A	2	461–462
3. Dependent; infinitely many solutions	3	8.1A	3	462

$(0, -1)$

Answer	If you missed:		Review:	
	Question	Section	Examples	Pages
4. $(-2, -3)$	4	8.1B	4	464
5. Inconsistent; no solution	5	8.1B	5	465
6. Dependent; infinitely many solutions	6	8.1B	6	465
7. $(-1, 2)$	7	8.1C	7	466
8. Inconsistent; no solution	8	8.1C	8	467
9. Dependent; infinitely many solutions	9	8.1C	9	468–469
10. $(-2, 2)$	10	8.1C	7–8	466–467
11. $(1, 2, -1)$; consistent	11	8.2A	1	477–478
12. Inconsistent; no solution	12	8.2A	2	478
13. Dependent; infinitely many solutions	13	8.2A	3	479
14. 30 nickels, 20 dimes	14	8.3A	1	485–486
15. 216 ft	15	8.3B	2	486
16. 5 mi/hr	16	8.3C	3	487
17. \$10,000 at 4%, \$20,000 at 6%, \$30,000 at 8%	17	8.3D	4	487–488
18. $(-1, 2, -3)$	18	8.4A	1, 2, 3	497–500
19. a. -14	19a	8.5A	1	506
b. 0	19b	8.5A	1	506
20. $\left(\frac{1}{10}, -\frac{8}{5}\right)$	20	8.5B	2	508
21. 60	21	8.5C	3	509–510
22. -7	22	8.5C	3	509–510
23. -14	23	8.5C	4	510–511
24. $(2, 3, -2)$; consistent	24	8.5C	5	512
25. Inconsistent; no solution	25	8.5C	6	513

Quadratic Functions and the Conic Sections

In this chapter we study curves called **conic sections**. To visualize conic sections, imagine two ice-cream cones placed vertically tip to tip (see Figure 1). Slice the top cone *parallel* to the ground; the cut will be a **circle**. If the cut is slightly *slanted* and *not* parallel to the side of the cone, the result is an **ellipse**. If the cut is parallel to the side of the cone, the exposed part is a **parabola**. Now suppose you slice through both cones but not through their tips. The exposed part of the cone will form a **hyperbola**. Because these curves can be described using a cone, circles, ellipses, parabolas, and hyperbolas are called conic sections. We begin this chapter by graphing parabolas and continue our study by graphing circles, ellipses, and hyperbolas and discussing the ways in which different conic sections can be identified. We end the chapter by discussing nonlinear systems of equations and inequalities.

Why do we study conic sections? The answer is a tangled tale based in antiquity. The first 300 years of Greek mathematics yielded three famous problems: the squaring of the circle, the trisection of an angle, and the duplication of the cube. The third problem was first mentioned by a Greek poet describing mythical king Minos as dissatisfied with the size of the tomb erected for his son Glaucus. "You have embraced too little space; quickly double it without spoiling its beautiful cubical form." When the construction was done incorrectly, legend says, local geometers were hastily summoned to solve the problem. A different version has the Athenians appealing to the oracle at Delos to get rid of pestilence. "Apollus' cubical altar must be doubled in size" was the reply. When workmen did the work, again incorrectly, "the indignant god made the pestilence even worse than before."

Hippocrates of Chios (ca. 440–380 B.C.) finally showed that the duplication of the cube could be done by using curves with certain properties, but it was the Greek mathematician Menaechmus (350 B.C.), the tutor of Alexander the Great, who is reputed to have discovered the curves that were later known as the ellipse, the parabola, and the hyperbola (the conics) to solve the problem.

9.1

QUADRATIC FUNCTIONS AND THEIR GRAPHS

To succeed, review how to:

1. Graph points in the Cartesian coordinate system (pp. 186–187).

2. Find x- and y-intercepts (pp. 390–391).

3. Use the quadratic formula (pp. 345–349).

4. Complete the square in a quadratic equation (pp. 338–341).

5. Find the discriminant of a quadratic equation (p. 355).

Objectives:

A   Graph a parabola of the form $y = ax^2 + k$.

B   Graph a parabola of the form $y = a(x - h)^2 + k$.

C   Graph a parabola of the form $y = ax^2 + bx + c$.

D   Graph parabolas of the form $x = a(y - k)^2 + h$ and $x = ay^2 + by + c$.

E   Solve applications involving parabolas.

getting started

The Fountain of Parabolas

Have you seen any parabolas lately? They are as near as your fountain: The streams of water follow the path of a quadratic function called a parabola. Parabolas, ellipses, circles, and hyperbolas are called **conic sections** because they can be obtained by intersecting (slicing) a cone with a plane, as shown in Figure 1. As you will see later, these conic sections occur in many practical applications. For example, your satellite dish, your flashlight lens and your telescope lens have a parabolic shape, while comets travel in orbits that are either elliptical (those may be seen from earth more than once) or hyperbolic (once in a lifetime viewing!). We shall begin our study of conic sections by examining the parabola.

| Circle | Ellipse | Parabola | Hyperbola |

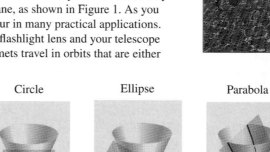

FIGURE 1

Graphing the Parabola $y = f(x) = ax^2 + k$

In Chapter 7, we studied *linear functions* like $f(x) = 2x + 5$ and $g(x) = -3x - 4$ whose graphs were straight lines. In this chapter we shall study equations (functions) defined by a quadratic (second-degree) polynomial of the form

$$f(x) = ax^2 + bx + c$$

These functions are **quadratic functions** and their graphs are **parabolas**.

Just as the "simplest" *line* to graph is $y = f(x) = x$, the "simplest" *parabola* to graph is $y = f(x) = x^2$. To draw this graph, we select values for x and find the corresponding values of y:

x-Value	$f(x) = y$-Value
$x = -2$	$f(-2) = (-2)^2 = 4$
$x = -1$	$f(-1) = (-1)^2 = 1$
$x = 0$	$f(0) = (0)^2 = 0$
$x = 1$	$f(1) = (1)^2 = 1$
$x = 2$	$f(2) = (2)^2 = 4$

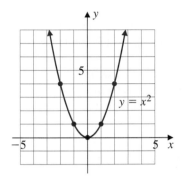

FIGURE 2

Then we make a table of ordered pairs, plot the ordered pairs on a coordinate system, and draw a smooth curve through the plotted points as in Figure 2.

x	y	(x, y)
-2	4	$(-2, 4)$
-1	1	$(-1, 1)$
0	0	$(0, 0)$
1	1	$(1, 1)$
2	4	$(2, 4)$

A very important feature of this parabola is its **symmetry** to the y-axis. This means that if you folded the graph of $y = x^2$ along the y-axis, the two halves of the graph would coincide because the same value of y is obtained for any value of x and its opposite $-x$. For instance, $x = 2$ and $x = -2$ both give $y = 4$. (See the preceding tables.) Because of this symmetry, the y-axis is called the **axis of symmetry** or simply the **axis** of the parabola. The point $(0, 0)$, where the parabola crosses its axis, is called the **vertex** of the curve. Note that the arrows on the curve in Figure 2 mean that the parabola goes on without end.

EXAMPLE 1 Graphing a parabola that opens downward

Graph: $y = -x^2$

SOLUTION We could make a table of x- and y-values as before. However, note that for any x-value, the y-value will be the *negative* of the y-value on the parabola $y = x^2$. (If you don't believe this, go ahead and make the table and check it, but it's easier to copy the table for $y = x^2$ with the negatives of the y-values entered as

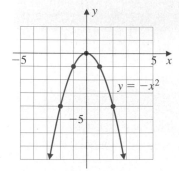

FIGURE 3

shown. Thus, the parabola $y = -x^2$ has the same shape as $y = x^2$, but it is turned in the *opposite* direction (opens *downward*). The graph of $y = -x^2$ is shown in Figure 3.

x	y	(x, y)
-2	-4	$(-2, -4)$
-1	-1	$(-1, -1)$
0	0	$(0, 0)$
1	-1	$(1, -1)$
2	-4	$(2, -4)$

■

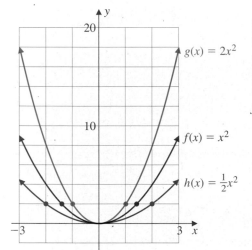

FIGURE 4

As you can see from the two preceding examples, when the coefficient of x^2 is positive (as in $y = x^2 = 1x^2$), the parabola opens upward (is **concave up**), but when the coefficient of x^2 is negative (as in $y = -x^2 = -1x^2$), the parabola opens downward (is **concave down**). In either case, the vertex is at $(0, 0)$. To see the effect of a in $f(x) = ax^2$ in general, let's plot some points and see how the graphs of $g(x) = 2x^2$ and $h(x) = \frac{1}{2}x^2$ compare to the graph of $f(x) = x^2$. All three graphs are shown in Figure 4.

x	$g(x) = 2x^2$	x	$h(x) = \frac{1}{2}x^2$
-2	$2(-2)^2 = 8$	-2	$\frac{1}{2}(-2)^2 = 2$
-1	$2(-1)^2 = 2$	-1	$\frac{1}{2}(-1)^2 = \frac{1}{2}$
0	$2(0)^2 = 0$	0	$\frac{1}{2}(0)^2 = 0$
1	$2(1)^2 = 2$	1	$\frac{1}{2}(1)^2 = \frac{1}{2}$
2	$2(2)^2 = 8$	2	$\frac{1}{2}(-2)^2 = 2$

Note that the graph of $g(x) = 2x^2$ is narrower than that of $f(x) = x^2$, while the graph of $h(x) = \frac{1}{2}x^2$ is wider. The vertex and line of symmetry is the same for the three curves. In general, we have the following.

PROPERTIES OF THE PARABOLA $g(x) = ax^2$

The graph of $g(x) = ax^2$ is a parabola with the vertex at the origin and the y-axis as its line of symmetry.

If a is *positive*, the parabola opens *upward*, if a is *negative*, the parabola opens *downward*.

If $|a|$ is greater than 1 ($|a| > 1$), the parabola is narrower than the parabola $f(x) = x^2$.

If $|a|$ is between 0 and 1 ($0 < |a| < 1$), the parabola is wider than the parabola $f(x) = x^2$.

Using this information, you can draw the graph of any parabola of the form $g(x) = ax^2$, as we illustrate in Example 2.

EXAMPLE 2   Graphing a parabola of the form $y = ax^2$

Graph:

a. $f(x) = 3x^2$ **b.** $g(x) = -3x^2$ **c.** $h(x) = \frac{1}{3}x^2$

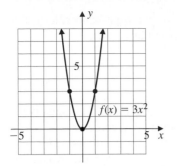

FIGURE 5

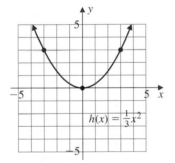

FIGURE 6

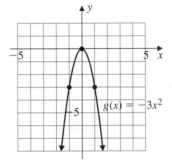

FIGURE 7

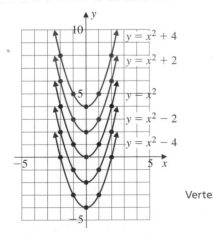

FIGURE 8

SOLUTION

a. By looking at the properties of the parabola $y = ax^2$, we know that the vertex of the parabola $f(x) = 3x^2$ is at the origin and that the y-axis is its line of symmetry. Since $3 > 0$, we also know that the parabola $f(x) = 3x^2$ opens upward and is narrower than the parabola $y = x^2$. We pick three easy points to complete our graph, which is shown in Figure 5.

x	$y = 3x^2$
-1	$y = 3(-1)^2 = 3$
0	$y = 3(0)^2 \quad = 0$
1	$y = 3(1)^2 \quad = 3$

b. The parabola $g(x) = -3x^2$ opens downward but is still narrower than the parabola $y = x^2$. As a matter of fact, the parabola $g(x) = -3x^2$ is the reflection of the parabola $f(x) = 3x^2$ across the x-axis. Again, we pick three points to complete the graph, which is shown in Figure 6.

x	$y = -3x^2$
-1	$y = -3(-1)^2 = -3$
0	$y = -3(0)^2 \quad = \quad 0$
1	$y = -3(1)^2 \quad = -3$

c. The parabola $h(x) = \frac{1}{3}x^2$ opens upward, since $\frac{1}{3} > 0$, but is wider than the parabola $y = x^2$. This time, instead of selecting $x = -1, 0$, and 1, we select $x = -3, 0$, and 3 for ease of computation; the completed graph is shown in Figure 7.

x	$y = \frac{1}{3}x^2$
-3	$y = \frac{1}{3}(-3)^2 = 3$
0	$y = \frac{1}{3}(0)^2 \quad = 0$
3	$y = \frac{1}{3}(3)^2 \quad = 3$

■

What do you think will happen if we graph the parabola $y = x^2 + 2$? Two things: First, the parabola opens upward since the coefficient of x^2 is understood to be $+1$. Second, all of the points will be 2 units higher than those for the same value of x on the parabola $y = x^2$. Thus we can make the graph of $y = x^2 + 2$ by following the pattern of $y = x^2$. The graphs of $y = x^2 + 2$, $y = x^2 + 4$, $y = x^2 - 2$, and $y = x^2 - 4$ are shown in Figure 8. The points used to make the graphs are listed in the following table.

For $y = x^2 + 2$		For $y = x^2 + 4$		For $y = x^2 - 2$		For $y = x^2 - 4$	
x	y	x	y	x	y	x	y
Vertex ⟶ 0	2	0	4	0	-2	0	-4
± 1	3	± 1	5	± 1	-1	± 1	-3
± 2	6	± 2	8	± 2	2	± 2	0

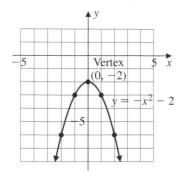

FIGURE 9

Note that adding or subtracting a positive number k on the right-hand side of the equation $y = x^2$ raises or lowers the graph (and the vertex) by k units.

EXAMPLE 3   Graphing a parabola opening downward

Graph: $y = -x^2 - 2$

SOLUTION   Since the coefficient of x^2 (which is understood to be -1) is negative, the parabola opens downward. It is also 2 units lower than the graph of $y = -x^2$. Thus the graph of $y = -x^2 - 2$ is a parabola opening downward with its vertex at $(0, -2)$. Letting $x = 1$, we get $y = -3$ and for $x = 2$, $y = -6$. Graph the two points $(1, -3)$ and $(2, -6)$ and, by symmetry, the points $(-1, -3)$ and $(-2, -6)$. The parabola passing through all these points is shown in Figure 9.   ■

Graphing a Parabola of Form $y = f(x) = a(x - h)^2 + k$

So far, we have graphed only parabolas of the form $y = ax^2 + k$. What do you think the graph of $y = (x - 1)^2$ looks like? As before, we make a table of values.

x	$y = (x - 1)^2$		x	y
$x = -1,$	$y = (-1 - 1)^2 = (-2)^2 = 4$		-1	4
$x = 0,$	$y = (0 - 1)^2 = (-1)^2 = 1$ ←—y-intercept		0	1
$x = 1,$	$y = (1 - 1)^2 = (0)^2 = 0$ ←—Vertex		1	0
$x = 2,$	$y = (2 - 1)^2 = 1^2 = 1$		2	1
$x = 3,$	$y = (3 - 1)^2 = 2^2 = 4$		3	4

or

The graph appears in Figure 10.

Note that the shape of the graph is identical to that of $y = x^2$ but it is shifted 1 unit to the *right*. Thus the vertex is at $(1, 0)$ and the line of symmetry is as shown in Figure 10. Similarly, the graph of $y = -(x + 1)^2$ is identical to that of $y = -x^2$ but shifted 1 unit to the *left*. Thus the vertex is at $(-1, 0)$ and the line of symmetry is as shown in Figure 11. Some easy points to plot are $x = 0$, $y = -(1)^2 = -1$ and $x = 1$, $y = -(1 + 1)^2 = -2^2 = -4$. When we plot the points $(0, -1)$ and $(1, -4)$, by symmetry the points $(-2, -1)$ and $(-3, -4)$ are also on the graph.

FIGURE 10

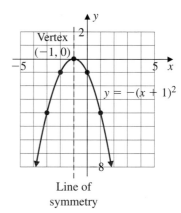

Line of symmetry

FIGURE 11

EXAMPLE 4   Graphing a parabola of the form $y = a(x - h)^2 + k$

Graph: $y = (x - 1)^2 - 2$

SOLUTION   The graph of this equation is identical to the graph of $y = x^2$ except for its position. The new parabola is shifted 1 unit to the right (because of the -1) and 2 units down (because of the -2). Thus the vertex is $(1, -2)$. Note the line

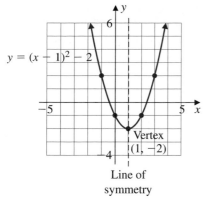

$y = (x - 1)^2 - 2$

Vertex (1, −2)

Line of symmetry

FIGURE 12

PROPERTIES OF THE PARABOLA $y = a(x - h)^2 + k$

of symmetry. Figure 12 indicates these two facts and shows the finished graph of $y = (x - 1)^2 - 2$.

$$y = (x - 1)^2 - 2$$

Opens upward (positive) — Shifted 1 unit right — Shifted 2 units down

Vertex (1, −2)

From these examples we can see that

1. The graph of $y = -x^2 - 2$ (Example 3) is exactly the same as the graph of $y = -x^2$ (Example 1) but moved 2 units *down*. In general, the graph of $y = ax^2 + k$ is the same as the graph of $y = ax^2$ but moved vertically k units. The vertex is at $(0, k)$.

2. The graph of $y = (x - 1)^2$ is the same as that of $y = x^2$ but moved 1 unit *right*. The vertex is at $(1, 0)$.

3. The graph of $y = (x - 1)^2 - 2$ (Example 4) is exactly the same as the graph of $y = (x - 1)^2$ but moved 2 units *down*. The vertex is at $(1, -2)$.

Here is the summary of this discussion.

> The graph of the parabola $y = a(x - h)^2 + k$ is the same as that of $y = ax^2$ but moved h units horizontally and k units vertically. The *vertex* is at the point (h, k), and the line of symmetry is $x = h$.

In conclusion, follow the given directions to graph an equation of the form

$$y = a(x - h)^2 + k \qquad \text{Vertex } (h, k)$$

Opens upward for $a > 0$, downward for $a < 0$ — Shifts the graph right or left — Moves the graph up or down

The graphs of $y = 2(x - 1)^2 + 1$, $y = 2(x - 1)^2 + 3$, $y = -2(x - 1)^2 - 1$, and $y = -2(x - 1)^2 - 3$ are shown in Figure 13.

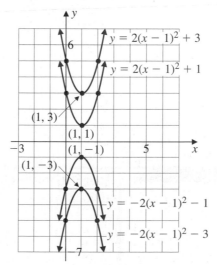

$y = 2(x - 1)^2 + 3$

$y = 2(x - 1)^2 + 1$

(1, 3)

(1, 1)

(1, −1)

(1, −3)

$y = -2(x - 1)^2 - 1$

$y = -2(x - 1)^2 - 3$

FIGURE 13

Graphing the Parabola $y = f(x) = ax^2 + bx + c$

How can we graph $f(x) = ax^2 + bx + c$? If we learn to write $f(x) = ax^2 + bx + c$ as $f(x) = a(x - h)^2 + k$, we can do it by using the techniques we just learned. We do this by completing the square. Here's how:

$f(x) = ax^2 + bx + c$	Given.
$= (ax^2 + bx) + c$	Group.
$= a\left[x^2 + \dfrac{b}{a}x + \phantom{\left(\dfrac{b}{2a}\right)^2}\right] + c$	Factor a.
$= a\left[x^2 + \dfrac{b}{a}x + \left(\dfrac{b}{2a}\right)^2 - \left(\dfrac{b}{2a}\right)^2\right] + c$	To complete the square, add and subtract $\left(\dfrac{1}{2} \cdot \dfrac{b}{a}\right)^2 = \left(\dfrac{b}{2a}\right)^2$ inside the brackets.
$= a\left(x + \dfrac{b}{2a}\right)^2 - a\left(\dfrac{b}{2a}\right)^2 + c$	Use the distributive property and factor inside the brackets.
$= a\left(x + \dfrac{b}{2a}\right)^2 - a \cdot \dfrac{b^2}{4a^2} + c$	Square $\dfrac{b}{2a}$.
$= a\left(x + \dfrac{b}{2a}\right)^2 - \dfrac{b^2}{4a} + c$	Multiply $-a \cdot \dfrac{b^2}{4a^2} = -\dfrac{b^2}{4a}$.
$= a\left(x + \dfrac{b}{2a}\right)^2 + \dfrac{4ac - b^2}{4a}$	Find the LCD of $\dfrac{-b^2}{4a}$ and c.

Thus to write

$$f(x) = a\left(x + \dfrac{b}{2a}\right)^2 + \dfrac{4ac - b^2}{4a}$$

as $f(x) = \underbrace{a(x - h)^2}_{} + \underbrace{k}_{}$

we must have

$$h = -\dfrac{b}{2a} \quad \text{and} \quad k = \dfrac{4ac - b^2}{4a}$$

the coordinates of the vertex. Note that you *do not* have to memorize the y-coordinate of the vertex. After you find the x-coordinate, substitute in the equation and find y.

    Here is a summary of our discussion.

PROCEDURE

Graphing the Parabola $y = f(x) = ax^2 + bx + c$

1. To find the vertex use one of the following methods:

    METHOD 1   Let $x = -\dfrac{b}{2a}$ in the equation and solve for y,

    or

METHOD 2 Complete the square and compare with $y = a(x - h)^2 + k$.

2. Let $x = 0$. The result, c, is the y-intercept.

3. Since the parabola is symmetric to its axis, use this symmetry to find additional points.

4. Let $y = 0$. Find x by solving $ax^2 + bx + c = 0$. If the solutions are real numbers, they are the x-intercepts. If not, the parabola does not intersect the x-axis.

5. Draw a smooth curve through the points found in steps $1-4$. Remember that if $a > 0$, the parabola opens *upward;* if $a < 0$, the parabola opens *downward.*

We demonstrate this procedure in Example 5.

EXAMPLE 5 Graphing a parabola of the form $y = ax^2 + bx + c$

Graph: $y = x^2 + 3x + 2$

SOLUTION

1. We first find the vertex using either of the two methods.

Method 1

Use the vertex formula for the x-coordinate. Since $a = 1, b = 3$, and $c = 2$,

$$x = -\frac{b}{2a} = -\frac{3}{2}$$

Substituting for x in the equation gives

$$y = x^2 + 3x + 2$$

$$= \left(-\frac{3}{2}\right)^2 + 3\left(-\frac{3}{2}\right) + 2$$

$$= \frac{9}{4} - \frac{9}{2} + 2$$

$$= \frac{9}{4} - \frac{18}{4} + \frac{8}{4} = -\frac{1}{4}$$

The vertex is at $\left(-\frac{3}{2}, -\frac{1}{4}\right)$.

Method 2

Complete the square.

$$y = [x^2 + 3x + \quad] + 2$$

$$= \left[x^2 + 3x + \left(\frac{3}{2}\right)^2\right] + 2 - \left(\frac{3}{2}\right)^2$$

$$= \left(x + \frac{3}{2}\right)^2 + 2 - \frac{9}{4}$$

$$= \left(x + \frac{3}{2}\right)^2 - \frac{1}{4}$$

The vertex is at $\left(-\frac{3}{2}, -\frac{1}{4}\right)$.

2. Let $x = 0$; then $y = x^2 + 3x + 2$ becomes $y = 2$. The y-intercept is 2.

3. By symmetry, the point $(-3, 2)$ is also on the graph.

4. Let $y = 0$; $y = x^2 + 3x + 2$ becomes

$$0 = x^2 + 3x + 2$$

$$= (x + 2)(x + 1)$$

Thus $x = -2$ or $x = -1$. The graph intersects the x-axis at $(-2, 0)$ and $(-1, 0)$.

5. Since the coefficient of x^2 is 1, $a > 0$ and the parabola opens upward.

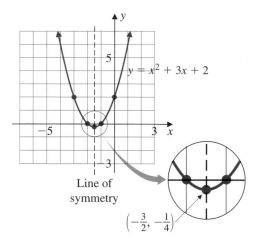

FIGURE 14

We draw a smooth curve through these points to obtain the graph of the parabola as shown in Figure 14. ∎

In Example 5, $x^2 + 3x + 2$ can be factored, and we can thus find the points at which the parabola crosses the x-axis. If the equation of the parabola cannot be factored, look at the discriminant $D = b^2 - 4ac$ and determine what kind of roots the equation has.

PROCEDURE

Using the Discriminant to Graph Quadratics

1. If $D < 0$, there are no real roots and the graph will *not* cross the x-axis.
2. If $D \geq 0$, either factor or use the quadratic formula to find x and approximate the answers so you can graph them.

The possibilities are shown in Figures 15–17.

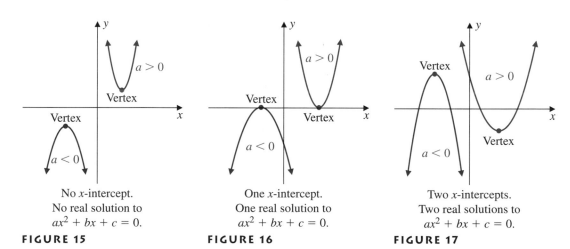

No x-intercept.
No real solution to
$ax^2 + bx + c = 0$.
FIGURE 15

One x-intercept.
One real solution to
$ax^2 + bx + c = 0$.
FIGURE 16

Two x-intercepts.
Two real solutions to
$ax^2 + bx + c = 0$.
FIGURE 17

EXAMPLE 6   Graphing a parabola whose equation cannot be factored

Graph: $y = -2x^2 + 4x - 3$

SOLUTION

1. To find the vertex, we can use either of these methods:

Method 1

Use the vertex formula.

Here $a = -2$ and $b = 4$, so

$$x = -\frac{b}{2a}$$

$$= \frac{-4}{2(-2)}$$

$$= 1$$

If we substitute $x = 1$ in $y = -2x^2 + 4x - 3$,

$$y = -2(1)^2 + 4(1) - 3$$

$$= -2 + 4 - 3$$

$$= -1$$

Thus the vertex is at $(1, -1)$.

Method 2

Complete the square.

$$y = -2x^2 + 4x - 3$$

$$= -2(x^2 - 2x + \quad) - 3$$

$$= -2(x^2 - 2x + 1) - 3 + 2$$

$$= -2(x - 1)^2 - 1$$

The vertex is at $(1, -1)$.

2. If $x = 0$, $y = -2x^2 + 4x - 3 = -3$, the y-intercept.

3. We graph the vertex $(1, -1)$ and the y-intercept -3. To make a more accurate graph, we need some more points. Since the parabola is symmetric, we can find a point across from the y-intercept by letting $x = 2$. Then

$$y = -2(2)^2 + 4(2) - 3$$

$$= -8 + 8 - 3 = -3$$

as expected.

4. For $y = 0$, $0 = -2x^2 + 4x - 3$. However, the right-hand side is not factorable. As a matter of fact, the discriminant of the equation is $4^2 - 4(-2)(-3) = 16 - 24 = -8$. This means that this equation has no solution and there are no x-intercepts. The graph does not cross the x-axis.

5. Since $a = -2 < 0$, the parabola opens downward. The completed graph is shown in Figure 18. ∎

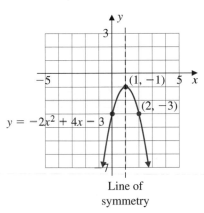

$y = -2x^2 + 4x - 3$

$(1, -1)$ $(2, -3)$

Line of symmetry

FIGURE 18

D Graphing Parabolas $x = a(y - k)^2 + h$ or $x = ay^2 + by + c$

Do you remember how the graph of $y = x^2$ looks? The graph of $x = y^2$ has a similar shape but opens *horizontally to the right*. Similarly, the graph of $x = y^2 + 1$ looks like that of $x = y^2$ but is shifted *right* 1 unit, as shown in Figure 19. Note that when $y = 0$, $x = 1$ so the vertex is at $(1, 0)$. The graph of $x = y^2 - 2$ is similar to that of $x = y^2$ but is shifted *left* 2 units. Here, when $y = 0$, $x = -2$ so the vertex is at $(-2, 0)$.

Similarly, the graphs of $x = -y^2$, $x = -y^2 + 1$, and $x = -y^2 - 2$ look like the graph of $x = -y^2$, opening *horizontally to the left* and then shifted right or left the correct number of units. Thus for $x = -y^2 + 1$, when $y = 0$, $x = 1$ and the vertex is at $(1, 0)$. For $x = -y^2 - 2$, when $y = 0$, $x = -2$ so the vertex is at $(-2, 0)$. The

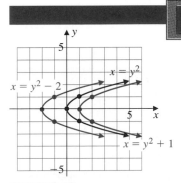

$x = y^2 - 2$ $x = y^2$ $x = y^2 + 1$

FIGURE 19

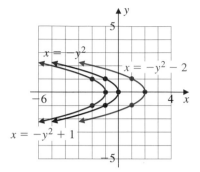

FIGURE 20

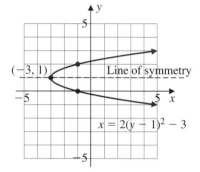

FIGURE 21

graphs are shown in Figure 20. Note that none of these are graphs of functions. (Remember the vertical line test? You can find several vertical lines that will intersect all of these graphs at more than one point.)

EXAMPLE 7　Graphing a parabola of the form $x = a(y - k)^2 + h$

Graph: $x = 2(y - 1)^2 - 3$

SOLUTION　In this problem, the roles of x and y are reversed, so the graph will look like that of $y = 2(x - 1)^2 - 3$ but opening horizontally.

The vertex of $x = 2(y - 1)^2 - 3$ is at $(-3, 1)$. The curve opens to the right, the positive x-direction. The graph is shown in Figure 21. You can verify that the graph is correct by letting $x = -1$, which gives $y = 0$ or $y = 2$.　∎

EXAMPLE 8　Graphing a parabola of the form $x = ay^2 + by + c$

Graph: $x = y^2 + 3y + 2$

SOLUTION　The graph is similar to that of $y = x^2 + 3x + 2$, but it opens horizontally (see Example 5). The vertex occurs where

$$y = -\frac{b}{2a} = -\frac{3}{2}$$

Substituting for y in the equation gives $x = \left(-\frac{3}{2}\right)^2 + 3\left(-\frac{3}{2}\right) + 2 = -\frac{1}{4}$. Thus the vertex is at $\left(-\frac{1}{4}, -\frac{3}{2}\right)$. The x-intercept is 2 and the y-intercepts are where $x = 0$. Thus,

$$0 = y^2 + 3y + 2$$
$$= (y + 2)(y + 1)$$

That is, $y = -2$ or $y = -1$. The parabola opens to the right since the coefficient of y^2 is $1 > 0$, and the completed graph is shown in Figure 22.　∎

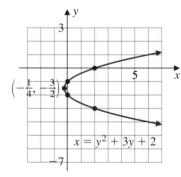

FIGURE 22

Solving Applications Involving Parabolas

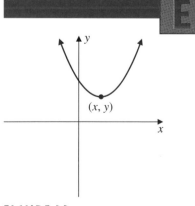

FIGURE 23

$f(x)$ has a minimum at the vertex (x, y).

Every parabola of the form $y = ax^2 + bx + c$ that we have graphed has its vertex at either its maximum (highest) or minimum (lowest) point on the graph. If the graph opens upward, the vertex is the minimum (Figure 23) and if the graph opens downward, the vertex is the maximum (Figure 24).

Thus if we are dealing with a quadratic function, we can find its maximum or minimum by finding the vertex of the corresponding parabola. This idea can be used to solve many real-world applications. For example, suppose that a CD company manufactures and sells x CDs per week.

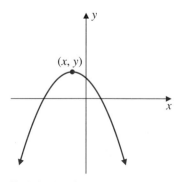

FIGURE 24

$f(x)$ has a maximum at the vertex (x, y).

If the revenue is given by $R = 10x - 0.01x^2$, we can use the techniques we've just studied to maximize the revenue. We do that next.

EXAMPLE 9   Recording a maximum revenue

If $R = 10x - 0.01x^2$, how many CDs does the company have to sell in order to obtain maximum revenue?

SOLUTION   We first write the equation as $R = -0.01x^2 + 10x$. Since the coefficient of x^2 is negative, the parabola opens downward (is concave down), and the vertex is its highest point. Letting

$$x = -\frac{b}{2a} = -\frac{10}{-0.02} = 500$$

$R = 10(500) - 0.01(500)^2 = 5000 - 2500 = 2500$. Thus when the company sells $x = 500$ CDs a week, the revenue is a maximum: $2500. ∎

We close this section by presenting in Table 1 a summary of the material we have studied.

TABLE 1 Summary of Parabola

Description	Graph when $a > 0$	Graph when $a < 0$
$f(x) = ax^2 + k$ A parabola with vertex at $(0, k)$. When $\lvert a \rvert > 1$, the graph is narrower than the graph of $y = x^2$. When $0 < \lvert a \rvert < 1$, it is wider.	Vertex at $(0, k)$ For $a > 0$, the parabola opens upward.	Vertex at $(0, k)$ For $a < 0$, the parabola opens downward.
$f(x) = a(x - h)^2 + k$ A parabola with vertex at (h, k). When $\lvert a \rvert > 1$, the graph is narrower than the graph of $y = x^2$. When $0 < \lvert a \rvert < 1$, it is wider.	Vertex at (h, k) For $a > 0$, the parabola opens upward.	Vertex at (h, k) For $a < 0$, the parabola opens downward.
$x = a(y - k)^2 + h$ A parabola with vertex at (h, k) and is not the graph of a function. When $\lvert a \rvert > 1$, the graph is narrower than the graph of $x = y^2$. When $0 < \lvert a \rvert < 1$, it is wider.	Vertex at (h, k) For $a > 0$, the parabola opens to the right.	Vertex at (h, k) For $a < 0$, the parabola opens to the left.

GRAPH IT

The graph of the function $y = f(x) = ax^2 + k$ can be easily obtained with a grapher. The only difficulty here is in selecting a window that will show the vertex $(0, k)$. Thus to graph $f(x) = 2x^2 + 6$ whose vertex is at $(0, 6)$, make sure that $(0, 6)$ is part of the graph by selecting a standard window. Of course, the graph opens upward as shown in Window 1.

If you were graphing $f(x) = 2x^2 + 10$, the vertex would be at $(0, 10)$, so you would have to adjust the window to see more of the graph. Try a $[-15, 15]$ by $[-15, 15]$ window. The resulting graph is shown in Window 2. Use this idea and a standard window to check the graphs in Examples 1, 2, 3, and 4. You can also use this same idea to check the graphs of Examples 5 and 6.

Now, can you use your grapher to check the graphs of $x = y^2$ or $x = -y^2$? It can be done, but it's almost easier to do it like we did in the text! Why? Because the grapher only graphs equations that are solved for y. Thus to graph $x = y^2$, you must **solve** this equation for y by extracting roots to obtain $y = \pm\sqrt{x}$. You would then graph $Y_1 = \sqrt{x}$ and $Y_2 = -\sqrt{x}$. The two graphs together correspond to the graph of $x = y^2$ as shown in Window 3.

To graph $x = 2(y - 1)^2 - 3$ in Example 7, solve for y as follows:

$$x + 3 = 2(y - 1)^2 \qquad \text{Add 3.}$$

$$\frac{x + 3}{2} = (y - 1)^2 \qquad \text{Divide by 2.}$$

$$\pm\sqrt{\frac{x + 3}{2}} = y - 1 \qquad \text{Extract roots.}$$

$$1 \pm \sqrt{\frac{x + 3}{2}} = y \qquad \text{Add 1.}$$

Now graph

$$Y_1 = 1 + \sqrt{\frac{x + 3}{2}} \qquad \text{and} \qquad Y_2 = 1 - \sqrt{\frac{x + 3}{2}}$$

The two graphs together correspond to $x = 2(y - 1)^2 - 3$ as shown in Window 4.

How would you graph $x = y^2 + 3y + 2$ (Example 8)? You have to complete the square and then solve for y! Try it and see whether you agree that it's easier to do it algebraically!

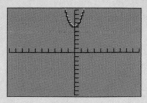

WINDOW 1
$y = 2x^2 + 6$
with a standard window

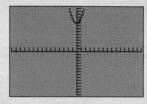

WINDOW 2
$y = 2x^2 + 10$ with a
$[-15, 15]$ by $[-15, 15]$ window

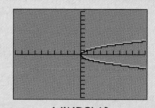

WINDOW 3
The graphs of $Y_1 = x$ and
$Y_2 = -x$ form the graph
of $x = y^2$.

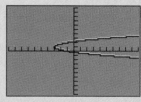

WINDOW 4
$x = 2(y - 1)^2 - 3$

EXERCISE 9.1

A In Problems 1–8, graph the given equations on the same coordinate axes.

1. a. $y = 2x^2$
   b. $y = 2x^2 + 2$
   c. $y = 2x^2 - 2$

2. a. $y = 3x^2 + 1$
   b. $y = 3x^2 + 3$
   c. $y = 3x^2 - 2$

3. a. $y = -2x^2$
   b. $y = -2x^2 + 1$
   c. $y = -2x^2 - 1$

4. a. $y = -4x^2$
   b. $y = -4x^2 + 1$
   c. $y = -4x^2 - 1$

5. a. $y = \dfrac{1}{4}x^2$

b. $y = -\dfrac{1}{4}x^2$

6. a. $y = \dfrac{1}{5}x^2$

b. $y = -\dfrac{1}{5}x^2$

7. a. $y = \dfrac{1}{3}x^2 + 1$

b. $y = -\dfrac{1}{3}x^2 + 1$

8. a. $y = \dfrac{1}{4}x^2 + 1$

b. $y = -\dfrac{1}{4}x^2 + 1$

B In Problems 9–16, graph the given equations on the same coordinate axes.

9. a. $y = (x + 2)^2 + 3$
b. $y = (x + 2)^2$
c. $y = (x + 2)^2 - 2$

10. a. $y = (x - 2)^2 + 2$
b. $y = (x - 2)^2$
c. $y = (x - 2)^2 - 2$

11. a. $y = -(x + 2)^2 - 2$
b. $y = -(x + 2)^2$
c. $y = -(x + 2)^2 - 4$

12. a. $y = -(x - 1)^2 + 1$
b. $y = -(x - 1)^2$
c. $y = -(x - 1)^2 + 2$

13. a. $y = -2(x + 2)^2 - 2$
b. $y = -2(x + 2)^2$
c. $y = -2(x + 2)^2 - 4$

14. a. $y = -2(x - 1)^2 + 1$
b. $y = -2(x - 1)^2$
c. $y = -2(x - 1)^2 + 2$

15. a. $y = 2(x + 1)^2 + \dfrac{1}{2}$

b. $y = 2(x + 1)^2$

16. a. $y = 2(x + 1)^2 - \dfrac{1}{2}$

b. $y = 2(x + 1)^2$

C In Problems 17–28, use the five-step procedure in the text to sketch the graph. Label the vertex and the intercepts.

17. $y = x^2 + 2x + 1$

18. $y = x^2 + 4x + 4$

19. $y = -x^2 + 2x + 1$

20. $y = -x^2 + 4x - 2$

21. $y = -x^2 + 4x - 5$

22. $y = -x^2 + 4x - 3$

23. $y = 3 - 5x + 2x^2$

24. $y = 3 + 5x + 2x^2$

25. $y = 5 - 4x - 2x^2$
(*Hint:* $\sqrt{56} = 7.5$)

26. $y = 3 - 4x - 2x^2$
(*Hint:* $\sqrt{40} = 6.3$)

27. $y = -3x^2 + 3x + 2$
(*Hint:* $\sqrt{33} = 5.7$)

28. $y = -3x^2 + 3x + 1$
(*Hint:* $\sqrt{21} = 4.6$)

D In Problems 29–34, graph on the same coordinate axes.

29. a. $x = (y + 2)^2 + 3$
b. $x = (y + 2)^2$

30. a. $x = (y - 2)^2 + 2$
b. $x = (y - 2)^2$

31. a. $x = -(y + 2)^2 - 2$
b. $x = -(y + 2)^2$

32. a. $x = -(y - 1)^2 + 1$
b. $x = -(y - 1)^2$

33. **a.** $x = -y^2 + 2y + 1$

    b. $x = -y^2 + 2y + 4$

34. **a.** $x = -y^2 + 4y - 5$

    b. $x = -y^2 + 4y - 3$

APPLICATIONS

35. The profit P (in dollars) for a company is $P = -5000 + 8x - 0.001x^2$, where x is the number of items produced each month. How many items does the company have to produce in order to obtain maximum profit? What is this profit?

36. The revenue R for Shady Glasses is given by $R = 1500p - 75p^2$, where p is the price of each pair of sunglasses (R and p in dollars). What should the price be to maximize revenue?

37. After spending x thousand dollars in an advertising campaign, the number of units N sold is given by $N = 50x - x^2$. How much should be spent in the campaign to obtain maximum sales?

38. The number N of units of a product sold after a television commercial blitz is $N = 40x - x^2$, where x is the amount spent in thousands of dollars. How much should be spent on television commercials to obtain maximum sales?

39. If a ball is batted up at 160 ft/sec, its height h feet after t seconds is given by $h = -16t^2 + 160t$. Find the maximum height reached by the ball.

40. If a ball is thrown upward at 20 ft/sec, its height h feet after t seconds is given by $h = -16t^2 + 20t$. How many seconds does it take for the ball to reach its maximum height, and what is this height?

41. If a farmer digs potatoes today, she will have 600 bushels worth $1 per bushel. Every week she waits, the crop increases by 100 bushels, but the price decreases 10¢ a bushel. Show that she should dig and sell her potatoes at the end of 2 weeks.

42. A man has a large piece of property along Washington Street. He wants to fence the sides and back of a rectangular plot. If he has 400 ft of fencing, what dimensions will give him the maximum area?

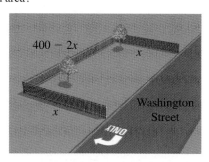

43. Have you read the story "The Jumping Frog of Calaveras County"? According to the *Guinness Book of Records,* the second greatest distance covered by a frog in a triple jump is 21 ft, $5\frac{3}{4}$ in. at the annual Calaveras Jumping Jubilee; this occurred on May 18, 1986.

    a. If Rosie the Ribiter's (the winner) path in her first jump is approximated by $R = -\frac{1}{98}x^2 + \frac{6}{7}x$ (where x is the horizontal distance covered in inches), what are the coordinates of the vertex of Rosie's path?

    b. Find the maximum height attained by Rosie in her first jump.

    c. Use symmetry to find the horizontal length of Rosie's first jump.

    d. Make a sketch for R showing the initial position $(0, 0)$, the vertex, and Rosie's ending position after her first jump.

44. Amazingly, Rosie's is not the best triple jump on record. That distinction belongs to Santjie, a South African frog who jumped 33 ft, $5\frac{1}{2}$ in. on May 21, 1977.

    a. If Santjie's path in his first jump is approximated by $S = -\frac{1}{200}x^2 + \frac{7}{10}x$ (where x is the distance covered in inches), what are the coordinates of the vertex of Santjie's path?

    b. Find Santjie's maximum height in his first jump.

    c. Use symmetry to find the horizontal length of Santjie's first jump.

    d. Make a sketch for S showing the initial position $(0, 0)$, the vertex, and Santjie's ending position after his first jump.

45. A baseball hit at an angle of 35° has a velocity of 130 mi/hr. Its trajectory can be approximated by the equation $d = -\frac{1}{400}x^2 + x$, where x is the distance the ball travels in feet.

    a. What are the coordinates of the vertex of the trajectory?

    b. Find the maximum height attained by the ball.

    c. Use symmetry to find how far the ball travels horizontally.

    d. Make a sketch for d showing the initial position $(0, 0)$, the vertex, and the ending position of the baseball.

46. From 1983 to 1990, the graph of the average men's SAT verbal scores is nearly a parabola.

    a. What is the maximum average verbal score for men in this period?

    b. What is the minimum average verbal score for men in this period?

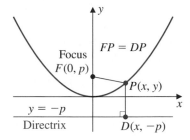

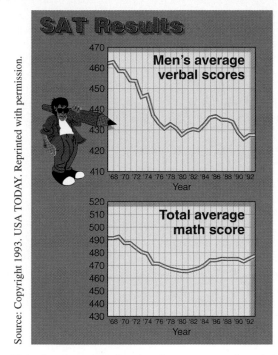

c. If the function approximating the average men's verbal score is

$$f(x) = ax^2 + bx + c$$

what can you say about a?

47. From 1976 to 1984, the graph of the total average SAT math scores is nearly a parabola.

   a. What is the maximum total average math score in this period?

   b. What is the minimum total average math score in this period?

   c. If the function approximating the total average math score is

$$f(x) = ax^2 + bx + c$$

what can you say about a?

SKILL CHECKER

Find the distance between each pair of points:

48. $A(3, 4)$ and $B(6, 8)$

49. $A(2, -3)$ and $B(4, 2)$

50. $A(3, 2)$ and $B(3, -4)$

USING YOUR KNOWLEDGE

Parabolas Revisited

Here is another way of defining a parabola.

PARABOLA

A **parabola** is the set of all points equidistant from a fixed point $F(0, p)$ (called the **focus**) and a fixed line $y = -p$ (called the **directrix**).

If $P(x, y)$ is a point on the parabola, this definition says that $FP = DP$; that is, the distance from F to P is the same as the distance from D to P:

51. Find FP.

52. Find DP.

53. Set $FP = DP$ and solve for x^2.

54. For the parabola $x^2 = 4y$,

   a. Locate the focus.

   b. Write the equation of the directrix.

Many applications of the parabola depend on an important focal property of the curve. If the parabola is a mirror, a ray of light parallel to the axis reflects to the focus, and a ray originating at the focus reflects parallel to the axis. (This can be proved by methods of calculus.)

If the parabola is revolved about its axis, a surface called a *paraboloid of revolution* is formed. This is the shape used for automobile headlights and searchlights that throw a parallel beam of light when the light source is placed at the focus; it's also the shape of a radar dish or a reflecting telescope mirror that collects parallel rays of energy (light) and reflects them to the focus.

We can find the equation of the parabola needed to generate a paraboloid of revolution by using the equation $x^2 = 4py$ as follows: Suppose a parabolic mirror has a diameter of 6 ft and a depth of 1 ft. Then, we find the value of p that makes the parabola pass through the point $(3, 1)$. This means that we substitute into the equation and solve for p. Thus we have

$$3^2 = 4p(1)$$

so that $4p = 9$ and $p = 2.25$. The equation of the parabola is $x^2 = 9y$ and the focus is at $(0, 2.25)$.

55. A radar dish has a diameter of 10 ft and a depth of 2 ft. The dish is in the shape of a paraboloid of revolution. Find an equation for a parabola that would generate this dish and locate the focus.

56. The cables of a suspension bridge hang very nearly in the shape of a parabola. A cable on such a bridge spans a distance of 1000 ft and sags 50 ft in the middle. Find an equation for this parabola.

WRITE ON . . .

Explain:

57. How you determine whether the graph of a quadratic function opens up or down.

58. What causes the graph of the function $f(x) = ax^2$ to be wider or narrower than the graph of $f(x) = x^2$.

59. The effect of the constant k on the graph of the function $f(x) = ax^2 + k$.

60. How a parabola that has two x-intercepts and vertex at $(1, 1)$ opens—that is, does it open up or down? Why?

61. Why the graph of a function never has two y-intercepts.

62. How can you tell if the vertex of a parabola is the maximum or the minimum point on the graph of a parabola with a vertical axis?

MASTERY TEST   If you know how to do these problems, you have learned your lesson!

Graph:

63. $x = y^2 + 2y - 3$

64. $x = -y^2 + 2y + 3$

65. $x = 2(y - 1)^2 + 3$

66. $x = -2(y - 1)^2 + 3$

67. $y = -2x^2 - 4x - 3$

68. $y = x^2 + 2x - 3$

69. $y = (x - 2)^2 + 3$

70. $y = -(x - 3)^2 + 2$

71. $f(x) = -x^2 + 4$

72. $f(x) = x^2 - 4$

73. $f(x) = 2x^2$

74. $g(x) = \dfrac{1}{2}x^2$

75. $h(x) = -2x^2$

76. $f(x) = -\dfrac{1}{2}x^2$

77. The revenue R for a company is $R = 300p - 15p^2$, where p is the price of each unit (R and p are in dollars). What should the price p be so that the revenue R is maximized?

9.2

CIRCLES AND ELLIPSES

To succeed, review how to:

1. Find the distance between two points (pp. 396–398).

2. Complete the square for a quadratic equation (pp. 338–341).

Objectives:

A Find the equation of a circle with a given center and radius.

B Find the center and radius and sketch the graph of a circle when its equation is given.

C Graph an ellipse when its equation is given.

getting started

Comets and Conics

This diagram shows the orbit of the comet Kohoutek with respect to the orbit of the Earth. The comet's orbit is an *ellipse,* whereas the Earth's orbit is nearly a perfect *circle.* In this section we continue our study of the conic sections by discussing circles and ellipses.

Source: From *The Nature of Comets* by Fred I. Whippler. © 1974 by Scientific American, Inc. All rights reserved.

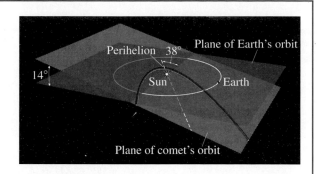

A Finding the Equation of a Circle

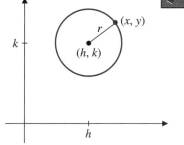

FIGURE 25

The distance from (h, k) to (x, y) is r.

Can you define a circle? A **circle** is defined as a set of points in a plane that are equidistant from a fixed point. The fixed point is called the *center* and the given distance is the *radius.* To find the equation of a circle of radius r, suppose the center is at a point $C(h, k)$; see Figure 25. The distance from C to any point $P(x, y)$ on the circle is found by the distance formula. Since r is the radius, this distance must be r. Thus

$$\sqrt{(x - h)^2 + (y - k)^2} = r$$

$$(x - h)^2 + (y - k)^2 = r^2 \qquad \text{Square both sides.}$$

We then have the following.

STANDARD FORM OF THE EQUATION OF A CIRCLE

> The equation of a circle with **radius** r and with **center** at $C(h, k)$ is
> $$(x - h)^2 + (y - k)^2 = r^2$$

EXAMPLE 1 Finding the equation of a circle

Find: The equation of the circle with center at $(3, -5)$ and radius 2

SOLUTION Here, the center $(h, k) = (3, -5)$ and $r = 2$. This means $h = 3$, $k = -5$, and $r = 2$. Using the formula, we have

$$(x - h)^2 + (y - k)^2 = r^2$$
$$(x - 3)^2 + [(y - (-5)]^2 = 2^2 \qquad \text{Substitute } h = 3, k = -5, r = 2.$$
$$(x - 3)^2 + (y + 5)^2 = 4 \qquad \text{Write in standard form.} \qquad \blacksquare$$

EXAMPLE 2 Finding the equation of a circle centered at the origin

Find: The equation of a circle of radius 3 and with center at the origin

SOLUTION The center is at $(h, k) = (0, 0)$. Thus $h = 0$, $k = 0$, and $r = 3$. Substituting $h = 0$, $k = 0$, $r = 3$ in $(x - h)^2 + (y - k)^2 = r^2$ gives

$$(x - 0)^2 + (y - 0)^2 = 3^2$$
$$x^2 + y^2 = 9 \qquad \blacksquare$$

In general, we have the following.

STANDARD FORM OF A CIRCLE CENTERED AT THE ORIGIN

The equation of a circle of radius r with center at the origin is

$$x^2 + y^2 = r^2$$

B | **Finding the Center and Radius of a Circle**

If we have the equation of a circle, we can write it in the standard form $(x - h)^2 + (y - k)^2 = r^2$ and find the center and radius. For example, if a circle has equation $(x - 3)^2 + (y - 4)^2 = 5^2$, then $h = 3$, $k = 4$, and $r = 5$. Thus the equation $(x - 3)^2 + (y - 4)^2 = 5^2$ is the equation of a circle of radius 5 with center at (3, 4).

EXAMPLE 3   Finding the center and radius of a circle

Find the center and radius and sketch the graph of the circle whose equation is

$$(x + 2)^2 + (y - 1)^2 = 9$$

SOLUTION   We first write the equation in the standard form

$$(x - h)^2 + (y - k)^2 = r^2$$

Thus

$$[x - (-2)]^2 + (y - 1)^2 = 3^2$$

We have $h = -2$, $k = 1$, and $r = 3$. The center is at $(h, k) = (-2, 1)$ and the radius is 3. The sketch is shown in Figure 26.   ■

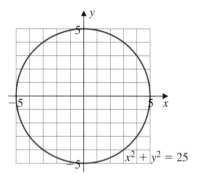

FIGURE 26

(−2, 1)

$(x + 2)^2 + (y - 1)^2 = 9$

EXAMPLE 4   Sketching the graph of a circle centered at the origin

Sketch: The graph of $x^2 + y^2 = 25$

SOLUTION   This equation can be written as $x^2 + y^2 = 5^2$, the equation of a circle of radius 5 centered at the origin. The graph is shown in Figure 27.   ■

FIGURE 27

$x^2 + y^2 = 25$

EXAMPLE 5   Sketching the graph of a circle not centered at the origin

Sketch: The graph of $x^2 - 4x + y^2 + 6y + 8 = 0$

SOLUTION   We must find the center and the radius by writing the equation in the standard form $(x - h)^2 + (y - k)^2 = r^2$. We can do this by subtracting 8 from both sides and then completing the square on x and y.

$$x^2 - 4x + y^2 + 6y + 8 = 0 \qquad \text{Given}$$

$$x^2 - 4x + \underline{\hspace{0.5cm}} + y^2 + 6y + \underline{\hspace{0.5cm}} = -8 \qquad \text{To complete the squares add } \left(\tfrac{1}{2} \cdot 4\right)^2 \text{ and } \left(\tfrac{1}{2} \cdot 6\right)^2.$$

$$x^2 - 4x + 4 + y^2 + 6y + 9 = -8 + 4 + 9$$

$$(x - 2)^2 + (y + 3)^2 = 5$$

$$(x - 2)^2 + (y + 3)^2 = (\sqrt{5})^2 \qquad \text{Remember that } (\sqrt{5})^2 = 5.$$

Thus the center is at $(2, -3)$, and the radius is $\sqrt{5} \approx 2.2$. The graph is shown in Figure 28.   ■

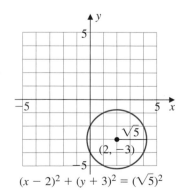

$(2, -3)$

$\sqrt{5}$

$(x - 2)^2 + (y + 3)^2 = (\sqrt{5})^2$

FIGURE 28

Graphing Ellipses

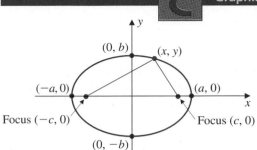

FIGURE 29

If you look at the diagram in the *Getting Started* at the beginning of this section, you will see the drawing of part of an ellipse. An *ellipse* is the set of points in a plane such that the sum of the distances of each point from two fixed points (called the **foci**; singular, *focus*) is a constant. If the coordinates of the foci are $(c, 0)$ and $(-c, 0)$, then the center of the ellipse is at the origin, and the x- and y-intercepts are given by $x = \pm a$ and $y = \pm b$. This is shown in Figure 29.

We can use the distance formula to find the equation of an ellipse (see *Using Your Knowledge* 7.2) but for the time being, we assume the following.

STANDARD FORM OF THE EQUATION OF AN ELLIPSE WITH FOCI ON THE AXIS AND CENTER AT THE ORIGIN

The equation of the ellipse whose x-intercepts are $x = \pm a$ and whose y-intercepts are $y = \pm b$ is

$$\frac{x^2}{a^2} + \frac{y^2}{b^2} = 1$$

If a and b are equal, the ellipse is a circle.

EXAMPLE 6   Graphing an ellipse

Graph: $4x^2 + 25y^2 = 100$

SOLUTION   To make sure we have an ellipse, we write the equation in the standard form

$$\frac{x^2}{a^2} + \frac{y^2}{b^2} = 1$$

If we divide each term by 100 (to make the right side 1), we have

$$\frac{x^2}{25} + \frac{y^2}{4} = 1$$

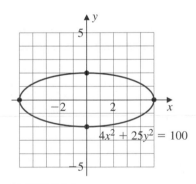

FIGURE 30

The x-intercepts are $x = \pm 5$, the y-intercepts are $y = \pm 2$, and the center is at the origin. We then pass the ellipse through the four intercepts, as shown in Figure 30.

Note that we can actually graph the original equation $4x^2 + 25y^2 = 100$ by letting $x = 0$ to obtain

$$25y^2 = 100$$
$$y^2 = 4$$
$$y = \pm\sqrt{4} = \pm 2$$

Letting $y = 0$ will yield $x = \pm 5$, as before.   ■

As in the case of circles, ellipses can be centered away from the origin, as Example 7 shows.

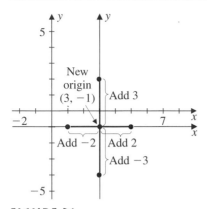

FIGURE 31

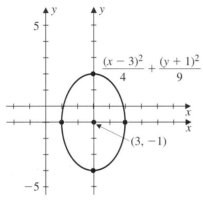

FIGURE 32

EXAMPLE 7   Graphing an ellipse not centered at the origin

Graph: $\dfrac{(x-3)^2}{4} + \dfrac{(y+1)^2}{9} = 1$

SOLUTION   The center of this ellipse is at $(3, -1)$. We construct a new coordinate system with origin at $(3, -1)$, as shown in Figure 31. The x-intercepts of the new system are at ± 2 units from 3, and the y-intercepts are at ± 3 units from -1.

     The graph of the ellipse is shown in Figure 32. ∎

Note that we can find the equation of an ellipse with foci on one of the axes and with center at the origin if we know its x- and y-intercepts. Thus the ellipse of Example 6 passes through $(\pm 5, 0)$ and $(0, \pm 2)$ and has equation

$$\frac{x^2}{5^2} + \frac{y^2}{2^2} = 1$$

Similarly, if the ellipse passes through $(\pm 3, 0)$ and $(0, \pm 4)$, its equation would be

$$\frac{x^2}{3^2} + \frac{y^2}{4^2} = 1$$

We will use this idea in Problems 52–55. Finally, make sure you know how to tell the difference between the graph of a circle and that of an ellipse. The equation

$$Ax^2 + By^2 = C \qquad (A, B, \text{ and } C \text{ positive})$$

has a **circle** as its graph when $A = B$ and an **ellipse** when $A \neq B$.

GRAPH IT

By now, you should be aware that most graphers only graph certain types of algebraic expressions. Do you know what type of expressions? Functions! Would your grapher be able to graph $x^2 + y^2 = 25$ from Example 4? The vertical line test should convince you that $x^2 + y^2 = 25$ does not represent a function. How can we graph it? We will do it by first solving for y, obtaining $y = \pm\sqrt{25 - x^2}$, and then graphing each of the two halves of the circle,

$$Y_1 = \sqrt{25 - x^2} \text{ (the top half)} \qquad \text{and} \qquad Y_2 = -\sqrt{25 - x^2} \text{ (the bottom half)}$$

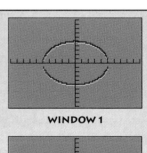

WINDOW 1

Unfortunately, the result shown in Window 1 looks like an ellipse. Do you know why? Because the units on the x-axis are longer than the units on the y-axis; that is, the graph is wider than it is high. To fix this, use a square window ([ZOOM] 5 on a TI-82) to obtain the graph shown in Window 2. (You can save time when you graph $y = \pm\sqrt{25 - x^2}$ by entering $Y_1 = \sqrt{25 - x^2}$ and then $Y_2 = -Y_1$. With a TI-82, place the cursor by $Y_2 =$ and press [(-)] [2nd] [Y-VARS] 1 1 [ENTER].)

WINDOW 2

Now let's try to graph $(x + 2)^2 + (y - 1)^2 = 9$ from Example 3. To solve for y, subtract $(x - 2)^2$ from both sides, take the square roots of both sides and add 1 to obtain $y = \pm\sqrt{9 - (x + 2)^2} + 1$. As we did before, graph the top and bottom of the circle using a square window, as shown in Window 3. It's easier to find the center of the circle from the equation itself, but can we find the center of the circle from the graph? First, note that the x-coordinate has a corresponding y-coordinate that is a *maximum* for the top half of the circle. Finding the coordinates of this maximum shows that the maximum occurs

(continued on facing page)

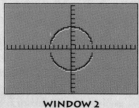

Maximum
X=-2   Y=4

WINDOW 3

when $y = 4$. (With a TI-82, press 2nd CALC 4 and follow the prompts.) The minimum of the lower half of the circle also occurs when $x = -2$ and corresponds to $y = -2$, so the radius is

$$\frac{4 - (-2)}{2} = 3$$

and thus the center is at $(-2, 1)$.

The ellipses in Examples 6 and 7 can be graphed similarly. Thus if we solve for y in Example 6, we have

$$y = \pm\sqrt{\frac{100 - 4x^2}{25}}$$

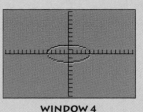

WINDOW 4

Graphing the top and bottom halves of the ellipse produces the graph of $4x^2 + 25y^2 = 100$ shown in Window 4. Now try Example 7.

EXERCISE 9.2

A In Problems 1–10, find the equation of a circle with the given center and radius.

1. Center $(3, 8)$, radius 2

2. Center $(2, 5)$, radius 3

3. Center $(-3, 4)$, radius 5

4. Center $(-5, 2)$, radius 5

5. Center $(-3, -2)$, radius 4

6. Center $(-1, -7)$, radius 9

7. Center $(2, -4)$, radius $\sqrt{5}$

8. Center $(3, -5)$, radius $\sqrt{7}$

9. Radius 3, center at the origin

10. Radius 4, center at the origin

B In Problems 11–32, find the center and radius of the circle and sketch the graph.

11. $(x - 1)^2 + (y - 2)^2 = 9$

12. $(x - 2)^2 + (y - 1)^2 = 4$

13. $(x + 1)^2 + (y - 2)^2 = 4$

14. $(x + 2)^2 + (y - 1)^2 = 9$

15. $(x - 1)^2 + (y + 2)^2 = 1$

16. $(x - 2)^2 + (y + 1)^2 = 4$

17. $(x + 2)^2 + (y + 1)^2 = 9$

18. $(x + 3)^2 + (y + 1)^2 = 4$

19. $(x - 1)^2 + (y - 1)^2 = 7$

20. $(x - 1)^2 + (y - 1)^2 = 3$

21. $x^2 - 6x + y^2 - 4y + 9 = 0$

22. $x^2 - 6x + y^2 - 2y + 9 = 0$

23. $x^2 + y^2 - 4x + 2y - 4 = 0$

24. $x^2 + y^2 + 2x - 4y - 4 = 0$

25. $x^2 + y^2 - 25 = 0$

26. $x^2 + y^2 - 9 = 0$

27. $x^2 + y^2 - 7 = 0$

28. $x^2 + y^2 - 3 = 0$

29. $x^2 + y^2 + 6x - 2y = -6$

30. $x^2 + y^2 + 4x - 2y = -4$

31. $x^2 + y^2 - 6x - 2y + 6 = 0$

32. $x^2 + y^2 - 4x - 6y + 12 = 0$

In Problems 33–46, graph the ellipse. Give the coordinates of the center and the values of a and b.

33. $25x^2 + 4y^2 = 100$

34. $9x^2 + 4y^2 = 36$

35. $x^2 + 4y^2 = 4$

36. $x^2 + 9y^2 = 9$

37. $x^2 + 4y^2 = 16$

38. $x^2 + 9y^2 = 25$

39. $\dfrac{x^2}{9} + \dfrac{y^2}{16} = 1$

40. $\dfrac{x^2}{4} + \dfrac{y^2}{1} = 1$

41. $\dfrac{(x-1)^2}{4} + \dfrac{(y-2)^2}{9} = 1$

42. $\dfrac{(x-2)^2}{9} + \dfrac{(y-1)^2}{4} = 1$

43. $\dfrac{(x-2)^2}{9} + \dfrac{(y+3)^2}{4} = 1$

44. $\dfrac{(x-1)^2}{4} + \dfrac{(y+2)^2}{9} = 1$

45. $\dfrac{(x-1)^2}{16} + \dfrac{(y-1)^2}{9} = 1$

46. $\dfrac{(x-2)^2}{9} + \dfrac{(y-1)^2}{16} = 1$

In Problems 47–51, find an equation of the circle with center at the origin and:

47. Passing through the point $(4, 3)$

48. Passing through the point $(3, 4)$

49. Passing through the point $(-5, -12)$

50. x-intercepts ± 5

51. y-intercepts ± 3

In Problems 52–55, find an equation of the ellipse centered at the origin and passing through:

52. Points $(\pm 7, 0)$ and $(0, \pm 2)$

53. Points $(\pm 2, 0)$ and $(0, \pm 6)$

54. Points $(\pm 3, 0)$ and $(0, \pm 7)$

55. Points $(\pm 6, 0)$ and $(0, \pm 4)$

APPLICATIONS

56. A circular arch for a bridge has a 100-ft span. If the height of the arch above the water is 25 ft, find an equation of the circle containing the arch if the center is at the origin as shown. [*Hint:* If the radius is r, $(50, r - 25)$ must satisfy the equation $x^2 + y^2 = r^2$.]

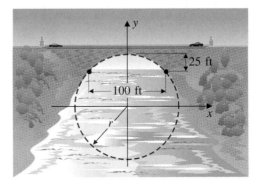

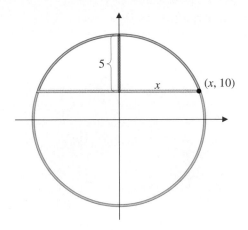

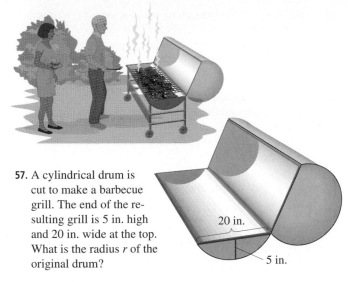

57. A cylindrical drum is cut to make a barbecue grill. The end of the resulting grill is 5 in. high and 20 in. wide at the top. What is the radius r of the original drum?

20 in.

5 in.

58. The larger semicircle has a radius of 15 ft.

 a. If the x-axis is placed at floor level, what is the equation of the circle?

 b. If the vertical bars inside the smaller semicircle are 1 ft apart, what is the length of the longest vertical bar?

 c. What is the length of the bar 1 ft to the right of the longest bar?

59. A portion of a circle with a 15-ft radius (blue) is shown. If the vertical bar (red) is 5 ft, how long is the horizontal bar (green)? (*Hint:* Look at the diagram.)

60. The elliptical drain pipe is 150 cm long and 100 cm high. If the origin is placed at the center of the pipe, what is the equation in standard form of the elliptical opening?

61. The elliptical portion of the sign is 8 ft wide and 5 ft high. What is the standard form of the equation of this ellipse if the origin is at the center of the sign?

62. The top half of a sign is half of an ellipse 7 ft high and 13 ft wide. If the origin is at the center of the sign, what is the equation of the complete ellipse in standard form?

63. The elliptical portion of the tanker truck is 8 ft wide and 6 ft high. What is the equation of the ellipse whose origin is at the center of the elliptical portion of the truck?

64. An elliptical running track is 100 yd long and 50 yd wide.
    a. If the origin is at its center, what is the equation of the ellipse?
    b. A running strip for pole vaulting is parallel to the y-axis 20 yd from the right-hand end of the track. Both ends of this strip are 5 yd from the running track. How long is the strip?

65. The orbit of the Earth around the Sun (one of the foci) is an ellipse as shown. The equation of the ellipse is written as

$$\frac{x^2}{a^2} + \frac{y^2}{b^2} = 1$$

where x and y are in millions of miles.

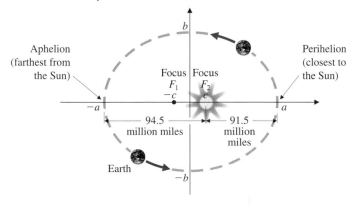

    a. Find a.   b. Find c.   c. If $b^2 = a^2 - c^2$, find b.

66. A semielliptic arch is supporting a bridge spanning a river 50 ft wide. The center of the arch is 20 ft above the center of the river. Write an equation of the ellipse in which the x-axis coincides with the water level and the y-axis passes through the center of the arch.

67. Have you eaten from plastic food plates lately? These plates are made using an elliptical mold 12 in. long and 9 in. wide.
    a. Write in standard form the equation of the outside ellipse of the mold.
    b. Find the width of the dish at a distance of 4 in. from the center. (This is the width in the direction perpendicular to the x-axis).

68. A semielliptic arch spanning a bridge has the dimensions shown. What is the height h of the arch at a distance of 10 ft from the center?

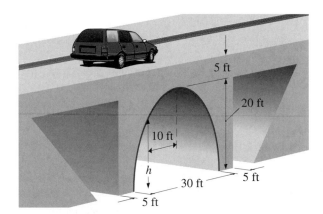

69. An 8-ft-wide boat with a mast whose top is 15 ft above the water is about to go under the bridge of Problem 66. How close can it get to the bank on the right side of the river and still fit under the bridge?

SKILL CHECKER

70. If you solve the equation $x^2 + y^2 = 25$ for x, you obtain two answers. What are these answers?

71. One of the answers in Problem 70 is always nonnegative. Look at the graph of $x^2 + y^2 = 25$ in Example 4. To what part of the graph does the positive answer correspond?

72. One of the answers in Problem 70 is always nonpositive. Look at the graph of $x^2 + y^2 = 25$ in Example 4. To what part of the graph does the negative answer correspond?

Ellipses Revisited

The definition of an ellipse is as follows.

ELLIPSE

An **ellipse** is the set of all points, the sum of whose distances from two fixed points $(c, 0)$ and $(-c, 0)$ is a constant $2a$ $(a > c)$. Each fixed point is called a **focus**.

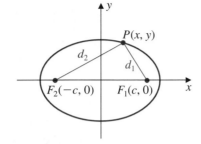

In Problems 73–78, we will prove that this definition leads to the equation of the ellipse that we gave in the text.

73. Suppose $P(x, y)$ is a point on the ellipse. Find the distance d_1 from P to F_1.

74. Find the distance d_2 from P to F_2.

75. The sum of the distances found in Problems 73 and 74 must be $2a$. Thus

$$\sqrt{(x - c)^2 + y^2} + \sqrt{(x + c)^2 + y^2} = 2a$$

Rewrite this equation with one radical on each side. Then square both sides and simplify. What is your answer?

76. Rewrite your answer for Problem 75 with the radical on one side and then square both sides again and simplify. What is your answer?

77. In your answer for Problem 76, since $a > c$, let $a^2 - c^2 = b^2$. Isolate all the variables on the left side. What is your answer?

78. Divide all terms of the answer you obtained in Problem 77 by $a^2 b^2$. What is your answer? If everything went well, you should have

$$\frac{x^2}{a^2} + \frac{y^2}{b^2} = 1$$

the equation of an ellipse.

In general, the equation for an ellipse centered at the origin and with foci on one of the axes is

$$\frac{x^2}{a^2} + \frac{y^2}{b^2} = 1$$

Discuss the resulting graph:

79. When $a > b$.   80. When $a < b$.   81. When $a = b$.

82. Can you explain why a circle is a special case of an ellipse?

If you know how to do these problems, you have learned your lesson!

Graph:

83. $\dfrac{(x + 3)^2}{4} + \dfrac{(y + 1)^2}{9} = 1$

84. $\dfrac{(x - 3)^2}{4} + \dfrac{(y - 1)^2}{9} = 1$

85. $4x^2 + 9y^2 = 36$ 86. $9x^2 + 4y^2 = 36$

87. $x^2 - 4x + y^2 - 6y + 9 = 0$

88. $x^2 + 6x + y^2 + 2y + 7 = 0$

89. $x^2 + y^2 = 4$ 90. $x^2 + y^2 - 6 = 0$

Find the center and radius and sketch the graph of:

91. The circle whose equation is $(x - 3)^2 + (y - 1)^2 = 4$.

92. The circle whose equation is $(x + 3)^2 + (y + 1)^2 - 5 = 0$.

Find the equation of:

93. A circle of radius 5 and with center at the origin.

94. A circle of radius 4 and with center at the origin.

95. A circle with center at $(-3, 6)$ and radius 3.

96. A circle with center at $(-2, -3)$ and radius 2.

9.3

HYPERBOLAS AND IDENTIFICATION OF CONICS

To succeed, review how to:

1. Graph points on the Cartesian plane (pp. 186–187).

2. Find the x- and y-intercepts of a curve (pp. 390–391).

Objectives:

 Graph hyperbolas.

B Identify conic sections by examining their equations.

getting started

Hyperbolas in the Night

Have you seen any hyperbolas lately? Next time you are outside at night, look at building lights. Many of the beams of light you see are hyperbolas. If you've studied chemistry or physics, you might also know that alpha particles (one of three types of radiation resulting from natural radioactivity) have trajectories (paths) that are hyperbolas. In this section we shall study hyperbolas by examining their equations and then graphing them.

A **Graphing Hyperbolas**

A **hyperbola** is the set of points in a plane such that the difference of the distances of each point from two fixed points (called the **foci**) is a constant (See *Using Your Knowledge* at the end of this section.) Consider the equation

$$\frac{x^2}{4} - \frac{y^2}{9} = 1$$

When $x = 0$, $y^2 = -9$, so there are no y-intercepts because $y^2 = -9$ has no real-number solution. When $y = 0$, $x^2 = 4$, and $x = \pm 2$ are the x-intercepts. The graph is shown in Figure 33.

Similarly, the graph of

$$\frac{y^2}{9} - \frac{x^2}{4} = 1$$

has no x-intercept since $y = 0$ yields $x^2 = -4$, which has no real-number solution. The y-intercepts are ± 3. The graph is shown in Figure 34.

The hyperbola

$$\frac{x^2}{4} - \frac{y^2}{9} = 1$$

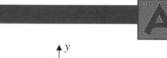

FIGURE 33

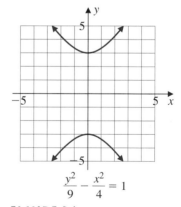

$$\frac{y^2}{9} - \frac{x^2}{4} = 1$$

FIGURE 34

has intercepts $x = \pm 2$. We can use the denominator of y^2 to help us with the graph. If we draw a rectangle with sides parallel to the x- and y-axes and passing through the x-intercept and the points on the y-axis corresponding to the square root of the denominator of $\frac{y^2}{9}$.

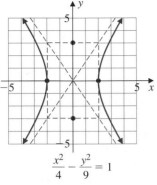

$$\frac{x^2}{4} - \frac{y^2}{9} = 1$$

FIGURE 35

in this case ± 3, and then connect opposite corners of the rectangle with a line, the graph of the hyperbola will approach these lines, called *asymptotes*. The asymptotes are *not* part of the hyperbola, but are used to help graph it. Note that the hyperbola never touches the asymptotes but gets closer and closer to them as x and y get larger and larger in absolute value. The graphs of the hyperbolas

$$\frac{x^2}{4} - \frac{y^2}{9} = 1 \qquad \text{and} \qquad \frac{y^2}{9} - \frac{x^2}{4} = 1$$

are shown in Figures 35 and 36. Here is a summary of this discussion.

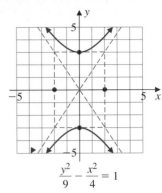

$$\frac{y^2}{9} - \frac{x^2}{4} = 1$$

FIGURE 36

STANDARD FORMS OF GRAPHS OF HYPERBOLAS

The graph of the equation

$$\frac{x^2}{a^2} - \frac{y^2}{b^2} = 1 \qquad \textbf{(1)}$$

is a **hyperbola** centered at the origin with x-intercepts $\pm a$.

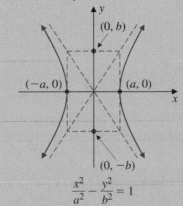

$$\frac{x^2}{a^2} - \frac{y^2}{b^2} = 1$$

The graph of the equation

$$\frac{y^2}{a^2} - \frac{x^2}{b^2} = 1 \qquad \textbf{(2)}$$

is a **hyperbola** centered at the origin with y-intercepts $\pm a$.

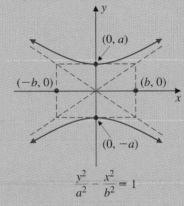

$$\frac{y^2}{a^2} - \frac{x^2}{b^2} = 1$$

The **asymptotes** of either hyperbola are the lines through opposite corners of the auxiliary rectangle whose sides pass through $(\pm a, 0)$ and $(0, \pm b)$ for hyperbola (1), and $(0, \pm a)$ and $(\pm b, 0)$ for hyperbola (2).

PROCEDURE

Graphing a Hyperbola

1. Find and graph the points

$$(\pm a, 0) \text{ and } (0, \pm b) \text{ for } \frac{x^2}{a^2} - \frac{y^2}{b^2} = 1$$

or

$$(0, \pm a) \text{ and } (\pm b, 0) \text{ for } \frac{y^2}{a^2} - \frac{x^2}{b^2} = 1$$

2. Connect the opposite corners of the resulting rectangle passing through $\pm a$ and $\pm b$ with lines called asymptotes.

3. Start the graph from the vertices $(\pm a, 0)$ or $(0, \pm a)$ and draw the hyperbola so that it approaches (but does not touch) the asymptotes.

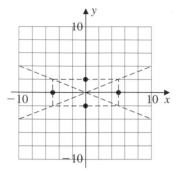

FIGURE 37

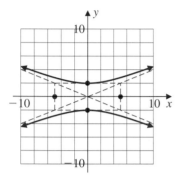

FIGURE 38

EXAMPLE 1   Graphing hyperbolas

Graph:

a. $\dfrac{y^2}{4} - \dfrac{x^2}{25} = 1$ 　　　　　　　　**b.** $25x^2 - 4y^2 = 100$

SOLUTION

a. Since the y^2 term is positive, the hyperbola is centered at the origin and has y-intercepts $y = \pm a = \pm 2$. (There are no x-intercepts.) Our auxiliary rectangle will pass through $y = \pm 2$ and through $x = \pm 5$, the square root of the denominator of x^2. We then connect opposite corners to complete our asymptotes, as shown in Figure 37. Since our hyperbola has y-intercepts $y = \pm 2$, we start our graph from the vertex $y = 2$ and approach the asymptotes, obtaining the top half of the hyperbola. The bottom half is obtained similarly by starting at the vertex $y = -2$. (See Figure 38.)

b. Divide each term by 100 to obtain a 1 on the right-hand side. We then write

$$\frac{x^2}{4} - \frac{y^2}{25} = 1$$

This time, we will show our auxiliary rectangle and the hyperbola on the same graph. Since the x^2 term is positive, the hyperbola has x-intercepts ± 2. Our auxiliary rectangle will pass through $x = \pm 2$ and through $y = \pm 5$. We then complete the auxiliary rectangle, the asymptotes, and the graph of the hyperbola with x-intercepts at ± 2, as shown in Figure 39. ■

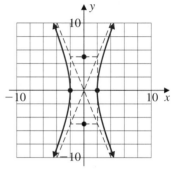

FIGURE 39

You have probably noticed that the equations whose graphs are hyperbolas are similar to the equations whose graphs are ellipses (when written in standard form, hyperbolas are written as differences). We examine this idea in more detail next.

B Identifying Conic Sections by Their Equations

How would you know the shape of the graph of a conic by studying its equation? Table 2 will help you with this.

EXAMPLE 2   Identifying conic sections

Identify:

a. $x^2 = 9 - y^2$ 　　**b.** $y = x^2 - 4$ 　　**c.** $4x^2 = 36 - 9y^2$ 　**d.** $9x^2 = 36 + 4y^2$

SOLUTION   If both variables appear to the second power, we write all variables on the left to make the identification easier.

a. In $x^2 = 9 - y^2$, both variables appear to the second power. Thus $x^2 = 9 - y^2$ is written as $x^2 + y^2 = 9$. The square terms have the same coefficient (1) and are added. The equation $x^2 = 9 - y^2$ represents a circle centered at the origin and with radius 3.

b. In this case, only one variable is squared, the x. Thus the conic is a parabola with the vertex at $(0, -4)$ and opening upward.

c. Both variables are squared. We then write $4x^2 = 36 - 9y^2$ as $4x^2 + 9y^2 = 36$. Here the square terms have different coefficients and are both positive. The equation corresponds to an ellipse centered at the origin. The x-intercepts are found by letting $y = 0$ to obtain $x = \pm 3$. Similarly, the y-intercepts are $y = \pm 2$.

TABLE 2 Identifying Conic Sections

Equation	Graph	Description	Identification
$y = a(x - h)^2 + k$		Parabola with vertex at (h, k). Opens upward for $a > 0$, downward for $a < 0$.	y is not squared.
$y = ax^2 + bx + c$		Parabola with vertex at $x = -\frac{b}{2a}$. Opens upward for $a > 0$, downward for $a < 0$.	y is not squared.
$x = a(y - k)^2 + h$		Parabola with vertex at (h, k). Opens right if $a > 0$, left if $a < 0$.	x is not squared.
$x = ay^2 + by + c$		Parabola with vertex at $y = -\frac{b}{2a}$. Opens right if $a > 0$, left if $a < 0$.	x is not squared.
$(x - h)^2 + (y - k)^2 = r^2$		Circle of radius r, centered at (h, k).	The coefficients of $(x - h)^2$ and $(y - k)^2$ are positive and equal when the variables are on the same side of the equation.
$\dfrac{x^2}{a^2} + \dfrac{y^2}{b^2} = 1$		Ellipse with x-intercepts $\pm a$, y-intercepts $\pm b$.	The coefficients of x^2 and y^2 are positive and not equal.
$\dfrac{x^2}{a^2} - \dfrac{y^2}{b^2} = 1$		Hyperbola with x-intercepts $\pm a$. Auxiliary rectangle passing through $(\pm a, 0)$ and $(0, \pm b)$. Asymptotes drawn through the corners of the auxiliary rectangle.	x^2 has positive coefficient, y^2 has negative coefficient.
$\dfrac{y^2}{a^2} - \dfrac{x^2}{b^2} = 1$		Hyperbola with y-intercepts $\pm a$. Auxiliary rectangle passing through $(0, \pm a)$ and $(\pm b, 0)$. Asymptotes drawn through the corners of the auxiliary rectangle.	y^2 has positive coefficient, x^2 has negative coefficient.

To confirm the fact that $4x^2 = 36 - 9y^2$ is an ellipse, we write the equation in standard form by dividing each term in the equation by 36 so we obtain a 1 on the right-hand side of the equation $4x^2 + 9y^2 = 36$. We then have

$$\frac{4x^2}{36} + \frac{9y^2}{36} = \frac{36}{36}$$

$$\frac{x^2}{9} + \frac{y^2}{4} = 1$$

$$\frac{x^2}{3^2} + \frac{y^2}{2^2} = 1$$

confirming that the conic is an ellipse.

d. Again, both variables are squared. Thus we write $9x^2 = 36 + 4y^2$ as $9x^2 - 4y^2 = 36$. The minus sign indicates that the conic is a hyperbola with x-intercepts $x = \pm 2$. To confirm that $9x^2 - 4y^2 = 36$ corresponds to a hyperbola, divide each term by 36 to obtain

$$\frac{9x^2}{36} - \frac{4y^2}{36} = \frac{36}{36}$$

$$\frac{x^2}{2^2} - \frac{y^2}{3^2} = 1$$

which confirms that the conic is a hyperbola. ■

GRAPH IT

As we mentioned in Section 9.2, most graphers only graph functions. Is a hyperbola the graph of a function? If you apply the vertical line test to the hyperbolas shown in Example 1a, you will see that the graphs are **not** graphs of functions. Thus to graph

$$\frac{y^2}{4} - \frac{x^2}{25} = 1$$

we first solve for y by adding

$$\frac{x^2}{25}$$

to both sides of the equation, multiplying both sides of the equation by 4, and taking the square root of both sides of the equation to obtain

$$y = \pm \sqrt{4 + \frac{4x^2}{25}}$$

We then graph

$$Y_1 = \sqrt{4 + \frac{4x^2}{25}} \qquad \text{and} \qquad Y_2 = -\sqrt{4 + \frac{4x^2}{25}}$$

The graphs for Y_1 (top part) and Y_2 (bottom part) are shown in Window 1. (Remember, you can save time when you enter the expressions to be graphed by entering

$$Y_1 = \sqrt{4 + \frac{4x^2}{25}}$$

and $Y_2 = -Y_1$). Now try to graph Example 1b.

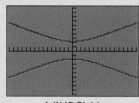

WINDOW 1

Graph of $\dfrac{y^2}{4} - \dfrac{x^2}{25} = 1$

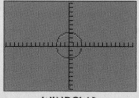

WINDOW 2

Graph of $x^2 = 9 - y^2$

You can easily identify the graphs corresponding to the equations in Example 2 if you solve each of the equations for y and graph the result. [To make sure you don't mistake a circle for an ellipse, use a "square" window (press ZOOM 5 on a TI-82) and check the intercepts.] For 2a, we obtain $y = \pm \sqrt{9 - x^2}$. Graphing $Y_1 = \sqrt{9 - x^2}$ and $Y_2 = -\sqrt{9 - x^2}$, we obtain the graph of $x^2 = 9 - y^2$, which can be easily identified as a circle. (See Window 2.) Now solve for y in parts b, c, and d, graph the results, and identify each of the graphs.

EXERCISE 9.3

A In Problems 1–12, draw the auxiliary rectangle, name the intercepts, and graph.

1. $\dfrac{x^2}{25} - \dfrac{y^2}{9} = 1$

2. $\dfrac{y^2}{9} - \dfrac{x^2}{25} = 1$

3. $\dfrac{y^2}{9} - \dfrac{x^2}{9} = 1$

4. $\dfrac{x^2}{9} - \dfrac{y^2}{9} = 1$

5. $\dfrac{x^2}{9} - \dfrac{y^2}{1} = 1$

6. $\dfrac{y^2}{16} - \dfrac{x^2}{1} = 1$

7. $\dfrac{x^2}{64} - \dfrac{y^2}{49} = 1$

8. $\dfrac{y^2}{49} - \dfrac{x^2}{64} = 1$

9. $\dfrac{y^2}{\frac{16}{9}} - \dfrac{x^2}{\frac{9}{16}} = 1$

10. $\dfrac{x^2}{\frac{9}{4}} - \dfrac{y^2}{\frac{4}{9}} = 1$

11. $x^2 - 9y^2 = 9$

12. $y^2 - 16x^2 = 16$

B In Problems 13–26, identify the conic and give the intercepts. If the conic is a parabola, give the vertex.

13. $x^2 + y^2 = 25$

14. $x^2 - y^2 = 25$

15. $x^2 - y^2 = 36$

16. $x^2 + y^2 = 36$

17. $x^2 - y = 9$

18. $x^2 + y = 9$

19. $y^2 - x = 4$

20. $y^2 + x = 4$

21. $9x^2 = 36 - 9y^2$

22. $4x^2 = 16 - 4y^2$

23. $9x^2 = 36 + 9y^2$

24. $4x^2 = 36 - 9y^2$

25. $x^2 = 9 - 9y^2$

26. $y^2 = 4 - 4x^2$

In Problems 27–30, sketch the hyperbola. (*Hint:* They are not centered at the origin.)

27. $\dfrac{(x-1)^2}{4} - \dfrac{(y+1)^2}{9} = 1$

28. $\dfrac{(x-2)^2}{9} - \dfrac{(y+1)^2}{4} = 1$

29. $\dfrac{(x-1)^2}{9} - \dfrac{(y-2)^2}{4} = 1$

30. $\dfrac{(x-2)^2}{4} - \dfrac{(y-1)^2}{9} = 1$

31. A semicircular plate of diameter D with a circular opening of diameter d is to be constructed. If the area of the plate is π square inches, the relationship between D and d is given by

$$\frac{D^2}{8} - \frac{d^2}{4} = 1$$

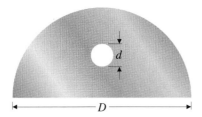

a. What type of conic corresponds to this equation?

b. Sketch the graph of

$$\frac{D^2}{8} - \frac{d^2}{4} = 1$$

(Use $\sqrt{8} \approx 2.8$.)

32. If three holes of diameter d were drilled in a semicircular plate of diameter D and the remaining area was π square inches, the relationship between D and d would be

$$\frac{D^2}{8} - \frac{3d^2}{4} = 1$$

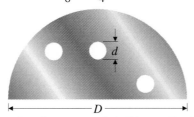

a. What type of conic corresponds to this equation?

b. Show that you can write the equation of the conic as

$$\frac{D^2}{8} - \frac{d^2}{\dfrac{4}{3}} = 1$$

c. Sketch the graph of

$$\frac{D^2}{8} - \frac{3d^2}{4} = 1$$

$\left(\text{Use } \sqrt{\tfrac{4}{3}} \approx 1.15.\right)$

33. The total kinetic energy of a spinning body moving through the air is 144 foot-pound (ft-lb). The velocity v through the air and the spinning velocity ω are related by the equation $4v^2 + 9\omega^2 = 144$. Graph this equation.

34. If the equation in Problem 33 is $16v^2 + 4\omega^2 = 256$, graph the equation.

Graph:

35. $y = x - 4$

36. $x^2 + y^2 = 4$

37. $y = x^2 + 1$

38. $y = x^2 - 1$

39. $4x^2 + 9y^2 = 36$

40. $9x^2 - 4y^2 = 36$

The definition for a hyperbola is as follows.

HYPERBOLA

A **hyperbola** is the set of points in a plane such that the difference of the distances of each point from two fixed points (called the **foci**) is a constant.

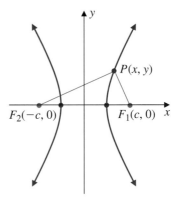

In Problems 41–46, we will prove that this definition leads to the equation of the hyperbola we gave in the text.

41. Suppose (x, y) is a point on the hyperbola. Find the distance from P to F_1.

42. Find the distance from P to F_2.

43. Let the difference of the distances found in Problems 41 and 42 be $2a$. Thus

$$\sqrt{(x-c)^2 + y^2} - \sqrt{(x+c)^2 + y^2} = \pm 2a$$

Rewrite this equation with one radical on each side. Then square both sides and simplify. What is your answer?

44. Rewrite your answer for Problem 43 with the radical on one side. Then square both sides again and simplify. What is your answer?

45. In your answer for Problem 44, let $c^2 - a^2 = b^2$, where $a < c$. Isolate all the variables on the left side. What is your answer?

46. Divide all terms of the answer you obtained in Problem 45 by $a^2 b^2$. What is your answer? If everything went well, you should have

$$\frac{x^2}{a^2} - \frac{y^2}{b^2} = 1$$

the equation of a hyperbola.

We have shown you how to graph the asymptotes of a hyperbola. We are now ready to use this knowledge to find the equation of these asymptotes. Consider the hyperbola

$$\frac{x^2}{a^2} - \frac{y^2}{b^2} = 1$$

47. The expression on the left is the difference of two squares. Factor it.

48. Isolate $\frac{x}{a} - \frac{y}{b}$ on the left. The expression on the right is a complex fraction with 1 as the numerator. What is it?

49. Look at the denominator of the complex fraction in Problem 48. If x and y are positive and very large, what happens to the denominator? What happens to the complex fraction?

50. If you answered that the complex fraction is very small, you are correct. In mathematics we say that $\frac{x}{a} - \frac{y}{b} \to 0$ (the expression approaches zero). Thus for very large positive x and y, $\frac{x}{a} - \frac{y}{b} \approx 0$. This means that $\frac{x}{a} - \frac{y}{b} = 0$ is an asymptote. Solve for y and find its equation.

51. We can show in the same way that $\frac{x}{a} + \frac{y}{b} = 0$ is an asymptote. Solve for y and find its equation.

52. a. In summary, what are the equations of the asymptotes for the hyperbola

$$\frac{x^2}{a^2} - \frac{y^2}{b^2} = 1$$

b. What about the asymptotes for the hyperbola

$$\frac{y^2}{a^2} - \frac{x^2}{b^2} = 1$$

WRITE ON . . .

53. Write an explanation of the procedure you would use to determine whether the graph of an equation is an ellipse or a hyperbola.

54. Write your own definition of the asymptote of a hyperbola.

55. Consider the equation $Ax^2 + By^2 = C$, where A, B, and C are real numbers. Under what conditions will the graph of this equation be
a. A circle?   **b.** An ellipse?   **c.** A hyperbola?

 MASTERY TEST | If you know how to do these problems, you have learned your lesson!

Identify:

56. $9x^2 = 36 - 4y^2$

57. $y^2 = 9 - x^2$

58. $y^2 = 9 - 4x^2$

59. $y = x^2 + 3$

60. $4x^2 = 36 + 9y^2$

Graph:

61. $\dfrac{y^2}{16} - \dfrac{x^2}{9} = 1$

62. $\dfrac{x^2}{16} - \dfrac{y^2}{9} = 1$

63. $9x^2 - 25y^2 = 225$

64. $25y^2 - 9x^2 = 225$

9.4

NONLINEAR SYSTEMS OF EQUATIONS

To succeed, review how to:

1. Graph lines and conics (pp. 387–389, 543–554).

2. Use the graphical and substitution methods to solve systems of equations (pp. 463–465, 466–469).

Objectives:

A Solve a nonlinear system by substitution.

B Solve a system with two second-degree equations by elimination.

C Solve applications involving nonlinear systems.

getting started   Consumer's Demand and Supply

How do supply and demand determine the market price and quantity of wheat available for sale? As the price of wheat *decreases,* the quantity demanded by consumers *increases*. If the price *increases,* the demand *decreases*. The point *C* of intersection of the two curves is called the **equilibrium point**. At this point, the price of a bushel of wheat is $3, and the amount demanded by the consumers (12 million bushels per month) exactly equals the amount supplied by producers. The equations that describe this system are an example of a nonlinear system of equations. Graphical results may be difficult to confirm so we use the substitution method instead. We show you how to do this next.

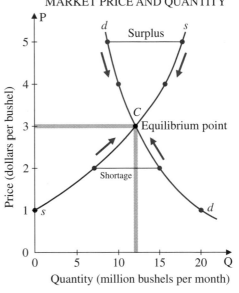

HOW SUPPLY AND DEMAND DETERMINE MARKET PRICE AND QUANTITY

Quantity (million bushels per month)

A   Solving Nonlinear Systems by Substitution

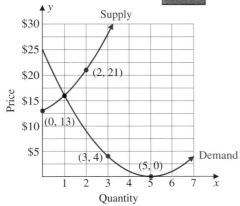

FIGURE 40

The systems of equations we studied in Chapter 8 consisted of linear equations. An equation in which some terms have more than one variable or have a variable of degree 2 or higher is a **nonlinear equation**. A system of equations including at least one nonlinear equation is called a **nonlinear system of equations**. Such systems may have **one, more than one**, or **no** real solutions. Let's start with a system consisting of two parabolas, solve them graphically, and confirm the results by substitution.

Suppose the demand curve for a product is given by $y = (x - 5)^2$ (a parabola), where x is the number of units produced and y is the price, and the supply curve is given by $y = x^2 + 2x + 13$ (another parabola), where y is the price and x the number of units available. To find the equilibrium point, we sketch both curves to find the intersection, as shown in Figure 40.

The equilibrium point seems to be the point (1, 16). How can we be sure? We use the substitution method to solve the system and confirm our result.

$$y = (x - 5)^2 \qquad \text{A parabola} \qquad \textbf{(1)}$$

$$y = x^2 + 2x + 13 \qquad \text{A parabola} \qquad \textbf{(2)}$$

We wish to find an ordered pair (x, y) that is a solution of *both* equations (1) and (2). Using the substitution method, we substitute $(x - 5)^2$ for y on the left-hand side of equation (2) to obtain

$$(x - 5)^2 = x^2 + 2x + 13$$

$$x^2 - 10x + 25 = x^2 + 2x + 13 \qquad \text{Expand.}$$

$$-10x + 25 = 2x + 13 \qquad \text{Subtract } x^2.$$

$$-12x = -12 \qquad \text{Subtract } 2x \text{ and } 25.$$

$$x = 1 \qquad \text{Divide by } -12.$$

If $x = 1$ in equation (2), then

$$y = (1)^2 + 2(1) + 13 = 16$$

Hence the solution for the given system is $(1, 16)$, as you can verify by substituting $x = 1$ and $y = 16$ in the two equations.

EXAMPLE 1   Solving a nonlinear system by substitution

Find the solution of the given system by the substitution method. Check the solution by sketching the graphs of the equations.

$$x^2 + y^2 = 25 \qquad \text{A circle} \qquad \textbf{(1)}$$

$$x + y = 5 \qquad \text{A line} \qquad \textbf{(2)}$$

SOLUTION   We first rewrite equation (2) in the equivalent form $y = 5 - x$ to obtain

$$x^2 + y^2 = 25 \qquad \textbf{(1)}$$

$$y = 5 - x \qquad \textbf{(3)}$$

Replacing y in equation (1) by $(5 - x)$, we get

$$x^2 + (5 - x)^2 = 25 \qquad \text{Substitute } (5 - x) \text{ in equation (1).} \qquad \textbf{(4)}$$

$$x^2 + 25 - 10x + x^2 = 25 \qquad \text{Expand.}$$

$$2x^2 - 10x = 0 \qquad \text{Simplify; subtract 25.}$$

$$x^2 - 5x = 0 \qquad \text{Divide by 2.}$$

$$x(x - 5) = 0 \qquad \text{Factor.}$$

$$x = 0 \quad \text{or} \quad x - 5 = 0$$

$$x = 0 \quad \text{or} \quad x = 5 \qquad \text{Solve for } x.$$

We now let $x = 0$ and $x = 5$ in equation (3) to obtain the corresponding y-values:

$$y = 5 - 0 = 5 \quad \text{and} \quad y = 5 - 5 = 0$$

Thus when $x = 0$, $y = 5$, and when $x = 5$, $y = 0$. Therefore, the solutions of the system are $(0, 5)$ and $(5, 0)$.

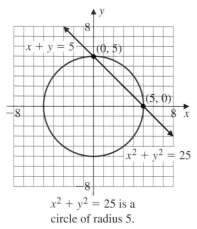

$x^2 + y^2 = 25$ is a circle of radius 5.

FIGURE 41

Note that if we had substituted $x = 0$ and $x = 5$ in equation (1) rather than in equation (4), we would have obtained

$$0^2 + y^2 = 25 \quad \text{and} \quad 5^2 + y^2 = 25$$

That is, $y = \pm 5$ and $y = 0$. In this case, the solutions obtained would have been $(0, 5)$, $(0, -5)$, and $(5, 0)$. However, $(0, -5)$ is *not* a solution of equation (3), since $-5 \neq 5 - 0$. Therefore, the only solutions are $(0, 5)$ and $(5, 0)$, as before.

> ⚠ **CAUTION** If the degrees of the equations are different, one component of a solution should be substituted in the *lower-degree* equation to find the ordered pairs satisfying *both* equations.

For this reason, we double-check our work by graphing the given system (see Figure 41) and verify that our solutions are correct. ∎

EXAMPLE 2 Solving nonlinear systems by substitution and checking with a graph

Find the solution of the given system by the substitution method. Check the solution by sketching the graphs of the equations.

$$x^2 + y^2 = 9 \tag{1}$$

$$x + y = 5 \tag{2}$$

SOLUTION Rewriting equation (2) in the form $y = 5 - x$, we obtain the equivalent system

$$x^2 + y^2 = 9 \tag{1}$$

$$y = 5 - x \tag{3}$$

Substituting $y = 5 - x$ in equation (1), we get

$$x^2 + (5 - x)^2 = 9$$

$$x^2 + 25 - 10x + x^2 = 9 \qquad \text{Expand.}$$

$$2x^2 - 10x + 25 = 9 \qquad \text{Simplify.}$$

$$2x^2 - 10x + 16 = 0 \qquad \text{Subtract 9.}$$

$$x^2 - 5x + 8 = 0 \qquad \text{Divide by 2.}$$

Using the quadratic formula with $a = 1, b = -5, c = 8$, we get

$$x = \frac{5 \pm \sqrt{25 - 4 \cdot 8}}{2} = \frac{5 \pm \sqrt{-7}}{2} = \frac{5 \pm \sqrt{7}i}{2}$$

Substituting these values in equation (3), we obtain

$$y = 5 - \frac{5 + \sqrt{7}i}{2} \quad \text{and} \quad y = 5 - \frac{5 - \sqrt{7}i}{2}$$

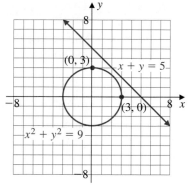

FIGURE 42

That is,

$$y = \frac{5}{2} - \frac{\sqrt{7}}{2}i \qquad \text{and} \qquad y = \frac{5}{2} + \frac{\sqrt{7}}{2}i$$

Hence the solutions of the system are

$$\left(\frac{5 + \sqrt{7}i}{2}, \frac{5 - \sqrt{7}i}{2}\right) \qquad \text{and} \qquad \left(\frac{5 - \sqrt{7}i}{2}, \frac{5 + \sqrt{7}i}{2}\right)$$

as can be checked in the original equations. The graphs of the two equations are shown in Figure 42. As you can see, the graphs *do not intersect.* When the solutions of a system of equations are imaginary numbers, there are no points of intersection for the graphs. This is because the coordinates of points in the real plane are *real* numbers. ∎

B | Solving Systems with Two Second-Degree Equations

When both equations in a system are of degree 2, it's easier to use the elimination method, as in Example 3.

EXAMPLE 3   Solving a system with two second-degree equations

Solve the system:

$$x^2 - 2y^2 = 1$$
$$x^2 + 4y^2 = 25$$

Verify the solution by graphing.

SOLUTION   To eliminate y^2, we multiply the first equation by 2 and add the result to the second equation:

$$\begin{aligned} 2x^2 - 4y^2 &= 2 \\ x^2 + 4y^2 &= 25 \\ \hline 3x^2 &= 27 \\ x^2 &= 9 \\ x &= \pm 3 \end{aligned}$$

The x-coordinates of the point of intersection are 3 and -3. Substituting in the second equation,

$$(\pm 3)^2 + 4y^2 = 25$$
$$4y^2 = 16$$
$$y^2 = 4$$
$$y = \pm 2$$

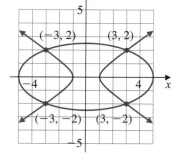

FIGURE 43

Thus the four points of intersection are $(3, 2)$, $(3, -2)$, $(-3, 2)$, and $(-3, -2)$, as you can check in the original equations. The graph for the two equations is shown in Figure 43. ∎

> NOTE   An independent system of equations with at least one being quadratic can have four, three, two, one, or no real-number solutions.

Solving Applications of Nonlinear Systems

The break-even point is the point at which enough units have been sold so that the cost C and the revenue R are equal. Example 4 shows how to use the methods we've studied to find the break-even point.

EXAMPLE 4   Breaking even

The total cost C for manufacturing and selling x units of a product each week is given by $C = 30x + 100$, whereas the revenue R is given by $R = 81x - 0.5x^2$. How many items must be manufactured and sold for the company to break even—that is, for C to equal R?

SOLUTION   We are asked to find the value of x for which $C = R$, that is,

$$\overbrace{C}^{} = \overbrace{R}^{}$$
$$30x + 100 = 81x - 0.5x^2$$

or, in standard form,

$$0.5x^2 - 51x + 100 = 0$$

where $a = 0.5$, $b = -51$, and $c = 100$. Using the quadratic formula, we get

$$x = \frac{51 \pm \sqrt{(-51)^2 - 4(0.5)(100)}}{2 \cdot 0.5}$$

$$= \frac{51 \pm \sqrt{2601 - 200}}{1}$$

$$= \frac{51 \pm \sqrt{2401}}{1}$$

$$= 51 \pm 49$$

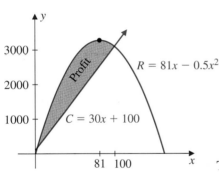

FIGURE 44

Thus x is $51 - 49 = 2$ or $x = 51 + 49 = 100$; that is, the company will break even when 2 or 100 items are sold (see Figure 44). Note that the company makes a profit when they sell between 2 and 100 items. ■

problem solving   **EXAMPLE 5**   A new dimension in desserts

Can you guess what the dimensions of the largest shortcake must have been? It covered 360 square feet and had a 106 ft perimeter! Use the RSTUV method to find the dimensions of this rectangular shortcake.

SOLUTION

1. Read the problem.

We are asked to find the dimensions, and we know that the area is 360 ft^2 and the perimeter is 106 ft.

2. Select the unknowns.

Let L be the length and W the width.

3. Think of a plan.

Translate the problem.

Area A of the rectangle: $A = LW$

Perimeter P of the rectangle: $P = 2L + 2W$

In our case $A = 360$ and $P = 106$, thus $360 = LW$

$106 = 2L + 2W$

4. Use the techniques we have studied to solve this system.

First, solve $360 = LW$ for W to obtain
$$W = \frac{360}{L}$$

Substitute $\frac{360}{L}$ for W in
$$106 = 2L + 2W$$

We then have
$$106 = 2L + 2\left(\frac{360}{L}\right)$$

$$106L = 2L^2 + 720 \qquad \text{Multiply each term by } L.$$

$$L^2 - 53L + 360 = 0 \qquad \text{Divide by 2 and write in standard form.}$$

$$(L - 45)(L - 8) = 0 \qquad \text{Factor.}$$

$$L = 45 \qquad \text{or} \qquad L = 8$$

If $L = 45$, $W = \frac{360}{L} = \frac{360}{45} = 8$. If $L = 8$, $W = \frac{360}{8} = 45$.

Since the length is usually longer than the width, we select the first case in which $L = 45$ and $W = 8$. Thus the shortcake is 45 ft long and 8 ft wide.

5. Verify the answer.

Since $L = 45$ and $W = 8$, the area is $A = 45 \cdot 8 = 360 \text{ ft}^2$, and the perimeter is $P = 2(45) + 2(8) = 106$ ft, as required. ■

GRAPH IT

You can use your grapher to check the solutions of systems of nonlinear equations. In Example 1, start by graphing $x^2 + y^2 = 25$ by solving for y to obtain $y = \pm\sqrt{25 - x^2}$, and then graph $y = 5 - x$ using a square window. As you can see, the graphs intersect at two points, so there are two solutions. Use the TRACE key to find them or, with a TI-82, press [2nd] [CALC] 5, and answer the grapher's questions: "First curve?" "Second curve?" and "Guess?" The grapher gives the points of intersection as shown in Window 1. Now try Example 2.

WINDOW 1

For Example 3, graph

$$y = \pm\sqrt{\frac{x^2 - 1}{2}} \qquad \text{and} \qquad y = \pm\sqrt{\frac{x^2 - 25}{4}}$$

Then press TRACE or [2nd] [CALC] 5 to find the points of intersection. The graph and one of the points of intersection, point $(-3, 2)$, are shown in Window 2.

WINDOW 2

Can you do Example 4 with your grapher? Of course, but first you must understand what the problem asks for. You need to find the points at which $C = 30x + 100$ is the same as $R = 81x - 0.5x^2$. To do this, let $C = Y_1 = 30x + 100$, $R = Y_2 = 81x - 0.5x^2$, and find the points at which $C = R$—that is, the points at which the graphs intersect. To obtain the complete graph, you must be careful with the window you use. Start with a $[-10, 100]$ by $[-10, 3500]$ window Note that when $x = 100$,

$$R = \frac{81(100) - (100)^2}{2} = 3500$$

Since you didn't get a complete graph (see Window 3), change the domain for x to $[-10, 200]$. You now get a complete graph, as shown in Window 4, and are able to use the ZOOM and TRACE keys to find the points of intersection. With a TI-82, use the intersect feature to find these points. The graph with one of the points of intersection, point $(100, 3100)$, is shown

WINDOW 3
This is not a complete graph.

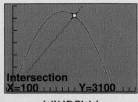

WINDOW 4

in Window 4. You should find the second point to convince yourself that the answers are as before. Now do Example 5 using your grapher. Of course, you have to follow the RSTUV method. The grapher is used when you solve the resulting system of nonlinear equations *after* the conditions of the problem have been translated into mathematical equations.

EXERCISE 9.4

A In Problems 1–16, solve the system and check by graphing.

1. $x^2 + y^2 = 16$
$x + y = 4$

2. $x^2 + y^2 = 9$
$x + y = 3$

3. $x^2 + y^2 = 25$
$y - x = 5$

4. $x^2 + y^2 = 9$
$y - x = 3$

5. $x^2 + y^2 = 25$
$y - x = 1$

6. $x^2 + y^2 = 5$
$y - x = 1$

7. $y = x^2 - 5x + 4$
$x - y = 1$

8. $y = x^2 - 2x + 1$
$x - y = 1$

9. $y = (x - 1)^2$
$y - x = 1$

10. $y = (x + 3)^2$
$x + y = -1$

11. $4x^2 + 9y^2 = 36$
$3y - 2x = 6$

12. $4x^2 + 9y^2 = 36$
$3y + 2x = 6$

13. $x^2 - y^2 = 16$
$x + 4y = 4$

14. $x^2 - y^2 = 9$
$x + 3y = 3$

15. $x^2 + y^2 = 4$
$y - x = 5$

16. $x^2 + y^2 = 4$
$y - x = 3$

B In Problems 17–30, solve the system and check by graphing.

17. $y = 4 - x^2$
$y = x^2 - 4$

18. $y = 2 - x^2$
$y = x^2 - 2$

19. $x^2 + y^2 = 25$
$x^2 - y^2 = 7$

20. $x^2 + y^2 = 20$
$x^2 - y^2 = 2$

21. $x^2 + y^2 = 16$
$x^2 + 16y^2 = 16$

22. $x^2 + y^2 = 9$
$x^2 + 9y^2 = 9$

23. $3x^2 - y^2 = 2$
$x^2 + 2y^2 = 3$

24. $x^2 - 4y^2 = 4$
$9x^2 + 4y^2 = 36$

25. $x^2 + 2y^2 = 11$
$2x^2 + y^2 = 19$

26. $4x^2 + 9y^2 = 52$
$9x^2 + 4y^2 = 52$

27. $x^2 + y^2 = 4$
$x^2 - y^2 = 9$

28. $x^2 + y^2 = 9$
$x^2 - y^2 = 16$

29. $x^2 + y^2 = 1$
$4x^2 + 9y^2 = 36$

30. $x^2 + y^2 = 25$
$4x^2 + 9y^2 = 36$

APPLICATIONS

31. The total cost C (in thousands of dollars) for manufacturing and selling x (thousand) items of a product each month is given by $C = x + 4$, and the revenue is $R = 6x - x^2$. How many items must be manufactured and sold for the company to break even?

32. The total cost C (in thousands of dollars) for manufacturing and selling x (thousand) items of a product each month is given by $C = 2x + 6$, and the revenue is $R = 150 + 20x - x^2$. How many items must be manufactured and sold for the company to break even?

33. There are two very interesting numbers. Their sum is 15 and the difference of their squares is also 15. Find the numbers.

34. Two squares differ in area by 108 ft^2 and their sides differ by 6 ft. Find the dimensions of these squares. (*Hint:* Make a picture!)

35. Find two numbers such that their product is 176 and the sum of their squares is 377.

36. Find the lengths of the sides of a rectangle whose area is 96 in.^{2} and whose perimeter is 44 in.

37. Have you written any big checks lately? The world's *physically* largest check had an area of 2170 ft^2 and a perimeter of 202 ft. If the check was rectangular, what were its dimensions?

38. The largest ancient carpet (A.D. 743) was a gold-enriched silk carpet that covered 54,000 ft^2. If the carpet was rectangular and its perimeter was 960 ft, what were the dimensions of the carpet?

39. A person receives $340 interest from an amount loaned at simple interest for 1 yr. If the interest rate had been 1% higher, she would have received $476. What was the amount loaned and what was the interest rate? (*Hint:* $I = Pr$, where I is the interest, P is the principal, and r is the rate.)

40. There are two interesting positive numbers such that their product, their difference, and the difference of their squares are all equal. Find the numbers. (*Hint:* They are irrational!)

SKILL CHECKER

Graph:

41. $x - y < 4$

42. $y - x < 4$

43. $2x - 3y \geq 6$

44. $3x - 2y \geq 6$

45. $y \geq 2x + 4$

46. $y \geq 3x + 6$

USING YOUR KNOWLEDGE   100-year-old Problems!

Do you think the algebra we have been studying is relatively new? What does it have to do with the price of eggs? Here are a couple of problems from *Elementary Algebra*, published by Macmillan in 1894. See if you have the knowledge to solve these problems.

47. "The number of eggs which can be bought for 25 cents is equal to twice the number of cents which 8 eggs cost. How many eggs can be bought for 25 cents?" (*Hint:* If you let p be the price of eggs, then $\frac{25}{p}$ is the number of eggs you can buy for 25¢.)

48. "One half of the number of cents which a dozen apples cost is greater by 2 than twice the number of apples which can be bought for 30 cents. How many can be bought for $2.50?" (*Hint:* If you let d be the cost of a dozen apples, then $\frac{d}{12}$ is the cost of one apple and

$$\frac{30}{\dfrac{d}{12}}$$

is the number of apples that can be bought for 30¢.

WRITE ON . . .

49. Consider the system

$$y = ax + b$$
$$y = cx + d$$

a. If $b \neq d$, what is the maximum number of solutions for this system? Write an explanation to support your answer. How many solutions are possible if b can be equal to d?

b. Write a set of conditions under which this system will have no solution. What will the relationship between a and c be then?

50. Consider the system

$$y = ax + b$$
$$x^2 + y^2 = r^2$$

a. What is the maximum number of real solutions for this system? Write an explanation to support your answer.

b. Make sketches showing a system similar to the given one and that has no real solution, one solution, and two solutions. What is the geometric interpretation for the number of solutions in a system?

Explain:

51. What is the maximum number of solutions you can have for a system of equations consisting of
   a. Two hyperbolas?   **b.** Two circles?
   c. Two ellipses?   **d.** A hyperbola and a circle?

MASTERY TEST   If you know how to do these problems, you have learned your lesson!

52. The total cost C for manufacturing and selling x units of a product is given by $C = 30x + 100$, whereas the revenue is given by $R = 57x - 0.5x^2$. How many items must be manufactured and sold for the company to break even?

53. The perimeter of a rectangle is 44 cm and its area is 120 cm². What are the dimensions of this rectangle?

Solve and check by graphing:

54. $x^2 - 9y^2 = 9$
    $x^2 + y^2 = 9$

55. $4y^2 - x^2 = 4$
    $x^2 + y^2 = 1$

56. $x^2 + y^2 = 3$
    $x + y = 3$

57. $x^2 + y^2 = 4$
    $x + y = 4$

58. $x^2 + 4y^2 = 16$
    $x + 2y = 4$

59. $x^2 + y^2 = 1$
    $x + y = 1$

9.5

NONLINEAR SYSTEMS OF INEQUALITIES

To succeed, review how to:

1. Graph linear inequalities in two variables (pp. 108–112).
2. Graph conic sections (pp. 543–554).

Objectives:

 Graph second-degree inequalities.

 Graph the solution set of a system of nonlinear inequalities.

getting started   **Profits and Losses**

This graph offers a wealth of information if you know how to read it. The line represents the cost C of a product, whereas the parabola represents the revenue R obtained when the product is sold. Here are some facts about the graph:

1. Where the graph representing the cost C lies above the graph of the revenue R, there is a loss.
2. The revenue equals the cost at a point called the break-even point.
3. Where the graph representing the revenue R lies above the graph of the cost C, there is a profit.

This is an example of a nonlinear system of inequalities, a concept we shall investigate in detail in this section.

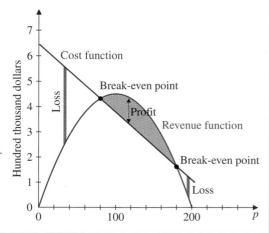

A Solving Second-Degree Inequalities

In Section 7.4, we graphed linear inequalities by graphing the boundary, selecting a test point not on the boundary, and determining the regions corresponding to the solution set. The region under the parabola in the graph is obtained by graphing a **second-degree inequality** using a similar procedure. Let's do an example.

EXAMPLE 1   Solving a second-degree inequality involving a parabola

Graph: $y \leq -x^2 + 3$

SOLUTION   The boundary of the required region is the parabola $y = -x^2 + 3$ with its vertex at $(0, 3)$, as shown in Figure 45. Since the inequality sign is $\leq$, the boundary *is included* in the solution set. To determine which region represents the solution set, select a test point not on the boundary and test the original inequality. A convenient point is $(0, 0)$. When $x = 0$ and $y = 0$, $y \leq -x^2 + 3$ becomes $0 \leq 0 + 3$, a true statement. Thus all points on the same side as $(0, 0)$ will be in the solution set. The graph of the solution set is shown shaded in Figure 46.

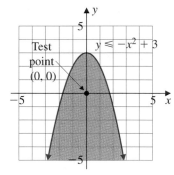

FIGURE 45

FIGURE 46

EXAMPLE 2   Solving a second-degree inequality involving a circle

Graph: $x^2 + y^2 > 4$

SOLUTION   This time, the boundary is the circle $x^2 + y^2 = 4$ with radius 2 and centered at the origin. The inequality sign is $>$, so that the boundary is *not* included in the solution set. It is shown dashed in Figure 47. For the point $(0, 0)$, $x = 0$ and $y = 0$, and $x^2 + y^2 > 4$ becomes $0 + 0 > 4$, a false statement. This means that the region containing $(0, 0)$ is *not* in the solution set. We then shade the region *outside* the circle to represent the solution set, as shown in Figure 47.

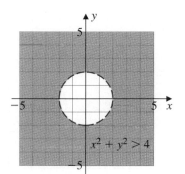

FIGURE 47

EXAMPLE 3   Solving a second-degree inequality involving a hyperbola

Graph: $9y^2 - 4x^2 \leq 36$

SOLUTION   Since all the variables are on the left and there is a minus sign between the terms, the boundary is a hyperbola with y-intercepts $y = \pm 2$. Since the inequality is $\leq$, the boundary *is* part of the solution set. If we use $(0, 0)$ for our test point, $9y^2 - 4x^2 \leq 36$ becomes $0 - 0 \leq 36$, which is true. Thus the point $(0, 0)$ is part of the complete solution set, which is shown shaded in Figure 48.

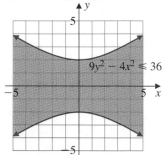

FIGURE 48

B Graphing Systems of Inequalities

We have already solved a linear system of inequalities in two variables using the graphical method. To solve a *nonlinear system of inequalities* graphically, we use the following procedure.

PROCEDURE

Graphing a Nonlinear System of Inequalities
1. Graph each of the inequalities on the same set of axes.
2. Find the region common to both graphs, that is, the intersection of both graphs. The result is the solution set of the system.

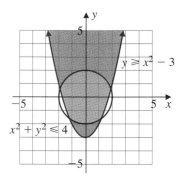

FIGURE 49

We illustrate this procedure next.

EXAMPLE 4   Graphing a nonlinear system of inequalities

Graph: The system $x^2 + y^2 \leq 4$ and $y \geq x^2 - 3$

SOLUTION   The boundary for the first inequality is a circle of radius 2 centered at the origin. The solution set of this inequality includes the boundary and the points inside the circle and is shown in the lighter color in Figure 49. The boundary for the second inequality is a parabola with vertex at $(0, -3)$, and the solution set includes the boundary and all the points above the parabola.

The solution set for the system is the intersection of the two solution sets, the region inside *both* the circle and the parabola and the points on the circle *above* or *on* the parabola.   ■

GRAPH IT

You can use your grapher to do Examples 1, 2, and 3 in this section by simply graphing the given expressions. Then, as we did in the text, you decide which points are in the solution set by examining the given inequalities.

Example 4 is a more challenging problem. First, solve for y in $x^2 + y^2 = 4$ to obtain $y = \pm\sqrt{4 - x^2}$. The corresponding inequality is $y \leq \pm\sqrt{4 - x^2}$, that is, the points *on* the boundary and *inside* the circle $x^2 + y^2 = 4$. [Remember to use a square window (ZOOM 5 on a TI-82) to make the circle appear round.] The solution set of $y \geq x^2 - 3$ are the points *on* the boundary and *above* the parabola $y = x^2 - 3$, as shown in the window. The intersection of these two sets are the points *inside* the circle $x^2 + y^2 = 4$ and *above* the parabola $y = x^2 - 3$, as before, and the points *on* the circle that are *above* or *on* the parabola.

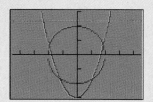

EXERCISE 9.5

A In Problems 1–12, graph the solution set of the inequality.

1. $x^2 + y^2 > 16$

2. $x^2 + y^2 < 16$

3. $x^2 + y^2 \leq 1$

4. $x^2 + y^2 \geq 1$

5. $y < x^2 - 2$

6. $y > x^2 - 2$

7. $y \leq -x^2 + 3$

8. $y \geq -x^2 + 3$

9. $4x^2 - 9y^2 > 36$

10. $4x^2 - 9y^2 < 36$

11. $x^2 - y^2 \geq 1$

12. $x^2 - y^2 \leq 1$

B **In Problems 13–24, graph the solution set of the system.**

13. $x^2 + y^2 \leq 25$
$\qquad y \geq x^2$

14. $x^2 + y^2 \leq 25$
$\qquad y \leq x^2$

15. $x^2 + y^2 \geq 25$
$\qquad y \leq x^2$

16. $x^2 + y^2 \geq 25$
$\qquad y \geq x^2$

17. $y < x^2 + 2$
$\qquad y > x^2 - 2$

18. $y < x^2 + 2$
$\qquad y < x^2 - 2$

19. $y \geq x^2 + 2$
$\qquad y \geq x^2 - 2$

20. $y \geq x^2 + 2$
$\qquad y \leq x^2 - 2$

21. $\dfrac{x^2}{4} - \dfrac{y^2}{4} \geq 1$
$\quad \dfrac{x^2}{25} + \dfrac{y^2}{4} \leq 1$

22. $\dfrac{x^2}{4} - \dfrac{y^2}{4} \geq 1$
$\quad \dfrac{x^2}{25} + \dfrac{y^2}{4} \geq 1$

23. $\dfrac{x^2}{36} + \dfrac{y^2}{16} < 1$
$\quad \dfrac{x^2}{16} + \dfrac{y^2}{36} < 1$

24. $\dfrac{x^2}{36} + \dfrac{y^2}{16} > 1$
$\quad \dfrac{x^2}{16} + \dfrac{y^2}{36} < 1$

25. Fill in each blank with an inequality symbol so that the system

$$\frac{x^2}{36} + \frac{y^2}{16} \underline{\quad} 1 \qquad \text{(ellipse)}$$

$$\frac{x^2}{16} + \frac{y^2}{16} \underline{\quad} 1 \qquad \text{(circle)}$$

will have
a. No solution. **b.** Two solutions.

26. Can you fill in each blank with an inequality symbol so that the system

$$\frac{x^2}{25} - \frac{y^2}{4} \underline{\quad} 1 \qquad \text{(hyperbola)}$$

$$\frac{x^2}{25} + \frac{y^2}{4} \underline{\quad} 1 \qquad \text{(ellipse)}$$

will have no solution?

SKILL CHECKER

Find the value of $y = 3x + 5$ when:

27. $x = -2$ **28.** $x = -5$

If $P(x) = x^2 - 9$ and $Q(x) = x + 3$, find:

29. The sum and the difference of $P(x)$ and $Q(x)$.

30. The product and the quotient of $P(x)$ and $Q(x)$.

USING YOUR KNOWLEDGE Give Me a Break-Even Point!

You may have wondered about the equations representing the cost C and the revenue R in the graph in the *Getting Started* at the beginning of this section. Note that both the cost C and the revenue R are given in terms of p, the price per unit. Can we find the break-even point—that is, the point at which the revenue R equals the cost C? Let's try.

31. The demand equation—that is, the number x of units retailers are likely to buy at p dollars per unit—is given by $x = 6000 - 30p$, and the cost C is given by $C = 72{,}000 + 60x$. Substitute $x = 6000 - 30p$ in $C = 72{,}000 + 60x$ and find C in terms of the price p.

32. The revenue is $R = xp$—that is, the revenue is the product of the number of units retailers are likely to buy and the price p per unit. To find R in terms of p, substitute $x = 6000 - 30p$ into $R = xp$. What is R in terms of p? What shape does the graph of R have?

33. The graph in the *Getting Started* on page 568 shows two points at which the revenue R equals the cost—that is, $R = C$. Substitute the expressions for R and C from Problems 31 and 32 in the equation $R = C$ and solve for p.

WRITE ON . . .

34. Write your own definition of a second-degree inequality.

35. How do you decide whether the boundary of a region should be included in the solution set?

36. Write the procedure you use to solve a system of nonlinear inequalities graphically.

37. The system of inequalities we solved in Example 4 has a solution set with infinitely many points. Can you find a system of nonlinear inequalities with

a. A solution set consisting of only one point? Make a sketch to show the system.

b. A solution set consisting of exactly two points? Make a sketch to show the system.

MASTERY TEST

If you know how to do these problems, you have learned your lesson!

Graph:

38. $x^2 + y^2 \leq 9$ and $y \geq x^2 + 2$

39. $x^2 + 4y^2 \geq 4$ and $y \geq x^2 + 1$

40. $4y^2 - 9x^2 \leq 36$

41. $x^2 - 9y^2 \geq 9$

42. $x^2 + y^2 > 9$

43. $x^2 > 4 - y^2$

44. $y \leq -x^2 - 3$

45. $y \geq -x^2 + 2$

research questions

Sources of information for these questions can be found in the *Bibliography* at the end of this book.

1. Write a paper detailing the origin and meaning of the words *ellipse, hyperbola,* and *parabola.*

2. Who was the first person that used the terminology "ellipse, hyperbola, and parabola" and from what book or paper were these terms adapted?

3. Who was "The Great Geometer"?

4. Write a short paper about Hippocrates of Chios, his discoveries, his works, and his connection to the conics.

5. Write a short paper about Menaechmus, his discoveries, his works, and his connection to the conics.

6. Write a short paper about Apollonius of Perga, his discoveries, his works, and his connection to the conics.

7. One of the three famous problems in the first 300 years of Greek mathematics was "the duplication of the cube." What were the other two?

8. How are conics used in mathematics, science, art, and other areas?

9. Write a paper detailing Girard Desargues' contribution to the study of the conic sections.

10. A famous French mathematician wrote *A Complete Work on Conics.* Who was this mathematician and what did the book contain?

SUMMARY

SECTION	ITEM	MEANING	EXAMPLE
9.1	Conic sections	Curves obtained by slicing a cone with a plane	Parabolas, circles, ellipses, and hyperbolas
9.1A	Vertex	The highest or lowest point of a parabola opening vertically	The vertex of $y = x^2 + 2$ is at $(0, 2)$.
9.1B	Parabola (vertical axis)	A curve with equation $y = a(x - h)^2 + k$ or $y = ax^2 + bx + c$	$y = x^2$, $y = 3(x - 1)^2 + 2$, and $y = -2x^2 + 3x - 7$
9.1D	Parabola (horizontal axis)	A curve with equation $x = a(y - k)^2 + h$ or $x = ay^2 + by + c$	$x = y^2$, $x = 3(y - 1)^2 + 2$, and $x = -4(y + 1)^2 - 3$
9.2A	Circle centered at (h, k)	A curve with the equation $(x - h)^2 + (y - k)^2 = r^2$	$(x - 3)^2 + (y + 4)^2 = 9$ is a circle with center at $(3, -4)$ and radius 3.
9.2C	Ellipse centered at the origin	A curve with equation $\dfrac{x^2}{a^2} + \dfrac{y^2}{b^2} = 1$	$\dfrac{x^2}{16} + \dfrac{y^2}{9} = 1$ is an ellipse.
9.3A	Hyperbola centered at the origin	A curve with equation $\dfrac{x^2}{a^2} - \dfrac{y^2}{b^2} = 1$ or $\dfrac{y^2}{a^2} - \dfrac{x^2}{b^2} = 1$	$\dfrac{x^2}{9} - \dfrac{y^2}{16} = 1$ is a hyperbola.
	Asymptotes	Lines through the opposite corners of the rectangle whose sides pass through $\pm a$ and $\pm b$ associated with a hyperbola	The asymptotes for $\dfrac{x^2}{a^2} - \dfrac{y^2}{b^2} = 1$ are shown
9.4A	Nonlinear systems	A system of equations containing at least one second-degree equation	$\begin{aligned} x^2 + y^2 &= 9 \\ x - y &= 3 \end{aligned}$ is a nonlinear system.
	Substitution method	A method of solving nonlinear systems in which substitution is made from one of the equations into the other	To solve $\begin{aligned} x^2 + y^2 &= 9 \\ x &= y + 3 \end{aligned}$ by substitution, replace x by $y + 3$ in $x^2 + y^2 = 9$.
9.5	Second-degree inequality	An inequality containing at least one second-degree term	$y \leq x^2 + 2$ and $x^2 + y^2 > 9$ are second-degree inequalities.

REVIEW EXERCISES

(If you need help with these exercises, look in the section indicated in brackets.)

1. [9.1A] Graph.
  a. $y = 9x^2$
  b. $y = -9x^2$

2. [9.1A] Find the vertex and graph.
  a. $y = (x - 1)^2 - 2$
  b. $y = -(x - 1)^2 + 2$

3. [9.1B] Find the vertex and graph.
  a. $y = x^2 - 4x + 2$
  b. $y = -x^2 + 6x - 5$

4. [9.1B] Find the vertex and graph.
  a. $y = 2x^2 - 4x + 3$
  b. $y = -2x^2 + 4x - 5$

5. [9.1C] Find the vertex and graph.
  a. $x = 2(y - 2)^2 - 2$
  b. $x = -2(y - 3)^2 + 1$

6. [9.1C] Find the vertex and graph.
  a. $x = y^2 - 4y + 1$
  b. $x = y^2 - 2y + 3$

7. [9.1E] Find the value of x that gives the maximum revenue R if
  a. $R = 20x - 0.01x^2$ **b.** $R = 10x - 0.02x^2$

8. [9.2A] Find an equation of the circle of radius 3 and with center at
  a. $(-2, 2)$ **b.** $(3, -2)$

9. [9.2A] Find an equation of the circle with center at the origin and of radius
  a. 5   **b.** 8

10. [9.2B] Find the center and the radius and sketch the graph.
  a. $(x + 2)^2 + (y - 1)^2 = 4$ **b.** $(x - 1)^2 + (y + 2)^2 = 9$

11. [9.2B] Sketch the graph.
  a. $x^2 + y^2 = 4$ **b.** $x^2 + y^2 = 25$

12. [9.2B] Find the center and the radius and sketch the graph.
  a. $x^2 + y^2 + 2x + 2y - 2 = 0$
  b. $x^2 + y^2 - 4x + 6y + 9 = 0$

13. [9.2C] Graph.
  a. $4x^2 + 9y^2 = 36$
  b. $9x^2 + y^2 = 9$

14. [9.2C] Graph.
  a. $\dfrac{(x - 1)^2}{4} + \dfrac{(y - 2)^2}{9} = 1$
  b. $\dfrac{(x + 2)^2}{9} + \dfrac{(y - 2)^2}{4} = 1$

15. [9.3A] Graph.
  a. $\dfrac{x^2}{9} - \dfrac{y^2}{16} = 1$
  b. $\dfrac{x^2}{16} - \dfrac{y^2}{9} = 1$

16. [9.3A] Graph.

a. $\dfrac{y^2}{9} - \dfrac{x^2}{16} = 1$

b. $\dfrac{y^2}{16} - \dfrac{x^2}{9} = 1$

17. [9.3B] Identify each of the curves.

a. $x = 1 - y^2$ b. $x^2 = 4y^2 - 4$

c. $y^2 = 9 - x^2$ d. $4y^2 = 36 - 9x^2$

18. [9.4A] Solve the system by the substitution method.

a. $x^2 + y^2 = 1$ b. $x^2 + y^2 = 10$
     $x + y = 1$ $x + y = 4$

19. [9.4A] Solve the system by the substitution method.

a. $x^2 - y^2 = 16$ b. $x^2 + y^2 = 4$
      $2x = y$ $x + y = 3$

20. [9.4B] Solve the system.

a. $x^2 - y^2 = 5$ b. $x^2 - y^2 = 3$
    $x^2 + 2y^2 = 17$ $2x^2 + y^2 = 9$

21. [9.4C]

a. The cost C of manufacturing and selling x units of a product is $C = 10x + 400$, and the corresponding revenue R is $R = x^2 - 200$. Find the break-even point.

b. Repeat part a if $C = 6x + 80$ and $R = x^2 - 200$.

22. [9.5A] Graph the inequality.

a. $y \le 1 - x^2$

b. $x \le 4 - y^2$

23. [9.5A] Graph the inequality.

a. $x^2 + y^2 \le 4$

b. $x^2 + y^2 > 9$

24. [9.5A] Graph the inequality.

a. $4x^2 - y^2 \le 4$

b. $x^2 - 4y^2 \le 4$

25. [9.5B] Graph the system.

a. $x^2 + y^2 \le 4$ and $y \le 2 - x^2$

b. $x^2 + y^2 \le 4$ and $y \ge 4x^2$

PRACTICE TEST

(Answers on pages 576–581.)

1. Graph the parabola $y = -x^2 - 4$.

2. Find the vertex and graph the parabola $y = (x - 2)^2 + 2$.

3. Find the vertex and graph the parabola $y = -x^2 - 4x - 1$.

4. Find the vertex and graph the parabola $y = -2x^2 + 4x + 1$.

5. Find the vertex and graph the parabola $x = 2(y + 1)^2 - 1$.

6. Find the vertex and graph the parabola $x = y^2 + 2y - 1$.

7. If the revenue is given by $R = 60x - 0.03x^2$, find the value of x that yields the maximum revenue.

8. Find an equation of the circle of radius 2 with its center at $(1, -2)$.

9. Find an equation of the circle of radius 4 with center at the origin.

10. Find the center and the radius and sketch the graph of $(x + 1)^2 + (y - 2)^2 = 9$.

11. Sketch the graph of $x^2 + y^2 = 16$.

12. Find the center and the radius and sketch the graph of $x^2 + y^2 + 4x - 2y - 4 = 0$.

13. Graph $x^2 + 9y^2 = 9$.

14. Graph $\dfrac{(x - 2)^2}{4} + \dfrac{(y + 1)^2}{9} = 1$.

15. Graph $\dfrac{y^2}{9} - \dfrac{x^2}{25} = 1$.   16. Graph $\dfrac{x^2}{9} - \dfrac{y^2}{25} = 1$.

17. Identify each of the following curves.

a. $x = y^2 - 4$ b. $16y^2 = 144 - 9x^2$

c. $9y^2 = 144 + 16x^2$ d. $x^2 = 16 - y^2$

18. Use the substitution method to solve the system.
$$x^2 + y^2 = 4$$
$$x + y = 2$$

19. Use the substitution method to solve the system.
$$x^2 + y^2 = 1$$
$$x + y = 2$$

20. Solve the system.
$$x^2 + y^2 = 20$$
$$x^2 - 2y^2 = 8$$

21. The cost C of manufacturing and selling x units of a product is $C = 20x + 50$, and the corresponding revenue R is $R = x^2 - 75$. Find the break-even value of x.

22. Graph the inequality $y \leq -x^2 - 1$.

23. Graph the inequality $x^2 + y^2 < 9$.

24. Graph the inequality $y^2 - 4x^2 \leq 4$.

25. Graph the system $x^2 + y^2 \leq 4$ and $y \leq 2 - x^2$.

ANSWERS TO PRACTICE TEST

Answer	If you missed: Question	Review: Section	Examples	Page
1.	1	9.1A	1,2	527–529
2. Vertex $(2, 2)$	2	9.1A	2	528–529

Answer	If you missed:		Review:	
	Question	Section	Examples	Page
3. Vertex $(-2, 3)$	3	9.1B	4	530–531
4. Vertex $(1, 3)$	4	9.1C	5	533–534
5. Vertex $(-1, -1)$	5	9.1C	6	534–535

3. Vertex $(-2, 3)$

4. Vertex $(1, 3)$

5. Vertex $(-1, -1)$

Answer	If you missed:		Review:	
	Question	Section	Examples	Page
6. Vertex $(-2, -1)$	6	9.1D	7	536

7. $x = 1000$	7	9.1E	9	537
8. $(x - 1)^2 + (y + 2)^2 = 4$	8	9.2A	1	543
9. $x^2 + y^2 = 16$	9	9.2A	2	543
10. Center $(-1, 2)$; $r = 3$	10	9.2B	3	544

11.

	11	9.2B	4	544

Answer	If you missed:		Review:	
	Question	Section	Examples	Page
12. Center $(-2, 1)$; $r = 3$	12	9.2B	5	544
13.	13	9.2C	6	545
14.	14	9.2C	7	546
15.	15	9.3A	1a	554

12. Center $(-2, 1)$; $r = 3$

13.

14.

15.

Answer	If you missed:		Review:	
	Question	Section	Examples	Page
16.	16	9.3A	1b	554
17a. Parabola	17a	9.3B	2	554–555
17b. Ellipse	17b	9.3B	2	554–555
17c. Hyperbola	17c	9.3B	2	554–555
17d. Circle	17d	9.3B	2	554–555
18. $(2, 0)$ and $(0, 2)$	18	9.4A	1	561–562
19. $\left(1 + \dfrac{\sqrt{2}}{2}i,\ 1 - \dfrac{\sqrt{2}}{2}i\right),$ $\left(1 - \dfrac{\sqrt{2}}{2}i,\ 1 + \dfrac{\sqrt{2}}{2}i\right)$	19	9.4A	2	562–563
20. $(4, 2), (4, -2), (-4, 2), (-4, -2)$	20	9.4B	3	563
21. $x = 25$	21	9.4C	4	564
22.	22	9.5A	1	569

Answer	If you missed:		Review:	
	Question	Section	Examples	Page
23.	23	9.5A	2	569
24.	24	9.5A	3	569
25.	25	9.5B	4	570

Inverse, Exponential, and Logarithmic Functions

In this chapter, we shall study ways to combine functions by finding their sum, difference, product, quotient, and composition (Section 10.1). In Section 10.2, we learn how to find the inverse of a function, graph it, and determine whether this inverse is also a function. We then study how to graph two special types of functions: exponential (Section 10.3) and logarithmic (Section 10.4). In Section 10.4, we establish that exponential and logarithmic equations are inverses of each other, study the properties of the logarithmic functions, and solve certain types of logarithmic equations. The last two sections (10.5 and 10.6) are devoted to finding logarithms and antilogarithms (common and natural) of numbers and to the techniques used to solve exponential and logarithmic equations and their applications.

Who invented logarithms? It's sometimes implied that the invention of logarithms was the work of one man: John Napier. Interestingly enough, Napier was not a professional mathematician but a Scottish laird (landowner) and Baron of Murchiston who dabbled in many controversial topics. (In a commentary on the Book of Revelations, he argued that the pope at Rome was the Antichrist!) Napier claimed that he worked on the invention of logarithms for 20 years before he published his *Mirifici logarithmorum canonis descriptio* (*A description of the Marvelous Rule of Logarithms*) in 1614.

In 1617, the year of Napier's death, Henry Briggs constructed the first table of Briggsian or common logarithms. But there is one more participant in the invention of logarithms. Jobst Bürgi of Switzerland independently developed the idea of logarithms as early as 1588. Unfortunately, his results were not published until 1620. In his scheme, Bürgi used "red" and "black" numbers, with the "red" numbers appearing on the side of the page and "black" numbers in the body of the table, thus creating what could now be described as an antilogarithm table.

10.1

THE ALGEBRA OF FUNCTIONS

To succeed, review how to:

1. Evaluate expressions (pp. 42–49).
2. Add, subtract, multiply, and divide polynomials (pp. 142–144, 151–153, 208–210).

Objectives:

A   Find the sum, difference, product, and quotient of two functions.

B   Find the composite of two functions.

C   Solve an application.

getting started   **A Lot of Garbage!**

How many pounds of garbage does each person generate each day? The average is more than 4 lb and growing! Of course, some of this waste is recovered and recycled into some other products. As a matter of fact, the amount of solid waste recovered (in millions of tons) is a **function** of time and is approximated by

$$R(t) = 0.04t^2 - 0.59t + 7.42$$

where t is the number of years after 1960.

On the other hand, the amount of solids *not* recovered is

$$N(t) = 2.93t + 82.58$$

Can we find the total amount of solid waste generated? This amount is the **sum** of $R(t)$ and $N(t)$, that is,

$$R(t) + N(t) = (0.04t^2 - 0.59t + 7.42) + (2.93t + 82.58)$$
$$= 0.04t^2 + 2.34t + 90$$

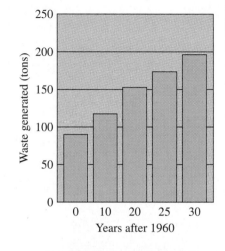

The items recovered most often are paper and paperboard. The amount of paper and paperboard is also a function of time and is given by $P(t) = 0.02t^2 - 0.25t + 6$. You may be surprised to learn what fraction of the total recovered waste $R(t)$ is actually paper and paperboard. That fraction is the **quotient**:

$$\frac{P(t)}{R(t)}$$

Now suppose you want to know what this fraction was in 1990. Since 1990 is 30 (1990 − 1960) years after 1960, this fraction for 1990 is

$$\frac{P(t)}{R(t)} = \frac{0.02(30)^2 - 0.25(30) + 6}{0.04(30)^2 - 0.59(30) + 7.42} = \frac{16.5}{25.72} = 0.64$$

Thus 64% of the total amount of recovered waste in 1990 was paper and paperboard.

As you can see, we can perform the fundamental operations of addition, subtraction, multiplication, and division using functions. We shall study such operations in this section.

Using Operations with Functions

Here are the definitions we need to add, subtract, multiply, and divide functions.

OPERATIONS WITH FUNCTIONS

If f and g are functions and x is in the domain of both functions, then

$$(f + g)(x) = f(x) + g(x) \qquad \text{The sum of } f \text{ and } g$$

$$(f - g)(x) = f(x) - g(x) \qquad \text{The difference of } f \text{ and } g$$

$$(fg)(x) = f(x) \cdot g(x) \qquad \text{The product of } f \text{ and } g$$

$$\left(\frac{f}{g}\right)(x) = \frac{f(x)}{g(x)}, \quad g(x) \neq 0 \qquad \text{The quotient of } f \text{ and } g$$

EXAMPLE 1   Performing operations with functions

If $f(x) = x^2 + 4$ and $g(x) = x + 2$, find:

a. $(f + g)(x)$ **b.** $(f - g)(x)$ **c.** $(fg)(x)$ **d.** $\left(\dfrac{f}{g}\right)(x)$

SOLUTION

a. By definition,

$$(f + g)(x) = \quad f(x) \quad + \quad g(x)$$
$$\qquad\qquad\quad \downarrow \qquad\qquad \downarrow$$
$$= (x^2 + 4) + (x + 2)$$
$$= x^2 + x + 6$$

b. $(f - g)(x) = \quad f(x) \quad - \quad g(x)$
$$\qquad\qquad\qquad\quad \downarrow \qquad\qquad \downarrow$$
$$= (x^2 + 4) - (x + 2) \qquad \text{Recall that } -(a + b) = -a - b.$$
$$= x^2 + 4 - x - 2$$
$$= x^2 - x + 2$$

c. $(fg)(x) = \quad f(x) \quad \cdot g(x)$
$$\qquad\qquad\quad \downarrow \qquad\quad \downarrow$$
$$= (x^2 + 4)(x + 2) \qquad \text{Multiply } x^2(x + 2) \text{ and } 4(x + 2).$$
$$= x^3 + 2x^2 + 4x + 8$$

d. $\left(\dfrac{f}{g}\right)(x) = \dfrac{f(x)}{g(x)}$
$$= \frac{x^2 + 4}{x + 2}$$

Note that the denominator can't be zero, so $x + 2 \neq 0$, that is, $x \neq -2$.   ∎

EXAMPLE 2   More practice in performing operations with functions

If $f(x) = 3x + 1$, find:

$$\frac{f(x) - f(a)}{x - a}, \quad x \neq a$$

SOLUTION   Since $f(x) = 3x + 1$, and $f(a) = 3a + 1$, we have

$$\frac{f(x) - f(a)}{x - a} = \frac{(3x + 1) - (3a + 1)}{x - a}$$

$$= \frac{3x + 1 - 3a - 1}{x - a}$$

$$= \frac{3x - 3a}{x - a}$$

$$= \frac{3(x - a)}{x - a}$$

$$= 3, \quad x \neq a$$ ∎

Finding Composite Functions

There are many instances in which some quantity depends on a variable that, in turn, depends on another variable. For instance, the tax you pay on your house depends on the assessed value and the millage rate. (A 1-mill rate means that you pay $1 for each $1000 of assessed value.) Many states offer a Homestead Exemption, exempting a certain amount of the house value from taxes. In Florida, this exemption is $25,000. Thus the function g giving the correspondence between the assessed value x of a home in Florida and its taxable value is $g(x) = x - 25,000$.

A house assessed at $150,000 will have a taxable value given by $g(150,000) = 150,000 - 25,000 = \$125,000$. When the tax rate is 5 mills, the function f that computes the tax on the house is

$$f(x) = \frac{5}{1000} \cdot V = 0.005V$$

where V is the taxable value of the house. Thus a house valued at $150,000 with a $25,000 Homestead Exemption will pay $0.005 \cdot (150,000 - 25,000) = 0.005 \cdot 125,000 = \$625$ in taxes. If you look at the table, you will see that a house assessed at $150,000 should indeed pay $625 in taxes. Can you find a function h that will find the tax? If you guessed $h(x) = 0.005(x - 25,000)$, you guessed correctly.

$g(x) = x - 25,000$ $f(x) = 0.005g(x)$

Assessed Value (A)	Taxable Value (V)	Tax T
$50,000	$25,000	0.005(25,000) = $125
$100,000	$75,000	0.005(75,000) = $375
$150,000	$125,000	0.005(125,000) = $625

$h(x)$

The taxable value of a house with a $25,000 exemption is $g(x) = x - 25,000$. To find the actual tax, we find

$$f(V) = f(g(x)) = f(x - 25,000) = 0.005(x - 25,000)$$

This gives a formula for h: $h(x) = 0.005(x - 25,000)$. The function h is called the composite of f and g and is denoted by $f \circ g$ (read "f circle g"). Here is the definition.

COMPOSITE FUNCTION

If f and g are functions, then

$$(f \circ g)(x) = f(g(x))$$

is the **composite of f with g**.

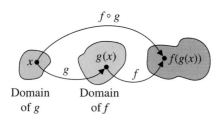

Domain of g Domain of f

NOTE The domain of $f \circ g$ is the set of all x in the domain of g such that $g(x)$ is in the domain of f.

Thus if $f(x) = x^2$ and $g(x) = x + 2$,

$$(f \circ g)(x) = f(g(x)) = f(x + 2)$$
$$= (x + 2)^2$$

EXAMPLE 3   Finding composite functions

If $f(x) = x^3$ and $g(x) = x - 1$, find:

a. $(f \circ g)(x)$ **b.** $(g \circ f)(x)$ **c.** $(f \circ g)(3)$

SOLUTION

a. Substituting $x - 1$ for $g(x)$ in $f(g(x))$, we have

$$(f \circ g)(x) = f(g(x)) = f(x - 1)$$
$$= (x - 1)^3$$

b. Substituting x^3 for $f(x)$ in $g(f(x))$, we have

$$(g \circ f)(x) = g(f(x)) = g(x^3) = x^3 - 1$$

NOTE $(f \circ g)(x) = (x - 1)^3$ and $(g \circ f)(x) = x^3 - 1$. Thus, $(f \circ g)(x) \neq (g \circ f)(x)$.

c. Since $(f \circ g)(x) = f(g(x)) = (x - 1)^3$,

$$(f \circ g)(3) = f(g(3)) = (3 - 1)^3 = 8$$ ∎

Solving an Application

Functions and their operations are used frequently in business. For example, suppose the cost C of making x items and the resulting revenue R are given as functions of x.

When does one make a profit? Since the profit (or loss) is the difference between the revenue and the cost, the profit P is

$$P(x) = R(x) - C(x)$$

EXAMPLE 4   The profit function

The revenue (in dollars) obtained from selling x units of a product is given by

$$R(x) = 200x - \frac{x^2}{30}$$

and the cost is given by $C(x) = 72{,}000 + 60x$.

a. Find the profit function, $P(x)$.

b. How many units must be made and sold to yield the maximum profit? Find this profit.

SOLUTION

a. $P(x) = R(x) - C(x)$

$$= \left(200x - \frac{x^2}{30}\right) - (72{,}000 + 60x)$$

$$= -\frac{x^2}{30} + 140x - 72{,}000$$

b. To find the value of x that gives the maximum profit, we note that the graph of the profit function is a parabola opening downward. Thus the maximum value of P is at the vertex. The vertex is at the point where

$$x = -\frac{b}{2a} = -\frac{140}{-\dfrac{2}{30}} = (15)(140) = 2100$$

Thus 2100 units must be made and sold to give the maximum profit. This profit is P dollars, where P is given by

$$P(2100) = -\frac{(2100)^2}{30} + 140(2100) - 72{,}000$$

$$= 75{,}000$$

So the maximum profit is \$75,000.   ■

GRAPH IT

You can use a grapher to find the maximum or minimum value of the functions discussed in this section. In Example 4, the profit function can be graphed using an appropriate window (try $[-1000, 4000]$ by $[-1000, 75{,}000]$ with Xscl = 1000, Yscl = 1000). You can now use the [ZOOM] and [TRACE] keys to find the maximum or, if you have a TI-82, enter

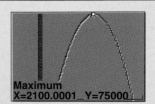

Maximum
X=2100.0001 Y=75000

$$Y_1 = -\frac{x^2}{30} + 140x - 72{,}000$$

Now press [2nd] [CALC] 4 and enter "Lower bound?" "Upper bound?" and "Guess?" The grapher then gives the coordinates of the maximum for the function, $X = 2100.0001$ and $Y = 75{,}000$, which you can approximate to 2100 and 75,000, as before.

EXERCISE 10.1

A If $f(x) = x + 4$, $g(x) = x^2 - 5x + 4$, and $h(x) = x^2 + 16$, find the following.

1. $f + g$

2. $f - g$

3. hf

4. $\dfrac{h}{f}$

5. $\dfrac{f}{h}$

6. $f + g - h$

B In Problems 7–14, find the following.

   a. $(f \circ g)(x)$ **b.** $(g \circ f)(x)$

7. $f(x) = x^2$, $g(x) = \sqrt{x}$, $x > 0$

8. $f(x) = x - 1$, $g(x) = x^2$

9. $f(x) = 3x - 2$, $g(x) = x + 1$

10. $f(x) = x^2$, $g(x) = x - 1$

11. $f(x) = \sqrt{x + 1}$, $g(x) = x^2 - 1$

12. $f(x) = \sqrt{x^2 + 1}$, $g(x) = 2x + 1$

13. $f(x) = 3$, $g(x) = -1$ 14. $f(x) = ax$, $g(x) = bx$

In Problems 15–20, find

$$\frac{f(x) - f(a)}{x - a}, \quad x \neq a$$

15. $f(x) = 3x - 2$

16. $f(x) = 5x - 1$

17. $f(x) = x^2$

18. $f(x) = x^3$

19. $f(x) = x^2 + 3x$

20. $f(x) = x^2 - 2x$

In Problems 21–30, evaluate the indicated combination of f and g for the given values of the independent variable. (If this is not possible, state the reason.)

21. $f(x) = \sqrt{x}$, $g(x) = x^2 - 1$

   a. $(f + g)(4)$ **b.** $(f - g)(4)$

22. $f(x) = \sqrt{x - 2}$, $g(x) = x^2 + 1$

   a. $(f + g)(1)$ **b.** $(f - g)(1)$

23. $f(x) = |x|$, $g(x) = 3$ 24. $f(x) = x^2 - 4$, $g(x) = x + 2$

   a. $\left(\dfrac{f}{g}\right)(3)$ **a.** $\left(\dfrac{f}{g}\right)(2)$

   b. $\left(\dfrac{f}{g}\right)(0)$ **b.** $\left(\dfrac{f}{g}\right)(-2)$

25. $f(x) = x - 3$, $g(x) = (x + 3)(x - 3)$

   a. $\left(\dfrac{f}{g}\right)(3)$ **b.** $\left(\dfrac{g}{f}\right)(3)$

26. $f(x) = x + 5$, $g(x) = (x + 5)(x - 5)$

   a. $\left(\dfrac{f}{g}\right)(5)$ **b.** $\left(\dfrac{g}{f}\right)(-5)$

27. $f(x) = \sqrt{x}$, $g(x) = x^2 + 1$

   a. $(f \circ g)(1)$ **b.** $(g \circ f)(-1)$

   c. $(f \circ g)(x)$ **d.** $(g \circ f)(x)$

28. $f(x) = \sqrt{x + 1}$, $g(x) = x^2$

   a. $(f \circ g)(-1)$ **b.** $(g \circ f)(-1)$

   c. $(f \circ g)(x)$ **d.** $(g \circ f)(x)$

29. $f(x) = \dfrac{1}{x^2 - 2}$, $g(x) = \sqrt{x}$

   a. $(f \circ g)(2)$ **b.** $(g \circ f)(2)$

   c. $(f \circ g)(x)$ **d.** $(g \circ f)(x)$

30. $f(x) = \dfrac{1}{x^2}$, $g(x) = \sqrt{x}$

   a. $(f \circ g)(-1)$ **b.** $(g \circ f)(-1)$

   c. $(f \circ g)(x)$ **d.** $(g \circ f)(x)$

APPLICATIONS

31. The dollar revenue obtained from selling x copies of a textbook is $R(x) = 40x - 0.0005x^2$. The production cost C is $C(x) = 120,000 + 6x$. Find the profit function $P(x)$.

32. Repeat Problem 31 if the production cost increases by $20,000.

33. The clam and crab catch (in millions of pounds) in New England between 1980 and 1990 can be approximated by $L(x) = -0.28x^2 + 2.8x + 15$ and $R(x) = 0.2x + 7$, respectively, where x is the number of years after 1980.

   a. What is the total catch of clams and crabs?

   b. How many pounds of clams and crabs were caught in 1980?

   c. How many pounds of clams and crabs were caught in 1990?

   d. How many more pounds of clams than of crabs were caught in 1990?

34. The total fish catch (in millions of pounds) in Hawaii from 1980 to 1990 can be approximated by $C(x) = 1.5x + 10.5$ where x is the number of years after 1980. The tuna catch during the same period was $T(x) = 0.7x + 7$.

a. How many pounds of fish other than tuna were caught?

b. How many pounds of tuna did they catch in Hawaii in 1990?

c. How many pounds of fish other than tuna were caught in Hawaii in 1990?

d. What fraction of the total catch in Hawaii in 1990 was tuna?

e. What percent of the total catch in Hawaii in 1990 was tuna?

35. The total Medicare costs (in billions of dollars) between 1990 and 1995 can be predicted by the function $C(t) = 2.5t^2 + 8.5t + 111$, where t is the number of years after 1990. The total U.S. population age 65 and older in the same period is

$$P(t) = -0.46t^2 + 1.14t + 31.08 \text{ (in millions)}$$

a. Find a function that represents the average cost of Medicare for persons 65 and older during the years 1990–1995.

b. Find the average cost of Medicare for persons 65 and older for 1990.

c. Do the same for 1995.

36. The reaction distance $R(v) = 0.75v$ (in feet) is the distance a car moving at v miles per hour travels while a driver with a reaction time of 0.5 sec is reacting to apply the brakes. If the braking distance for the car is $B(v) = 0.06v^2$, find a function that gives the total distance (in feet) a car moving at v miles per hour travels during a panic stop. What is this distance if the car is moving at 30 mi/hr?

37. The function $C(F) = \frac{5}{9}(F - 32)$ converts the temperature F in degrees Fahrenheit to C in degrees Celsius. The function $K(C) = C + 273$ converts degrees Celsius to kelvins.

a. Find a composite function that converts degrees Fahrenheit to kelvins.

b. If the temperature is 41°F, what is the Celsius temperature?

c. If water boils at 212°F, at what Kelvin temperature does water boil?

38. A company manufactures skates for $50 a pair. For sales of 100 or more pairs, the regular $90 price is discounted and given by $(90 - 0.2x)$, where x is the number of pairs sold. The profit function is:

$$P(x) = \text{pairs sold} \cdot \text{profit per pair} = x(90 - 0.2x - 50)$$

a. What is the profit when 80 pairs are sold?

b. If the number of pairs sold is $s(x) = x + 100$, find a composite function that gives the profit.

c. What is the profit when 200 pairs are sold?

39. The function F giving the correspondence between dress sizes in the United States and France is $F(x) = x + 32$, where x is the U.S. size and $F(x)$ the French size. The function $E(F) = F - 30$ gives the correspondence between dress sizes in France and those in England.

a. What French size corresponds to a U.S. size 6?

b. Find a function that will give a correspondence between U.S. and English dress sizes.

c. What English size corresponds to a U.S. size 8?

40. The correspondence between dress sizes in the United States and Italy is $I(x) = 2x + 22$ where x is the U.S. dress size and the correspondence between Italian and English dress sizes is $f(I) = \frac{1}{2}I - 9$.

a. What Italian size corresponds to a U.S. size 8?

b. What English size corresponds to a U.S. size 10?

c. Find a composite function that will give the correspondence between U.S. and English dress sizes.

41. There are many interesting functions that can be defined using the ideas of this section. For example, we have mentioned that the frequency with which a cricket chirps is a function of the temperature. The table shows the number of chirps per minute and the temperature in degrees Fahrenheit. If f is the function that relates the number of chirps per minute, c, and the temperature x, find

a. $f(40)$ **b.** $f(42)$ **c.** $f(44)$

Temperature (°F)	40	41	42	43	44
Chirps per minute	0	4	8	12	16

42. The function relating the number of chirps per minute of the cricket and the temperature x (in degrees Fahrenheit) is given by $f(x) = 4(x - 40)$. If the temperature is 80°F, how many chirps per minute will you hear from your friendly house cricket?

43. The function $C(x) = \frac{5}{9}(x - 32)$ converts the temperature from degrees Fahrenheit to degrees Celsius:

a. Find a composite function that would relate the number of chirps per minute a cricket makes (Problem 42) to the temperature in degrees Celsius.

b. How many chirps per minute would a cricket make when the temperature is 10°C?

44. The distance (in centimeters per second) traveled by a certain type of ant when the temperature is C (in degrees Celsius) is $d(C) = \frac{1}{6}(C - 4)$. The function $F(C) = \frac{9}{5}C + 32$ converts the temperature from degrees Celsius to Fahrenheit.

a. Find a composite function that would relate the distance the ant travels to the temperature in degrees Fahrenheit.

b. How fast is the ant traveling when the temperature is 50°F?

SKILL CHECKER

Graph:

45. $x + y = 3$

46. $2x - y = 2$

47. $2x + \dfrac{1}{2}y = 2$

48. $y = -x - 3$

49. $y = -4x + 4$

50. $y = -3x + 6$

USING YOUR KNOWLEDGE   Odd and Even Functions

In this section we have discussed several types of functions. As is the case with numbers, functions can be classified as *odd* or *even*. Here is the definition.

ODD AND EVEN FUNCTIONS

The function f is **even** if $f(-x) = f(x)$ for every value of x.

The function f is **odd** if $f(-x) = -f(x)$ for every value of x.

For example, the function $f(x) = x^2$ is even. Let's see why.

$$f(-x) = (-x)^2 = x^2 = f(x)$$

Thus $f(-x) = f(x)$. On the other hand, $f(x) = x^3$ is odd because $f(-x) = (-x)^3 = -x^3 = -f(x)$.

51. Is $f(x) = x^4$ odd or even?   **52.** Is $f(x) = x^5$ odd or even?

53. Is $f(x) = x^2 + x^3$ odd, even, or neither?

54. Is $f(x) = 0$ odd, even, or both?

WRITE ON . . .

Explain:

55. Why $(f + g)(x) = (g + f)(x)$ for every value of x but $(f - g)(x) \neq (g - f)(x)$.

56. Why

$$\left(\frac{f}{g}\right)(x) \neq \left(\frac{g}{f}\right)(x)$$

57. Is $(f \circ g)(x) = (g \circ f)(x)$ for every value of x? Explain why or why not.

58. If $f(x) = \frac{1}{2}x$ and $g(x) = 2x$,
   a. Find $(f \circ g)(x)$ and $(g \circ f)(x)$. Are they the same?

   b. Can you find other functions f and g so that $(f \circ g)(x) = (g \circ f)(x)$?

MASTERY TEST   If you know how to do these problems, you have learned your lesson!

59. The revenue (in dollars) obtained by selling x units of a product is given by

$$R(x) = 200x - \frac{x^2}{30}$$

and the cost $C(x)$ of making x units is given by $C(x) = 72{,}000 + 60x$. Find the profit function $P(x)$.

If $f(x) = x^3$ and $g(x) = x + 1$, find:

60. $(f \circ g)(x)$

61. $(g \circ f)(x)$

62. $(f \circ g)(3)$

63. $(g \circ f)(-3)$

64. Find $\dfrac{f(x) - f(a)}{x - a}$ if $f(x) = 2x + 1$.

65. Find $\dfrac{f(x) - f(a)}{x - a}$ if $f(x) = x^2 + 1$.

If $f(x) = x^2 + 4$ and $g(x) = x - 2$, find (if possible):

66. $(f + g)(x)$

67. $(f - g)(x)$

68. $(g - f)(x)$

69. $\left(\dfrac{f}{g}\right)(-2)$

70. $\left(\dfrac{g}{f}\right)(x)$

71. $\left(\dfrac{g}{f}\right)(-2)$

72. $\left(\dfrac{f}{g}\right)(2)$

10.2

INVERSE FUNCTIONS

To succeed, review how to:

1. Find the range and domain of a relation (pp. 431–433).

2. Solve an equation for a specified variable (pp. 76–80).

3. Graph linear and quadratic equations (pp. 387–390, 528–530).

Objectives:

A Find the inverse of a function when the function is given as a set of ordered pairs.

B Find the equation of the inverse of a function.

C Graph a function and its inverse and determine whether the inverse is a function.

D Solve applications involving functions.

getting started

Money, Money, Money

FOREIGN EXCHANGE

Country	Foreign Currency in U.S. $ Fri.	U.S. $ in Foreign Currency Fri.	Foreign Currency in U.S. $ Last Fri.	U.S. $ in Foreign Currency Last Fri.
Argentina (Austral)	.0005587	1790.03	.0006803	1470.00
Australia (Dollar)	.7600	1.3158	.7937	1.2599
Austria (Schilling)	.08427	11.87	.08316	12.03
Bahrain (Dinar)	2.5971	.3851	2.6525	.3770
Belgium (Franc)				
Commercial rate	.02837	35.25	.02798	35.74
Financial rate	.02836	35.26	.2791	35.83
Brazil (New Cruzado)	.06090	16.42	.06805	14.70
Britain (Pound)	1.6640	.6010	1.6465	.6073
30-Day Forward	1.6552	.6042	1.6372	.6108
90-Day Forward	1.6374	.6107	1.6202	.6172
180-Day Forward	1.6133	.6198	1.6044	.6233
Canada (Dollar)	.8370	1.1947	.8452	1.1832
30-Day Forward	.8341	1.1989	.8423	1.1872
90-Day Forward	.8289	1.2064	.8377	1.1938
180-Day Forward	.8226	1.2156	.8328	1.2008

If you're planning on traveling somewhere, it's a good idea to know how much your dollars are worth in the country you plan to visit. The second column in the table shows the value of $1 (U.S.) in different currencies in a recent year. If you look at the first column, the value of a Canadian dollar in U.S. dollars is $0.8370. This means that the number of dollars D you get for C Canadian dollars is given by

$$D = 0.8370C$$

To find how many Canadian dollars you get for a U.S. dollar, solve for C in $D = 0.8370C$ to obtain

$$C = \frac{D}{0.8370} = 1.1947D$$

that is, you get 1.1947 Canadian dollars for every U.S. dollar. You can check this in the table! In this section we learn how to find the inverse of a function, which is a similar procedure.

The Inverse of a Function

Let's look again at how we found the exchange rate for Canadian dollars in the *Getting Started*. If we think of D as the definition for the function $f(C)$, we can make a table and write the function f as

$$f = \{(1, 0.8370), (10, 8.370), (100, 83.70), (100, 837)\}$$

Canadian Dollars	U.S. Dollars
1	0.8370
10	8.37
100	83.70
1000	837.00

On the other hand, if you exchange U.S. currency for Canadian, the number of Canadian dollars you get for \$0.8370, \$8.370, \$83.70, and \$837 (U.S.), respectively, corresponds to the set of ordered pairs (D, C) and is given by

$$g = \{(0.8370, 1), (8.370, 10), (83.70, 100), (837, 1000)\}$$

The relation g obtained by reversing the order of the coordinates in each ordered pair in f is called the *inverse of f*. As you can see, the domain of f is the range of g and the range of f is the domain of g. Here is the definition we need.

INVERSE OF A FUNCTION

> If f is a relation, then the **inverse of** f, denoted by f^{-1} (read "f inverse," or "the inverse of f") is the relation obtained by reversing the order of the coordinates in each ordered pair in f.

NOTE The -1 in f^{-1} is *not* an exponent. Here it denotes the inverse of f.

EXAMPLE 1   Finding the inverse of a function

Let $S = \{(1, 2), (3, 4), (5, 4)\}$ and find:
a. The domain and range of S
b. S^{-1}
c. The domain and range of S^{-1}
d. The graphs of S and S^{-1} on the same coordinate axes

SOLUTION
a. The domain of S is $\{1, 3, 5\}$. The range is $\{2, 4\}$.
b. $S^{-1} = \{(2, 1), (4, 3), (4, 5)\}$
c. The domain of S^{-1} is $\{2, 4\}$; the range is $\{1, 3, 5\}$.
d. The graphs of S (in blue) and S^{-1} (in red) are shown in Figure 1. As you can see, the two graphs are symmetric with respect to the line $y = x$ (shown dashed in Figure 1).   ■

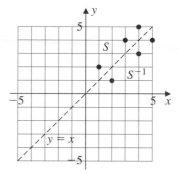

FIGURE 1

Finding the Equation of the Inverse Function

The equation $D = 0.8370C$ gives the U.S. dollar value of C Canadian dollars. To find the inverse of

$$f = \{(1, 0.8370), (10, 8.370), (100, 83.70), (1000, 837)\}$$

we reverse the *order* of the coordinates in each ordered pair in *f*. To find the inverse of $D = 0.8370C$, we interchange the *variables* in $D = 0.8370C$ to obtain

$$C = 0.8370D$$

$$D = \frac{1}{0.8370}C \qquad \text{Solve for } D.$$

$$D = 1.1947C \qquad \text{Divide 1 by 0.8370.}$$

This means that \$1 (U.S.) is worth \$1.1947 (Canadian). You can verify this by looking at the second column in the *Getting Started*.

PROCEDURE

> **Finding the equation of an inverse function**
>
> 1. Interchange the roles of *x* and *y* in the equation for *f*.
> 2. Solve for *y*.

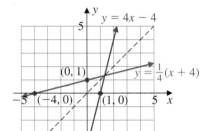

FIGURE 2

For example, consider the relation

$$y = 4x - 4 \qquad \textbf{(1)}$$

The inverse of this relation is obtained by first interchanging the *x*- and *y*-coordinates—that is, by writing $x = 4y - 4$ and then solving for *y*:

$$y = \frac{1}{4}(x + 4) \qquad \textbf{(2)}$$

The graphs of equation (1) (in blue) and its inverse, equation (2) (in red), are shown in Figure 2. Clearly, the graphs are symmetric to each other with respect to the line $y = x$, shown dashed. This is to be expected, since one relation was obtained from the other by interchanging *x* and *y*.

EXAMPLE 2   Finding and graphing an inverse function

Let $f(x) = y = 4x - 2$:

a. Find $f^{-1}(x)$.   **b.** Graph *f* and its inverse.

SOLUTION

a. Since $y = 4x - 2$, we interchange the variables *x* and *y* and solve for *y* to obtain

$$x = 4y - 2$$

$$x + 2 = 4y$$

$$y = \frac{x + 2}{4}$$

Thus the inverse of *f* is

$$f^{-1}(x) = \frac{x + 2}{4}$$

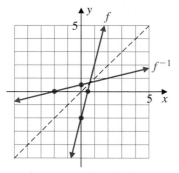

FIGURE 3

b. We can graph $y = 4x - 2$ in the usual way. For $x = 0$, $y = -2$. For $y = 0$, $x = \frac{1}{2}$. The graph is shown in blue in Figure 3. We then graph

$$y = \frac{x + 2}{4}$$

in a similar manner. This graph is shown in red and is symmetric to the graph of $y = 4x - 2$ with respect to the line $y = x$, which is shown dashed. Note that we could have obtained the graph of the inverse function by reflecting the graph of $y = 4x - 2$ about the line $y = x$. ∎

Graphing Functions Whose Inverses Are Also Functions

Since every function is a set of ordered pairs (a relation), every function has an inverse. Is this inverse always a function? We can see very quickly that it is not. For example, if S is the function defined by $S = \{(1, 2,), (3, 2)\}$, the inverse of S is $S^{-1} = \{(2, 1), (2, 3)\}$, which is *not* a function, since two distinct ordered pairs have the same first component, 2. On the other hand, if G is the function defined by $G = \{(3, 4), (5, 6)\}$, the inverse is $G^{-1} = \{(4, 3), (6, 5)\}$, which *is* a function. The reason that the inverse of S is not a function is that S has two ordered pairs with the same *second* component. A function in which no two distinct ordered pairs have the same second component is called a **one-to-one function**. The inverse of such a function is always a function. We summarize this discussion as follows.

ONE-TO-ONE FUNCTION

> If the function $y = f(x)$ is one-to-one, then the inverse of f is also a function and is denoted by $y = f^{-1}(x)$.

Thus to determine whether the inverse of a function is a function, we must ascertain whether the original function is one-to-one. In order to do this, we must return to the definition of a one-to-one function—that is, a one-to-one function *cannot* have two ordered pairs with the same second component. Since any two points with the same second coordinate will be on a *horizontal* line parallel to the x-axis, if any horizontal line intersects the graph of a function more than once, the function will not be one-to-one. Its inverse will not be a function. Thus we have the following **horizontal line test**.

THE HORIZONTAL LINE TEST

> If a horizontal line intersects the graph of a function more than once, the inverse of the function is not a function.

Thus linear functions have inverses that are functions (horizontal lines will intersect the graph only once), but quadratic functions do not have inverses that are functions. This is because quadratic functions have graphs that can be intersected by a horizontal line more than once.

Here are the steps we need to find the inverse of a function.

PROCEDURE

Finding the Inverse of a Function $y = f(x)$

1. Replace $f(x)$ by y, if necessary.
2. Interchange the roles of x and y.
3. Solve the resulting equation for y, if possible.
4. Replace y by $f^{-1}(x)$. (If the original function was one-to-one, f^{-1} is a function.)

EXAMPLE 3   Finding inverses and using the horizontal line test

Find the inverse of: $f(x) = x^2$. Is the inverse a function?

SOLUTION   We use the four-step procedure.

1. Replace $f(x)$ by y. $y = x^2$
2. Interchange the roles of x and y. $x = y^2$
3. Solve for y. $y = \pm\sqrt{x}$
4. Replace y by $f^{-1}(x)$. $f^{-1}(x) = \pm\sqrt{x}$

The inverse of $f(x)$ is *not* a function, since we can draw a horizontal line that intersects the graph of $y = x^2$ at more than one point.

The function $f(x) = x^2$ (in blue) and its inverse (in red) are shown in the graph in Figure 4.   ■

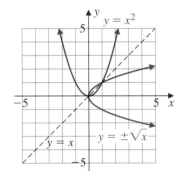

FIGURE 4

EXAMPLE 4   More practice using the horizontal line test

Graph: The function $f(x) = y = 3^x$. Is the inverse a function?

SOLUTION   We make a table of values to graph $y = 3^x$ (on the left). If we examine the graph of $y = 3^x$ (in blue in Figure 5), we can see that any horizontal line will intersect the graph only once. Thus the inverse is a function. To try to find the inverse, we interchange the x and y in $y = 3^x$ to obtain $x = 3^y$.

Unfortunately, we are not able to solve for y at this time (we will have to wait until later in this chapter to do this). However, we can still graph the inverse $x = 3^y$ by giving values to y and finding the corresponding x-values, as shown in the table on the right. The graph of the inverse relation $x = 3^y$ is in red.   ■

FIGURE 5

x	$y = 3^x$		$x = 3^y$	y
-1	$3^{-1} = \frac{1}{3}$		$3^{-1} = \frac{1}{3}$	-1
0	$3^0 = 1$		$3^0 = 1$	0
1	$3^1 = 3$		$3^1 = 3$	1

   Applications Involving Functions

Can you tell yet what the temperature is by listening to crickets? Let's return to our cricket problem one more time and solve it yet another way.

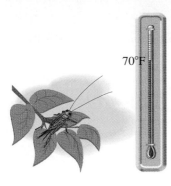

70°F

EXAMPLE 5   "Functional" cricket chirping

The relationship between the temperature F (in degrees Fahrenheit) and the number of chirps c a cricket makes in 1 minute is given by the function

$$c = n(F) = 4(F - 40)$$

a. Find the inverse of n.

b. If a cricket is chirping 120 times a minute, what is the temperature in degrees Fahrenheit?

SOLUTION

a. We again use our four-step procedure, noting that the inverse of n will give the temperature F as a function of the number of chirps c. To find $n^{-1}(F)$,

1. Replace $n(F)$ by c. $c = 4(F - 40)$

2. Interchange the roles of F and c. $F = 4(c - 40)$

3. Solve for c. $c = \dfrac{F}{4} + 40$ Divide by 4, add 40.

4. Replace c by $n^{-1}(c)$. $n^{-1}(c) = \dfrac{c}{4} + 40$

b. If the cricket is chirping 120 times a minute, the temperature is

$$n^{-1}(120) = \frac{120}{4} + 40 = 70°F$$

CHECK: Does the cricket make 120 chirps per minute when the temperature is 70°F? Using the equation $n(F) = 4(F - 40)$, $n(70) = 4(70 - 40) = 120$ as expected. ∎

GRAPH IT

With some graphers (the TI-82, for example), you can graph ordered pairs by entering the domain and range using the list feature. To do Example 1a, press $\boxed{\text{STAT}}$ 1 and enter 1, 3, and 5 (the domain) under L1 and 2, 4, and 4 (the range) under L2. The result is shown in Window 1. Now you have to tell the grapher to plot these points. Press $\boxed{\text{2nd}}$ $\boxed{\text{STAT PLOT}}$ 1 and select $\boxed{\text{ON}}$, $\boxed{\text{⋅⋅}}$, L1, L2 and □ by moving the cursor and pressing $\boxed{\text{ENTER}}$ five times. Finally, press $\boxed{\text{GRAPH}}$. The result is shown in Window 2.

Follow this procedure to graph S^{-1} using lists L3 (2, 4, 4) and L4 (1, 3, 5) for the domain and range of S^{-1}. Press $\boxed{\text{2nd}}$ $\boxed{\text{STAT PLOT}}$ 2 and select $\boxed{\text{ON}}$ $\boxed{\text{⋅⋅}}$, L3, L4, and $\boxed{+}$. Now press $\boxed{\text{GRAPH}}$ to obtain the results in Window 3. (Note that we used a square window so that S and S^{-1} appear as reflections of each other across the line $y = x$.) Examples 2 and 3 can be done using any grapher. To do Example 4, you need a grapher with a draw feature. First, let's get rid of the two plots you used in Example 1. Enter $\boxed{\text{2nd}}$ $\boxed{\text{STAT PLOT}}$ 1 and select "OFF" then enter $\boxed{\text{2nd}}$ $\boxed{\text{STAT PLOT}}$ 2 and select "OFF." This way the two plots won't appear on your next graph. Now enter $Y_1 = 3^x$ and graph it using a square window. To graph the inverse using the draw feature on a TI-82, press $\boxed{\text{2nd}}$ $\boxed{\text{DRAW}}$ 8 to tell the grapher you want an inverse. Then press $\boxed{\text{2nd}}$ $\boxed{\text{Y-VARS}}$ 1 1 $\boxed{\text{ENTER}}$ to tell the grapher you specifically want the inverse of Y_1. The result is shown in Window 4. Note that the grapher still doesn't tell you how to find the inverse, or what the equation for this inverse is. To find that out, you have to learn the techniques in the next section.

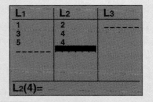

L1	L2	L3
1	2	----
3	4	
5	4	
----	▬▬	

L2(4)=

WINDOW 1

WINDOW 2

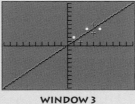

WINDOW 3

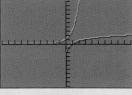

WINDOW 4

EXERCISE 10.2

A B **In Problems 1–4, find f^{-1}, draw the graphs of f and f^{-1} on the same axes, and determine whether f^{-1} is a function.**

1. $f = \{(1, 3), (2, 4), (3, 5)\}$ 2. $f = \{(2, 3), (3, 4), (4, 5)\}$

3. $f = \{(-1, 5), (-3, 4), (-4, 4)\}$

4. $f = \{(-2, 4), (-3, 3), (-5, 3)\}$

C D **In Problems 5–14, find the equation of the inverse, graph it, and state whether the inverse is a function.**

5. $\{(x, y) \mid y = 3x + 3\}$

6. $\{(x, y) \mid y = 2x + 4\}$

7. $\{(x, y) \mid y = 2x - 4\}$

8. $\{(x, y) \mid y = 3x - 3\}$

9. $\{(x, y) \mid y = 2x^2\}$

10. $\{(x, y) \mid y = x^2 + 1\}$

11. $\{(x, y) \mid y = x^2 - 1\}$

12. $\{(x, y) \mid y = x^3 - 1\}$

13. $\{(x, y) \mid y = -x^3\}$

14. $\{(x, y) \mid y = -2x^3\}$

E **In Problems 15–20, make a table of values (see Example 4). Graph the function and its inverse and determine whether the inverse is a function.**

15. $y = f(x) = 2^{x+1}$

16. $y = f(x) = 3^{x+1}$

17. $y = f(x) = \left(\dfrac{1}{3}\right)^x$

18. $y = f(x) = \left(\dfrac{1}{2}\right)^x$

19. $y = f(x) = 2^{-x}$

20. $y = f(x) = 3^{-x}$

21. If $f(x) = 4x + 4, f^{-1}(x) = \dfrac{x - 4}{4}$. Find:

     a. $f((f^{-1}(3))$ b. $f^{-1}(f(-1))$

22. If $f(x) = 2x - 2, f^{-1}(x) = \dfrac{x + 2}{2}$. Find:

     a. $f(f^{-1}(-1))$ b. $f^{-1}(f(x))$

23. If $y = f(x) = \dfrac{1}{x}$, find $f^{-1}(x)$.

24. If $y = f(x) = \dfrac{2}{x}$, find $f^{-1}(x)$.

In Problems 25–35, determine whether the given function has an inverse. If so, graph it by reflecting the given function along the line $y = x$ (shown dashed).

25. $y = 2x$

26. $y = -3x$

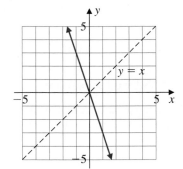

30. $y = -\sqrt{4 - x^2}$

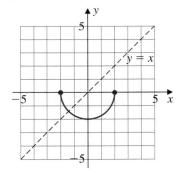

27. $y = -x^2 + 2$

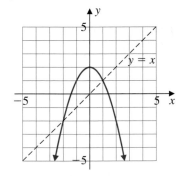

31. $y = \sqrt{9 - x^2}$

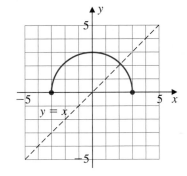

28. $y = x^2 + 1$

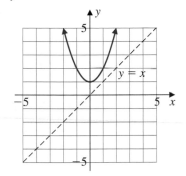

32. $y = -\sqrt{9 - x^2}$

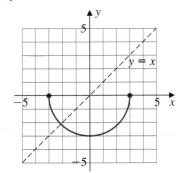

29. $y = \sqrt{4 - x^2}$

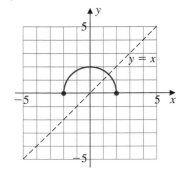

33. $y = x^3$

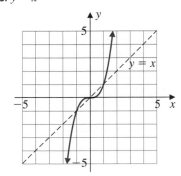

34. $y = -x^3$

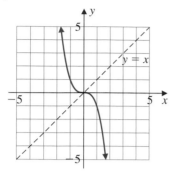

35. $y = |x|$

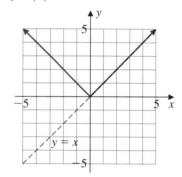

APPLICATIONS

---

36. Your shoe size S is a function of the length of your foot L in inches. For men, the function giving this correspondence is

$$S = f(L) = 3L - 22$$

a. Find $f^{-1}(S)$.

b. If a man's shoe size is 7, what is the length of his foot?

37. For women, the function giving the correspondence between S and L is

$$S = f(L) = 3L - 21$$

a. Find $f^{-1}(S)$.

b. If a woman's shoe size is 7, what is the length of her foot?

38. There is a correspondence between dress sizes in the United States and France. If the U.S. dress size is x, the corresponding size in France is

$$f(x) = x + 32$$

a. Find $f^{-1}(x)$.

b. What U.S. dress size corresponds to a French size 40?

39. In the United States, dress sizes d are a function of waist sizes w (in inches) and are given by

$$d = f(w) = w - 16$$

a. What dress size corresponds to a 32-in. waist?

b. Find $f^{-1}(d)$.

c. If a woman wears a size 12, what is her waist size?

40. There is also a relationship between bust size b (in inches) and dress size d. The correspondence is defined by the function

$$d = f(b) = b - 24$$

a. If $b = 38$, what is d?   **b.** Find $f^{-1}(d)$.

c. If $d = 16$, what is b?

41. Can you predict Olympic outcomes? You can come close if you start with the right function! The winning time in the Women's Olympic 400-m track relay is a function of the year x in which the event was run and is given by

$$f(x) = -0.12x + 280 \quad \text{(seconds)}$$

a. Predict the winning time for the 1988 Olympics (the actual time was 41.98 sec).

b. Find $f^{-1}(x)$.

c. Use $f^{-1}(x)$ to predict in what year the winning time will be 40 sec.

42. The winning time in the Women's Olympic 200-m dash is a function of the year x in which the event was run, starting with 1948, and is given by

$$f(x) = -0.0661x + 152.8 \quad \text{(seconds)}$$

a. Predict the winning time for the 1988 Olympics. (The actual time was 21.34 sec.)

b. Find $f^{-1}(x)$.

c. Use $f^{-1}(x)$ to predict in what year the winning time will be 20 sec.

SKILL CHECKER

Evaluate:

43. 3^2 **44.** 3^{-2}

45. $\left(\dfrac{1}{3}\right)^{-3}$ **46.** $\left(\dfrac{1}{2}\right)^{x/3}$ when $x = 6$

47. $\left(\dfrac{1}{2}\right)^{x/3}$ when $x = -6$ **48.** $\left(\dfrac{1}{3}\right)^{x/2}$ when $x = -4$

The "Undoer"

What does the inverse of a function do? You can think of the
inverse as "undoing" the operations performed on the variable
by the function. We will demonstrate this with the function
from Example 2, $f(x) = 4x - 2$. The following table shows how.

To construct the inverse, we start with x at the bottom of column (4) and work our way up. Columns (2) and (3) respectively
show the operations performed on x and their "undoing."

	Operations		
Function (1)	Function (2)	Inverse Operation (3)	Inverse Function (4)
$f(x) = 4x - 2$	Multiply by 4. $\xrightarrow{\text{undo}}$ Subtract 2.	Divide by 4. Add 2.	Finish here. $f^{-1}(x) = \dfrac{x+2}{4}$ ↑ $\dfrac{x+2}{x}$ Start here.

Now let's examine the function from Example 5,
$cn(F) = 4(F - 40)$, which as you will remember became
$F = 4(c - 40)$. So we shall "undo" the operations on c. We
proceed in a similar manner.

	Operations		
Function (1)	Function (2)	Inverse Operation (3)	Inverse Function (4)
$c = n(F) = 4(F - 40)$	Subtract 40. $\xrightarrow{\text{undo}}$ Multiply. $\xrightarrow{\text{undo}}$	Add 40. Divide by 4.	Finish here. $n^{-1}(c) = \dfrac{c}{4} + 40$ ↑ $\dfrac{c}{4}$ $\dfrac{c}{4}$ Start here.

Use this method to construct the inverse for the given functions.

49. $f(x) = 3x - 2$

50. $f(x) = 5x + 2$

51. $f(x) = \dfrac{x+1}{2}$

52. $f(x) = \dfrac{x-1}{2}$

53. $f(x) = x^3 + 1$

54. $f(x) = x^3 - 1$

55. $f(x) = \sqrt{x}$

56. $f(x) = \sqrt[3]{x}$

57. Explain why the function $f(x) = ax + b$ ($a \neq 0$) *always* has an
inverse that is a function.

58. Explain why the function $f(x) = ax^2 + bx + c$ ($a \neq 0$) *never*
has an inverse that is a function.

59. If f and f^{-1} are functions that are inverses of each other,
explain the result of the composition of f and f^{-1}; that is,
what happens when you take $(f \circ f^{-1})(x)$?

60. Under the same conditions as in Problem 59, what happens
when you take $(f^{-1} \circ f)(x)$?

61. In view of your answers for Problems 59 and 60, how could
you verify that f and f^{-1} are inverses of each other?

If you know how to do these problems, you have learned your lesson!

62. Graph $y = f(x) = 2^x$. Is the inverse a function?

63. Find the inverse of $f(x) = x^3$. Is the inverse a function?
Graph it.

64. Let $y = f(x) = 2x - 4$.
  a. Find $f^{-1}(x)$.   **b.** Graph f and its inverse.

Let $S = \{(4, 3), (3, 2), (2, 1)\}$:

65. Find the domain and range of S.

66. Find S^{-1}.

67. Find the domain and range of S^{-1}.

68. Graph S and S^{-1} on the same coordinate axes.

10.3

EXPONENTIAL FUNCTIONS

To succeed, review how to:

1. Understand the concept of base and exponent (pp. 28–29).

2. Interpret and evaluate expressions containing rational exponents (pp. 281–286).

Objectives:

A Graph exponential functions of the form a^x or a^{-x} $(a > 0)$.

B Determine whether an exponential function is increasing or decreasing.

C Solve applications involving exponential functions.

started

What Do Cells Know about Exponents!

Are you taking biology? Have you studied cell reproduction? The photographs show a cell reproducing by a process called *mitosis*. In mitosis, a single cell or bacterium divides and forms two identical daughter cells. Each daughter cell then doubles in size and divides. As you can see, the number of bacteria present is a function of time. If we start with one cell and assume that each cell divides after 10 min, then the number of bacteria present at the end of the first 10-min period $(t = 10)$ is

$$2 = 2^1 = 2^{10/10}$$

At the end of the second 10-min period $(t = 20)$, the two cells divide, and the number of bacteria present is

$$4 = 2^2 = 2^{20/10}$$

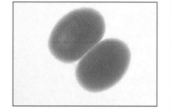
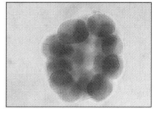

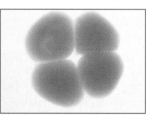

Similarly, at the end of the third 10-min period $(t = 30)$, the number is

$$8 = 2^3 = 2^{30/10}$$

Thus we can see that the number of bacteria present at the end of t minutes is given by the function

$$f(t) = 2^{t/10}$$

Note that this also gives the correct result for $t = 0$, because $2^0 = 1$.

The function $f(t) = 2^{t/10}$ is called an *exponential function* because the variable t is in the exponent. In this section we shall learn more about graphing exponential functions, and we shall see how such functions can be used to solve real-world problems.

---

z

A   Graphing Exponential Functions

Exponential functions take many forms. For example, the following functions are also exponential functions:

$$f(x) = 3^x, \qquad F(y) = \left(\frac{1}{2}\right)^y, \qquad H(z) = (1.02)^{z/2}$$

In general, we have the following definition.

EXPONENTIAL FUNCTION

> An **exponential function** is a function defined for all real values of x by
> $$f(x) = b^x \qquad (b > 0, b \neq 1)$$

> **NOTE** The variable *b* *must not* equal 1 because $f(x) = 1^x = 1$ is a constant function, *not* an exponential function.

In this definition, *b* is a constant called the **base**, and the **exponent** *x* is the variable. It is proved in more advanced courses that b^x for $b > 0$ has a unique real value for each real value of *x*. We assume this in all the following work.

An exponential function is frequently not in the form given in our definition, but it can be put in that form. For example, $f(t) = 2^{t/10}$ can be written as $(2^{1/10})^t$ so that the base is $2^{1/10}$ and the exponent is *t*.

The exponential function defined by $f(t) = 2^{t/10}$ can be graphed and used to predict the number of bacteria present after a period of time *t*. To make this graph, we first construct a table giving the value of the function for certain convenient times:

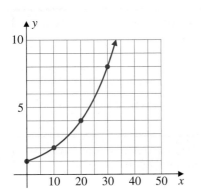

FIGURE 6

t	0	10	20	30
$f(t) = 2^{t/10}$	1	2^1	2^2	2^3

The corresponding points can then be graphed and joined with a smooth curve, as shown in Figure 6. In general, we graph an exponential function by plotting several points calculated from the function and then drawing a smooth curve through these points.

EXAMPLE 1   Graphing exponential functions

Graph on the same coordinate system:

a. $f(x) = 2^x$ **b.** $g(x) = \left(\dfrac{1}{2}\right)^x$

SOLUTION

a. We first make a table with convenient values for *x* and find the corresponding values for $f(x)$. We then graph the points and connect them with a smooth curve, as shown in blue in Figure 7.

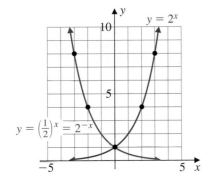

FIGURE 7

x	−2	−1	0	1	2
$f(x) = 2^x$	$2^{-2} = \frac{1}{4}$	$2^{-1} = \frac{1}{2}$	$2^0 = 1$	$2^1 = 2$	$2^2 = 4$

b. If we let $x = -2$,

$$g(-2) = \left(\frac{1}{2}\right)^{-2} = \frac{1}{\left(\frac{1}{2}\right)^2} = 4$$

Similarly, for $x = -1$,

$$g(-1) = \left(\frac{1}{2}\right)^{-1} = \frac{1}{\left(\frac{1}{2}\right)^1} = 2$$

For $x = 0$, 1, and 2, the function values are $\left(\frac{1}{2}\right)^0 = 1$, $\left(\frac{1}{2}\right)^1 = \frac{1}{2}$, and $\left(\frac{1}{2}\right)^2 = \frac{1}{4}$, as shown in the table.

x	−2	−1	0	1	2
$g(x) = \left(\frac{1}{2}\right)^x$	4	2	1	$\frac{1}{2}$	$\frac{1}{4}$

Note that we can save time if we realize that $\left(\frac{1}{2}\right)^x = (2^{-1})^x$, whose values are shown in the table for $f(x)$ in part a. The graph of $g(x) = \left(\frac{1}{2}\right)^x = 2^{-x}$ is shown in red in Figure 7. ∎

Notice that the two graphs in Figure 7 are symmetric to each other with respect to the y-axis. In general, we have the following fact.

SYMMETRIC GRAPHS

> The graphs of $y = b^x$ and $y = b^{-x}$ are **symmetric** with respect to the y-axis.

 Determining Whether Functions Are Increasing or Decreasing

As you have seen, the graphs of some functions increase and some decrease. We'll make the idea more precise next.

INCREASING AND DECREASING FUNCTIONS

> If the graph of a function goes *up* from left to right, the function is an **increasing function**. If the graph goes *down* from left to right, the function is a **decreasing function**.

Thus we see that $f(x) = 2^x$ is an increasing function and $g(x) = \left(\frac{1}{2}\right)^x$ is a decreasing function. (See Figure 7.)

In our definition of the function $y = b^x$, it was required only that $b > 0$ and $b \neq 1$. For many practical applications, however, there is a particularly important base, the irrational number e. The value of e is approximately 2.7182818. The reasons for using this base are made clear in more advanced mathematics courses, but for our purposes we need only note that e is defined as the value that the quantity $\left(1 + \frac{1}{n}\right)^n$ approaches as n increases indefinitely. In symbols,

$$\left(1 + \frac{1}{n}\right)^n \to e \approx 2.7182819 \qquad \text{as} \qquad n \to \infty$$

To show this, we use increasing values of n (1000, 10,000, 100,000, 1,000,000) and evaluate the expression $\left(1 + \frac{1}{n}\right)^n$ using a calculator with a $\boxed{x^y}$ key or your grapher. (See the *Graph It* section to find out how to do this with your grapher.)

For $n = 1000, \left(1 + \dfrac{1}{n}\right)^n \qquad = (1.001)^{1000} \qquad = 2.7169239$ Enter 1.001
$\boxed{x^y}$ 1000 $\boxed{=}$.

For $n = 10,000, \left(1 + \dfrac{1}{n}\right)^n \qquad = (1.0001)^{10,000} \qquad = 2.7181459$

For $n = 100,000, \left(1 + \dfrac{1}{n}\right)^n \qquad = (1.00001)^{100,000} \qquad = 2.7182682$

For $n = 1,000,000, \left(1 + \dfrac{1}{n}\right)^n = (1.000001)^{1,000,000} = 2.7182805$

As you can see, the value of $\left(1 + \frac{1}{n}\right)^n$ is indeed getting closer to $e \approx 2.7182819$.

To graph the functions $f(x) = e^x$ and $g(x) = e^{-x}$, we make a table giving x different values (say $-2, -1, 0, 1, 2$) and finding the corresponding $y = e^x$ and $y = e^{-x}$ values. This can be done with a calculator with an $\boxed{e^x}$ key. On such calculators, you usually have to enter $\boxed{\text{INV}}$ or $\boxed{\text{2nd}}$ to find the value of e^x. [Enter 1 $\boxed{\text{2nd}}$ (or $\boxed{\text{INV}}$) $\boxed{e^x}$, and the calculator will give the value 2.7182818.] We use these ideas next.

EXAMPLE 2 Graphing increasing and decreasing functions

Use the values in the table to graph: $f(x) = e^x$ and $g(x) = e^{-x}$ and determine which of these is increasing and which is decreasing. (Use the same coordinate system.)

x	-2	-1	0	1	2
e^x	0.1353	0.3679	1	2.7183	7.3891

x	-2	-1	0	1	2
e^{-x}	7.3891	2.7183	1	0.3679	0.1353

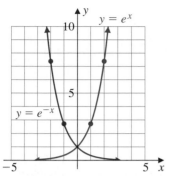

SOLUTION Plotting the given values, we obtain the graphs of $f(x) = e^x$ and $g(x) = e^{-x}$ shown in Figure 8. Note that e^x and e^{-x} are symmetric with respect to the y-axis. Also, e^x is increasing and e^{-x} is decreasing. ■

FIGURE 8

Applications Involving Exponential Functions

Do you have some money invested? Is it earning interest compounded annually, quarterly, monthly, or daily? Does the frequency of compounding make a difference? Some banks have instituted what is called *continuous interest compounding*. We can compare continuous compounding and n compoundings per year by examining their formulas:

$$\text{Continuous compounding:} \quad A = Pe^{rt}$$

$$n \text{ compoundings per year:} \quad A = P\left(1 + \frac{r}{n}\right)^{nt}$$

where

$A = $ compound amount

$P = $ principal

$r = $ interest rate

$t = $ time in years

and $n = $ periods per year (second formula only)

Note that as the number of times the interest is compounded increases, n increases, and it can be shown that

$$\left(1 + \frac{r}{n}\right)^{nt}$$

gets closer to e^{rt}. Let's see what the two formulas earn.

EXAMPLE 3 Maximizing your investment

Find the compound amount:
a. For $100 compounded continuously for 18 months at 6%.
b. For $100 compounded quarterly for 18 months at 6%.

SOLUTION
a. Here, $P = 100$, $r = 0.06$, and $t = 1.5$, so

$$A = 100e^{(0.06)(1.5)}$$

$$= 100e^{0.09}$$

A calculator gives the value

$$e^{0.09} = 1.0942$$

Thus

$$A = (100)(1.0942) = 109.42$$

and the compound amount is $109.42.

b. As before, $P = 100$, $r = 0.06$, $t = 1.5$, and $n = 4$, so

$$1 + \frac{r}{n} = 1 + \frac{0.06}{4} = 1.015 \qquad \text{and} \qquad nt = 4 \cdot 1.5 = 6$$

Thus the compound amount for $100 at the same rate, compounded quarterly, is given by $A = 100(1.015)^6 = 109.34$. Note that at 18 months, the difference between continuous and quarterly compounding is only 8¢! For more comparisons, see the *Using Your Knowledge* at the end of this section. ∎

EXAMPLE 4   Calculating radioactive decay

A radioactive substance decays so that G, the number of grams present, is given by

$$G = 1000e^{-1.2t}$$

where t is the time in years. Find, to the nearest gram, the amount of the substance present:

a. At the start   **b.** In 2 yr

SOLUTION

a. Here, $t = 0$, so $G = 1000e^0 = 1000(1) = 1000$—that is, 1000 g of the substance are present at the start.

b. Since $t = 2$, $G = 1000e^{-2.4}$. To evaluate G, we use a calculator to obtain

$$e^{-2.4} = 0.090718$$

so that

$$G = (1000)(0.090718)$$

$$= 90.718$$

So there are about 91 g present in 2 yr. ∎

EXAMPLE 5   Predicting U.S. population

Thomas Robert Malthus invented a model for predicting population based on the idea that, when the birth rate (B) and the death rate (D) are constant and no other factors are considered, the population P is given by

$$P = P_0 e^{kt}$$

where P = population at any time t

P_0 = initial population

k = annual growth rate ($B - D$)

t = time in years after 1980

According to the *Statistical Abstract of the United States,* the population in 1980 was 226,546,000, the birth rate was 0.016, and the death rate 0.0086. Use this information to predict the number of people in the United States in the year

a. 1990   **b.** 2000

SOLUTION

a. The initial population is $P_0 = 226{,}546{,}000$, $k = 0.016 - 0.0086 = 0.0074$, and the number of years from 1980 to 1990 is $t = 10$.

$$P = 226{,}546{,}000 e^{0.0074 \cdot 10}$$

$$= 226{,}546{,}000 e^{0.074}$$

$$= 243{,}946{,}000$$

Thus, the predicted population is 243,946,000. The actual population in 1990 was 248,143,000. Why do you think there is a discrepancy?

b. To predict the population in the year 2000 ($t = 20$), we compute

$$P = 226{,}546{,}000 e^{0.0074 \cdot 20}$$

$$= 226{,}546{,}000 e^{0.148}$$

$$= 262{,}683{,}000$$

The predicted population for 2000 is 262,683,000. ∎

GRAPH IT

Exponential functions can be graphed using a grapher. To do Example 1, enter $Y_1 = 2^x$ (Recall that to enter 2^x with a TI-81 or TI-82, you press 2 [^] [XIT], or [X,T,θ].) Enter $Y_2 = \left(\frac{1}{2}\right)^x$ by pressing [(] 1 [÷] 2 [)] [^] [XIT] (or [X,T,θ]). Now press [GRAPH]. The result is shown in Window 1.

The numerical work preceding Example 2 can also be done using a grapher. For example, to find $(1.001)^{1000}$, press [2nd] [QUIT] [CLEAR] (to clear your home screen) and then 1 [.] 0 0 1 [^] 1 0 0 0 [ENTER]. For a more dramatic approximation for $e \approx 2.718281828$, let's tell the grapher to graph the sequence of numbers $\left(1 + \frac{1}{n}\right)^n$ as n increases. Press [MODE], move the cursor three lines down to the line starting with "FUNC"; select "SEQ." Go down to the next line and select "DOT" and then press [ENTER]. (We have told the grapher we are about to graph a sequence.) Now press [Y=]. The symbols U_n and V_n are on your screen. Let's define the sequence U_n by entering $\left(1 + \frac{1}{n}\right)^n$. (The n is entered by pressing [2nd] 9.) To adjust the window, press [WINDOW] and select Xmin $= -1$, Xmax $= 10$, Ymin $= -1$, and Ymax $= 3$. Now press [GRAPH]. The result is shown in Window 2. To show you that the sequence of dots are getting closer to e, let's use the draw feature to graph $y = e$. [Since we have entered a sequence, we can't enter the function $f(x) = e$.] Press [2nd] [DRAW] 6 and then enter [2nd] [e^x] 1 [ENTER]. The result is shown in Window 3. Do you now see how the sequence of dots representing $\left(1 + \frac{1}{n}\right)^n$ approaches e?

You can also use your grapher to obtain an understanding of the nature of a problem. In Example 4, we have the function $G = 1000e^{-1.2t}$. Let $Y_1 = 1000e^{-1.2t}$ using a window $[-1, 10]$ by $[-100, 1000]$ and Yscl $= 100$. Do you now see that the substance is *decaying*? Now let's calculate the value of the function when $x = 2$ as required in the example. Press [2nd] [CALC] 1. Answer the grapher question by entering $X = 2$ and press [ENTER]. The result is $Y = 90.717953$, as shown in Window 4. Rounding the answer to the nearest gram, the answer is 91, as before.

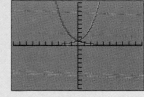

WINDOW 1

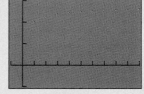

WINDOW 2

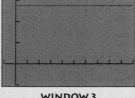

WINDOW 3

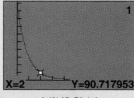

WINDOW 4

EXERCISE 10.3

A **In Problems 1–6, find the value of the given exponential for the indicated values of the variable.**

1. 5^x
   a. $x = -1$
   b. $x = 0$
   c. $x = 1$

2. 5^{-x}
   a. $x = -1$
   b. $x = 0$
   c. $x = 1$

3. 3^t
   a. $t = -2$
   b. $t = 0$
   c. $t = 2$

4. 3^{-t}
   a. $t = -2$
   b. $t = 0$
   c. $t = 2$

5. $10^{t/2}$
   a. $t = -2$
   b. $t = 0$
   c. $t = 2$

6. $10^{-t/2}$
   a. $t = -2$
   b. $t = 0$
   c. $t = 2$

B **In Problems 7–16, graph the functions given in parts a and b on the same coordinate system. State whether the function is increasing or decreasing.**

7. a. $f(x) = 5^x$
   b. $g(x) = 5^{-x}$

8. a. $f(t) = 3^t$
   b. $g(t) = 3^{-t}$

9. a. $f(x) = 10^x$
   b. $g(x) = 10^{-x}$

10. a. $f(t) = 10^{t/2}$
   b. $g(t) = 10^{-t/2}$

11. a. $f(x) = e^{2x}$
   b. $g(x) = e^{-2x}$

12. a. $f(t) = e^{t/4}$
   b. $g(t) = e^{-t/4}$

13. a. $f(x) = 3^x + 1$
   b. $g(x) = 3^{-x} + 1$

14. a. $f(x) = \left(\dfrac{1}{3}\right)^x + 1$
   b. $g(x) = \left(\dfrac{1}{3}\right)^{-x} + 1$

15. a. $f(x) = 2^{x+1}$
   b. $g(x) = 2^{-x+1}$

16. a. $f(x) = \left(\dfrac{1}{2}\right)^{x+1}$
   b. $g(x) = \left(\dfrac{1}{2}\right)^{-x+1}$

APPLICATIONS

In Problems 17–20, find the compound amount if the compounding is (a) continuous or (b) quarterly.

17. $1000 at 9% for 10 yr

18. $1000 at 9% for 20 yr

19. $1000 at 6% for 10 yr

20. $1000 at 6% for 20 yr

21. The population of a town is given by the equation $P = 2000(2^{0.2t})$, where t is the time in years from 1985. Find the population in
   a. 1985   **b.** 1990   **c.** 1995

22. A colony of bacteria grows so that their number, B, is given by the equation $B = 1200(2^t)$, where t is in days. Find the number of bacteria
   a. At the start ($t = 0$).   **b.** In 5 days.   **c.** In 10 days.

23. A radioactive substance decays so that G, the number of grams present, is given by $G = 2000e^{-1.05t}$, where t is the time in years. Find the amount of the substance present
   a. At the start.   **b.** In 1 yr.   **c.** In 2 yr.

24. Solve Problem 23 where the equation is $G = 2000e^{-1.1t}$.

In Problems 25–32, you may use a calculator with $\boxed{e^x}$ **and** $\boxed{x^y}$ **keys, or you can use your grapher.**

25. In 1980, the number of persons of Hispanic origin living in the United States was 14,609,000. If their birth rate is 0.0232 and their death rate is 0.004, predict the number of persons of Hispanic origin living in the United States for the year 2000. (Use a calculator.) (*Hint:* See Example 5.)

26. In 1980, the number of African-Americans living in the United States was 26,683,000. If their birth rate is 0.0221 and their death rate is 0.0088, predict the number of African-Americans living in the United States in the year 2000. (*Hint:* See Example 5.)

27. The number of compact discs (CDs) sold in the United States (in millions) since 1985 can be approximated by the exponential function $S(t) = 32(10)^{0.19t}$, where t is the number of years after 1985.

a. Predict the number of CDs sold in 1990.

b. Predict the number that will be sold in the year 2000.

28. The number of cellular phones sold in the United States (in thousands) since 1989 can be approximated by the exponential function $S(t) = 900(10)^{0.27t}$, where t is the number of years after 1989.
   a. Predict the number of cellular phones sold in 1990.

   b. Predict the number that will be sold in the year 2000.

29. According to the *Statistical Abstract of the United States*, about $\frac{2}{3}$ of all aluminum cans distributed are recycled. If a company distributes 500,000 cans, the number still in use after t years is given by the exponential function $N(t) = 500,000\left(\frac{2}{3}\right)^t$. How many cans are still in use after
   a. 1 yr?   **b.** 2 yr?   **c.** 10 yr?

30. If the value of an item each year is about 60% of its value the year before, after t years the salvage value of an item costing C dollars is given by $S(t) = C(0.6)^t$. Find the salvage value of a computer costing $10,000
   a. 1 yr after it was bought.   **b.** 10 yr after it was bought.

31. The atmospheric pressure A (in pounds per square inch) can be approximated by the exponential function $A(a) = 14.7(10)^{-0.000018a}$, where a is the altitude in feet.
   a. The highest mountain in the world is Mount Everest, which is about 29,000 ft high. Find the atmospheric pressure at the top of Mount Everest.

   b. In the United States, the highest mountain is Mount McKinley in Alaska, whose highest point is about 20,000 ft. Find the atmospheric pressure at the top of Mount McKinley.

32. The atmospheric pressure A (in pounds per square inch) can also be approximated by $A(a) = 14.7e^{-0.21a}$, where a is the altitude in miles.
   a. If we assume that the altitude of Mount Everest is about 6 mi, what is the atmospheric pressure at the top of Mount Everest?

   b. If we assume that the altitude of Mount McKinley is about 4 mi, what is the atmospheric pressure at the top of Mount McKinley?

SKILL CHECKER

State the property of exponents that is being applied:

33. $(10^x)(10^y) = 10^{x+y}$ 34. $\dfrac{10^x}{10^y} = 10^{x-y}$

35. $(10^x)^3 = 10^{3x}$ 36. $[(2)(10^x)]^y = (2^y)(10^{xy})$

USING YOUR KNOWLEDGE — **Compounding Your Money**

In this section you learned that for continuous compounding, the compound amount is given by $A = Pe^{rt}$. For ordinary compound interest, the compound amount is given by $A = P\left(1 + \frac{r}{n}\right)^{nt}$, where r is the annual interest rate and n is the number of periods per year. Suppose you have $1000 to put into an account where the interest rate is 6%. How much more would you have at the end of 2 yr for continuous compounding than for monthly compounding?

For continuous compounding, the amount is given by

$$A = 1000e^{(0.06)(2)} = 1000e^{0.12}$$

$$= (1000)(1.1275) \qquad \text{Use a calculator.}$$

$$= 1127.50$$

Thus the amount is $1127.50. For monthly compounding,

$$\frac{r}{n} = \frac{0.06}{12} = 0.005 \qquad \text{and} \qquad nt = 24$$

Thus the amount is given by $A = 1000(1 + 0.005)^{24} = 1000(1.005)^{24}$. Compound interest tables or your calculator will give the value

$$(1.005)^{24} = 1.1271598$$

so that $A = 1127.16$ (to the nearest hundredth). Thus the amount is $1127.16. Continuous compounding earns you only 34¢ more! But, see what happens when the period t is extended in Problems 37 and 38!

37. Make the same comparison where the time is 10 yr.

38. Make the same comparison where the time is 20 yr.

WRITE ON . . .

39. The definition of the exponential function $f(x) = b^x$ does not allow $b = 1$.
   a. What type of graph will $f(x)$ have when $b = 1$?
   b. Is $f(x) = b^x$ a function when $b = 1$? Explain.
   c. Does $f(x) = b^x$ have an inverse when $b = 1$? Explain.

40. List some reasons to justify the condition $b > 0$ in the definition of the exponential function $f(x) = b^x$.

41. Discuss the relationship between the graphs of $f(x) = b^x$ and $g(x) = b^{-x}$.

42. In Example 5, we predicted the U.S. population for the year 1990 to be 243,946,000. The actual population in 1990 was 248,143,000. Can you give some reasons for this discrepancy?

MASTERY TEST

If you know how to do these problems, you have learned your lesson!

43. A radioactive substance decays so that the number of grams present, G, is given by $G = 1000e^{-1.2t}$, where t is the time in years. Find, to the nearest gram, the amount of substance present in 18 months.

44. The compound amount A for a principal P at rate r compounded continuously for t years is $A = Pe^{rt}$. Find the compound amount for \$100 compounded continuously for 18 months where the rate is 6%.

Graph:

45. $f(x) = e^{x/2}$

46. $f(x) = e^{-x/2}$

47. $f(x) = 6^x$

48. $f(x) = 6^{-x}$

10.4

LOGARITHMIC FUNCTIONS AND THEIR PROPERTIES

To succeed, review how to:

1. Use the rules of exponents in multiplication, division, and raising variables to a power (pp. 31–37).

2. Simplify algebraic expressions (pp. 149–151).

Objectives:

A Graph logarithmic functions.

B Write an exponential equation in logarithmic form and a logarithmic equation in exponential form.

C Solve logarithmic equations.

D Use the properties of logarithms to simplify logarithms of products, quotients, and powers.

E Solve applications involving logarithmic functions.

getting started   **Shake, Rattle, and Roll**

The damage done by the 1989 earthquake in San Francisco was tremendous, as can be seen in the photo. If I_0 denotes the minimum intensity of an earthquake (used for comparison purposes), the intensity of this quake was $10^{7.1} \cdot I_0$—that is, $10^{7.1}$ times as intense as the minimum intensity.

The magnitude of an earthquake is usually measured on the Richter scale. The Richter scale is logarithmic. In this scale the magnitude R of an earthquake of intensity $I = 10^{7.1} \cdot I_0$ is reported as

$$R = \log_{10} \frac{I}{I_0} = \log_{10} \frac{10^{7.1} \cdot \cancel{I_0}}{\cancel{I_0}} = \log_{10} 10^{7.1} = 7.1$$

Thus the magnitude was reported as 7.1 on the Richter scale. Note that 7.1 is the *exponent* of 10. The exponent 7.1 is called the logarithm, base 10, of $10^{7.1}$ and is written as

$$\log_{10} 10^{7.1} = 7.1$$

In this section we shall study logarithmic functions and their properties.

Graphing Logarithmic Functions

In Section 10.2, we graphed the function $f(x) = 3^x$ and its inverse $x = 3^y$ and promised to show how to find $f^{-1}(x)$. Here are the steps we shall use:

$$f(x) = 3^x \qquad \text{Given.}$$

1. Replace $f(x)$ by y. $\qquad y = 3^x$

2. Interchange x and y. $\qquad x = 3^y$

3. Solve for y. $\qquad y = $ the exponent to which we raise 3 to get x

4. Replace y by $f^{-1}(x)$. $\quad f^{-1}(x) = $ the exponent to which we raise 3 to get x

Since a logarithm is an exponent, we define "the exponent to which we raise 3 to get x" as follows.

DEFINITION OF LOG$_3 x$

> $\log_3 x$ ("the logarithm base 3 of x") means "the exponent to which we raise 3 to get x."

If we use this definition in step 4, $f^{-1}(x) = \log_3 x$. Note that $f^{-1}(9) = \log_3 9 = 2$ because 2 is the exponent to which we raise the base 3 to get 9; that is, $\log_3 9 = 2$ means $3^2 = 9$.

In general, for any exponential function $f(x) = b^x$, the inverse $f^{-1}(x) = \log_b x$. The graph of $f^{-1}(x) = \log_b x$ can be drawn by reflecting the graph of $f(x) = b^x$ across the line $y = x$ (see Figure 9). (Recall that the definition of exponential function requires $b > 0$ and $b \neq 1$.) Using these ideas, we have the following definition.

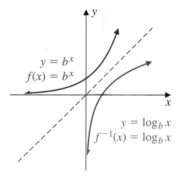

FIGURE 9

DEFINITION OF LOGARITHM

> $y = \log_b x$ is equivalent to $b^y = x$ $\quad (b > 0, b \neq 1)$

Note that this definition simply means that the logarithm of a number is an exponent. For example,

$4 = \log_2 16$	is equivalent to	$2^4 = 16$	The logarithm is the exponent.
$2 = \log_5 25$	is equivalent to	$5^2 = 25$	The logarithm is the exponent.
$-3 = \log_{10} 0.001$	is equivalent to	$10^{-3} = 0.001$	The logarithm is the exponent.

EXAMPLE 1 Graphing a logarithmic function

Graph: $y = f(x) = \log_4 x$

SOLUTION   By the definition of logarithm, $y = \log_4 x$ is equivalent to $4^y = x$. We can graph $4^y = x$ by first assigning values to y and calculating the corresponding x-values.

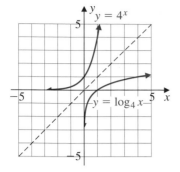

FIGURE 10

y	$x = 4^y$	Ordered Pair
0	$x = 4^0 = 1$	$(1, 0)$
1	$x = 4^1 = 4$	$(4, 1)$
2	$x = 4^2 = 16$	$(16, 2)$
-1	$x = 4^{-1} = \frac{1}{4}$	$\left(\frac{1}{4}, -1\right)$
-2	$x = 4^{-2} = \frac{1}{16}$	$\left(\frac{1}{16}, -2\right)$

We then graph the ordered pairs and connect them with a smooth curve, the graph of $y = \log_4 x$. To confirm that our graph is correct, graph $y = 4^x$ on the same coordinate axes (see Figure 10). As expected, the graphs are reflections of each other across the line $y = x$. ∎

B Converting Exponential to Logarithmic Equations and Vice Versa

The definition of logarithm states "$y = \log_b x$ is equivalent to $b^y = x$." We can use this definition to write logarithmic equations as exponential equations, and vice versa. The fact to remember is that *the logarithm is simply an exponent for a given base.* Thus

$8 = 2^3$	is equivalent to	$\log_2 8 = 3$	The logarithm 3 is the exponent for the base 2.
$25 = 5^2$	is equivalent to	$\log_5 25 = 2$	The logarithm 2 is the exponent for the base 5.
$9 = b^a$	is equivalent to	$\log_b 9 = a$	The logarithm a is the exponent for the base b.

Similarly,

$y = \log_5 3$	is equivalent to	$5^y = 3$	The logarithm y is the exponent for the base 5.
$-2 = \log_b 4$	is equivalent to	$b^{-2} = 4$	The logarithm -2 is the exponent for the base b.
$a = \log_b c$	is equivalent to	$b^a = c$	The logarithm a is the exponent for the base b.

EXAMPLE 2   Converting an exponential equation to logarithmic form

Write in logarithmic form: $125 = 5^x$

SOLUTION   By the definition of logarithm,

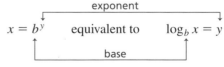

Thus $125 = 5^x$ is equivalent to $\log_5 125 = x$. ∎

EXAMPLE 3   Converting a logarithmic equation to exponential form

Write in exponential form and check its accuracy: $\log_{32} 64 = \frac{6}{5}$

SOLUTION   By the definition of logarithm, $\log_b x = y$ means $x = b^y$. Hence $\log_{32} 64 = \frac{6}{5}$ means $32^{6/5} = 64$. Because $32^{6/5} = (\sqrt[5]{32})^6 = 2^6 = 64$, the equation $\log_{32} 64 = \frac{6}{5}$ is correct. ∎

Solving Logarithmic Equations

In Example 2, we pointed out that $125 = 5^x$ is equivalent to $\log_5 125 = x$. Can we actually find x? Yes, if we write the equation $125 = 5^x$ *as exponentials with the same base*, set the exponents equal, and then solve the resulting equation. This can be done because the exponential function $f(x) = b^x$ is *one-to-one*, so when $b^x = b^y$, $x = y$. Thus if we rewrite the equation $125 = 5^x$ as $5^3 = 5^x$, it's easy to see that $3 = x$. In doing this, we have used the following property.

EQUIVALENCE PROPERTY

For any $b > 0$, $b \neq 1$, $b^x = b^y$ is equivalent to $x = y$.

EXAMPLE 4   Solving logarithmic equations

Solve:

a. $\log_3 x = -2$ **b.** $\log_x 9 = 2$

SOLUTION

a. The equation $\log_3 x = -2$ is equivalent to

$$3^{-2} = x$$

$$\frac{1}{9} = x$$

CHECK: Substitute $x = \frac{1}{9}$ in the original equation to obtain $\log_3 \frac{1}{9} = -2$, which is equivalent to $3^{-2} = \frac{1}{9}$, a true statement.

b. The equation $\log_x 9 = 2$ is equivalent to

$$x^2 = 9$$

$$x = \pm 3 \qquad \text{Extract roots.}$$

But x represents the base, which must be positive, so $x = 3$ is the only answer. (We discard $x = -3$, which is negative and cannot be the base.)

CHECK: Substitute $x = 3$ in the original equation to obtain $\log_3 9 = 2$, which is equivalent to $3^2 = 9$, a true statement. ■

EXAMPLE 5   Finding logarithms

Find:

a. $\log_2 32$ **b.** $\log_2\left(\dfrac{1}{4}\right)$

c. $\log_{10} 1000$ **d.** $\log_{10} 0.01$

SOLUTION

a. We know that $\log_2 32$ means the exponent to which we must raise the base 2 to get 32. Since $2^5 = 32$, the exponent is 5 and $\log_2 32 = 5$. Alternatively, since we are looking for $\log_2 32$, we can write

$$\log_2 32 = x$$

which is equivalent to $\qquad\qquad 2^x = 32$

or $\qquad\qquad\qquad\qquad 2^x = 2^5$

Since the exponents must be equal, $x = 5$, as before.

b. Since $\log_2\left(\frac{1}{4}\right)$ is the exponent to which we must raise the base 2 to get $\frac{1}{4}$ and since $2^{-2} = \frac{1}{4}$, the exponent is -2 and $\log_2\left(\frac{1}{4}\right) = -2$. Alternatively, since we are looking for $\log_2\left(\frac{1}{4}\right)$, we can write

$$\log_2\left(\frac{1}{4}\right) = x$$

which is equivalent to $\qquad\qquad 2^x = \dfrac{1}{4}$

or $\qquad\qquad\qquad\qquad 2^x = 2^{-2}$

Since the exponents must be equal, $x = -2$, as before.

c. Since $10^3 = 1000$, $\log_{10} 1000 = 3$. Alternatively, let $\log_{10} 1000 = x$. Then

$$10^x = 1000 = 10^3$$

and $\qquad\qquad\qquad\qquad x = 3$

d. Since $10^{-2} = \frac{1}{100} = 0.01$, $\log_{10} 0.01 = -2$. Alternatively, let $\log_{10} 0.01 = x$ or equivalently

$$10^x = 0.01 = 10^{-2}$$

$$x = -2 \qquad\qquad \text{Since the exponents must be equal} \qquad ■$$

Properties of Logarithms

Logarithms have three important properties that are the counterparts of the corresponding properties of exponents. By the definition of logarithm, if $x = \log_b M$ and $y = \log_b N$, then $M = b^x$ and $N = b^y$. From the properties of exponents, it follows that

$$MN = b^x b^y = b^{x+y}$$

so that

$$\log_b MN = x + y = \log_b M + \log_b N$$

This means that the *logarithm of a product is the sum of the logarithms of its factors.* Similarly,

$$\frac{M}{N} = \frac{b^x}{b^y} = b^{x-y}$$

which shows that

$$\log_b \frac{M}{N} = x - y = \log_b M - \log_b N$$

Thus the *logarithm of M divided by N is the logarithm of M minus the logarithm of N.* If we have a power of a number such as M^r, then the fact that

$$M^r = (b^x)^r = b^{rx}$$

means that $\log_b M^r = rx = r \log_b M$. In words, the *logarithm of M^r is r times the logarithm of M.* We summarize these results as follows.

PROPERTIES OF LOGARITHMS

$\log_b MN = \log_b M + \log_b N$	Product property
$\log_b \dfrac{M}{N} = \log_b M - \log_b N$	Quotient property
$\log_b M^r = r \log_b M$	Power property

EXAMPLE 6   Using the power and quotient properties

Show that

$$\log_b \sqrt{\frac{61}{37}} = \frac{1}{2} (\log_b 61 - \log_b 37)$$

SOLUTION

$$\log_b \sqrt{\frac{61}{37}} = \log_b \left(\frac{61}{37}\right)^{1/2}$$

$$= \frac{1}{2} \log_b \frac{61}{37} \qquad \text{Power property}$$

$$= \frac{1}{2} (\log_b 61 - \log_b 37) \qquad \text{Quotient property} \qquad \blacksquare$$

EXAMPLE 7   Using the product and quotient properties

Use the properties of logarithms to show that

$$\log \left(x - \frac{1}{x}\right) = \log(x + 1) + \log(x - 1) - \log x$$

where base 10 is understood throughout.

SOLUTION   Since

$$x - \frac{1}{x} = \frac{x^2 - 1}{x} = \frac{(x + 1)(x - 1)}{x}$$

we can apply the product and the quotient properties to write

$$\log \left(x - \frac{1}{x}\right) = \log \frac{(x + 1)(x - 1)}{x}$$

$$= \log(x + 1) + \log(x - 1) - \log x \qquad \blacksquare$$

There are three more properties of logarithms worth mentioning. We have stated why these properties are true by writing the given logarithmic equation as an equivalent exponential equation.

OTHER PROPERTIES OF LOGARITHMS

$\log_b 1 = 0$	Because $b^0 = 1$
$\log_b b = 1$	Because $b^1 = b$
$\log_b b^x = x$	Because $b^x = b^x$

Solving Applications Involving Logarithmic Functions

We can use logarithmic functions to solve a variety of applications, from measuring earthquake intensity to compounding interest.

EXAMPLE 8   An earthquake of logarithmic proportion

In the *Getting Started* we mentioned the 1989 San Francisco earthquake, but earthquakes are commonplace in California. A recent California earthquake was $10^{6.4}$ times as intense as that of a quake of minimum intensity, I_0. What was the magnitude of that quake on the Richter scale?

SOLUTION   We are given that $I = 10^{6.4} I_0$, so the magnitude on the Richter scale is

$$R = \log_{10} \frac{I}{I_0} = \log_{10} 10^{6.4} = 6.4 \qquad \blacksquare$$

EXAMPLE 9   Compounding the purchase price of Manhattan

In 1626, Peter Minuit bought the island of Manhattan from the Native Americans for the equivalent of about \$24. If this money had been invested at 5% compounded annually, then in 2001 the money would have been worth $\$24(1.05)^{375}$. If it's known that $\log_{10} 24 = 1.3802$ and $\log_{10} 1.05 = 0.0212$, find $\log_{10} 24(1.05)^{375}$.

SOLUTION

$$
\begin{aligned}
\log_{10} 24(1.05)^{375} &= \log_{10} 24 + \log_{10} (1.05)^{375} \\
&= \log_{10} 24 + 375(\log_{10} 1.05) \\
&= 1.3802 + 375(0.0212) \\
&= 1.3802 + 7.9500 \\
&= 9.3302 \qquad \text{Log of the accumulated amount} \qquad \blacksquare
\end{aligned}
$$

Now suppose the accumulated amount is A. We have

$$\log_{10} A = 9.3302$$

or $\qquad 10^{9.3302} = A = 2{,}139{,}000{,}000 \qquad$ Rounded from 2,138,946,884

Over two billion dollars! Of course, you can find $24(1.05)^{375}$ with your calculator. What answer do you get then?

GRAPH IT

If your grapher has a $\boxed{\text{LOG}}$ key, you can certainly graph functions of the form $f(x) = \log x$, where the base is understood to be 10. If this is the case, how would you graph Example 1, $f(x) = \log_4 x$? To do so, you need the following property, which we cover in the next section:

$$\log_b x = \frac{\log x}{\log b}$$

Letting $b = 4$,

$$Y_1 = \log_4 x = \frac{\log x}{\log 4}$$

(continued on facing page)

The graph, using a $[-5, 5]$ by $[-5, 5]$ window, is shown in Window 1.

You can solve logarithmic equations with the same techniques used to solve other equations; that is, graph Y_1 (the left-hand side of the equation) and Y_2 (the right-hand side of the equation) and find the point of intersection by using the ZOOM and TRACE keys or the intersection feature of your grapher. The point of intersection is the solution of the equation. Thus in Example 4a, graph

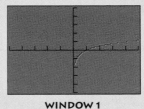

WINDOW 1

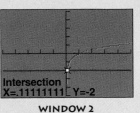

Intersection
X=.11111111 Y=-2

WINDOW 2

$$Y_1 = \log_3 x = \frac{\log x}{\log 3} \quad \text{and} \quad Y_2 = -2$$

If you have the intersection feature, press 2nd CALC 5. The answer is $x = 0.11111111$, as shown in Window 2. How do you know this is the same answer we obtained before—that is, $\frac{1}{9}$? Divide 1 by 9, and you get $0.11111111\ldots$. Same answer! Now try Example 4b.

EXERCISE 10.4

A In Problems 1–8, graph the equation.

1. $y = \log_2 x$

2. $y = \log_3 x$

3. $y = \log_5 x$

4. $f(x) = \log_6 x$

5. $f(x) = \log_{1/2} x$

6. $f(x) = \log_{1/3} x$

7. $y = \log_{1.5} x$

8. $y = \log_{2.5} x$

B In Problems 9–18, write the equation in logarithmic form.

9. $2^x = 128$

10. $3^x = 81$

11. $10^t = 1000$

12. $10^{-t} = 0.001$

13. $81^{1/2} = 9$

14. $16^{1/2} = 4$

15. $216^{1/3} = 6$

16. $64^{1/6} = 2$

17. $e^3 = t$

18. $e^2 = 7.389056$

In Problems 19–32, write the equation in exponential form and check its accuracy (if possible).

19. $\log_9 729 = 3$

20. $\log_7 343 = 3$

21. $\log_2 \frac{1}{256} = -8$

22. $\log_5 \frac{1}{125} = -3$

23. $\log_{81} 27 = \frac{3}{4}$

24. $\log_{64} 32 = \frac{5}{6}$

25. $x = \log_4 16$

26. $t = \log_5 10$

27. $-2 = \log 0.01$

28. $-3 = \log 0.001$

29. $\log_e 30 = 3.4012$

30. $\log_e 40 = 3.6889$

31. $\log_e 0.3166 = -1.15$

32. $\log_e 0.2592 = -1.35$

C In Problems 33–50, solve the equation.

33. $\log_3 x = 2$

34. $\log_2 x = 3$

35. $\log_3 x = -3$

36. $\log_2 x = -3$

37. $\log_4 x = \frac{1}{2}$

38. $\log_9 x = \frac{1}{2}$

39. $\log_8 x = \frac{1}{3}$

40. $\log_x 4 = 2$

41. $\log_x 16 = 4$

42. $\log_x 8 = 3$

43. $\log_x 27 = 3$

44. $\log_8 \frac{1}{8} = x$

45. $\log_2 \dfrac{1}{4} = x$

46. $\log_3 1 = x$

47. $\log_{16} \dfrac{1}{2} = x$

48. $\log_{32} \dfrac{1}{2} = x$

49. $\log_3 \dfrac{1}{9} = x$

50. $\log_2 \dfrac{1}{8} = x$

In Problems 51–74, find the value of the logarithm.

51. $\log_2 256$

52. $\log_2 128$

53. $\log_3 81$

54. $\log_3 243$

55. $\log_2 \dfrac{1}{8}$

56. $\log_3 \dfrac{1}{27}$

57. $\log_{10} 1,000,000$

58. $\log_{10} 0.001$

59. $\log_3 1$

60. $\log_{10} 1$

61. $\log_{10} 10$

62. $\log_2 2$

63. $\log_e e$

64. $\log_e \dfrac{1}{e}$

65. $\log_5 \dfrac{1}{5}$

66. $\log_{27} 3$

67. $\log_8 2$

68. $\log_e e^2$

69. $\log_e e^{-3}$

70. $\log_e e^{-5}$

71. $\log_{10} 10^t$

72. $\log_e e^x$

73. $\log_4 4^t$

74. $\log_5 5^t$

In Problems 75–84, use the properties of logarithms to transform the left-hand side into the right-hand side of the stated equation. Assume the logarithms are all to base 10.

75. $\log \dfrac{26}{7} - \log \dfrac{15}{63} + \log \dfrac{5}{26} = \log 3$

76. $\log 9 - \log 8 - \log \sqrt{75} + \log \sqrt{\dfrac{25}{27}} = -3 \log 2$

77. $\log b^3 + \log 2 - \log \sqrt{b} + \log \dfrac{\sqrt{b^3}}{2} = 4 \log b$

78. $\log k^2 - \log k^{-2} - \log \sqrt{k} - \log k^{-1} = \dfrac{9}{2} \log k$

79. $\log k^{3/2} + \log r - \log k - \log r^{3/4} = \dfrac{1}{4} (\log k^2 r)$

80. $\log a - \dfrac{1}{6} \log b - \dfrac{1}{2} \log a + \dfrac{1}{3} \log b = \dfrac{1}{6} \log a^3 b$

81. $\log \left(y - \dfrac{1}{y^2} \right)^3 = 3 \log(y - 1) + 3 \log(y^2 + y + 1) - 6 \log y$

82. $\log \dfrac{x^2(x + 5)^{3/2}}{x - 5} = 2 \log x + \dfrac{3}{2} \log(x + 5) - \log(x - 5)$

83. $\log \dfrac{(x^2 - 4) \sqrt{x^2 + 2x + 4}}{(x^3 - 8)^2} =$

$\log(x + 2) - \log(x - 2) - \dfrac{3}{2} \log(x^2 + 2x + 4)$

84. $\log \left[\dfrac{1}{12(z - 3)^2} - \dfrac{1}{12(z + 3)^2} \right] =$

$\log z - 2 \log(z + 3) - 2 \log(z - 3)$

APPLICATIONS

85. The worst earthquake ever recorded occurred in the Pacific Ocean near Colombia in 1906. The intensity of this earthquake was $10^{8.9}$ as great as that of an earthquake of minimum intensity I_0. What was the magnitude of this earthquake on the Richter scale?

86. The San Francisco earthquake of 1906 was $10^{8.3}$ times as intense as an earthquake of minimum intensity I_0. What was the magnitude of the San Francisco earthquake on the Richter scale?

87. When Johnny was born, his father deposited $5000 in an account that paid 6% compounded monthly. This was to help pay Johnny's college expenses starting on his eighteenth birthday. The compound amount in this account was $A = \$5000(1.005)^{216}$. If $\log 5000 = 3.69897$ and $\log 1.005 = 0.00217$, find $\log A$ to five decimal places using the properties of logarithms. Check your answer with a calculator.

88. Suppose that the account in Problem 87 paid 10% compounded quarterly. The compound amount would then be given by $A = \$5000(1.025)^{72}$. If $\log 1.025 = 0.01072$, find $\log A$ to five decimal places using the properties of logarithms. Check your answer with a calculator.

89. By the age of 2, most children have reached about 50% of their mature height. Thus if you measure the height of a 2-yr-old child and double this height, the result should be close to the child's mature height. The percent P of adult height attained by a boy can be approximated by the function

$$P(A) = 29 + 50 \log(A + 1)$$

where A is the age in years $(0 < A < 17)$.

Use a calculator to answer the following questions. (Approximate the answers to the nearest percent.)

a. What percent of his mature height is a 2-yr-old boy?

b. What percent of his mature height is an 8-yr-old boy?

c. What percent of his mature height is a 16-yr-old boy?

d. Graph $P(A)$.

e. If a 12-yr-old boy is 60 in. tall, how tall would you expect him to be when he is an adult?

90. The height H of a girl (in inches) can be approximated by the function $H(A) = 11 + 19.44 \log_e A$, where A is the age in years and $6 < A < 15$.

a. What should the height of a 7-yr-old girl be?

b. What should the height of an 11-yr-old girl be?

c. What should the height of a 13-yr-old girl be?

d. Graph $H(A)$. Remember that the domain is $\{A \mid 6 < A < 15\}$.

SKILL CHECKER

Write in scientific notation:

91. 32.68

92. 326.8

93. 0.002387

94. 0.0004392

95. 0.0000569

96. 0.000006731

USING YOUR KNOWLEDGE **Which Is Better?**

One way of comparing the compound amounts in Problems 87 and 88 is to find their ratio. If we denote this ratio by R, then

$$R = \frac{5000(1.005)^{216}}{5000(1.025)^{72}} = \frac{(1.005)^{216}}{(1.025)^{72}}$$

so that

$$\log R = \log(1.005)^{216} - \log(1.025)^{72}$$
$$= 216 \log 1.005 - 72 \log 1.025$$

Using the values of the logarithms given in Problems 87 and 88, we get

$$\log R = (216)(0.00217) - (72)(0.01072) = -0.30312$$

The negative value for $\log R$ means that R is less than 1. Can you see why? This means that the compound amount in Problem 87 is less than that in Problem 88. (A calculator gives the value of R as 0.49760 so that the first amount is less than half the second!)

97. Compare an investment of $1000 at 8% compounded quarterly with the same amount invested at 8.5% compounded annually for a period of 10 yr. You will need the values $\log 1.02 = 0.00860$ and $\log 1.085 = 0.03543$.

98. Compare an investment of $1000 at 6% compounded quarterly with the same amount invested at 6.25% compounded annually for a period of 10 yr. You will need the values $\log 1.015 = 0.00647$ and $\log 1.0625 = 0.02633$.

WRITE ON ...

Explain:

99. The relationship between the exponential function $f(x) = b^x$ and the logarithmic function $g(x) = \log_b x$.

100. Why the logarithm of a negative number is not defined.

101. Why the number 1 is not allowed as a base for the logarithmic function $f(x) = \log_b x$.

102. Why $\log_b 1 = 0$.

MASTERY TEST **If you know how to do these problems, you have learned your lesson!**

103. A recent California earthquake was $10^{6.4}$ times as intense as that of an earthquake of minimum intensity I_0. What was the magnitude of this earthquake on the Richter scale?

Show that:

104. $\log\left(x^2 + \dfrac{1}{x}\right) = \log(x + 1) + \log(x^2 - x + 1) - \log x$

105. $\log_b \sqrt[3]{\dfrac{4}{5}} = \dfrac{1}{3}(\log_b 4 - \log_b 5)$

Write the equation:

106. $\log_4 32 = \dfrac{5}{2}$ in exponential form

107. $2^{10} = 1024$ in logarithmic form

Find:

108. $\log_2 64$ 109. $\log_2 \dfrac{1}{8}$ 110. $\log_{10} 100$ 111. $\log_{10} 0.1$

Solve:

112. $\log_3 x = 2$ 113. $\log_3 x = -3$

114. $\log_x 16 = 2$ 115. $\log_3 \dfrac{1}{27} = x$

116. $\log_{16} x = \dfrac{1}{4}$

Graph:

117. $y = \log_7 x$

118. $f(x) = \log_8 x$

10.5

COMMON AND NATURAL LOGARITHMS

To succeed, review how to:

1. Write numbers in scientific notation (pp. 38–39).
2. Write a radical expression using the exponential form (p. 283).

Objectives:

A Find logarithms and antilogarithms base 10.

B Find logarithms and antilogarithms base e.

C Change the base of a logarithm.

D Graph exponential and logarithmic functions base e.

E Solve applications involving common and natural logarithms.

---

getting started

How Many Decibels Was That?

The scale for measuring the loudness of sound is called the *decibel* scale. When the loudness L of a sound of intensity I is measured in decibels (dB), it is expressed as

$$L = 10 \log_{10} \frac{I}{I_0}$$

where I_0 is the minimum intensity detectable by the human ear. For example, the sound of a riveting machine 30 ft away is 10^{10} times as intense as the minimum intensity I_0, and hence its loudness in decibels is expressed as

$$L = 10 \log_{10} \frac{10^{10} I_0}{I_0} = 10 \log_{10} 10^{10} = (10 \cdot 10) \log_{10} 10$$

$$= 100 \text{ dB}$$

The decibel scale uses logarithms to the base 10; these are called *common* logarithms and we shall study them and another type of logarithm called *natural* logarithms in this section.

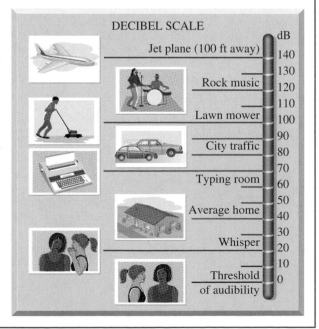

DECIBEL SCALE

	dB
Jet plane (100 ft away)	140
	130
Rock music	120
	110
Lawn mower	100
	90
City traffic	80
	70
Typing room	60
	50
Average home	40
	30
Whisper	20
	10
Threshold of audibility	0

Finding Common Logarithms

The decibel scale is just one example of the many applications of logarithms that use the base 10. Logarithms to base 10 are called **common logarithms** because they use the same base as our "common" decimal system; it's customary to omit the base when working with these logarithms. Be sure to keep in mind that when the base is omitted,

$$\log M = \log_{10} M$$

$\uparrow$ no base (10 is understood)

If you have a scientific calculator, you can find the logarithm of a number by using the $\boxed{\log}$ key. To find log 396 using a scientific calculator, enter 396 and press $\boxed{\log}$. The result is 2.5976952 or, to four decimal places, 2.5977. Before calculators were widely used, the logarithm of a number was found using a table and the properties we studied in Section 10.4. Any number M can be written in scientific notation as

$$M = N \times 10^p$$

where $1 \leq N < 10$ and p is an integer. Then the following is true:

$$\log M = \log(N \times 10^p)$$

$= \log N + \log 10^p$	Product property
$= \log N + p \log 10$	Power property
$= \log N + p$	Since log 10 = 1
$= p + \log N$	Commutative property of addition

The integer p is called the **characteristic**, and the decimal part, $\log N$, is called the **mantissa**. Using the fact that $\log (N \times 10^p) = p + \log N$ we can find log 396 by writing

$$\log 396 = \log 3.96 \times 10^2$$
$$= 2 + \log 3.96$$
$$= 2 + 0.5976952$$
$$= 2.5976952 \qquad \text{As before}$$

EXAMPLE 1   Finding common logarithms

Find:

a. log 42,500   **b.** log 0.000425

SOLUTION

a. Enter 42,500 and press $\boxed{\log}$ to obtain

$$\log 42{,}500 \approx 4.6284 \qquad \text{Rounded to four decimal places}$$

Note that $42{,}500 = 4.25 \times 10^4$, so the characteristic is 4 and the mantissa is 0.6284.

b. Enter 0.000425 and press $\boxed{\log}$ to obtain

$$\log 0.000425 \approx -3.3716$$

Here, $0.000425 = 4.25 \times 10^{-4}$, so the characteristic is -4 and the mantissa is 0.6284. Since $-4.0000 + 0.6284 \approx -3.3716$, the result is the same. ∎

As you recall, the inverse of a logarithmic function is an exponential function. The inverse of a logarithm is an **antilogarithm** or **inverse logarithm**. To find the antilogarithm, we find the power of the base. Thus

$$\text{if} \quad f(x) = \log x, \qquad f^{-1}(x) = \text{antilog } x = 10^x$$

For example, if $\log x = 2.5105$, then antilog $2.5105 = 10^x = 10^{2.5105}$. Most calculators don't have an antilog key, so you have to *know* that to find the inverse, you must use the $\boxed{10^x}$ key, if your calculator has one. If it doesn't you can use the $\boxed{x^y}$ key. Sometimes, the $\boxed{\log}$ key is used as the $\boxed{10^x}$ key after using the $\boxed{\text{2nd}}$ or $\boxed{\text{INV}}$ key. In either case, to find antilog 2.5105, enter 2.5105 and press $\boxed{\text{2nd}}$ (or $\boxed{\text{INV}}$) $\boxed{10^x}$. The result is 324.0.

EXAMPLE 2 Finding antilogarithms

Find:

a. antilog 0.8176 **b.** antilog(-2.1824)

SOLUTION

a. Enter 0.8176 and press $\boxed{\text{2nd}}$ (or $\boxed{\text{INV}}$) $\boxed{10^x}$ to obtain

$$\text{antilog } 0.8176 \approx 6.5705 \qquad \text{Rounded to four decimal places}$$

b. Enter -2.1824 $\boxed{\text{2nd}}$ (or $\boxed{\text{INV}}$) $\boxed{10^x}$ to obtain

$$\text{antilog}(-2.1824) \approx 0.0066 \qquad \text{Rounded to four decimal places}$$

Note that to enter -2.1824, you have to enter 2.1824 and press the $\boxed{+/-}$ key (if available) to change the sign. ■

B Finding Natural Logarithms

The irrational number $e \approx 2.718281828$ was discussed in Section 10.3. This number is used as the base of an important system of logarithms called **natural logarithms** or **Napierian logarithms** in honor of John Napier (1550–1617), the discoverer of logarithms (see *The Human Side of Algebra* at the beginning of this chapter). To distinguish natural from common logarithms, the abbreviation ln (pronounced *el-en*) is used instead of $\log_e$.

NATURAL LOGARITHM

$$\ln x \text{ means } \log_e x$$

The calculator key $\boxed{\ln}$ or $\boxed{\ln x}$ is used for natural logarithms.

EXAMPLE 3 Finding a natural logarithm

Find: $\ln 3252$

SOLUTION Enter 3252 and press $\boxed{\ln}$ to obtain

$\ln 3252 \approx 8.0870$ Rounded to four decimal places ■

Antilogs base e can be found using the $\boxed{e^x}$ key, if your calculator has one. If it doesn't, use the $\boxed{x^y}$ key and approximate e. Sometimes, the $\boxed{\ln}$ key is used as the $\boxed{e^x}$ key after using the $\boxed{\text{2nd}}$ or $\boxed{\text{INV}}$ key.

EXAMPLE 4 Finding the antilog of a natural logarithm

Find: Antilog$_e(-3.4865)$

SOLUTION Enter -3.4865 (you may have to enter 3.4865 and press $\boxed{+/-}$). Now, to find the antilog$_e$, press $\boxed{\text{2nd}}$ (or $\boxed{\text{INV}}$) $\boxed{e^x}$ to obtain

$$\text{antilog}_e(-3.4865) \approx 0.0306 \qquad \text{Rounded to four decimal places} \qquad \blacksquare$$

Changing Logarithmic Bases

Calculators are used to find $\log x$ and $\ln x$, but can we find $\log_4 x$ or $\log_5 x$? To do this, we need the following conversion formula.

CHANGE-OF-BASE FORMULA

For any logarithms with base a and b and any number $M > 0$,

$$\log_b M = \frac{\log_a M}{\log_a b}$$

This fact can be proved as follows: Let $x = \log_b M$. By the definition of logarithm,

$$b^x = M$$

$$\log_a b^x = \log_a M \qquad \text{Take the logarithm base } a \text{ of both sides.}$$

$$x \log_a b = \log_a M \qquad \text{Use the power rule on the left side.}$$

$$x = \frac{\log_a M}{\log_a b} \qquad \text{Solve for } x.$$

$$\log_b M = \frac{\log_a M}{\log_a b} \qquad \text{Since we let } x = \log_b M$$

EXAMPLE 5 Using the change-of-base formula with common logarithms

Using common logarithms, find: $\log_4 20$

SOLUTION Using the change-of-base formula with $b = 4$, $a = 10$,

$$\log_4 20 = \frac{\log_{10} 20}{\log_{10} 4}$$

We then use a calculator to find $\log_{10} 20 \approx 1.3010$ and $\log_{10} 4 \approx 0.6021$ and substitute the values in the equation. Thus

$$\log_4 20 = \frac{\log_{10} 20}{\log_{10} 4}$$

$$\approx \frac{1.3010}{0.6021}$$

$$\approx 2.1608 \qquad \blacksquare$$

EXAMPLE 6 Using the change-of-base formula with natural logarithms

Using natural logarithms, find: $\log_4 20$.

SOLUTION Using the change-of-base formula with $b = 4$, $a = e$,

$$\log_4 = \frac{\ln 20}{\ln 4}$$

We then use a calculator to find ln 20 ≈ 2.9957 and ln 4 ≈ 1.3863 and substitute the values in the equation. Thus

$$\log_4 20 = \frac{\ln 20}{\ln 4}$$

$$\approx \frac{2.9957}{1.3863}$$

$$\approx 2.1609$$

Note that because of rounding, the answer differs from the previous one in the last decimal digit.

To avoid rounding errors, **do not** round the intermediate values $\log_{10} 20$ and $\log_{10} 4$ in Example 5, or ln 20 and ln 4 in Example 6; that is, in Example 5 enter

$$2\ 0\ \boxed{\log}\ \boxed{\div}\ 4\ \boxed{\log}\ \boxed{=}$$

and *then* round the answer 2.190964047 to 2.1910. In Example 6, enter

$$2\ 0\ \boxed{\ln}\ \boxed{\div}\ 4\ \boxed{\ln}\ \boxed{=}$$

and you will obtain the same answer as before.   ■

> NOTE It's best to wait until the final step to round off the answer.

D   Graphing Exponential and Logarithmic Functions Base e

The logarithmic function $f(x) = \ln x$ is the **inverse** of the exponential function $f(x) = e^x$. Because of that, they are reflections across the line $y = x$ as shown in Figure 11. What will the graphs of $f(x) = e^{ax}$, $f(x) = -e^{ax}$, and $f(x) = e^x + b$ look like? The answers are given in Table 1.

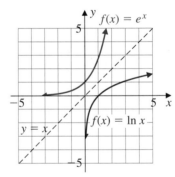

FIGURE 11

EXAMPLE 7   Graphing exponential functions

Graph:

a. $f(x) = e^{(1/3)x}$
b. $f(x) = -e^{(1/3)x}$
c. $f(x) = e^x + 3$

SOLUTION

a. We use convenient values for x (say $x = -3$, 0, and 3) and find the $f(x) = y$-values using a calculator; then we plot the resulting points and draw the graph (Figure 12). For example, when $x = 3$, $f(3) = e^1 \approx 2.72$.

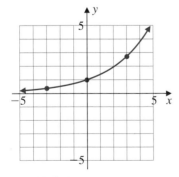

FIGURE 12

x	$e^{(1/3)x}$
-3	0.37
0	1
3	2.72

b. Before you draw the graph, recall that $f(x) = e^{(1/3)x}$ and $f(x) = -e^{(1/3)x}$ are reflections of each other across the x-axis. As before, use convenient points such

TABLE 1 Graphs of $f(x) = e^{ax}$, $f(x) = -e^{ax}$, and $f(x) = e^x + b$

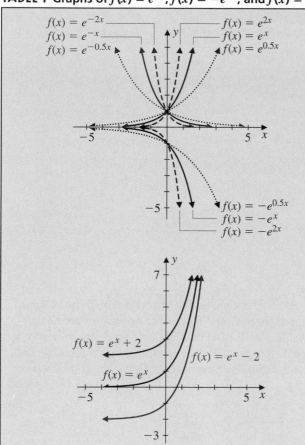

1. The graph of $f(x) = e^{ax}$, where a is positive, looks like that of $f(x) = e^x$. The larger the value of a, the "steeper" the graph. Thus the graph of $f(x) = e^{2x}$ is steeper than the graph of $f(x) = e^{0.5x}$.

2. The graph of $f(x) = -e^{ax}$, where a is positive, is the reflection across the x-axis of the graph of $f(x) = e^{ax}$. Thus the graph of $f(x) = -e^{2x}$ is the reflection across the x-axis of the graph of $f(x) = e^{2x}$.

3. The graph of $f(x) = e^{-ax}$, where a is positive, is the reflection across the y-axis of the graph of $f(x) = e^{ax}$. Thus the graph of $f(x) = e^{-2x}$ is the reflection across the y-axis of the graph of $f(x) = e^{2x}$.

The graph of $f(x) = e^x + b$ is identical to that of $f(x) = e^x$ shifted b units (upward when b is positive, downward when b is negative).

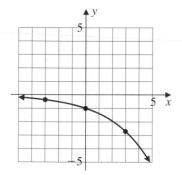

FIGURE 13

as $x = -3, 0$, and 3 but note that the y-values are the *negatives* of the y-values obtained in part a. Plot the resulting points and draw the graph (Figure 13).

x	$e^{(1/3)x}$
-3	-0.37
0	-1
3	-2.72

c. The graph of $f(x) = e^x + 3$ is identical to that of $f(x) = e^x$ but shifted 3 units up. Using $x = -1, 0$, and 1 and a calculator, we find the points shown in the table. We plot these points to obtain the graph shown in Figure 14.

x	$e^x + 3$
-1	3.37
0	4
1	5.72

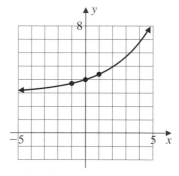

FIGURE 14

■

The functions $f(x) = \ln(x + 1)$ and $f(x) = \ln x + 1$ have graphs that are similar to that of $f(x) = \ln x$. Thus the graph of $f(x) = \ln(x + 1)$ is the graph of $f(x) = \ln x$ shifted 1 unit to the *left*, while the graph of $f(x) = \ln x + 1$ is the graph of $f(x) = \ln x$ shifted 1 unit *up*.

> **CAUTION** Remember that $\ln(x + 1)$ and $\ln x + 1$ are **different**. The 1 added inside the parentheses in $\ln(x + 1)$ shifts the curve $\ln x$ 1 unit *left*. The 1 added to $\ln x$ in $\ln x + 1$ shifts the curve 1 unit *up*. In general, the a in $\ln(x + a)$ shifts the curve a units *right* or *left* (right if a is *negative*, left if a is *positive*.) The a in $\ln x + a$ shifts the curve *up* or *down* (up if a is *positive*, down if a is *negative*).

We use these ideas next.

EXAMPLE 8   Graphing logarithmic functions

Graph:

a. $f(x) = \ln(x + 2)$

b. $f(x) = \ln x + 2$

SOLUTION

a. The graph of $f(x) = \ln(x + 2)$ is the same as that of $f(x) = \ln x$ shifted 2 units left. Since $\ln x$ is only defined for positive values of x, we must select x's that make $x + 2$ positive so that $\ln(x + 2)$ is defined. Using convenient values for x, say $x = 0$, 1, and 2 and a calculator to find the corresponding y-values, we construct a table and then plot the resulting points to obtain the graph shown in Figure 15.

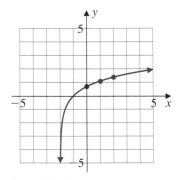

FIGURE 15

x	$\ln(x + 2)$
0	0.69
1	1.10
2	1.39

b. The graph of $f(x) = \ln x + 2$ is the same as that of $f(x) = \ln x$ shifted 2 units up. Using the values $x = 1$, 2, and 3 and a calculator, we obtain the numbers in the table, plot the resulting points, and draw the graph (Figure 16).

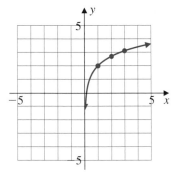

FIGURE 16

x	$\ln x + 2$
1	2
2	2.69
3	3.10

■

  Applications Involving Common and Natural Logarithms

Here are some examples of the many problems that involve logarithms for their solution.

EXAMPLE 9   Finding the pH of a solution

In chemistry, the pH (a measure of acidity) of a solution is defined by the formula $pH = -\log[H^+]$, where $[H^+]$ is the hydrogen ion concentration of the solution in moles per liter. Find the pH of a solution for which $[H^+] = 8 \times 10^{-6}$.

SOLUTION   The pH is defined to be $-\log[H^+]$, so for this solution

$$\begin{aligned}
pH &= -\log(8 \times 10^{-6}) \\
&= -(\log 8 + \log 10^{-6}) \\
&= -(\log 8 - 6) \\
&= 6 - \log 8 \\
&= 6 - 0.9031 \\
&= 5.0969
\end{aligned}$$

The pH of the solution is 5.0969. ■

EXAMPLE 10   Diastolic pressure

Have you taken your blood pressure lately? Normal blood pressures are stated as 120/80, where 120 is the *systolic* pressure (when the heart is contracting) and 80 is the *diastolic* pressure (when the heart is relaxed). Over short periods of time, the diastolic pressure in the aorta of a normal adult is a function of time and can be approximated by the equation

$$P = 90e^{-0.5t}$$

a. What is the diastolic pressure when the aortic valve is closed ($t = 0$)?
b. What is the diastolic pressure when $t = 0.3$ sec?

SOLUTION
a. At $t = 0$, $P = 90e^{-0.5 \cdot 0} = 90$.
b. When $t = 0.3$,

$$\begin{aligned}
P &= 90e^{-0.5 \cdot 0.3} \\
&= 90e^{-0.15} \\
&= 90(0.8607) \qquad \text{Use a calculator.} \\
&= 77.46
\end{aligned}$$

■

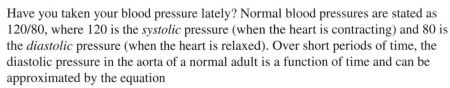

GRAPH IT

The numerical and graphical examples in this section can be done using a grapher. To do Example 1a, go to your home screen. To find log 42,500, press [LOG]; then enter 42,500 and press [ENTER] to obtain 4.62838893, which can be rounded to the desired number of decimal places. Now do Example 1b.

For Example 2a, remember that to find the antilog$_{10}$, we use the [10^x] key. Thus to find antilog$_{10}$ 0.8176, press [2nd] [10^x] [.] 8 1 7 6 and [ENTER] to obtain 6.57052391. Now try Example 2b.

Examples 3 and 4 can be done using the [LN] and [e^x] keys.

Examples 7 and 8 are easily done using the [e^x] and [LN] keys and convenient windows, say $[-5, 5]$ by $[-5, 5]$. Thus to do Example 7a with a TI-82, press [WINDOW] and set Xmin $= -5$, Xmax $= 5$, Xscl $= 1$, Ymin $= -5$, Ymax $= -5$, and Yscl $= 1$. To select the function to be entered, press [Y=] and enter [2nd] [e^x] [(] 1 [÷] 3 [)] [X,T,θ]. Now press [GRAPH]. The result is shown in Window 1. You can graph Example 7b, $f(x) = -e^{(1/3)x}$ by entering [(-)] 1 [2nd] [Y-VARS] 1 1, that is, by making $Y_2 = -Y_1$, to obtain the result in Window 2. Note that $f(x) = e^{(1/3)x}$ and $f(x) = -e^{(1/3)x}$ are reflections of each other across the *x*-axis as expected. Now do Example 7c.

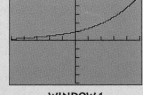
WINDOW 1

To do Example 8a, enter $Y_1 = \ln(x + 2)$ making sure you use parentheses as written. The result is Window 3. Note that the domain consists of all real numbers *x* greater than or equal to -2. Actually, the line $x = -2$ is an *asymptote* for the function. Now do Example 8b.

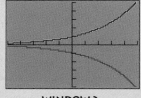
WINDOW 2

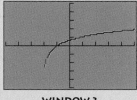
WINDOW 3

A In Problems 1–12, find the common logarithm of the given number.

1. 74.5

2. 952

3. 1840

4. 3.05

5. 0.0437

6. 0.0673

7. 50.18

8. 94.44

9. 0.01238

10. 0.01004

11. 0.008606

12. 0.0004632

In Problems 13–20, find the antilogarithm base 10.

13. antilog 1.2672

14. antilog 2.4409

15. antilog(7.7672 − 10)

16. antilog(−3.4045)

17. antilog 1.4630

18. antilog 2.9408

19. antilog(−0.134)

20. antilog(−1.2275)

B In Problems 21–30, find the natural logarithm.

21. $\ln 3$

22. $\ln 4$

23. $\ln 52$

24. $\ln 62$

25. $\ln 2356$

26. $\ln 208.3$

27. $\ln 0.054$

28. $\ln 0.049$

29. $\ln 0.00062$

30. $\ln 0.00132$

In Problems 31–40, find the antilogarithm.

31. $\text{antilog}_e 1.2528$

32. $\text{antilog}_e 2.2925$

33. $\text{antilog}_e 4.1744$

34. $\text{antilog}_e 4.6052$

35. $\text{antilog}_e 0.0392$

36. $\text{antilog}_e 0.0198$

37. $\text{antilog}_e(-2.3025)$

38. $\text{antilog}_e(-0.3566)$

39. $\text{antilog}_e(-4.6051)$

40. $\text{antilog}_e(-4.8281)$

C In Problems 41–50, use the change-of-base formula to find the logarithm using (a) common logarithms and (b) natural logarithms.

41. $\log_3 20$

42. $\log_3 40$

43. $\log_{100} 40$

44. $\log_{200} 50$

45. $\log_{0.2} 3$

46. $\log_{0.1} 4$

47. $\log_4 0.4$

48. $\log_2 0.16$

49. $\log_{\sqrt{2}} 0.8$

50. $\log_{\sqrt{2}} 0.16$

D In Problems 51–70, graph the function.

51. $f(x) = e^{3x}$

52. $f(x) = e^{4x}$

53. $f(x) = -e^{3x}$

54. $f(x) = -e^{4x}$

55. $f(x) = e^{-3x}$

56. $f(x) = e^{-4x}$

57. $f(x) = e^{(1/2)x}$

58. $f(x) = -e^{(1/2)x}$

59. $f(x) = e^x + 1$

60. $f(x) = e^x - 1$

61. $f(x) = e^{0.5x} + 1$

62. $f(x) = e^{0.5x} - 1$

63. $f(x) = 2e^x$

64. $f(x) = -2e^x$

65. $f(x) = \ln(x + 1)$

66. $f(x) = \ln x + 1$

67. $f(x) = \ln x + 4$

68. $f(x) = \ln(x + 4)$

69. $f(x) = \ln(x - 1)$

70. $f(x) = \ln x - 1$

APPLICATIONS

In Problems 71–76, find the pH of a solution with the given [H⁺]. (*Hint:* pH = −log[H⁺].)

71. $[H^+] = 7 \times 10^{-7}$ **72.** $[H^+] = 1.5 \times 10^{-9}$

73. Eggs whose $[H^+]$ is 1.6×10^{-8}

74. Tomatoes whose $[H^+]$ is 6.3×10^{-5}

75. Milk whose $[H^+]$ is 4×10^{-7}

76. $[H^+] = 5 \times 10^{-8}$

77. The number of Hispanic persons in the United States can be approximated by

$$H = 15{,}000{,}000 e^{0.02t}$$

where t is the number of years after 1980.

a. What was the number of Hispanic persons in 1980?

b. Using this approximation, what will be the number of Hispanic persons in the United States in the year 2000?

78. The number of African-Americans in the United States can be approximated by

$$B = 27{,}000{,}000 e^{0.013t}$$

where t is the number of years after 1980.

a. What was the number of African-Americans in 1980?

b. Using this approximation, what will be the number of African-Americans in the United States in the year 2000?

79. The number of bacteria present in a certain culture can be approximated by

$$B = 50{,}000 e^{0.2t}$$

where t is measured in hours and $t = 0$ corresponds to 12 noon. Find the number of bacteria present at

a. noon   **b.** 2 P.M.   **c.** 6 P.M.

80. If a bactericide (a bacteria killer) is introduced into a bacteria culture, the number of bacteria can be approximated by

$$B = 50{,}000 e^{-0.1t}$$

where t is measured in hours. Find the number of bacteria present

a. When $t = 0$.   **b.** When $t = 1$.   **c.** When $t = 10$.

81. Sales begin to decline d days after the end of an advertising campaign and can be approximated by

$$S = 1000 e^{-0.1d}$$

a. How many sales will be made on the last day of the campaign—that is, when $d = 0$?

b. How many sales will be made 10 days after the end of the campaign?

82. The demand function for a certain commodity is approximated by

$$p = 100 e^{-q/2}$$

where q is the number of units demanded at a price of p dollars per unit.

a. If there is a 100-unit demand for the product, what will be its price?

b. If there is no demand for the product, what will its price be?

83. The concentration C of a drug in the bloodstream at time t (in hours) can be approximated by

$$C = 100(1 - e^{-0.5t})$$

a. What will the concentration be when $t = 0$?

b. What will the concentration be after 1 hr?

84. The number of people $N(t)$ reached by a particular rumor at time t is approximated by

$$N(t) = \frac{5050}{1 + 100 e^{-0.06t}}$$

a. Find $N(0)$.   **b.** Find $N(10)$.

85. The stellar magnitude M of a star is defined by

$$M = -2.5 \log\left(\frac{B}{B_0}\right)$$

where B is the brightness of the star and B_0 the minimum of brightness.

a. Find the stellar magnitude of the North Star, which is 2.1 times as bright as B_0.

b. Find the stellar magnitude of Venus, which is 36.2 times as bright as B_0.

86. The percent P of adult height a male has reached at age A $(13 \le A \le 18)$ is

$$P = 16.7 \log(A - 12) + 87$$

   a. What percent of adult height has a 13-yr-old male reached?

   b. What percent of adult height has an 18-yr-old male reached?

Solve for k:

87. $\log 3 = k \log 2$

88. $\log 25 = 0.15k$

89. $25 = 10^k$

90. $100 = 5^k$

Exponential and Logarithmic Functions

Is there a relationship between the graph of an exponential function and the graph of the logarithm of the function? Let's use what we learned in this section to find out.

91. a. Graph $f(x) = 2^x$.   b. Graph $g(x) = \log 2^x$.

   c. What is the slope of $g(x) = \log 2^x$? (*Hint:* Use the properties of logarithms to simplify $\log 2^x$.)

92. a. Graph $f(x) = 3^x$.   b. Graph $g(x) = \log 3^x$.

   c. What is the slope of $g(x) = \log 3^x$?

93. Compare the base used for the function f in Problems 91 and 92 with the slope of the function g. What is the relationship between the two?

Explain:

94. The difference between **common** logarithms and **natural** logarithms.

95. What the antilog$_{10}$ of a number is.

96. The usefulness of the change-of-base formula.

97. The relationship between the graphs of $f(x) = 10^x$ and $g(x) = \log x$.

98. The relationship between the graphs of $f(x) = e^x$ and $g(x) = \ln x$.

99. The meaning of "the graph of $f(x) = e^{2x}$ is steeper than the graph of $g(x) = e^{0.5x}$."

If you know how to do these problems, you have learned your lesson!

Find:

100. $\ln 3120$

101. antilog$_e(-1.5960)$

102. $\log_4 40$

103. $\log_5 20$

104. $\log 41,500$

105. $\log 0.000415$

106. antilog 0.8432

107. antilog(-2.4683)

Graph:

108. $f(x) = e^{(1/4)x}$

109. $f(x) = -e^{(1/4)x}$

110. $f(x) = e^{(-1/4)x}$

111. $f(x) = e^{(1/4)x} + 1$

112. $f(x) = \ln(x + 3)$

113. $f(x) = \ln x + 3$

114. Over short periods of time, the systolic pressure of a normal adult can be approximated by

$$P = 130e^{-0.5t}$$

   a. What is the systolic pressure when $t = 0$?

   b. What is the systolic pressure when $t = 0.5$ sec?

10.6

EXPONENTIAL AND LOGARITHMIC EQUATIONS AND APPLICATIONS

To succeed, review how to:

1. Use the properties of logarithms (p. 615).

2. Use the laws of exponents (pp. 31–37).

3. Solve linear equations (pp. 69–73).

4. Evaluate logarithms (p. 621).

Objectives:

 Solve exponential equations.

 Solve logarithmic equations.

C Solve applications involving exponential or logarithmic equations.

---

getting started

Don't Drink and Drive!

Is there a relationship between blood alcohol level (BAC) and the probability of having an accident? Absolutely! As the chart shows, the probability $P(b)$ of having an accident, written as a percent, is a function of your blood alcohol level b. The formula is given by

$$P(b) = e^{kb} \qquad \textbf{(1)}$$

As you can see from the chart, this probability is 25% when the BAC is 0.15%. Can we find k using this information? To do this, let $P(b) = 25$ and $b = 0.15$ in equation (1). We then have

$$25 = e^{0.15k} \qquad \textbf{(2)}$$

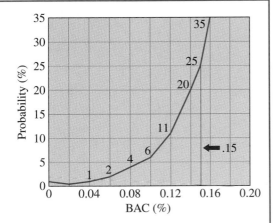

Equation (2) is an example of an *exponential* equation, since the variable k occurs in the exponent. We shall solve this equation in Examples 3 and 4. We will even be able to predict what BAC theoretically leads to certain disaster; that is, what alcohol level b corresponds to a 100% probability of an accident!

A Solving Exponential Equations

In all the equations we have solved, we have seldom used variables as exponents. When this happens, we have an exponential equation.

EXPONENTIAL EQUATION

> An **exponential equation** is an equation in which the variable occurs in an exponent.

Thus $6^x = 14$, $3^{5x} = 20$, and $2^{6x} = 32$ are exponential equations. The equation $2^{6x} = 32$ can be rewritten using powers of the base 2, that is, as

$$2^{6x} = 2^5$$

Because we are using the same base, the exponents must be equal; thus,

$$6x = 5 \qquad \text{and} \qquad x = \frac{5}{6}$$

Remember the equivalence property from Section 10.4? We restate it here for your convenience.

EQUIVALENCE PROPERTY

> For $b > 0$, $b \neq 1$, $b^x = b^y$ is equivalent to $x = y$.

EXAMPLE 1   Solving exponential equations

Solve:

a. $3^{2x-1} = 81$

b. $2^{x+1} = 8^{x-1}$

SOLUTION

a. Since $3^4 = 81$, we write each side of $3^{2x-1} = 81$ as a power of 3.

$$3^{2x-1} = 3^4$$

Since the base is the same (3), the exponents must be equal. Thus

$$2x - 1 = 4 \qquad \text{Equate the exponents.}$$

$$2x = 5 \qquad \text{Add 1.}$$

$$x = \frac{5}{2} \qquad \text{Divide by 2.}$$

CHECK: Letting $x = \frac{5}{2}$ in $3^{2x-1} = 81$, we obtain

$$3^{2 \cdot (5/2) - 1} = 81$$

or $3^4 = 81 \qquad$ A true statement

Thus the solution is $\frac{5}{2}$.

b. The idea is to write both sides of the equation using the same base. Since $8 = 2^3$, the equation can be rewritten as

$$2^{x+1} = (2^3)^{x-1} = 2^{3x-3}$$

Since the base is the same, 2, the exponents must be equal. Thus

$$x + 1 = 3x - 3$$

$$-2x + 1 = -3 \qquad \text{Subtract } 3x.$$

$$-2x = -4 \qquad \text{Subtract 1.}$$

$$x = 2 \qquad \text{Divide by } -2.$$

CHECK: If we let $x = 2$ in the original equation, we obtain

$$2^{2+1} = 8^{2-1}$$

$$2^3 = 8^1 \qquad \text{A true statement}$$

Thus the solution is 2.   ■

How can we solve $6^x = 14$? Since it isn't possible to write each side of $6^x = 14$ as a power of the same base, we make use of a fundamental property of logarithms.

EQUIVALENCE PROPERTY FOR LOGARITHMS

> If M, N, and b are all positive numbers and $b \neq 1$, then
>
> $$\log_b M = \log_b N \qquad \text{is equivalent to} \qquad M = N$$

This means that

$$\text{if } \log_b M = \log_b N, \quad \text{then} \quad M = N$$

and conversely

$$\text{if } M = N, \quad \text{then} \quad \log_b M = \log_b N$$

EXAMPLE 2   More practice solving exponential equations

Solve: $6^x = 14$

SOLUTION

$$6^x = 14 \qquad \text{Given.}$$

$$\log 6^x = \log 14 \qquad \text{Take the log of both sides.}$$

$$x \log 6 = \log 14 \qquad \text{Since } \log 6^x = x \log 6$$

$$x = \frac{\log 14}{\log 6} \qquad \text{Divide by } \log 6.$$

Note that this is the *exact* answer. If we wish to approximate it, use a calculator to obtain

$$x = \frac{\log 14}{\log 6} \approx 1.4729$$

Remember to wait until the final step to round off your answer!   ■

Now let's return to the *Getting Started* and solve equation (2), $25 = e^{0.15k}$. Keep in mind that the property "$\log_b M = \log_b N$ is equivalent to $M = N$" works when the base $b = e$, that is

$$M = N \qquad \text{is equivalent to} \qquad \ln M = \ln N$$

EXAMPLE 3   Using an equivalency property
                    to solve an exponential equation

Solve: $25 = e^{0.15k}$

SOLUTION

$$25 = e^{0.15k} \qquad \text{Given.}$$

$$\ln 25 = \ln e^{0.15k} \qquad \text{Take the natural logarithm of both sides.}$$

$$\ln 25 = 0.15k \qquad \text{Since } \log_b b^x = x, \ln e^{0.15k} = 0.15k.$$

$$\frac{\ln 25}{0.15} = k \qquad \text{Divide both sides by 0.15.}$$

Thus

$$k = \frac{\ln 25}{0.15} \approx \frac{3.2189}{0.15} \approx 21.5 \qquad \text{(to the nearest tenth)} \qquad ■$$

problem solving   **EXAMPLE 4**   Accident rate and blood alcohol level

If we substitute $k = 21.5$ into equation (2) in the *Getting Started*, the formula for $P(b)$ becomes $P(b) = e^{21.5b}$. At what blood alcohol level b will the probability of having an accident be 100%?

1. Read the problem.

2. Select the unknown.

SOLUTION   We use the RSTUV method.

We are asked to find b so that $P(b) = 100$. To do this we must solve the equation

$$100 = e^{21.5b}$$

3. Think of a plan.   First, we take natural logarithms of both sides of the equation, then solve for b.

4. Use algebra to solve the problem.
$$\ln 100 = \ln e^{21.5b}$$

$$= 21.5b \ln e \qquad \text{Use the power property.}$$

$$= 21.5b \qquad \ln e = 1$$

$$b = \frac{\ln 100}{21.5} \qquad \text{Solve for } b.$$

$$= \frac{2 \ln 10}{21.5} \qquad 100 = 10^2$$

The value of $\ln 10$ can be found with a calculator. To four decimal places, $\ln 10 = 2.3026$, so

$$b = \frac{4.6052}{21.5} = 0.214$$

5. Verify the answer.   The verification that $100 = e^{21.5(0.214)}$ is left to you (use your calculator!)

Thus when the blood alcohol level is about 0.21%, the probability of an accident is 100%. Of course, if your blood alcohol level is 0.21%, you are probably too drunk to even get in your car!!   ■

B Solving Logarithmic Equations

We have already solved certain types of **logarithmic equations**—that is, equations containing *logarithmic expressions*—in Section 10.4. We used the definition of a logarithm and converted the given equation into an exponential equation. We use this technique in Example 5.

EXAMPLE 5   Solving logarithmic equations

Solve: $\log_5(2x - 3) = 2$

SOLUTION
$$\log_5(2x - 3) = 2 \qquad \text{Given.}$$

$$2x - 3 = 5^2 \qquad \text{Since } \log_b x = y \text{ is equivalent to } b^y = x$$

$$2x - 3 = 25$$

$$2x = 28 \qquad \text{Add 3.}$$

$$x = 14 \qquad \text{Divide by 2.}$$

Thus the solution is 14. You can check this by letting $x = 14$ in $\log_5(2x - 3) = 2$ to obtain

$$\log_5(2 \cdot 14 - 3) = 2$$

$$\log_5(25) = 2$$

$$5^2 = 25 \qquad \text{A true statement}$$ ■

In general, we use the following procedure to solve logarithmic equations.

PROCEDURE

Solving Logarithmic Equations

1. Write the equation as an equivalent one with a single logarithmic expression on one side; that is, write the equation in the form $\log_b M = N$.
2. Write the equivalent exponential equation $b^N = M$ and solve.
3. Always check your answer.

EXAMPLE 6 More practice solving logarithmic equations

Solve:

a. $\log(x + 3) + \log x = 1$ **b.** $\log_3(x - 1) - \log_3(x - 3) = 1$

SOLUTION

a.

$\log(x + 3) + \log x = 1$	Given.
$\log(x + 3)x = 1$	Use the product rule.
$(x + 3)x = 10$	Write as an equivalent equation.
$x^2 + 3x = 10$	Simplify.
$x^2 + 3x - 10 = 0$	Subtract 10.
$(x + 5)(x - 2) = 0$	Factor.
$(x + 5) = 0$ or $(x - 2) = 0$	By the zero-factor property
$x = -5$ or $x = 2$	

CHECK: For $x = -5$, the expression $\log x$ becomes $\log(-5)$, which isn't defined because we cannot find the logarithm of a negative number. For $x = 2$,

$$\log(x + 3) + \log x = 1$$

becomes $\log 5 + \log 2 = 1$

or

$\log(5 \cdot 2) = 1$	Use the product rule.
$10^1 = 5 \cdot 2$	Use the definition of logarithm.

Since the result is a true statement, the solution is 2.

b.

$\log_3(x - 1) - \log_3(x - 3) = 1$	Given.
$\log_3\left(\dfrac{x - 1}{x - 3}\right) = 1$	Use the quotient rule.
$\dfrac{x - 1}{x - 3} = 3^1$	Write as an equivalent equation.
$x - 1 = 3(x - 3)$	Multiply both sides by $(x - 3)$.
$x - 1 = 3x - 9$	By the distributive property
$-1 = 2x - 9$	Subtract x from both sides.
$8 = 2x$	Add 9 to both sides.
$4 = x$	Divide both sides by 2.

CHECK: For $x = 4$,

$$\log_3(x - 1) - \log_3(x - 3) = 1$$

becomes $\log_3(4 - 1) - \log_3(4 - 3) = 1$

or $\log_3 3 - \log_3 1 = 1$

                                                $1 - 0 = 1$ Since $\log_3 3 = 1$ and $\log_3 1 = 0$

The resulting statement $1 - 0 = 1$ is true, so the solution is 4.   ■

 Applications Involving Exponential or Logarithmic Equations

Exponential and logarithmic equations have many applications in such areas as business, engineering, social science, psychology, and science. The following examples will give you an idea of the variety and range of their use.

problem solving   **EXAMPLE 7**   Doubling your money at 6% interest

With continuous compounding, a principal of P dollars accumulates to an amount A given by the equation

$$A = Pe^{rt}$$

where r is the interest rate and t is the time in years. If the interest rate is 6%, how long would it take for the money in your bank account to double?

1. Read the problem.   **SOLUTION**   We use the RSTUV method.

2. Select the unknown.   With $A = 2P$ and $r = 0.06$, the equation becomes

$$2P = Pe^{0.06t}$$

3. Think of a plan.   or

$$2 = e^{0.06t}$$ Divide by P.

4. Use algebra to solve the problem.   We want to solve this equation for t, so we take natural logarithms of both sides:

$$\ln 2 = \ln e^{0.06t} = 0.06t \ln e = 0.06t$$

Thus

$$t = \frac{\ln 2}{0.06}$$

Using a calculator, we find $\ln 2 = 0.69315$ so that

$$t = \frac{0.69315}{0.06} = 11.6$$

This means that it would take about 11.6 yr for your money to double.

5. Verify the solution.   The verification is left to you.   ■

 problem solving   **EXAMPLE 8**   World population in the year 2000

In 1984, the population of the world was about 4.8 billion and the yearly growth rate was 2%. The equation giving the population P in terms of the time t is

$$P = 4.8e^{0.02t}$$

Estimate the world population P in the year 2000.

1. Read the problem.	**SOLUTION** We use the RSTUV method.
2. Select the unknown.	We are looking for the population P in the year 2000.
3. Think of a plan.	Since $P = 4.8$ for $t = 0$, the equation shows that t is measured from the year 1984. To estimate the population in 2000, we use $t = 16$ in the equation:

$$P = 4.8e^{(0.02)(16)} = 4.8e^{0.32}$$

4. Use arithmetic to solve the problem.	The value $e^{0.32} = 1.3771$ can be found with a calculator. Hence

$$P = (4.8)(1.3771) = 6.6$$

Thus our estimate for the population in 2000 is about 6.6 billion.

5. Verify the answer.	The verification is left to you. ■

 problem solving   **EXAMPLE 9**   Population explosion in a bacteria culture

If B is the number of bacteria present in a laboratory culture after t minutes, then, under ideal conditions,

$$B = Ke^{0.05t}$$

where K is a constant. If the initial number of bacteria is 1000, how long would it take for there to be 50,000 bacteria present?

1. Read the problem.	**SOLUTION** We use the RSTUV method.
2. Select the unknown.	Since $B = 1000$ for $t = 0$, we have

$$1000 = Ke^0 = K$$

The equation for B is then

$$B = 1000e^{0.05t}, \text{ where } t \text{ is the unknown.}$$

3. Think of a plan.	Now let $B = 50{,}000$:

$$50{,}000 = 1000e^{0.05t}$$
$$50 = e^{0.05t}$$

4. Use algebra to solve the problem.	To solve for t, take natural logarithms of both sides:

$$\ln 50 = \ln e^{0.05t}$$
$$\ln 50 = 0.05t \qquad \ln e = 1$$

Thus

$$t = \frac{\ln 50}{0.05} = \frac{3.9120}{0.05} = 78.2 \text{ min}$$

It will take 78.2 min for there to be 50,000 bacteria present.

5. Verify the solution.	The verification is left to you. ■

 problem solving   **EXAMPLE 10**   Half-life of cesium-137

The element cesium-137 decays at the rate of 2.3% per year. Find the half-life of this element.

1. Read the problem.	**SOLUTION** We use the RSTUV method.
2. Select the unknown.	The half-life of a substance is found by using the equation

$$A(t) = A_0 e^{-kt}$$

where $A(t)$ is the amount present at time t (years), k is the decay rate, and A_0 is the initial amount of the substance present. In this problem,

$$k = 2.3\% = 0.023, \qquad A(t) = \frac{1}{2}A_0$$

and we want to find t.

3. Think of a plan.

With this information, the basic equation becomes

$$\frac{1}{2}A_0 = A_0 e^{-0.023t}$$

$$\frac{1}{2} = e^{-0.023t} \qquad \text{Divide by } A_0.$$

4. Use algebra to solve the problem.

Now take natural logarithms of both sides:

$$\ln \frac{1}{2} = \ln e^{-0.023t}$$

$$\ln 1 - \ln 2 = -0.023t$$

$$t = \frac{\ln 2}{0.023} \qquad \text{Since } \ln 1 = 0$$

$$= \frac{0.69315}{0.023} = 30.14$$

Thus the half-life of cesium-137 is about 30.14 yr.

5. Verify the solution.

The verification is left to you.   ■

GRAPH IT

Do you remember how to solve equations using your grapher? The idea is to graph both sides of the equation, call them Y_1 and Y_2 and use $\boxed{\text{TRACE}}$ and $\boxed{\text{ZOOM}}$ (or the intersect feature) to find the point of intersection (the solution). In Example 1a, graph $Y_1 = 3^{(2x-1)}$ (note the parentheses around the $2x - 1$) and $Y_2 = 81$ using a $[-3, 3]$ by $[-10, 100]$ window with the Yscl = 10. On the TI-82, press $\boxed{\text{2nd}}$ $\boxed{\text{CALC}}$ 5 and $\boxed{\text{ENTER}}$ three times to answer the grapher questions "First Curve?" "Second Curve?" "Guess?" The intersection is $X = 2.5$, as shown in Window 1. Now try Example 1b.

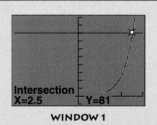

WINDOW 1

For Example 2, let $Y_1 = 6^x$ and $Y_2 = 14$ and follow the procedure of Example 1a using a $[-1, 10]$ by $[-1, 20]$ window. The intersection is $X = 1.4728859$, as shown in Window 2. To do Example 3, let $Y_1 = 25$ and $Y_2 = e^{0.15x}$ with a $[-5, 30]$ by $[-5, 30]$ window and Xscl = YScl = 5. As usual, press $\boxed{\text{2nd}}$ $\boxed{\text{CALC}}$ 5 and $\boxed{\text{ENTER}}$ three times to find the intersection $X = 21.459172$, as shown in Window 3.

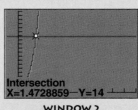

WINDOW 2

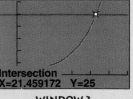

WINDOW 3

(continued on facing page)

What type of window do you need for Example 4? Note that $Y_1 = 100$. After you find the appropriate window, you can check the results of Example 4.

In Example 5, we have to graph $Y_1 = \log_5(2x - 3)$, but graphers don't have a $\boxed{\log_5}$ key. Here, *you* have to remember that the change-of-base formula states that

$$Y_1 = \log_5(2x - 3) = \frac{\log(2x - 3)}{\log 5}$$

Then let $Y_2 = 2$. If you use a standard window, the point of intersection **does not** show (try it!), so we suggest a [0, 20] by [−1, 3] window. Press $\boxed{\text{2nd}}$ $\boxed{\text{CALC}}$ 5 and $\boxed{\text{ENTER}}$ three times to obtain the intersection $X = 14$, as shown in Window 4.

To do Example 6, use the $\boxed{\text{LOG}}$ function in your grapher. Let $Y_1 = \log(x + 3) + \log x$ (there is no need to simplify Y_1!) and $Y_2 = 1$ and use a standard window. Press $\boxed{\text{2nd}}$ $\boxed{\text{CALC}}$ 5 and move the cursor so that it shows a positive value for *x*. Then press $\boxed{\text{ENTER}}$ three times. The intersection $X = 2$ is shown in Window 5.

The equations of Examples 7, 8, and 9 can be solved with your grapher using the same techniques as those employed to solve Examples 3 and 4. Of course, you must know how to set up the equations first!

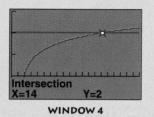

WINDOW 4

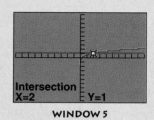

WINDOW 5

EXERCISE 10.6

A **In Problems 1–26, solve the equation.**

1. $5^x = 25$ **2.** $3^x = 81$

3. $2^x = 32$ **4.** $3^{2x} = 81$

5. $5^{3x} = 625$ **6.** $6^{-2x} = 216$

7. $7^{-3x} = 343$ **8.** $5^x = 4$

9. $7^x = 512$ **10.** $3^{x+1} = 729$

11. $5^{x-2} = 625$ **12.** $2^{3x+1} = 128$

13. $3^x = 2$ **14.** $3^x = 20$

15. $2^{3x-2} = 32$ **16.** $3^{4x-3} = 27$

17. $5^{3x} \cdot 5^{x^2} = 25$ **18.** $3^{4x} \cdot 3^{x^2} = 243$

19. $e^x = 10$ **20.** $e^x = 100$

21. $e^{-x} = 0.1$ **22.** $e^{-x} = 0.01$

23. $30 = e^{2k}$ **24.** $40 = e^{3k}$

25. $10 = e^{-2k}$ **26.** $20 = e^{-3k}$

B **In Problems 27–50, solve the equation.**

27. $\log_2 x = 3$ **28.** $\log_3 x = 2$

29. $\log_2 x = -3$ **30.** $\log_3 x = -2$

31. $\ln x = 1$ **32.** $\ln x = -1$

33. $\ln x = 3$ **34.** $\ln x = -3$

35. $\log_2(3x - 5) = 1$ **36.** $\log_3(2x - 1) = 4$

37. $\log_4(3x - 1) = 2$ **38.** $\log_6(2x + 1) = 2$

39. $\log_5(3x + 1) = 2$ **40.** $\log_2(4x - 1) = 4$

41. $\log x + \log(x - 3) = 1$ **42.** $\log x + \log(x - 11) = 1$

43. $\log_2(x + 1) + \log_2(x + 3) = 3$

44. $\log_3(x + 4) + \log_3(x - 2) = 3$

45. $\log(x + 1) - \log x = 1$ **46.** $\log(x - 1) - \log x = 1$

47. $\log_2(3 + x) - \log_2(7 - x) = 2$

48. $\log_3(2 + x) - \log_3(8 - x) = 2$

49. $\log_2(x^2 + 4x + 7) = 2$

50. $\log_2(x^2 + 4x + 3) = 3$

APPLICATIONS

In Problems 51–54, assume continuous compounding and follow the procedure in Example 7 to find how long it takes a given amount to double at the given interest rate.

51. $r = 5\%$ **52.** $r = 7\%$

53. $r = 6.5\%$ **54.** $r = 7.5\%$

55. Suppose that the population of the world grows at the rate of 1.5% and that the population in 1984 was about 4.8 billion. Follow the procedure of Example 8 to estimate the population in the year 2000.

56. Repeat Problem 55 for growth rate of 1.75%.

In Problems 57–60, assume that the number of bacteria present in a culture after t minutes is given by $B = 1000e^{0.04t}$. Find the time it takes for the number of bacteria present to be

57. 2000   **58.** 5000

59. 25,000   **60.** 50,000

61. When a bacteria-killing solution is introduced into a certain culture, the number of live bacteria is given by the equation $B = 100,000e^{-0.2t}$, where t is the time in hours. Find the number of live bacteria present at the following times.
   a. $t = 0$ **b.** $t = 2$ **c.** $t = 10$ **d.** $t = 20$

62. The number of honey bees in a hive is growing according to the equation $N = N_0e^{0.015t}$, where t is the time in days. If the bees swarm when their number is tripled, find how many days till this hive swarms.

In Problems 63–66, follow the procedure of Example 10 to find the half-life of the substance.

63. Plutonium, whose decay rate is 0.003% per year

64. Krypton, whose decay rate is 6.3% per year

65. A radioactive substance whose decay rate is 5.2% per year

66. A radioactive substance whose decay rate is 0.2% per year

67. The atmospheric pressure P in pounds per square inch at an altitude of h feet above the Earth is given by the equation $P = 14.7e^{-0.00005h}$. Find the pressure at an altitude of
   a. 0 ft   **b.** 5000 ft   **c.** 10,000 ft

68. If the atmospheric pressure in Problem 67 is measured in inches of mercury, then $P = 30e^{-0.207h}$, where h is the altitude in miles. Find the pressure
   a. At sea level.   **b.** At 5 mi above sea level.

69. According to the National Football League Players Association (NFLPA), average NFL salaries (in thousands of dollars) are as shown in the graph and can be approximated by

$$S = 163e^{0.18t}$$

where t is the number of years after 1985.

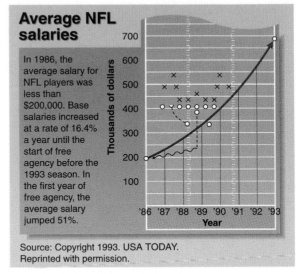

Average NFL salaries

In 1986, the average salary for NFL players was less than $200,000. Base salaries increased at a rate of 16.4% a year until the start of free agency before the 1993 season. In the first year of free agency, the average salary jumped 51%.

Source: Copyright 1993. USA TODAY. Reprinted with permission.

   a. In how many years will average salaries be $800,000? (Answer to the nearest year.)

   b. Based on this graph, in how many years will salaries reach the 1 million dollar mark? (Answer to the nearest year.)

70. According to the *Statistical Abstract of the United States*, about $\frac{2}{3}$ of all aluminum cans distributed are recycled. If a company distributes 500,000 cans, the number in use after t years is

$$N(t) = 500,000 \left(\frac{2}{3}\right)^t$$

How many years will it take for the number of cans to reach 100,000? (Answer to the nearest year.)

71. Do you have a fear of flying? The U.S. Department of Transportation has good news for nervous fliers: The number of general aviation accidents A has gone down significantly in the last 20 years! It can be approximated by

$$A = 5000e^{-0.04t}$$

where t is the number of years after 1970.
   a. How many accidents were there in 1970?

   b. How many accidents were there in 1990?

   c. In what year do you predict the number of accidents to be 1000? (Answer to the nearest year.)

72. After exercise the diastolic blood pressure of normal adults is a function of time and can be approximated by

$$P = 90e^{-0.5t}$$

where t is the time in minutes. How long would it be before the diastolic pressure comes down to 80? (Answer to two decimal places.)

73. How much do you spend on your credit cards? According to the Federal Reserve Board and the Bankcard Holders of America, credit card spending (in millions of dollars) is on the increase and can be approximated by

$$S = 54e^{0.15t}$$

where t is the number of years after 1980.
a. How many millions of dollars were spent in 1980?

b. In what year did the amount spent on credit cards reach 500 million dollars? (Answer to the nearest year.)

74. The percentage of adult height P attained by a boy can be approximated by

$$P = h(A) = 29 + 50 \log(A + 1)$$

where P is the percentage of adult height and A is the age in years ($0 < A < 17$). At what age will a boy reach 89% of his adult height?

75. The height H of a girl (in inches) can be approximated by the function

$$H = h(A) = 11 + 19.44 \ln A$$

where A is the age in years ($6 < A < 15$). To the nearest year, at what age would you expect a girl to be
a. 60 in. tall?   b. 50 in. tall?

76. According to the National Football League, paid attendance (in millions) is as shown in the graph and can be approximated by

$$A = 17.4 + 0.32 \ln t$$

where t is the number of years after 1988.

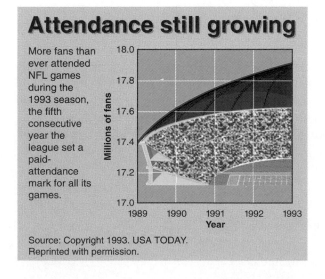

Attendance still growing

More fans than ever attended NFL games during the 1993 season, the fifth consecutive year the league set a paid-attendance mark for all its games.

Millions of fans

1989 1990 1991 1992 1993
Year

Source: Copyright 1993. USA TODAY. Reprinted with permission.

a. Based on this graph, in what year did attendance reach 18 million? (Answer to the nearest year.)

b. Based on this graph, in what year will attendance reach 18.5 million? (Answer to the nearest year.)

77. If $1000 is invested in an account earning 1% interest each month, the number of months t it takes the account to grow to an amount A can be approximated by

$$t = -694.2 + 231.4 \log A$$

where $A \geq 1000$. How long would it take for this account to grow to 1 million dollars?

78. Evaluate the expression $\frac{1}{2}n(n - 1)$ when:
a. $n = 1$ b. $n = 3$ c. $n = 10$

79. Find the value of the function $f(n) = 2n + 1$ when:
a. $n = 1$ b. $n = 5$ c. $n = 10$

USING YOUR KNOWLEDGE   Just the Fax, Ma'am

Have you ever wondered how the approximations in this exercise set are done? To construct a growth equation that is based on time t, we start with the basic equation

$$A = A_0 e^{kt}$$

When $t = 0$, $A = A_0 e^0 = A_0$. Now, if we know the value A for a certain time t, we can solve the resulting equation and find k. Use this idea and the fact that the sale of fax machines soared from 580,000 in 1987 to 11 million in 1992 to solve Problems 80–85.

80. If the number N of fax machines sold (in thousands) is given by $N = N_0 e^{kt}$, where t is the number of years after 1987, what is N_0?

81. Using the N_0 obtained in Problem 80, we can write $N = 580e^{kt}$. We know that $N = 11,000$ in 1992, that is, when $t = 5$ ($1992 - 1987 = 5$). Thus $11,000 = 580e^{5k}$. Solve for k (to two decimal places) in this equation.

82. Use the values obtained in Problems 80 and 81 to find an exponential equation $N = N_0 e^{kt}$ for the number of fax machines sold from 1987 to 1992.

83. According to the equation obtained in Problem 82, how many fax machines were sold in 1990?

84. a. According to the equation obtained in Problem 82, how many fax machines were sold in 1992?

b. What was the actual number of fax machines sold in 1992?

c. What was the percent error between your approximation and actual number of fax machines sold?

85. How many fax machines would you predict to be sold in the year 2000?

WRITE ON . . .

In Problems 80–85, you constructed your own mathematical model, $A = A_0 e^{kt}$, based on a given set of data. How do you know what type of model to use for a particular set of data? Two of the possibilities are logarithmic or exponential:

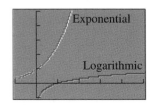

86. The number of cellular phones sold from 1985 to 1994 is shown in the graph. What type of mathematical model would you use to approximate the number of phones sold based on the information provided in the graph? Explain why.

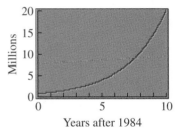

Years after 1984

87. If you use the exponential model $S = S_0 e^{kt}$ to approximate the information in the graph for Problem 86, explain the procedure you would use to find S_0.

88. Explain how you would find k in $S = S_0 e^{kt}$.

89. The number of cigarettes produced per capita from 1969 to 1990 is shown in the graph. What type of model would you use to approximate this information? Explain.

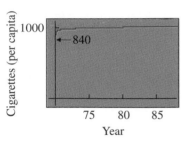

Year

90. If you assume that the equation modeling the graph in Problem 89 is of the form $N = a \ln x + b$, what information do you need in order to find b? Explain.

91. For the equation in Problem 90, what other information do you need to find a?

92. Explain the procedure you would use to find a in the equation in Problem 90.

MASTERY TEST If you know how to do these problems, you have learned your lesson!

93. The number of bacteria B present in a laboratory culture after t minutes is given by $B = K e^{0.05t}$. If the initial number of bacteria is 1000, how long would it take for there to be 20,000 bacteria present?

94. After a bactericide is introduced, the number of bacteria present in a laboratory culture after t minutes is given by $B = K e^{-0.02t}$. If the initial number of bacteria is 50,000, how long would it be before this number is reduced to 10,000?

Solve:

95. $\log(x + 9) + \log x = 1$ 96. $\log_2(x + 3) - \log_2 x = 1$

97. $\log_6(4x - 4) = 2$ 98. $50 = e^{0.15k}$

99. $5^x = 10$ 100. $2^{3x-1} = 4$

101. $2^{2x+1} = 8^{x-1}$

research questions

Sources of information for these questions can be found in the *Bibliography* at the end of this book.

1. Who was the first person to publish a book describing the rules of logarithms, what was the name of the book he published, and what does "logarithm" mean?

2. Who wrote the book from which our words "mantissa" and "characteristic" are derived, and what is the meaning of each of these words?

3. Write a few paragraphs about Jobst Bürgi and describe how he used "red" and "black" numbers in his logarithmic table.

4. Name the author and the title of his book published in 1748 that "gave his approval for e to represent the base of natural logarithms."

5. The symbol e was first used in 1731. Name the circumstances under which the symbol e was used.

SUMMARY

SECTION	ITEM	MEANING	EXAMPLE
10.1A	$(f + g)(x)$	$f(x) + g(x)$	If $f(x) = x^2$ and $g(x) = x + 1$, then $(f + g)(x) = x^2 + x + 1$.
	$(f - g)(x)$	$f(x) - g(x)$	If $f(x) = x^2$ and $g(x) = x + 1$, then $(f - g)(x) = x^2 - x - 1$.
	$(fg)(x)$	$f(x) \cdot g(x)$	If $f(x) = x^2$ and $g(x) = x + 1$, then $(fg)(x) = x^3 + x^2$.
	$\left(\dfrac{f}{g}\right)(x)$	$\dfrac{f(x)}{g(x)}, \quad g(x) \neq 0$	If $f(x) = x^2$ and $g(x) = x + 1$, then $\left(\dfrac{f}{g}\right)(x) = \dfrac{x^2}{x + 1}$.
10.1B	$(f \circ g)(x)$	$f(g(x))$, the composite of f with g	If $f(x) = x^2$ and $g(x) = x + 1$, then $f \circ g = f[g(x)]$ $= (x + 1)^2$
10.2A	f^{-1}	The inverse of a relation, obtained by reversing the order of the coordinates in each ordered pair in f	If $f = \{(1, 2), (4, 6), (6, 9)\}$, then $f^{-1} = \{(2, 1), (6, 4), (9, 6)\}$
10.2C	Horizontal line test	If any horizontal line intersects the graph of $f(x)$ more than once, then f^{-1} is not a function.	$f(x) = x^2$. The graph is a parabola that opens upward. Thus any horizontal line $y = b > 0$ cuts the graph in two points. The inverse, $f^{-1} = \pm\sqrt{x}$, is not a function.
10.3A	Exponential function	A function defined for all real values of x by $f(x) = b^x, b > 0, b \neq 1$	$f(x) = 2^x$
10.3B	Increasing function	A function whose graph goes up to the right	$f(x) = 2^x$
	Decreasing function	A function whose graph goes down to the right	$f(x) = 2^{-x}$
10.4A	Logarithm	$\log_b x = y$ means $x = b^y$.	$\log_2 8 = 3$ because $2^3 = 8$.
10.4D	Logarithm of a product	$\log_b MN = \log_b M + \log_b N$	$\log_2(4 \times 8) = \log_2 4 + \log_2 8$
	Logarithm of a quotient	$\log_b \dfrac{M}{N} = \log_b M - \log_b N$	$\log_2 \dfrac{4}{8} = \log_2 4 - \log_2 8$
	Logarithm of a power	$\log_b M^r = r \log_b M$	$\log_2 4^3 = 3 \log_2 4$
10.5A	Common logarithm	The logarithm to base 10	$\log 10 = 1$, $\log 100 = 2$
	Characteristic	The integer part of the common log	$\log 200 = 2.3010$ has characteristic 2. $\log 0.002 = 7.3010 - 10$ has characteristic $7 - 10$, or -3.
	Mantissa	The decimal part of the common log	Both $\log 2$ and $\log 0.002$ have mantissa 0.3010.
	Antilogarithm	The number that corresponds to a given logarithm	antilog $2.3010 = 200$

SECTION	ITEM	MEANING	EXAMPLE
10.5B	Natural logarithms	Logarithms to base e	$\ln 2 = 0.69315$
10.5C	Change-of-base formula	$\log_b M = \dfrac{\log_a M}{\log_a b}$	$\log_5 20 = \dfrac{\log 20}{\log 5} = \dfrac{1.30103}{0.69897}$ $= 1.86135$
10.6A	Exponential equation	An equation in which the variable occurs in an exponent	$10^{2x} = 5$
10.6B	Logarithmic equation	Equations containing logarithmic expressions	$\log(2x - 1) = 3$ and $\log_2(3x - 1) = 2$ are logarithmic equations.

REVIEW EXERCISES

(If you need help with these exercises, look in the section indicated in brackets.)

1. **[10.1A]** Let $f(x) = 2 - x^2$ and $g(x) = 2 + x$ and find
   a. $(f + g)(x)$ b. $(f - g)(x)$

   c. $(fg)(x)$ d. $\left(\dfrac{f}{g}\right)(x)$

2. **[10.1A]** Let $f(x) = 3 - x^2$ and $g(x) = 3 + x$ and find
   a. $(f + g)(x)$ b. $(f - g)(x)$

   c. $(fg)(x)$ d. $\left(\dfrac{f}{g}\right)(x)$

3. **[10.1A]** If $f(x) = 6x + 1$, find

   $$\dfrac{f(x) - f(a)}{x - a}, \qquad x \neq a$$

4. **[10.1A]** If $f(x) = 7x + 1$, find

   $$\dfrac{f(x) - f(a)}{x - a}, \qquad x \neq a$$

5. **[10.1B]** If $f(x) = x^3$ and $g(x) = 2 - x$, find
   a. $(f \circ g)(x)$ b. $(g \circ f)(x)$ c. $(g \circ f)(2)$

6. **[10.1B]** If $f(x) = x^3$ and $g(x) = 3 - x$, find
   a. $(f \circ g)(x)$ b. $(g \circ f)(x)$ c. $(g \circ f)(2)$

7. **[10.1C]** The lower limit L (heartbeats per minute) of a person's target zone is given by $L(a) = -\frac{2}{3}a + 150$, where a is age (in years). Find L for a person who is
   a. 21 years old   b. 30 years old

8. **[10.1C]** The revenue (in dollars) obtained from selling x units of a product is $R(x) = 100x - 0.02x^2$ and the cost is $C(x) = 30{,}000 + 30x$.
   a. Find the profit function, $P(x) = R(x) - C(x)$.

   b. How many units must be made and sold to yield the maximum profit?

9. **[10.1C]** The revenue (in dollars) obtained from selling x units of a product is $R(x) = 100x - 0.02x^2$ and the cost is $C(x) = 40{,}000 + 40x$.
   a. Find the profit function, $P(x) = R(x) - C(x)$.

   b. How many units must be made and sold to yield the maximum profit?

10. **[10.2A]** Let $S = \{(4, 4), (6, 6), (8, 8)\}$ and find
    a. The domain and range of S
    b. S^{-1}
    c. The domain and range of S^{-1}
    d. The graphs of S and S^{-1}

11. **[10.2A]** Let $S = \{(4, 5), (6, 7), (8, 9)\}$ and find
    a. The domain and range of S
    b. S^{-1}

c. The domain and range of S^{-1}

d. The graphs of S and S^{-1}

12. [10.2B] Let $f(x) = y = 3x - 3$.
  a. Find $f^{-1}(x)$. b. Graph f and its inverse.

13. [10.2B] Let $f(x) = y = 4x - 4$.
  a. Find $f^{-1}(x)$. b. Graph f and its inverse.

14. [10.2C] Find the inverse of $f(x) = y = 4x^2$. Is the inverse a function?

15. [10.2C] Find the inverse of $f(x) = y = 5x^2$. Is the inverse a function?

16. [10.2C] Graph $y = 3^x$. Is the inverse a function?

17. [10.2C] Graph $y = 4^x$. Is the inverse a function?

18. [10.2D] The relationship between dress size d and waist size w (in inches) is given by $d = f(w) = w - 16$.
  a. What dress size corresponds to a 28-in. waist?
  b. Find $f^{-1}(d)$.
  c. If a woman wears a size 10, what is her waist size?

19. [10.3A] For $-6 \le x \le 6$, graph the function:
  a. $f(x) = 2^{x/2}$ b. $f(x) = 2^{-x/2}$

20. [10.3A] For $-6 \le x \le 6$, graph the function:
  a. $g(x) = \left(\dfrac{1}{2}\right)^{x/2}$ b. $g(x) = \left(\dfrac{1}{2}\right)^{-x/2}$

21. [10.3C] A radioactive substance decays so that G, the number of grams present, is given by

$$G = 1000e^{-1.4t}$$

where t is the time in years. Find, to the nearest gram, the amount of the substance present
  a. At the start. b. In 2 yr.

22. [10.4A] Graph $f(x) = \log_5 x$.

23. [10.4B] Write in logarithmic form.
  a. $243 = 3^5$ b. $\dfrac{1}{8} = 2^{-3}$

24. [10.4B] Write in exponential form.
  a. $\log_2 32 = 5$ b. $\log_3 \dfrac{1}{81} = -4$

25. [10.4C] Solve.
  a. $\log_4 x = -2$ b. $\log_x 16 = 2$

26. [10.4C] Find.
  a. $\log_2 16$ b. $\log_3 \dfrac{1}{27}$

27. [10.4D] Fill in the blank with the correct expression.
  a. $\log_b MN = $ _____ b. $\log_b M - \log_b N = $ _____
  c. $\log_b M^r = $ _____

[10.5A] In Problems 28–30, use a calculator.

28. a. $\log 975$ b. $\log 837$

29. a. $\log 0.00759$ b. $\log 0.000648$

30. a. antilog 2.8215 b. antilog -3.3904

[10.5B] In Problems 31–32, use a calculator.

31. a. $\ln 2850$ b. $\ln 0.345$

32. a. $\text{antilog}_e\, 20.0855$ b. $\text{antilog}_e\, 2.7183$

33. [10.5C] Use common logarithms to find
  a. $\log_3 10$ b. $\log_3 100$

34. [10.5D] Graph.
  a. $f(x) = e^{(1/2)x}$
  b. $g(x) = -e^{(1/2)x}$

35. [10.5D] Graph.
  a. $f(x) = \ln(x + 1)$
  b. $g(x) = \ln x + 1$

36. [10.5E] The pH of a solution is defined by $pH = -\log[H^+]$, where $[H^+]$ is the hydrogen ion concentration of the solution in moles per liter. Find the pH of a solution for which $[H^+] = 4 \times 10^{-6}$.

37. [10.6A] Solve.
  a. $2^{2x-1} = 32$ **b.** $3^{x+1} = 9^{x-1}$

38. [10.6A] Solve.
  a. $2^x = 3$ **b.** $5^{2x} = 2.5$

39. [10.6A] Solve.
  a. $e^{5.6x} = 2$ **b.** $e^{-0.33x} = 2$

40. [10.6B] Solve.
  a. $\log x + \log(x - 10) = 1$
  b. $\log_3(x + 1) - \log_3(x - 1) = 1$

41. [10.6C] The compound amount with continuous compounding is given by $A = Pe^{rt}$, where P is the principal, r the rate, and t the time in years. Find how long it takes for the money to double—that is, for A to equal $2P$—if $\ln 2 = 0.69315$ and the rate is
  a. 5%   **b.** 8.5%

42. [10.6C] The number of bacteria in a culture after t minutes is given by $N = 1000e^{kt}$. If there are 1804 bacteria after the given time, find k.
  a. 2 min   **b.** 4 min

43. [10.6C] A radioactive substance decays so that the amount present in t years is given by $A = A_0 e^{-kt}$. Use $\ln 2 = 0.69315$ and find the half-life if
  a. $k = 0.5$ **b.** $k = 0.02$

PRACTICE TEST

1. Let $f(x) = 1 - x^2$ and $g(x) = 1 - x$ and find
  a. $(f + g)(x)$ **b.** $(f - g)(x)$
  c. $(fg)(x)$ **d.** $\left(\dfrac{f}{g}\right)(x)$

2. If $f(x) = 5x + 1$, find
$$\frac{f(x) - f(a)}{x - a}, \qquad x \ne a$$

3. If $f(x) = x^3$ and $g(x) = 1 + x$, find
  a. $(f \circ g)(x)$ **b.** $(g \circ f)(x)$ **c.** $(g \circ f)(2)$

4. The revenue (in dollars) obtained from selling x units of a product is $R(x) = 100x - 0.02x^2$ and the cost is $C(x) = 10{,}000 + 20x$.
  a. Find the profit function, $P(x) = R(x) - C(x)$.
  b. How many units must be made and sold to yield the maximum profit?

5. Let $S = \{(3, 5), (5, 7), (7, 9)\}$ and find
  a. The domain and range of S.
  b. S^{-1}
  c. The domain and range of S^{-1}.
  d. The graph of S and S^{-1}

6. Let $f(x) = y = 4x - 4$.
  a. Find $f^{-1}(x)$.   **b.** Graph f and its inverse.

7. Find the inverse of $f(x) = y = 3x^2$. Is the inverse a function?

8. Graph $y = 3^x$.
  a. Is the inverse a function?
  b. Is $y = 3^x$ increasing or decreasing?

9. A radioactive substance decays so that the number of grams present after t years is
$$G = 1000e^{-1.4t}$$
Find, to the nearest gram, the amount of the substance present
  a. At the start.   **b.** In 2 yr.

10. Graph on the same coordinate axes.
  a. $f(x) = 2^x$ **b.** $f(x) = \log_2 x$

11. Write the equation
  a. $27 = 3^x$ in logarithmic form.
  b. $\log_5 25 = x$ in exponential form.

12. Solve.
  a. $\log_4 x = -1$ **b.** $\log_x 16 = 2$

13. Show that $\log_b \sqrt{\dfrac{53}{79}} = \dfrac{1}{2}(\log_b 53 - \log_b 79)$.

14. Show that $\log\left(1 + \dfrac{1}{x^3}\right) =$
$$\log(x + 1) + \log(x^2 - x + 1) - 3 \log x.$$

15. Find.
    a. log 325   b. antilog 3.5502

16. Find.
    a. ln 325   b. antilog_e 1.1618

17. a. Use the change-of-base formula to fill in the blank:
    $\log_3 10 =$ _____
    b. Use the result of part a to find a numerical approximation for $\log_3 10$.

18. Graph.
    a. $f(x) = e^{(1/2)x}$ b. $g(x) = -e^{(1/2)x}$

19. Graph.
    a. $f(x) = \ln(x + 1)$ b. $g(x) = \ln x + 1$

20. Solve.
    a. $5^{2x+1} = 25$ b. $3^{x+1} = 9^{2x-1}$

21. Solve.
    a. $3^x = 2$ b. $50 = e^{0.20k}$

22. Solve.
    a. $\log(x + 2) + \log(x - 7) = 1$
    b. $\log_3(x + 5) - \log_3(x - 1) = 1$

23. The compound amount with continuous compounding is given by $A = Pe^{rt}$, where P is the principal, r is the interest rate, and t is the time in years. If the rate is 8%, find how long it takes for the money to double—that is, for A to equal $2P$ ($\ln 2 = 0.69315$).

24. The number of bacteria in a culture after t minutes is given by $N = 1000e^{kt}$. Use the following table to find k if there are 2117 bacteria after 4 min.

x	e^x	e^{-x}
0.55	1.7333	0.5769
0.60	1.8221	0.5488
0.65	1.9155	0.5220
0.70	2.0138	0.4966
0.75	2.1170	0.4724

25. A radioactive substance decays so that the amount A present at time t (years) is $A = A_0 e^{-0.5t}$. Find the half-life (time for half to decay) of this substance ($\ln 2 = 0.69315$).

ANSWERS TO PRACTICE TEST

Answer	If you missed: Question	Review: Section	Examples	Page
1a. $-x^2 - x + 2$	1a	10.1A	1a	585
1b. $-x^2 + x$	1b	10.1A	1b	585
1c. $x^3 - x^2 - x + 1$	1c	10.1A	1c	585
1d. $1 + x, x \neq 1$	1d	10.1A	1d	585
2. 5	2	10.1A	2	585–586
3a. $(1 + x)^3$	3a	10.1B	3a	587
3b. $1 + x^3$	3b	10.1B	3b	587
3c. 9	3c	10.1B	3c	587
4a. $P(x) = -0.02x^2 + 80x - 10,000$	4a	10.1C	4a	588
4b. 2000	4b	10.1C	4b	588

Answer	If you missed:	Review:		
	Question	Section	Examples	Page
5a. $D = \{3, 5, 7\}$; $R = \{5, 7, 9\}$	5a	10.2A	1a	593
5b. $\{(5, 3), (7, 5), (9, 7)\}$	5b	10.2A	1b	593
5c. $D = \{5, 7, 9\}$; $R = \{3, 5, 7\}$	5c	10.2A	1c	593
5d.	5d	10.2A	1d	593

6a. $f^{-1}(x) = \dfrac{x + 4}{4}$	6a	10.2B	2a	594
6b.	6b	10.2B	2b	595

7. $y = \pm\sqrt{\dfrac{x}{3}}$ or $\pm\dfrac{\sqrt{3x}}{3}$; no	7	10.2C	3	596

Answer	If you missed:		Review:	
	Question	Section	Examples	Page
8a. yes	8a	10.3A	1a	603
8b. Increasing	8b	10.3B	2	605
9a. 1000 g	9a	10.3C	4a	605–606
9b. 61 g	9b	10.3C	4b	606
10.	10a	10.4A	1	611–612
	10b	10.4A	1	611–612
11a. $\log_3 27 = x$	11a	10.4B	2	612
11b. $5^x = 25$	11b	10.4B	3	612
12a. $\dfrac{1}{4}$	12a	10.4C	4a	613
12b. 4	12b	10.4C	4b	613

8a.

$f(x)$ $y = 3^x$

10.

$y = 2^x$

$y = \log_2 x$

Answer	If you missed:		Review:	
	Question	Section	Examples	Page
13. $\log_b \sqrt{\dfrac{53}{79}} = \dfrac{1}{2}\log_b \dfrac{53}{79}$ $= \dfrac{1}{2}(\log_b 53 - \log_b 79)$	13	10.4D	6	615
14. $1 + \dfrac{1}{x^3} = \dfrac{x^3 + 1}{x^3}$ $= \dfrac{(x+1)(x^2 - x + 1)}{x^3}$ $\log\left(1 + \dfrac{1}{x^3}\right) = \log(x+1)$ $+ \log(x^2 + x + 1) - 3\log x$	14	10.4D	7	615
15a. 2.5119	15a	10.5A	1	621
15b. 3550	15b	10.5A	2	622
16a. 5.7838	16a	10.5B	3	622
16b. 3.1957	16b	10.5B	4	622–623
17a. $\dfrac{\log_{10} 10}{\log_{10} 3}$	17a	10.5C	5	623
17b. 2.0959	17b	10.5C	5	623
18a. and 18b.	18a	10.5D	7a	624
	18b	10.5D	7b	624–625

Answer	If you missed:		Review:	
	Question	Section	Examples	Page
19a. and 19b.	19a	10.5D	8a	626
	19b	10.5D	8b	626
$y = \ln x + 1$ $y = \ln (x + 1)$				
20a. $\dfrac{1}{2}$	20a	10.6A	1a	632
20b. 1	20b	10.6A	1b	632
21a. $\dfrac{\log 2}{\log 3} = 0.6309$	21a	10.6A	2	633
21b. 19.56	21b	10.6A	3	633
22a. 8	22a	10.6B	6a	635
22b. 4	22b	10.6B	6b	635
23. About 8.66 yr	23	10.6C	7	636
24. 0.1875	24	10.6C	9	637
25. About 1.386 yr	25	10.6C	10	637–638

Sequences and Series

In this chapter we study sequences of numbers and their applications. The two main types of sequences are arithmetic, in which the difference between any term and the preceding term is a constant, and geometric, in which each term is a constant times the preceding term. We also look at some sequences that are neither arithmetic nor geometric, such as the Fibonacci sequence. We learn how to sum the terms of finite arithmetic and geometric sequences and find conditions under which the sum of an infinite geometric sequence can be found. We end the chapter with a discussion of binomial expansion using Pascal's triangle for the coefficients. In this regard, we show you an interesting application to probability involving binomial coefficients.

In our study of sequences and series, we frequently encounter the contributions of two giants of mathematics: Karl Friedrich Gauss and Blaise Pascal. In the late eighteenth century, a precocious boy of ten was enrolled in an arithmetic class. "It was easy for the heroic teacher (Buttner) to give out a long problem in addition whose answer he could find by a formula involving arithmetic series." The problem was of this sort:

$$1 + 2 + 3 + 4 + \cdots + 97 + 98 + 99 + 100$$

Buttner had barely finished stating the problem when the boy flung his slate on the table. "*Liggett se* (there it lies)," he said in his peasant dialect. The rest of the hour, while the class worked on the problem, the boy sat with his hands folded, favored now and then by a sarcastic glance from Buttner, who imagined the boy to be just another blockhead. At the end of the period, Buttner looked over the answers. On the boy's slate there appeared but a single number, and it was the correct answer, 5050! How did the boy get it? He figured that the first and last numbers,

1 and 100, added to 101. The second and next to the last number, 2 and 99, also added to 101, and so on. Since there are 50 pairs of numbers, the total sum would be, you guessed it, 50×101 or 5050. The boy was Karl Friedrich Gauss, and he became one of the most renowned mathematicians of his time.

11.1

SEQUENCES AND SERIES

To succeed, review how to:

1. Simplify algebraic expressions (pp. 45–48, 149–151).
2. Evaluate a formula (pp. 76–80).

Objectives:

A Find specific and general terms in a sequence.

B Find specified terms in a sequence when the general term is given.

C Use summation notation to find partial sums of a series.

D Solve an application involving sequences.

getting started

Rabbity Numbers

Leonardo Fibonacci was one of the greatest mathematicians of the Middle Ages. Here is a problem that greatly interested him.

> Let's suppose you have a 1-month-old pair of rabbits. Assume that in the second month, and every month thereafter, they produce a new pair. If the new pair does the same and none of the rabbits die, how many pairs of rabbits will there be at the beginning of each month?

You can check the solution just by counting:

Beginning month	m_1	m_2	m_3	m_4	m_5	m_6	m_7	...
Number of pairs	1	1	2	3	5	8	13	...

The number of pairs 1, 1, 2, 3, 5, 8, 13, . . . form an *infinite sequence* called the **Fibonacci sequence**. The three dots (called an ellipsis) indicate that the sequence continues without stopping. If we stop after a certain number of months, say 6, we obtain the *finite sequence* 1, 1, 2, 3, 5, 8.

In this section we shall study sequences like this one and learn how to find terms in them.

 ### Finding Specific and General Terms in a Sequence

We can think of the Fibonacci sequence as a function that pairs 1 with 1, 2 with 1, 3 with 2, 4 with 3, and so on. In general, we have this definition.

SEQUENCE

An infinite **sequence** is a function whose domain is the set of natural numbers.

Intuitively, you can think of a sequence as a set of numbers arranged according to some pattern. The numbers in a sequence are called the *first term, second term, third term,* and so on. These **terms** are usually denoted by subscripts. Thus, a_1 (read "a sub-one"), a_2, and a_3 are the first three terms in a sequence. The expression a_n, which defines a sequence, is called the **general term**. Note that the notation a_n means the same as $a(n)$. Here are some examples of sequences.

$$\text{Positive multiples of 2:} \quad 2, 4, 6, \ldots$$

$$\text{Powers of 10:} \quad 10^1, 10^2, 10^3, \ldots$$

$$-1 \text{ and } 1 \text{ alternating:} \quad -1, 1, -1, \ldots$$

To find the first three terms and the general term in the sequence consisting of the positive multiples of 2, write the terms a_1, a_2, a_3, and so on in one line and the sequence $2, 4, 6, \ldots$ on the next line and find the pattern associating (linking) them:

$$
\begin{array}{ccccc}
a_1 & a_2 & a_3 & \ldots & a_n \quad \ldots \\
\downarrow & \downarrow & \downarrow & & \downarrow \\
2 & 4 & 6 & \ldots & ? \quad \ldots \\
a_1 = 2 \cdot 1 & a_2 = 2 \cdot 2 & a_3 = 2 \cdot 3 & \ldots & a_n = 2 \cdot n \quad \ldots
\end{array}
$$

The first three terms are $a_1 = 2$, $a_2 = 4$, $a_3 = 6$, and the general term is $a_n = 2n$.

Similarly, the first three terms of the sequence $10^1, 10^2, 10^3, \ldots$ are $a_1 = 10$, $a_2 = 10^2 = 100$, and $a_3 = 10^3 = 1000$. The general term is $a_n = 10^n$.

The first three terms in the sequence $-1, 1, -1, \ldots$ are $a_1 = -1$, $a_2 = 1$, and $a_3 = -1$.

$$
\begin{array}{ccccc}
a_1 & a_2 & a_3 & \ldots \; a_n & \ldots \\
\downarrow & \downarrow & \downarrow & \downarrow & \\
-1 & 1 & -1 & \ldots \quad ? & \ldots \\
a_1 = (-1)^1 = -1 & a_2 = (-1)^2 = 1 & a_3 = (-1)^3 = -1 & \ldots \; a_n = (-1)^n & \ldots
\end{array}
$$

From the pattern, you can see that $a_n = (-1)^n$.

EXAMPLE 1 Finding specific terms in a sequence

For the sequence $2, 4, 8, \ldots$, where each term after a_1 is double the preceding term, find: $a_2, a_5,$ and a_n

SOLUTION As before, we write

$$
\begin{array}{ccccc}
a_1 & a_2 & a_3 & \ldots & a_n \quad \ldots \\
\downarrow & \downarrow & \downarrow & & \downarrow \\
2 & 4 & 8 & \ldots & ? \quad \ldots \\
a_1 = 2^1 & a_2 = 2^2 & a_3 = 2^3 & \ldots & a_n = 2^n \quad \ldots
\end{array}
$$

Following the pattern, the second term is $a_2 = 2^2 = 4$, the fifth term is $a_5 = 2^5 = 32$, and the general term is $a_n = 2^n$. ∎

656   CHAPTER 11 · SEQUENCES AND SERIES

EXAMPLE 2   More practice finding terms

For the sequence of even integers with alternating signs $2, -4, 6, -8, \ldots$, find: a_{10} and a_n

SOLUTION   The only difference between this problem and the sequence $2, 4, 6, 8, \ldots$ is the sign of the terms of the sequence. To get the correct sign, we use the factor $(-1)^{n-1}$, which gives alternately $+1$ and -1. Then we write

$$
\begin{array}{cccc}
a_1 & a_2 & a_3 & \cdots \\
\downarrow & \downarrow & \downarrow & \\
2 & -4 & 6 & \cdots
\end{array}
$$

$$a_1 = (-1)^{1-1}2 \cdot 1 = 2 \quad a_2 = (-1)^{2-1}2 \cdot 2 = -4 \quad a_3 = (-1)^{3-1}2 \cdot 3 = 8 \ldots$$

$$
\begin{array}{c}
a_n \quad \cdots \\
\downarrow \\
? \quad \cdots
\end{array}
$$

$$a_n = (-1)^{n-1}2n \ldots$$

Thus $a_{10} = (-1)^{10-1}2 \cdot 10 = (-1)^9 2 \cdot 10 = (-1)20 = -20$, and $a_n = (-1)^{n-1}2n$. ∎

EXAMPLE 3   Sequential folios

In the publishing industry large printed pages are folded to make the pages of a book. If sheets are folded once, this makes 2 pages or a *folio*. If the sheets are folded twice, this makes 4 pages or a *quarto*. If sheets are folded three times, this makes 8 pages, or an *octavo*. Further folding produces units called 16mo, 32mo, and so on.
a. Write the sequence that gives the number of pages after each fold.
b. If a sheet is folded 6 times, how many pages are there?
c. How many pages result after n folds?

SOLUTION   Since 1 fold produces 2 pages, $a_1 = 2$. Then 2 folds produce 4 pages, so $a_2 = 4$, and 3 folds produce 8 pages, so $a_3 = 8$. We then write

$$
\begin{array}{ccccc}
a_1 & a_2 & a_3 & \cdots & a_n & \cdots \\
\downarrow & \downarrow & \downarrow & & \downarrow & \\
2 & 4 & 8 & \cdots & ? & \cdots
\end{array}
$$

$$a_1 = 2^1 \quad a_2 = 2^2 \quad a_3 = 2^3 \quad \cdots \quad a_n = 2^n \quad \cdots$$

a. The sequence is $2, 4, 8, \ldots, 2^n \ldots$.
b. If a sheet is folded 6 times, the number of pages is $2^6 = 64$.
c. After n folds, there are 2^n pages. ∎

B Finding Specified Terms When the General Term Is Given

In Examples 1 and 2, we found the general term of a given sequence. However, if only a finite number of successive terms are given *without* a rule that defines the general term, a *unique* general term cannot be obtained. Let's see why.

Consider two sequences with general terms:

$$a_n = 2n \quad \text{and} \quad a_n = 2n + \tfrac{1}{2}(n-1)(n-2)(n-3)(n-4)$$

$$a_1 = 2 \cdot 1 = 2 \quad \text{and} \quad a_1 = 2 \cdot 1 + \tfrac{1}{2}(1-1)(1-2)(1-3)(1-4) = 2$$

$$a_2 = 2 \cdot 2 = 4 \quad \text{and} \quad a_2 = 2 \cdot 2 + \tfrac{1}{2}(2-1)(2-2)(2-3)(2-4) = 4$$

$$a_3 = 2 \cdot 3 = 6 \quad \text{and} \quad a_3 = 2 \cdot 3 + \tfrac{1}{2}(3-1)(3-2)(3-3)(3-4) = 6$$

$$a_4 = 2 \cdot 4 = 8 \quad \text{and} \quad a_4 = 2 \cdot 4 + \tfrac{1}{2}(4-1)(4-2)(4-3)(4-4) = 8$$

$$a_5 = 2 \cdot 5 = 10 \quad \text{but} \quad a_5 = 2 \cdot 5 + \tfrac{1}{2}(5-1)(5-2)(5-3)(5-4)$$

$$= 10 + \tfrac{1}{2}(24) = 10 + 12 = 22$$

Thus examining the first four terms in the sequence $2, 4, 6, 8, \ldots$ may lead you to use $a_n = 2n$ or $a_n = 2n + \tfrac{1}{2}(n-1)(n-2)(n-3)(n-4)$ as the general term. They are both correct! However, the fifth terms are not equal. You cannot find a unique general term from a finite number of terms.

> **NOTE** There may be more than one general term that produces the same first three or four terms in a sequence and, consequently, there are no rules for finding the general term of a sequence from the first few terms.

EXAMPLE 4 Finding specified terms when the general term is given

Find the first three terms and the ninth term of the sequence whose general term is

$$a_n = \frac{1}{2}n(n-1)$$

SOLUTION We find the required terms by substituting the corresponding values of n into the given formula. Thus

$$a_1 = \frac{1}{2}(1)(1-1) = 0$$

$$a_2 = \frac{1}{2}(2)(2-1) = 1$$

$$a_3 = \frac{1}{2}(3)(3-1) = 3$$

$$a_9 = \frac{1}{2}(9)(9-1) = 36 \qquad \blacksquare$$

Sometimes function notation rather than subscript notation is used to define the terms of a sequence. For instance, in Example 4, we could use function notation and write:

$$a(n) = \frac{1}{2}n(n-1) \qquad \text{Instead of } a_n = \tfrac{1}{2}n(n-1)$$

so that

$$a(1) = \frac{1}{2}(1)(1-1) = 0$$

$$a(2) = \frac{1}{2}(2)(2-1) = 1, \qquad \text{and so on}$$

EXAMPLE 5   Finding the sequence given a function

Consider the function

$$a(n) = 2n + 1, \qquad n = 1, 2, 3, \ldots$$

Find: The sequence corresponding to this function

SOLUTION

$$\text{for } n = 1, a(1) = 2(1) + 1 = 3$$
$$\text{for } n = 2, a(2) = 2(2) + 1 = 5$$
$$\text{for } n = 3, a(3) = 2(3) + 1 = 7$$
$$\text{for } n = 4, a(4) = 2(4) + 1 = 9$$

The sequence is $3, 5, 7, 9, \ldots.$ ■

 Summation Notation

There is an Old English rhyme that reads as follows:

> As I was going to St. Ives
> I met a man with seven wives;
> Every wife had seven sacks;
> Every sack had seven cats;
> Every cat had seven kits,
> Kits, cats, sacks and wives,
> How many were going to St. Ives?

The answer is really 1. If you don't believe this, then read the rhyme again. A less misleading question would be, How many were *leaving* St. Ives? The sequence of numbers involved in this second question is

1	$1 \cdot 7 = 7$	$7 \cdot 7 = 49$	$7 \cdot 7 \cdot 7 = 343$	$7 \cdot 7 \cdot 7 \cdot 7 = 2401$
↓	↓	↓	↓	↓
Man	Wives	Sacks	Cats	Kits

Thus the number leaving St. Ives is $1 + 7 + 49 + 343 + 2401 = 2801$.

The sequence 1, 7, 49, 343, 2401 is a *finite sequence,* and the indicated sum for the sequence $1 + 7 + 49 + 343 + 2401$ is called a *series.*

INFINITE AND FINITE SERIES

> Given the infinite sequence
>
> $$a_1, a_2, a_3, \ldots, a_n, \ldots$$
>
> the sum of the terms is called an **infinite series**:
>
> $$a_1 + a_2 + a_3 + \cdots + a_n + \cdots$$
>
> The **partial sum**
>
> $$a_1 + a_2 + a_3 + \cdots + a_n$$
>
> is called a **finite series** and is denoted by S_n.

Thus the sequence $7, 7^2, 7^3, 7^4, \ldots, 7^n \ldots$ has the following partial sums:

$S_1 = 7$ The **first** term of the sequence

$S_2 = 7 + 49 = 56$ The sum of the first **two** terms

$S_3 = 7 + 49 + 343 = 399$ The sum of the first **three** terms

$S_4 = 7 + 49 + 343 + 2401 = 2800$ The sum of the first **four** terms

EXAMPLE 6 Finding partial sums

For the sequence $-1, 3, -5, 7, -9, 11, -13$, find the sums:

a. S_4 **b.** S_7

SOLUTION

a. $S_4 = -1 + 3 + (-5) + 7 = 4$

b. $S_7 = -1 + 3 + (-5) + 7 + (-9) + 11 + (-13) = -7$ ■

If we know the general term of a sequence, we can represent a sum of terms using *summation (sigma) notation.* In this notation, the Greek letter Σ (capital sigma), which corresponds to the English letter S, indicates that we are to add the given terms. Thus

$$\sum_{i=1}^{n} a_i \qquad \text{Read "the sum of } a_i \text{ from } i = 1 \text{ to } n.\text{"}$$

is defined by

$$\sum_{i=1}^{n} a_i = a_1 + a_2 + a_3 + \cdots + a_n$$

Of course, the sum need not start at $i = 1$ and end at $i = n$. Furthermore, any letter may be used in place of the *index i.* The following examples illustrate some of the possibilities.

$$\sum_{i=3}^{6} a_i = a_3 + a_4 + a_5 + a_6$$

$$\sum_{j=1}^{5} j = 1 + 2 + 3 + 4 + 5$$

$$\sum_{k=1}^{6} k^2 = 1^2 + 2^2 + 3^2 + 4^2 + 5^2 + 6^2$$

$$\sum_{n=1}^{5} (-1)^n na_n = -a_1 + 2a_2 - 3a_3 + 4a_4 - 5a_5$$

EXAMPLE 7 Evaluating sums given in summation notation

Find and evaluate the sum:

a. $\displaystyle\sum_{n=1}^{4} n^2$ **b.** $\displaystyle\sum_{k=0}^{3} (2k + 1)$

SOLUTION

a. $\displaystyle\sum_{n=1}^{4} n^2 = 1^2 + 2^2 + 3^2 + 4^2 = 1 + 4 + 9 + 16 = 30$

Evaluate n^2 for $n = 1, 2, 3,$ and 4 and then add.

b. $\displaystyle\sum_{k=0}^{3} (2k + 1) = (2 \cdot 0 + 1) + (2 \cdot 1 + 1) + (2 \cdot 2 + 1) + (2 \cdot 3 + 1)$

$\qquad\qquad = \quad 1 \quad + \quad 3 \quad + \quad 5 \quad + \quad 7 \quad = 16 \quad$ ■

EXAMPLE 8   More practice using summation notation

Write using summation notation:

a. $2 + 4 + 6 + \cdots + 20$

b. $\dfrac{1}{2} + \dfrac{1}{3} + \dfrac{1}{4} + \cdots + \dfrac{1}{30}$

SOLUTION

a. The finite series $2 + 4 + 6 + \cdots + 20$ is a sum of even numbers with general term $2n$ starting with $n = 1$ and ending with $n = 10$. The summation notation is

$$\sum_{n=1}^{10} 2n$$

b. The finite series

$$\frac{1}{2} + \frac{1}{3} + \frac{1}{4} + \cdots + \frac{1}{30}$$

is a sum of fractions whose denominators are consecutive numbers starting with 2 and ending with 30. The summation notation is

$$\sum_{n=2}^{30} \frac{1}{n}$$

■

 An Application Involving Sequences

Did you know that sequences were used to find some of the planets of our solar system? In 1772, a German astronomer named Johann Bode discovered a pattern in the distances of the planets from the Sun. We examine his sequence in Example 9.

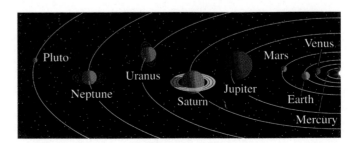

EXAMPLE 9　Finding missing planets

Bode's sequence is as follows:

p_1	p_2	p_3	p_4	p_5	p_6	p_7	p_8
Mercury	Venus	Earth	Mars	?	Jupiter	Saturn	?
↓	↓	↓	↓	↓	↓	↓	
$0 + 4 = 4$	$3 + 4 = 7$	$6 + 4 = 10$	$12 + 4 = 16$	?	$48 + 4 = 52$	$96 + 4 = 100$	?

a. What number corresponds to the fifth missing planet?
b. What number corresponds to the eighth planet?

SOLUTION

a. After each term starting with p_3, the number in front of 4 is twice the number in front of 4 in the preceding term. Thus the number for the unknown planet is $2 \cdot 12 + 4 = 28$. (It turns out that this planet is really Ceres, a planetoid or asteroid.)

b. The number for the planet after Saturn is $2 \cdot 96 + 4 = 196$, which corresponds to Uranus, which was discovered by William Herschel in 1781. ■

GRAPH IT

If you know the general term of a sequence, most graphers will give you any term you wish. In Example 1, the general term is $a_n = 2^n$. Go to the home screen (press [2nd] [QUIT] [CLEAR]). Tell the grapher that the general term is 2^n (press [ALPHA] [+] 2 [^] [2nd] 9 [ALPHA] [+]). Next, define your sequence U_n by storing 2^n into Un. This is done by pressing [STO] [2nd] [Y-VARS] 4 1. Finally, tell the grapher you want the second term by pressing [2nd] [·] [2nd] [Y-VARS] 4 1 [(] 2 [)]. Now press [ENTER] and the answer 4 will be on the next line, as shown in Window 1. Fortunately, you don't have to reenter all the information to find the fifth term. Simply press [2nd] [ENTER] and move the cursor left until it's on top of the 2. Press 5 [ENTER] and the answer 32 will appear. Now try Examples 2, 3, 4, and 5.

To do sums (as in Example 6), press [2nd] [STAT], move the cursor to the "MATH" (second) column and press 5. Enter the numbers to be added (separate them by commas and enclose them in braces). To find S_7, enter $\{-1, 3, -5, 7, -9, 11, -13\}$; pressing [ENTER] gives the answer, as shown in Window 2.

Many graphers use the summation (sigma) notation to find the sum of a sequence. To do Example 7a, you must tell the grapher you want the sum of a sequence. Pressing [2nd] [STAT], moving the cursor to the "MATH" column, and pressing 5 tells the grapher you want a sum. Then press [2nd] [STAT] 5 to make it clear that you want the sum of a sequence. Enter the sequence as N^2 by pressing [ALPHA] [LOG] [^] 2 [,]; the variable you are using (press [ALPHA] [LOG] [,]); the starting point (press 1 [,]); the ending point (press 4 [,]); and the increments you want for N (press 1). Close the parentheses and press [ENTER]. The answer is 30, as shown in Window 3. Now try Examples 7b and 8.

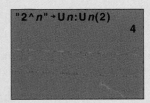

WINDOW 1

WINDOW 2

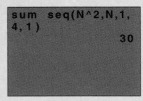

WINDOW 3

EXERCISE 11.1

[A] **In Problems 1–22, find a tenth term and an nth term to fit the given sequence.**

1. $1, 4, 7, 10, \ldots$ 2. $5, 7, 9, 11, \ldots$

3. $5, 8, 11, 14, \ldots$ 4. $3, 8, 13, 18, \ldots$

5. $20, 25, 30, 35, \ldots$ 6. $15, 18, 21, 24, \ldots$

7. $50, 45, 40, 35, \ldots$ 8. $30, 28, 26, 24, \ldots$

9. $\dfrac{1}{2}, \dfrac{1}{3}, \dfrac{1}{4}, \dfrac{1}{5}, \ldots$ 10. $\dfrac{1}{2}, \dfrac{2}{3}, \dfrac{3}{4}, \dfrac{4}{5}, \ldots$

11. $1, -1, 1, -1, \ldots$ 12. $-1, 2, -4, 8, \ldots$

13. $x, x^2, x^3, x^4, \ldots$ 14. $x^2, x^4, x^6, x^8, \ldots$

15. $x, -x^3, x^5, -x^7, \ldots$ 16. $-x, x^2, -x^4, x^8, \ldots$

17. $x, -x, x, -x, \ldots$ 18. $-x, x, -x, x, \ldots$

19. $x, \dfrac{x^2}{2}, \dfrac{x^3}{3}, \dfrac{x^4}{4}, \ldots$ 20. $\dfrac{x}{5}, \dfrac{x^2}{10}, \dfrac{x^3}{15}, \dfrac{x^4}{20}, \ldots$

21. $\dfrac{x}{2}, -\dfrac{x^2}{4}, \dfrac{x^3}{8}, -\dfrac{x^4}{16}, \ldots$

22. $\dfrac{x}{2}, -\dfrac{x^3}{4}, \dfrac{x^5}{8}, -\dfrac{x^7}{16}, \ldots$

[B] **In Problems 23–36, find the first three terms of the sequence with the given general term.**

23. $a_n = 2n - 3$ 24. $a_n = 2n + 3$

25. $a_n = \dfrac{n(n-2)}{2}$ 26. $a_n = \dfrac{n(n+2)}{2}$

27. $a(n) = 1 - \dfrac{1}{n}$ 28. $a(n) = 1 + \dfrac{2}{n}$

29. $a_n = n^2$ 30. $a_n = -n^3$

31. $a(n) = \dfrac{n}{2n+1}$ 32. $a(n) = \dfrac{n}{3n-1}$

33. $a_n = (-1)^n$ 34. $a_n = (-2)^{n-1}$

35. $a(n) = (-1)^n 2^{-n}$ 36. $a(n) = (-1)^{n-1} 3^n$

[C] **In Problems 37–48, compute the indicated sums.**

37. $\displaystyle\sum_{k=1}^{6} k^2$ 38. $\displaystyle\sum_{i=1}^{8} i$ 39. $\displaystyle\sum_{k=1}^{4} k^3$

40. $\displaystyle\sum_{n=1}^{5} 2n$ 41. $\displaystyle\sum_{i=1}^{7} 3$ 42. $\displaystyle\sum_{k=3}^{8} 5$

43. $\displaystyle\sum_{j=1}^{4} \dfrac{1}{2j}$ 44. $\displaystyle\sum_{j=0}^{6} \dfrac{1}{j+1}$ 45. $\displaystyle\sum_{k=1}^{7} \dfrac{k+1}{k}$

46. $\displaystyle\sum_{n=1}^{6} (-1)^n$ 47. $\displaystyle\sum_{k=1}^{5} (-1)^{k+1}$ 48. $\displaystyle\sum_{k=1}^{6} (-1)^k 3^{k+1}$

In Problems 49–56, write each expression using the sigma notation.

49. $1 + 2 + 3 + \cdots + 200$ 50. $1 + 4 + 9 + 16 + \cdots + 49$

51. $1 + \dfrac{1}{2} + \dfrac{1}{3} + \dfrac{1}{4} + \cdots + \dfrac{1}{50}$

52. $x_1^2 + x_2^2 + x_3^2 + \cdots + x_{100}^2$

53. $1 - 2 + 3 - 4 + 5 - 6 + \cdots - 50$

54. $2 - 4 + 8 - 16 + 32$

55. $1 + 6 + 11 + 16 + 21$ 56. $\dfrac{1}{2} + 1 + \dfrac{3}{2} + 2 + \dfrac{5}{2} + 3$

APPLICATIONS

57. A property valued at \$30,000 will depreciate \$1380 the first year, \$1340 the second year, \$1300 the third year, and so on. What will be the depreciation during
   a. The eighth year?   **b.** The tenth year?

58. Strikers at a plant were ordered to return to work and were told they would be fined \$50 the first day they failed to do so, \$75 the second day, \$100 the third day, and so on. If the strikers stayed out for 6 days, what was the fine for the sixth day?

59. When dropped on a hard surface, a Super Ball takes a sequence of bounces, each one about $\frac{9}{10}$ as high as the preceding one. If a Super Ball is dropped from a height of 10 ft, find how high it will bounce on the
   a. First bounce   **b.** Third bounce   **c.** nth bounce

60. An ancient legend says that the Shah of Persia offered the inventor of chess anything he wished as a reward for his invention. The man asked for 1 grain of wheat to be placed on the first square of the chessboard, 2 grains on the second, 4 grains on the third, and so on. How many grains would there be on
   a. The fifth square?   **b.** The ninth square?
   c. The nth square?

61. A colony of bacteria starts with 100 members and doubles every hour. How many bacteria are there at the end of
   a. 2 hr?   **b.** 4 hr?   **c.** n hr?

62. A free-falling body falls about 16 ft the first second, 48 ft the next second, 80 ft the third second, 112 ft the fourth second, and so on. How far does it fall during
    a. The eighth second?   b. The nth second?

63. A salesman sold $100 worth of goods on Monday and doubled his sales each day thereafter for a week. What was the amount of sales on Saturday?

64. A sprinter runs 6 meters in the first second of a certain race and increases her speed by 25 cm/sec in each succeeding second. (This means that she goes 6 m 25 cm the second second, 6 m 50 cm the third second, and so on.) How far does she go during
    a. The eighth second?   b. The nth second?

SKILL CHECKER

Simplify:

65. $7 + (n - 1)(3)$ 66. $16 + (n - 1)(2)$

67. $\frac{1}{2}n(16 + 32n - 16)$ 68. $\frac{n}{2}(3 + 5n - 2)$

Factor:

69. $5n^2 + n - 328$ 70. $7n^2 - n - 336$

USING YOUR KNOWLEDGE   **The Fibonacci Rabbits**

Let's return to the rabbit problem introduced in the *Getting Started*. Here is what happens in the first 5 months:

NA = New Adult   OA = Old Adult   B = Babies

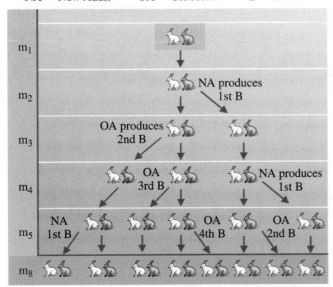

As we can see, the number of rabbit pairs at the beginning of each month are the terms of the Fibonacci sequence 1, 1, 2, 3, 5, 8, 13, ...

Notice that, starting with the third term, each term is the sum of the two preceding terms: $2 = 1 + 1$, $3 = 1 + 2$, $5 = 2 + 3$, and so on. This leads us to the general formula

$$a_n = a_{n-2} + a_{n-1}$$

Use these ideas to write the following terms of the Fibonacci sequence.

71. The eighth term   72. The ninth term

73. The tenth term   74. The eleventh term

75. The twelfth term

WRITE ON . . .

76. Write your own definition of a sequence.

77. What is the difference between a finite sequence and an infinite sequence?

78. Given the sequence 1, 3, 5, . . . can you find a *unique* general term a_n? Explain why or why not.

79. If the general term a_n for a sequence is known, is the resulting sequence unique? Explain why or why not.

MASTERY TEST   If you know how to do these problems, you have learned your lesson!

80. A famous painting doubles in value every 50 yr. Find the value of the painting in the year 2000, if it was worth $1000 in the year 1500.

81. Find the first three terms and the tenth term of the sequence whose general term is $a_n = \frac{1}{2}(n^2 - 1)$.

82. Suppose each card in a pack of five index cards is cut in half and all the halves are put in a single pile and cut again. This procedure is repeated again and again.
    a. Write the sequence that gives the number of cards in the pack after each cut.
    b. How many cards are in the pack after the fourth cut?
    c. How many cards are in the pack after the nth cut?

83. For the sequence of multiples of 3 with alternating signs, $-3, 6, -9, 12, \ldots$ find a_{10} and a_n.

84. For the sequence of positive multiples of 3, that is, 3, 6, 9, 12, . . . , find:
    a. a_2 and a_4 b. a_{10} c. a_n

85. Write using summation (sigma) notation:
    a. $6 + 12 + 18 + 24$ b. $2 - 4 + 8 - 16 + 32 - 64$

86. Find and evaluate the sum:
    a. $\sum_{k=1}^{6} 2^k + 1$ b. $\sum_{k=1}^{5} \frac{1}{k^2}$

11.2

ARITHMETIC SEQUENCES AND SERIES

To succeed, review how to:

1. Recognize the terms of a sequence (pp. 655–656).
2. Find the *n*th term of a sequence when the general term is given (pp. 657–658).

Objectives:

A Find the common difference and general term in an arithmetic sequence.

B Find the sum of an arithmetic sequence.

C Solve applications involving arithmetic sequences.

getting started

Falling Sequences

A skydiver plunges toward the ground. Do you know how far he will fall in the first 5 sec? A free-falling body travels about 16 ft in the first second, 48 ft in the next second, 80 ft in the third second, and so on. The number of feet traveled in each successive second is

$$16, 48, 80, 112, 144, \ldots$$

Can you find the next number in the sequence? First, note that the second term (48) is obtained by adding 32 to the first term (16). Similarly, the third term (80) is obtained by adding 32 to the second term (48), and so on. Thus the term after 144 is found by adding 32 to 144 to obtain 176. Sequences in which successive terms are found by adding a constant to the preceding term are called *arithmetic sequences* or *arithmetic progressions,* and we shall discuss them in this section.

A   **Arithmetic Sequences**

The *Getting Started* gives an application of arithmetic sequences. You will find other applications in the problem set.

ARITHMETIC SEQUENCE

An **arithmetic sequence** or **arithmetic progression** is a sequence in which each term after the first is obtained by adding a quantity *d,* called the *common difference,* to the preceding term.

The sequence 16, 48, 80, 112, 144, . . . mentioned in the *Getting Started* is an arithmetic sequence in which each term is obtained by adding the common difference 32 to the preceding term. This means that the common difference for an arithmetic sequence is just the difference between any two consecutive terms.

COMMON DIFFERENCE

The **common difference** *d* is defined by

$$d = a_{n+1} - a_n$$

EXAMPLE 1 Finding the common difference

Find the common difference in each sequence:

a. $7, 37, 67, 97, \ldots$ **b.** $10, 5, 0, -5, \ldots$

SOLUTION

a. The common difference is $37 - 7 = 30$ (or $67 - 37$, or $97 - 67$).
b. The common difference is $5 - 10 = -5$ (or $0 - 5$, or $-5 - 0$). ■

It's customary to denote the first term of an arithmetic sequence by a_1 (read, "a sub 1"), the common difference by d, and the nth term by a_n. Thus in the sequence $16, 48, 80, 112, 144, \ldots$, we have $a_1 = 16$ and $d = 32$. The second term of the sequence, a_2, is

$$a_2 = a_1 + 32 = 16 + 32 = 48$$

Since each term is obtained from the preceding one by adding 32,

$$a_3 = a_2 + 32 = (a_1 + 32) + 32 = a_1 + 2 \cdot 32 = 80$$

$$a_4 = a_3 + 32 = (a_1 + 2 \cdot 32) + 32 = a_1 + 3 \cdot 32 = 112$$

$$a_5 = a_4 + 32 = (a_1 + 3 \cdot 32) + 32 = a_1 + 4 \cdot 32 = 144$$

By following this pattern, we make the following definition.

GENERAL TERM OF AN ARITHMETIC SEQUENCE

$$a_n = a_1 + (n - 1) \cdot d$$

EXAMPLE 2 Working with an arithmetic sequence

Consider the sequence $7, 10, 13, 16, \ldots$ and find:
a. a_1, the first term **b.** d, the common difference
c. a_{11}, the eleventh term **d.** a_n, the nth term

SOLUTION

a. The first term a_1 is 7.
b. The common difference d is $10 - 7 = 3$.
c. The eleventh term is $a_{11} = 7 + (11 - 1) \cdot 3 = 7 + 10 \cdot 3 = 37$.
d. $a_n = a_1 + (n - 1) \cdot d = 7 + (n - 1) \cdot 3 = 7 + 3n - 3 = 4 + 3n$ ■

Finding the Sum of an Arithmetic Sequence

Let's return to the problem of the skydiver. How far does he fall in 5 sec? The first five terms of the sequence are 16, 48, 80, 112, and 144; thus we need to find the sum

$$16 + 48 + 80 + 112 + 144$$

Since successive terms of an arithmetic sequence are obtained by adding the common difference d, the sum S_n of the first n terms is

$$S_n = a_1 + (a_1 + d) + (a_1 + 2d) + (a_1 + 3d) + \cdots + a_n \qquad \textbf{(1)}$$

We can also start with a_n and obtain successive terms by subtracting the common difference d. Thus with the terms written in reverse order,

$$S_n = a_n + (a_n - d) + (a_n - 2d) + \cdots + a_1 \qquad \textbf{(2)}$$

Adding equations (1) and (2), we find that the d's drop out, and we obtain

$$2S_n = (a_1 + a_n) + (a_1 + a_n) + \cdots + (a_1 + a_n)$$
$$= n(a_1 + a_n)$$

We thus find the following formula.

SUM OF AN ARITHMETIC SEQUENCE

> The sum S_n of the first n terms starting with a_1 and ending with a_n is given by
>
> $$S_n = \frac{n}{2}(a_1 + a_n)$$

We are now able to determine the sum S_5, the distance the skydiver dropped in 5 sec:

$$S_5 = \frac{5}{2}(16 + 144) = 400 \text{ ft}$$

EXAMPLE 3   Finding the sum of an arithmetic sequence

Find the distance the skydiver falls in:
a. 10 sec   **b.** n sec

SOLUTION   To do this problem, we first find the distance fallen in the nth second. Since $a_1 = 16$ and $d = 32$, we get

$$a_n = a_1 + (n - 1)d$$
$$= 16 + (n - 1)(32)$$
$$= 16 + 32n - 32$$

Thus

$$a_n = 32n - 16$$

a. For $n = 10$, $a_{10} = (32)(10) - 16 = 320 - 16 = 304$ so that

$$S_{10} = \frac{10}{2}(16 + 304) = 5(320) = 1600$$

Thus the skydiver falls 1600 ft in 10 sec.
b. Here, we use the general formulas for S_n and a_n.

$$S_n = \frac{n}{2}(a_1 + a_n) \qquad \text{and} \qquad a_n = 32n - 16$$

Now we substitute $a_1 = 16$ and $a_n = 32n - 16$ into the formula for S_n to get

$$S_n = \frac{n}{2}(16 + 32n - 16) = \frac{n}{2}(32n)$$

or

$$S_n = 16n^2$$

You can check that this surprisingly simple formula gives the same answers that we found in part a. ∎

EXAMPLE 4   Using the sum to find a term and the common difference

The sum of the first 10 terms of an arithmetic sequence is 205 and the tenth term is 34. Find:

a. a_1, the first term   **b.** d, the common difference

SOLUTION

a. We use the formula for the sum with $n = 10$:

$$S_{10} = \frac{10}{2}(a_1 + a_{10}) = 5(a_1 + a_{10})$$

We then substitute $S_{10} = 205$ and $a_{10} = 34$ to get

$$205 = 5(a_1 + 34)$$

$$41 = a_1 + 34 \qquad \text{Divide by 5.}$$

$$7 = a_1 \qquad \text{Subtract 34.}$$

Thus, the first term a_1 is 7.

b. Now we use the formula for the nth term,

$$a_n = a_1 + (n - 1)d$$

with $n = 10$, $a_1 = 7$, and $a_{10} = 34$, to get

$$34 = 7 + 9d$$

$$27 = 9d \qquad \text{Subtract 7.}$$

$$3 = d \qquad \text{Divide by 9.} \quad ∎$$

Thus, the common difference d is 3.

Applications of Arithmetic Sequences

EXAMPLE 5   Depreciating a truck

A heavy-duty truck valued at $50,000 depreciates $5000 the first year, $4800 the second year, $4600 the third year, and so on. What will be the value of the truck at the end of 8 yr?

SOLUTION   The yearly depreciations form an arithmetic sequence

$$5000, 4800, 4600, \ldots$$

so that $a_1 = 5000$, $d = -200$, and $n = 8$. Since

$$a_n = a_1 + (n - 1)d$$

$$a_8 = 5000 + (7)(-200)$$

$$= 3600$$

The total depreciation will be the sum of the first eight terms of the sequence:

$$S_8 = \frac{8}{2}(a_1 + a_8)$$

$$= 4(5000 + 3600)$$

$$= 4(8600)$$

$$= 34,400$$

So the total depreciation is $34,400 and the remaining value is

$$\$50,000 - \$34,400 = \$15,600$$ ∎

problem solving   **EXAMPLE 6**   A long time saving

Alice started a savings campaign. She put aside 3¢ the first day, 8¢ the second day, 13¢ the third day, and so on in arithmetic sequence. After a few days, Alice found that she had saved $1.64. How many days had she been saving?

1. Read the problem.

SOLUTION   Here, we know that $a_1 = 3$, $d = 5$, and $S_n = 164$, and we want to find n. So we use the formula for the sum

$$S_n = \frac{n}{2}(a_1 + a_n)$$

and the formula for the nth term

2. Select the unknown.

$$a_n = a_1 + (n - 1)d$$

Substituting the known values into the preceding two formulas, we get

$$164 = \frac{n}{2}(3 + a_n)$$

$$a_n = 3 + (n - 1)(5)$$

$$= 5n - 2$$

3. Think of a plan.

We next substitute for a_n in the first equation to obtain

$$164 = \frac{n}{2}(3 + 5n - 2)$$

$$328 = n(5n + 1)$$ Multiply by 2 and simplify.

4. Use algebra to solve the problem.

We rewrite this equation in standard quadratic form:

$$5n^2 + n - 328 = 0$$

$$(5n + 41)(n - 8) = 0$$ Factor.

5. Verify the solution.

Since n must be positive, the solution is $n = 8$, so it took 8 days for Alice to get up to $1.64. ∎

As we mentioned in Section 11.1, if you know the general term of a sequence, most graphers can find any term you wish. Moreover, they can also find the sum of the terms of an arithmetic sequence (that is, a series). In Example 3b, the general term is $a_n = 32n - 16$. To find a_{10}, tell the grapher what the general term is. On a TI-82, press ALPHA + 3 2 2nd 9 − 1 6 ALPHA + . Define the sequence U_n by storing $3n - 16$ into Un by pressing STO 2nd Y-VARS 4 1 and then tell the grapher you want the tenth term by pressing 2nd . 2nd Y-VARS 4 1 [(1 0)]. Now press ENTER , and the answer 304 will appear on the next line, as shown in the window.

> ```
> "32n−16"→Un:Un(1
> 0)
>                  304
> ```

What about the sum S_{10}? Since you know that

$$S_{10} = \frac{10}{2}(a_1 + a_n) \qquad \text{and} \qquad a_1 = 16$$

and we found $a_{10} = 304$, you can use the calculator feature of your grapher to find the answer.

What about the rest of the problems? There's not much a grapher can do for you in those. Remember, you still have to know your algebra!

EXERCISE 11.2

A **In Problems 1–10, an arithmetic sequence is given. Find:**
a. a_1, the first term
b. d, the common difference
c. a_n, the nth term

1. $5, 8, 11, 14, \ldots$

2. $5, 10, 15, 20, \ldots$

3. $11, 6, 1, -4, \ldots$

4. $43, 32, 21, 10, \ldots$

5. $3, -1, -5, -9, \ldots$

6. $0.6, 0.2, -0.2, -0.6, \ldots$

7. $\dfrac{1}{2}, \dfrac{1}{4}, 0, -\dfrac{1}{4}, \ldots$

8. $\dfrac{2}{3}, \dfrac{5}{6}, 1, \dfrac{7}{6}, \ldots$

9. $-\dfrac{5}{6}, -\dfrac{1}{3}, \dfrac{1}{6}, \dfrac{2}{3}, \ldots$

10. $-\dfrac{1}{4}, \dfrac{1}{4}, \dfrac{3}{4}, \dfrac{5}{4}, \ldots$

B **In Problems 11–20, some values for an arithmetic progression are given. Find the other indicated values.**

11. $a_1 = 7, n = 15, d = 6; a_{15}, S_{15}$

12. $a_1 = -2, d = -5, a_n = -72; n, S_n$

13. a_1, d, S_8 for the sequence $4, 10, 16, 22, \ldots$

14. a_1, d, S_n for the sequence $3, -1, -5, -9, \ldots$

15. $a_1 = 3, a_6 = 8; d, S_6$

16. $a_1 = -1, a_{10} = -4; d, S_{10}$

17. $a_1 = 6, S_{14} = -280; d, a_{14}$

18. $a_1 = 15, a_n = -25, S_n = -85; d, n$

19. $d = 40, S_{40} = 40; a_1, a_{40}$

20. $a_1 = 4, d = 2, a_n = 30; n, S_n$

APPLICATIONS

21. A certain property valued at \$30,000 will depreciate \$1380 the first year, \$1340 the second year, \$1300 the third year, and so on, the annual depreciation decreasing \$40 per year. What will the property be worth at the end of 20 yr?

22. Strikers at a certain plant were ordered to return to work and told that their union would be fined \$50 the first day they refused to do so, \$60 the second day, \$70 the third day, and so on. If their union paid a \$680 fine, after how many days did they go back to work?

23. The diagram shows a sequence of crosses. Starting with a single square, a cross is constructed by adding a square to each side. Then the crosses are extended by adding a square to both ends of the vertical and the horizontal pieces. Find the total number of squares in the first 10 elements of this sequence.

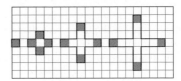

24. The diagram shows a sequence of staircases. Starting with a single square, more squares are added at each stage, as shown by the black squares. Find the number of squares in the tenth staircase.

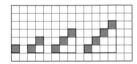

25. Show that the sum of the first n natural numbers is

$$\frac{n(n + 1)}{2}$$

26. Show that the sum of the first n odd natural numbers is n^2.

27. Show that the sum of the first n even natural numbers is $n^2 + n$.

28. Find the sum of the natural numbers between 50 and 100 that are divisible by 3.

SKILL CHECKER

Evaluate:

29. r^{n-1} for $r = 2, n = 6$ 30. r^{n-1} for $r = 3, n = 4$

31. $\dfrac{1 - r^n}{1 - r}$ for $r = 2, n = 6$ 32. $\dfrac{1 - r^n}{1 - r}$ for $r = 3, n = 4$

33. $\dfrac{10}{1 - r}$ for $r = \dfrac{1}{2}$ 34. $\dfrac{10}{1 - r}$ for $r = \dfrac{1}{4}$

USING YOUR KNOWLEDGE   Depressing Depreciation

The ideas in this section can be applied to many areas. For example, the IRS code lists many types of depreciation so we will do some depreciation problems here.

35. You have a piece of property valued at $35,000, which for tax purposes is to be depreciated to a value of $5000 in 5 yr. Suppose that the first year's depreciation is $10,000, and the depreciation for the successive years decreases by a fixed amount each year. Find the depreciation for each of the remaining 4 yr.

36. In Problem 35, suppose the first year's depreciation is $5000 and the depreciation for the successive years increases by a fixed amount. Find the depreciation for each of the remaining 4 yr.

37. The book value after t years for an asset depreciated using the **straight-line method**, that is, decreasing in value by a fixed amount each year, is an arithmetic sequence given by

$$b_t = C - t\left(\frac{C - S}{N}\right)$$

where C is the cost of the asset, S is the salvage value (trade-in) of the asset, and N is the years of expected life for the asset. A $12,000 machine has an expected life of 5 yr and a salvage value of $6000 at the end of 5 yr.
a. Find the formula for b_t.

b. Find the salvage value of the machine after 0, 1, 2, 3, 4, and 5 yr.

WRITE ON . . .

Write your own definition of:

38. An arithmetic sequence   39. An arithmetic series

40. What is the difference between an arithmetic sequence and an arithmetic series?

Describe a situation you have encountered that can be modeled by:

41. An arithmetic sequence   42. An arithmetic series

MASTERY TEST   If you know how to do these problems, you have learned your lesson!

43. Consider the sequence 5, 9, 13, 17, . . . and find:
a. a_1, the first term   b. d, the common difference
c. a_{10}, the tenth term   d. a_n, the nth term

44. Do you remember the song "The Twelve Days of Christmas"? On the first day, you get 1 gift. On the second day, you get $1 + 2$ gifts, on the third day you get $1 + 2 + 3$ gifts, and so on.
a. How many gifts would you get on the tenth day?

b. How many gifts would you get on the nth day? (*Hint:* You are adding the terms in the sequence 1, 2, 3, . . . , n.)

45. The sum of the first eight terms in an arithmetic sequence is 136 and the eighth term is 24. Find:
a. The first term   b. The common difference

46. A heavy-duty truck valued at $50,000 depreciates $6000 the first year, $5800 the second year, $5600 the third year, and so on. What will be the value of the truck at the end of 8 yr?

47. Pedro started a bank account with $30. He then saved $40 the next month, $50 the next, and so on in an arithmetic sequence. How many months did it take for Pedro to accumulate $750?

11.3

GEOMETRIC SEQUENCES AND SERIES

To succeed, review how to:

1. Define a sequence (p. 664).

2. Recognize arithmetic sequences (p. 665).

Objectives:

A Find the common ratio and the general term in a geometric sequence.

B Find the sum of a geometric sequence.

C Find the sum of an infinite geometric series if it exists.

D Solve applications involving geometric sequences.

getting started

Games, Games, Games

The game of chess is said to have originated in Persia. Legend has it that the Shah of Persia offered the inventor of the game anything he wanted as a reward. The inventor asked that 1 grain of wheat be placed on the first square of the chessboard, 2 grains on the second, 4 on the third, and so on. The sequence enumerating the number of grains in each square is given by

$$1, 2, 4, 8, 16, \ldots$$

and the inventor was to receive the sum of the first 64 terms in this sequence (there are 64 squares on a chessboard). As you can see, each *term* after the first in this sequence is obtained by *doubling* (multiplying by 2) the preceding term.

Clearly, the sequence 1, 2, 4, 8, 16, . . . is **not** an arithmetic sequence. It's a different type of sequence called a *geometric sequence* or *geometric progression*. How many grains of wheat should the man have received? More than is produced in the entire world in 1 year! But for a better answer, you have to learn how to sum geometric sequences, so we'll wait until Example 7 to find the exact answer.

Geometric Sequences

Many everyday applications involve geometric sequences. We will discuss some of them in the exercises.

GEOMETRIC SEQUENCE

A **geometric sequence** or **geometric progression** is a sequence in which each term after the first is obtained by *multiplying* the preceding term by a constant r called the **common ratio**.

Multiplying each term by the number r produces a fixed ratio between any two consecutive terms. Thus we can find r by finding the ratio of two successive terms. For example, in the sequence 1, 2, 4, 8, 16, ... The ratio r is given by

$$r = \frac{2}{1} = \frac{4}{2} = \frac{8}{4} = \frac{16}{8} = 2$$

COMMON RATIO

The **common ratio** r is defined by

$$r = \frac{a_{n+1}}{a_n}$$

Since in a geometric sequence each term after the first is obtained by multiplying the preceding term by r, the first n terms in such a sequence are

$$a_1, a_1r, a_1r^2, a_1r^3, \ldots, a_1r^{n-1}$$
$$\begin{array}{ccccccc} \uparrow & \uparrow & \uparrow & \uparrow & & & \uparrow \\ 1 & 2 & 3 & 4 & & & n \end{array}$$

GENERAL TERM OF A GEOMETRIC SEQUENCE

The nth term of the geometric sequence is expressed as

$$a_n = a_1r^{n-1}$$

EXAMPLE 1   Finding terms in a geometric sequence

For the geometric sequence $1, 3, 9, 27, \ldots$, find:

a. a_1 **b.** r **c.** a_6 **d.** a_n

SOLUTION

a. By inspection, we see that the first term is $a_1 = 1$.

b. The common ratio r can be found by taking the ratio of any term to the preceding term. Thus using the ratio of the second term to the first, we find

$$r = \frac{3}{1} = 3$$

c. The formula $a_n = a_1r^{n-1}$ gives, for $n = 6$,

$$a_6 = (1)(3^{6-1}) = 3^5 = 243$$

d. The general term is obtained with $r = 3$.

$$a_n = (1)(3^{n-1}) = 3^{n-1}$$ ∎

B   **Finding the Sum of a Geometric Sequence**

Can we find the sum S_n of the first n terms in a geometric sequence?

$$S_n = a_1 + a_1r + a_1r^2 + a_1r^3 + \cdots + a_1r^{n-1} \qquad \text{By definition}$$

$$rS_n = a_1r + a_1r^2 + a_1r^3 + \cdots + a_1r^{n-1} + a_1r^n \quad \text{Multiply by } r.$$

$$S_n - rS_n = a_1 - a_1r^n = a_1(1 - r^n) \qquad\qquad \text{Subtract.}$$

$$S_n(1 - r) = a_1(1 - r^n) \qquad\qquad\qquad \text{By the distributive property}$$

$$S_n = \frac{a_1(1 - r^n)}{1 - r} \qquad\qquad\qquad\qquad \text{Divide by } 1 - r.$$

Thus we have the following result.

SUM OF A GEOMETRIC SEQUENCE

The sum S_n of the first n terms starting with a_1 and ending with a_n is given by

$$S_n = \frac{a_1(1 - r^n)}{1 - r}$$

EXAMPLE 2   Finding the sum of a geometric sequence

For the geometric sequence $4, -8, 16, -32, \ldots$, find:

a. r **b.** a_{10} **c.** S_{10}

SOLUTION

a. Since r is the common ratio,

$$r = \frac{-8}{4} = -2$$

b. We have $a_1 = 4$ and $r = -2$, so

$$a_{10} = a_1 r^{n-1}$$
$$= (4)(-2)^9 = (4)(-512)$$
$$= -2048$$

c. Using the formula for S_n with $n = 10$, we get

$$S_{10} = \frac{4[1 - (-2)^{10}]}{1 - (-2)} = \frac{4(1 - 2^{10})}{1 + 2}$$
$$= \frac{4(1 - 1024)}{3} = -1364 \qquad\blacksquare$$

EXAMPLE 3   Finding a specific term and the common ratio of a geometric sequence

Given a geometric sequence with $a_1 = 2$ and $S_3 = 26$, find: a_3 and r

SOLUTION   Using the formulas for the nth term and the sum of n terms, we first find r and use it to find a_3. So

$$a_3 = a_1 r^2 = 2r^2 \quad\text{and}\quad S_3 = \frac{a_1(1 - r^3)}{1 - r} = \frac{2(1 - r^3)}{1 - r}$$

Since $1 - r^3 = (1 - r)(1 - r + r^2)$, the formula for S_3 can be simplified to

$$S_3 = \frac{2(1 - r)(1 + r + r^2)}{1 - r}$$
$$= 2(1 + r + r^2)$$

Thus we have

$$26 = 2(1 + r + r^2)$$

$$13 = 1 + r + r^2$$

which gives a quadratic equation for r:

$$r^2 + r - 12 = 0$$

By factoring, we get

$$(r + 4)(r - 3) = 0$$

so that $r = -4$ or $r = 3$. If $r = -4$, then $a_3 = 2(-4)^2 = 32$, and if $r = 3$, then $a_3 = 2(3^2) = 18$.

We can check these results by writing the three terms of the sequence.

For $r = -4$:   2, -8, 32, which add to 26

For $r = 3$:   2, 6, 18, which add to 26

Hence the correct answers are $a_3 = 32$, $r = -4$ or $a_3 = 18$, $r = 3$. ■

Finding the Sum of an Infinite Geometric Series

In the preceding examples, we found the sum of the first n terms of a geometric sequence. We now consider what happens if n is allowed to increase without bound. We indicate that the number of terms is *infinite*—that is, unlimited—by writing

$$\sum_{n=1}^{\infty} a_1 r^{n-1} = a_1 + a_1 r + a_1 r^2 + \cdots + a_1 r^{n-1} + \cdots$$

This expression is called an *infinite geometric series.* We can find the sum of the first n terms of this series by using the formula

$$S_n = \frac{a_1(1 - r^n)}{1 - r} = \frac{a_1}{1 - r}(1 - r^n)$$

If r is less than 1 in absolute value, that is, $|r| < 1$, or, equivalently, $-1 < r < 1$, then r^n becomes smaller and smaller as n increases. For example, if $r = 0.6$, then a calculation with logarithms or with a calculator gives the following approximate results:

$$r^{10} = (0.6)^{10} = 6.05 \times 10^{-3}$$

$$r^{100} = (0.6)^{100} = 6.53 \times 10^{-23}$$

$$r^{1000} = (0.6)^{1000} = 1.4 \times 10^{-222}$$

(The first of these has 2 zeros before the first significant digit, and the second and third have 22 and 221 zeros, respectively, before the first significant digit.) So we can make r^n as small as we like by taking n large enough.

You can see that if $|r| < 1$, the factor $(1 - r^n)$ in the sum formula can be made as close to 1 as we wish by taking n large enough. Hence as n becomes greater and greater, the sum

$$S_n = \frac{a_1}{1 - r}(1 - r^n)$$

is more and more closely approximated by the expression

$$\frac{a_1}{1 - r}$$

We summarize this discussion by making the following definition.

SUM OF AN INFINITE GEOMETRIC SERIES

> If $|r| < 1$, then the sum S of the geometric series with first term a_1 and common ratio r is defined to be
>
> $$S = a_1 + a_1 r + a_1 r^2 + \cdots + a_1 r^{n-1} + \cdots$$
>
> $$= \frac{a_1}{1 - r}$$

If $|r| \geq 1$, the sum of n terms does not get closer and closer to any number as n becomes larger and larger without bound. In this case, we say that S *does not exist*.

 NOTE If $|r| \geq 1$, the sum of an infinite geometric series **does not exist**.

For $r = \frac{1}{2}$ and $a_1 = 1$, we can give a graphical interpretation of the behavior of S_n as n increases without bound. We consider the series

$$\sum_{n=1}^{\infty} \left(\frac{1}{2}\right)^{n-1} = 1 + \frac{1}{2} + \frac{1}{4} + \frac{1}{8} + \cdots + \left(\frac{1}{2}\right)^{n-1} + \cdots$$

For this series,

$$S_1 = 1$$

$$S_2 = 1 + \frac{1}{2} = 1\frac{1}{2}$$

$$S_3 = 1 + \frac{1}{2} + \frac{1}{3} = 1\frac{3}{4}$$

$$S_4 = 1 + \frac{1}{2} + \frac{1}{3} + \frac{1}{4} = 1\frac{7}{8}$$

The sums are as shown here:

As you can see, each step cuts in half the remaining distance to the point marked 2. Thus by making n sufficiently large, we can make the value of S_n as close as we like to 2. For the series $1 + \frac{1}{2} + \frac{1}{4} + \frac{1}{8} + \cdots + \left(\frac{1}{2}\right)^{n-1} + \cdots$

$$S = \frac{a_1}{1 - r} = \frac{1}{1 - \frac{1}{2}} = 2$$

EXAMPLE 4 Finding the sum of an infinite geometric series

Find the sum of the geometric series $4 - 2 + 1 - \frac{1}{2} + \cdots$.

SOLUTION In this series, $a_1 = 4$ and $r = -\frac{1}{2}$, so

$$S = \frac{a_1}{1 - r} = \frac{4}{1 - \left(-\frac{1}{2}\right)} = \frac{8}{2 + 1} = \frac{8}{3} \qquad \blacksquare$$

EXAMPLE 5 More practice finding the sum of an infinite geometric series

If the following geometric series has a sum, find it.

$$\sum_{n=1}^{\infty} (1.01)^n = (1.01) + (1.01)^2 + (1.01)^3 + \cdots$$

SOLUTION For this series the ratio r is

$$\frac{(1.01)^2}{(1.01)} = 1.01$$

which is greater than 1. So the sum of this series does not exist. $\qquad \blacksquare$

Applications of Geometric Series

Geometric series can be used to express nonterminating repeating decimals as fractions. For example, the decimal

$$0.333\ldots = 0.\overline{3}$$

can be written as

$$\frac{3}{10} + \frac{3}{100} + \frac{3}{1000} + \cdots$$

which is an infinite geometric series with

$$a_1 = \frac{3}{10} \quad \text{and} \quad r = \frac{1}{10}$$

Thus

$$S = \frac{\frac{3}{10}}{1 - \frac{1}{10}} = \frac{3}{9} = \frac{1}{3}$$

EXAMPLE 6   Finding fraction equivalents

Find the fraction equivalent to the repeating decimal $0.414141\ldots$.

SOLUTION   We can write this decimal as

$$\frac{41}{100} + \frac{41}{(100)^2} + \frac{41}{(100)^3} + \cdots$$

which is an infinite geometric series with

$$a_1 = \frac{41}{100} \quad \text{and} \quad r = \frac{1}{100}$$

Thus the sum of this series is

$$S = \frac{a_1}{1 - r} = \frac{\frac{41}{100}}{1 - \frac{1}{100}}$$

$$= \frac{\frac{41}{100}}{\frac{99}{100}} = \frac{41}{99} \qquad \blacksquare$$

problem solving

EXAMPLE 7   Job offers and series

Suppose you have two job offers for a 2-week (14-day) trial period. Job A starts at $50 per day with a $50 raise each day. Job B starts at 50¢ per day and your salary is doubled every day. Find the total amount paid by each of the jobs at the end of the 14 days.

1. Read the problem.

SOLUTION   Note that we must find the total amount paid by each of the jobs by the end of the 14 days.

2. Select the unknown.

We are asked to find the total amount paid, S_{14}, for each job.

3. Think of a plan.

Let's find the amount job A pays at the end of 14 days. Then we'll find the amount job B pays at the end of 14 days.

The pay for job A starts at \$50 ($a_1 = 50$) and increases by \$50 each day ($d = 50$). The salary for the fourteenth day is $a_{14} = 50 + 13 \cdot 50 = 700$. The pay for job B starts at \$0.50 ($a_1 = 0.50$) and doubles every day ($r = 2$).

4. Use the formula for the sum of an arithmetic and a geometric sequence to solve the problem.

For job A, the sum of the arithmetic sequence for 14 days is

$$S_{14} = \frac{n(a_1 + a_n)}{2} = \frac{14 \cdot 750}{2} = \$5250$$

For job B, the sum of the geometric sequence for 14 days is

$$S_{14} = \frac{a_1(1 - r^n)}{1 - r} = \frac{0.50(1 - 2^{14})}{1 - 2}$$

$$= \frac{0.50(1 - 2^{14})}{-1}$$

$$= 0.50(2^{14} - 1)$$

$$= 0.50\,(16{,}383)$$

$$= \$8191.50$$

Job B pays much more!

5. Verify the solution.

We leave the verification to you.   ■

EXAMPLE 8   Take it with a grain of wheat

As you recall from the *Getting Started,* the Shah of Persia so liked the game of chess that he offered its inventor anything he wanted. The inventor asked that 1 grain of wheat be placed on the first square of the chessboard, 2 grains on the second, 4 on the third, and so on. If there are 64 squares on a chessboard:
a. How many grains were to be placed on the 64th square?
b. What is the total number of grains the inventor should have received?

SOLUTION
a. The number of grains in each square is

Square 1	Square 2	Square 3	Square 4	...	Square 64
1	2	$4 = 2^{3-1}$	$8 = 2^{4-1}$	...	$2^{63} = 2^{64-1}$

b. The sum of the geometric sequence $1, 2, 4, \ldots, 2^{63}$ where $a_1 = 1$ and $r = 2$ (since the number of grains is doubled on each succeeding square) is

$$S_{64} = \frac{1(1 - 2^{64})}{1 - 2} = \frac{1 - 2^{64}}{-1}$$

$$= 2^{64} - 1$$

By the way, since 2^{10} is about 1000, $2^{60} = (2^{10})^6$ is about $(1000)^6$ or 1,000,000,000,000,000,000 (one quintillion)! Actually, if you do this with a calculator, you will find that the answer is closer to 18 quintillion!   ■

GRAPH IT

In Example 1, you can find a_n for any n if you do the algebra involved in parts a, b, and c. After that, graph $Y_1 = 3^{x-1}$ using an integer window. If you want to find a_6, simply use the $\boxed{\text{TRACE}}$ key until you get to $X = 6$. (See Window 1.) The grapher shows the corresponding Y-value even though the X-value *is not shown* in the window!

If you understand the terminology of sequences, you can find the sum of a geometric sequence using a grapher. First, recall that

$$S_n = a_1 + a_2r + a_3r^2 + \cdots + a_nr^{n-1} = \frac{a_1(1 - r^n)}{1 - r}$$

In Example 2, we want to find S_n. Let's graph

$$Y_1 = \frac{a_1(1 - r^n)}{1 - r}$$

by replacing a_1 by 4, n by x, r by -2, and using the dot mode with an integer window. Thus we graph

$$Y_1 = \frac{4(1 - (-2)^x)}{1 - (-2)} = \frac{4(1 - (-2)^x)}{3}$$

Now use the $\boxed{\text{TRACE}}$ key to find S_{10}, as shown in Window 2.

Next, let's do some exploration with Example 4. The series in this example can be written as

$$\sum_{n=1}^{N} (-1)^{n+1}2^{3-n}$$

What happens as N increases? Let's try some sums with a TI-82 by pressing $\boxed{\text{2nd}}$ $\boxed{\text{STAT}}$, moving the cursor to the "MATH" column and pressing 5, and then pressing $\boxed{\text{2nd}}$ $\boxed{\text{STAT}}$ 5 to add the terms of the given sequence. Now enter $(-1)^{N+1}2^{3-N}$ followed by $\boxed{,}$; the variable being used (which is N—you can enter N by pressing $\boxed{\text{ALPHA}}$ $\boxed{\text{log}}$); the starting point for N, say, 1, followed by $\boxed{,}$; how far you want the N to go (enter 1 5); and the increment for N, say, 1. Now close the parentheses and press $\boxed{\text{ENTER}}$. The answer is shown in Window 3. Try to change the 15 to 30 (press $\boxed{\text{2nd}}$ $\boxed{\text{ENTER}}$ and replace the 15 by 30 and press $\boxed{\text{ENTER}}$). You get 2.666666664, which is very close to the answer we obtained before, $\frac{8}{3}$.

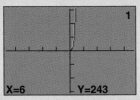

X=6   Y=243

WINDOW 1

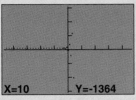

X=10   Y=-1364

WINDOW 2

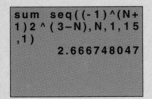

```
sum seq((-1)^(N+
1)2^(3-N),N,1,15
,1)
        2.666748047
```

WINDOW 3

EXERCISE 11.3

$\boxed{\text{A}}$ $\boxed{\text{B}}$ **In Problems 1–10, a geometric sequence is given. Find the indicated values.**

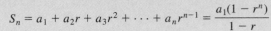

    a. a_1 **b.** r **c.** a_n **d.** S_n

1. 3, 6, 12, 24, ...

2. $\frac{1}{3}$, 1, 3, 9, ...

3. 8, 24, 72, 216, ...

4. $\frac{1}{5}, \frac{1}{10}, \frac{1}{20}, \frac{1}{40}, \ldots$

5. 16, -4, 1, $-\frac{1}{4}$, ...

6. 3, -1, $\frac{1}{3}$, $-\frac{1}{9}$, ...

7. $-\frac{3}{5}, \frac{3}{2}, -\frac{15}{4}, \frac{75}{8}, \ldots$

8. 60, -6, $\frac{6}{10}$, $-\frac{6}{100}$, ...

9. $-\frac{3}{4}, -\frac{1}{4}, -\frac{1}{12}, -\frac{1}{36}, \ldots$

10. $-\frac{5}{6}, -\frac{1}{3}, -\frac{2}{15}, -\frac{4}{75}, \ldots$

In Problems 11–20, some values for a geometric sequence are given. Find the remaining indicated values.

11. $a_1 = 1, S_3 = \dfrac{7}{4}; a_3, r$

12. $a_1 = 4, S_3 = 7; a_3, r$

13. $a_1 = 3, S_3 = 21; a_3, r$

14. $a_1 = \dfrac{1}{2}, S_3 = \dfrac{39}{50}; a_3, r$

15. $r = 2, S_8 = 1785; a_1, a_8$

16. $a_6 = -\dfrac{16}{27}, r = -\dfrac{1}{3}; a_1, S_6$

17. $a_1 = -4, a_n = 108, S_n = 80; r, n$

18. $a_1 = \dfrac{3}{4}, a_n = 192, S_n = 255\dfrac{3}{4}; r, n$

19. $a_1 = \dfrac{16}{125}, r = \dfrac{5}{2}, a_n = \dfrac{25}{2}; n, S_n$

20. $a_1 = 7, r = 2, a_n = 896; n, S_n$

In Problems 21–30, an infinite geometric series is given. Find the sum if it exists.

21. $6 + 3 + 1\dfrac{1}{2} + \cdots$

22. $12 + 4 + 1\dfrac{1}{3} + \cdots$

23. $(-6) + (-3) + \left(-\dfrac{3}{2}\right) + \cdots$

24. $(-8) + (-4) + (-2) + \cdots$

25. $\displaystyle\sum_{n=1}^{\infty} (-1)^{n-1} 2^{2-n} = 2 - 1 + \dfrac{1}{2} - \dfrac{1}{4} + \cdots$

26. $\displaystyle\sum_{n=1}^{\infty} (-1)^{n-1} 3^{3-n} = 9 - 3 + 1 - \dfrac{1}{3} + \cdots$

27. $4 - 8 + 16 - 32 + \cdots$ **28.** $(-5) + (-10) + (-20) + \cdots$

29. $\dfrac{1}{10} + \dfrac{1}{5} + \dfrac{2}{5} + \cdots$ **30.** $0.0001 - 0.001 + 0.01 - \cdots$

In Problems 31–40, find a fraction equivalent to the given repeating decimal.

31. $0.555\ldots$ **32.** $0.666\ldots$

33. $0.181818\ldots$ **34.** $0.242424\ldots$

35. $4.050505\ldots$ **36.** $2.313131\ldots$

37. $2.3161616\ldots$ **38.** $4.1272727\ldots$

39. $0.140140140\ldots$ **40.** $1.123123123\ldots$

41. The population of a certain town increases at the rate of 4% per year.
   a. Write the sequence associated with this population.
   b. If the present population is 20,000, what will the population be at the end of 5 yr?

42. The number of bacteria in a culture increased from 320,000 at the beginning of the first day to 2,430,000 at the end of the fifth day. Find the daily rate of increase if this rate is assumed to be constant—that is, if the starting number and the numbers at the ends of the successive days form the geometric sequence

$$a_1, a_1 r, a_1 r^2, a_1 r^3, a_1 r^4, a_1 r^5$$

43. The distance traveled in any swing by a point on a compound pendulum is 20% less than in the preceding swing. If the length of the first swing is 62.5 cm, find the total distance the point has traveled at the end of the fourth swing. (*Hint:* Write the sequence associated with the four swings.)

44. A small business makes a net profit of $10,000 in its first year. If the net profit increases by 25% each year for the next 4 yr, what is the total net profit for these 5 yr?

45. If the rate of increase in Problem 44 is 50%, what is the total net profit for the 5 yr?

46. A polluted tank holds 100 gal of a poisonous chemical that mixes readily with water. After 25 gal of the chemical are drawn off, the tank is refilled with water. Then 25 gal of the mixture are drawn off, and the tank is again refilled with water. If this operation is performed until five batches have been drawn from the tank, how much of the original chemical remains?

47. If the first three terms in an arithmetic sequence are increased by 1, 3, and 13, respectively, the resulting numbers are in geometric sequence. Find the original terms if their sum is 9.

48. If the first three terms in an arithmetic sequence are increased by 9, 7, and 9, respectively, the resulting numbers are in geometric sequence. Find the original terms if their sum is 3.

49. Roberto and Jimmy are golf pals. Yesterday, Roberto persuaded Jimmy to bet 1¢ on the first hole, 2¢ on the second hole, 4¢ on the third hole, and so on (doubling the bet on each successive hole). Roberto did not have very good luck! He won the first hole, lost the second hole, and continued winning a hole and losing a hole in that order for the remainder of the game. How much did Roberto lose in all? (*Hint:* You will need to use the number $2^{18} = 262,144$.)

50. Refer to Problem 49. Suppose Roberto and Jimmy play again, making the same bets as before. This time Roberto wins the first two holes, loses the next two holes, and continues winning two holes and losing two holes in that order for the remainder of the game. How did Roberto come out this time?

51. A rubber ball is dropped from a height of 8 ft. It makes a sequence of bounces, each three-fourths the height of the preceding bounce. About how far does the ball travel before coming to rest?

52. A pendulum on each separate swing describes an arc whose length is 98% of the length of the preceding swing. If the first arc is 12 in. long, about how far does the pendulum travel before it comes to rest?

SKILL CHECKER

Expand:

53. $(a + b)^2$

54. $(x - y)^2$

55. $(x + y)^3$

56. $(a - b)^3$

57. $(y - 2z)^3$

58. $(2a + b)^3$

59. $\left(\dfrac{1}{x} - \dfrac{1}{y}\right)^3$

60. $\left(\dfrac{2}{y} - \dfrac{1}{2}\right)^3$

USING YOUR KNOWLEDGE   What Is the Sequence?

In the preceding sections you have learned what we mean by a sequence, an arithmetic sequence, and a geometric sequence. You can use this knowledge in the following problems.

In Problems 61–72, the first four terms of a series are given. Determine whether these terms form an *arithmetic* sequence, a *geometric* sequence, or *neither*. If the terms form a geometric sequence and the series is extended to form an infinite geometric series, find the sum if it has one.

61. $1 + \dfrac{2}{5} + \dfrac{4}{25} + \dfrac{8}{125} + \cdots$ **62.** $1 - \dfrac{1}{2} + \dfrac{1}{3} - \dfrac{1}{4} + \cdots$

63. $1 + \dfrac{2}{5} - \dfrac{1}{5} - \dfrac{4}{5} - \cdots$ **64.** $1 + 3 + 4 + 7 + \cdots$

65. $2 + \dfrac{7}{4} + \dfrac{49}{32} + \dfrac{343}{256} + \cdots$ **66.** $2 + 1\dfrac{3}{4} + 1\dfrac{1}{2} + 1\dfrac{1}{4} + \cdots$

67. $5 + 4 + 2 - 1 - \cdots$ **68.** $4 + 2 + 0 - 2 - \cdots$

69. $\dfrac{1}{2} + \dfrac{2}{3} + \dfrac{3}{4} + \dfrac{4}{5} + \cdots$ **70.** $-12 + 4 - \dfrac{4}{3} + \dfrac{4}{9} - \cdots$

71. $4 - 2 + 1 - \dfrac{1}{2} + \cdots$ **72.** $-6 - 3 + 0 + 3 + \cdots$

WRITE ON . . .

73. Explain the difference between an arithmetic sequence and a geometric sequence.

74. The Fibonacci sequence is 1, 1, 2, 3, 5, 8, ... Is this sequence an arithmetic sequence, a geometric sequence, or neither? Explain your answer.

75. Explain the difference between an infinite geometric series and an infinite geometric sequence.

76. What conditions do you need for

$$S_n = \frac{a_1}{1 - r}(1 - r^n)$$

to be very close to

$$\frac{a_1}{1 - r}?$$

MASTERY TEST   If you know how to do these problems, you have learned your lesson!

77. For the geometric sequence $2, 1, \frac{1}{2}, \frac{1}{4}, \ldots$, find:
a. a_1 **b.** r **c.** a_6 **d.** a_n

78. For the geometric sequence $4, -8, 16, -32, \ldots$, find:
a. a_n **b.** S_n

79. Given a geometric sequence with $a_1 = 2$, $S_3 = 42$, find a_3 and r.

80. Find the sum of the geometric series $9 - 3 + 1 - \frac{1}{3} + \cdots$.

81. If the geometric series defined by

$$\sum_{n=1}^{\infty} (-1)^{n+1}(1.01)^n = (1.01) - (1.01)^2 + (1.01)^3 - \cdots$$

has a sum, find it.

82. Find the fraction equivalent to the repeating decimal $0.432432432\ldots$.

83. An investment firm claims that you can double your money every year if you invest with them. Suppose you invest $100.
a. Write the first five terms of the sequence associated with this claim, starting with $a_1 = 100$.
b. How much money would you have at the end of 6 years?
c. How much money would you have at the end of the nth year?
d. How many years would it take for you to have over $100,000? (Answer to the nearest year.)

11.4

THE BINOMIAL EXPANSION

To succeed, review how to:

1. Simplify an algebraic expression (pp. 149–151).
2. Expand binomials of the form $(x + y)^n$, where $n < 6$ (pp. 153–154).

Objectives:

A Use the binomial expansion to expand and simplify a power of a binomial.

B Find the coefficient of a term in a binomial expansion.

C Solve applications involving binomial expansions.

getting started

It's Chinese to Me

This drawing made in A.D. 1303 by Chu Shi-kie turns out to be a nice depiction of Pascal's triangle (Pascal having lived in the 1600s). As you can see, every row starts and ends with the symbol ⊖. If we make ⊖ = 1, we can say the triangle starts with the number 1. The next row has the symbols ⊖ and ⊖, or in our translation, the numbers 1 and 1. The next row has the symbols ⊖, ⊜, and ⊖—that is, the numbers 1, 2, and 1. What are the "numbers" in the next row? Can you see a pattern emerging? It looks like this:

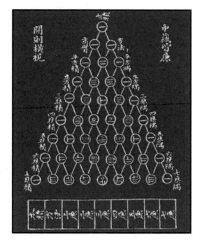

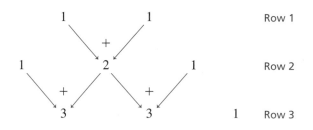

Note that the 2 in the second row is the sum of the two elements in the first row. The first 3 in the third row is the sum of the first two elements in the second row, and so on. If we continue to construct rows in this way, we get the following triangular array

$$
\begin{array}{ccccccccccc}
& & & & 1 & & 1 & & & & \\
& & & 1 & & 2 & & 1 & & & \\
& & 1 & & 3 & & 3 & & 1 & & \\
& 1 & & 4 & & 6 & & 4 & & 1 & \\
1 & & 5 & & 10 & & 10 & & 5 & & 1 \\
& & & & & \vdots & & & & &
\end{array}
$$

This array is known as Pascal's triangle, in honor of the French mathematician Blaise Pascal. We shall examine Pascal's triangle as our introduction to binomial expansions.

Pascal's Triangle and Binomial Expansions

The seventeenth-century mathematician Blaise Pascal wrote a book about the triangle presented in the *Getting Started*. In the book he stated that if we have 5 coins, the relationship between the number of heads that can come up and the number of ways in which these heads can occur is as follows:

Number of heads	5	4	3	2	1	0
Number of occurrences	1	5	10	10	5	1

For example, if 5 coins are tossed, 5 heads can only occur one way (HHHHH), but 4 heads can occur in 5 ways (THHHH, HTHHH, HHTHH, HHHTH, HHHHT). What do these results have to do with algebra? As it turns out, there is a somewhat different way to arrive at the numbers in Pascal's triangle. From our previous work, you can verify that

$$(x + y)^1 = x + y$$
$$(x + y)^2 = x^2 + 2xy + y^2$$
$$(x + y)^3 = x^3 + 3x^2y + 3xy^2 + y^3$$
$$(x + y)^4 = x^4 + 4x^3y + 6x^2y^2 + 4xy^3 + y^4$$
$$(x + y)^5 = x^5 + 5x^4y + 10x^3y^2 + 10x^2y^3 + 5xy^4 + y^5$$

Look at the coefficients (the numbers that multiply x and y) of the terms in the expansion of $(x + y)^5$. They are 1, 5, 10, 10, 5, and 1. Now compare this with the fifth row in Pascal's triangle. The numbers are identical!

Can we predict the expansion of $(x + y)^6$ by studying these expansions? The pattern for the powers of x and y is easy. In going from left to right, the powers of x decrease from n to 0 and the powers of y increase from 0 to n. Ignoring the coefficients, the expansion of $(x + y)^n$ has the form

$$(x + y)^n = x^ny^0 + \Box x^{n-1}y^1 + \Box x^{n-2}y^2 + \cdots + \Box x^{n-k}y^k + \cdots + y^nx^0$$

where $\Box$ represents the missing coefficients. (Note that the sum of the powers of x and y is always n.) Thus to write the expansion of $(x + y)^n$, we need only know how to find these missing coefficients. In the case of $(x + y)^6$, the coefficients should be 1, 6, 15, 20, 15, 6, and 1, and they are obtained by constructing the sixth row of Pascal's triangle, starting and ending with a 1 and finding each coefficient by addition in the pattern we have just discussed. In going from left to right, the powers of x decrease from $n = 6$ to 0, the powers of y increase from 0 to 6, and the sum of the exponents of x and y is always 6. Thus we have

$$(x + y)^6 = 1x^6y^0 + 6x^5y^1 + 15x^4y^2 + 20x^3y^3 + 15x^2y^4 + 6x^1y^5 + 1x^0y^6$$
$$= x^6 + 6x^5y + 15x^4y^2 + 20x^3y^3 + 15x^2y^4 + 6xy^5 + y^6$$

Here is the generalization for the expression of $(x + y)^n$.

PROCEDURE

The Expansion of $(x + y)^n$

1. The first term is x^n and the last term is y^n.

2. The powers of x *decrease* by 1 from term to term, the powers of y *increase* by 1 from term to term, and the sum of the powers of x and y is always n.

3. The coefficients for the expansion correspond to those in the nth row of Pascal's triangle and can be written as $\binom{n}{r}$ where r is the power of x.

Using this notation,

$$(x + y)^6 = \binom{6}{6}x^6 + \binom{6}{5}x^5y + \binom{6}{4}x^4y^2 + \binom{6}{3}x^3y^3 + \binom{6}{2}x^2y^4 + \binom{6}{1}xy^5 + \binom{6}{0}y^6$$

$$= x^6 \quad + 6x^5y \quad + 15x^4y^2 \quad + 20x^3y^3 \quad + 15x^2y^4 \quad + 6xy^5 \quad + y^6$$

where

$$\binom{6}{6} = 1, \binom{6}{5} = 6, \binom{6}{4} = 15, \binom{6}{3} = 20, \binom{6}{2} = 15, \binom{6}{1} = 6, \text{ and } \binom{6}{0} = 1$$

To facilitate defining $\binom{n}{r}$, we first introduce a special symbol, $n!$ (read "n factorial"), for which we have the following definition.

FACTORIAL

For any natural number n,

$$n! = n(n - 1)(n - 2) \cdots (1)$$

Thus

$$5! = 5 \cdot 4 \cdot 3 \cdot 2 \cdot 1 = 120$$

$$7! = 7 \cdot 6 \cdot 5 \cdot 4 \cdot 3 \cdot 2 \cdot 1 = 5040$$

This notation can also be used to denote the product of consecutive integers beginning with integers different from 1. For example,

$$\frac{7!}{4!} = 7 \cdot 6 \cdot 5$$

since

$$\frac{7!}{4!} = \frac{7 \cdot 6 \cdot 5 \cdot \cancel{4} \cdot \cancel{3} \cdot \cancel{2} \cdot \cancel{1}}{\cancel{4} \cdot \cancel{3} \cdot \cancel{2} \cdot \cancel{1}}$$

Similarly,

$$\frac{9!}{5!} = 9 \cdot 8 \cdot 7 \cdot 6 \qquad \text{and} \qquad \frac{10!}{6!} = 10 \cdot 9 \cdot 8 \cdot 7$$

Also,

$$n! = n[(n - 1)(n - 2)(n - 3) \cdots (1)]$$

$$= n(n - 1)!$$

Thus we can write

$$n! = n(n - 1)!$$

With this notation,

$$9! = 9 \cdot 8!$$

$$13! = 13 \cdot 12!$$

If we let $n = 1$ in $n! = n(n - 1)!$, we obtain

$$1! = 1 \cdot 0!$$

Hence we define 0! as follows.

ZERO FACTORIAL

$$0! = 1$$

EXAMPLE 1   Calculations with factorials

Find:

a. 4!   **b.** $\dfrac{8!}{5!}$

SOLUTION

a. Since $n! = n(n - 1)(n - 2) \cdots (1)$

$$4! = 4 \cdot 3 \cdot 2 \cdot 1 = 24$$

b. $$8! = 8 \cdot 7 \cdot 6 \cdot 5 \cdot 4 \cdot 3 \cdot 2 \cdot 1$$

$$\text{and } 5! = 5 \cdot 4 \cdot 3 \cdot 2 \cdot 1$$

Thus

$$\frac{8!}{5!} = \frac{8 \cdot 7 \cdot 6 \cdot \cancel{5} \cdot \cancel{4} \cdot \cancel{3} \cdot \cancel{2} \cdot \cancel{1}}{\cancel{5} \cdot \cancel{4} \cdot \cancel{3} \cdot \cancel{2} \cdot \cancel{1}} = 336$$

∎

We are now ready to use factorial notation to define the binomial coefficient.

BINOMIAL COEFFICIENT

The **binomial coefficient**, denoted $\binom{n}{r}$, is defined as

$$\binom{n}{r} = \frac{n!}{r!(n - r)!}$$

Thus

$$\binom{5}{2} = \frac{5!}{2!(5 - 2)!} = \frac{5!}{2!3!} = \frac{5 \cdot 4 \cdot \cancel{3!}}{2!\cancel{3!}} = 10$$

$$\binom{5}{5} = \frac{5!}{5!(5 - 5)!} = \frac{5!}{5!0!} = 1$$

EXAMPLE 2   Evaluating a binomial coefficient

Evaluate:

$$\binom{6}{4}$$

SOLUTION By definition,

$$\binom{n}{r} = \frac{n!}{r!(n-r)!}$$

$$\binom{6}{4} = \frac{6!}{4!2!} = \frac{6 \cdot 5 \cdot 4!}{4!2!} = \frac{6 \cdot 5}{2} = 15 \quad\blacksquare$$

Now that we know the meaning of $\binom{n}{r}$, we can use it to write the coefficients of a binomial expansion.

EXAMPLE 3   Writing the coefficients of a binomial expansion

Given that $(a + b)^4 = a^4 + 4a^3b + 6a^2b^2 + 4ab^3 + b^4$, write each coefficient in the expansion using the $\binom{n}{r}$ notation.

SOLUTION

$$(a + b)^4 = a^4 + 4a^3b + 6a^2b^2 + 4ab^3 + b^4$$

$$= \binom{4}{4}a^4 + \binom{4}{3}a^3b + \binom{4}{2}a^2b^2 + \binom{4}{1}ab^3 + \binom{4}{0}b^4$$

You can easily verify that

$$\binom{4}{4} = 1, \binom{4}{3} = 3, \binom{4}{2} = 6, \binom{4}{1} = 4, \text{ and } \binom{4}{0} = 1 \quad\blacksquare$$

Following the pattern in Example 3, the general formula for $(x + y)^n$ can be written using the $\binom{n}{r}$ notation, where n represents the exponent of the binomial being expanded and r is the power of x. Note that the powers of x *decrease* by 1 from term to term, while those of y *increase* by 1 from term to term. Thus we have

THE GENERAL BINOMIAL EXPANSION

$$(x + y)^n = \binom{n}{n}x^n + \binom{n}{n-1}x^{n-1}y^1 + \binom{n}{n-2}x^{n-2}y^2 + \cdots$$

$$+ \binom{n}{n-k}x^{n-k}y^k + \cdots + \binom{n}{0}y^n$$

Using this formula, we have

$$(x + y)^5 = \binom{5}{5}x^5 + \binom{5}{4}x^4y^1 + \binom{5}{3}x^3y^2 + \binom{5}{2}x^2y^3 + \binom{5}{1}x^1y^4 + \binom{5}{0}y^5$$

$$= x^5 + 5x^4y + 10x^3y^2 + 10x^2y^3 + 5xy^4 + y^5$$

To avoid confusion, we first write the expansion completely, and *then* substitute the coefficients

$$\binom{5}{5} = 1, \binom{5}{4} = 5, \binom{5}{3} = 10, \binom{5}{2} = 10, \binom{5}{1} = 5, \text{ and } \binom{5}{0} = 1$$

in the expansion. Note that

$$\binom{5}{5} = \binom{5}{0}, \binom{5}{4} = \binom{5}{1}, \text{ and } \binom{5}{3} = \binom{5}{2}$$

In general,

$$\binom{n}{k} = \binom{n}{n-k}$$

This tells you that the coefficients of the first and last terms, second and next to last, and so on, are equal, thus saving you time and computation when you expand binomials.

EXAMPLE 4   Using the binomial expansion

Expand: $(a - 2b)^5$

SOLUTION   We use the binomial expansion with $x = a$, $y = -2b$, and $n = 5$. This gives

$$(a - 2b)^5 = \binom{5}{5}a^5 + \binom{5}{4}a^4(-2b)^1 + \binom{5}{3}a^3(-2b)^2 + \binom{5}{2}a^2(-2b)^3 + \binom{5}{1}a^1(-2b)^4 + \binom{5}{0}(-2b)^5$$

We know that

$$\binom{5}{5} = \binom{5}{0} = 1, \binom{5}{4} = \binom{5}{1} = \frac{5!}{1!4!} = 5 \text{ and } \binom{5}{3} = \binom{5}{2} = \frac{5!}{2!3!} = 10$$

Thus

$$(a - 2b)^5 = a^5 + 5a^4(-2b)^1 + 10a^3(-2b)^2 + 10a^2(-2b)^3 + 5a^1(-2b)^4 + (-2b)^5$$

$$= a^5 - 10a^4b + 40a^3b^2 - 80a^2b^3 + 80ab^4 - 32b^5 \quad \blacksquare$$

B   Finding the Coefficient of a Term in a Binomial Expansion

As you recall, when we expand $(x + y)^n$, in the coefficient $\binom{n}{r}$ n represents the exponent of the binomial being expanded and r is the power of x. We shall use this fact to find the coefficient of a particular term in an expansion without writing out the whole expansion.

EXAMPLE 5   Finding the coefficient of a term in a binomial expansion

Find: The coefficient of a^4b^3 in the expansion of $(a - 3b)^7$

SOLUTION In our case, $n = 7$, $x = a$, $y = -3b$, and the power of a is $r = 4$. Thus the coefficient of $a^4 b^3$ can be obtained by simplifying

$$\binom{7}{4} a^4 (-3b)^3 = \frac{7!}{4!3!} a^4 (-3b)^3 = \frac{7 \cdot 6 \cdot 5}{3 \cdot 2 \cdot 1} a^4 (-27b^3) = -945a^4 b^3$$

Hence the coefficient of $a^4 b^3$ is -945. ∎

 Applications Involving Binomial Expansions

The binomial coefficients for $(x + y)^n$ furnish us with a row of Pascal's triangle. As you recall, the numbers in such a row also tell us how many ways a stated number of heads can come up if n coins are tossed. For example, the first coefficient $\binom{n}{n} = 1$ tells us that the number of ways that n heads can occur when n coins are tossed is 1. The second coefficient,

$$\binom{n}{n-1} = n$$

tells us that the number of ways that $n - 1$ heads can occur when n coins are tossed is n, the third coefficient

$$\binom{n}{n-2} = \frac{n(n-1)}{1 \cdot 2}$$

tells us that the number of ways that $n - 2$ heads can occur when n coins are tossed is

$$\frac{n(n-1)}{1 \cdot 2}$$

and so on. We use this idea in Example 6.

EXAMPLE 6 Using the binomial expansion

Six coins are tossed. In how many ways can exactly 2 heads come up?

SOLUTION In this case, $n = 6$ and $r = 2$. Thus the number of ways in which exactly 2 heads can come up when 6 coins are tossed is

$$\binom{6}{2} = \frac{6!}{2!4!} = \frac{6 \cdot 5}{2 \cdot 1} = 15$$ ∎

It is shown in probability theory that if p is the probability of an event occurring favorably in *one* trial, then the probability of its occurring favorably r times in n trials is

$$\binom{n}{r} p^r (1 - p)^{n-r}$$

For example, if a fair coin is tossed, the probability of its coming up heads is $\frac{1}{2}$. So, if the coin is tossed 6 times, the probability of getting exactly 4 heads is

$$\binom{6}{4} \left(\frac{1}{2}\right)^4 \left(1 - \frac{1}{2}\right)^2 = \left(\frac{6 \cdot 5}{2 \cdot 1}\right) \left(\frac{1}{16}\right) \left(\frac{1}{4}\right) = \frac{30}{128} = \frac{15}{64}$$

EXAMPLE 7   More practice using binomial expansion

A fair coin is tossed 8 times. Find the probability of getting exactly 4 heads.

SOLUTION   Here $n = 8$, $r = 4$, and $p = \frac{1}{2}$. Substituting in the probability formula, we get

$$\binom{8}{4}\left(\frac{1}{2}\right)^4\left(1 - \frac{1}{2}\right)^4 = 70\left(\frac{1}{2}\right)^8 = \frac{70}{256} = \frac{35}{128}$$ ■

GRAPH IT

You can use the numerical capabilities of your grapher in this section. Suppose you want to do Example 1. With a TI-82, go to the home screen by pressing [2nd] [QUIT] and enter the number 4. To get 4!, press [MATH], move the cursor three places to the right so you are in the PRB column and then press 4. The home screen now shows 4!. Press [ENTER] and the answer 24 appears, as shown in Window 1.

Now suppose you want to find $\binom{8}{4}$. Most graphers use the notation nCr instead of $\binom{n}{r}$. Thus to find $\binom{8}{4}$ start at the home screen, enter 8, then press [MATH], move to the PRB column, press 3 4 and [ENTER]. The answer 70 is obtained, as shown in Window 2.

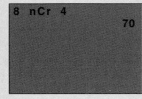

WINDOW 1   **WINDOW 2**

EXERCISE 11.4

A **In Problems 1–14, evaluate the given expression.**

1. $3!$
2. $6!$
3. $10!$
4. $2!$
5. $\frac{6!}{4!}$
6. $\frac{11!}{10!}$
7. $\frac{9!}{6!}$
8. $\frac{3!}{0!}$
9. $\binom{6}{2}$
10. $\binom{6}{5}$
11. $\binom{11}{1}$
12. $\binom{11}{0}$
13. $\binom{4}{0}$
14. $\binom{7}{3}$

B **In Problems 15–26, use the binomial expansion to expand these expressions.**

15. $(a + 3b)^4$
16. $(a - 3b)^4$
17. $(x + 4)^4$
18. $(4x - 1)^4$
19. $(2x - y)^5$
20. $(x + 2y)^5$
21. $(2x + 3y)^5$
22. $(3x - 2y)^5$
23. $\left(\frac{1}{x} - \frac{y}{2}\right)^4$
24. $\left(\frac{x}{2} + \frac{3}{y}\right)^4$

25. $(x + 1)^6$
26. $(y - 1)^6$

In Problems 27–34, find the coefficient of the indicated term in the expansion.

27. $(x - 3)^6$; x^3
28. $(x + 2)^6$; x^3
29. $(x + 2y)^7$; x^2y^5
30. $(y - 2z)^7$; y^2z^5
31. $(2x - 1)^8$; x^4
32. $(y + 2z)^8$; y^4z^4
33. $\left(\frac{a}{2} - 1\right)^5$; a^2
34. $\left(y + \frac{z}{2}\right)^5$; y^3z^2

---**APPLICATIONS**---

35. Six coins are tossed. In how many ways can exactly 3 heads come up?

36. Six coins are tossed. In how many ways can exactly 2 heads come up?

37. Nine coins are tossed. In how many ways can exactly 3 heads come up?

38. Nine coins are tossed. In how many ways can exactly 4 heads come up?

39. A fair coin is tossed 6 times. Find the probability of getting exactly 2 heads.

40. A fair coin is tossed 6 times. Find the probability of getting exactly 3 heads.

41. A fair coin is tossed 9 times. Find the probability of getting exactly 3 heads.

42. A fair coin is tossed 9 times. Find the probability of getting exactly 4 heads.

43. A single die (plural, dice) is rolled 4 times. Find the probability that a 4 comes up exactly twice. ($Note:$ The probability that a specified number, 1, 2, 3, 4, 5, or 6, comes up in a single throw is $\frac{1}{6}$.)

USING YOUR KNOWLEDGE Follow the Pattern

How do you relate the pattern used to construct Pascal's triangle with the $\binom{n}{r}$ notation? As you recall, the numbers in Pascal's triangle are obtained by adding the numbers above and to the left and right of the number in question. Using $\binom{n}{r}$ notation, this fact is written as

$$\binom{n}{k} + \binom{n}{k-1} \to \binom{n+1}{k}$$

44. Prove that

$$\binom{n}{k} + \binom{n}{k-1} = \binom{n+1}{k}$$

45. In Example 4, we used the fact that $\binom{5}{5} = \binom{5}{0}$, $\binom{5}{4} = \binom{5}{1}$ and so on. Prove that

$$\binom{n}{k} = \binom{n}{n-k}$$

where $k \le n$.

WRITE ON . . .

46. Write an explanation of how Pascal's triangle is constructed.

47. Give your own definition of $n!$

48. In the binomial expansion, which part is the "binomial"?

MASTERY TEST If you know how to do these problems, you have learned your lesson!

Evaluate:

49. $6!$

50. $\dfrac{7!}{3!}$

51. $\dbinom{6}{2}$

52. $\dbinom{7}{3}$

Expand:

53. $(2a - b)^4$ **54.** $(3a - 2b)^5$

Find the coefficient of:

55. $a^2 b^4$ in the expansion of $(2a - 3b)^6$

56. ab^3 in the expansion of $(3a - b)^4$

57. Five coins are tossed. In how many ways can exactly 4 heads come up?

58. A fair coin is tossed 7 times. Find the probability of getting exactly 5 tails.

research questions	**Sources of information for these questions can be found in the *Bibliography* at the end of this book.**

1. Write a short essay about Gauss's childhood.

2. Find out and write a report about Gauss's inventions.

3. Aside from being a superb mathematician, Gauss did some work in the field of astronomy. Report on some of his discoveries in this field.

4. In 1807, a famous French mathematician made Gauss pay an involuntary 2000-franc contribution to the French government. Find out who this famous mathematician was and the circumstances of the payment.

5. Write a report on the correspondence between Blaise Pascal and Pierre Fermat and its influence on the development of the theory of probability.

6. The binomial coefficients studied in this chapter date back to the year A.D. 1050. Write a report about the binomial coefficients and their connection to the Chinese mathematicians Chu Shih-Chien and Chia Hsien.

7. Write a report about the first triangular arrangement of binomial coefficients printed in Europe. (In what book did they appear and who was the author?)

SUMMARY

SECTION	ITEM	MEANING	EXAMPLE
11.1A	Sequence	A set of numbers arranged according to some given law	$2, 4, 6, \ldots$
	Terms $a_1, a_2, a_3, \ldots$	The numbers of a sequence	The terms of $2, 4, 6, \ldots$ are $a_1 = 2$, $a_2 = 4$, $a_3 = 6$, and so on.
	General term	The formula that generates the terms of the sequence	In the sequence $2, 4, 6, \ldots$, the general term is $a_n = 2n$.
11.1C	Σ	Summation notation	$\displaystyle\sum_{i=1}^{4} 2^i = 2 + 2^2 + 2^3 + 2^4$
11.2A	Arithmetic sequence	A sequence in which each term after the first is obtained by adding a quantity d to the preceding term	$2, 4, 6, \ldots$ is an arithmetic sequence.
	Common difference d	The difference between two successive terms of an arithmetic sequence	In the arithmetic sequence $3, 6, 9$, the common difference is $d = 3$.
	General term of an arithmetic sequence	$a_n = a_1 + (n-1)d$	In the sequence $8, 12, 16, \ldots$, the general term is $a_n = 8 + (n-1)(4)$.
11.2B	Sum S_n of an arithmetic sequence	$S_n = \dfrac{n}{2}(a_1 + a_n)$	In the sequence $2, 4, 6, \ldots$, $S_6 = \dfrac{6(2 + 12)}{2} = 42$.

SECTION	ITEM	MEANING	EXAMPLE
11.3A	Geometric sequence	A sequence in which each term after the first is formed by multiplying the preceding term by a constant r	3, 6, 12, . . . is a geometric sequence.
	Common ratio r	The ratio of two successive terms of a geometric sequence	In the geometric sequence 3, 6, 12, . . . , $r = \frac{6}{3} = 2$.
	General term of a geometric sequence	$a_n = a_1 r^{n-1}$	In the geometric sequence 3, 6, 12, . . . , $a_n = 3 \cdot 2^{n-1}$.
11.3B	Sum S_n of a geometric sequence	$S_n = \dfrac{a_1(1 - r^n)}{1 - r}$	For the geometric sequence 3, 6, 12, . . . , $$S_6 = \frac{3(1 - 2^6)}{1 - 2}$$ $$= \frac{3 \cdot (-63)}{-1}$$ $$= 189$$
11.3C	Infinite geometric series	A series of the form $a_1 + a_1 r + a_1 r^2 + \cdots + a_1 r^{n-1} + \cdots$	$2 + 1 + \frac{1}{2} + \cdots$, where $a_1 = 2$ and $r = \frac{1}{2}$
	Sum S of a geometric series with $|r| < 1$	$S = \dfrac{a_1}{1 - r}$	For $2 + 1 + \frac{1}{2} + \cdots$, $$S = \frac{2}{1 - \dfrac{1}{2}} = 4$$
11.4A	Pascal's triangle	An arrangement of numbers of the form $$\begin{array}{ccccccc} & & 1 & & 1 & & \\ & 1 & & 2 & & 1 & \\ 1 & & 3 & & 3 & & 1 \end{array}$$ $$\vdots$$	
	Binomial coefficient	$\dbinom{n}{r} = \dfrac{n!}{r!(n - r)!}$	$\dbinom{5}{2} = \dfrac{5!}{2!3!} = \dfrac{5 \cdot 4 \cdot 3 \cdot 2 \cdot \cancel{1}}{2 \cdot 1 \cdot 3 \cdot 2 \cdot \cancel{1}} = 10$
	Binomial expansion $(x + y)^n$	An expansion in which the general term is $\dbinom{n}{r} x^{n-r+1} y^{r-1}$	$(x + y)^3 = x^3 + \dfrac{3}{1} x^2 y + \dfrac{3 \cdot 2}{1 \cdot 2} x y^2$ $\qquad + \dfrac{3 \cdot 2 \cdot 1}{1 \cdot 2 \cdot 3} y^3$ $= x^3 + 3x^2 y + 3xy^2 + y^3$

REVIEW EXERCISES

(If you need help with these exercises, look in the section indicated in brackets.)

1. [11.1A] For the given sequence, find a_6 and a_n.
  a. The sequence of odd counting numbers: 1, 3, 5, 7, . . .

  b. The sequence of multiples of 3: 3, 6, 9, 12, . . .

2. [11.1A] Each card in a pack of four index cards is cut in half, and the halves are put in a single pile. This step is repeated again and again.

  a. How many cards are in the pack after the fourth cut?

  b. How many cards are in the pack after the nth cut?

3. [11.1B] Find the first three terms, the tenth term, and S_3 for the sequence whose general term is

    a. $\dfrac{1}{3}(n^2 + 1)$ **b.** $\dfrac{1}{4}(n^2 + n)$

4. [11.1B, C] Find the sequence that corresponds to the function.

    a. $a(n) = n^2 + 2, n = 1, 2, 3, \ldots$; then find $\displaystyle\sum_{n=1}^{4} (n^2 + 2)$

    b. $a(n) = 2n^2 - 1, n = 1, 2, 3, \ldots$, then find $\displaystyle\sum_{n=1}^{3} (2n^2 - 1)$

5. [11.1D] A painting doubles in value every 100 yr. Find the value of this painting in the year 2000 if it was worth $1000 in the year

    a. 1400   **b.** 1600

6. [11.2A] For the given arithmetic sequence, find d and a_{10}.

    a. $3, 6, 9, 12, \ldots$ **b.** $4, 8, 12, 16, \ldots$

7. [11.2A] For the given arithmetic sequence, find a_n.

    a. $3, 6, 9, 12, \ldots$ **b.** $4, 8, 12, 16, \ldots$

8. [11.2B] The distance (in feet) that a free-falling body falls in each second, starting with the first second, is given by the arithmetic sequence $16, 48, 80, 112, \ldots$. Find what distance the body falls in the

    a. Sixth second   **b.** Eighth second

9. [11.2B] The sixth term of an arithmetic sequence is 24. Find the first term and the common difference if the sum of the first six terms is

    a. 114   **b.** 99

10. [11.2B] The sum of the first six terms of an arithmetic sequence is 54. Find the first term and the common difference if the sixth term is

    a. 14   **b.** 16.5

11. [11.2C] A machine valued at $80,000 depreciates $8000 the first year, $7800 the second year, $7600 the third year, and so on. Find the value of the machine at the end of

    a. 8 yr   **b.** 10 yr

12. [11.2C] Juan is saving his pennies. He saved 5¢ the first week, 9¢ the second week, 13¢ the third week, and so on, in arithmetic sequence. How many weeks will it take Juan to save

    a. $2.30?   **b.** $3.24?

13. [11.3A] For the given geometric sequence, find r and a_6.

    a. $3, 6, 12, 24, \ldots$ **b.** $4, 2, 1, \dfrac{1}{2}, \ldots$

14. [11.3A] For the given geometric sequence, find r and a_n.

    a. $1, 4, 16, 64, \ldots$ **b.** $4, -2, 1, -\dfrac{1}{2}, \ldots$

15. [11.3B] For the given geometric sequence, find a_n and S_n.

    a. $2, -4, 8, -16, \ldots$ **b.** $1, -\dfrac{1}{2}, \dfrac{1}{4}, -\dfrac{1}{8}, \ldots$

16. [11.3C] Find the sum (if it exists) of the geometric series.

    a. $32 - 16 + 8 - 4 + \cdots$ **b.** $18 - 6 + 2 - \dfrac{2}{3} + \ldots$

17. [11.3C] Find the sum (if it exists) of the geometric series.

    a. $1 + (1.005) + (1.005)^2 + (1.005)^3 + \cdots$

    b. $1 - (1.001) + (1.001)^2 - (1.001)^3 + \cdots$

18. [11.3D] Find the fraction equivalent to the repeating decimal.

    a. $0.313131\ldots$ **b.** $0.324324324\ldots$

19. [11.3D] A rubber ball is dropped and takes a sequence of bounces, each one 0.5 as high as the preceding one. If the ball continues bouncing indefinitely, find the total distance it travels, given that it was dropped from a height of

    a. 8 ft   **b.** 6 ft

20. [11.4A] Find:

    a. $8!$ **b.** $\dfrac{8!}{4!}$

21. [11.4A] Find:

    a. $\dbinom{9}{3}$ **b.** $\dbinom{8}{8}$

22. [11.4A] Expand:

    a. $(a - 3b)^4$ **b.** $(3a + 2b)^4$

23. [11.4B] Find the coefficient of x^4 in the expansion of

    a. $(x + 2)^8$ **b.** $(3x - 1)^7$

24. [11.4C] Seven coins are tossed. In how many ways can exactly

    a. 4 heads come up?   **b.** 3 heads come up?

25. [11.4C] A fair coin is tossed 7 times. Find the probability of getting exactly

    a. 4 heads   **b.** 3 heads

PRACTICE TEST

(Answers are on pages 695–696.)

1. For the sequence of odd natural numbers 1, 3, 5, 7, . . . , find
   a. a_{10}, the tenth term   b. a_n, the general term

2. Each card in a pack of three index cards is cut in half, and the halves are put in a single pile. This step is repeated again and again.
   a. Write the sequence that gives the number of cards in the deck after each step.
   b. How many cards are in the pack after the fourth step?
   c. How many cards are in the pack after the nth step?

3. Find the first three terms and the eighth term of the sequence whose general term is
$$a_n = \frac{n(n + 1)}{2}$$

4. Find the sequence that corresponds to the function $a(n) = 3n - 1$, $n = 1, 2, 3, \ldots$; then find
$$\sum_{n=1}^{5} (3n - 1)$$

5. A sculpture by a famous artist doubles in value every 50 yr. Find the value of the sculpture in the year 2000 if it was worth $500 in the year 1500.

6. For the arithmetic sequence 5, 8, 11, 14, . . . , find
   a. d b. a_{10}

7. For the arithmetic sequence 5, 8, 11, 14, . . . , find a_n.

8. Sally draws a sequence of circles. She starts with a row of three circles, then adds two circles to get the second row, and adds two more circles to get the third row shown. Find how many circles she would have in the
   a. Fifth row   b. Tenth row   c. nth row

9. The distance (in feet) that a free-falling body falls in each second, starting with the first second, is given by the arithmetic sequence 16, 48, 80, 112, ... Find the distance that the body falls in 7 sec.

10. The sum of the first eight terms of an arithmetic sequence is 172 and the eighth term is 32. Find:
    a. a_1, the first term   b. d, the common difference

11. A piece of machinery valued at $50,000 depreciates $8000 the first year, $7500 the second year, $7000 the third year, and so on. Find the value of this piece of machinery at the end of 6 yr.

12. Natasha is trying to save her pennies. She saved 5¢ the first week, 8¢ the second week, 11¢ the third week, and so on, in arithmetic sequence. After a few weeks, Natasha has saved a total of $1.85. For how many weeks has she been saving?

13. For the geometric sequence 3, 1, $\frac{1}{3}$, $\frac{1}{9}$, . . . , find:
    a. r b. a_6

14. For the sequence in Problem 13, find a_n.

15. For the geometric sequence 2, -4, 8, -16, . . . , find:
    a. r b. a_8 c. S_8

16. Find the sum of the geometric series $8 - 4 + 2 - 1 + \cdots$.

17. Find the sum of the geometric series $\sum_{n=1}^{\infty} (1.001)^n = 1.001 + (1.001)^2 + (1.001)^3 + \cdots$, if the sum exists.

18. Find the fraction that is equivalent to the repeating decimal 0.312312312 . . . .

19. A rubber ball, dropped on a hard surface, takes a sequence of bounces, each one-half as high as the preceding one. The ball is dropped from a height of 12 ft, and it is assumed to continue bouncing indefinitely. Find the total distance it would travel.

20. Find:
    a. 9!   b. $\dfrac{9!}{7!}$

21. Find:
    a. $\dbinom{9}{8}$ b. $\dbinom{9}{9}$

22. Expand $(2a - b)^4$.

23. Find the coefficient of $a^4 b^3$ in the expansion of $(a - 2b)^7$.

24. Seven coins are tossed. In how many ways can exactly 4 heads turn up?

25. A fair coin is tossed 12 times. Find the probability of getting exactly 6 heads.

ANSWERS TO PRACTICE TEST

Answer	If you missed: Question	Review: Section	Examples	Page
1a. $a_{10} = 19$	1a	11.1A	1	655
1b. $a_n = 2n - 1$	1b	11.1A	1	655
2a. $6, 12, 24, \ldots$	2a	11.1A	2, 3	656
2b. 48	2b	11.1A	2, 3	656
2c. $3 \cdot 2^n$	2c	11.1A	2, 3	656
3. $a_1 = 1, a_2 = 3, a_3 = 6, a_8 = 36$	3	11.1B	4	657
4. $2, 5, 8, 11, 14, \ldots ; 40$	4	11.1B	5	657
5. \$512,000	5	11.1C	6	659
6a. $d = 3$	6a	11.2A	1	665
6b. $a_{10} = 32$	6b	11.2A	2	665
7. $a_n = 3n + 2$	7	11.2A	2	665
8a. 11	8a	11.2A	2	665
8b. 21	8b	11.2A	2	665
8c. $2n + 1$	8c	11.2A	2	665
9. 784 ft	9	11.2B	3	666–667
10a. $a_1 = 11$	10a	11.2B	4	667
10b. $d = 3$	10b	11.2B	4	667
11. \$9500	11	11.2C	5	667–668
12. 10	12	11.2C	6	668
13a. $r = \dfrac{1}{3}$	13a	11.3A	1	672
13b. $a_6 = \dfrac{1}{81}$	13b	11.3A	1	672
14. $a_n = \left(\dfrac{1}{3}\right)^{n-2}$	14	11.3A	1	672
15a. $r = -2$	15a	11.3B	2	673
15b. $a_8 = -256$	15b	11.3B	2	673
15c. $S_8 = -170$	15c	11.3B	2	673

Answer	If you missed:		Review:	
	Question	Section	Examples	Page
16. $\dfrac{16}{3}$, or $5\dfrac{1}{3}$	16	11.3C	4	676
17. Sum does not exist.	17	11.3C	5	676
18. $\dfrac{104}{333}$	18	11.3D	6	677
19. 36 ft	19	11.3D	7, 8	677–678
20a. 362,880	20a	11.4A	1	685
20b. 72	20b	11.4A	1	685
21a. 9	21a	11.4A	2	686
21b. 1	21b	11.4A	2	686
22. $16a^4 - 32a^3b + 24a^2b^2 - 8ab^3 + b^4$	22	11.4A	4	687
23. $-280a^4b^3$	23	11.4B	5	687–688
24. 35	24	11.4C	6	688
25. $\dfrac{231}{1024}$	25	11.4C	7	689

Answers to Odd-Numbered Exercises

CHAPTER 1

Exercise 1.1

A **1.** $\{1, 2\}$ **3.** $\{5, 6, 7\}$ **5.** $\{-1, -2, -3\}$
7. $\{0, 1, 2, 3\}$ **9.** $\{1, 2, 3, \ldots\}$
11. $\{x \mid x$ is an integer between 0 and 4$\}$
13. $\{x \mid x$ is an integer between -3 and 3$\}$
15. $\{x \mid x\}$ is an even integer between 19 and 78$\}$
17. False   **19.** True   **21.** True   **23.** True   **25.** True
B **27.** $0.\overline{6}$ **29.** 0.875 **31.** 2.5 **33.** $1.1\overline{6}$
C **35.** Check Rational, Real   **37.** Check Irrational, Real
39. Check Rational, Real   **41.** Check Rational, Real
43. Check Rational, Real
D **45.** -8 **47.** 7 **49.** $-\frac{3}{4}$ **51.** $\frac{1}{5}$ **53.** -0.5
55. $-0.\overline{2}$ **57.** $1.\overline{36}$ **59.** $-\pi$
E **61.** 10 **63.** 17 **65.** $\frac{3}{5}$ **67.** $0.\overline{5}$ **69.** $3.\overline{61}$
71. $-\sqrt{2}$ **73.** π
F **75.** $<$ **77.** $>$ **79.** $>$ **81.** $<$ **83.** $<$
USING YOUR KNOWLEDGE **85.** 0.11; the Burger King
87. 0.625; the Yankees   **89.** Monjane
91. False. 5 is a number, not a set of numbers. The symbol $\in$ would make the statement correct.
MASTERY TEST **97.** 0.125   **99.** -8 **101.** $+2$
103. $\{x \mid x = 3$ or an integer obtained by adding 3, 4, 5, . . . in succession$\}$ **105.** $>$

Exercise 1.2

A **1.** $\frac{2}{5}$ **3.** -0.1 **5.** 2 **7.** -0.8 **9.** $\frac{1}{5}$ **11.** -7
13. -0.3 **15.** $-\frac{13}{56}$ **17.** -14 **19.** -0.6 **21.** $-\frac{41}{63}$
23. -1 **25.** -4 **27.** -0.1 **29.** $\frac{22}{21}$ **31.** -40
33. -12 **35.** 50 **37.** 60 **39.** 40 **41.** 30 **43.** -40
45. -7.26 **47.** 2.86 **49.** $-\frac{25}{42}$ **51.** $\frac{1}{4}$ **53.** $-\frac{15}{4}$
55. -2 **57.** -4 **59.** 2 **61.** 0 **63.** Not defined
65. -2 **67.** 9 **69.** 3 **71.** 1 **73.** -4 **75.** -7
77. $-\frac{21}{20}$ **79.** $\frac{4}{7}$ **81.** $-\frac{5}{7}$ **83.** $-\frac{1}{2}$ **85.** $\frac{1}{6}$
B **87.** The commutative property of addition
89. The commutative property of multiplication
91. The commutative property of multiplication
93. The multiplicative identity   **95.** $A = a(b + c) = ab + ac$
APPLICATIONS **97.** 3500°C   **99.** $46   **101.** 14°C
SKILL CHECKER **103.** Additive inverse -7 Reciprocal $\frac{1}{7}$
105. Additive inverse 0   Reciprocal not defined
USING YOUR KNOWLEDGE **107.** -1.35 **109.** -3.045
111. 3.875
MASTERY TEST **115.** $\frac{16}{25}$ **117.** 2 **119.** $-\frac{59}{56}$
121. -12.8 **123.** 16 **125.** The sum of opposites is zero.
127. The associative property of multiplication
129. The commutative property of multiplication

Exercise 1.3

A 1. -16 3. 25 5. -125 7. 1296 9. -32

B 11. $\frac{1}{16}$ 13. $\frac{1}{125}$ 15. $\frac{1}{81}$ 17. $\frac{1}{x^6}$ 19. $\frac{1}{a^8}$

C 21. $\frac{1}{64}$ 23. $12x^2$ 25. $-15y^2$ 27. $\frac{20}{a^5}$ 29. $-\frac{30y^3}{x^2}$

31. $-\frac{24y^6}{x^4}$ 33. $-\frac{40}{a^2b^3}$ 35. -30 37. $2x^4$ 39. $\frac{a^2}{2}$

41. $-2x^3y^2$ 43. $-\frac{x}{2}$ 45. $\frac{2}{3a^3}$ 47. $\frac{3}{4}$ 49. $\frac{3b^3}{2a^6}$

D 51. $\frac{8x^9}{y^6}$ 53. $\frac{4y^6}{x^4}$ 55. $-\frac{1}{27x^9y^6}$ 57. $\frac{1}{x^{12}y^6}$

59. $x^{12}y^{12}$

E 61. $\frac{a^2}{b^6}$ 63. $-\frac{8b^6}{27a^3}$ 65. a^8b^4 67. $\frac{1}{x^{15}y^6}$

69. $x^{27}y^6$

F 71. 2.68×10^8 73. 2.4×10^{-4} 75. 8,000,000
77. 0.23 79. 2×10^3 81. 3×10^8 83. 31 yr

SKILL CHECKER 85. a. -9.856 b. -2.772
USING YOUR KNOWLEDGE 87. 3.34×10^5
89. a. 7.3 10 b. 1.23 -07

MASTERY TEST 97. $\frac{8}{x^4y^2}$ 99. $-\frac{1}{5x^{14}}$ 101. -64

103. $\frac{x^8}{9y^{10}}$ 105. $-\frac{1}{x^5}$ 107. $\frac{49x^{10}}{25y^8}$ 109. 2.4×10^{-2}

Exercise 1.4

A 1. a -26 b. -70 3. a. 27 b. 3 5. -47 7. 20
9. -15 11. -13 13. -36 15. -10 17. -4
19. 0 21. 1 23. 57 25. 3 27. -6 29. 1
B 31. -20 33. 8 35. -24 37. -33 39. 11,800
C 41. $4x - 4y$ 43. $-9a + 9b$ 45. $1.2x - 0.6$

47. $-\frac{3a}{2} + \frac{6}{7}$ or $\frac{-21a + 12}{14}$ 49. $-2x + 6y$

51. $-2.1 - 3y$ 53. $-4a - 20$ 55. $-6x - xy$
57. $-8x + 8y$ 59. $-6a + 21b$ 61. $0.5x + 0.5y - 1.0$
63. $-\frac{6}{5}a + \frac{6}{5}b - 6$ 65. $-2x + 2y - 6z - 10$
67. $-0.3x - 0.3y + 0.6z + 1.8$ 69. $-\frac{5}{2}a + 5b - \frac{5}{2}c - 5d + 5$
D 71. $9x - 6$ 73. $-9x - 8$ 75. $11L - 4W$ 77. $-2x$
79. $\frac{x}{9} + 2$ 81. $6a + 2b$ 83. $3x - 4y$ 85. $x - 5y + 36$
E 87. $-2x^3 + 9x^2 - 3x + 12$ 89. $\frac{2}{7}x^2 + \frac{4}{5}x - \frac{3}{4}$
91. $4a - 11$ 93. $-7a + 10b - 3$ 95. $-4.8x + 3.4y + 5$
SKILL CHECKER 97. 16 99. -2
USING YOUR KNOWLEDGE 101. $v_a = \frac{1}{2}v_1 + \frac{1}{2}v_2$
103. $KE = \frac{1}{2}mv_1^2 + \frac{1}{2}mv_2^2$
MASTERY TEST 107. -3 109. $-2x + 6y - 4z + 8$
111. $\frac{x}{4} + 5$ 113. $-2a^3 + 2a^2 - a - 17$ 115. 6

Review Exercises

1. a. $\{4, 5, 6, 7, 8\}$ b. $\{5, 6, 7\}$ 2. a. 0.2 b. 0.4
3. a. $0.\overline{1}$ b. $0.\overline{2}$
4. 0.3 Rational, Real
   0 Whole, Integer, Rational, Real
   $-\frac{3}{4}$ Rational, Real
   -5 Integer, Rational, Real
   $\sqrt{3}$ Irrational, Real
5. a. 3.5 b. $-\frac{3}{4}$ 6. a. 9 b. 4.2 7. a. $\frac{1}{8}$ b. $0.\overline{4}$
8. a. $<$ b. $<$ 9. a. $>$ b. $=$ 10. a. $<$ b. $=$
11. a. -11 b. -3 12. a. $-\frac{2}{7}$ b. -0.6 13. a. 12 b. 4
14. a. $\frac{19}{20}$ b. $\frac{13}{12}$ 15. a. -36 b. -14.4 16. a. $-\frac{21}{32}$ b. $\frac{5}{21}$
17. a. 0 b. Not defined 18. a. $-\frac{5}{3}$ b. $\frac{1}{0.3}$ or $\frac{10}{3}$
19. a. $-\frac{9}{4}$ b. -3 20. a. Commutative property of addition
b. Associative property of addition

21. a. 81 b. -81 22. a. 1 b. 1 23. a. $-\frac{1}{512}$ b. $\frac{1}{x^{10}}$

24. a. $-\frac{15y^{10}}{x^4}$ b. $-\frac{24}{x^{11}y^8}$ 25. a. $\frac{3}{x^2}$ b. $-4x^{11}$

26. a. $-\frac{x}{3}$ b. $-\frac{2}{x^{11}}$ 27. a. $-\frac{8x^{21}}{y^{18}}$ b. $\frac{16}{x^{24}y^{24}}$

28. a. $\frac{1}{x^{24}y^{12}}$ b. $x^{25}y^{15}$ 29. a. 3.4×10^5 b. 4.7×10^{-5}

30. a. 37,000 b. 0.0078 31. a. -75 b. 50
32. a. 9 b. 20 33. a. -7 b. -73 34. a. $-3x + 21$
b. $2x + 17$ 35. a. $-x + 10y - 16$ b. $3x^2 + 3x + 8$

CHAPTER 2
Exercise 2.1

A 1. Yes 3. Yes 5. Yes 7. No 9. No
B 11. 4 13. 1 15. 2 17. 2 19. 7 21. -1
23. -8 25. -4 27. 0 29. 0 31. $\frac{1}{13}$ 33. 8
35. $-\frac{11}{80}$
C 37. 24 39. 1 41. 15 43. 10 45. 4 47. 10
49. -5 51. 0 53. -1 55. $-\frac{10}{11}$ 57. $\frac{5}{2}$
59. An identity
D 61. 4 63. -6 65. 57 67. 0 69. 16
71. $\frac{43}{2}$ or 21.5 73. 1500 75. 6
SKILL CHECKER 77. $-3x - 7$ 79. 2 81. 50 83. 5
USING YOUR KNOWLEDGE 85. 11 in. 87. 8 89. 12 in.
91. a. 47 b. 48
MASTERY TEST 95. 10 97. 16 99. -1 101. No

Exercise 2.2

A 1. $h = \frac{V}{\pi r^2}$ 3. $W = \frac{V}{LH}$ 5. $b = P - s_1 - s_2$

7. $s = \frac{A}{\pi r} - r = \frac{A - \pi r^2}{\pi r}$ 9. $V_2 = \frac{P_1 V_1}{P_2}$ 11. $P_2 = \frac{P_1 V_1}{V_2}$

13. a. $T = \dfrac{D}{R}$ **b.** 4 hr **15. a.** $A = 34 - 2H$ **b.** 18 yr

17. a. CGS = (OPM)(NS) − OE **b.** $29,500

19. a. $U = \dfrac{F - PF}{P}$ **b.** 1000

21. a. $L = \dfrac{H - 32}{1.88}$ **b.** No **c.** Yes **d.** $L = \dfrac{H - 29}{1.95}$ **e.** Yes

23. 68.4 in.

B 25. a. $C = 0.36 + 0.28(t - 1), t \geq 1$ **b.** 14 min

27. a. $F = 20 + 10m$ **b.** $m = 12$

C 29. $x = -5, (15 - 4x)° = (25 - 2x)° = 35°$

31. $x = 20, (80 + 3x)° = (40 + 5x)° = 140°$

33. $x = 3, (14x + 8)° = (16x + 2)° = 50°$

35. $x = -5.5, (50 - 10x)° = 105°, (42 - 6x)° = 75°$

SKILL CHECKER 37. $3n + 6$ **39.** $p = 25,000$

41. $m = 70$ **43.** $x = 18$

USING YOUR KNOWLEDGE 45. 2025 **47.** 3.04 ft

MASTERY TEST 51. $x = -20, (90 - 4x)° = (50 - 6x)° = 170°$

53. a. $C = 32 + 23(w - 1), w = 1, 2, 3, \ldots$

b. $w = 1 + \dfrac{C - 32}{23}$ or $w = \dfrac{C - 9}{23}$ **c.** $2.39

55. a. $C = 6S + 4$ **b.** 16°C

Exercise 2.3

A 1. $4m = m + 18$ **3.** $x + (x - 3) = 7$

5. $4x + 5 = 29, x = 6$ **7.** $3x + 8 = 29, x = 7$

9. $3x - 2 = 16, x = 6$ **11.** $5x = 12 + 2x, x = 4$

13. $\frac{1}{3}x - 2 = 10, x = 36$

B 15. 2.71 million lb

17. Russia has 6575 ships. Japan has 8851 ships. **19.** 130

C 21. 44, 46, 48 **23.** −11, −9, −7 **25.** 87 and 92

27. 47 **29.** 937 **31.** 168 lb

D 33. 12 ft by 5000 ft **35.** 518 ft by 716 ft **37.** 72°

39. 135° **41.** $x = 19$ $(3x + 10)° = 67°$ $(2x - 15)° = 23°$

43. $x = 39$ $(3x + 10)° = 127°$ $(2x - 25)° = 53°$

45. 30°, 60°, and 90°

SKILL CHECKER 47. $n = 20$ **49.** $W = 20$ **51.** $x = 1200$

53. $T = 8$

USING YOUR KNOWLEDGE 55. 5

MASTERY TEST 63. 31, 33, and 35 **65.** $\dfrac{x - 5}{x + 8} = 2$

67. $x = 27$ $(3x + 4)° = 85°$ $(4x - 18)° = 95°$

69. $n = 12.60$ million students

Exercise 2.4

A 1. a. 64.5 **b.** 9.75 **3.** $10.81 **5.** 98.8 million homes

7. $23.13 **9.** Markup, $24, 44.4̄% of selling price **11.** 40%

13. 50 **15.** 20% **17.** $275,000 **19.** $95.04 billion

B 21. $6000 at 7%, $9000 at 5%

23. $8000 at 7.5%, $17,000 at 6%

C 25. 360 km **27.** 60 mi **29.** 300 mi

D 31. 6 **33.** 30 lb copper, 40 lb zinc **35.** 20

SKILL CHECKER 37. a. > **b.** > **c.** >

USING YOUR KNOWLEDGE 39. $229.005 billion

MASTERY TEST 47. 10 gal **49.** 175 mi

Exercise 2.5

A 1. $x > 3$ $(3, \infty)$

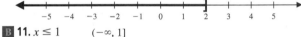

3. $x \leq -3$ $(-\infty, -3]$

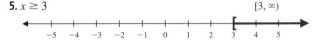

5. $x \geq 3$ $[3, \infty)$

7. $x \geq -1$ $[-1, \infty)$

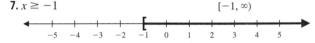

9. $x \leq 2$ $(-\infty, 2]$

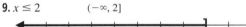

B 11. $x \leq 1$ $(-\infty, 1]$

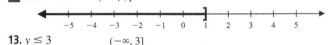

13. $y \leq 3$ $(-\infty, 3]$

15. $x \geq 5$ $[5, \infty)$

17. $a \leq 2$ $(-\infty, 2]$

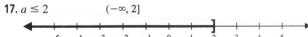

19. $z \leq -4$ $(-\infty, -4]$

21. $x \leq -1$ $(-\infty, -1]$

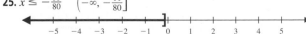

23. $x \leq 0$ $(-\infty, 0]$

25. $x \leq -\frac{11}{80}$ $\left(-\infty, -\frac{11}{80}\right]$

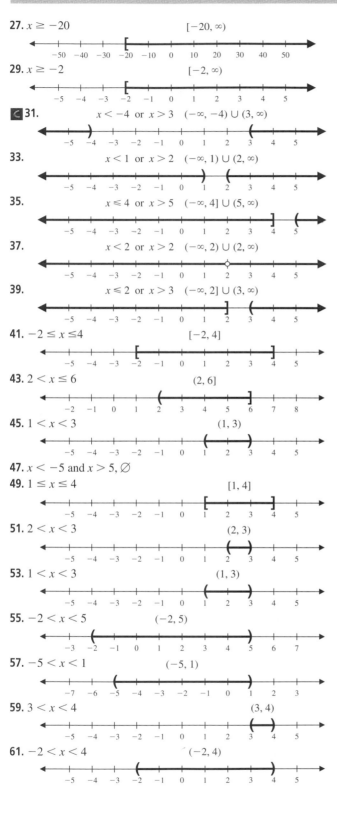

27. $x \geq -20$ $[-20, \infty)$

29. $x \geq -2$ $[-2, \infty)$

31. $x < -4$ or $x > 3$ $(-\infty, -4) \cup (3, \infty)$

33. $x < 1$ or $x > 2$ $(-\infty, 1) \cup (2, \infty)$

35. $x \leq 4$ or $x > 5$ $(-\infty, 4] \cup (5, \infty)$

37. $x < 2$ or $x > 2$ $(-\infty, 2) \cup (2, \infty)$

39. $x \leq 2$ or $x > 3$ $(-\infty, 2] \cup (3, \infty)$

41. $-2 \leq x \leq 4$ $[-2, 4]$

43. $2 < x \leq 6$ $(2, 6]$

45. $1 < x < 3$ $(1, 3)$

47. $x < -5$ and $x > 5$, $\varnothing$

49. $1 \leq x \leq 4$ $[1, 4]$

51. $2 < x < 3$ $(2, 3)$

53. $1 < x < 3$ $(1, 3)$

55. $-2 < x < 5$ $(-2, 5)$

57. $-5 < x < 1$ $(-5, 1)$

59. $3 < x < 4$ $(3, 4)$

61. $-2 < x < 4$ $(-2, 4)$

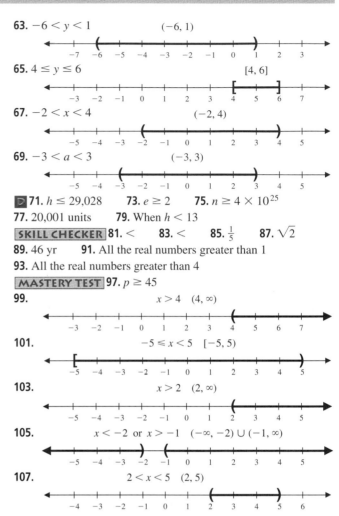

63. $-6 < y < 1$ $(-6, 1)$

65. $4 \leq y \leq 6$ $[4, 6]$

67. $-2 < x < 4$ $(-2, 4)$

69. $-3 < a < 3$ $(-3, 3)$

71. $h \leq 29{,}028$ **73.** $e \geq 2$ **75.** $n \geq 4 \times 10^{25}$

77. 20,001 units   **79.** When $h < 13$

SKILL CHECKER **81.** $<$ **83.** $<$ **85.** $\frac{1}{5}$ **87.** $\sqrt{2}$

89. 46 yr   **91.** All the real numbers greater than 1

93. All the real numbers greater than 4

MASTERY TEST **97.** $p \geq 45$

99. $x > 4$ $(4, \infty)$

101. $-5 \leq x < 5$ $[-5, 5)$

103. $x > 2$ $(2, \infty)$

105. $x < -2$ or $x > -1$ $(-\infty, -2) \cup (-1, \infty)$

107. $2 < x < 5$ $(2, 5)$

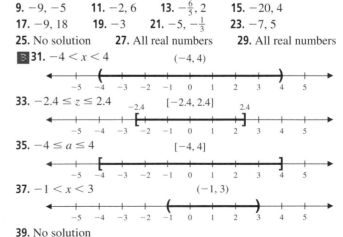

Exercise 2.6

1. $-13, 13$ **3.** $-2.3, 2.3$ **5.** 0 **7.** No solution

9. $-9, -5$ **11.** $-2, 6$ **13.** $-\frac{6}{5}, 2$ **15.** $-20, 4$

17. $-9, 18$ **19.** -3 **21.** $-5, -\frac{1}{3}$ **23.** $-7, 5$

25. No solution   **27.** All real numbers   **29.** All real numbers

31. $-4 < x < 4$ $(-4, 4)$

33. $-2.4 \leq z \leq 2.4$ $[-2.4, 2.4]$

35. $-4 \leq a \leq 4$ $[-4, 4]$

37. $-1 < x < 3$ $(-1, 3)$

39. No solution

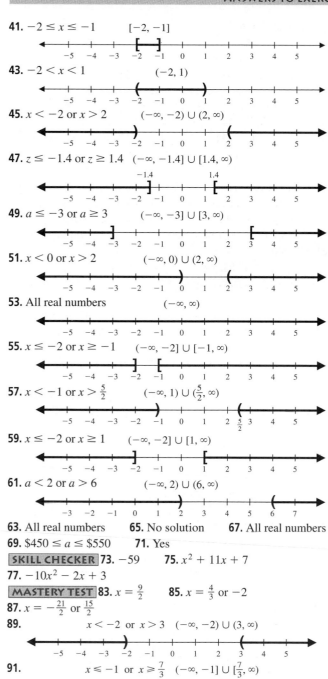

41. $-2 \le x \le -1$ $\qquad$ [−2, −1]

43. $-2 < x < 1$ $\qquad$ (−2, 1)

45. $x < -2$ or $x > 2$ $\qquad$ $(-\infty, -2) \cup (2, \infty)$

47. $z \le -1.4$ or $z \ge 1.4$ $\quad$ $(-\infty, -1.4] \cup [1.4, \infty)$

49. $a \le -3$ or $a \ge 3$ $\qquad$ $(-\infty, -3] \cup [3, \infty)$

51. $x < 0$ or $x > 2$ $\qquad$ $(-\infty, 0) \cup (2, \infty)$

53. All real numbers $\qquad$ $(-\infty, \infty)$

55. $x \le -2$ or $x \ge -1$ $\quad$ $(-\infty, -2] \cup [-1, \infty)$

57. $x < -1$ or $x > \frac{5}{2}$ $\qquad$ $(-\infty, 1) \cup (\frac{5}{2}, \infty)$

59. $x \le -2$ or $x \ge 1$ $\qquad$ $(-\infty, -2] \cup [1, \infty)$

61. $a < 2$ or $a > 6$ $\qquad$ $(-\infty, 2) \cup (6, \infty)$

63. All real numbers $\qquad$ **65.** No solution $\qquad$ **67.** All real numbers

69. $\$450 \le a \le \$550$ $\qquad$ **71.** Yes

SKILL CHECKER **73.** -59 $\qquad$ **75.** $x^2 + 11x + 7$

77. $-10x^2 - 2x + 3$

MASTERY TEST **83.** $x = \frac{9}{2}$ $\qquad$ **85.** $x = \frac{4}{3}$ or -2

87. $x = -\frac{21}{2}$ or $\frac{15}{2}$

89. $\qquad$ $x < -2$ or $x > 3$ $\quad$ $(-\infty, -2) \cup (3, \infty)$

91. $\qquad$ $x \le -1$ or $x \ge \frac{7}{3}$ $\quad$ $(-\infty, -1] \cup [\frac{7}{3}, \infty)$

Review Exercises

1. a. No $\quad$ **b.** No $\quad$ **c.** Yes $\qquad$ **2. a.** 12 $\quad$ **b.** 15 $\quad$ **c.** 18

3. a. 2 $\quad$ **b.** 3 $\quad$ **c.** 4 $\qquad$ **4. a.** 14 $\quad$ **b.** 21 $\quad$ **c.** 28

5. a. 3 $\quad$ **b.** 3 $\quad$ **c.** 3 $\qquad$ **6. a.** $P = 500$ $\quad$ **b.** $P = 2000$ $\quad$ **c.** $P = 4000$

7. a. $h = \dfrac{H - 72.48}{2.5}; 4$ $\quad$ **b.** $h = \dfrac{H - 77.48}{2.5}; 2$

c. $h = \dfrac{H - 84.98}{2.5}; -1$

8. a. $A = \dfrac{7B + 14}{2}$ $\quad$ **b.** $A = \dfrac{7B + 21}{3}$ $\quad$ **c.** $A = \dfrac{7B + 28}{4}$

9. $L = \dfrac{P - 2W}{2}$

a. 40 ft by 50 ft $\quad$ **b.** 50 ft by 60 ft $\quad$ **c.** 60 ft by 70 ft

10. a. $x = 10; 50°$ $\quad$ **b.** $x = 15; 57°$ $\quad$ **c.** $x = 20; 58°$

11. a. 100 mi $\quad$ **b.** 150 mi $\quad$ **c.** 200 mi

12. a. 49, 51, 53 $\quad$ **b.** 51, 53, 55 $\quad$ **c.** 67, 69, 71

13. a. $\$20,000$ $\quad$ **b.** $\$30,000$ $\quad$ **c.** $\$15,000$

14. a. $\$5000$ in bonds, $\$15,000$ in CDs $\quad$ **b.** $\$17,000$ in bonds, $\$3000$ in CDs $\quad$ **c.** $\$10,000$ in bonds, $\$10,000$ in CDs

15. a. 200 mi from town $\quad$ **b.** 300 mi from town

c. 120 mi from town $\qquad$ **16. a.** 100 L $\quad$ **b.** 25 L $\quad$ **c.** 0 L

17. a. $x \ge -2$ $\qquad$ [−2, ∞)

b. $x \ge -3$ $\qquad$ [−3, ∞)

c. $x \ge -4$ $\qquad$ [−4, ∞)

18. a. $x \ge -3$ $\qquad$ [−3, ∞)

b. $x \ge -2$ $\qquad$ [−2, ∞)

c. $x \ge -2$ $\qquad$ [−2, ∞)

19. a. $x > 3$ $\qquad$ (3, ∞)

b. $x > 3$ $\qquad$ (3, ∞)

c. $x > 3$ $\qquad$ (3, ∞)

20. a. $-1 < x < 2$ $\qquad$ (−1, 2)

b. $-2 < x < 3$ $\qquad$ (−2, 3)

c. $-3 < x < 4$ $\qquad$ (−3, 4)

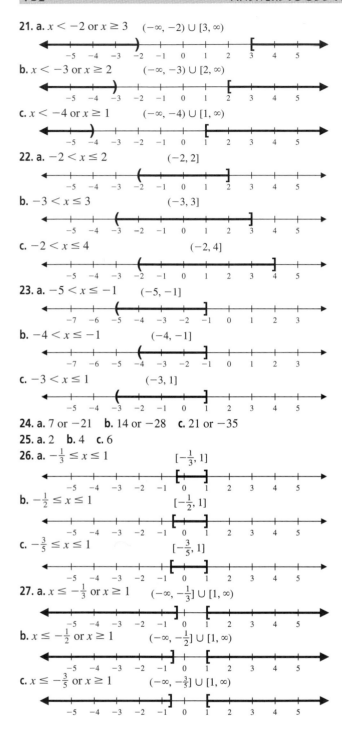

21. a. $x < -2$ or $x \geq 3$ $(-\infty, -2) \cup [3, \infty)$

b. $x < -3$ or $x \geq 2$ $(-\infty, -3) \cup [2, \infty)$

c. $x < -4$ or $x \geq 1$ $(-\infty, -4) \cup [1, \infty)$

22. a. $-2 < x \leq 2$ $(-2, 2]$

b. $-3 < x \leq 3$ $(-3, 3]$

c. $-2 < x \leq 4$ $(-2, 4]$

23. a. $-5 < x \leq -1$ $(-5, -1]$

b. $-4 < x \leq -1$ $(-4, -1]$

c. $-3 < x \leq 1$ $(-3, 1]$

24. a. 7 or -21 **b.** 14 or -28 **c.** 21 or -35

25. a. 2   **b.** 4   **c.** 6

26. a. $-\frac{1}{3} \leq x \leq 1$ $[-\frac{1}{3}, 1]$

b. $-\frac{1}{2} \leq x \leq 1$ $[-\frac{1}{2}, 1]$

c. $-\frac{3}{5} \leq x \leq 1$ $[-\frac{3}{5}, 1]$

27. a. $x \leq -\frac{1}{3}$ or $x \geq 1$ $(-\infty, -\frac{1}{3}] \cup [1, \infty)$

b. $x \leq -\frac{1}{2}$ or $x \geq 1$ $(-\infty, -\frac{1}{2}] \cup [1, \infty)$

c. $x \leq -\frac{3}{5}$ or $x \geq 1$ $(-\infty, -\frac{3}{5}] \cup [1, \infty)$

CHAPTER 3
Exercise 3.1

A B **1.** Monomial; 4   **3.** Binomial; 3   **5.** Trinomial; 3
7. Trinomial; 5   **9.** Zero polynomial; no degree
B C **11.** $x^3 + y^4$; 4   **13.** $x^4 + y^4 + z^4$; 4

15. $3x^2 - 4y^3 + 5x^2y^2$; 4
D **17.** -8 **19.** 4   **21.** -1 **23.** 16   **25.** -136
27. a. 0   **b.** -9 **c.** -9 **29. a.** 8   **b.** 2   **c.** 6
E **31.** $6x^2 - 5$ **33.** $x^2 - 2x - 4$ **35.** $3x^2 + 4x - 9$
37. $-11y^2 - y - 3$ **39.** $4x^3 - 12x^2 + 9x - 6$
41. $7y^2 + 12y - 11$ **43.** $3v^3 - 2v^2 + v - 7$
45. $-5u^3 - 5u^2 - 3u + 10$ **47.** $4x^3 + 2xy - 1$
49. $x^3 - x^2 - 4$ **51.** $-2a^2 + 5a$ **53.** $4y$ **55.** $3x^2 + 7y$
57. $P(0) = 3$ $Q(0) = -1$ $P(0) + Q(0) = 3 + (-1) = 2$
59. $P(x) - P(x) = 0$ **61.** Commutative property $(+)$
63. Distributive property   **65.** Associative property $(+)$
67. Commutative property $(+)$ **69.** Distributive property
F **71. a.** 48 ft   **b.** 64 ft   **73. a.** \$42,500   **b.** \$70,000
75. a. \$50,000   **b.** \$0 (No value)   **77.** \$20   **79.** \$1000
81. \$30,000

SKILL CHECKER **83.** $18x^3y^2$ **85.** $14x^4y^2$
USING YOUR KNOWLEDGE **87.** $2x^2 + 6x$ **89.** $3x^2 + 4x$
91. $27x^2 + 10x$
MASTERY TEST **95. a.** $P = -0.2x^2 + 47x - 100$ **b.** \$2600
97. 4   **99. a.** Binomial   **b.** Polynomial   **c.** Monomial
d. Trinomial   **101.** $P(2) - Q(2) = -3$
103. $P(x) - Q(x) = -5x + 7$ **105.** $y^3 + 5xy - xyz^2$; degree 4

Exercise 3.2

A **1.** $12x^2 - 6x$ **3.** $-3x^3 + 9x^2$ **5.** $-24x^3 + 16x^2 - 8x$
7. $-18x^3y^2 - 9xy^4 + 21xy^2$ **9.** $6x^3y^6 - 10x^2y^5 + 2x^2y^4$
B **11.** $x^3 + 4x^2 + 8x + 15$ **13.** $x^3 + 3x^2 - x + 12$
15. $x^3 + 2x^2 - 5x - 6$ **17.** $x^3 - 8$
19. $x^4 - x^3 + x^2 + x - 2$
C **21.** $9x^2 + 9x + 2$ **23.** $5x^2 + 11x - 12$
25. $3a^2 + 14a - 5$ **27.** $2y^2 + 7y - 15$ **29.** $x^2 - 8x + 15$
31. $6x^2 - 7x + 2$ **33.** $4x^2 + 4ax - 15a^2$
35. $x^2 + 15x + 56$ **37.** $4a^2 + 10ab + 4b^2$
D E **39.** $16u^2 + 8uv + v^2$ **41.** $4y^2 + 4yz + z^2$
43. $9a^2 - 6ab + b^2$ **45.** $a^2 - b^2$ **47.** $25x^2 - 4y^2$
49. $b^2 - 9a^2$ **51.** $3x^3 + 9x^2 + 6x$ **53.** $-3x^3 + 12x^2 - 9x$
55. $x^3 + 6x^2 + 9x$ **57.** $-2x^3 + 4x^2 - 2x$ **59.** $4x^2y^2 - y^4$
61. $x^2 + \frac{3}{2}x + \frac{9}{16}$ **63.** $4y^2 - \frac{4}{5}y + \frac{1}{25}$
65. $\frac{9}{16}p^2 + \frac{3}{10}pq + \frac{1}{25}q^2$
67. $9x^2 + 24xy + 16y^2 + 6x + 8y + 1$
69. $9x^2 - 24xy + 16y^2 - 6x + 8y + 1$
71. $9x^2 + 12xy + 4y^2 - 6x - 4y + 1$
73. $16p^2 - 24pq + 9q^2 + 8p - 6q + 1$
F **75. a.** $R = 1000p - 30p^2$ **b.** \$8000   **77.** $T_1^4 - T_2^4$
79. $Kt_n^2 - 2Kt_nt_a + Kt_a^2$
SKILL CHECKER **81.** $5(x + y)$ **83.** $3a(b + c)$
USING YOUR KNOWLEDGE **85. a.** 9   **b.** 5   **c.** No
87. a. x^2 **b.** xy **c.** y^2 **d.** xy **89.** $(x + y)^2 = x^2 + 2xy + y^2$
MASTERY TEST **95.** $12x^2 + 17xy + 6y^2$
97. $25x^2 - 30xy + 9y^2$ **99.** $9x^2 - y^2$

101. $x^3 - 6x^2 + 5x + 6$ **103.** $20x^6 + 12x^5 - 8x^4 - 20x^3$
105. $9x^2 - 24xy + 16y^2 + 6x - 8y + 1$

Exercise 3.3

1. $8(x + 2)$ **3.** $9(y - 2)$ **5.** $-5(y - 5)$
7. $-8(x + 3)$ **9.** $4x(x + 9)$ **11.** $6x(1 - 7x^2)$
13. $-5x^2(1 + 7x^2)$ **15.** $3x(x^2 + 2x + 13)$
17. $9y(7y^2 - 2y + 3)$ **19.** $6x^2(6x^4 + 2x^3 - 3x^2 + 5)$
21. $8y^3(6y^5 + 2y^2 - 3y + 1)$ **23.** $\frac{1}{7}(4x^3 + 3x^2 - 9x + 3)$
25. $\frac{1}{8}y^2(7y^7 + 3y^4 - 5y^2 + 5)$
27. $(x + 2)(x^2 + 1)$ **29.** $(y - 3)(y^2 + 1)$
31. $(2x + 3)(2x^2 + 1)$ **33.** $(3x - 1)(2x^2 + 1)$
35. $(y + 2)(4y^2 + 1)$ **37.** $(2a^2 + 3)(a^4 + 1)$
39. $(x^2 + 4)(3x^3 + 1)$ **41.** $(2y^2 + 3)(3y^3 + 1)$
43. $y^2(y^2 + 3)(4y^3 + 1)$ **45.** $a^2(a^2 - 2)(3a^3 - 2)$
47. $a^2(4a^2 - 5)(2a - 3)$ **49.** $x^3(x - 2)(x^2 + 2)$
51. $(x - 4)(2x + 5)$ **53.** $\alpha L(t_2 - t_1)$ **55.** $(R - 1)(R - 1)$
SKILL CHECKER **57.** $x^2 + 7x + 12$ **59.** $x^2 + 3x - 10$
61. $25x^2 + 20xy + 4y^2$ **63.** $25x^2 - 20xy + 4y^2$
65. $u^2 - 36$
USING YOUR KNOWLEDGE **67.** $-w(l - z)$ **69.** $a(a + 2s)$
71. $-16(t^2 - 5t - 15)$
MASTERY TEST **77.** $x(3x^3 + 1)(2x^2 - 3)$
79. $(2x - 3)(3x^2 - 1)$ **81.** $3x^2(4x^5 + x^4 - 2x^3 + 9)$
83. $6(x - 8)$ **85.** $-\frac{1}{5}x^2(x^3 + 4x^2 - 2)$ or $\frac{1}{5}x^2(2 - 4x^2 - x^3)$

Exercise 3.4

1. $(x + 2)(x + 3)$ **3.** $(a + 2)(a + 5)$ **5.** $(x + 4)(x - 3)$
7. $(x + 2)(x - 1)$ **9.** $(x + 1)(x - 2)$ **11.** $(x + 2)(x - 5)$
13. $(a - 7)(a - 9)$ **15.** $(y - 2)(y - 11)$
17. $(9x + 1)(x + 4)$ **19.** $(3a + 1)(a - 2)$
21. $(2y + 5)(y - 4)$ **23.** $(4x - 3)(x - 2)$
25. $(2x + 3)(3x - 4)$ **27.** $(3a + 2)(7a - 1)$
29. $(2x + 3y)(3x - y)$ **31.** $x^2(7x - 3y)(x - y)$
33. $y^3(3x + y)(5x - 2y)$ **35.** $xy^2(15x^2 - 2xy - 2y^2)$
37. $-(2b - 5)(b - 4)$ **39.** $-(4y - 3)(3y + 4)$
41. $(2y + 7)(y + 1)$ **43.** $(2x - 3)(x - 3)$ **45.** $-(a + 1)^4$
47. $(2g + 7)(g - 3)$ **49.** $(2R - 1)(R - 1)$
SKILL CHECKER **51.** $4a^2 + 4ab + b^2$ **53.** $a^2 - 4ab + 4b^2$
55. $a^2 - b^2$ **57.** $4x^2 - 9y^2$
USING YOUR KNOWLEDGE **59.** $(L - 3)(2L - 3)$
61. $(5t - 7)(t - 1)$
MASTERY TEST **65.** $(3x + 2)(2x + 3)$ **67.** $(x - 5)(x + 2)$
69. Not factorable   **71.** $2x^2y(2x - y)(3x + 2y)$
73. Not factorable   **75.** $-(4x - 7)(2x + 3)$
77. $(2y - 5)(y - 1)$

Exercise 3.5

1. $(x + 1)^2$ **3.** $(y + 11)^2$ **5.** $(1 + 2x)^2$ **7.** $(3x + 5y)^2$
9. $(6a + 4)^2 = 4(3a + 2)^2$ **11.** $(y - 1)^2$ **13.** $(7 - x)^2$

15. $(7a - 2x)^2$ **17.** $(4x - 3y)^2$ **19.** $(3x^2 + 2)^2$
21. $(4x^2 - 3)^2$ **23.** $(1 + x^2)^2$
25. $(y + 8)(y - 8)$ **27.** $\left(a + \frac{1}{3}\right)\left(a - \frac{1}{3}\right)$
29. $(8 + b)(8 - b)$ **31.** $(6a + 7b)(6a - 7b)$
33. $\left(\frac{x}{3} + \frac{y}{4}\right)\left(\frac{x}{3} - \frac{y}{4}\right)$ **35.** $(a + 2b + c)(a + 2b - c)$
37. $(2x - y + 1)(2x - y - 1)$
39. $(3y - 2x + 5)(3y - 2x - 5)$
41. $(4a + x + 3y)(4a - x - 3y)$ **43.** $(y + a - b)(y - a + b)$
45. $(x + 5)(x^2 - 5x + 25)$ **47.** $(1 + a)(1 - a + a^2)$
49. $(2x + y)(4x^2 - 2xy + y^2)$ **51.** $(x - 1)(x^2 + x + 1)$
53. $(5a - 2b)(25a^2 + 10ab + 4b^2)$
55. $(x + 2)(x - 2)(x^2 + 2x + 4)(x^2 - 2x + 4)$
57. $\left(x + \frac{1}{2}\right)\left(x - \frac{1}{2}\right)\left(x^2 + \frac{1}{2}x + \frac{1}{4}\right)\left(x^2 - \frac{1}{2}x + \frac{1}{4}\right)$
59. $\left(\frac{x}{2} + 1\right)\left(\frac{x}{2} - 1\right)\left(\frac{x^2}{4} + \frac{1}{2}x + 1\right)\left(\frac{x^2}{4} - \frac{1}{2}x + 1\right)$
61. $(x - y + 1)(x^2 - 2xy + y^2 - x + y + 1)$
63. $(1 + x + 2y)(1 - x - 2y + x^2 + 4xy + 4y^2)$
65. $(y - 2x - 1)(y^2 - 4xy + 4x^2 + y - 2x + 1)$
67. $(3 - x - 2y)(9 + 3x + 6y + x^2 + 4xy + 4y^2)$
69. $(4 + x^2 - y^2)(16 - 4x^2 + 4y^2 + x^4 - 2x^2y^2 + y^4)$
SKILL CHECKER **71.** $x^2 - 2x - 15$ **73.** $x^2 - 6x - 16$
75. $4x^2 - 12xy + 9y^2$
USING YOUR KNOWLEDGE **77.** $(10 + x)(10 - x)$
79. $(2x + 1)(4x^2 - 2x + 1)$
MASTERY TEST **87.** $(2x + 3y)(4x^2 - 6xy + 9y^2)$
89. $(x^2 + 1)(x^4 - x^2 + 1)$
91. $(2x - 3y)(2x + 3y)(4x^2 + 9y^2)$
93. $(x + 5y - 4)(x + 5y + 4)$ **95.** $(2x - 3y)^2$
97. $(x + 9)^2$ **99.** $4xy$
101. $(5 + x + y)(25 - 5x - 5y + x^2 + 2xy + y^2)$

Exercise 3.6

1. $3x^2(x + 2)(x - 3)$ **3.** $5x^2(x + 4y)(x - 2y)$
5. $-3x^4(x^2 + 2x + 7)$ **7.** $2x^4y(x^2 - 2xy - 5y^2)$
9. $-2x^4(2x^2 + 6xy + 9y^2)$ **11.** $2y^2(3x^2 + 1)(x + 2)$
13. $-3xy(3x^2 + 2)(x + 1)$ **15.** $-2x(2x^2 - y)(x + y)$
17. $3y^2(x + 4y)^2$ **19.** $-2k(3x + 2y)^2$ **21.** $4xy^2(2x - 3y)^2$
23. Not factorable   **25.** $3x^3(x + 2y)^2$ **27.** $2x^4(3x + y)^2$
29. $3x^2y^2(2x - 3y)^2$ **31.** $6(x + 2)(x + 1)(x - 1)$
33. $7(x^2 + y^2)(x + y)(x - y)$
35. $2x^2(x^2 + 4y^2)(x + 2y)(x - 2y)$ **37.** $-2(x + 3)^2$
39. $-3(x + 2)^2$ **41.** $-x^2(2x + y)^2$ **43.** $-y^2(3x + 2y)^2$
45. $-2y^2(2x - 3y)^2$ **47.** $-2x(3x + 2y)^2$
49. $-2x(3x + 5y)^2$ **51.** $-x(x + y)(x - y)$
53. $-x^2(x + 2y)(x - 2y)$ **55.** $-x^2(2x + 3y)(2x - 3y)$
57. $-2x(2x + 3y)(2x - 3y)$ **59.** $-2x^2(3x + 2y)(3x - 2y)$
61. $x^2(3 - x)(9 + 3x + x^2)$ **63.** $x^4(x - 2)(x^2 + 2x + 4)$

65. $x^4(3 + 2x)(9 - 6x + 4x^2)$

67. $x^4(3x + 4y)(9x^2 - 12xy + 16y^2)$

69. $(x - y + 2)(x + y + 2)$ **71.** Prime (the sum of two squares)

73. $(x - y - 2)(x + y + 2)$ **75.** $-(3x - 5y)^2$

77. $2x(3x - 5y)^2$

SKILL CHECKER **79.** $(2x + 1)(3x - 2)$ **81.** $(3x - 1)(4x + 1)$

USING YOUR KNOWLEDGE **83.** $\dfrac{2\pi A}{360}(R + Kt)$

85. $\dfrac{3S}{2bd^3}(d + 2z)(d - 2z)$

MASTERY TEST **91.** $(x - 2)(3x - 1)(3x + 1)$

93. $x^2(2x^2 + xy + 2y^2)$ **95.** $(x - y - 5)(x + y - 5)$

97. $-(4y - 7)(2y + 3)$ **99.** $4x^3(3x + y)^2$ **101.** $8x^3(x^2 + 9)$

Exercise 3.7

A **1.** $-1, -2$ **3.** $1, -4, -3$ **5.** $\frac{1}{2}, \frac{1}{3}$ **7.** $0, 3$

9. $8, -8$ **11.** $9, -9$ **13.** $0, -6$ **15.** $0, 3$ **17.** $3, 9$

19. $-1, -5$ **21.** $5, -3$ **23.** $-\frac{2}{3}, -1$ **25.** $1, \frac{1}{2}$

27. $1, -\frac{1}{2}$ **29.** $2, -6$ **31.** $1, \frac{1}{2}$ **33.** $-2, -4$ **35.** $4, 1$

37. $-2, -\frac{5}{2}$ **39.** 1 **41.** $2, -2, -4$ **43.** $5, 3, -3$

45. $2, -2, -1$

B **47.** $6, 8, 10$ **49.** 10 in., 24 in., 26 in.

C **51.** 1 sec   **53.** 1 sec   **55.** 40   **57.** 10

59. a. $5.50 - x$ **b.** $550 + 450x - 100x^2$ **c.** 50¢ ($0.50) or $4

d. $0.50

SKILL CHECKER **61.** $\dfrac{1}{5x^2}$ **63.** $2x^{10}$ **65.** $\dfrac{2}{x^{12}}$

USING YOUR KNOWLEDGE **67.** 4.25 sec   Recall that t is the time 1 sec after the ball leaves the bat.   **69.** 400 ft

MASTERY TEST **75.** $-\frac{1}{4}, \frac{1}{2}$ **77.** $1, 5$ **79.** $-4, 4$

81. $5, 12, 13$

Review Exercises

1. a. Binomial, degree 6   **b.** Monomial, degree 8

c. Trinomial, degree 5   **2. a.** $8x^3y^3 - 3x^3y^2 + 7z^3$, degree 6

b. $3x^2y^2z^3 + 4x^2y^2 - 4z^5$, degree 7

c. $6x^2y^2z^2 + 3xyz^3 - 4x^2y^3z^3$, degree 8   **3. a.** 9   **b.** 3   **c.** 23

4. a. $x^3 + 8x^2 - 10x + 7$ **b.** $5x^3 - 3x^2 - 3x + 8$

c. $3x^3 + 5x^2 - 7x + 6$ **5. a.** $-3x^3 + 7x^2 - 3x + 3$

b. $10x^2 + x - 6$ **c.** $-7x^3 + 13x^2 - 12x + 4$

6. a. $-2x^4y - 6x^3y^2 + 4x^2y^4$ **b.** $-3x^4y^2 - 9x^3y^3 + 6x^2y^5$

c. $-4x^3y^2 - 12x^2y^3 + 8xy^5$ **7. a.** $x^3 - 4x^2 + x + 2$

b. $x^3 - 5x^2 + 4x + 4$ **c.** $x^3 - 6x^2 + 7x + 6$

8. a. $8x^2 + 2xy - 15y^2$ **b.** $6x^2 + 5xy - 6y^2$

c. $10x^2 + 9xy - 9y^2$ **9. a.** $4x^2 + 28xy + 49y^2$

b. $9x^2 + 42xy + 49y^2$ **c.** $16x^2 + 56xy + 49y^2$

10. a. $9x^2 - 42xy + 49y^2$ **b.** $16x^2 - 56xy + 49y^2$

c. $25x^2 - 70xy + 49y^2$ **11. a.** $9x^2 - 4y^2$ **b.** $16x^2 - 9y^2$

c. $25x^2 - 9y^2$ **12. a.** $5x^2(3x^3 - 4x^2 + 2x + 5)$

b. $3x^2(3x^3 - 4x^2 + 2x + 5)$ **c.** $2x^2(3x^3 - 4x^2 + 2x + 5)$

13. a. $x(2x^3 + 5)(3x^2 - 1)$ **b.** $x(2x^3 + 5)(3x^2 - 4)$

c. $x(2x^3 + 3)(3x^2 - 2)$ **14. a.** $(x - 6y)(x + 3y)$

b. $(x - 6y)(x + 2y)$ **c.** $(x - 6y)(x + y)$

15. a. $(2x + 5y)(x - 6y)$ **b.** $(2x + 5y)(x - 4y)$

c. $(2x + 5y)(x - 5y)$ **16. a.** $-3x^2y(3x + 2y)(2x - y)$

b. $-5x^2y(3x + 2y)(2x - y)$ **c.** $-6x^2y(3x + 2y)(2x - y)$

17. a. $(2x - 7y)^2$ **b.** $(3x - 7y)^2$ **c.** $(4x - 7y)^2$

18. a. $(3x + 4y)^2$ **b.** $(3x + 5y)^2$ **c.** $(3x + 6y)^2 = 9(x + 2y)^2$

19. a. $(9x^2 + y^2)(3x + y)(3x - y)$

b. $(x^2 + 4y^2)(x + 2y)(x - 2y)$

c. $(9x^2 + 4y^2)(3x + 2y)(3x - 2y)$

20. a. $(x - 2 + y)(x - 2 - y)$ **b.** $(x - 3 + y)(x - 3 - y)$

c. $(x - 4 + y)(x - 4 - y)$ **21. a.** $(3x + 2y)(9x^2 - 6xy + 4y^2)$

b. $(3x + 4y)(9x^2 - 12xy + 16y^2)$

c. $(4x + 3y)(16x^2 - 12xy + 9y^2)$

22. a. $(3x - 2y)(9x^2 + 6xy + 4y^2)$

b. $(3x - 4y)(9x^2 + 12xy + 16y^2)$

c. $(4x - 3y)(16x^2 + 12xy + 9y^2)$

23. a. $x^3(3x - 2y)(9x^2 + 6xy + 4y^2)$

b. $x^4(3x - 4y)(9x^2 + 12xy + 16y^2)$

c. $x^5(4x - 3y)(16x^2 + 12xy + 9y^2)$

24. a. $3x^4(9x^2 + 1)$ **b.** $4x^4(x^2 + 16)$ **c.** $2x^4(x^2 + 9)$

25. a. $3x^2(3x + 2y)^2$ **b.** $4x^2(3x + 2y)^2$ **c.** $5x^2(3x + 2y)^2$

26. a. $3x^2(3x - 2y)^2$ **b.** $4x^2(3x - 2y)^2$ **c.** $5x^2(3x - 2y)^2$

27. a. $4xy(3x + y)(x - 4y)$ **b.** $5xy(3x + y)(x - 4y)$

c. $6xy(3x + y)(x - 4y)$

28. a. $(x + 1)(x - 1)(2x - 1)$ **b.** $(3x + 1)(3x - 1)(2x - 1)$

c. $(4x + 1)(4x - 1)(2x - 1)$

29. a. 3 or -4 **b.** 4 or -5 **c.** 4 or -6

30. a. $\frac{1}{3}$ or $-\frac{1}{2}$ **b.** $\frac{1}{4}$ or $-\frac{1}{2}$ **c.** $\frac{1}{5}$ or $-\frac{1}{2}$

31. a. 1 or -1 or -2 **b.** 1 or -1 or -4 **c.** 3 or -2 or -3

32. a. 6 units, 8 units, 10 units   **b.** 9 units, 12 units, 15 units

c. 12 units, 16 units, 20 units

CHAPTER 4

Exercise 4.1

A **1.** $x = -3$ **3.** Defined for all values of m

5. $m = -1$ and 2   **7.** $p = -3$ and 3   **9.** $a = -\frac{1}{2}$ and 6

11. Defined for all real values of v

B **13.** $\dfrac{4xy^2}{6y^3}$ **15.** $\dfrac{x(x - y)}{x^2 - y^2}$ or $\dfrac{x^2 - xy}{x^2 - y^2}$

17. $\dfrac{-x(y + x)}{y^2 - x^2}$ or $\dfrac{-(xy + x^2)}{y^2 - x^2}$ or $\dfrac{-xy - x^2}{y^2 - x^2}$

19. $\dfrac{-x(2x + 3y)}{4x^2 - 9y^2}$ or $\dfrac{-(2x^2 + 3xy)}{4x^2 - 9y^2}$

21. $\dfrac{4x(x - 2)}{x^2 - x - 2}$ or $\dfrac{4x^2 - 8x}{x^2 - x - 2}$

23. $\dfrac{-5x(x - 2)}{x^2 + x - 6}$ or $\dfrac{-5x^2 + 10x}{x^2 + x - 6}$

25. $\dfrac{3(x^2 - xy + y^2)}{x^3 + y^3}$ or $\dfrac{3x^2 - 3xy + 3y^2}{x^3 + y^3}$

27. $\dfrac{x(x^2 + xy + y^2)}{x^3 - y^3}$ or $\dfrac{x^3 + x^2y + xy^2}{x^3 - y^3}$

29. $\dfrac{x(x + y)}{x^3 + y^3}$ or $\dfrac{x^2 + xy}{x^3 + y^3}$

31. $\dfrac{y}{2}$ **33.** $\dfrac{x}{5 - x}$ **35.** $\dfrac{-2x}{5y}$ **37.** $\dfrac{x + y}{x - y}$

39. $\dfrac{1}{x - 2}$

41. $\dfrac{x^3}{y^3}$ **43.** 3   **45.** $\dfrac{1}{3x + 2y}$ **47.** $\dfrac{(x - y)^2}{x + y}$

49. $y - 1$ **51.** $\dfrac{x + y}{x - y}$ **53.** $\dfrac{y - 5}{y + 6}$ **55.** -1

57. $-(x + 3)$ **59.** $-(y^2 + 2y + 4)$ **61.** -1

63. $-(x + 5)$ **65.** $2 - x$ **67.** $\dfrac{-1}{x + 6}$ **69.** $\dfrac{1}{x - 2}$

SKILL CHECKER **71.** $2x^2$ **73.** $-2x^2$ **75.** $(x + 4)(x - 3)$
77. $(3x + 2y)(3x - 2y)$ **79.** $(x - 1)(x^2 + x + 1)$
81. $(2x + 1)(4x^2 - 2x + 1)$ **83.** $\frac{2}{3}$ **85.** $\frac{3}{2}$ **87.** $-\frac{3}{2}$

USING YOUR KNOWLEDGE **89. a.** $525 million
b. $1400 million (or 1.4 billion)   **c.** $3150 million (or 3.15 billion)
d. No. The denominator $= 0$ for $p = 100$. As p increases toward
100, the fraction increases without bound.
91. a. $5(x - 1)$ **b.** $45   **c.** $495

MASTERY TEST **99.** $x^4 y^4$ **101.** $-\dfrac{x + y}{x^2 + xy + y^2}$

103. $4 - x$ **105.** $-1, 1$ **107.** $\frac{4}{x}$ **109.** $\frac{14}{16}$

111. $\dfrac{4x^2 - 11x - 3}{x^2 - x - 6}$

Exercise 4.2

1. $\frac{3}{10}$ **3.** $\dfrac{2x}{3}$ **5.** $\dfrac{2y}{21x^2 z^4}$ **7.** $\dfrac{4}{x + 1}$ **9.** 1

11. $\dfrac{-5}{x(x + y)}$ or $\dfrac{-5}{x^2 + xy}$ **13.** $\dfrac{3y - 2}{y - 1}$ **15.** 1

17. $1 - x$ **19.** $\dfrac{a + b}{a - b}$

21. $\frac{27}{50}$ **23.** $\dfrac{5x}{3}$ **25.** $\dfrac{9ad}{c}$ **27.** $\dfrac{3x}{x + 1}$

29. $\dfrac{4(y + 5)}{3(y + 2)}$ or $\dfrac{4y + 20}{3y + 6}$ **31.** $\dfrac{a + b}{a - b}$

33. $\dfrac{-(4a^2 + 2a + 1)}{2u^2 w^2}$ or $\dfrac{-4a^2 - 2a - 1}{2u^2 w^2}$

35. $\dfrac{-(1 + x)}{5x}$ or $\dfrac{-1 - x}{5x}$ **37.** $\dfrac{2y + 3}{3y - 5}$ **39.** $\dfrac{2x - 3}{5 - 2x}$

41. $\dfrac{1}{x + 1}$ **43.** $\dfrac{x - 1}{x - 4}$ **45.** -1 **47.** $\dfrac{x}{x + y}$

49. $\dfrac{xy + 2y^2 + 2x + 4y}{xy - 3y^2 - 2x + 6y}$ **51.** $\dfrac{x^2 + 4x + 3}{x^2 - x + 1}$

53. $\dfrac{1}{x^3 + 2x^2 + 4x}$

55. $\dfrac{450}{x}$ **57.** $\dfrac{4t(t + 3)}{t^2 + 9}$ or $\dfrac{4t^2 + 12t}{t^2 + 9}$

SKILL CHECKER **59.** $\dfrac{x^2 + x - 2}{x^2 - x - 6}$ **61.** $\dfrac{x - 2}{x^3 - 8}$

USING YOUR KNOWLEDGE **63.** $\dfrac{RR_T}{R - R_T}$

65. $R = \dfrac{60{,}000 + 9000x}{x}$ **67.** $\dfrac{2W^2 - LW}{6}$

MASTERY TEST **71.** $\dfrac{-(x^2 - 3x + 9)}{x^2 + 2x + 4}$

73. $\dfrac{3(x - 1)}{x + 2}$ or $\dfrac{3x - 3}{x + 2}$ **75.** -1

77. $\dfrac{x - 4}{x^2 - x + 1}$ **79.** $\dfrac{x(x - 3)}{3(3x + 2y)}$ or $\dfrac{x^2 - 3x}{9x + 6y}$

Exercise 4.3

1. $\dfrac{3x}{5}$ **3.** $\dfrac{5x}{3}$ **5.** $\dfrac{3 + 2x}{5(x + 2)}$ **7.** $\dfrac{x + 2}{2(x + 1)}$

9. $\dfrac{2x + 5}{3(x - 1)}$

11. $\dfrac{2x^2 - 5x}{(x + 4)(x - 4)(x - 1)}$ **13.** $\dfrac{5x^2 + 19x}{(x + 5)(x - 2)(x + 3)}$

15. $\dfrac{6x - 4y}{(x + y)^2(x - y)}$ **17.** $\dfrac{10 - x}{x^2 - 25}$

19. $\dfrac{-6x - 10}{(x + 1)(x + 2)(x + 3)}$ **21.** $\dfrac{2x - 5}{(x - 2)(x - 3)}$

23. $\dfrac{2}{x + 3}$ **25.** $\dfrac{2a + 6}{(a + 2)(a + 4)}$ **27.** $\dfrac{2}{a + 4}$

29. $\dfrac{15 - a^2}{(a + 5)(a + 3)}$ **31.** $\dfrac{y^2 + 2y}{(y + 1)(y - 1)}$

33. $\dfrac{-2y^2 - 4y + 5}{(y + 4)(y - 4)}$ **35.** $\dfrac{50x^2 + 8y^2}{(5x - 2y)(5x + 2y)}$

37. $\dfrac{5x^2}{(2x - y)(3x + y)}$ **39.** $\dfrac{2x^2 + x + 5}{(x - 2)(x + 1)^2}$

41. $\dfrac{4 - 3x}{(x - 2)(x - 4)(x - 1)}$ **43.** $\dfrac{-45}{(x + 5)(x - 5)}$

45. $\dfrac{-x}{(x - y)(x - 2)}$ or $\dfrac{x}{(y - x)(x - 2)}$ or $\dfrac{-x}{(y - x)(2 - x)}$

47. $\dfrac{18b^2 + 12ab + 12a - 18b}{(2a + 3b)(2a - 3b)}$ **49.** $\dfrac{1}{x^2 - 5x + 25}$

C **51.** $\dfrac{-w_0x^3 + 3w_0L^2x - 2w_0L^3}{6L}$ **53.** $\dfrac{p^2 - 2gm^2Mr}{2mr^2}$

SKILL CHECKER **55.** 20 **57.** $24x + 18y$ **59.** $x^2 - 1$

USING YOUR KNOWLEDGE **61.** $P(x + h) = x^2 + 2xh + h^2$

63. $\dfrac{P(x + h) - P(x)}{h} = 2x + h$

MASTERY TEST **69.** $\dfrac{-(5x + 9)}{(x + 1)(x + 2)(x - 2)}$ **71.** $\dfrac{1}{x + 3}$

73. $\dfrac{2x}{x - 2}$ **75.** $\dfrac{1}{x + 3}$

Exercise 4.4

1. $\frac{178}{33}$ **3.** $\dfrac{a}{c}$ **5.** $\dfrac{z}{xy}$ **7.** $\dfrac{2z}{5y}$ **9.** $\frac{1}{3}$ **11.** $\dfrac{ab - a}{b + a}$

13. $\dfrac{x}{xy - 2}$ **15.** $\dfrac{x - y}{x^2y^2}$ **17.** $\frac{9}{5}$ **19.** $\dfrac{2a^2 - a}{2a + 1}$

21. $\dfrac{x^2}{2x - 1}$ **23.** $\frac{30}{43}$ **25.** $\dfrac{-(x^2 + 1)}{2x}$ **27.** $\dfrac{x}{y}$ **29.** -1

31. $\dfrac{1 - x}{1 + x}$ **33.** $\dfrac{y + 7}{y - 2}$ **35.** $\dfrac{x^2 - 1}{5x^2 - 4x - 2}$ **37.** $\dfrac{c + d}{c - d}$

39. $\dfrac{ab}{a^2 + b^2}$ **41.** $R = \dfrac{R_1R_2}{R_1 + R_2}$ **43.** f static $\sqrt{\dfrac{c + v}{c - v}}$

SKILL CHECKER **45.** $4x$ **47.** $-5x^2$ **49.** $-\dfrac{3}{x^2}$

51. $5x^3 + 5x^2$ **53.** $9x^5 - 15x^4$ **55.** $(3x - 1)(2x + 3)$

57. $x^2 + 2x - 8$

USING YOUR KNOWLEDGE **59.** $\dfrac{288(NM - P)}{12PN + N^2M}$

61. $\frac{6}{25}$ yr **63.** $11\frac{43}{50}$ yr

MASTERY TEST **69.** $\dfrac{a^2 + 6a + 12}{2a + 6}$

71. $\dfrac{x(x - 2)}{x - 3}$ or $\dfrac{x^2 - 2x}{x - 3}$ **73.** $\dfrac{4(2a - 3b)}{2a + 3b}$

75. $\dfrac{(x + 1)(x^2 + 1)}{x(x^2 + x + 1)}$ or $\dfrac{x^3 + x^2 + x + 1}{x^3 + x^2 + x}$ **77.** -1

Exercise 4.5

A **1.** $x^2 + 3x - 2$ **3.** $-2x^2 + x - 3$ **5.** $-2y^2 + 8y - 3$

7. $5x^2 + 4x - 8 + \dfrac{3}{x}$ **9.** $3xy - 2 + \dfrac{3}{xy}$

B **11.** $x + 3$ **13.** $y + 5$ **15.** $x^2 - x - 1$

17. $x^2 + 5x + 6$ **19.** $x^2 - 2x - 8$

21. $x^2 + 4x + 3$ R 1 or $x^2 + 4x + 3 + \dfrac{1}{2x - 1}$

23. $y^2 - y - 1$

25. $4x^2 + 3x + 7$ R 12 or $4x^2 + 3x + 7 + \dfrac{12}{2x - 3}$

27. $x^2 - 2x + 4$ **29.** $4y^2 + 8y + 16$ **31.** $a^2 - a - 1$

33. $x^3 + 2x^2 - x$ **35.** $2x^3 - 3x^2 + x - 4$ R -1

C **37.** $(x + 1)(x - 3)(x - 2)$

39. $(x - 2)(x - 2)(x + 1)(x - 1)$

41. $(x + 5)(x + 4)(x^2 - 3x + 7)$

D **43.** $v^2 + 3v + 1$ R 0 **45.** $x^2 + 6x + 5$ R 15

47. $z^2 - 6z + 4$ R 0 **49.** $3y^3 + 12y^2 + 7y + 15$ R 52

51. $2y^3 - y^2 + 5$ R 0

E **53.** R = 0, so 4 is a solution. **55.** R = 0, so −4 is a solution.

57. R = 0, so 5 is a solution. **59.** R = 0, so −1 is a solution.

F **61.** $\dfrac{500}{x} + 4$

SKILL CHECKER **63.** 4 **65.** 3

USING YOUR KNOWLEDGE **67.** $x + 1, x + 2, x + 3, x + 6$

MASTERY TEST **73.** $2x^3 - 3x^2 + x - 4$ R 0

75. R = 0, so −2 is a solution. **77.** $(z + 2)(z - 2)(z - 3)$

79. $3x^2 - 8x + 15$ R -36 **81.** $2x - \dfrac{1}{2} + \dfrac{1}{x} - \dfrac{2}{x^2} + \dfrac{5}{x^3}$

Exercise 4.6

1. 6 **3.** −5 **5.** $\frac{1}{3}$ **7.** −4 **9.** $\frac{26}{9}$ **11.** $-\frac{1}{12}$

13. No solution **15.** Any real number ($x \neq -\frac{1}{3}$) **17.** −11

19. 7 **21.** 2 **23.** −3 **25.** $-\frac{3}{2}$ **27.** No solution

29. 0 **31.** $\frac{4}{5}$ **33.** 2 **35.** −4

SKILL CHECKER **37.** 21, 23, 25 **39.** 60 gal

USING YOUR KNOWLEDGE **41.** $h = \dfrac{2A}{b_1 + b_2}$

43. $Q_1 = \dfrac{PQ_2}{1 + P}$ **45.** $f = \dfrac{ab}{a + b}$

MASTERY TEST **51.** −7 (8 is extraneous.) **53.** 3

55. −4, 1 **57.** 5

Exercise 4.7

A **1.** 8 **3.** 5 and 10 **5.** 6 and 8 **7.** $\frac{2}{7}$ **9.** $20,000

B **11.** $1\frac{7}{8}$ hr **13.** $4\frac{14}{19}$ hr **15.** 6 hr and 3 hr **17.** $15\frac{3}{4}$ hr

19. $5\frac{1}{4}$ hr **21.** 36 sec **23.** 18 hr

C **25.** 150 mi/hr **27.** 30 mi/hr

29. auto: 25 mi/hr   plane: 125 mi/hr

SKILL CHECKER **31.** $\dfrac{1}{x^8}$ **33.** x^{20} **35.** $-8x^3y^6$ **37.** $\dfrac{1}{x^2}$

39. $\dfrac{1}{a^8b^6}$

USING YOUR KNOWLEDGE **41.** $F = \dfrac{f_1 f_2}{f_1 + f_2}$

43. $R = \dfrac{2E - 2ri}{i}$ **45.** 12 tablets per day

MASTERY TEST **49.** 36 mi/hr   **51.** 4 and 6

Review Exercises

1. a. $\dfrac{8x^2 y^3}{36y^7}$ **b.** $\dfrac{10x^2 y^4}{45y^8}$ **c.** $\dfrac{12x^2 y^5}{54y^9}$

2. a. $\dfrac{2x^2 + 9x + 4}{x^2 + 5x + 4}$ **b.** $\dfrac{2x^2 + 11x + 5}{x^2 + 6x + 5}$ **c.** $\dfrac{2x^2 + 13x + 6}{x^2 + 7x + 6}$

3. a. $\dfrac{6}{y}$ **b.** $\dfrac{7}{y}$ **c.** $\dfrac{8}{y}$

4. a. $\dfrac{y - x}{6}$ **b.** $\dfrac{y - x}{7}$ **c.** $\dfrac{y - x}{8}$

5. a. $x^3 y^5$ **b.** $x^3 y^6$ **c.** $x^3 y^7$

6. a. $\dfrac{y^2}{x - y}$ **b.** $\dfrac{y^3}{x - y}$ **c.** $\dfrac{y^4}{x - y}$

7. a. $\dfrac{-(x - 2y)}{x^2 - 2xy + 4y^2}$ or $\dfrac{2y - x}{x^2 - 2xy + 4y^2}$
b. $\dfrac{-(x + 2y)}{x^2 + 2xy + 4y^2}$ **c.** $\dfrac{-(x + 3y)}{x^2 + 3xy + 9y^2}$

8. a. $\dfrac{3x + 2y}{3x + 1}$ **b.** $\dfrac{3x + 2y}{4x + 1}$ **c.** $\dfrac{3x + 2y}{5x + 1}$

9. a. $\dfrac{1}{(x - 2)(x + 4)}$ **b.** $\dfrac{1}{(x - 2)(x + 5)}$ **c.** $\dfrac{1}{(x - 2)(x + 6)}$

10. a. $\dfrac{-(x^2 - 3x + 9)}{x^2 + 2x + 4}$ **b.** $\dfrac{-(x^2 - 3x + 9)}{x^2 + 2x + 4}$
c. $\dfrac{-(x^2 - 3x + 9)}{x^2 + 2x + 4}$

11. a. $\dfrac{1}{x - 2}$ **b.** $\dfrac{1}{x - 3}$ **c.** $\dfrac{1}{x - 4}$

12. a. $\dfrac{1}{x + 3}$ **b.** $\dfrac{1}{x + 4}$ **c.** $\dfrac{1}{x + 5}$

13. a. $\dfrac{2x^2 + 9x + 11}{(x + 1)(x - 1)(x + 2)}$ **b.** $\dfrac{2x^2 + 10x + 13}{(x + 1)(x - 1)(x + 2)}$
c. $\dfrac{2x^2 + 11x + 15}{(x + 1)(x - 1)(x + 2)}$

14. a. $\dfrac{-4x - 14}{(x + 3)(x - 3)(x + 2)}$ **b.** $\dfrac{-3x - 11}{(x + 3)(x - 3)(x + 2)}$
c. $\dfrac{-x - 5}{(x + 3)(x - 3)(x + 2)}$

15. a. $\dfrac{x(x^2 - x + 1)}{(x^2 + 1)(x - 1)}$ **b.** $\dfrac{x(x^2 - x + 1)}{(x^2 + 1)(x - 1)}$
c. $\dfrac{x(x^2 - x + 1)}{(x^2 + 1)(x - 1)}$

16. a. $\dfrac{80 + 20a + a^2}{20 + 4a}$ **b.** $\dfrac{150 + 30a + a^2}{30 + 5a}$ **c.** $\dfrac{252 + 42a + a^2}{42 + 6a}$

17. a. $3x^3 - 2x + 1$ **b.** $3x^2 - 2 + \dfrac{1}{x}$ **c.** $3x - \dfrac{2}{x} + \dfrac{1}{x^2}$

18. a. $x^2 - x - 1$ R (-6) **b.** $x^2 - x - 1$ R (-7)
c. $x^2 - x - 1$ R (-8)

19. a. $(x - 1)(x - 2)(x - 3)$ **b.** $(x - 1)(x - 2)(x - 3)$
c. $(x - 1)(x - 2)(x - 3)$

20. a. $x^3 + 9x^2 + 26x + 24$ R 4   **b.** $x^3 + 8x^2 + 19x + 12$ R 4
c. $x^3 + 7x^2 + 14x + 8$ R 4

21. a. R $= 0$, so -1 is a solution.   **b.** R $= 0$, so -2 is a solution.
c. R $= 0$, so -3 is a solution.

22. a. No solution   **b.** No solution   **c.** No solution

23. a. -9 **b.** -11 **c.** -13

24. a. 10 and 12   **b.** 12 and 14   **c.** 14 and 16

25. a. $2\frac{2}{9}$ hr   **b.** $2\frac{2}{5}$ hr   **c.** $2\frac{6}{11}$ hr

26. a. 225 mi/hr   **b.** 275 mi/hr   **c.** 350 mi/hr

27. a. $a = 2A - 2b - 3c$

b. $b = A - \dfrac{a}{2} - \dfrac{3c}{2}$ or $b = \dfrac{2A - a - 3c}{2}$

c. $c = \dfrac{2A}{3} - \dfrac{a}{3} - \dfrac{2b}{3}$ or $c = \dfrac{2A - a - 2b}{3}$

CHAPTER 5
Exercise 5.1

A 1. 2   **3.** 2   **5.** -2 **7.** $-\frac{1}{4}$ **9.** 2   **11.** 2
B 13. 3   **15.** Not a real number   **17.** 3   **19.** 3
21. $-\frac{1}{2}$ **23.** Not a real number   **25.** 9   **27.** 25   **29.** $\frac{1}{4}$
31. 16   **33.** 16   **35.** -16 **37.** $\frac{1}{16}$ **39.** 7

C 41. $x^{3/7}$ **43.** $\dfrac{1}{x^{5/9}}$ **45.** $x^{2/5}$ **47.** z **49.** x^2

51. $\dfrac{1}{z^2}$ **53.** $\dfrac{1}{b^{4/5}}$ **55.** $\dfrac{1}{a^8 b^9}$ **57.** $\dfrac{b^9}{a^{10}}$ **59.** $x^{16} y^{30}$

61. $x + x^{1/3} y^{1/2}$ **63.** $x^{1/2} y^{3/4} - y^{5/4}$ **65.** $\dfrac{1}{x}$ **67.** $\dfrac{x^2}{y}$

69. $x^3 y^{11}$

D 71. $v = 15$ m/sec   **73.** $v = 28$ ft/sec   **75.** 3   **77.** 13
79. $\frac{2}{3}$

SKILL CHECKER **81.** $\dfrac{16x^3 y}{8x^3 y^3}$ **83.** $\dfrac{48x^3 y^2}{16x^4 y^4}$ **85.** $\dfrac{20x}{32x^5}$

USING YOUR KNOWLEDGE **87. a.** 2   **b.** 8   **89.** $-\frac{1}{2}$
91. $\frac{1}{2}$

MASTERY TEST **97.** $\dfrac{1}{x^5 y^{12}}$ **99.** $\dfrac{y^{1/3}}{x^{1/2}}$ **101.** $x^{1/5} + x^{2/5} y^{1/4}$

103. $x^{8/15}$ **105.** $\dfrac{1}{x^{11/12}}$ **107.** $\dfrac{1}{y^{3/20}}$ **109.** $\frac{1}{27}$ **111.** $\frac{1}{36}$

113. $\frac{1}{16}$ **115.** 7   **117.** $\frac{1}{3}$ **119.** -5

Exercise 5.2

A 1. 5 **3.** -4 **5.** $|-x| = |x|$ **7.** $|x + 6|$
9. $|3x - 2|$ **11.** $4|xy|\sqrt{xy}$ **13.** $2x\sqrt[3]{5xy}$ **15.** $|xy|\sqrt[4]{xy^3}$

17. $-3a^2b^3\sqrt[5]{b^2}$ **19.** $\dfrac{\sqrt{13}}{7}$ **21.** $\dfrac{\sqrt{17}}{2|x|}$ **23.** $\dfrac{\sqrt[3]{3}}{4x}$

B 25. $\dfrac{\sqrt{6}}{3}$ **27.** $\dfrac{-\sqrt{14}}{7}$ **29.** $\dfrac{\sqrt{10a}}{2a}$ **31.** $\dfrac{\sqrt{10ab}}{8ab}$

33. $-\dfrac{\sqrt{6ab}}{2a^2b^2}$ **35.** xy **37.** $-\dfrac{\sqrt[3]{21}}{3}$ **39.** $\dfrac{\sqrt[3]{12x}}{4x}$

C 41. $\sqrt[3]{3}$ **43.** $\sqrt{2a}$ **45.** $x\sqrt{5xy}$ **47.** $x^2y\sqrt{7xy}$

49. $\sqrt{2ab}$ **51.** $\dfrac{\sqrt[3]{a^2b^2}}{b^2}$ **53.** $\dfrac{2\sqrt{6ab}}{3b^2}$ **55.** $\dfrac{\sqrt{2abx}}{2x}$

D 57. a. $\dfrac{\sqrt[3]{6\pi^2V}}{2\pi}$ **b.** 3 ft **59.** $m = \dfrac{m_0c\sqrt{c^2 - v^2}}{c^2 - v^2}$

SKILL CHECKER **61.** $a^2 - b^2$ **63.** $x^3 - y^3$ **65.** $1 + 2x$
67. 7 **69.** 8

USING YOUR KNOWLEDGE **71.** $m = \dfrac{m_0c(c^2 - v^2)^{1/2}}{c^2 - v^2}$

73. $T = \dfrac{(2\pi Lg)^{1/2}}{g}$ **75.** $v = \dfrac{\sqrt{3kTm}}{m}$

MASTERY TEST **81.** $\dfrac{\sqrt{33}}{6}$ **83.** $\dfrac{\sqrt[4]{6x^2}}{2x^2}$ **85.** $2\sqrt[3]{4}$

87. $\dfrac{\sqrt{2ax}}{2x}$ **89.** 10 **91.** $-x$ **93.** $\dfrac{\sqrt{3}}{2}$ **95.** $\dfrac{\sqrt{35}}{5}$

97. $\dfrac{\sqrt{30x}}{6x}$

Exercise 5.3

A 1. $15\sqrt{2}$ **3.** $9\sqrt{5a}$ **5.** $\sqrt{2}$ **7.** $-5a\sqrt{2}$
9. $-26\sqrt{3}$ **11.** $7\sqrt[3]{5}$ **13.** $-12\sqrt[3]{3}$ **15.** $3\sqrt[3]{3}$

17. $4\sqrt[3]{3a}$ **19.** $\dfrac{7\sqrt[3]{3}}{6}$ **21.** $\dfrac{3\sqrt{2} + 2\sqrt{3} + \sqrt{6}}{6}$

23. $\dfrac{15\sqrt{3} + 28\sqrt{15}}{30}$ **25.** $\dfrac{\sqrt{xy}(x + y - 1)}{xy}$ **27.** $3\sqrt[3]{75}$

29. $\dfrac{4\sqrt[3]{18x^2} + \sqrt[3]{6x^2}}{12x}$

B 31. $15 - 3\sqrt{2}$ **33.** $2 + 3\sqrt[3]{2}$ **35.** $14\sqrt{15} + 30$
37. $6\sqrt[3]{15} - 15$ **39.** $-8\sqrt{21} + 20\sqrt{14}$ **41.** $55 + 13\sqrt{15}$
43. $42 + 21\sqrt{2}$ **45.** $-441 + \sqrt{35}$ **47.** -1 **49.** -23
51. $5 + 2\sqrt{6}$ **53.** $a^2 + 2a\sqrt{b} + b$ **55.** $5 - 2\sqrt{6}$
57. $a^2 - 2a\sqrt{b} + b$ **59.** $a - 2\sqrt{ab} + b$ **61.** $1 + \sqrt{2}$

63. $\dfrac{2 - \sqrt{3}}{4}$

C 65. $\dfrac{3\sqrt{2} + \sqrt{6}}{2}$ **67.** $\dfrac{6 + 2\sqrt{2}}{7}$ **69.** $3a + a\sqrt{5}$

71. $\dfrac{9a - 3a\sqrt{2} + 6b - 2b\sqrt{2}}{7}$ **73.** $\dfrac{a + 2b\sqrt{a} + b^2}{a - b^2}$

75. $\dfrac{a + 2\sqrt{2ab} + 2b}{a - 2b}$

SKILL CHECKER **77.** 11 **79.** 2 or 1

USING YOUR KNOWLEDGE **81.** $\dfrac{1}{\sqrt{5} - \sqrt{2}}$

83. $\dfrac{x - 2}{5(\sqrt{x} + \sqrt{2})}$ **85.** $\dfrac{x - y}{x(\sqrt{x} - \sqrt{y})}$ or $\dfrac{x - y}{x\sqrt{x} - x\sqrt{y}}$

87. $\dfrac{x - y}{\sqrt{x}(\sqrt{x} - \sqrt{y})}$ or $\dfrac{x - y}{x - \sqrt{xy}}$

89. $\dfrac{x - y}{\sqrt{x}(\sqrt{x} + \sqrt{y})}$ or $\dfrac{x - y}{x + \sqrt{xy}}$

MASTERY TEST **97.** $\dfrac{y + \sqrt{xy}}{y - x}$ **99.** $5 + \sqrt{2}$

101. $11 - 2\sqrt{21}$ **103.** $\sqrt{5x} - \sqrt{15x}$ **105.** $\dfrac{3\sqrt{2}}{4}$

107. $2\sqrt[3]{2x}$ **109.** $2\sqrt{2}$

Exercise 5.4

A 1. 16 **3.** 43 **5.** 18 **7.** No solution **9.** 3
11. 0 (-3 is *not* a solution.) **13.** 6 **15.** -16 **17.** 11
19. 9 **21.** 0 **23.** 1 **25.** 0 **27.** 1 **29.** 4

31. $a + b^2$ **33.** $\dfrac{a - c^3}{b}$ **35.** ab^2 **37.** $\dfrac{b^2 - a}{b}$ **39.** $\dfrac{b}{3}$

B 41. $5\sqrt{3}$ ft **43. a.** $d = \dfrac{gt^2}{2}$ **b.** 144.9 ft

45. a. $L = \dfrac{gt^2}{4\pi^2}$ **b.** $L = \dfrac{392}{121}$ ft ≈ 3.2 ft

SKILL CHECKER **47.** $11 + 6x$ **49.** $-1 + 8x$

51. $\dfrac{1 + 5\sqrt{2}}{7}$ **53.** $\dfrac{4 + \sqrt{2}}{7}$ **55.** $\dfrac{x + y - 2\sqrt{xy}}{x - y}$

USING YOUR KNOWLEDGE **57.** $r = 100$ ft
59. $\dfrac{1600}{9}$ ft or $177.\overline{7}$ ft
MASTERY TEST **65.** No solution **67.** -1 and 0
69. No solution **71.** 0 (4 is an extraneous solution.)

Exercise 5.5

A 1. $5i$ **3.** $5i\sqrt{2}$ **5.** $24i\sqrt{2}$ **7.** $-12i\sqrt{2}$
9. $8i\sqrt{7} + 3$
B 11. $6 + 4i$ **13.** $-2 - 6i$ **15.** $-5 - 6i$ **17.** $-2 + 5i$
19. $-7 + 4i$ **21.** $8 - i$ **23.** $10 + 5i$ **25.** $-3 - 2i\sqrt{2}$

27. $-1 + 2i\sqrt{2}$ **29.** $-7 + 3i\sqrt{5}$

C 31. $12 + 6i$ **33.** $-12 + 20i$ **35.** $-4 + 6i$

37. $-3 + 3i\sqrt{3}$ **39.** $-6 + 9i$ **41.** $28 + 12i$

43. $-20 + 20i$ **45.** $3 + 11i$ **47.** 13 **49.** $24 + 7i$

51. 31 **53.** $-3i$ **55.** $6i$ **57.** $\frac{2}{5} + \frac{1}{5}i$ **59.** $-\frac{6}{5} + \frac{3}{5}i$

61. $-1 + 2i$ **63.** $\frac{17}{13} - \frac{6}{13}i$ **65.** $-\frac{3}{2}i$

67. $\frac{12 + \sqrt{10}}{18} + \frac{4\sqrt{5} - 3\sqrt{2}}{18}i$ **69.** $\frac{3 + \sqrt{6}}{12} + \frac{3\sqrt{2} - \sqrt{3}}{12}i$

D 71. 1 **73.** $-i$ **75.** i **77.** 1 **79.** i

APPLICATIONS **81.** $Z_1 + Z_2 = (8 + i)$ ohms

83. $Z_T = \frac{167}{65} - \frac{29}{65}i$

SKILL CHECKER **85.** 2 or -2 **87.** 5 or -5

USING YOUR KNOWLEDGE **89.** 5 **91.** $\sqrt{13}$

MASTERY TEST **97.** $-i$ **99.** -1 **101.** $\frac{27}{25} - \frac{11}{25}i$

103. $-10\sqrt{2} + 30i$ **105.** $28 + 4i$ **107.** $3 - 2i$

109. $-1 + 7i$ **111.** $5\sqrt{2}i$

Review Exercises

1. a. Not a real number   **b.** -4

2. a. Not a real number   **b.** -5

3. a. -3 **b.** -4 **4. a.** $\frac{1}{2}$ **b.** $\frac{1}{4}$ **5. a.** 25 **b.** 16

6. a. Not a real number   **b.** Not a real number   **7. a.** $\frac{1}{4}$ **b.** $\frac{1}{16}$

8. a. $\frac{1}{4}$ **b.** $\frac{1}{16}$ **9. a.** $x^{8/15}$ **b.** $x^{9/20}$ **10. a.** $\frac{1}{x^{9/20}}$ **b.** $\frac{1}{x^{8/15}}$

11. a. $\frac{1}{x^5 y^6}$ **b.** $\frac{1}{x^5 y^{12}}$

12. a. $x^{2/5} + x^{3/5}y^{3/5}$ **b.** $x^{3/5} + x^{4/5}y^{3/5}$ **13. a.** 7 **b.** 6

14. a. $|-x|$ **b.** $|-x|$ **15. a.** $2\sqrt[3]{6}$ **b.** $2\sqrt[3]{7}$

16. a. $2xy^2 \sqrt[3]{2x}$ **b.** $2x^2 y^5 \sqrt[3]{2x^2}$ **17. a.** $\frac{\sqrt{15}}{27}$ **b.** $\frac{\sqrt{5}}{32}$

18. a. $\frac{1}{x}$ **b.** $\frac{\sqrt[3]{5}}{x}$ **19. a.** $\frac{\sqrt{55}}{11}$ **b.** $\frac{\sqrt{65}}{13}$

20. a. $\frac{\sqrt{10x}}{5x}$ **b.** $\frac{\sqrt{15x}}{5x}$ **21. a.** $\frac{\sqrt[3]{25x^2}}{5x}$ **b.** $\frac{\sqrt[3]{49x^2}}{7x}$

22. a. $\frac{\sqrt[5]{2x^2}}{2x}$ **b.** $\frac{\sqrt[5]{10x^2}}{2x}$ **23. a.** $\frac{4}{3}$ **b.** $\frac{5}{3}$

24. a. $\sqrt[3]{9c^2 d^2}$ **b.** $\sqrt[3]{25c^2 d^2}$ **25. a.** $\frac{\sqrt[3]{3ac}}{3c^3}$ **b.** $\frac{\sqrt[3]{9ac}}{3c^3}$

26. a. $6\sqrt{2}$ **b.** $7\sqrt{2}$ **27. a.** $\sqrt{7}$ **b.** $\sqrt{7}$

28. a. $\frac{3\sqrt{2}}{4}$ **b.** $\frac{5\sqrt{2}}{4}$ **29. a.** $\frac{9\sqrt[3]{6x^2}}{4x}$ **b.** $\frac{11\sqrt[3]{6x^2}}{4x}$

30. a. $6 + \sqrt{6}$ **b.** $8 + \sqrt{6}$

31. a. $2x\sqrt[3]{6} - 3\sqrt[3]{6x^2}$ **b.** $2x\sqrt[3]{6} - 3\sqrt[3]{6x^2}$

32. a. $30 + 12\sqrt{6}$ **b.** $24 + 10\sqrt{6}$

33. a. $19 + 8\sqrt{3}$ **b.** $28 + 10\sqrt{3}$

34. a. $12 - 6\sqrt{3}$ **b.** $19 - 8\sqrt{3}$ **35. a.** 3 **b.** 4

36. a. $4 - \sqrt{2}$ **b.** $6 - \sqrt{2}$ **37. a.** $\frac{\sqrt{xy} + y}{x - y}$ **b.** $\frac{x + \sqrt{xy}}{x - y}$

38. a. No real number solution   **b.** No real number solution

39. a. 4 **b.** 10 **40. a.** 4 or 3 **b.** 5 or 4 **41. a.** 4 **b.** 9

42. a. 30 **b.** 67 **43. a.** No solution   **b.** No solution

44. a. $I = \frac{k}{d^2}$ **b.** $k = d^2 I$ **45. a.** $10i$ **b.** $11i$

46. a. $6i\sqrt{2}$ **b.** $5i\sqrt{2}$ **47. a.** $10 + 3i$ **b.** $6 + 3i$

48. a. $-4 + 7i$ **b.** $2 + 11i$ **49. a.** $21 + i$ **b.** $23 - 2i$

50. a. $24\sqrt{2} + 16i$ **b.** $36\sqrt{2} + 24i$

51. a. $\frac{17 + 6i}{25}$ or $\frac{17}{25} + \frac{6}{25}i$ **b.** $\frac{27 - 11i}{25}$ or $\frac{27}{25} - \frac{11}{25}i$

52. a. -1 **b.** $-i$ **53. a.** -1 **b.** i

CHAPTER 6
Exercise 6.1

A 1. ± 8 **3.** $\pm 11i$ **5.** ± 13 **7.** $\pm 2i$ **9.** $\pm \frac{7}{6}$

11. $\pm \frac{9}{2}i$ **13.** $\pm \frac{5\sqrt{3}}{3}$ **15.** $\pm \frac{6\sqrt{5}}{5}i$ **17.** $\pm \frac{10\sqrt{3}}{3}$

19. $\pm \frac{9\sqrt{13}}{13}i$

B 21. -3 or -7 **23.** $-2 \pm 5i$ **25.** $6 \pm 3\sqrt{2}$

27. $1 \pm 2\sqrt{7}i$ **29.** $1 \pm 5\sqrt{2}$ **31.** $5 \pm 4\sqrt{2}$ **33.** $9 \pm 8i$

35. $-1 \pm 4\sqrt{2}$ **37.** $2 \pm 5\sqrt{2}$ **39.** $5 \pm 3\sqrt{3}i$

C 41. -1 or -5 **43.** -5 or -3 **45.** $-3 \pm i$

47. 4 or 6 **49.** 3 or 7 **51.** $4 \pm i$

53. $-1 \pm \frac{\sqrt{2}}{2}i$ or $\frac{-2 \pm \sqrt{2}i}{2}$ **55.** $-1 \pm 5i$ **57.** $\frac{3}{5}$ or $\frac{2}{5}$

59. $\frac{1}{2} \pm i$ **61.** $\frac{1}{2} \pm \sqrt{2}$ or $\frac{1 \pm 2\sqrt{2}}{2}$

63. $1 \pm \frac{\sqrt{2}}{2}$ or $\frac{2 \pm \sqrt{2}}{2}$ **65.** 1 or -2 **67.** $\frac{-5 \pm 3\sqrt{3}}{2}$

69. $\frac{-3 \pm \sqrt{29}}{2}$ **71.** 2 sec   **73.** 10%

SKILL CHECKER **75.** 0 or 9 **77.** $1 \pm \sqrt{3}$ **79.** $\frac{1}{8}$ or -1

81. $\frac{4}{3} \pm \frac{\sqrt{5}}{3}i$ or $\frac{4 \pm \sqrt{5}i}{3}$ **83.** $-1 \pm \sqrt{5}i$

USING YOUR KNOWLEDGE **85. a.** 2 thousand   **b.** $\$2$

87. 3 days

MASTERY TEST **91.** $\frac{5 \pm 3\sqrt{5}}{10}$ **93.** $-6 \pm 2\sqrt{11}$

95. $\dfrac{-6 \pm 3\sqrt{6}}{2}$ **97.** $3 \pm \dfrac{6\sqrt{5}}{5}i$ or $\dfrac{15 \pm 6\sqrt{5}i}{5}$

99. $2 \pm 2\sqrt{6}$ **101.** $\pm\dfrac{4\sqrt{3}}{3}$ **103.** $\pm 2\sqrt{3}i$

Exercise 6.2

A **1.** 1 or -2 **3.** $-2 \pm \sqrt{3}$ **5.** $\dfrac{3 \pm \sqrt{17}}{2}$

7. $\frac{5}{7}$ or 1   **9.** $\dfrac{-4}{5} \pm \dfrac{3}{5}i$ or $\dfrac{-4 \pm 3i}{5}$ **11.** $-\frac{3}{2}$ or -2

13. $\frac{3}{2}$ or 1   **15.** $-\frac{1}{2}$ or -3 **17.** $\dfrac{3 \pm \sqrt{5}}{4}$ **19.** -1

21. $\dfrac{3 \pm \sqrt{5}}{4}$ **23.** $\dfrac{-3 \pm \sqrt{21}}{6}$ **25.** ± 2 **27.** $\pm 3\sqrt{5}$

29. $\pm\dfrac{2\sqrt{3}}{3}$

B **31.** 2 or $-1 \pm \sqrt{3}i$ **33.** $\dfrac{1}{2}, -\dfrac{1}{4} \pm \dfrac{\sqrt{3}}{4}i$ or $\dfrac{-1 \pm \sqrt{3}i}{4}$

C **35. a.** Yes **b.** 1987   **37.** \$$\dfrac{5 + \sqrt{37}}{6} \approx \$1.85$

39. $10 \pm 2\sqrt{15}$ **41.** 1

SKILL CHECKER **43.** 4   **45.** $\sqrt{23}i$ **47.** $6x^2 - 5x - 4$
49. $12x^2 - 19x - 21$

USING YOUR KNOWLEDGE **51.** Multiply both sides by $4a$.
53. To complete the square on the left side, add b^2 to both sides.
55. Take the square root of each side.
57. Divide both sides by $2a$.

MASTERY TEST **61.** $\dfrac{2}{3}, \dfrac{-1}{3} \pm \dfrac{\sqrt{3}}{3}i$ or $\dfrac{-1 \pm \sqrt{3}i}{3}$

63. $\dfrac{-1}{3} \pm \dfrac{\sqrt{2}}{3}i$ or $\dfrac{-1 \pm \sqrt{2}i}{3}$

65. $-\frac{1}{2}, 2$ **67.** 0, 6   **69.** $2 \pm 2\sqrt{2}$ **71.** $-\frac{5}{3}, 1$ **73.** \$1

Exercise 6.3

A **1.** $D = 49$ The two solutions are rational numbers.
3. $D = 0$ There is one rational solution.
5. $D = 0$ There is one rational solution.
7. $D = -23$ The two solutions are imaginary numbers.
9. $D = 57$ The two solutions are irrational numbers.
11. ± 4 **13.** -5 **15.** ± 8 **17.** ± 20 **19.** -16
B **21.** Not factorable   **23.** $(4x - 3)(3x - 2)$
25. $(9x - 7)(3x + 8)$ **27.** $(5x - 6)(3x + 14)$
29. $(4x - 15)(3x - 4)$

C **31.** $x^2 - 7x + 12 = 0$ **33.** $x^2 + 12x + 35 = 0$
35. $3x^2 - 7x - 6 = 0$ **37.** $4x^2 - 1 = 0$ **39.** $5x^2 + x = 0$
D **41. a.** Sum $= \frac{6}{4} = \frac{3}{2}$ **b.** Product $= \frac{5}{4}$ **c.** No
43. a. Sum $= -\frac{13}{5}$ **b.** Product $= -\frac{6}{5}$ **c.** Yes
45. a. Sum $= -\frac{5}{2}$ **b.** Product $= 1$ **c.** No   **47.** $\frac{8}{3}$ **49.** 15
SKILL CHECKER **51.** 1   **53.** $-1, -5$
USING YOUR KNOWLEDGE **55.** Occurs twice
57. Does not occur ($b^2 - 4ac$ is negative).
MASTERY TEST **63.** Yes   **65.** Factorable. $(3x - 5)(4x + 7)$
67. $3x^2 + x - 2 = 0$

Exercise 6.4

A **1.** 0 or $-\frac{5}{2}$ **3.** 7 or -3 **5.** 0   **7.** 2 or 1
9. $-\frac{16}{5}$ or -1
B **11.** ± 3 or ± 2 **13.** $\pm\frac{1}{2}$ or $\pm 3i$ **15.** $\pm\sqrt{2}$ or $\pm\dfrac{\sqrt{3}}{3}i$

17. $1, -2, 1 \pm \sqrt{3}i, -\dfrac{1}{2} \pm \dfrac{\sqrt{3}}{2}i$ or $\dfrac{-1 \pm \sqrt{3}i}{2}$

19. 7 or -6 **21.** $\dfrac{1 \pm \sqrt{37}}{3}, \dfrac{1}{2} \pm \dfrac{\sqrt{3}}{2}i$ or $\dfrac{1 \pm \sqrt{3}i}{2}$ **23.** 16

25. 8 or 27   **27.** 4   **29.** $2 \pm \sqrt{29}, 2 \pm \sqrt{13}$ **31.** 6

33. $\frac{16}{81}$ or 1   **35.** $-\frac{1}{4}$ or $\frac{1}{2}$ **37.** $\pm\dfrac{\sqrt{2}}{2}$ or $\pm\sqrt{3}i$

39. $\pm\sqrt{2}$ or $\pm\dfrac{\sqrt{3}i}{2}$

C **41.** 10 hr, 15 hr
SKILL CHECKER **43.** $x \geq -2$
USING YOUR KNOWLEDGE **45.** 10
MASTERY TEST **51.** 1 or 16   **53.** 16 or 81

55. $\pm\dfrac{\sqrt{2}}{2}$ or $\pm\dfrac{\sqrt{7}}{7}$ **57.** $-1, 2,$ or $\dfrac{1 \pm \sqrt{3}i}{2}$

59. -4 (3 is an extraneous solution.)   **61.** ± 1 or ± 2

Exercise 6.5

A **1.**
$x < -1$ or $x > 3$ $(-\infty, 1) \cup (3, \infty)$

3.
$-4 \leq x \leq 0$ $[-4, 0]$

5.
$-1 \leq x \leq 2$ $[-1, 2]$

7.
$x \leq 0$ or $x \geq 3$ $(-\infty, 0] \cup [3, \infty)$

9.
$1 < x < 2$ $(1, 2)$

11. $-3 < x < 1$ $(-3, 1)$

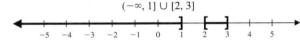

13. $x = -5$

15. All real values $(-\infty, \infty)$

17. $x = 1.6$ and -0.6
19. $x = 2.55$ and -1.55
21. $x \geq 2$ or $-3 \leq x \leq -1$
$[-3, -1] \cup [2, \infty)$

23. $x \leq 1$ or $2 \leq x \leq 3$
$(-\infty, 1] \cup [2, 3]$

25. $x > 2$ $(2, \infty)$

27. $1 < x < 7$ $(1, 7)$

29. $\frac{1}{2} < x < 3$ $\left(\frac{1}{2}, 3\right)$

31. $-2 < x < 1$ $(-2, 1)$

33. $x < -\frac{1}{2}$ or $x > 0$ $\left(-\infty, -\frac{1}{2}\right) \cup (0, \infty)$

35. $x \leq -3$ or $x \geq 3$ $(-\infty, -3] \cup [3, \infty)$
37. $x \leq 1$ or $x \geq 5$ $(-\infty, 1] \cup [5, \infty)$
APPLICATIONS 39. $R > 4$ **41.** $10 < T < 100$
43. $1 < t < 2$
SKILL CHECKER 45. 12 **47.** 2 **49.** -2
USING YOUR KNOWLEDGE 51. $23.2 \leq v \leq 26.1$
53. $v = 128.2$ mi/hr
MASTERY TEST
59. $x > 0$ $(0, \infty)$

61. $x \leq -3$ or $1 \leq x \leq 4$ $(-\infty, -3] \cup [1, 4]$

63. $-3 \leq x \leq 2$ $[-3, 2]$

65. $x \leq -1$ or $x \geq 3$ $(-\infty, -1] \cup [3, \infty)$

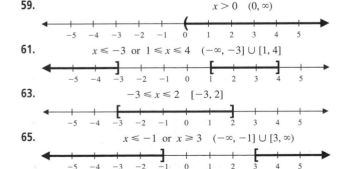

Review Exercises

1. a. $\pm \frac{7}{4}$ **b.** $\pm \frac{4}{5}$ **2. a.** $\pm\sqrt{6}i$ **b.** $\pm\sqrt{7}i$
3. a. $3 \pm 4\sqrt{2}$ **b.** $5 \pm 5\sqrt{2}$

4. a. $2 \pm \frac{5\sqrt{2}}{2}i$ or $\frac{4 \pm 5\sqrt{2}i}{2}$ **b.** $3 \pm \frac{8\sqrt{3}}{3}i$ or $\frac{9 \pm 8\sqrt{3}i}{3}$

5. a. $1 \pm \frac{2\sqrt{15}}{5}$ or $\frac{5 \pm 2\sqrt{15}}{5}$

b. $-\frac{1}{2} \pm \frac{2\sqrt{3}}{3}$ or $\frac{-3 \pm 4\sqrt{3}}{6}$

6. a. 9 or -1 **b.** -4 or -8 **7. a.** $\frac{1}{2}, -\frac{3}{2}$ **b.** $\frac{3 \pm \sqrt{2}}{4}$

8. a. $\frac{1}{3}$ or -2 **b.** 2 or $-\frac{1}{5}$ **9. a.** $\frac{1 \pm \sqrt{13}}{3}$ **b.** $\frac{3 \pm \sqrt{21}}{4}$

10. a. 0 or 16 **b.** 0 or 12 **11. a.** $\frac{-1 \pm \sqrt{301}}{30}$ **b.** $\frac{1}{5}$ or -2

12. a. $\frac{1}{3} \pm \frac{\sqrt{2}}{3}i$ or $\frac{1 \pm \sqrt{2}i}{3}$ **b.** $\frac{1}{5} \pm \frac{\sqrt{19}}{5}i$ or $\frac{1 \pm \sqrt{19}i}{5}$

13. a. $\frac{5}{2}, -\frac{5}{4} \pm \frac{5\sqrt{3}}{4}i$ or $\frac{-5 \pm 5\sqrt{3}i}{4}$

b. $\frac{2}{5}, -\frac{1}{5} \pm \frac{\sqrt{3}}{5}i$ or $\frac{-1 \pm \sqrt{3}i}{5}$

14. a. 3 **b.** 1
15. a. $D = k^2 - 64; k = \pm 8$ **b.** $D = k^2 - 64; k = \pm 8$
16. a. Not factorable **b.** $(2x + 1)(9x + 2)$
17. a. $(6x - 5)(3x + 1)$ **b.** Not factorable
18. a. $x^2 - x - 6 = 0$ **b.** $12x^2 + 5x - 2 = 0$
19. a. Sum $= -\frac{4}{15}$, Product $= -\frac{1}{5}$ **b.** Sum $= \frac{4}{3}$, Product $= -\frac{5}{9}$
20. a. Yes **b.** No **21. a.** -5 **b.** 0 or 2

22. a. $1, -2, \frac{-1 \pm \sqrt{15}i}{2}$ **b.** $1, 2, \frac{3 \pm \sqrt{33}}{2}$

23. a. 1 or 81 **b.** 16 **24. a.** 27 or -64 **b.** 216 or -1
25. a. 9 **b.** 25 **26. a.** $\pm 1, \pm\sqrt{3}$ **b.** $\pm 1, \pm\sqrt{3}i$
27. a. $-3 < x < 2$ $(-3, 2)$

b. $-2 < x < 3$ $(-2, 3)$

28. a. $x \geq 0$ or $x \leq -4$
$(-\infty, -4] \cup [0, \infty)$

b. $x \geq 3$ or $x \leq 0$ $(-\infty, 0] \cup [3, \infty)$

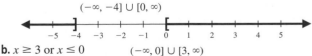

29. a. $x \le -2 - 2\sqrt{3}$ or $x \ge -2 + 2\sqrt{3}$
$(-\infty, -2 - 2\sqrt{3}] \cup [-2 + 2\sqrt{3}, \infty)$
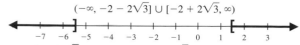

b. $x \le -3 - 3\sqrt{3}$ or $x \ge -3 + 3\sqrt{3}$
$(-\infty, -3 - 3\sqrt{3}] \cup [-3 + 3\sqrt{3}, \infty)$
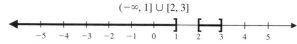

30. a. $x \le 1$ or $2 \le x \le 3$
$(-\infty, 1] \cup [2, 3]$
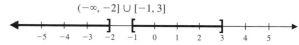

b. $x \le -2$ or $-1 \le x \le 3$
$(-\infty, -2] \cup [-1, 3]$
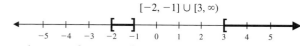

31. a. $x \ge 3$ or $-2 \le x \le -1$
$[-2, -1] \cup [3, \infty)$
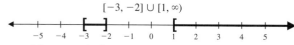

b. $x \ge 1$ or $-3 \le x \le -2$
$[-3, -2] \cup [1, \infty)$

32. a. $x < 2$ or $x \ge 6$ $(-\infty, 2) \cup [6, \infty)$

b. $-4 \le x < -2$
$[-4, -2)$
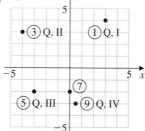

CHAPTER 7
Exercise 7.1

A 1-9
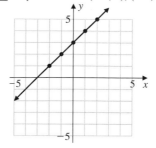

11. $(2, 1)$
13. $(3, 5)$
15. $(-2, 3)$
17. $(-4, 0)$
19. $(0, -3)$

B 21. $y = x + 3$ $(-2, 1), (-1, 2), (0, 3), (1, 4), (2, 5)$
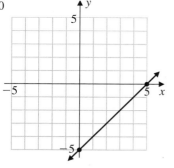

23. $x - y = 4$
$(-1, -5), (0, -4), (1, -3)$
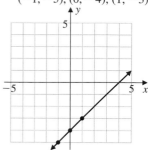

25. $2x - y - 3 = 0$
$(-1, -5), (0, -3), (1, -1)$
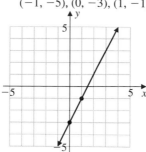

C 27. $y = x - 5$
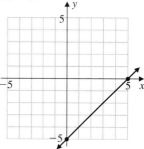

29. $2x + 3y = 6$
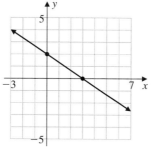

31. $2x - y = 4$
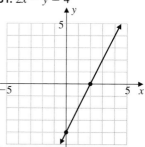

33. $2x + y - 4 = 0$
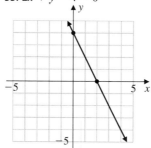

35. $y + 4x = 0$
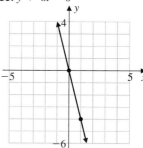

37. $2x - 5y = -10$
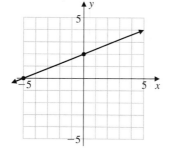

39. $x - y - 5 = 0$

41. $-\frac{7}{2}x = 14$

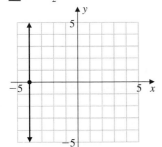

43. $\frac{3}{2}x = 6$

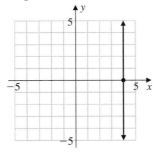

45. $-\frac{3}{4}x = 3$

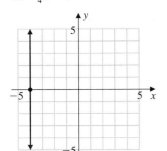

47. $-\frac{1}{3} + y = \frac{2}{3}$

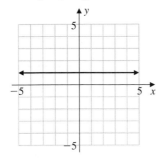

49. $\frac{2}{3} = x - \frac{4}{3}$

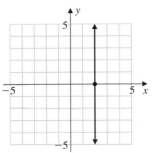

APPLICATIONS **51. a.** Yes **b.** 80 **c.** 160
53. (20, 140) **55.** (45, 148)
SKILL CHECKER **57.** $-\frac{3}{2}$ **59.** -3 **61.** $\frac{2}{3}$
USING YOUR KNOWLEDGE **63.** 86°F **65.** Less than 10%
67. 11°F

MASTERY TEST
77. $y = -3$

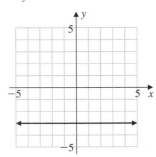

79. $-3x - y = -6$

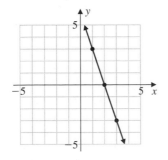

81. $3x - 2y = -6$

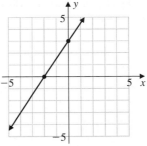

83. $A(-2, 3)$ $B(-3, 0)$ $C(4, -2)$

Exercise 7.2

A **B** **1.** Distance 5; slope $\frac{4}{3}$

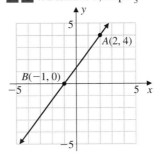

3. Distance $\sqrt{73}$; slope $\frac{8}{3}$

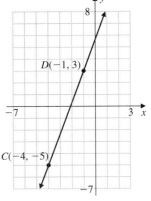

5. Distance $3\sqrt{10}$; slope 3

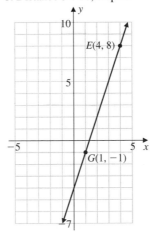

7. Distance 5; slope 0

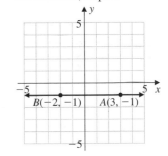

9. Distance 6; slope undefined

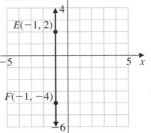

11. Parallel   **13.** Perpendicular   **15.** Neither

17. Perpendicular   **19.** Parallel   **21.** 2   **23.** 4   **25.** $-\frac{16}{3}$

27. -8 **29.** 3

31. 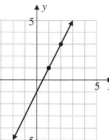   **33.**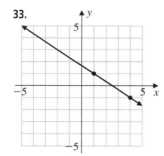

35. 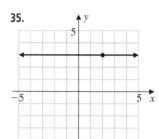   **37.**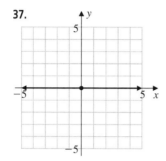

39. 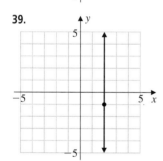   **41.** $(5, 3)$
43. $(-4, -4)$
45. $(-5.5, 0)$
47. $(-4, -2)$
49. $(83, 35)$

APPLICATIONS **51.** No $(584 \neq 449 + 13)$
53. Yes $(650 = 637 + 13)$ **55.** Yes $(584 = 292 + 292)$
57. a. 224 million   **b.** 246 million   **c.** 268 million   **d.** 2020
e.

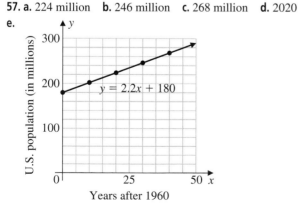

SKILL CHECKER **59.** $y = -2x + 4$ **61.** $y = \frac{2}{3}x + 4$

USING YOUR KNOWLEDGE **63.** $\frac{3}{2}$ **65.** $-2\sqrt{3}$

67. $\dfrac{\sqrt{2}}{4}$ **69.** 2, -1 **71.** 2 or -1

MASTERY TEST

77. 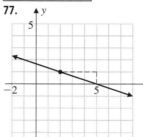   **79.** $y = 4$
81. a. Perpendicular
   b. Parallel
83. a. 10
   b. $2\sqrt{17}$
   c. 4
   d. 4

Exercise 7.3

A **1.** $3x - y = 4$ 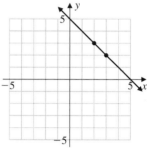   **3.** $x + y = 5$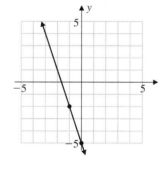

B **5.** $y - 5 = 2(x + 3)$ **7.** $y + 2 = -3(x + 1)$

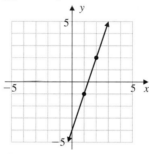   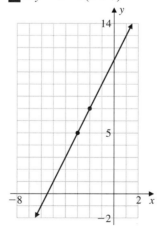

C **9.** $y = 5x + 2,\ 5x - y = -2$
11. $y = -\frac{1}{5}x - \frac{1}{3},\ 3x + 15y = -5$
D **13.** $y = 2x - 4$ **15.** $y = -3x - 12$ **17.** $y = -2x + 3$
19. $y = -x - 6$ **21.** $y = \frac{1}{3}x - \frac{5}{3}$ **23.** $y = -\frac{5}{2}x + 3$
25. $y = -\frac{1}{8}x + \frac{1}{2}$ **27.** $y = 4$ **29.** $x = -2$ **31.** $x = -2$
33. $y = 2$ **35.** $y = -2x + 4$ **37.** $y = \frac{2}{3}x + 2$
39. $y = -4$ **41.** $2x - 3y = -6$

43.

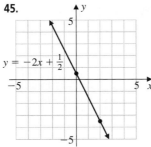

$y = \frac{2}{3}x + 2$

45.

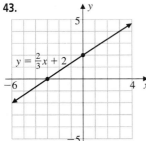

$y = -2x + \frac{1}{2}$

1. $x + 2y > 4$

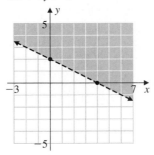

3. $-2x - 5y \le -10$

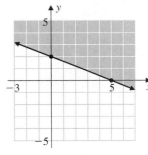

47.

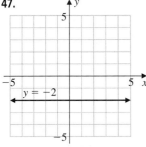

$y = -2$

5. $y \ge 2x - 2$

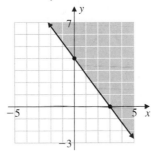

7. $6 < 3x - 2y$

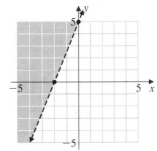

APPLICATIONS **49. a.** $d + 2p = 14$ **b.** 10
51. a. $s - 2p = -20$ **b.** $10 **c.** 60 **53.** $4
SKILL CHECKER **55.** x-intercept 4; y-intercept -2
57. x-intercept -3; y-intercept 6 **59.** $x = -2$
USING YOUR KNOWLEDGE **61.** $y = 2x + 50$
63. a. $2 **b.** $75
MASTERY TEST **69.** $m = -2, b = 4$

9. $4x + 3y \ge 12$

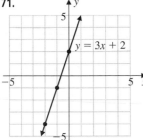

11. $10 < -5x + 2y$

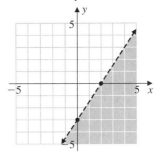

71.

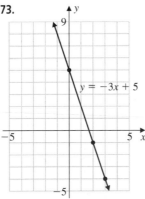

$y = 3x + 2$

73.

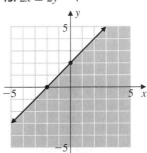

$y = -3x + 5$

13. $2x \ge 2y - 4$

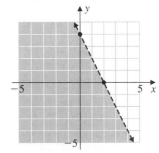

15. $2y < -4x + 8$

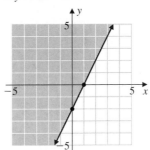

75.

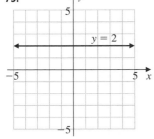

$y = 2$

17. $x \geq -3$

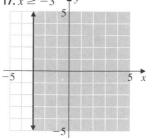

19. $y < 3$

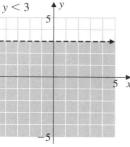

37. $|y - 2| < 1$

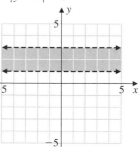

 21. $|x| < 1$

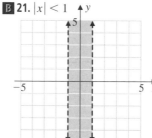

23. $|y| < 4$

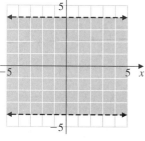

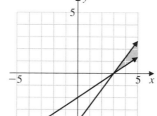 **39.** $x - y \geq 2$
and $x + y \leq 6$

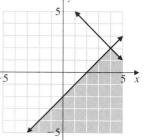

41. $2x - 3y \leq 6$
and $4x - 3y \geq 12$

25. $|x| \geq 1$

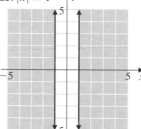

27. $|y| \geq 2$

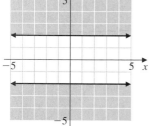

43. $2x - 3y \leq 5$ and $x \geq y$

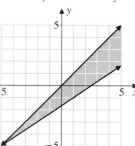

45. $x + 3y \leq 6$ and $x \geq y$

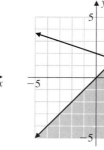

29. $|x + 2| < 1$

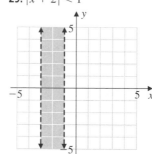

31. $|y + 2| < 1$

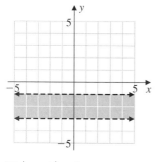

47. $x - y \leq 1$ and $3x - y < 3$

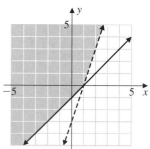

33. $|x + 1| \geq 3$

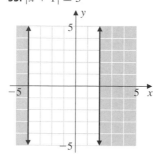

35. $|x - 1| \leq 2$

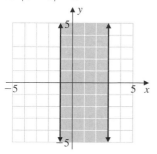

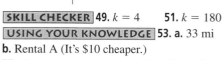 **SKILL CHECKER** **49.** $k = 4$ **51.** $k = 180$
USING YOUR KNOWLEDGE **53. a.** 33 mi
b. Rental A (It's $10 cheaper.)
55. If you plan to drive more than 33 mi, Rental A is the cheaper.

MASTERY TEST

61. $|x + 2| > 3$

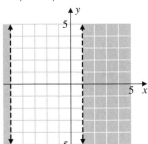

63. $|y| \le 1$

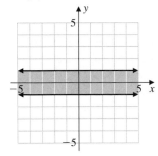

65. $x \ge -2$

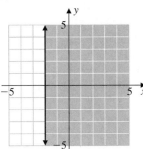

67. $3x - 2y < -6$

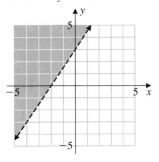

69. $x > 2$ and $y < -1$

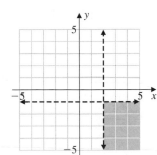

71. $x - 2y \le -2$ and $2y - x < 4$

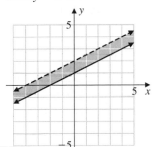

Exercise 7.5

A **1.** $T = ks$ **3.** $W = kh^3$ **5.** $W = kB$

B **7.** $R = \dfrac{k}{D^2}$

C **9.** $I = kPr$ **11.** $A = ksv^2$ **13.** $V = kdw$

15. $I = \dfrac{ki}{d^2}$ **17.** $R = \dfrac{kL}{A}$ **19.** $W = \dfrac{k}{d^2}$

APPLICATIONS **21. a.** $I = km$ **b.** $k = 0.055$ or 5.5% **c.** $\$41.25$

23. a. $d = ks^2$ **b.** $k = 0.06$ **c.** 216 ft

25. a. $S = \dfrac{k}{y}$ **b.** 30 songs

27. a. $W = \dfrac{k}{d^2}$ **b.** $k = 121(3960)^2$ **c.** 81 lb

29. a. $d = ks$ **b.** $k = 17.63$

c. k is the number of hours to travel the distance d at the speed s.

31. a. $C = 4(F - 37)$ **b.** 212

33. a. $I = kn$ **b.** $k = \dfrac{31.4}{23} \approx 1.365$

c. $319.9 + 47.8 = 367.7$ ppm

35. $\$208.33$ **37.** $C = 102.9$

SKILL CHECKER

39. $2x - y = 2$

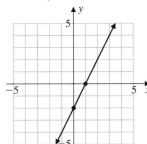

41. $y = -x - 3$

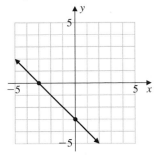

USING YOUR KNOWLEDGE

43.

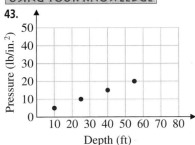

45. $p = kd$

47. They are equal.

MASTERY TEST **53. a.** $F = kAV^2$ **b.** $k = 0.0045$ **c.** 32.4 lb

55. a. $P = \dfrac{k}{r}$ **b.** $k = 10$

Exercise 7.6

A **1. a.** $D = \{-3, -2, -1\}$ **b.** $R = \{0, 1, 2\}$ **c.** A function
3. a. $D = \{3, 4, 5\}$ **b.** $R = \{0\}$ **c.** A function
5. a. $D = \{1, 2\}$ **b.** $R = \{2, 3\}$ **c.** Not a function
7. a. $D = \{1, 3, 5, 7\}$ **b.** $R = \{-1\}$ **c.** A function
9. a. $D = \{2\}$ **b.** $R = \{1, 0, -1, -2\}$ **c.** Not a function

B **11.**

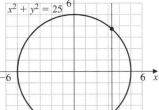

$D = \{x \mid -5 \le x \le 5\}$
$R = \{y \mid -5 \le y \le 5\}$
Not a function

13.

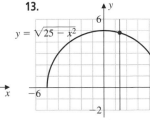

$D = \{x \mid -5 \le x \le 5\}$
$R = \{y \mid 0 \le y \le 5\}$
A function

15.

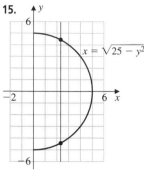

$D = \{x \mid 0 \le x \le 5\}$
$R = \{y \mid -5 \le y \le 5\}$
Not a function

17.

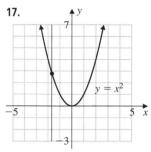

$D = \{x \mid x \text{ is a real number}$
$R = \{y \mid y \ge 0\}$
A function

19.

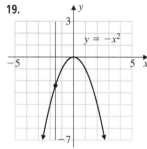

$D = \{x \mid x \text{ is a real number}\}$
$R = \{y \mid y \le 0\}$
A function

21.

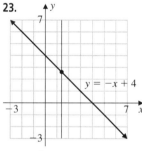

$D = \{x \mid x \ge 0\}$
$R = \{y \mid y \text{ is a real number}\}$
Not a function

23.

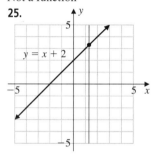

$D = \{x \mid x \text{ is a real number}\}$
$R = \{y \mid y \text{ is a real number)}$
A function

25.

$D = \{x \mid x \text{ is a real number}\}$
$R = \{y \mid y \text{ is a real number}\}$
A function

27.

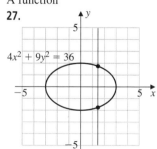

$D = \{x \mid -3 \le x \le 3\}$
$R = \{y \mid -2 \le y \le 2\}$
Not a function

29.

$D = \{x \mid x \ge 2 \text{ or } x \le -2\}$
$R = \{y \mid y \text{ is a real number}\}$
Not a function

31. Yes   **33.** Yes   **35.** Yes

37. $D = \{x \mid x \ge 5\}$ **39.** $D = \{x \mid x \le 2\}$
41. $D = \{x \mid x \text{ is a real number}\}$
43. $D = \{x \mid x \text{ is a real number and } x \ne 5\}$
45. $D = \{x \mid \text{ is a real number and } x \ne -5\}$
47. $D = \{x \mid x \text{ is a real number and } x \ne -1 \text{ or } -2\}$
49. $D = \{x \mid x \text{ is a real number and } x \ne \pm 4\}$
51. a. $f(0) = 1$ **b.** $f(2) = 7$ **c.** $f(-2) = -5$
53. a. $F(1) = 0$ **b.** $F(5) = 2$ **c.** $F(26) = 5$

55. a. $f(1) = \frac{1}{4}$ **b.** $f(1) - f(2) = \frac{3}{28}$ **c.** $\dfrac{f(1) - f(2)}{3} = \dfrac{1}{28}$

57. a. $f(3) = 5$ **b.** $g(3) = 19$ **c.** $f(3) + g(3) = 24$
59. a. $f(-2) = -10$ **b.** $g(-3) = 7$ **c.** $f(-2) \cdot g(-3) = -70$
61. a. $f(1) = 3$ **b.** $g(-2) = 4$ **c.** $f(1) + g(-2) = 7$
63. a. $f(-3) = 7$ **b.** $g(2) = 8$ **c.** $f(-3) \cdot g(2) = 56$
65. a. $P(x) = -0.0005x^2 + 24x - 100,000$
b. Maximum profit is $90,000.
67. a. $U(50) = 140$ **b.** $U(60) = 130$
69. a. 160 lb **b.** 78 in.   **71. a.** 639 lb/ft^2 **b.** 6390 lb/ft^2
73. a. L **b.** S **c.** Size 11 **d.** Size 12
75. a. 262 robberies per 100,000 population
b. 619.2 robberies per 100,000 population
77. a. Yes   **b.** $D = \{x \mid 1 \le x \le 6\}, R = \{y \mid 0 \le y \le 0.10\}$
c. $D = \{x \mid 1 \le x \le 4\}, R = \{y \mid 0.0 \le y \le 0.09\}$
d. $m = 0.0225$ **e.** $C = 0.0225D - 0.035$

USING YOUR KNOWLEDGE **85.** Yes   **87.** Yes   **89.** Yes
MASTERY TEST **91. a.** 3 **b.** -1 **c.** 4
93. $D = \{x \mid x \text{ is a real number and } x \ne 2\}$ **95.** Not a function
97. $D = \{x \mid -3 \le x \le 3\}, R = \{y \mid -3 \le y \le 3\}$
99. $D = \{x \mid -3 \le x \le 3\}, R = \{y \mid -3 \le y \le 0\}$
101. $D = \{7, 8, 9\}, R = (8, 9, 10)$

Review Exercises

1. a.

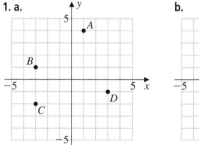

b.

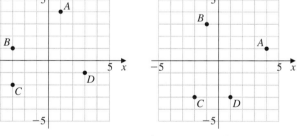

2. a.

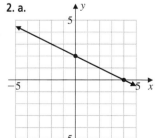

b.

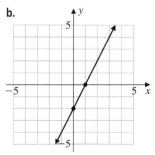

3. a. *x*-intercept -1; *y*-intercept 3   **b.** *x*-intercept 2; *y*-intercept -4

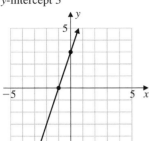
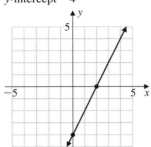

4.
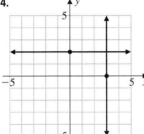

5.
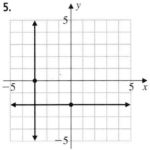

6. a. 13   **b.** $2\sqrt{5}$ **c.** 4   **7. a.** $-\frac{1}{2}$ **b.** Undefined

8. a. -1 **b.** $-\frac{3}{2}$ **9. a.** Perpendicular   **b.** Parallel

10. a. Parallel   **b.** Neither   **11. a.** $y = 2$ **b.** $y = \frac{11}{2}$

12. a.
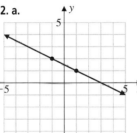

b.
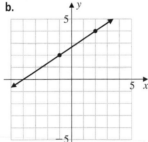

13. a. $x - y = -3$ **b.** $5x + 2y = -14$

14. a. $2x - y = 1$ **b.** $2x - y = 10$

15. a. $y = 3x + 2$ **b.** $y = -3x + 4$

16. a. Slope 2; *y*-intercept -4 **b.** Slope $-\frac{1}{2}$; *y*-intercept 2

17. a. Slope -2; *y*-intercept 4   **b.** Slope $\frac{4}{3}$; *y*-intercept 4

18. a. $2x + y = 5$ **b.** $3x - y = 5$

19. a. $3x - 2y = 4$ **b.** $2x + 3y = 7$

20. a.
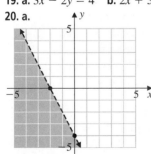

b.
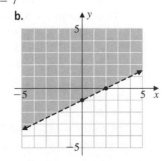

21. a.
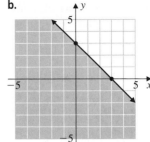

b.
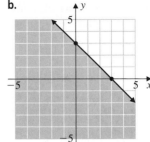

22. a.
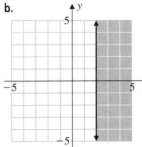

b.
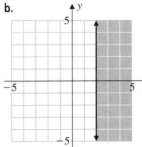

23. a.
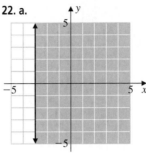

b.
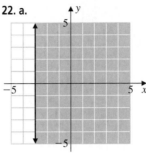

24. a.
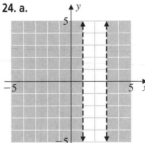

b.
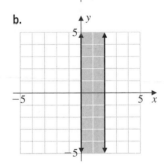

25. a. $P = kT$ **b.** $k = \frac{1}{120}$ **26. a.** $P = \dfrac{k}{V}$ **b.** $k = 3200$

27. a. $F = \dfrac{k}{d^2}$ **b.** $k = 1.92 \times 10^9$ **28.** 76.8 lb

29. a. $h = kd^3r$ **b.** $k = \frac{1}{20}$ **30.** 600 hp

31. a. $D = \{0, 2, 3, 5\}$ $R = (5, 8, 9, 10)$

b. $D = (0, 2, 3, 5)$ $R = \{6, 9, 10, 11\}$

32. a. $D = \{x \mid x \text{ is a real number}\}$
$R = \{y \mid y \text{ is a real number}\}$ **b.** $D = \{x \mid x \text{ is a real number}\}$
$R = \{y \mid y \text{ is a real number}\}$

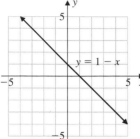

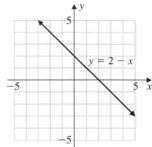

c. $D = \{x \mid x \text{ is a real number}\}$ $R = \{y \mid y \text{ is a real number}\}$

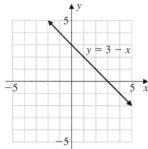

33. a. $D = \{x \mid -2 \le x \le 2\}$ $R = \{y \mid -2 \le y \le 2\}$
b. $D = \{x \mid -3 \le x \le 3\}$ $R = \{y \mid -3 \le y \le 3\}$
34. a. $D = \{x \mid -2 \le x \le 2\}$ $R = \{y \mid 0 \le y \le 2\}$
b. $D = \{x \mid -3 \le x \le 3\}$ $R = \{y \mid 0 \le y \le 3\}$
35. a. $D = \{x \mid -2 \le x \le 2\}$ $R = \{y \mid -2 \le y \le 0\}$
b. $D = \{x \mid -3 \le x \le 3\}$ $R = \{y \mid -3 \le y \le 0)$
36. a. A function   **b.** Not a function

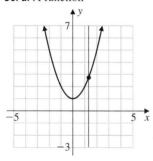

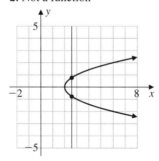

37. a. $D = \{x \mid x \text{ is a real number and } x \ne 1\}$
b. $D = \{x \mid x \text{ is a real number and } x \ne 2\}$
38. a. $D = \{x \mid x \ge 3\}$ **b.** $D = \{x \mid x \ge 4\}$
39. a. $f(2) = -2$ **b.** $f(1) = -3$ **c.** $f(2) - f(1) = 1$
40. a. $f(2) = 1$ **b.** $f(1) = -2$ **c.** $f(2) - f(1) = 3$
41. a. $f(2) = 0$ **b.** $f(1) = -1$ **c.** $f(2) - f(1) = 1$
42. a. $f(2) = 1$ **b.** $f(1) = 0$ **c.** $f(2) - f(1) = 1$

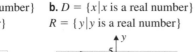

A 1. Consistent   **3.** Consistent

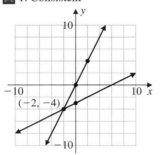

5. Consistent   **7.** Inconsistent

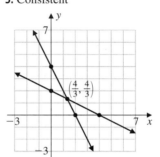

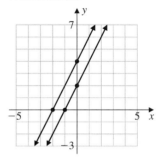

9. Dependent

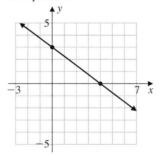

B **11.** $(2, 0)$; consistent   **13.** $\left(1, \frac{3}{2}\right)$; consistent
15. No solution; inconsistent   **17.** No solution; inconsistent
19. Infinitely many solutions; dependent
21. $(-1, -1)$; consistent   **23.** No solution; inconsistent
25. $(4, 1)$; consistent
C **27.** $(5, 3)$; consistent   **29.** $\left(0, \frac{1}{2}\right)$; consistent
31. $(0, 1)$; consistent   **33.** $(2, 3)$; consistent
35. $\left(\frac{5}{2}, -\frac{1}{2}\right)$; consistent   **37.** No solution; inconsistent
39. $(3, 2)$; consistent   **41.** Dependent; $\left\{(x, y) \mid y = \frac{2}{5}x - \frac{9}{5}\right\}$
43. $\left(\frac{1}{3}, 2\right)$; consistent   **45.** $(5, -2)$; consistent
47. $\left(-\frac{1}{2}, -\frac{2}{3}\right)$; consistent   **49.** $(8, -12)$; consistent
51. $(6, 8)$; consistent   **53.** $(4, -3)$; consistent
55. $(4, 2)$; consistent
D **57.** $p = 14$ **59.** $p = 200$ **61.** $p = 40$
63. $p = 28, D(28) = 512$ **65.** $p = 15, D(15) = S(15) = 435$
67. a. $x + y = 90, x = y + 15$ **b.** $x = 52.5°, y = 37.5°$

69. a. $x + y = 180$, $y = 4x$ **b.** $x = 36°$, $y = 144°$
71. a. $x + y = 465$, $x = y + 15$ **b.** 240 lb and 225 lb
73. a. $x + y = 4637$, $x = y + 437$ **b.** Japan: 2537; U.S.: 2100
75. 8.5 shekels **77.** 36 and 34
79. Antenna = 222 ft; building = 1250 ft **81.** 660 lb and 640 lb
SKILL CHECKER 83. x-intercept = 4, y-intercept = 6
85. x-intercept = $-\frac{4}{3}$, y-intercept = 2
USING YOUR KNOWLEDGE 87. Tweedledee = $120\frac{2}{3}$ lb,
Tweedledum = $119\frac{2}{3}$ lb
MASTERY TEST 93. $(-1, 3)$
95. Inconsistent; no solution **97.** $(3, 1)$
99. Inconsistent; no solution
101. Consistent; $(2, 2)$ **103.** Consistent; $(2, 1)$

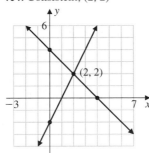

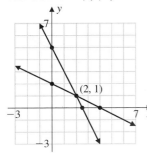

105. Inconsistent; no solution

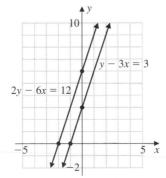

Exercise 8.2

A B 1. $(5, 3, 4)$; consistent **3.** $(-1, 1, 4)$; consistent
5. $(3, 4, 1)$; consistent **7.** No solution; inconsistent
9. $\left(\frac{1}{2}, \frac{1}{4}, \frac{1}{3}\right)$; consistent **11.** No solution; inconsistent
13. No solution; inconsistent **15.** $\left(\frac{9}{2}, \frac{1}{2}, \frac{5}{2}\right)$; consistent
17. $(6, 3, -1)$; consistent **19.** $(-2, -3, -4)$; consistent
21. $a = 1$, $b = 2$, $c = -3$ **23.** $\{(-1, 0, 2)\}$
25. $(-1, 0, 2)$, $k = -2$
APPLICATIONS 27. 8, 16, 25 **29.** 30°, 50°, 100°
31. corns: 70, heel pain: 41, ingrown toenail: 39
33. Pizza Hut: 24%; Domino's: 13%; and Little Caesar's: 11%
35. Gates: $6.3 billion; Kluge: $5.5 billion; and Waltons:
$5.1 billion
SKILL CHECKER 37. $12,000 at 5%; $8000 at 7%
39. 32 gal **41.** 250 mi/hr

USING YOUR KNOWLEDGE 43. $d = \dfrac{5n + 16}{216}$
MASTERY TEST 51. $(2, -1, 3)$
53. Inconsistent; no solution
55. Dependent; infinitely many solutions
$$\left\{(x, y, z)\,\middle|\,x = \frac{12 - 3z}{5}, y = \frac{4 - z}{5}, z = \text{any real number}\right\}$$
For example, $(0, 0, 4)$ and $(3, 1, -1)$ are two solutions.

Exercise 8.3

A 1. 25 nickels, 50 dimes **3.** 10 nickels, 20 quarters
5. 5 pennies, 5 nickels **7.** 20 tens, 5 twenties
9. 13 nickels, 9 dimes, 22 quarters
B 11. 43 and 59 **13.** 21 and 105 **15.** 4 and 20
17. Longs Peak is 14,255 ft high; Pikes Peak is 14,110 ft high
19. Butterscotch: $2033\frac{1}{3}$ lb; caramel: $2033\frac{1}{3}$ lb; and chocolate:
$2633\frac{1}{3}$ lb
C 21. Plane: 210 mi/hr, wind: 30 mi/hr
23. Boat: 12 mi/hr, current: 3 mi/hr
25. Plane: 450 mi/hr, wind: 50 mi/hr
D 27. $6000 at 8%: $4000 at 6%
29. $10,000 at 6%; $5000 at 8%; and $10,000 at 10%
E 31. $l = 505$ ft, $w = 255$ ft **33.** $l = 134$ ft, $w = 85$ ft
SKILL CHECKER 35. $(1, -2, -1)$
USING YOUR KNOWLEDGE 37. $15a + 5b + 20c = 170$
39. $10a + 15b + 10c = 110$
MASTERY TEST 43. $l = 50$ cm, $w = 35$ cm
45. Wind: 50 mi/hr; plane: 350 mi/hr **47.** 723 lb and 743 lb

Exercise 8.4

A 1. $(1, 2, 0)$ **3.** $(-1, -1, 3)$
5. $(-2, -1, 3)$ **7.** Inconsistent; no solution
9. Dependent; $(1 - k, 2, k)$, k any real number
C 13. Type I: 8; type II: 10; type III: 12
15. Type I: 50%; type II: 25%; type III: 25%
17. a. Singular **b.** Nonsingular **c.** Nonsingular
SKILL CHECKER 19. -150 **21.** -473 **23.** $\frac{1}{8}$
MASTERY TEST 29. $(6, -3, 4)$
31. Inconsistent; no solution
33. Dependent; $(1, k, k)$, k any real number

Exercise 8.5

A 1. 2 **3.** 7 **5.** 6 **7.** $\frac{1}{2}$ **9.** $-\frac{7}{40}$
B 11. $(2, 3)$ **13.** $(4, 5)$ **15.** $(3, -1)$ **17.** $(4, 5)$
19. Dependent; infinitely many solutions, $\left(x, \dfrac{-2x - 13}{3}\right)$
21. $(-2, -3)$ **23.** Inconsistent; no solution
25. $(10, 1)$ **27.** $(5, 2)$ **29.** $(-1, -1)$

C **31.** -7 **33.** 0 **35.** -1 **37.** -4 **39.** -9
D **41.** $(1, 2, 3)$ **43.** $(3, -1, -2)$ **45.** $(3, 0, 4)$
47. $(-5, 1, 5)$ **49.** $(-6, 2, 5)$

51.
$$\begin{vmatrix} a & b & 0 \\ c & d & 0 \\ e & f & 0 \end{vmatrix}$$

$$= a\begin{vmatrix} d & 0 \\ f & 0 \end{vmatrix} - b\begin{vmatrix} c & 0 \\ e & 0 \end{vmatrix} + 0\begin{vmatrix} c & d \\ e & f \end{vmatrix}$$

$$= a \cdot 0 - b \cdot 0 + 0 = 0$$

53.
$$\begin{vmatrix} a & b & c \\ 1 & 2 & 3 \\ a & b & c \end{vmatrix} = a\begin{vmatrix} 2 & 3 \\ b & c \end{vmatrix} - b\begin{vmatrix} 1 & 3 \\ a & c \end{vmatrix} + c\begin{vmatrix} 1 & 2 \\ a & b \end{vmatrix}$$

$$= a(2c - 3b) - b(c - 3a) + c(b - 2a)$$

$$= 2ac - 3ab - bc + 3ab + bc - 2ac$$

$$= 0$$

55.
$$\begin{vmatrix} 1 & 2 & 3 \\ 3 & 1 & 2 \\ k & 2k & 3k \end{vmatrix} \overset{?}{=} k\begin{vmatrix} 1 & 2 & 3 \\ 3 & 1 & 2 \\ 1 & 2 & 3 \end{vmatrix}$$

$$1\begin{vmatrix} 1 & 2 \\ 2k & 3k \end{vmatrix} - 2\begin{vmatrix} 3 & 2 \\ k & 3k \end{vmatrix} + 3\begin{vmatrix} 3 & 1 \\ k & 2k \end{vmatrix}$$

$$\overset{?}{=} k\left[1\begin{vmatrix} 1 & 2 \\ 2 & 3 \end{vmatrix} - 2\begin{vmatrix} 3 & 2 \\ 1 & 3 \end{vmatrix} + 3\begin{vmatrix} 3 & 1 \\ 1 & 2 \end{vmatrix} \right]$$

$$1(3k - 4k) - 2(9k - 2k) + 3(6k - k)$$

$$\overset{?}{=} k[1(3 - 4) - 2(9 - 2) + 3(6 - 1)]$$

$$- k - 14k + 15k$$

$$\overset{?}{=} k[-1 - 14 + 15]$$

$$0 \cdot k \overset{?}{=} k \cdot 0$$

$$0 = 0$$

57.
$$\begin{vmatrix} kb_1 & b_1 & 1 \\ kb_2 & b_2 & 2 \\ kb_3 & b_3 & 3 \end{vmatrix} = kb_1\begin{vmatrix} b_2 & 2 \\ b_3 & 3 \end{vmatrix} - b_1\begin{vmatrix} kb_2 & 2 \\ kb_3 & 3 \end{vmatrix} + 1\begin{vmatrix} kb_2 & b_2 \\ kb_3 & b_3 \end{vmatrix}$$

$$= kb_1(3b_2 - 2b_3) - b_1(3kb_2 - 2kb_3) + 1(kb_2b_3 - kb_2b_3)$$

$$= 3kb_1b_2 - 2kb_1b_3 - 3kb_1b_2 + 2kb_1b_3 + 0$$

$$= 0$$

59.
$$\begin{vmatrix} 1 & 1 & 1 \\ 2 & a & a \\ 3 & b & b \end{vmatrix} = 1\begin{vmatrix} a & a \\ b & b \end{vmatrix} - 1\begin{vmatrix} 2 & a \\ 3 & b \end{vmatrix} + 1\begin{vmatrix} 2 & a \\ 3 & b \end{vmatrix}$$

$$= 1(ab - ab) - (2b - 3a) + (2b - 3a) = 0$$

SKILL CHECKER **61.** $(x + 3)(x + 1)$ **63.** $(x + 3)(x - 1)$
65. $-1(x - 3)(x - 1)$

USING YOUR KNOWLEDGE **67.** $2x - y + 3 = 0$
69. $2x + 9y - 34 = 0$ **71.** $bx + ay - ab = 0$

MASTERY TEST

79. $2\begin{vmatrix} 1 & 0 \\ -1 & 1 \end{vmatrix} + 1\begin{vmatrix} 1 & 0 \\ 0 & 1 \end{vmatrix} - 3\begin{vmatrix} 1 & 1 \\ 0 & -1 \end{vmatrix}$

$$= 2 \cdot 1 + 1 \cdot 1 - 3 \cdot (-1) = 6$$

81. Inconsistent, no solution   **83.** -19 **85.** 8

Review Exercises

1. a. 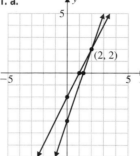 **b.**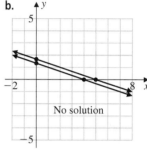

2. a. 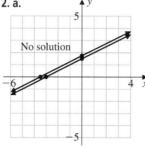 No solution **b.** 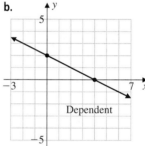 No solution

3. a. 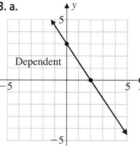 Dependent **b.** Dependent

4. a. $(3, 2)$ **b.** $\left(\frac{7}{5}, \frac{12}{5}\right)$
5. a. Inconsistent; no solution   **b.** Inconsistent; no solution
6. a. Dependent; infinitely many solutions, $\left(x, \frac{1}{2}x + \frac{5}{2}\right)$
b. Dependent; infinitely many solutions, $\left(x, 1 - \frac{1}{5}x\right)$
7. a. $(4, -1)$ **b.** $(-1, 2)$
8. a. Inconsistent; no solution   **b.** Inconsistent; no solution
9. a. Dependent; infinitely many solutions, $\left(x, \dfrac{2 - 2x}{5}\right)$

b. Dependent; infinitely many solutions, $\left(x, \frac{4}{3}x - 8\right)$
10. a. $(2, -3)$ **b.** $(1, 1)$
11. a. Dependent; infinitely many solutions, $(2, k, k + 2)$,
k is any real number
b. Dependent; infinitely many solutions, $(k, -1, -2k)$,
k is any real number
12. a. Inconsistent; no solution   **b.** Inconsistent; no solution
13. a. Dependent; infinitely many solutions, $(2 - k, 2 - k, k)$,
k is any real number
b. Dependent; infinitely many solutions, $(4k - 8, 6 - 2k, k)$,
k is any real number

14. a. 30 nickels   25 dimes   **b.** 10 nickels   15 dimes
15. a. 180 ft   **b.** 160 ft   **16. a.** 6 mi/hr   **b.** 3 mi/hr
17. a. $10,000 at 4% $10,000 at 6% $20,000 at 8%
b. $10,000 at 4% $15,000 at 6% $20,000 at 8%
18. a. 10 in. by 40 in.   **b.** 10 in. by 30 in.
19. a. $(3, -2, 5)$ **b.** Dependent; infinitely many solutions,

$\left(\dfrac{20 + 11k}{7}, \dfrac{2 + 6k}{7}, k\right)$, k is any real number

20. a. Inconsistent; no solution   **b.** $(1, 2, -1)$
21. a. -22 **b.** 14
22. a. $D = -23, D_x = -23, D_y = 46$ $x = 1, y = -2$
b. $D = -28, D_x = -14, D_y = 14$ $x = \frac{1}{2}, y = -\frac{1}{2}$
23. a. 15 **b.** -18
24. a. 9 **b.** 45 **25. a.** -10 **b.** -20
26. a. -1 **b.** -4
27. a. $D = -10, D_x = -20, D_y = -30, D_z = 20,$ $(2, 3, -2)$
b. $D = 6, D_x = 6, D_y = -12, D_z = -6,$ $(1, -2, -1)$
28. a. Inconsistent; no solution   **b.** Inconsistent; no solution

CHAPTER 9
Exercise 9.1

A 1.

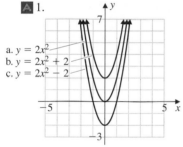

a. $y = 2x^2$
b. $y = 2x^2 + 2$
c. $y = 2x^2 - 2$

3.

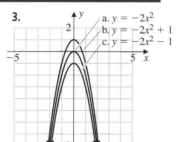

a. $y = -2x^2$
b. $y = -2x^2 + 1$
c. $y = -2x^2 - 1$

5.

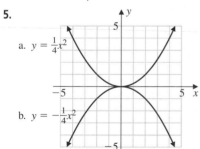

a. $y = \frac{1}{4}x^2$
b. $y = -\frac{1}{4}x^2$

7.

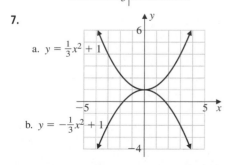

a. $y = \frac{1}{3}x^2 + 1$
b. $y = -\frac{1}{3}x^2 + 1$

B 9.

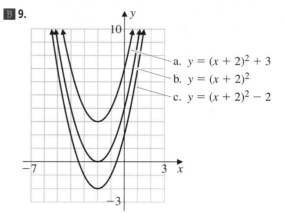

a. $y = (x + 2)^2 + 3$
b. $y = (x + 2)^2$
c. $y = (x + 2)^2 - 2$

11.

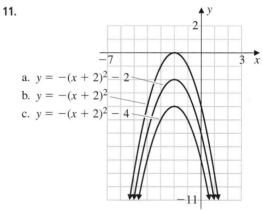

a. $y = -(x + 2)^2 - 2$
b. $y = -(x + 2)^2$
c. $y = -(x + 2)^2 - 4$

13.

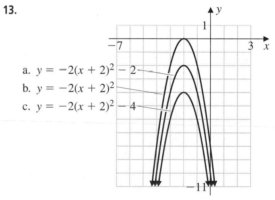

a. $y = -2(x + 2)^2 - 2$
b. $y = -2(x + 2)^2$
c. $y = -2(x + 2)^2 - 4$

15.

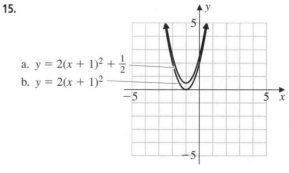

a. $y = 2(x + 1)^2 + \frac{1}{2}$
b. $y = 2(x + 1)^2$

C **17.**

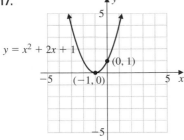

$y = x^2 + 2x + 1$

$(0, 1)$

$(-1, 0)$

19.

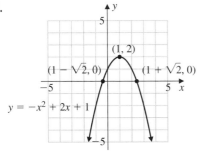

$(1, 2)$

$(1 - \sqrt{2}, 0)$ $(1 + \sqrt{2}, 0)$

$y = -x^2 + 2x + 1$

21.

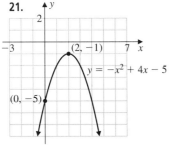

$(2, -1)$

$y = -x^2 + 4x - 5$

$(0, -5)$

23.

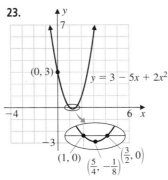

$(0, 3)$

$y = 3 - 5x + 2x^2$

$(1, 0)$ $\left(\frac{5}{4}, -\frac{1}{8}\right)$ $\left(\frac{3}{2}, 0\right)$

25.

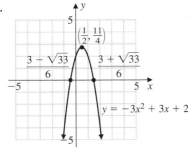

$(-1, 7)$

$y = 5 - 4x - 2x^2$ $(0, 5)$

$\dfrac{-4 - \sqrt{56}}{4} = \dfrac{-2 - \sqrt{14}}{2}$ $\dfrac{-4 + \sqrt{56}}{4} = \dfrac{-2 + \sqrt{14}}{2}$

27.

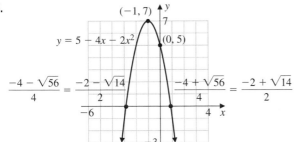

$\left(\frac{1}{2}, \frac{11}{4}\right)$

$\dfrac{3 - \sqrt{33}}{6}$ $\dfrac{3 + \sqrt{33}}{6}$

$y = -3x^2 + 3x + 2$

D **29.**

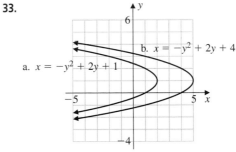

a. $x = (y + 2)^2$

b. $x = (y + 2)^2 + 3$

31.

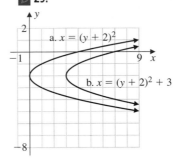

b. $x = -(y + 2)^2$

a. $x = -(y + 2)^2 - 2$

33.

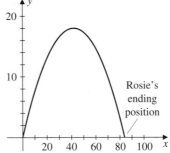

b. $x = -y^2 + 2y + 4$

a. $x = -y^2 + 2y + 1$

35. $x = 4000$, $P = \$11{,}000$ **37.** \$25(thousand) or \$25,000

39. 400 ft

41. $P = (600 + 100W)(1 - 0.10W)$; $P =$ price, $W =$ the weeks elapsed. The maximum for P occurs when $W = 2$ (at end of two weeks).

43. a. $(42, 18)$ **b.** 18 in.   **c.** 84 in.

d.

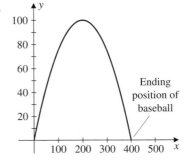

Rosie's ending position

45. a. $(200, 100)$ **b.** 100 ft   **c.** 400 ft

d.

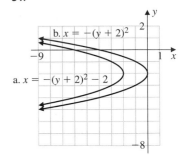

Ending position of baseball

47. a. about 472   **b.** about 465   **c.** $a > 0$

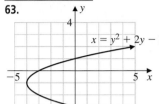

SKILL CHECKER **49.** $\sqrt{29} \approx 5.4$ units

USING YOUR KNOWLEDGE **51.** $FP = \sqrt{x^2 + (y - p)^2}$

53. $x^2 = 4py$

55. $y^2 = 12.5x$; focus $(3.125, 0)$ or $x^2 = 12.5y$; focus $(0, 3.125)$

MASTERY TEST

63.

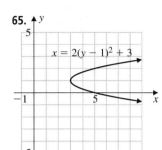

65.

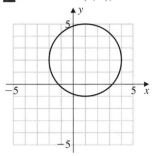

67.

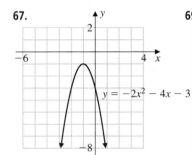

69.

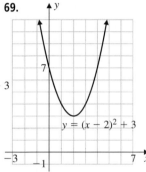

71.

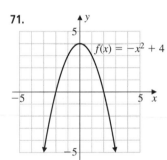

73.

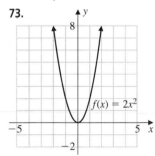

75.

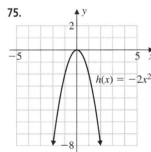

77. $10

Exercise 9.2

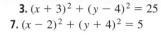

1. $(x - 3)^2 + (y - 8)^2 = 4$ **3.** $(x + 3)^2 + (y - 4)^2 = 25$

5. $(x + 3)^2 + (y + 2)^2 = 16$ **7.** $(x - 2)^2 + (y + 4)^2 = 5$

9. $x^2 + y^2 = 9$

11. Center at $(1, 2)$, radius 3   **13.** Center at $(-1, 2)$, radius 2

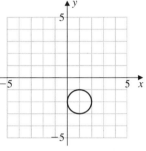
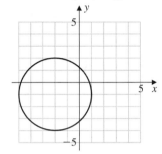

15. Center at $(1, -2)$, radius 1   **17.** Center at $(-2, -1)$, radius 3

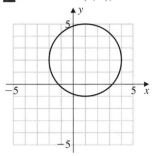
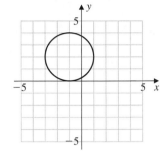

19. Center at $(1, 1)$, radius $\sqrt{7} \approx 2.6$ **21.** Center at $(3, 2)$, radius 2

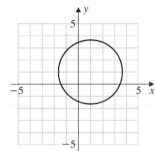
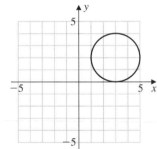

23. Center at $(2, -1)$, radius 3   **25.** Center at $(0, 0)$, radius 5

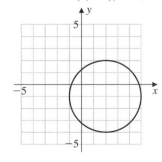
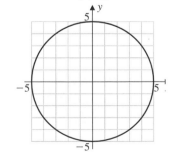

27. Center at $(0, 0)$, radius $\sqrt{7} \approx 2.6$

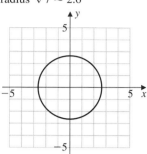

29. Center at $(-3, 1)$, radius 2

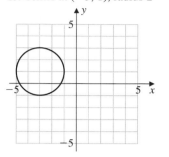

45.

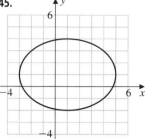

47. $x^2 + y^2 = 25$ **49.** $x^2 + y^2 = 169$ **51.** $x^2 + y^2 = 9$

53. $\dfrac{x^2}{4} + \dfrac{y^2}{36} = 1$ **55.** $\dfrac{x^2}{36} + \dfrac{y^2}{16} = 1$ **57.** 12.5 in.

59. 22.4 ft **61.** $\dfrac{x^2}{16} + \dfrac{y^2}{6.25} = 1$ **63.** $\dfrac{x^2}{16} + \dfrac{y^2}{9} = 1$

65. a. $a = 93$ **b.** $c = 1.5$ **c.** $b \approx \sqrt{8647} \approx 92.99$

67. a. $\dfrac{x^2}{36} + \dfrac{y^2}{20.25} = 1$ **b.** $3\sqrt{5}$ in. $= 6.71$ in.

69. About 4.5 ft from the side of the boat

SKILL CHECKER **71.** Right half of the circle

USING YOUR KNOWLEDGE **73.** $PF_1 = \sqrt{(x - c)^2 + y^2}$

75. $4xc = 4a^2 - 4a\sqrt{(x - c)^2 + y^2}$

or $a^2 - cx = a\sqrt{(x - c)^2 + y^2}$

77. $b^2x^2 + a^2y^2 = a^2b^2$

MASTERY TEST

31. Center at $(3, 1)$, radius 2

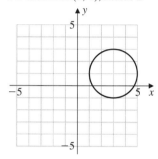

83. $\dfrac{(x + 3)^2}{4} + \dfrac{(y + 1)^2}{9} = 1$ **85.** $4x^2 + 9y^2 = 36$

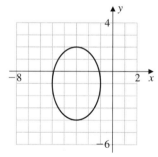

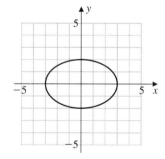

33.

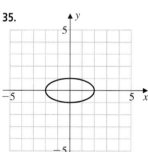

35.

37.

39.

87. $x^2 - 4x + y^2 - 6y + 9 = 0$ **89.** $x^2 + y^2 = 4$

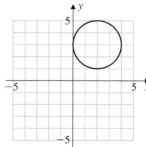

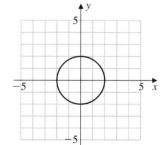

41.

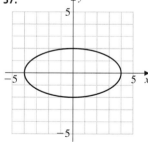

43.

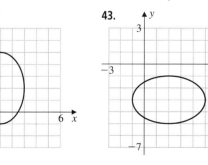

91. $(x - 3)^2 + (y - 1)^2 = 4$ Center (3, 1), radius 2

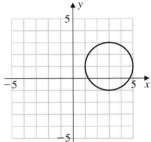

93. $x^2 + y^2 = 25$
95. $(x + 3)^2 + (y - 6)^2 = 9$

Exercise 9.3

A 1. $\dfrac{x^2}{25} - \dfrac{y^2}{9} = 1$;

intercepts: $(\pm 5, 0)$

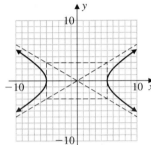

3. $\dfrac{y^2}{9} - \dfrac{x^2}{9} = 1$;

intercepts: $(0, \pm 3)$

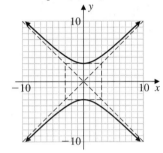

5. $\dfrac{x^2}{9} - \dfrac{y^2}{1} = 1$;

intercepts: $(\pm 3, 0)$

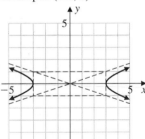

7. $\dfrac{x^2}{64} - \dfrac{y^2}{49} = 1$;

intercepts: $(\pm 8, 0)$

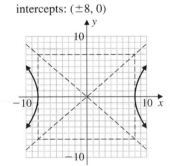

9. $\dfrac{y^2}{\frac{16}{9}} - \dfrac{x^2}{\frac{9}{16}} = 1$;

intercepts: $(0, \pm\frac{4}{3})$

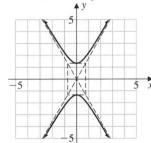

11. $x^2 - 9y^2 = 9$;

intercepts: $(\pm 3, 0)$

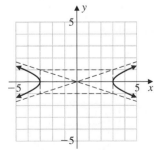

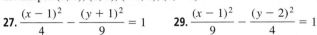 **13.** Circle (5, 0), (0, 5), (−5, 0), (0, −5)
15. Hyperbola (6, 0), (−6, 0)   **17.** Parabola (0, −9); [(±3, 0) int]
19. Parabola (−4, 0); [(0, ±2) int]
21. Circle (2, 0), (0, 2), (−2, 0), (0, −2)
23. Hyperbola (2, 0), (−2, 0)
25. Ellipse (3, 0), (0, 1), (−3, 0), (0, −1)

27. $\dfrac{(x - 1)^2}{4} - \dfrac{(y + 1)^2}{9} = 1$

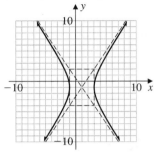

29. $\dfrac{(x - 1)^2}{9} - \dfrac{(y - 2)^2}{4} = 1$

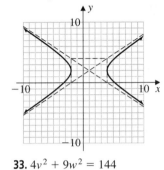

31. a. A hyperbola

b. $\dfrac{D^2}{8} - \dfrac{d^2}{4} = 1$

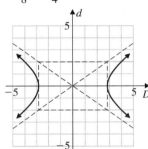

33. $4v^2 + 9w^2 = 144$

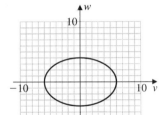

SKILL CHECKER

35. $y = x - 4$

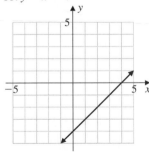

37. $y = x^2 + 1$

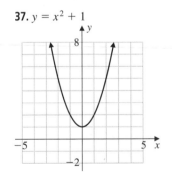

39. $4x^2 + 9y^2 = 36$

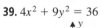
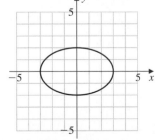

41. $PF_1 = \sqrt{(x-c)^2 + y^2}$

43. $cx + a^2 = \pm a\sqrt{(x+c)^2 + y^2}$

45. $b^2x^2 - a^2y^2 = a^2b^2$

47. $\left(\dfrac{x}{a} + \dfrac{y}{b}\right)\left(\dfrac{x}{a} - \dfrac{y}{b}\right) = 1$

49. The denominator becomes very large and the complex fraction approaches zero.

51. $y = -\dfrac{b}{a}x$

57. A circle, center at the origin, radius 3

59. A parabola opening upward, vertex at (0, 3)

61. $\dfrac{y^2}{16} - \dfrac{x^2}{9} = 1$

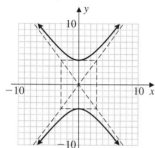

63. $9x^2 - 25y^2 = 225$

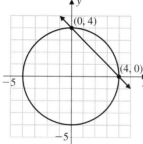

Exercise 9.4

A **1.** (0, 4), (4, 0)

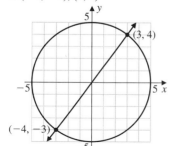

3. (−5, 0), (0, 5)

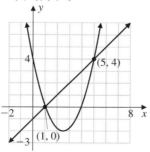

5. (−4, −3), (3, 4)

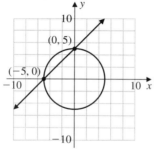

7. (1, 0), (5, 4)

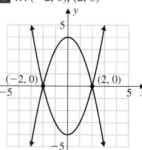

9. (0, 1), (3, 4)

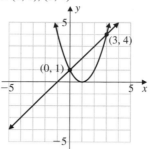

11. (−3, 0), (0, 2)

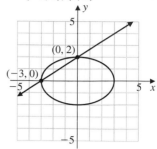

13. $(4, 0), \left(-\dfrac{68}{15}, \dfrac{32}{15}\right)$

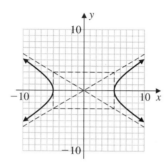

15. There is no real-number solution:

$$\left(\dfrac{-5 + \sqrt{17}i}{2}, \dfrac{5 + \sqrt{17}i}{2}\right)$$

$$\left(\dfrac{-5 - \sqrt{17}i}{2}, \dfrac{5 - \sqrt{17}i}{2}\right)$$

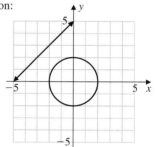

B **17.** (−2, 0), (2, 0)

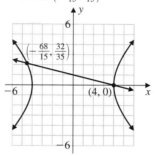

19. (−4, ±3), (4, ±3)

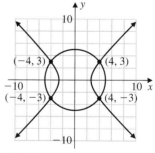

21. (−4, 0), (4, 0)

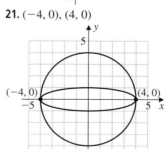

23. (±1, ±1)

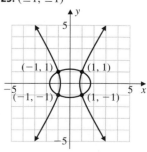

25. $(\pm 3, \pm 1)$

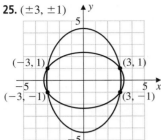

27. There is no real solution.

$$\left(\pm\frac{\sqrt{26}}{2}, \pm\frac{\sqrt{10}i}{2}\right)$$

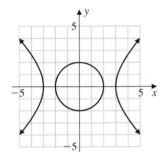

29. There is no real solution.

$$\left(\pm\frac{3\sqrt{15}i}{5}, \pm\frac{4\sqrt{10}}{5}\right)$$

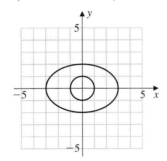

31. 1000 or 4000   **33.** 8 and 7

35. 11 and 16   or -11 and -16 **37.** 70 ft by 31 ft

39. $P = \$13,600, r = 2.5\%$

SKILL CHECKER

41. $x - y < 4$

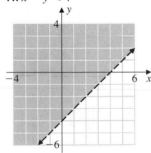

43. $2x - 3y \geq 6$

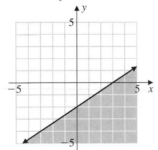

45. $y \geq 2x + 4$

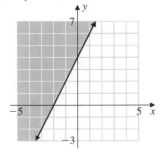

USING YOUR KNOWLEDGE   **47.** 20 eggs

MASTERY TEST   **53.** 12 cm by 10 cm

55. The solutions are $(0, \pm 1)$.

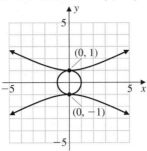

57. There is no real number solution. $(2 + \sqrt{2}i, 2 - \sqrt{2}i)$, $(2 - \sqrt{2}i, 2 + \sqrt{2}i)$

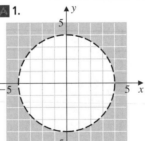

59. The solutions are $(0, 1), (1, 0)$

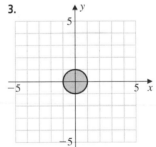

Exercise 9.5

1.

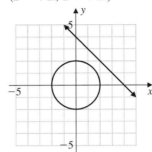

3.

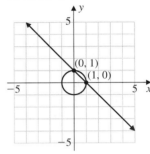

5.

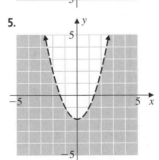

7.

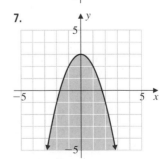

9.

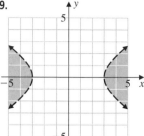

11.

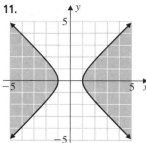

13.

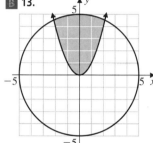

15.

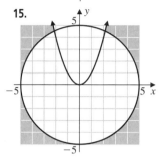

17.

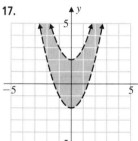

19.

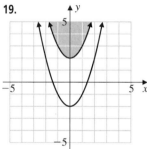

21.

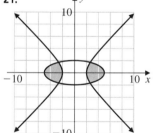

23.

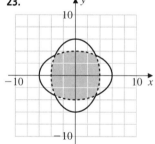

25. a. $\dfrac{x^2}{36} + \dfrac{y^2}{16} > 1$ $\dfrac{x^2}{16} + \dfrac{y^2}{16} < 1$

b. $\dfrac{x^2}{36} + \dfrac{y^2}{16} \geq 1$ $\dfrac{x^2}{16} + \dfrac{y^2}{16} \leq 1$

SKILL CHECKER **27.** -1

29. $P(x) + Q(x) = x^2 + x - 6$ $P(x) - Q(x) = x^2 - x - 12$

USING YOUR KNOWLEDGE **31.** $C = 432{,}000 - 1800p$

33. $p = 180$ or $p = 80$

MASTERY TEST

39. $x^2 + 4y^2 \geq 4$
and $y \geq x^2 + 1$

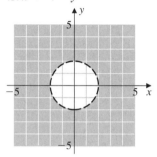

41. $x^2 - 9y^2 \geq 9$

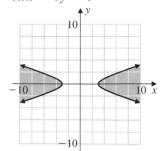
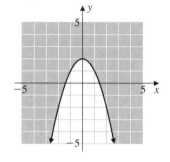

43. $x^2 > 4 - y^2$

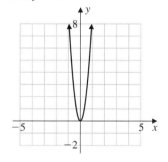

45. $y \geq -x^2 + 2$

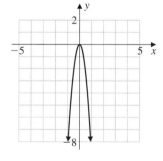

Review Exercises

1. a. $y = 9x^2$

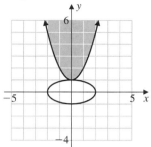

b. $y = -9x^2$

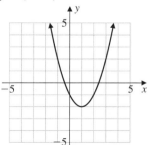

2. a. Vertex at $(1, -2)$
$y = (x - 1)^2 - 2$

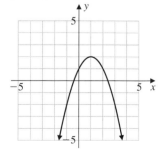

b. Vertex at $(1, 2)$
$y = -(x - 1)^2 + 2$

3. a. Vertex at $(2, -2)$
$y = x^2 - 4x + 2$

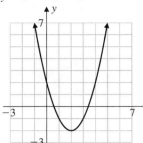

b. Vertex at $(3, 4)$
$y = -x^2 + 6x - 5$

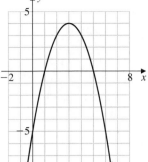

7. a. $x = 1000$ **b.** $x = 250$
8. a. $(x + 2)^2 + (y - 2)^2 = 9$ **b.** $(x - 3)^2 + (y + 2)^2 = 9$
9. a. $x^2 + y^2 = 25$ **b.** $x^2 + y^2 = 64$
10. a. Center at $(-2, 1)$, radius 2
$(x + 2)^2 + (y - 1)^2 = 4$

b. Center at $(1, -2)$, radius 3
$(x - 1)^2 + (y + 2)^2 = 9$

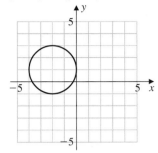

4. a. Vertex at $(1, 1)$
$y = 2x^2 - 4x + 3$

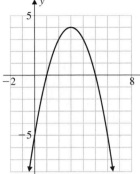

b. Vertex at $(1, -3)$
$y = -2x^2 + 4x - 5$

11. a. $x^2 + y^2 = 4$

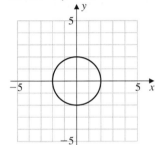

b. $x^2 + y^2 = 25$

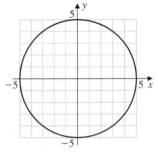

5. a. Vertex at $(-2, 2)$
$x = 2(y - 2)^2 - 2$

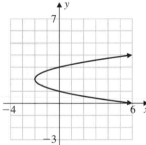

b. Vertex at $(1, 3)$
$x = -2(y - 3)^2 + 1$

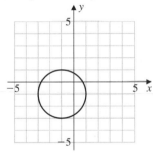

12. a. Center at $(-1, -1)$,
radius 2
$x^2 + y^2 + 2x + 2y - 2 = 0$

b. Center at $(2, -3)$, radius 2
$x^2 + y^2 - 4x + 6y + 9 = 0$

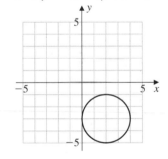

6. a. Vertex at $(-3, 2)$
$x = y^2 - 4y + 1$

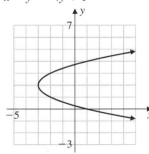

b. Vertex at $(2, 1)$
$x = y^2 - 2y + 3$

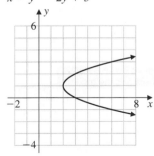

13. a. $4x^2 + 9y^2 = 36$

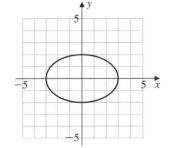

b. $9x^2 + y^2 = 9$

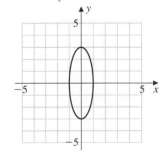

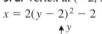
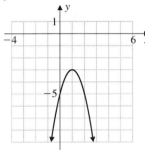
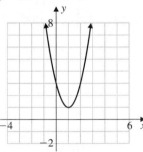
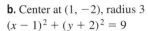

14. a. $\dfrac{(x-1)^2}{4} + \dfrac{(y-2)^2}{9} = 1$ **b.** $\dfrac{(x+2)^2}{9} + \dfrac{(y-2)^2}{4} = 1$

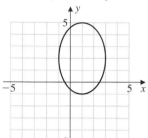   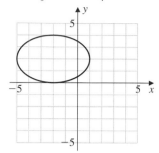

15. a. $\dfrac{x^2}{9} - \dfrac{y^2}{16} = 1$ **b.** $\dfrac{x^2}{16} - \dfrac{y^2}{9} = 1$

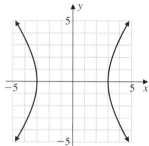   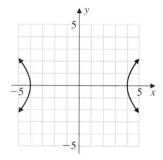

16. a. $\dfrac{y^2}{9} - \dfrac{x^2}{16} = 1$ **b.** $\dfrac{y^2}{16} - \dfrac{x^2}{9} = 1$

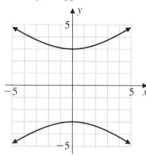   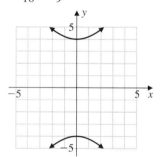

17. a. A parabola   **b.** A hyperbola   **c.** A circle   **d.** An ellipse
18. a. (0, 1) and (1, 0)   **b.** (1, 3) and (3, 1)

19. a. $\left(\dfrac{4\sqrt{3}}{3}i, \dfrac{8\sqrt{3}}{3}i\right), \left(-\dfrac{4\sqrt{3}}{3}i, -\dfrac{8\sqrt{3}}{3}i\right)$

b. $\left(\dfrac{3+i}{2}, \dfrac{3-i}{2}\right), \left(\dfrac{3-i}{2}, \dfrac{3+i}{2}\right)$

20. a. $(\pm 3, \pm 2)$ **b.** $(\pm 2, \pm 1)$ **21. a.** $x = 30$ **b.** $x = 20$

22. a. $y \leq 1 - x^2$ **b.** $x \leq 4 - y^2$

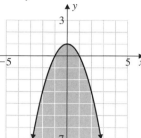   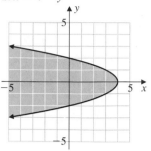

23. a. $x^2 + y^2 \leq 4$ **b.** $4x^2 + y^2 > 9$

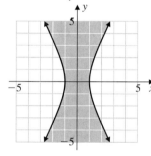   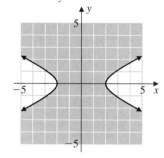

24. a. $4x^2 - y^2 \leq 4$ **b.** $x^2 - 4y^2 \leq 4$

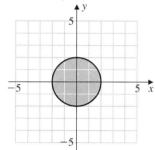   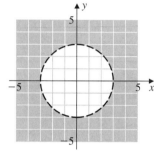

25. a. $x^2 + y^2 \leq 4$ $y \leq 2 - x^2$ **b.** $x^2 + y^2 \leq 4$ $y \geq 4x^2$

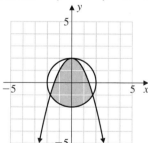   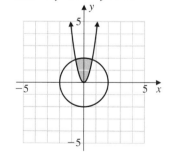

CHAPTER 10
Exercise 10.1

1. $f + g = x^2 - 4x + 8$ **3.** $hf = x^3 + 4x^2 + 16x + 64$

5. $\dfrac{f}{h} = \dfrac{x+4}{x^2+16}$

B **7. a.** $f \circ g(x) = x$ **b.** $g \circ f(x) = x$
9. a. $f \circ g = 3x + 1$ **b.** $g \circ f = 3x - 1$
11. a. $f \circ g = x$ **b.** $g \circ f = x$ **13. a.** $f \circ g = 3$ **b.** $g \circ f = -1$
15. 3   **17.** $x + a$ **19.** $x + a + 3$
21. a. $(f + g)(4) = 17$ **b.** $(f - g)(4) = -13$
23. a. $\left(\dfrac{f}{g}\right)(3) = 1$ **b.** $\left(\dfrac{f}{g}\right)(0) = 0$
25. a. Not defined   **b.** Not defined
27. a. $(f \circ g)(1) = \sqrt{2}$ **b.** $(g \circ f)(-1) = 0$
c. $(f \circ g)(x) = \sqrt{x^2 + 1}$ **d.** $(g \circ f)(x) = x + 1$

29. a. $(f \circ g)(2)$ is not defined.   **b.** $(g \circ f)(2) = \dfrac{1}{\sqrt{2}} = \dfrac{\sqrt{2}}{2}$

c. $(f \circ g)(x) = \dfrac{1}{x - 2}$ **d.** $(g * f)(x) = \dfrac{1}{\sqrt{x^2 - 2}} = \dfrac{\sqrt{x^2 - 2}}{x^2 - 2}$

31. $-0.0005x^2 + 34x - 120,000$
33. a. $-0.28x^2 + 3x + 22$ **b.** 22 million   **c.** 24 million
d. 6 million
35. a. $\dfrac{2.5t^2 + 8.5t + 111}{-0.46t^2 + 1.14t + 31.08}$ thousands   **b.** \$3571   **c.** \$8544
37. a. $(K \circ C)(F) = \frac{5}{9}(F - 32) + 273$ **b.** $C = 5°$ **c.** $K = 373°$
39. a. 38   **b.** $E(x) = x + 2$ **c.** 10   **41. a.** 0   **b.** 8   **c.** 16
43. a. $(C \circ f)(C) = 4\left(\frac{9}{5}C - 8\right)$ **b.** 40

 **45.**

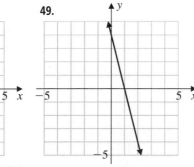

47. 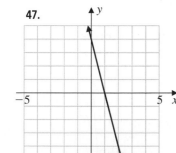   **49.**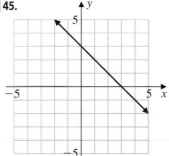

USING YOUR KNOWLEDGE   **51.** Even   **53.** Neither

MASTERY TEST   **59. a.** $P(x) = -\dfrac{x^2}{30} + 140x - 72,000$

61. $(g \circ f)(x) = x^3 + 1$
63. $(g \circ f)(-3) = -26$ **65.** $x + a, x \neq a$

67. $(f - g)(x) = x^2 - x + 6$ **69.** $\left(\dfrac{f}{g}\right)(-2) = -2$

71. $\left(\dfrac{g}{f}\right)(-2) = -\dfrac{1}{2}$

Exercise 10.2

A **B** **1.** $f^{-1} = \{(3, 1), (4, 2), (5, 3)\}$ Yes, f^{-1} is a function.

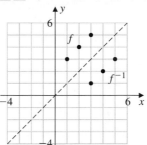

3. $f^{-1} = \{(5, -1), (4, -3), (4, -4)\}$ No, f^{-1} is not a function.

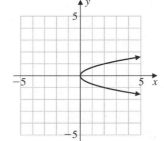

B **C** **5.** $y = \dfrac{x - 3}{3}$ Yes   **7.** $y = \dfrac{x + 4}{2}$ Yes

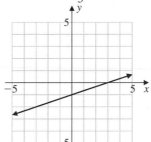   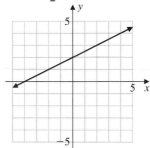

9. $y = \pm\dfrac{\sqrt{2x}}{2}$ No   **11.** $y = \pm\sqrt{x + 1}$ No

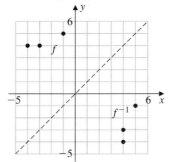   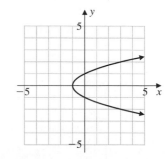

13. $y = -\sqrt[3]{x}$ Yes

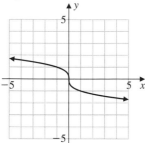

© **15.** The inverse $x = 2^{y+1}$ is a function.

17. The inverse $x = \left(\frac{1}{3}\right)^y$ is a function.

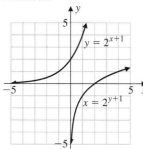

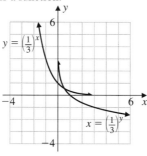

19. The inverse $x = 2^{-y}$ is a function.

21. a. 3   **b.** -1

23. $f^{-1}(x) = \dfrac{1}{x}$

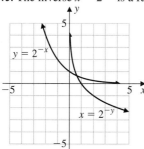

25.

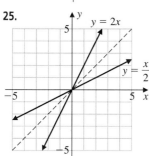

27.

Inverse not a function

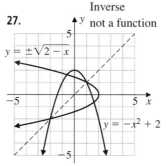

29.   Inverse not a function

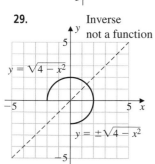

31.   Inverse not a function

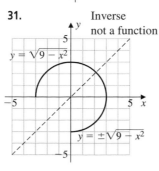

33.

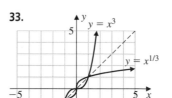

35.   Inverse not a function

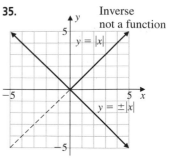

37. a. $L = f^{-1}(S) = \dfrac{S + 21}{3}$ **b.** $9\frac{1}{3}$ in.

39. a. $d = 16$ **b.** $f^{-1}(d) = d + 16$ **c.** 28 in.

41. a. $f(1988) = 41.44$ sec   **b.** $f^{-1}(x) = \dfrac{280 - x}{0.12}$

c. In the year 2000

SKILL CHECKER **43.** 9   **45.** 27   **47.** 4

USING YOUR KNOWLEDGE **49.** $f^{-1}(x) = \dfrac{x + 2}{3}$

51. $f^{-1}(x) = 2x - 1$ **53.** $f^{-1}(x) = \sqrt[3]{x - 1}$

55. $f^{-1}(x) = x^2, x \geq 0$

MASTERY TEST **63.** Yes

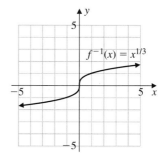

65. $D = \{2, 3, 4\}, R = \{1, 2, 3\}$

67. $D = \{1, 2, 3\}, R = \{2, 3, 4\}$

Exercise 10.3

Ⓐ **1. a.** $\frac{1}{5}$ **b.** 1   **c.** 5   **3. a.** $\frac{1}{9}$ **b.** 1   **c.** 9

5. a. $\frac{1}{10}$ **b.** 1   **c.** 10

Ⓑ **7. a.** $f(x)$ increasing   **9. a.** $f(x)$ increasing

b. $g(x)$ decreasing   **b.** $g(x)$ decreasing

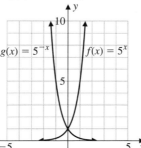

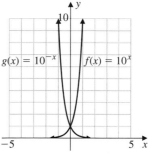

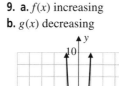

11. a. $f(x)$ increasing
b. $g(x)$ decreasing

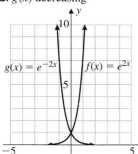

13. a. $f(x) = 3^x + 1$, increasing
b. $g(x) = 3^{-x} + 1$, decreasing

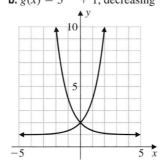

15. a. $f(x) = 2^{x+1}$, increasing **b.** $g(x) = 2^{-x+1}$, decreasing

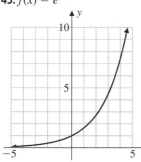

17. a. \$2459.60 **b.** \$2435.19
19. a. \$1822.12 **b.** \$1814.02 **21. a.** 2000 **b.** 4000 **c.** 8000
23. a. 2000 g **b.** 699.9 g **c.** 244.9 g
25. 21,448,000 **27. a.** 285.2 million **b.** 22,654 million
29. a. 333,333 **b.** 222,222 **c.** 8671
31. a. 4.42 lb/in.^{2} **b.** 6.42 lb/in.^{2}

SKILL CHECKER **33.** Product property **35.** Power property
USING YOUR KNOWLEDGE **37.** Continuous: \$1822.12
Monthly: \$1819.40 Continuous compounding gives about
\$2.72 more.

MASTERY TEST **43.** 165 g
45. $f(x) = e^{x/2}$

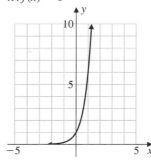

47. $f(x) = 6^x$

Exercise 10.4

A 1. $y = \log_2 x$

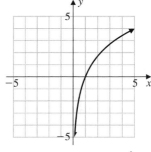

3. $y = \log_5 x$

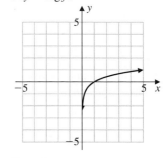

5. $f(x) = \log_{1/2} x$

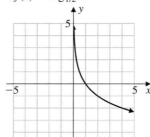

7. $y = \log_{1.5} x$

B 9. $\log_2 128 = x$ **11.** $\log_{10} 1000 = t$ **13.** $\log_{81} 9 = \frac{1}{2}$
15. $\log_{216} 6 = \frac{1}{3}$ **17.** $\log_e t = 3$ **19.** $9^3 = 729$, OK
21. $2^{-8} = \frac{1}{256}$, OK **23.** $81^{3/4} = 27$, OK **25.** $4^x = 16$
27. $10^{-2} = 0.01$, OK **29.** $e^{3.4012} = 30$, OK
31. $e^{-1.15} = 0.3166$, OK
C 33. $x = 9$ **35.** $x = 3^{-3} = \frac{1}{27}$ **37.** $x = 4^{1/2} = 2$
39. $x = 8^{1/3} = 2$ **41.** $x = 2$ **43.** $x = 3$ **45.** $x = -2$
47. $x = -\frac{1}{4}$ **49.** $x = -2$ **51.** 8 **53.** 4 **55.** -3
57. 6 **59.** 0 **61.** 1 **63.** 1 **65.** -1 **67.** $\frac{1}{3}$
69. -3 **71.** t **73.** t
D 75. $\log\left(\frac{26}{7} \times \frac{63}{15} \times \frac{5}{26}\right) = \log 3$

77. $\log\left(b^3 \times 2 \times \dfrac{1}{\sqrt{b}} \times \dfrac{\sqrt{b^3}}{2}\right) = \log b^4 = 4 \log b$

79. $\log\left(k^{3/2} \times r \times \dfrac{1}{k} \times \dfrac{1}{r^{3/4}}\right)$

$= \log k^{1/2} r^{1/4} = \log k^{2/4} r^{1/4}$
$= \log(k^2 r)^{1/4} = \frac{1}{4} \log k^2 r$

81. $3 \log \dfrac{y^3 - 1}{y^2}$

$= 3 \log \dfrac{(y - 1)(y^2 + y + 1)}{y^2}$

$= 3 \log(y - 1) + 3 \log(y^2 + y + 1) - 6 \log y$

83. $\log \dfrac{(x^2 - 4)\sqrt{x^2 + 2x + 4}}{(x^3 - 8)^2}$

$\quad = \log \dfrac{(x - 2)(x + 2)(x^2 + 2x + 4)^{1/2}}{(x - 2)^2(x^2 + 2x + 4)^2}$

$\quad = \log(x + 2) - \log(x - 2) - \frac{3}{2}\log(x^2 + 2x + 4)$

E **85.** $R = 8.9$ **87.** $\log A = 4.16769$

89. a. 53%   **b.** 77%   **c.** 91%

d. $h(A) = 29 + 50\log(A + 1)$ **e.** 70.84 in.

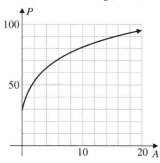

SKILL CHECKER **91.** 3.268×10^1 **93.** 2.387×10^{-3}
95. 5.69×10^{-5}

USING YOUR KNOWLEDGE

97. $R = \dfrac{(1.02)^{40}}{(1.085)^{10}}$, $\log R = -0.01030$

The 8.5% annually is better.

MASTERY TEST **103.** 6.4

105. $\log_b \sqrt[3]{\frac{4}{5}} = \frac{1}{3}\log_b \frac{4}{5} = \frac{1}{3}(\log_b 4 - \log_b 5)$

107. $\log_2 1024 = 10$ **109.** -3 **111.** -1

113. $\frac{1}{27}$ **115.** -3

117. $y = \log_7 x$

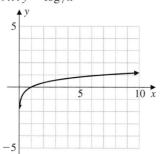

Exercise 10.5

A **1.** 1.8722   **3.** 3.2648   **5.** $8.6405 - 10 = -1.3595$
7. 1.7005   **9.** $8.0927 - 10 = -1.9073$
11. $7.9348 - 10 = -2.0652$ **13.** 18.5   **15.** 0.00585
17. 29.04   **19.** 0.7345
B **21.** 1.0986   **23.** 3.9512   **25.** 7.7647   **27.** -2.9188

29. -7.3858 **31.** 3.5   **33.** 65   **35.** 1.04   **37.** 0.1
39. 0.01
C **41.** 2.7268   **43.** 0.8010   **45.** -0.6826 **47.** -0.6610
49. -0.6439

D **51.** $f(x) = e^{3x}$

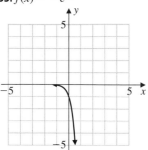

53. $f(x) = -e^{3x}$

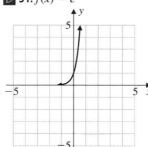

55. $f(x) = e^{-3x}$

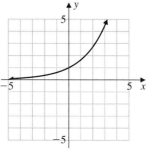

57. $f(x) = e^{(1/2)x}$

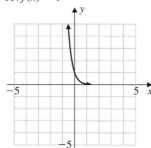

59. $f(x) = e^x + 1$

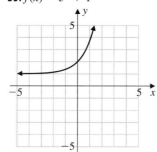

61. $f(x) = e^{0.5x} + 1$

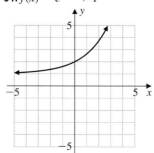

63. $f(x) = 2e^x$

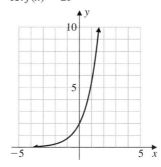

65. $y = \ln(x + 1)$

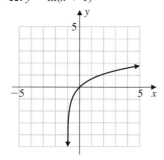

67. $y = \ln x + 4$

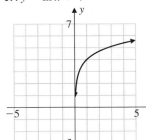

69. $y = \ln(x - 1)$

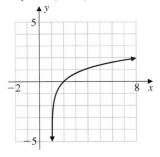

113. $y = \ln x + 3$

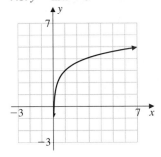

71. 6.2   **73.** 7.8   **75.** 6.4

77. a. 15,000,000   **b.** 22,377,000 (rounded from 22,377,370)

79. a. 50,000   **b.** 74,591   **c.** 166,006

81. a. 1000   **b.** 368   **83. a.** 0   **b.** 39.3

85. a. -0.8055 **b.** -3.8968

SKILL CHECKER **87.** $k = \dfrac{\log 3}{\log 2} \approx 1.585$

89. $k = \log 25 \approx 1.3979$

USING YOUR KNOWLEDGE

91. a. $f(x) = 2^x$

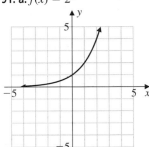

b. $g(x) = \log 2^x$

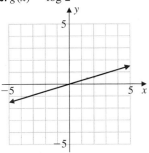

c. The slope is $\log 2$.

93. The slope of $g(x)$ is the common logarithm of the base of $f(x)$.

MASTERY TEST **101.** 0.2027   **103.** 1.8614   **105.** -3.3820

107. 0.003402

109. $y = -e^{(1/4)x}$

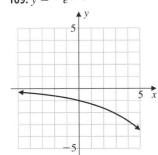

111. $y = e^{(1/4)x} + 1$

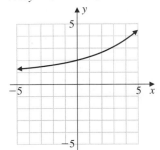

Exercise 10.6

A **1.** $x = 2$ **3.** $x = 5$ **5.** $x = \frac{4}{3}$ **7.** $x = -1$

9. $x = 3.2059$ **11.** $x = 6$ **13.** $x = \dfrac{\log 2}{\log 3} = 0.6309$

15. $x = \frac{7}{3}$ **17.** $x = \dfrac{-3 \pm \sqrt{17}}{2}$ **19.** $x = \ln 10 = 2.3026$

21. $x = -\ln 0.1 = 2.3026$ **23.** $k = 1.7006$

25. $k = -1.1513$

B **27.** $x = 8$ **29.** $x = \frac{1}{8}$ **31.** $x = e \approx 2.7184$

33. $x = e^3 \approx 20.0855$

35. $x = \frac{7}{3}$ **37.** $x = \frac{17}{3} = 5\frac{2}{3}$ **39.** $x = 8$ **41.** $x = 5$

43. $x = 1$ **45.** $x = \frac{1}{9}$ **47.** $x = 5$ **49.** $x = -1$ or -3

C **51.** 13.86 yr   **53.** 10.66 yr   **55.** About 6.10 billion

57. About 17.3 min   **59.** About 80.5 min

61. a. 100,000   **b.** 67,032   **c.** 13,534   **d.** 1832

63. About 23,105 yr   **65.** About 13.3 yr

67. a. 14.7 lb/in.^{2} **b.** 11.4 lb/in.^{2} **c.** 8.92 lb/in.^{2}

69. a. 9 yr   **b.** 10 yr   **71. a.** 5000   **b.** 2247   **c.** 2010

73. a. $54 million   **b.** 1995   **75. a.** 12   **b.** 7

77. About 694 months

SKILL CHECKER **79. a.** 3   **b.** 11   **c.** 21

USING YOUR KNOWLEDGE **81.** $k = 0.59$ **83.** 3,405,000

85. 1,243,000,000 (rounded from 1,242,987,000)

MASTERY TEST **93.** 59.9 min   **95.** $x = 1$ **97.** $x = 10$

99. $x = \dfrac{\log 10}{\log 5} = 1.4307$ **101.** $x = 4$

Review Exercises

1. a. $f + g = 4 + x - x^2$ **b.** $f - g = -x - x^2$

c. $fg = 4 + 2x - 2x^2 - x^3$ **d.** $\dfrac{f}{g} = \dfrac{2 - x^2}{2 + x}$

2. a. $f + g = 6 + x - x^2$ **b.** $f - g = -x - x^2$

c. $fg = 9 + 3x - 3x^2 - x^3$ **d.** $\dfrac{f}{g} = \dfrac{3 - x^2}{3 + x}$

3. 6   **4.** 7

5. a. $f \circ g = (2 - x)^3$ **b.** $g \circ f = 2 - x^3$ **c.** $(g \circ f)(2) = -6$
6. a. $f \circ g = (3 - x)^3$ **b.** $g \circ f = 3 - x^3$ **c.** $(g \circ f) = -5$
7. a. 136 beats per minute   **b.** 130 beats per minute
8. a. $P(x) = -0.02x^2 + 70x - 30,000$ **b.** 1750
9. a. $P(x) = -0.02x^2 + 60x - 40,000$ **b.** 1500
10. a. $D = \{4, 6, 8\}$ $R = \{4, 6, 8\}$
b. $S^{-1} = \{(4, 4), (6, 6), (8, 8)\}$
c. $D = \{4, 6, 8\}$ $R = \{4, 6, 8\}$
d.

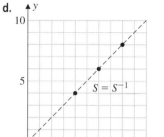

11. a. $D = \{4, 6, 8\}$ $R = \{5, 7, 9\}$
b. $S^{-1} = \{(5, 4), (7, 6), (9, 8)\}$
c. $D = \{5, 7, 9\}$ $R = \{4, 6, 8\}$
d.

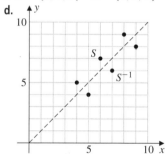

12. a. $f^{-1}(x) = \dfrac{x + 3}{3}$

13. a. $f^{-1}(x) = \dfrac{x + 4}{4}$

b.

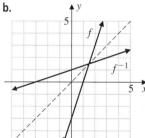

b.

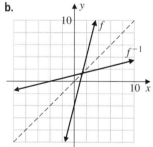

14. $y = \pm\dfrac{\sqrt{x}}{2}$; no

15. $y = \pm\dfrac{\sqrt{5x}}{5}$; no

16.

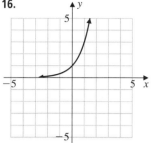

17.

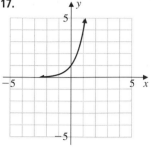

The inverse $x = 3^y$ is a function.

The inverse $x = 4^y$ is a function.

18. a. 12   **b.** $f^{-1}(d) = d + 16$ **c.** 26 in.
19. a. $f(x) = 2^{x/2}$ **b.** $f(x) = 2^{-x/2}$

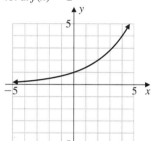

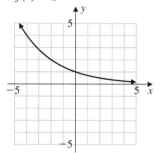

20. a. $g(x) = \left(\dfrac{1}{2}\right)^{x/2}$ **b.** $g(x) = \left(\dfrac{1}{2}\right)^{-x/2}$

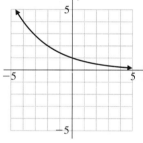

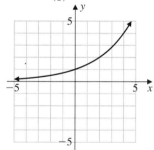

21. a. 1000   **b.** 61 g
22. $f(x) = \log_5 x$

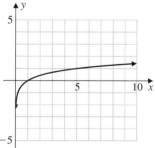

23. a. $\log_3 243 = 5$ **b.** $\log_2 \frac{1}{8} = -3$
24. a. $2^5 = 32$ **b.** $3^{-4} = \frac{1}{81}$
25. a. $x = \frac{1}{16}$ **b.** $x = 4$ **26. a.** 4   **b.** -3

27. a. $\log_b M + \log_b N$ **b.** $\log_b \dfrac{M}{N}$ **c.** $r \log_b M$

28. a. 2.9890 **b.** 2.9227
29. a. $7.8802 - 10 = -2.1198$ **b.** $6.8116 - 10 = -3.1884$
30. a. 663 **b.** 0.000407 **31. a.** 7.9551 **b.** -1.0642
32. a. 528,500,000 (rounded from 528,472,000) **b.** 15.155
33. a. 2.0959 **b.** 4.1918
34.

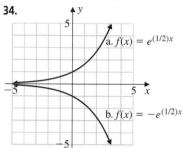

a. $f(x) = e^{(1/2)x}$
b. $f(x) = -e^{(1/2)x}$

35.

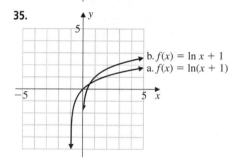

b. $f(x) = \ln x + 1$
a. $f(x) = \ln(x + 1)$

36. 5.398 **37. a.** $x = 3$ **b.** $x = 3$
38. a. $x = 1.5850$ **b.** $x = 0.2847$
39. a. $x = 0.1238$ **b.** $x = -2.1004$
40. a. $x = 10.9161$ **b.** $x = 2$
41. a. About 13.9 yr **b.** About 8.2 yr
42. a. 0.2950 **b.** 0.1475
43. a. About 1.39 yr **b.** About 34.7 yr

CHAPTER 11

Exercise 11.1

1. a. $a_{10} = 28;\ a_n = 3n - 2$ **3.** $a_{10} = 32;\ a_n = 3n + 2$
5. $a_{10} = 65;\ a_n = 5n + 15$ **7.** $a_{10} = 5;\ a_n = 55 - 5n$

9. $a_{10} = \dfrac{1}{11};\ a_n = \dfrac{1}{n + 1}$ **11.** $a_{10} = -1;\ a = (-1)^{n-1}$

13. $a_{10} = x^{10};\ a_n = x^n$ **15.** $a_{10} = -x^{19};\ a_n = (-1)^{n+1}(x)^{2n-1}$

17. $a_{10} = -x;\ a_n = (-1)^{n-1}(x)$ **19.** $a_{10} = \dfrac{x^{10}}{10};\ a_n = \dfrac{x^n}{n}$

21. $a_{10} = \dfrac{x^{10}}{1024};\ a_n = (-1)^{n-1}\left(\dfrac{x}{2}\right)^n$

23. $-1, 1, 3$ **25.** $-\dfrac{1}{2}, 0, \dfrac{3}{2}$ **27.** $0, \dfrac{1}{2}, \dfrac{2}{3}$ **29.** $1, 4, 9$
31. $\dfrac{1}{3}, \dfrac{2}{5}, \dfrac{3}{7}$ **33.** $-1, 1, -1$ **35.** $-\dfrac{1}{2}, \dfrac{1}{4}, -\dfrac{1}{8}$ **37.** 91

39. 100 **41.** 21 **43.** $\dfrac{25}{24}$ **45.** $9\dfrac{83}{140}$ **47.** 1

49. $\displaystyle\sum_{n=1}^{200} n$ **51.** $\displaystyle\sum_{n=1}^{50} \dfrac{1}{n}$ **53.** $\displaystyle\sum_{n=1}^{50} (-1)^{n-1}n$ **55.** $\displaystyle\sum_{n=1}^{5} (5n - 4)$

57. a. $a_8 = \$1100$ **b.** $a_{10} = \$1020$

59. a. 9 ft **b.** $\dfrac{729}{100}$ ft or 7.29 ft **c.** $\dfrac{9^n}{10^{n-1}}$

61. a. 400 **b.** 1600 **c.** $100(2)^n$ **63.** $\$3200$
SKILL CHECKER **65.** $4 + 3n$ **67.** $16n^2$
69. $(5n + 41)(n - 8)$
USING YOUR KNOWLEDGE **71.** $a_8 = 21$ **73.** $a_{10} = 55$
75. $a_{12} = 144$
MASTERY TEST **81.** $0, \dfrac{3}{2}, 4;\ a_{10} = \dfrac{99}{2}$

83. $a_{10} = 30,\ a_n = (-1)^n 3n$ **85. a.** $\displaystyle\sum_{n=1}^{4} 6n$ **b.** $\displaystyle\sum_{n=1}^{6} (-1)^{n-1}2^n$

Exercise 11.2

1. a. $a_1 = 5$ **b.** $d = 3$ **c.** $a_n = 3n + 2$
3. a. $a_1 = 11$ **b.** $d = -5$ **c.** $a_n = 16 - 5n$
5. a. $a_1 = 3$ **b.** $d = -4$ **c.** $a_n = 7 - 4n$

7. a. $a_1 = \dfrac{1}{2}$ **b.** $d = -\dfrac{1}{4}$ **c.** $a_n = \dfrac{3}{4} - \dfrac{1}{4}n$ or $\dfrac{3 - n}{4}$

9. a. $a_1 = -\dfrac{5}{6}$ **b.** $d = \dfrac{1}{2}$ **c.** $a_n = \dfrac{1}{2}n - \dfrac{4}{3}$ or $\dfrac{3n - 8}{6}$

11. $a_{15} = 91;\ S_{15} = 735$ **13.** $a_1 = 4;\ d = 6;\ S_8 = 200$
15. $d = 1;\ S_6 = 33$ **17.** $d = -4;\ a_{14} = -46$
19. $a_1 = -779;\ a_{40} = 781$
21. $\$10,000$ **23.** 190
25. Natural numbers are $1, 2, 3 \ldots$
$a_1 = 1;\ a_n = n$
$S_n = \dfrac{n}{2}(1 + n) = \dfrac{n(n + 1)}{2}$

27. Even natural numbers are $2, 4, 6 \ldots$
$a_1 = 2;\ a_n = 2n$
$S_n = \dfrac{n}{2}(2 + 2n) = n^2 + n$

SKILL CHECKER **29.** 32 **31.** 63 **33.** 20
USING YOUR KNOWLEDGE **35.** $\$8000, \$6000, \$4000, \$2000$
37. a. $b_t = 12,000 - 1200t$
b. $b_0 = \$12,000,\ b_1 = \$10,800,\ b_2 = \$9,600,\ b_3 = \$8,400,$
$b_4 = \$7,200,\ b_5 = \$6,000$
MASTERY TEST **43. a.** $a_1 = 5$ **b.** $d = 4$ **c.** $a_{10} = 41$
d. $a_n = 4n + 1$ **45. a.** $a_1 = 10$ **b.** $d = 2$ **47.** 10 months

Exercise 11.3

1. a. $a_1 = 3$ **b.** $r = 2$ **c.** $a_n = 3(2^{n-1})$ **d.** $S_n = 3(2^n - 1)$
3. a. $a_1 = 8$ **b.** $r = 3$ **c.** $a_n = 8(3^{n-1})$ **d.** $S_n = 4(3^n - 1)$
5. a. $a_1 = 16$ **b.** $r = -\dfrac{1}{4}$ **c.** $a_n = (-4)^{3-n}$

d. $S_n = \frac{64}{5}\left[1 - \left(-\frac{1}{4}\right)^n\right]$

7. a. $a_1 = -\frac{3}{5}$ **b.** $r = -\frac{5}{2}$ **c.** $a_n = \left(-\frac{3}{5}\right)\left(-\frac{5}{2}\right)^{n-1}$

d. $S_n = -\frac{6}{35}\left[1 - \left(-\frac{5}{2}\right)^n\right] = \frac{6}{35}\left[\left(-\frac{5}{2}\right)^n - 1\right]$

9. a. $a_1 = -\frac{3}{4}$ **b.** $r = \frac{1}{3}$ **c.** $a_n = -\frac{1}{4}(3^{2-n})$

d. $S_n = \left(-\frac{9}{8}\right)\left[1 - \left(\frac{1}{3}\right)^n\right]$ or $S_n = \left(-\frac{9}{8}\right)\left(\frac{3^n - 1}{3^n}\right)$

11. $a_3 = \frac{1}{4}, r = \frac{1}{2}$ or $a_3 = \frac{9}{4}, r = -\frac{3}{2}$

13. $a_3 = 12, r = 2$ or $a_3 = 27, r = -3$ **15.** $a_1 = 7, a_8 = 896$

17. $r = -3; n = 4$ **19.** $n = 6, S_n = S_6 = \frac{5187}{250}$

21. 12   **23.** -12 **25.** $\frac{4}{3}$ or $1\frac{1}{3}$

27. Sum does not exist, $|r| > 1$ **29.** Sum does not exist, $|r| > 1$

31. $\frac{5}{9}$ **33.** $\frac{2}{11}$ **35.** $\frac{401}{99}$ **37.** $\frac{2293}{990}$ **39.** $\frac{140}{999}$

41. a. $P_0, (1.04P_0), (1.04)^2P_0, \ldots$ **b.** 24,333

43. $S_5 = 184.5$ cm   **45.** \$131,875

47. 1, 3, 5 or 17, 3, -11 **49.** 87,381¢ or \$873.81   **51.** 56 ft

SKILL CHECKER   **53.** $a^2 + 2ab + b^2$

55. $x^3 + 3x^2y + 3xy^2 + y^3$ **57.** $y^3 - 6y^2z + 12yz^2 - 8z^3$

59. $\frac{1}{x^3} - \frac{3}{x^2y} + \frac{3}{xy^2} - \frac{1}{y^3}$

USING YOUR KNOWLEDGE   **61.** Geometric sequence, $|r| < 1$, $\therefore S = \frac{5}{3}$ **63.** Arithmetic sequence

65. Geometric sequence, $|r| < 1, \therefore S = 16$ **67.** Neither

69. Neither   **71.** Geometric sequence, $|r| < 1, \therefore S = \frac{8}{3}$

MASTERY TEST   **77. a.** $a_1 = 2$ **b.** $r = \frac{1}{2}$ **c.** $a_6 = \frac{1}{16}$

d. $a_n = \frac{1}{2^{n-2}}$ **79.** $a_3 = 32, r = 4$, or $a_3 = 50, r = -5$

81. Sum does not exist.   **83. a.** \$100, \$200, \$400, \$800, \$1600
b. \$6400   **c.** \$100(2^n)   **d.** 10 yr

Exercise 11.4

1. 6   **3.** 3,628,800   **5.** 30   **7.** 504   **9.** 15   **11.** 11
13. 1

15. $a^4 + 12a^3b + 54a^2b^2 + 108ab^3 + 81b^4$

17. $x^4 + 16x^3 + 96x^2 + 256x + 256$

19. $32x^5 - 80x^4y + 80x^3y^2 - 40x^2y^3 + 10xy^4 - y^5$

21. $32x^5 + 240x^4y + 720x^3y^2 + 1080x^2y^3 + 810xy^4 + 243y^5$

23. $\frac{1}{x^4} - \frac{2y}{x^3} + \frac{3y^2}{2x^2} - \frac{y^3}{2x} + \frac{y^4}{16}$

25. $x^6 + 6x^5 + 15x^4 + 20x^3 + 15x^2 + 6x + 1$ **27.** -540

29. 672   **31.** 1120   **33.** $-\frac{5}{2}$

35. 20   **37.** 84   **39.** $\frac{15}{64}$ **41.** $\frac{21}{128}$ **43.** $\frac{25}{216}$

USING YOUR KNOWLEDGE

45.
$$\binom{n}{k} = \frac{n!}{k!(n-k)!}$$

$$\binom{n}{n-k} = \frac{n!}{(n-k)!\,k!}$$

Thus, $$\binom{n}{k} = \binom{n}{n-k}$$

MASTERY TEST   **49.** 720   **51.** 15

53. $(2a - b)^4 = 16a^4 - 32a^3b + 24a^2b^2 - 8ab^3 + b^4$

55. 4860   **57.** 5

Review Exercises

1. a. $a_6 = 11, a_n = 2n - 1$ **b.** $a_6 = 18, a_n = 3n$

2. a. 64   **b.** 2^{n+2} **3. a.** $\frac{2}{3}, \frac{5}{3}, \frac{10}{3}; a_{10} = \frac{101}{3}; S_3 = \frac{17}{3}$
b. $\frac{1}{2}, \frac{3}{2}, 3; a_{10} = \frac{55}{2}; S_3 = 5$

4. a. $3, 6, 11, \ldots; S_4 = 38$

b. $1, 7, 17, \ldots; S_3 = 25$ **5. a.** \$64,000   **b.** \$16,000

6. a. $d = 3; a_{10} = 30$ **b.** $d = 4; a_{10} = 40$

7. a. $a_n = 3n$ **b.** $a_n = 4n$ **8. a.** 176 ft   **b.** 240 ft

9. a. $a_1 = 14, d = 2$ **b.** $a_1 = 9, d = 3$

10. a. $a_1 = 4, d = 2$ **b.** $a_1 = 1.5, d = 3$

11. a. \$21,600   **b.** \$9000   **12. a.** 10   **b.** 12

13. a. $r = 2, a_6 = 96$ **b.** $r = \frac{1}{2}, a_6 = \frac{1}{8}$

14. a. $r = 4, a_n = 4^{n-1}$ **b.** $r = -\frac{1}{2}, a_n = \frac{(-1)^{n-1}}{2^{n-3}}$

15. a. $a_n = (-1)^{n-1}2^n, S_n = \frac{2}{3}[1 - (-2)^n]$
b. $a_n = \left(-\frac{1}{2}\right)^{n-1}, S_n = \frac{2}{3}\left[1 - \left(-\frac{1}{2}\right)^n\right]$ **16. a.** $\frac{64}{3}$ **b.** $\frac{27}{2}$

17. a. Sum does not exist.   **b.** Sum does not exist.

18. a. $\frac{31}{99}$ **b.** $\frac{12}{37}$ **19. a.** 24 ft   **b.** 18 ft

20. a. 40,320   **b.** 1680   **21. a.** 84   **b.** 1

22. a. $a^4 - 12a^3b + 54a^2b^2 - 108ab^3 + 81b^4$
b. $81a^4 + 216a^3b + 216a^2b^2 + 96ab^3 + 16b^4$

23. a. 1120   **b.** -2835 **24. a.** 35   **b.** 35

25. a. $\frac{35}{128}$ **b.** $\frac{35}{128}$

Research Bibliography

The entries in this bibliography provide a first resource for investigating the Research Questions that appear at the end of each chapter in the text. Many of these books will also contain their own bibliographies that you can use for an even more thorough information search. Also, using your library's card catalog system, be sure to look under subject as well as title listings to gain a better idea of the resources that your library has available even beyond the specific titles listed here.

Bell, E. T. *Men of Mathematics.* New York: Simon and Schuster, 1965.

Billstein, R. et al. *A Problem Solving Approach to Mathematics,* 4th ed. Redwood City, CA: Benjamin Cummings Publishing Company, 1990.

Brewer, James W. and Martha K. Smith, ed. "Emmy Noether, A Tribute to Her Life and Work," *Pure and Applied Mathematics,* 69, Marcel Dekker, 1981.

Burton, David. *The History of Math: An Introduction,* 2d ed. Dubuque, IA: Wm. C. Brown Publishers, 1991.

Cajori, Florian. *A History of Mathematical Notation.* New York: Dover Publications, 1993.

Calinger, Ronald, ed. *Classics of Mathematics.* Englewood Cliffs: Prentice-Hall, 1995.

Copi, I. *Introduction to Logic,* 6th ed. New York: Macmillan, 1982.

Eves, Howard. *An Introduction to the History of Mathematics,* 4th ed. New York: Holt, Rinehart and Winston, 1976.

Hogben, Lancelot. *Mathematics in the Making.* London: Galahad Books, 1960.

Kahane, Howard. *Logic and Philosophy: A Modern Introduction,* 6th ed. Belmont, CA: Wadsworth Publishing, 1990.

Katz, Victor J. *A History of Mathematics.* New York: Harper Collins, 1993.

Klein, Morris. *Mathematical Thought: From Ancient to Modern Times,* 4 Vols. York York: Oxford University Press, 1990.

Krause, Eugene. *Mathematics for Elementary Teachers,* 2d ed. Lexington, MA: D.C. Heath and Company, 1991.

Merzbach, Roger. *A History of Mathematics,* 2d ed. New York: John Wiley & Sons, Inc., 1991.

Newman, James. *The World of Mathematics,* 4 Vols. New York: Simon and Schuster, 1956.

Osen, Lynn M. *Women in Mathematics.* Cambridge, MA: MIT Press, 1974.

Pedoe, Don. *The Gentle Art of Mathematics.* New York: Collier Books, 1963.

Perl, Teri H. *Math Equals.* Reading, MA: Addison-Wesley, 1978.

Photo Credits

Index

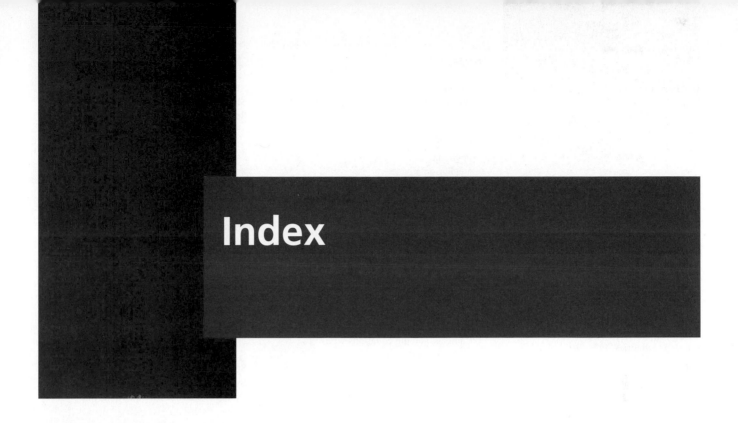

INTERMEDIATE ALGEBRA

THE REAL NUMBER SYSTEM

Natural numbers $\qquad N = \{1, 2, 3, \ldots\}$

Whole numbers $\qquad W = \{0, 1, 2, 3, \ldots\}$

Integers $\qquad I = \{\ldots, -3, -2, -1, 0, 1, 2, 3, \ldots\}$

Rational numbers $\qquad = \left\{ \dfrac{a}{b} \,\middle|\, a \in J, b \in J, b \neq 0 \right\}$

Irrational numbers $\qquad = \{\text{nonterminating, nonrepeating decimals}\}$

Real numbers $\qquad = \{\text{rational numbers}\} \cup \{\text{irrational numbers}\}$

Absolute value $\qquad |x| = \begin{cases} x & \text{if } x \geq 0 \\ -x & \text{if } x < 0 \end{cases}$

PROPERTIES OF REAL NUMBERS

If a, b, and c are real numbers, the following properties are assumed to be true.

	ADDITION	MULTIPLICATION
Commutative	$a + b = b + a$	$a \cdot b = b \cdot a$
Associative	$a + (b + c) = (a + b) + c$	$a(b \cdot c) = (a \cdot b)c$
Identity	$a + 0 = 0 + a = a$	$a \cdot 1 = 1 \cdot a = a$
Inverse	$a + (-a) = (-a) + a = 0$	$a \cdot \dfrac{1}{a} = \dfrac{1}{a} \cdot a = 1, \quad a \neq 0$
Distributive		$a(b + c) = ab + ac$

OPERATIONS WITH EQUATIONS

Addition property of equality $\qquad$ If $a = b$, then $a + c = b + c$.

Subtraction property of equality $\qquad$ If $a = b$, then $a - c = b - c$.

Multiplication property of equality $\qquad$ If $a = b$, then $ac = bc, \quad c \neq 0$.

Division property of equality $\qquad$ If $a = b$, then $\dfrac{a}{c} = \dfrac{b}{c}, \quad c \neq 0$.